Lecture Notes in Computer Science 13989

Founding Editors

Gerhard Goos
Juris Hartmanis

Editorial Board Members

Elisa Bertino, *Purdue University, West Lafayette, IN, USA*
Wen Gao, *Peking University, Beijing, China*
Bernhard Steffen , *TU Dortmund University, Dortmund, Germany*
Moti Yung , *Columbia University, New York, NY, USA*

The series Lecture Notes in Computer Science (LNCS), including its subseries Lecture Notes in Artificial Intelligence (LNAI) and Lecture Notes in Bioinformatics (LNBI), has established itself as a medium for the publication of new developments in computer science and information technology research, teaching, and education.

LNCS enjoys close cooperation with the computer science R & D community, the series counts many renowned academics among its volume editors and paper authors, and collaborates with prestigious societies. Its mission is to serve this international community by providing an invaluable service, mainly focused on the publication of conference and workshop proceedings and postproceedings. LNCS commenced publication in 1973.

João Correia · Stephen Smith · Raneem Qaddoura
Editors

Applications of Evolutionary Computation

26th European Conference, EvoApplications 2023
Held as Part of EvoStar 2023
Brno, Czech Republic, April 12–14, 2023
Proceedings

 Springer

Editors
João Correia 🄳
University of Coimbra
Coimbra, Portugal

Stephen Smith 🄳
University of York
York, UK

Raneem Qaddoura 🄳
Al Hussein Technical University
Amman, Jordan

ISSN 0302-9743   ISSN 1611-3349 (electronic)
Lecture Notes in Computer Science
ISBN 978-3-031-30228-2   ISBN 978-3-031-30229-9 (eBook)
https://doi.org/10.1007/978-3-031-30229-9

Preface

This volume contains the proceedings of EvoApplications 2023, the International Conference on the Applications of Evolutionary Computation. The conference is part of Evo*, the leading event on bio-inspired computation in Europe, and was held in Brno, Czech Republic, as a hybrid event, on April 12–14, 2023.

EvoApplications, formerly known as EvoWorkshops, aims to bring together high-quality research focusing on applied domains of bio-inspired computing. At the same time, under the Evo* umbrella, EuroGP focused on the technique of genetic programming, EvoCOP targeted evolutionary computation in combinatorial optimization, and EvoMUSART was dedicated to evolved and bio-inspired music, sound, art, and design. The proceedings for these co-located events are available in the LNCS series.

EvoApplications 2023 received 201 high-quality submissions distributed among the main session on Applications of Evolutionary Computation and 9 additional special sessions chaired by leading experts on the different areas: Analysis of Evolutionary Computation Methods: Theory, Empirics, and Real-World Applications (Thomas Bartz-Beielstein, Carola Doerr, and Christine Zarges); Applications of Bio-inspired Techniques on Social Networks (Giovanni Iacca and Doina Bucur); Computational Intelligence for Sustainability (Valentino Santucci and Fabio Caraffini); Evolutionary Computation in Edge, Fog, and Cloud Computing (Diego Oliva, Seyed Jalaleddin Mousavirad, and Mahshid Helali Moghadam); Evolutionary Computation in Image Analysis, Signal Processing, and Pattern Recognition (Pablo Mesejo and Harith Al-Sahaf); Machine Learning and AI in Digital Healthcare and Personalized Medicine (Stephen Smith and Marta Vallejo); Resilient bio-inspired Algorithms (Carlos Cotta and Gustavo Olague); Soft Computing Applied to Games (Alberto P. Tonda, Antonio M. Mora, and Pablo García-Sánchez); and Surrogate-Assisted Evolutionary Optimisation (Tinkle Chugh and Alma Rahat). We selected 37 of these papers for full oral presentation, while 14 works were presented in short oral presentations and as posters. Morevoer, these proceedings also include contributions from the Evolutionary Machine learning (EML) joint track, a combined effort of the International Conference on the Applications of Evolutionary Computation (EvoAPPS) and the European Conference on Genetic Programming (EuroGP), organized by Penousal Machado and Wolfgang Banzhaf. EML received 27 high-quality submissions. After careful review, seven were selected for oral presentation and five in short oral presentations and as posters. Since EML is a joint track these papers are shared among the proceedings, and this volume's part "Evolutionary Machine Learning" contains 7 of them. The remaining ones are published in the EuroGP proceedings. All accepted contributions, regardless of the presentation format, appear as full papers in this volume.

An event of this kind would not be possible without the contribution of a large number of people:

- We express our gratitude to the authors for submitting their works and to the members of the Program Committee for devoting selfless effort to the review process.
- We would also like to thank Nuno Lourenço (University of Coimbra, Portugal) for his dedicated work with the submission and registration system and Sérgio Rebelo and Tiago Martins (University of Coimbra, Portugal) for their important graphic design work.
- We are grateful to Francisco Chicano (University of Málaga, Spain) and João Correia (University of Coimbra, Portugal) for their impressive work managing and maintaining the Evo* website and handling the publicity, respectively.
- We credit the invited keynote speakers, Marek Vácha (Charles University, Prague, Czech Republic) and Evelyne Lutton (INRAE, France), for their fascinating and inspiring presentations.
- We would like to express our gratitude to the Steering Committee of EvoApplications for helping organize the conference.
- We are grateful for the support provided by SPECIES, the Society for the Promotion of Evolutionary Computation in Europe and its Surroundings, for the coordination and financial administration.
- Special thanks to Jiri Jaros (Brno University of Technology, Czech Republic) and Lukas Sekanina (Brno University of Technology, Czech Republic) as local organizers and to the Brno University of Technology, Czech Republic, for organizing and providing an enriching conference venue.

Finally, we express our continued appreciation to Anna I. Esparcia-Alcázar, from SPECIES, Europe, whose considerable efforts in managing and coordinating Evo* helped build a unique, vibrant, and friendly atmosphere.

April 2023

João Correia
Stephen Smith
Raneem Qaddoura

Organization

EvoApplications Conference Chair

João Correia   University of Coimbra, Portugal

EvoApplications Conference Co-chair

Stephen Smith   University of York, UK

EvoApplications Publication Chair

Raneem Qaddoura   Al Hussein Technical University, Jordan

Analysis of Evolutionary Computation Methods: Theory, Empirics, and Real-World Applications Chairs

Thomas Bartz-Beielstein   TH Köln, Germany
Carola Doerr   CNRS and Sorbonne Université, France
Christine Zarges   Aberystwyth University, UK

Applications of Bio-inspired Techniques on Social Networks Chairs

Giovanni Iacca   Università di Trento, Italy
Doina Bucur   University of Twente, The Netherlands

Computational Intelligence for Sustainability Chairs

Valentino Santucci   Università per Stranieri di Perugia, Italy
Fabio Caraffini   Swansea University, UK

Evolutionary Computation in Edge, Fog, and Cloud Computing Chairs

Diego Oliva   Universidad de Guadalajara, México
Seyed Jalaleddin Mousavirad   Hakim Sabzevari University, Iran
Mahshid Helali Moghadam   RISE Research Institutes of Sweden, Sweden

Evolutionary Computation in Image Analysis, Signal Processing and Pattern Recognition Chairs

Pablo Mesejo   Universidad de Granada, Spain
Harith Al-Sahaf   Victoria University of Wellington, New Zealand

Machine Learning and AI in Digital Healthcare and Personalized Medicine Chairs

Stephen Smith   University of York, UK
Marta Vallejo   Heriot-Watt University, UK

Resilient Bio-inspired Algorithms Chairs

Carlos Cotta   University of Málaga, Spain
Gustavo Olague   CICESE, Mexico

Soft Computing Applied to Games Chairs

Alberto P. Tonda   INRAE, France
Antonio M. Mora   Universidad de Granada, Spain
Pablo García-Sánchez   Universidad de Granada, Spain

Surrogate-Assisted Evolutionary Optimisation Chairs

Tinkle Chugh   University of Exeter, UK
Alma Rahat   Swansea University, UK

Evolutionary Machine Learning Chairs

Penousal Machado   University of Coimbra, Portugal
Wolfgang Banzhaf   Michigan State University, USA

EvoApplications Steering Committee

Stefano Cagnoni   University of Parma, Italy
Pedro A. Castillo   University of Granada, Spain
Anna I. Esparcia-Alcázar   Universitat Politècnica de València, Spain
Mario Giacobini   University of Torino, Italy
Paul Kaufmann   University of Mainz, Germany
Antonio Mora   University of Granada, Spain
Günther Raidl   Vienna University of Technology, Austria
Franz Rothlauf   Johannes Gutenberg University Mainz, Germany
Kevin Sim   Edinburgh Napier University, Scotland
Giovanni Squillero   Politecnico di Torino, Italy
Cecilia di Chio   King's College London, UK
   (Honorary Member)

Program Committee

Harith Al-Sahaf   Victoria University of Wellington, New Zealand
Jacopo Aleotti   University of Parma, Italy
Anca Andreica   Babes-Bolyai University, Romania
Claus Aranha   University of Tsukuba, Japan
Kehinde Babaagba   Edinburgh Napier University, UK
Jaume Bacardit   Newcastle University, UK
Marco Baioletti   Universitá degli Studi di Perugia, Italy
Illya Bakurov   NOVA IMS, Portugal
Wolfgang Banzhaf   Michigan State University, USA
Thomas Bartz-Beielstein   TH Koeln, Germany
János Botzheim   Eötvös Loránd University, Hungary
Doina Bucur   University of Twente, Netherlands
Maxim Buzdalov   ITMO University, Russia
Stefano Cagnoni   University of Parma, Italy
Francisco Carlos   Pontificia Universidad Javeriana, Colombia
   Calderon Bocanegra
Fabio Caraffini   De Montfort University, UK
Oscar Castillo   Tijuana Institute of Technology, Mexico

Pedro Castillo	University of Granada, Spain
Josu Ceberio	University of the Basque Country, Spain
Ying-Ping Chen	National Yang Ming Chiao Tung University, Taiwan
Francisco Chicano	University of Málaga, Spain
Anders Christensen	University of Southern Denmark, Denmark
Tinkle Chugh	University of Exeter, UK
Anthony Clark	Pomona College, California, USA
José Manuel Colmenar	Universidad Rey Juan Carlos, Spain
Feijoo Colomine D.	Universidad Nacional Experimental del Tachira, Venezuela
Stefano Coniglio	University of Bergamo, Italy
João Correia	University of Coimbra, Portugal
Carlos Cotta	Universidad de Málaga, Spain
Fabio D'Andreagiovanni	Sorbonne University, France
Gregoire Danoy	University of Luxembourg, Luxembourg
George De Ath	University of Exeter, UK
Amir Dehsarvi	Ludwig Maximilian University of Munich, Germany
Bilel Derbel	CRIStAL (Univ. Lille), France
Travis Desell	Rochester Institute of Technology, USA
Federico Divina	Pablo de Olavide University, Spain
Carola Doerr	Sorbonne University, France
Bernabe Dorronsoro	University of Cadiz, Spain
Tome Eftimov	Jozef Stefan Institute, Slovenia
Abdelrahman Elsaid	Rocheser Institute of Technology, USA
Edoardo Fadda	Politecnico di Torino, Italy
Andres Faina	IT University of Copenhagen, Denmark
Francisco Fernandez De Vega	Universidad de Extremadura, Spain
Johana Maria Florez Lozano	Pontificia Universidad Javeriana, Colombia
Francesco Fontanella	Università di Cassino e del Lazio Meridionale, Italy
Marcus Gallagher	University of Queensland, Australia
Pablo García Sánchez	Universidad de Granada, Spain
Mario Giacobini	University of Torino, Italy
Finley Gibson	University of Exeter, UK
Antonio Gonzalez-Pardo	Universidad Rey Juan Carlos, Spain
Michael Guckert	THM, Germany
Mahshid Helali Moghadam	RISE Research Institutes of Sweden, Sweden
Daniel Hernandez	Tecnológico Nacional de México/Instituto Tecnológico de Tijuana, Mexico
Rolf Hoffmann	TU Darmstadt, Germany

David Pelta	University of Granada, Spain
Diego Perez Liebana	Queen Mary University of London, UK
Arkadiusz Poteralski	Silesian University of Technology, Poland
Petr Pošík	Czech Technical University in Prague, Czechia
Raneem Qaddoura	Al Hussein Technical University, Jordan
Elena Raponi	Technical University of Munich, Germany
José Carlos Ribeiro	Polytechnic Institute of Leiria, Portugal
Jose Santos	University of A Coruña, Spain
Valentino Santucci	University for Foreigners of Perugia, Italy
Enrico Schumann	University of Basel, Switzerland
Lennart Schäpermeier	TU Dresden, Germany
Sevil Sen	University of York, UK
Roman Senkerik	Tomas Bata University in Zlin, Czechia
Chien-Chung Shen	University of Delaware, USA
Kevin Sim	Edinburgh Napier University, UK
Anabela Simões	Coimbra Institute of Engineering, Portugal
Stephen Smith	University of York, UK
Yanan Sun	Sichuan University, China
Shamik Sural	Indian Institute of Technology Kharagpur, India
Ernesto Tarantino	ICAR-CNR, Italy
Andrea Tettamanzi	Univ. Nice Sophia Antipolis, France
Renato Tinós	USP, Brazil
Marco Tomassini	University of Lausanne, Switzerland
Alberto Tonda	INRA, France
Jamal Toutouh	Massachusetts Institute of Technology, USA
Heike Trautmann	University of Münster, Germany
Koen van der Blom	Leiden Institute of Advanced Computer Science, Netherlands
Frank Veenstra	University of Oslo, Norway
Diederick Vermetten	Leiden Institute of Advanced Computer Science, Netherlands
Marco Villani	University of Modena and Reggio Emilia, Italy
Markus Wagner	University of Adelaide, Australia
Hao Wang	Leiden University, Netherlands
Thomas Weise	University of Science and Technology of China, China
Simon Wells	Edinburgh Napier University, UK
Phoenix Williams	University of Exeter, UK
Dennis Wilson	ISAE-Supaero, France
Anil Yaman	Vrije Universiteit Amsterdam, Netherlands
Furong Ye	Leiden Institute of Advanced Computer Science, Netherlands

Ales Zamuda   University of Maribor, Slovenia
Christine Zarges   Aberystwyth University, UK
Mengjie Zhang   Victoria University of Wellington, New Zealand

Contents

Analysis of Evolutionary Computation Methods: Theory, Empirics, and Real-World Applications

Computational Intelligence for Sustainability

Machine Learning and AI in Digital Healthcare and Personalized Medicine

Resilient Bio-inspired Algorithms

Soft Computing Applied to Games

Applications of Evolutionary Computation

An Evolutionary Approach for Scheduling a Fleet of Shared Electric Vehicles

Steffen Limmer[1]($\boxtimes$) (iD), Johannes Varga[2] (iD), and Günther R. Raidl[2] (iD)

[1] Honda Research Institute Europe GmbH, 63073 Offenbach, Germany
steffen.limmer@honda-ri.de
[2] Institute of Logic and Computation, TU Wien, 1040 Vienna, Austria
{jvarga,raidl}@ac.tuwien.ac.at

Abstract. In the present paper, we investigate the management of a fleet of electric vehicles. We propose a hybrid evolutionary approach for solving the problem of simultaneously planning the charging of electric vehicles and the assignment of electric vehicles to a set of reservations. The reservation assignment is optimized with an evolutionary algorithm while linear programming is used to compute optimal charging schedules. The evolutionary algorithm uses an indirect encoding and a problem-specific crossover operator. Furthermore, we propose the use of a surrogate fitness function. Experimental results on problem instances with up to 100 vehicles and 1600 reservations show that the proposed approach is able to notably outperform two approaches based on mixed integer linear programming.

Keywords: Electric vehicles · Scheduling · Evolutionary algorithm · Mixed integer linear programming · eMaaS

1 Introduction

There is an increasing trend towards electric, shared, and multimodal mobility in order to tackle environmental issues and to respond to current and future transportation needs of users [7]. Services, which integrate multiple transportation modes and shared electric mobility are commonly summarized under the term electric Mobility as a Service (eMaaS) [4]. The operation of an eMaaS service, like, for example, a dial-a-ride platform with electric vehicles (EVs), typically requires planning ahead the usage and recharging of a fleet of electric vehicles [2]. This in turn requires to solve an optimization problem. The runtime of the optimization is a critical aspect, especially in a dynamic setting, in which fast responses to newly arriving transportation requests are required.

In the present paper, we investigate the use of evolutionary computation to accelerate the optimization for such type of scheduling problem: The EV Fleet Charging and Allocation Problem (EVFCAP) [9]. In this problem, a fleet of EVs (e.g., a company car fleet) is considered. The EVs can charge energy at a

J. Correia et al. (Eds.): EvoApplications 2023, LNCS 13989, pp. 3–18, 2023.
https://doi.org/10.1007/978-3-031-30229-9_1

common site (e.g., a company building), where the energy can be either drawn from the power grid for time-varying electricity prices or from photovoltaics (PV) overproduction, if available, for free. Users (e.g., company employees) can make reservations of vehicles. For each reservation, a period of time, when a vehicle is required, and an estimated energy consumption is given. The problem consists in planning the charging of the EVs simultaneously with an assignment of the individual EVs to the reservations with the objective to maximize the usage of EVs for serving reservations while keeping the charging cost low.

The problem can be formulated as a mixed integer linear programming (MILP) problem, which can be reasonably well solved as long as the number of EVs and reservations is not too high. Betz et al. [1] use a MILP approach to solve a variant of the EVFCAP where only a limited number of heterogeneous charging stations are available for charging the EVs. They report that the largest problem instance, which could be solved to proven optimality, comprises eight EVs and about 30 reservations. Sassi and Oulamara [8] consider a problem variant where the total charging power is limited and reservations, which are not assigned to EVs, have to be assigned to a limited number of combustion engine vehicles. They propose a problem-specific heuristic for the solution of that problem and compare it to a MILP approach on problem instances with up to 200 vehicles and 320 reservations. Especially on larger problem instances, the MILP approach is clearly outperformed by the heuristic approach. Varga et al. [9] consider the same problem variant as considered in the present paper. They use a Benders decomposition approach to speed up the optimization with a MILP approach. The optimization is further accelerated by applying in early stages a general variable neighborhood search heuristic to solve the master problem of the Benders decomposition. The authors compare the proposed approach with a standard MILP approach on problem instances with up to 100 EVs and 1600 reservations.

We propose and evaluate an evolutionary approach for the solution of the EVFCAP. The assignment of reservations to EVs is optimized with an evolutionary algorithm (EA), which is hybridized with a linear programming approach in order to compute optimal charging schedules. To ensure feasibility of solution candidates, the EA uses an indirect encoding that does not encode the reservation assignment directly but only the order in which reservations are passed to an insertion operator. The optimization is supported by a problem-specific crossover operator. Furthermore, a surrogate fitness function is employed in order to accelerate the optimization. In experiments, the proposed approach is evaluated and compared to a standard MILP approach and the improved MILP approach proposed in [9] on publicly available problem instances from [9].

The rest of the paper is organized as follows: Sect. 2 provides a description of the problem. In Sect. 3, the proposed approach is explained in detail. Section 4 describes the experiments and discusses their results. Finally, Sect. 5 provides conclusions.

2 Problem Description

We consider a planning horizon $\mathcal{T} = \{1, \ldots, T\}$ of T discrete time steps of an equal length of Δt hours. There is a set $\mathcal{N} = \{1, \ldots, N\}$ of N EVs, each with a maximum charging power of $P^{\max}$ kW and with a battery capacity of $E^{\max}$ kWh. At the beginning of the planning horizon, each EV $n \in \mathcal{N}$ has a certain initial battery level of E_n^{init} kWh. The EVs can be charged at a common site. It is assumed that at this site, there is a certain electrical base load (consumption) and energy production by a PV system. If the PV production exceeds the base load, there is a surplus energy, which can be used for EV charging. Let Sur_t denote the amount of surplus energy in time step t. For time steps t, in which the base load is higher than the PV production, Sur_t is zero. In addition to the surplus energy, energy from the power grid can be used for EV charging. It is assumed that the price for grid energy varies over time. Let p_t denote the electricity price per kWh in time step t. There is a set $\mathcal{R} = \{1, \ldots, R\}$ of R reservations of vehicles by users. For each reservation $r \in \mathcal{R}$, there is a time period, in which a vehicle is required. Let t_r^s and t_r^e denote the first and last time step, respectively, of this period for reservation r. It is assumed that EVs are not charged externally while they are used for a reservation. Thus, it has to be ensured that EVs are sufficiently charged before they are used for a reservation. Let E_r^{res} denote the amount of energy, which is required for serving reservation r.

An operator of the fleet has to decide for each reservation whether it is assigned to an EV or not and if assigned to an EV, to which one exactly. Unassigned reservations might, e.g., be served by a fleet of combustion engine cars inducing additional cost. In addition to the reservation assignment, the charging of the EVs has to be planned. It is assumed that the operator of the fleet is interested in three objectives:

1. Minimizing the amount of energy required for reservations that are not assigned to an EV, which corresponds to maximizing the usage of EVs.
2. Minimizing the electricity cost incurred by EV charging.
3. Minimizing the amount of energy missing in the batteries of the EVs at the end of the planning horizon in order to increase the number of reservations that can be served by EVs in the time after the planning horizon.

We introduce binary variables y_r and $x_{n,r}$, where y_r indicates whether a reservation r is assigned to an EV or not and $x_{n,r}$ indicates whether reservation r is assigned to EV n or not. Furthermore, we introduce variables $P_{n,t}$ for the power by which EV n is charged in time step t. Let E_t^{grid} and E_t^{sur} denote the amount of charging energy consumed from the grid and from surplus energy, respectively, in time step t, and let $E_{n,t}$ denote the battery level of EV n in time step t. The scheduling problem can be expressed as the following MILP problem:

$$\min \alpha \sum_{r \in \mathcal{R}} E_r^{\text{res}} \cdot y_r + \sum_{t \in \mathcal{T}} p_t \cdot E_t^{\text{grid}} + \beta \sum_{n \in \mathcal{N}} (E^{\max} - E_{n,T}), \tag{1}$$

subject to:

$$\sum_{n\in\mathcal{N}} x_{n,r} + y_r = 1 \qquad\qquad \forall r \in \mathcal{R}, \ (2)$$

$$P_{n,t} \leq P^{\mathrm{max}} \cdot (1 - \sum_{r\in\mathcal{R}|t_r^s \leq t \leq t_r^e} x_{n,r}) \qquad\qquad \forall n \in \mathcal{N}, \forall t \in \mathcal{T}, \ (3)$$

$$E_{n,1} = E_n^{\mathrm{init}} + \Delta t \cdot P_{n,1} - \sum_{r\in\mathcal{R}|t_r^s=1} x_{n,r} \cdot E_r^{\mathrm{res}} \qquad\qquad \forall n \in \mathcal{N}, \ (4)$$

$$E_{n,t} = E_{n,t-1} + \Delta t \cdot P_{n,t} - \sum_{r\in\mathcal{R}|t_r^s=t} x_{n,r} \cdot E_r^{\mathrm{res}} \qquad\qquad \forall n \in \mathcal{N}, \forall t \in \mathcal{T} \setminus \{1\}, \ (5)$$

$$\sum_{n\in\mathcal{N}} \Delta t \cdot P_{n,t} = E_t^{\mathrm{grid}} + E_t^{\mathrm{sur}} \qquad\qquad \forall t \in \mathcal{T}, \ (6)$$

$$0 \leq P_{n,t} \leq P^{\mathrm{max}} \qquad\qquad \forall n \in \mathcal{N}, \forall t \in \mathcal{T}, \ (7)$$

$$0 \leq E_{n,t} \leq E^{\mathrm{max}} \qquad\qquad \forall n \in \mathcal{N}, \forall t \in \mathcal{T}, \ (8)$$

$$0 \leq E_t^{\mathrm{grid}} \qquad\qquad \forall t \in \mathcal{T}, \ (9)$$

$$0 \leq E_t^{\mathrm{sur}} \leq Sur_t \qquad\qquad \forall t \in \mathcal{T}, \ (10)$$

$$x_{n,r} \in \{0,1\}, y_r \in \{0,1\} \qquad\qquad \forall n \in \mathcal{N}, \forall r \in \mathcal{R}. \ (11)$$

The objective function is a weighted sum of the three objectives, with weights α and β. The weight α could be, for example, set to the cost arising from traveling the average distance corresponding to the consumption of one kW with a combustion engine vehicle. Constraint (2) ensures that each reservation $r \in \mathcal{R}$ is either assigned to exactly one EV or to no EV. Constraint (3) ensures that an EV is not charged during time steps in which it is used for a reservation. Furthermore, it ensures that an EV is not assigned to two or more temporarily overlapping reservations, because for time steps in which the reservations overlap, the right-hand side of the constraint would be negative, making the problem infeasible. Constraints (4) and (5) set the battery levels of the EVs after the first and the following time steps, respectively. It is assumed that the energy required by a reservation is consumed in the first time step of the reservation. Together with the lower and upper bounds for the battery levels (8), these constraints ensure that an EV has always a sufficient amount of energy before it is used for a reservation and that EVs cannot be charged higher than technically possible. Constraint (6) ensures that the energy charged in a time step t is consumed from the grid and/or from the available surplus energy.

3   Evolutionary Algorithm

We propose an evolutionary algorithm for the solution of the described EV fleet scheduling problem. More precisely, we apply a hybrid approach, where the assignment of reservations to EVs is optimized with an EA and the charging scheduling is determined via linear programming. Furthermore, a surrogate fitness function is used in order to accelerate the optimization. The following subsections provide a detailed description of the approach.

3.1 Encoding

An individual has to encode the assignment of reservations to EVs. An obvious encoding would be a list of R integer variables $i_1, \ldots, i_R$ between zero and N, where reservation r is assigned to no EV if i_r is zero and to EV i_r, otherwise. However, with this encoding it is hard to ensure that the encoded assignment is feasible. Thus, we use an indirect encoding, where the genotype is a permutation of the numbers $1, \ldots, R$. To compute the phenotype (the actual reservation assignment), the reservations are passed in the encoded order to an insertion operator. The insertion operator computes a (feasible) reservation assignment in the form of a list $A = [A_1, \ldots, A_N]$ of N lists, where the list A_n contains the reservations assigned to EV $n = 1, \ldots, N$. The EA makes use of two insertion operators: *basic insertion* and *random insertion*. The basic insertion is outlined in Algorithm 1. It starts with a list of empty lists and then iterates over the reservations in the order in which they were passed to the operator. For each reservation r it goes through the EVs and inserts r in the list belonging to the first EV for which the insertion does not lead to an infeasible reservation assignment. The assignment of reservations to an EV is infeasible if two or more of the assigned reservations overlap or if it is not possible to satisfy the energy requirements of the reservations with the EV. The latter can be easily determined by checking if the energy requirements are satisfied in the case of uncontrolled charging (i.e., charging the EV in all time steps in which it is not used for a reservation with the maximum possible power until the battery is full). The random insertion operator works similar to the basic insertion operator with the difference that it does not assign a reservation to the first possible EV it finds but to an EV which is randomly selected from all possible EVs.

Algorithm 1: Basic insertion operator.

Input: list P of reservations
Output: reservation assignment A
1 $A = [[\,], \ldots, [\,]]$;
2 **for** r in P **do**
3   **for** $n = 1, \ldots, N$ **do**
4          $A' = A[n] + [r]$;
5   **if** feasible(A') **then**
6              $A[n] = A'$;
7   break; // go to next reservation
8   **end**
9   **end**
10 **end**
11 **return** A;

It is obvious that the order in which the reservations are passed to the insertion operators has a big influence on the resulting reservation assignment. Reservations at the beginning of the list have a high chance of being inserted, while reservations at the end of the list often cannot be inserted since they overlap with already inserted reservations or lead to infeasible energy requirements. The basic insertion operator is the main insertion operator used in the EA, since it is deterministic. The random insertion operator is only used as part of the initialization process (see Sect. 3.2). As already stated, the used encoding in form of permutations, which are passed to the basic insertion operator in order to compute the phenotype, has the advantage that an individual always encodes a feasible reservation assignment. However, it has also a disadvantage: It cannot encode all possible feasible reservation assignments. For example, if there are two EVs and three reservations $r1$, $r2$, $r3$, which can be all assigned to the first EV, then the basic insertion operator will always assign the reservations to the first EV, no matter in which order they are passed. However, assigning one or more of the reservations to the second EV or to no EV at all might yield a better objective value. But as we will see from the experimental results, the encoding's advantage seems to outweigh its disadvantage.

3.2   Initialization

As already described in the previous subsection, reservations at the beginning of the list, which is passed to the insertion operator, have a higher chance of being inserted than reservations at the end of the list. Thus, it is preferable to have promising reservations at the beginning of the list. Reservations with high energy requirements and low durations can be considered to be promising. They contribute much to the reduction of the first term of the objective function while retaining a high flexibility in the charging schedule and/or in the insertion of further reservations. Hence, in the initialization of the EA's population, we first sort the reservations in decreasing order of required energy per time step of duration. The resulting list is then used to initialize the individuals as exemplary illustrated in Fig. 1 for a population of two individuals. For each individual, the sorted list of reservations is passed to the random insertion operator. The resulting reservation assignment is then flattened. The resulting flattened list is then filled up with the reservations that are not already in the list in decreasing order of energy requirement per time step of duration. However, in the fill-up, not all reservations are considered but only reservations which were inserted for at least one individual. In the example, reservation 6 is not considered in the individuals. This means, the optimization will never assign this reservation to an EV. It can be assumed that reservations which were not inserted at least once are likely to be also not assigned to an EV in the optimal solution.

Fig. 1. Example of the initialization of a population of two individuals Ind_1 and Ind_2.

By excluding them from the optimization (and considering them as unassigned from the beginning), a notable reduction of the search space can be achieved. For the sake of simplicity, we assume in the rest of the paper that always the whole set $\mathcal{R}$ of reservations is considered in the individuals.

3.3 Fitness Evaluation

In order to evaluate an individual, the encoded permutation of reservations is passed to the basic insertion operator as described in Sect. 3.1. Given the resulting reservation assignment, the charging of the EVs is optimized with respect to the second and third term of the objective function (1). Optimizing the charging for a fixed reservation assignment is a purely continuous problem, which can be efficiently solved with linear programming. The resulting charging schedule and the reservation assignment are then used to compute the overall objective.

3.4 Crossover

The crossover operator produces an offspring individual from two parent individuals. There are different standard crossover operators for permutations, like OX, PMX or alternating position crossover, which are popular for applications like the traveling salesman problem [6]. However, in preliminary experiments we noticed that such operators do not work well for the given EV fleet scheduling problem. Instead, we use the following approach for the crossover: We select four random integers s_1, e_1, s_2, e_2 with $1 \leq s_1 < e_1 < s_2 < e_2 \leq R$. The reservations between the positions s_1 and e_1 in the first parent are copied into the offspring. Then, the reservations between the positions s_2 and e_2 in the second parent, which are not already in the offspring, are copied into the offspring. Finally, the gaps in the offspring are filled up with the reservations which are not already in the offspring in decreasing order of energy requirement per time step of duration. This is illustrated in Fig. 2, where it is assumed that the energy requirement per time step of duration of a reservation r_1 is greater than or equal to that of an reservation r_2 iff $r_1 < r_2$. First, the reservations 4, 6, and 5 are copied from the first parent into the offspring. Then reservation 10 is copied from the second parent into the offspring. Reservation 6 is not copied from the second parent since

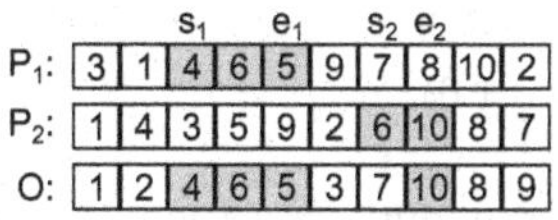

			s_1		e_1		s_2	e_2		
P_1:	3	1	4	6	5	9	7	8	10	2
P_2:	1	4	3	5	9	2	6	10	8	7
O:	1	2	4	6	5	3	7	10	8	9

Fig. 2. Example for the crossover of two parents P_1, P_2 to an offspring O. It is assumed that the energy requirement per time step of duration of a reservation r_1 is greater than or equal to that of an reservation r_2 iff $r_1 < r_2$.

it is already in the offspring. Then the first so far unset position of the offspring is set to reservation 1, since from the reservations, which are not already in the offspring, this is the reservation with the highest energy requirement per time step of duration. Finally, analogously, the remaining unset positions are set to reservations 2, 3, 7, 8, and 9.

3.5   Mutation

As mutation, two basic operations are applied: An exchange of two random reservations in the permutation and a shift of a random reservation to another position in the permutation. The complete mutation operator is outlined in Algorithm 2. With a probability of p_{swap}, reservations are swapped and otherwise reservations are shifted. If reservations are swapped, the number of swaps is determined randomly. A certain minimum number $n_{\text{swap\_start}}$ of swaps is always executed. With probability $p_{\text{n\_swap}}$, $n_{\text{swap\_start}} + 1$ or more swaps are executed, with probability $p_{\text{n\_swap}}^2$, $n_{\text{swap\_start}} + 2$ or more swaps are executed, and so on. Analogously, the number of shifts is randomly determined based on parameters $n_{\text{shift\_start}}$ and $p_{\text{n\_shift}}$.

Algorithm 2: Mutation operator.

Input: list P of reservations
Output: modified list P' of reservations

1 $P' = P$;
2 $p = \text{uniform\_rand}(0.0, 1.0)$;
3 **if** $p < p_{\text{swap}}$ **then**
4 $\quad$ $n_{\text{swap}} = n_{\text{swap\_start}}$;
5 $\quad$ $p = \text{uniform\_rand}(0.0, 1.0)$;
6 $\quad$ **while** $p < p_{\text{n\_swap}}$ **do**
7 $\quad\quad$ $n_{\text{swap}} = n_{\text{swap}} + 1$;
8 $\quad\quad$ $p = \text{uniform\_rand}(0.0, 1.0)$;
9 $\quad$ **end**
10 $\quad$ perform n_{swap} swaps of random pairs of reservations in P';
11 **else**
12 $\quad$ $n_{\text{shift}} = n_{\text{shift\_start}}$;
13 $\quad$ $p = \text{uniform\_rand}(0.0, 1.0)$;
14 $\quad$ **while** $p < p_{\text{n\_shift}}$ **do**
15 $\quad\quad$ $n_{\text{shift}} = n_{\text{shift}} + 1$;
16 $\quad\quad$ $p = \text{uniform\_rand}(0.0, 1.0)$;
17 $\quad$ **end**
18 $\quad$ perform n_{shifts} shifts of random reservations to random positions in P';
19 **end**
20 **return** P';

3.6   Optimization Process

We apply a steady-state generational scheme, where in each generation one off-spring is generated and is considered for insertion in the population. With a certain crossover probability pc, the offspring is produced by crossover of two parent individuals and with probability $1 - pc$ the offspring is produced by copying a parent individual. Parent individuals are selected with binary tournament selection without replacement. The offspring is then mutated. If one or both of the parameters $n_{\text{swap\_start}}$ and $n_{\text{shift\_start}}$ of the mutation operator are set to zero, it might happen that the offspring is not changed by the mutation operator. Hence, if an offspring was produced by copying a parent, we repeat its mutation until at least one swap or shift was executed. After the mutated offspring is evaluated, it replaces the so far worst individual in the population, if the offspring is better and if there is not already another individual with the same fitness in the population (to increase the diversity in the population).

3.7   Surrogate-Assisted Optimization

As outlined in Sect. 3.3, in the fitness evaluation the EV charging is optimized. Although this optimization can be done very efficiently with the help of linear programming, the fitness evaluation is responsible for a large fraction of the runtime of the whole optimization process. We apply a surrogate model – i.e., a fast approximation of the fitness function to accelerate the optimization. Popular surrogate models used in the context of evolutionary optimization are data-driven models like the Kriging model, radial basis functions or support vector machines [3]. However, in our case we do not have to rely on a data-driven approach. Instead, we can use a fast heuristic to set the charging powers and use the objective with the resulting charging schedule as a surrogate for the real fitness with globally optimal charging scheduling. We apply the following simple heuristic: We assume that the EVs are charged uncontrolled, i.e. with maximum possible power, when they are not used for reservations and are not fully charged already. The objective value with uncontrolled charging should be already a reasonable indicator for the real objective value with optimized charging. More precisely, it provides an upper bound for the real objective value.

There are different options for integrating the surrogate in the evolutionary optimization. Popular strategies are the generation-based strategy and the individual-based strategy [5]. In the generation-based strategy, the surrogate is used in some generations and in the rest of the generations, the real fitness function is used. In the individual-based strategy, in each generation it is determined with help of the surrogate, which offspring individuals are evaluated with the real fitness function. In the experiments described later, we compare the standard version of the EA without any surrogate (*EA*) with two surrogate-assisted versions (*EA-SI* and *EA-SG*). The *EA-SI* variant uses an individual-based strategy. In each generation, N^{inter} intermediate offspring individuals are generated and are evaluated with the surrogate fitness function. Only the best of these individuals (in terms of the surrogate fitness) is then evaluated with the real fitness function and is considered for insertion in the population. In the *EA-SG* variant, an extreme case of the generation-based strategy is applied: The whole optimization works only on the surrogate fitness and only the best individual (in terms of the surrogate fitness) of the final population is evaluated with the real fitness.

4   Experiments

4.1   Experimental Setup

We executed the experiments on a compute cluster. Each process was run on a separate node with an Intel(R) Xeon(R) E5-2623@3.00GHz 8-core CPU and 64 GB RAM. We use a single-threaded C/C++ implementation of the evolutionary algorithm. We compare the EA to a standard MILP approach, which solves the complete model (1–11), and to the improved MILP approach proposed in [9]. The improved approach (denoted as BDH) applies a Benders decomposition to split the problem into a master problem and a subproblem, which are iteratively solved in an alternating manner. In early stages of the optimization process, the master problem is solved with a general variable neighborhood search heuristic. For the solution of the subproblem as well as the master problem in later stages, MILP is used. For all MILP optimizations and for the charging optimization in the fitness evaluation of the EA, version 9.1 of the Gurobi solver is used and the number of threads for the solver is set to one. For the BDH approach, we use the same Julia implementation as in [9]. The parameters of the BDH approach are set to the same values used in the experiments in [9]. To achieve a fair comparison between the EA and the MILP approaches, which do not exhibit a number of fitness evaluations, we use a time limit as termination condition for the optimizations.

Depending on the concrete use case, there might be different runtime requirements. Hence, we observe the optimization results after 5 min, 15 min and 1 h of runtime. We evaluate the different approaches on artificially created problem instances from [9][1]. The instances have the naming scheme tmax$T$_nN_rmax$R$_I, where T is the number of time steps of the planning horizon, N is the number of EVs, R is the number of reservations and I is the index of the problem instance between 1 and 30. The length Δt of a time step is assumed to be 15 min. EVs can be charged with a maximum power of 3.3 kW and have a battery capacity of 20 kWh. See [9] for more details to the problem instances. For different problem sizes, i.e. different values of T, N, and R, there are 30 instances per size. In the experiments we use the first instance, i.e. with index $I = 1$, for each problem size with $T = 768$ time steps. These are the largest problem sizes considered in [9]. We are not interested in instances of small size, since these can be efficiently solved with a standard MILP approach. In the experiments we execute 21 optimization trials per problem instance with each EA variant. The setting of the parameters of the EA is discussed in the following subsection.

---

[1] The problem instances are publicly available at https://www.ac.tuwien.ac.at/research/problem-instances/#evfcap.

4.2 Parameter Setting and Analysis

Based on preliminary experimental results, we set the population size to 100 and the number N^{inter} of intermediate offspring per generation in the *EA-SI* variant to 20. In order to investigate the influence of different settings of the crossover rate pc and of the parameters of the mutation operator on the optimization performance, we drew 2200 random settings of these parameters and performed five optimization trials on the problem instance tmax768_n020_rmax0320_30 with each of these parameter settings. Please note that this problem instance is not part of the evaluation instances used in the experiments described later. The optimizations were done with the standard *EA* variant with a time limit of 1 min per trial. The minimum and maximum of the considered parameter ranges can be seen in Table 1. The ranges of continuous parameters are equidistantly discretized with a step size of 0.1. For each of the random parameter settings we computed the mean objective value over the five trials. The lowest mean objective value was obtained with the parameter setting show in Table 1. It yielded a mean objective value of 206,553.86. We applied this parameter setting in all EA variants in the subsequent experiments. The distribution of the mean objective over all evaluated parameter settings can be seen in Fig. 3(a). There are two peaks, one around 208,000 and one around 214,000. The first arises from parameter settings with a crossover rate $pc > 0$ and the second from parameter settings with $pc = 0$. This can be seen from the boxplot of the mean objective with different values for pc in Fig. 3(b). Thus, the crossover has a clearly positive effect on the optimization. The best results are achieved with crossover rates between 0.3 and 0.5. The probability p_{swap} is zero in the best found parameter

Table 1. Parameter ranges considered in the sampling and parameter setting yielding the lowest mean objective value.

Parameter	pc	n_{swap_start}	n_{shift_start}	p_{n_swap}	p_{n_shift}	p_{swap}
Range (min,max)	0.1, 0.9	0, 4	0, 4	0.2, 0.8	0.2, 0.8	0.1, 1.0
Setting	0.3	2	0	0.3	0.4	0.0

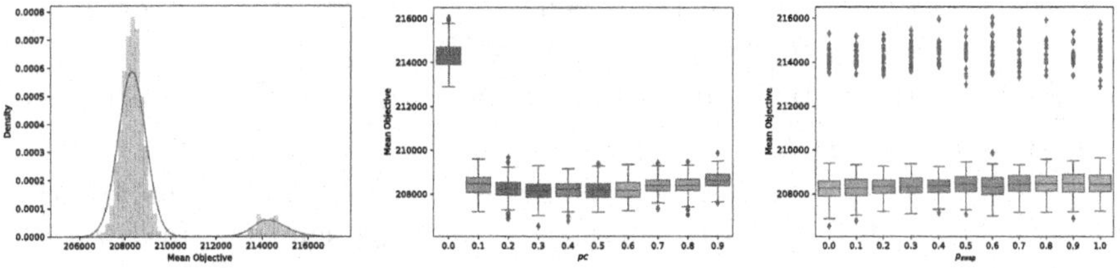

(a)   (b)   (c)

Fig. 3. (a) Distribution of mean objective with evaluated parameter settings. (b) Box-plot of mean objective values with different settings for the crossover rate pc. (c) Box-plot of mean objective value with different settings for the probability p_{swap}.

setting. That means, only shifts and no swaps are performed in the mutation operator. However, in the results of the parameter tuning there is no clear trend towards any best setting for p_{swap} as can be seen from the boxplot in Fig. 3(c). The same holds for the other parameters of the mutation operator. Thus, the EA seems not very sensitive regarding the setting of these parameters.

4.3   Experimental Results

Table 2 shows the objective values achieved with the standard MILP approach (*MILP*), the improved MILP approach from [9] (*BDH*) and the EA variants (*EA*, *EA-SI*, and *EA-SG*) on the different problem instances after a runtime of 5 min, 15 min, and 60 min.

Table 2. Objective values obtained by the different approaches on different problem instances after 5 min, 15 min, and 60 min runtime. The results of the EA variants are averages over 21 trials.

Time Limit [min]	*MILP*	*BDH*	*EA*	*EA-SI*	*EA-SG*
tmax768_n020_rmax0080_01					
5	**11,365.78**	11,580.85	11,445.63[2,3]	11,521.20[3]	11,990.64
15	**11,358.29**	11,417.99	11,402.05[2,3]	11,480.91[3]	12,003.68
60	**11,358.29**	11,396.97	11,381.07[2,3]	11,441.68[3]	12,009.36
tmax768_n020_rmax0160_01					
5	55,066.20	45,734.72	44,794.47	**43,685.21**[1]	43,749.39[1]
15	50,240.80	44,684.07	43,964.40	**43,414.14**[1,3]	43,627.42[1]
60	43,716.90	44,462.71	43,495.71	**43,192.74**[1,3]	43,568.40
tmax768_n020_rmax0320_01					
5	N/A	215,110.85	208,850.95	205,906.86[1]	**205,210.12**[1,2]
15	223,243.76	210,163.65	206,598.62	205,109.22[1]	**204,693.99**[1]
60	217,205.15	207,150.76	205,208.35	**204,219.44**[1]	204,301.36[1]
tmax768_n050_rmax0200_01					
5	N/A	30,193.53	**28,007.51**[2,3]	28,138.48[3]	28,762.28
15	**27,222.70**	30,131.23	27,778.85[2,3]	27,967.79[3]	28,735.96
60	**27,069.16**	27,913.11	27,506.53[2,3]	27,727.19[3]	28,789.72
tmax768_n050_rmax0400_01					
5	N/A	N/A	113,234.86	110,101.28[1]	**109,974.88**[1]
15	N/A	110,061.76	111,412.08	**108,826.50**[1]	109,407.65[1]
60	N/A	110,028.34	109,559.97	**107,634.57**[1,3]	108,990.14[1]
tmax768_n050_rmax0800_01					
5	N/A	N/A	514,038.25	508,647.80[1]	**506,261.09**[1,2]
15	N/A	N/A	510,606.16	505,195.21[1]	**503,542.10**[1,2]
60	N/A	510,663.74	505,904.61	502,171.33[1]	**501,239.37**[1,2]

(*continued*)

Table 2. (*continued*)

Time Limit [min]	*MILP*	*BDH*	*EA*	*EA-SI*	*EA-SG*
tmax768_n100_rmax0400_01					
5	N/A	N/A	56,426.29	**56,208.45**[1,3]	56,451.73
15	N/A	59,518.77	55,974.64	**55,765.65**[1,3]	56,314.65
60	N/A	57,847.22	55,394.08[3]	**55,150.19**[1,3]	56,296.16
tmax768_n100_rmax0800_01					
5	N/A	N/A	242,928.36	237,856.31[1]	**235,740.16**[1,2]
15	N/A	N/A	239,679.03	234,452.57[1]	**233,053.13**[1,2]
60	N/A	232,561.19	236,538.11	230,798.55[1]	**230,196.62**[1]
tmax768_n100_rmax1600_01					
5	N/A	N/A	1,014,829.75	1,009,492.52[1]	**1,005,598.19**[1,2]
15	N/A	N/A	1,010,564.52	1,004,557.84[1]	**1,001,765.96**[1,2]
60	N/A	N/A	1,005,903.36	998,538.39[1]	**996,443.36**[1,2]

For the EA variants, the shown objectives are averages over the 21 trials per problem instance. The best results per time limit and problem instance are highlighted in bold. An "N/A" denotes that no feasible solution was found. The superscripts 1, 2, and 3 at the results of the EA variants indicate that the results are statistically significantly lower (better) than the corresponding results of the *EA*, *EA-SI*, and *EA-SG* variant, respectively. This was determined with pairwise Wilcoxon rank sum tests with a significance level of 0.05.

On the smallest problem instance, the MILP approach performs very well. In 5 min, it achieves an objective value, which is not achieved by the EA variants in 60 min. On the instance with 50 EVs and 200 reservations, the MILP is not able to find a feasible solution in 5 min, but it yields better results than the EA variants after 15 min and 60 min. On the other instances, the MILP approach is outperformed by the EA. On the five largest instances the MILP approach is not even able to find a feasible solution within 60 min[2]. The BDH approach scales better than the standard MILP approach and finds feasible solutions within one hour for all problem instances except the largest one. However, in most cases it is also outperformed by the EA variants.

On most of the problem instances, the standard EA variant is outperformed by at least one of the surrogate-assisted variants. With the *EA-SG* variant, the results might get worse with a higher runtime since it uses only the surrogate

---

[2] Please note that we use a slightly different MILP problem formulation than [9] (we use helper variables for the battery levels), since we noticed that this yields a better performance. With the formulation from [9] the MILP approach is able to find feasible solutions for the larger instances within 60 min, but only the trivial solutions, where no reservation is assigned to an EV.

fitness to determine the quality of a solution. One can see a certain trend in the results: On instances with a high total number of reservations or a high number of reservations per EV, the *EA-SG* variant performs best. On instances with a low total number of reservations or a low number of reservations per EV, the standard EA variant yields better results than the other EA variants. However, on these small instances, the MILP approach is also still efficient. On instances with a medium total number of reservations or a medium number of reservations per EV, the *EA-SI* variant performs best. An explanation for this can be derived from Table 3, which shows detailed results on three problem instances computed with the three EA variants with a runtime of one hour. The table lists the three parts of the objective separately (cost for reservations not assigned to EVs, energy cost, and cost for missing energy in the batteries at the end of the planning horizon), the number of reservations assigned to EVs and the number of executed generations. The results are averages over the 21 trials. One can see that with an increasing number of reservations per EV, the rate of unassigned reservations increases and thus the cost for unassigned reservations contributes more to the objective value. These cost are computed exactly in the surrogate fitness and thus the surrogate becomes more accurate with an increasing number of reservations per EV. Furthermore, the larger the instance, the lower the number of generations, which can be executed and thus the higher the benefit from using a surrogate. This explains why the standard EA is better on smaller instances while the surrogate-assisted variants are better on the larger instances.

Table 3. Detailed results on three problem instances after a runtime of one hour. The results are averages over 21 trials.

EA Variant	Unserved Cost	Energy Cost	Battery Cost	#Served	Generations
tmax768_n020_rmax0080_01					
EA	62.3	11,318.8	0.0	78.0	33,956.7
EA-SI	62.3	11,379.4	0.0	78.0	51,032.6
EA-SG	62.3	11,947.1	0.0	78.0	4,445,403.3
tmax768_n050_rmax0400_01					
EA	41,456.8	65,164.8	2938.4	329.1	20,585.3
EA-SI	39,454.3	65,312.1	2868.2	333.6	16,440.7
EA-SG	38,717.8	67,376.4	2896.0	336.3	476,646.6
tmax768_n100_rmax1600_01					
EA	790,889.7	198,101.4	16,912.3	887.9	6,516.4
EA-SI	782,347.2	200,970.6	15,220.6	897.4	2,828.3
EA-SG	779,589.8	201,540.0	15,313.6	897.8	72,332.8

One can see in Table 3 that on the instance with 20 EVs and 80 reservations, more generations are executed with the *EA-SI* variant than with the *EA* variant. This appears counter-intuitive. However, we found that in the *EA-SI* variant the (real) fitness evaluation tends to be faster than in the *EA* variant. This is probably because the *EA-SI* variant tends to evaluate individuals with a higher number of assigned reservations, which makes the charging optimization easier since there are less time steps in which the EVs can charge. At the same time, the computation of the surrogate fitness costs nearly no time on the smaller problem instances. Figure 4 shows for the different EA variants the average and standard deviation of the optimization progress of the 21 trials on the different problem instances. From this one can see again the advantage of the surrogate-assisted variants on the larger instances and the poor performance of the *EA-SG* variant on instances with a small number of reservations per EV. There is also a comparatively high variance in the results of *EA-SG* on such instances.

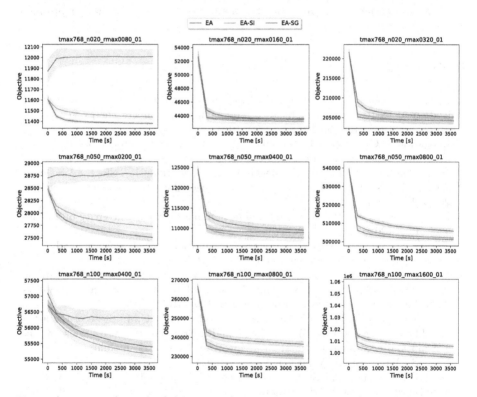

Fig. 4. Average and standard deviation of optimization progress of the EA variants on different problem instances.

5 Summary and Conclusion

We proposed a hybrid evolutionary algorithm (EA) for the electric vehicle fleet scheduling problem, which employs an indirect encoding in order to improve the handling of constraints. Furthermore, a problem-specific crossover operator is used. An analysis of the influence of the parameters of the EA revealed that this crossover is clearly beneficial for the optimization. We further propose the use of a problem-specific surrogate fitness. In experiments we evaluated two variants, *EA-SI* and *EA-SG*, of the surrogate-assisted EA and compared it to the standard EA and to two MILP-based approaches. The MILP approaches are clearly outperformed by the EA variants on the considered problem instances. Furthermore, the experimental results show that the use of the surrogate fitness is beneficial. There is no clear winner among the *EA-SI* and *EA-SG* variants, but the *EA-SI* variant appears to be a good compromise between the standard EA without surrogate and the completely surrogate-based *EA-SG* variant – it yields a reasonable performance on the smaller as well as on the larger problem instances.

References

1. Betz, J., Werner, D., Lienkamp, M.: Fleet disposition modeling to maximize utilization of battery electric vehicles in companies with on-site energy generation. Transport. Res. Procedia **19**, 241–257 (2016)
2. Bongiovanni, C., Kaspi, M., Geroliminis, N.: The electric autonomous dial-a-ride problem. Transport. Res. Part B: Methodol. **122**, 436–456 (2019)
3. Díaz-Manríquez, A., Toscano Pulido, G., Barron-Zambrano, J., Tello, E.: A review of surrogate assisted multiobjective evolutionary algorithms. Comput. Intell. Neurosci. **2016**, 1–14 (2016)
4. Haveman, S., et al.: eMaaS project public summary report. University of Twente, Tech. Rep. (2020)
5. Jin, Y.: Surrogate-assisted evolutionary computation: recent advances and future challenges. Swarm Evol. Comput. **1**(2), 61–70 (2011)
6. Larrañaga, P., Kuijpers, C., Murga, R., Inza, I., Dizdarevic, S.: Genetic algorithms for the travelling salesman problem: A review of representations and operators. Artif. Intell. Rev.: Int. Surv. Tutor. J. **13**(2), 129–170 (1999)
7. Reyes García, J.R., Lenz, G., Haveman, S.P., Bonnema, G.M.: State of the art of mobility as a service (MaaS) ecosystems and architectures - An overview of, and a definition, ecosystem and system architecture for electric mobility as a service (eMaaS). World Electr. Vehicle J. **11**(1), 7 (2020)
8. Sassi, O., Oulamara, A.: Electric vehicle scheduling and optimal charging problem: complexity, exact and heuristic approaches. Int. J. Prod. Res. **55**(2), 519–535 (2017)
9. Varga, J., Raidl, G.R., Limmer, S.: Computational methods for scheduling the charging and assignment of an on-site shared electric vehicle fleet. IEEE Access **10**, 105786–105806 (2022)

The Specialized Threat Evaluation and Weapon Target Assignment Problem: Genetic Algorithm Optimization and ILP Model Solution

Ahmet Burak Baraklı[1,2](✉), Fatih Semiz[1], and Emre Atasoy[1]

[1] Aselsan Inc., Ankara, Turkey
{abarakli,fatihsemiz,eatasoy}@aselsan.com.tr
[2] Middle East Technical University, Ankara, Turkey
ahmet.barakli@metu.edu.tr

Abstract. In this study, we developed an algorithm that provides automatic protection against swarm drones by using directional jammers in anti-drone systems. Directional jammers are a special type of jammers that can be rotated to a certain angle and do jamming around only that angle. This feature is useful for jamming particular targets and not jamming areas where it is not desired. We worked on a specialized version of the threat evaluation and weapon allocation (TEWA) problem, in which weapons (jammers for this problem) should be assigned to angles to cover threats at a maximum rate by satisfying their priorities. In this problem, it is aimed at keeping the threat within jamming signals at the maximum rate by turning the directional jammers to appropriate angles, taking into account the threat priorities. We have presented an algorithm that solves this problem by meeting the physical constraints of the jammers and tactical constraints defined by the user. The solutions created include: using as few jammers as possible, minimizing the angle changes jammers make in each new plan, prioritizing threats according to their characteristics (type, direction, speed, and distance), and preventing jammers from returning to the physical constraints defined for them. We solved the threat evaluation problem with the help of genetic programming and the jammer angle assignment problem by transforming it into an integer linear programming (ILP) formulation. We also handled physical constraints unsuitable for ILP formulation with post-processing. Since there are few studies directly dealing with this version of the problem, we compared our study with the study that was claimed to be the first to solve this particular version of this problem. Furthermore, we compared our study with the different versions of the algorithm we created. Experiments have shown that threat coverage percentage is vastly increased, achieving this without a significant drop in problem-solving speed.

Keywords: Threat Evaluation and Weapon Allocation (TEWA) ·
Integer Linear Programming (ILP) · Set Covering Problem · Genetic
Algorithm

J. Correia et al. (Eds.): EvoApplications 2023, LNCS 13989, pp. 19–34, 2023.
https://doi.org/10.1007/978-3-031-30229-9_2

1   Introduction

Anti-Drone systems aim to protect a certain area against unmanned aerial vehicles. These systems generally include detecting, tracking, identifying, engaging
with, and evaluating threats [17,18]. Anti-Drone systems perform two different
methods, soft-kill and hard-kill, to neutralize threats. Soft-kill mechanisms, also
called electronic countermeasures (ECM), are applied mainly in urban environments where the civilian population is dense. The most common soft-kill method
is to perform jamming with jammers [21]. In anti-drone systems, the automatic
assignment of weapons to threats and automatic neutralization of threats with
hard-kill or soft-kill methods are done with command and control (C2) applications. The weapon-target assignment problem (WTA) [9] is an optimization
problem that is frequently studied in C2 automation and is frequently seen in
military applications [2]. The WTA problem is an NP-Hard problem in terms of
complexity [12]. When the weapon assignment is performed taking the priorities
of the threats into account, that variation is called the threat evaluation and
weapon assignment (TEWA) problem. This study proposes new threat evaluation and weapon assignment methods on a customized version of TEWA for
soft-kill mechanisms. In this version, multiple jammers are assigned to specific
angles to maximize threat coverage. To solve this problem, we formulate the
problem as a static TEWA problem. By solving a series of TEWA problems, we
aim to provide continuous automatic protection of determined areas. In Fig. 1
an example problem setting is presented. In this setting, the circles represent
the jammers, the green areas represent jammer coverages, and the drone illustrations represent the threats. The threats are colored according to their types;
red for drones, orange for suspected drones, yellow for important, and gray for
insignificant. The arrows show the directions of the threats; we have only drawn
the covered threats for simplicity.

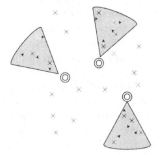

Fig. 1. An Example Jammer-Threat assignment problem. (Color figure online)

2   Related Work

Researchers have done extensive studies on WTA problems. WTA problem has two subcategories the Static WTA (SWTA) [10] and the Dynamic WTA (DWTA) [3,4] problems. Since new exact solution methods are presented in this article, we have reviewed the exact methods in the literature. Among the exact algorithms proposed for solving the WTA problem, there are algorithms such as Maximum Marginal Return (MMR) [23], exhaustive search [20], branch and bound algorithm approaches [1,14], linear integer programming techniques [15,19] and genetic algorithm [11]. Threat Evaluation (TE) and Weapon Assignment (WA), together the TEWA problem is a variation of the weapon-target assignment problem considering threat priorities [7,16]. For the threat evaluation part, Johansson and Falkman in 2008 [8] stated in their study that there are three categories of a threat that can be evaluated, proximity, capability, and intent. These parameters are used to measure the importance of a threat to be able to take precautions to defend against that threat. They also provided a Bayesian network-based algorithm to make threat evaluations. Lötter et al. [13] in 2014 reviewed the studies about the threat evaluation aspect of the problem. The early attempts include human operators that perform threat evaluation for the systems. Most preceding studies use rule-based algorithms, formulations, or mathematical calculations. In this study, we worked on the jammer-threat assignment problem, which can be summarized as a problem of maximizing the amount of threat coverage, taking into account the threat priority situations. This problem can be seen as a specialized version of the TEWA problem [6]. Since the weapon to be optimized in this specialized version is the jammer, it is necessary to work on maximum angle coverage. This problem can be expressed with the set covering formulation or ILP formulation in the literature [5]. Although set-covering formulations are not frequently encountered in TEWA problems in the literature, this formulation is frequently used for different problems [24]. In this problem, we used the genetic algorithm solution in the threat evaluation. While there are studies that use a genetic algorithm in the TEWA literature, they generally use this in the weapon assignment part of the problem [22,25]. In our case, once a system is deployed into an area, it is likely to be used for a long time. We can run a threat evaluation technique once and just use those parameters for evaluating threat weights after deployment. Because of this situation, we have the chance to use a more complex and time-consuming technique.

3   Problem Description

In this problem, the aim is to create a solution that will cover the maximum amount of threats, taking into account the weight of the threats. An instance of the jammer-threat assignment problem consists of N jammers and M threats. Each threat has its own attribute values, such as threat type, speed, distance, and information about whether the threat is approaching or moving away. For each threat i, a weight that indicates how important the corresponding threat is, w_i, is calculated based on the values of these attributes. A genetic algorithm method is used to optimize how much these features affect the weights. The information of how many degrees (azimuth for this problem since we model the system as 2D) each jammer can turn is given by the set g. For this problem it is predetermined as $g = \{5°, 10°, 15° \ldots 360°\}$. The processing power and the running time of each instance were effective in determining the increase in angles by five by five. Since the coverage angle of the jammer is $30°$, we determined the five-by-five increment as a value that works fast enough and tries enough possibilities.

K represents the number of angles at which a jammer is forbidden to rotate. The set of angles at which the jammer j is prohibited from turning is denoted by C_j, which represents one or more forbidden zones. $C_j = \{c_{j1}, c_{j2}, c_{j3} \ldots c_{jK}\}$ where $(c_{ji} \in g, 1 \leq i \leq K)$. Let p be the angle a jammer is facing. The jammer will cover threats in the $[p - \gamma, p + \gamma]$ range, where γ is the range of the jammer. The maximum distance the jammer can cover is also a feature of the device and is determined at the beginning of the problem. The information on whether a jammer is used or not is decided on a list named y. The y_j value for the jammer j is set to 1 if the jammer is used in that solution and 0 if the jammer is not used. Information about whether a jammer covers a threat at some angle is kept in the a list. If a jammer j covers the threat of i at the angle of g, the value a_{jgi} is set to 1, otherwise a_{jgi} is set to 0.

There are three decision variables for this ILP model. The first is the value x_{jg}; it checks if the jammer j is looking at the angle g. This value is used to enforce the constraint that a jammer can only rotate one angle at a time. Another decision variable is z. The value z_i checks whether the i threat is covered. This value is desired to be maximized in this problem with the coefficients of w. The last decision variable in the problem is d. This value holds the difference in angle change from the previous solution for each jammer. In this ILP model, it is aimed to minimize this value.

4   Method

The developed approach consists of three main steps:

1. Determination of the parameters of threat features by genetic algorithm (threat evaluation).
2. Finding the recommended angles to rotate the jammers by means of the mathematical model.
3. Arranging the rotation of the jammers from their previous angles to their next angles according to physical and tactical constraints and making the angle assignment proposal over the newly arranged angles.

The main flow of the proposed approach is presented in Fig. 2.

4.1   Threat Evaluation with Genetic Algorithm

In this section, the operations performed with the genetic algorithm to create the list of threat weights to be used in the mathematical model will be explained.

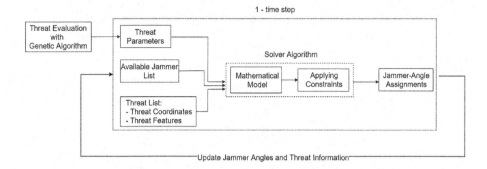

Fig. 2. Jammer-Threat assignment algorithm working scheme.

In our workflow, the genetic algorithm does not need to be run every time the jammer-threat assignment approach is run. It can be run with a certain data set as a preliminary calculation at desired times and used to create an optimized weight sequence. When not running, the mathematical model will work with pre-computed threat weights or the threat weights obtained with user input. In this section, an appropriate weight is determined for each threat using the information we have about the threats. The weight calculated for each threat indicates the priority of that threat.

While designing what parameters to use, we wanted to have some parameters in each category that Johansson and Falkman [8] used to classify the parameters. So we included proximity parameters (threat distance, threat speed), capability parameters (threat type), and intent parameters (threat direction). The first decision we make to create a genetic algorithm structure is to decide what information will be kept in each individual as a gene to create an individual. The first idea that emerged was that all the parameters we determined were included in individuals as genes. However, since it was predicted that this would slow down the computation when the number of individuals increased, instead of keeping a gene for each parameter, we could obtain the same information with a smaller array by keeping a gene for the ratios of both parameters to each other. So in the end we had individuals with 9 real number value genes. As seen in Fig. 3, when we moved from the first design to the second design, the number of genes in an individual decreased from 13 to 9. This drop will provide a serious acceleration (about 30%) during calculations as it requires fewer parameters to calculate.

In the general operation of our genetic algorithm structure, when each new generation is produced, after crossover and mutations, a mathematical model is run for each individual forming the population, and the UAV coverage ratio found is assigned as the fitness value of that individual. Afterward, a new generation is created according to the fitness values, and this structure continues until the stop condition is met. In our study, we followed the strategy of continuing until there was no improvement in the coverage ratios, which was the stop condition for our algorithm.

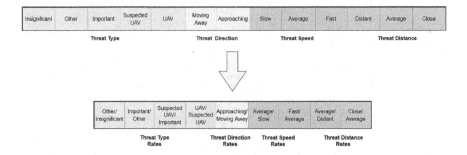

Fig. 3. Design of individuals in genetic algorithm.

4.2   Mathematical Model

The problem to be solved here can be seen as a set covering problem since the coverage process will be done over the scanning angles of the jammers. In 2018, Kıvanç Gül [6] modeled this problem mathematically with integer linear programming (ILP). We took that model further to make it better describe our problem design.

$$max \left(\sum_{i=1}^{M} w_i z_i - \epsilon \sum_{j=1}^{N} y_j - \epsilon \sum_{k=0}^{K} d_k \right) \tag{1}$$

$$\sum_{g=5}^{360} x_{jg} \leq y_j \quad \forall\, j, g = \{5°, 10°, 15° \ldots 360°\} - C_j \tag{2}$$

$$\sum_{j=1}^{N} \sum_{g=5}^{360} a_{jgi} x_{jg} \geq z_i \quad \forall\, j, g = \{5°, 10°, 15° \ldots 360°\} - C_j \tag{3}$$

$$x_{jg}, y_j, z_i \in \{0,1\} \quad d_j \in [0°, 360°] \tag{4}$$

In the integer linear programming model, Eq. 1 shows the model's objective function. Here, the weight of the track i based on the genetic algorithm module is represented by w_i. The amount of threats covered in this equation is maximized by taking their weights into account. The second sum symbol shows the process of minimizing the number of jammers used, and the last sum symbol minimizes the angle change of the jammers in each problem solution. Thus, when sufficient coverage is provided, it is ensured that jammers that do not need to be used are turned off. The devices on which the jammers are located may wear out by changing the angle in each calculation. Thus, providing as much coverage as possible with as little angle change as possible is an important constraint for the problem.

In Eq. 2, the constraint for each jammer, if used, must be facing only one angle is defined to the system. From all these possible angles, the angles in the subset formed by subtracting the angles determined as physical and tactical constraints on that jammer were input into the mathematical model as possible angles that jammer could turn. Equation 3 defines the restriction: for each threat, if that threat is covered, at least one jammer must be facing the angle at which that threat is found (provided that the jammer does not rotate to angles where it is prohibited to rotate). Equation 4 shows that the decision variables x_{jg}, y_j, z_i can take binary values and that the decision variable d_j can take values in the range of $0°, 360°$. We used IBM CPLEX library to solve this ILP model.

4.3   Application of Physical and Tactical Constraints on Jammer Angle Changes

In this section, we talked about the changes we made to adapt the solutions produced by the mathematical model to the physical and tactical characteristics of the device that will perform the rotation. Before we talk about these changes, we want to talk about the physical and tactical constraints of the problem. There are 3 constraints in this problem:

1. The jammer can rotate at most $\alpha°$ in a one-time step (per second). This restriction must be satisfied when assigning angles to jammers.
2. Results should be produced in such a way that the jammer is not rotated to angles where it is not physically possible to rotate (the jammer cannot turn to that angle and do jamming, and it will not be able to pass instantly over the angle where the physical constraint is).
3. If there is a tactical restriction, the jammer will not be able to turn to the angle with that restriction and do jamming. However, it can use those angles momentarily (for rotational movement) without jamming.

Ensuring the Application of the Maximum Angle of Rotation Constraint. There is a decision variable to minimize the maximum angle change in the mathematical model. In Eq. 1, the d variable in this formula is the angle that is used to minimize the change and has as much weight as the coefficient in front of it. In fact, thanks to this variable, it is expected that the mathematical model will not result in angle changes greater than the maximum rotational speed of the jammer. Suppose the weight coefficient of the decision variable is more than a certain threshold. In that case, the model will give more importance to angle minimization while optimizing, and the jammers will not want to move. For this reason, it is possible to determine only a certain amount of weight, which in some cases causes results that include more angle changes than the desired angle change.

In addition, although this formulation provides a reduction in the angle change, it does not set a limit to this change. In such cases, if an angle change exceeds the maximum rotational speed at that time step, we restrict the angle change to the determined maximum angle change by doing an extra operation on the solution. If the same assignment continues in the next time step, the jammer will complete this change in several time steps by complying with the rotation speed constraint. The duration of this difference can be increased or decreased by changing the rotational speed of the jammer. If another assignment is made to the jammer while it continues its movement, it will stop the rotation it was continuing to make and start moving toward the new assignment.

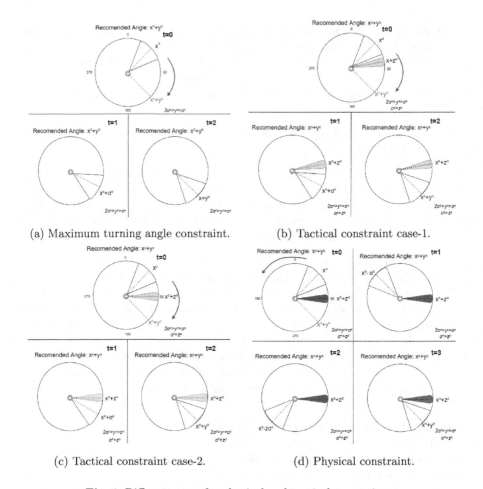

(a) Maximum turning angle constraint.   (b) Tactical constraint case-1.

(c) Tactical constraint case-2.   (d) Physical constraint.

Fig. 4. Different cases for physical and tactical constraints.

A visual showing the working logic of the method developed here is shared in Fig. 4. In this example, a jammer looking at the x angle at $t = 0$ is given. Then a proposal has been made to turn from the mathematical model to the y angle $(2\alpha > y > \alpha)$ at $t = 1$ where α is the maximum turning angle for a time step. Since the maximum angle of rotation of the jammer was α degrees, the jammer was rotated to $x + \alpha$ degrees at $t = 2$ and to $x + y$ degrees at $t = 3$.

In the mathematical model, the tactically restricted angles are subtracted from the angles that the jammer can rotate, which ensures applying this constraint. The angle set determined by the tactical constraint is removed from the list of all angles that a jammer j can rotate $C_j = \{c_{j1}, c_{j2}, c_{j3} \ldots c_{jk}\}(c_{ji} \in g, 1 \leq i \leq K))$. In tactical constraints, it is ensured that the jammer does not look at the final constraint angle but can look at that point momentarily during the turn. Also, when it completes the final turn, it will not look at the specified angles, but its coverage may intersect with tactical areas.

The cases where this intersection is not desired are discussed in the physical constraints. Possible situations that may occur in this design are given in Fig. 4.

In these examples, the jammer will rotate from the x angle to the $x + y$ angle. An area of $\pm 5°$ is given as a tactical constraint, with the $x + z$ angle being the midpoint. In Fig. 4b, the jammer jumps over the tactical area and advances to the turning point. In Fig. 4c, when the jammer makes the first turning move, the tactical area, and the scanning area intersect, then it turns to its final angle. The essential condition here is that the jammer does not turn to the determined tactical restricted angles. In this case, even if the coverage created from the angle of the jammer's point of view intersects with the restricted tactical areas, this will not pose a problem.

Applying the Physical Constraints of the Jammer. In order to fulfill the physical constraints of the jammer, it is not sufficient that the mathematical model does not assign those angles. Since this restriction indicates a physical feature of the device, the device will not be able to return to physically restricted angles in any case. To solve this problem, we created an algorithm as follows: we calculated which direction we should go to reach the angle of assignment from the mathematical model. When we consider all angles that can be rotated ($360°$), it will be possible to reach the determined angle by moving either clockwise or counterclockwise. The direction with the shortest one of these distances was determined as the direction in which the jammer would be directed. If the physical constraint is not between the jammer and the angle it will rotate, the jammer will be able to rotate to the assigned angle with the specified direction without any extra operation. If it is between the angle that needs to be turned and the jammer, then the jammer will turn in the opposite direction of the determined direction and turn to the angle it should turn. Figure 4d shows the application where the physical constraint is between the jammer and the angle it will rotate. The jammer will rotate from the x angle to the $x + y$ angle. An area of $\pm 5°$ is given as the physical constraint, with the $x + z$ angle being the midpoint. Since the physical constraint of the jammer is in the direction of the angle it will rotate, it has moved in the opposite direction. Algorithm 1 shows the calculations for adjusting the jammers' rotation angle according to the physical constraint.

Algorithm 1. Applying Physical Constraints

1: **function** PROCEDURE UPDATEJAMMERANGLE(suggestedAngle, suggestedDirection, constraintDirection)
2:     **if** suggestedDirection = constraintDirection **then**
3:         suggestedDirection.reverse()
4:     angleDifference = suggestedAngle − jammer.angle ∗ suggestedDirection
5:     **if** angleDifference > maxRotateAngle **then**
6:         jammer.angle + = maxRotateAngle ∗ suggestedDirection
7:     **else**
8:         jammer.angle + = angleDifference ∗ suggestedDirection
9:     **return** jammer.angle

---

5 Experimental Study

In this section, we will share our experiments to measure the quality of the developed algorithms. First, we will discuss the experiments we ran using a genetic algorithm for the threat evaluation part. Then we will discuss our experiments on the combined threat evaluation and jammer-angle assignment solution. Since we could not find any comparable paper working on this particular problem and Gül [6] stated that this is the first article in this field, we compared our work with it and the different versions we created in our experiments. We first compared the algorithm we developed with Gül's study. Then, for the solutions to the parts of the problem that have not been studied before, we will compare the different versions we have created for the comparison and discuss the results.

5.1 Data Sets

We did not find a benchmark dataset to test our results when we reviewed the literature and the previous study working on this problem. For this reason, we created our own datasets to carry out the experiments. We created four datasets based on the field data we have to use in the experiments of this study. According to what we observed in real-life data, we decided to set the radar range to 4 km, the jammer range to 5 km, and the threat speeds between 1 km/h to 30 km/h. Using randomized discrete test cases (each consisting of one time-step snapshot of the problem) seems to be an efficient method to see the performance of the mathematical model in different problem instances. However, it is observed that real-life problems arise by solving problem instances that are the continuation of each other. The main reason for this is that threats can move in specific directions at certain speeds and can only change place a certain amount at any time. For this reason, testing with threats that move at certain speeds, rather than threats located in entirely different locations, will better reflect the real-world problem. We decided to create test data in both structures for these reasons.

Table 1. Summary of datasets generated.

Datasets	Scenario Type	Number of Instances	Tests Used
Dataset 1	Independent	1840 instances	Genetic Algorithm
Dataset 2	Independent	310 instances	Genetic Algorithm
Dataset 3	Independent	16000 instances	Comparison with Previous Work
Dataset 4	Continuous	8000 instances	Tactical and Physical Constraints

A summary of the datasets we created is shown in Table 1. In these datasets, the number of jammers was increased from 1 to 4, and the number of threats was 10, 50, 100, and 500. There we an equal number of scenarios for each of these jammer-count and threat-count combinations. In all datasets created here, features of the threats and coordinates of the jammers are provided inside the feature list. The distance calculation is considered as the distance to the critical region. In these tests, the mid-point coordinate of the jammers was determined as the critical region coordinate. All the created datasets were created randomly but with a 50% probability of including drone-type threats. Of the drone type threats, 50% are created to be approaching and close type. In order to increase the number of generations produced in genetic algorithm experiments, we produced datasets containing fewer instances. Each independent instance of the dataset represents an entirely different problem. On the other hand, in continuous datasets, the successive instances for the same jammer-count and threat-count couples are the continuation of each other.

5.2   Genetic Algorithm Tests

We have tried several combinations of hyper-parameters for genetic algorithms. Since we kept the threat coverage values obtained by solving the problem with the current parameter values in the genetic algorithm, we continued with parameter sets that could reach higher values. During these experiments, we selected parameters in a way that could eliminate premature maturation of results and satisfy our CPU constraints. Due to lack of space, we will not be able to share all these experiments in this part, but the parameters we reached after our tests can be examined in Table 2:

Table 2. Genetic algorithm parameters used to get best results.

Genetic Algorithm Parameters	Values
Initial population	created with random small numbers
Rate of Mutation	0.9
Magnitude of Mutation	In [5,100] interval, random, decreasing with time
Size of Population	20
Crossover Type	two-point
How the Next Generation Produced	20% fittest individuals, 20% randomly selected from the previous generation, 20% crossover, 40% crossover + mutation

We found that in test cases where everything could mutate and crossover, in general some individuals suppressed the population. Therefore, we have handled mutation and crossover in a more separate manner.

We ran our genetic algorithm with the hyper-parameters in Table 2 on two different datasets. The first of these was dataset 1; although the learning rate was reasonable since this dataset is a large dataset, the number of generations we could produce would be low due to the memory constraints of the system we tested. For this reason, we produced more generations using a smaller dataset, dataset 2, and compared the learning rates in these two cases. The results of these experiments are shown in Fig. 5. When we examine the results, we see that while we see a more extended learning period in the first dataset, the learning starts very quickly in the second dataset and then stops. We obtained the parameter set, which gives better coverage than the parameters in the results, as a result of the tests performed in dataset 2.

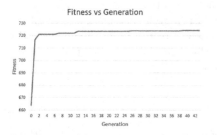
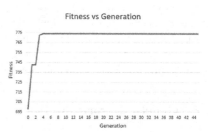

(a) Fitness by each generation using dataset with 1840 cases

(b) Fitness by each generation using dataset with 310 cases

Fig. 5. Fitness by each generation

5.3   Experiments with the Linear Programming Model

In this section, we will talk about the tests of the developed model as a whole. In these experiments, we used the parameter values that we mentioned about the tests in the previous section within the w value in the linear programming model. For results, coverage percentages for all threats, UAVs, Fast Close Approaching UAVs (FCA UAVs), duration (and for relevant cases, angle difference with previous time frame) information are presented.

Comparison with the Previous Work: While comparing the method developed in this study with the method developed by Kıvanç Gül [6], we conducted tests without adding constraints such as angle minimization and physical and tactical constraints. The reason for this is that such constraints are not included in that study. We used dataset 3 for the experiments in this section. The results are shared in Table 3. When we examine the results, although it is seen that Gül's model in general coverage is relatively better, we observe that our model in coverage gives much better coverage results in terms of covering important threats. We observed that the difference in coverage increased even more as the importance of the threat increased. In addition, we observed that although Gül's work is faster in terms of working speed, this difference is at a negligible level.

Table 3. Comparison of the algorithm developed in this study with Gül's Model. Bold font indicates the better result.

| | | Comparison with Gül's Model [6] | | | | | | | |
| | | Our Model | | | | Gül's Model [6] | | | |
# of Jammers	# of Threats	All Threats %	UAV %	FCA UAV %	Duration (ms)	All threats %	UAV %	FCA UAV %	Duration (ms)
1	10	38.62	**52.53**	**100**	20.03	41.94	42.63	77.78	**17.88**
	50	25.32	**30.42**	**94.44**	25.07	28.77	28.79	33.33	**22.63**
	100	22.16	25.29	**80.49**	30.5	25.5	25.41	34.15	**31.05**
	500	19.73	20.83	**48.61**	**50.81**	21.81	21.91	19.1	53.33
2	10	68.08	**83.98**	**100**	19.27	71.66	71.95	100	**16.81**
	50	49.76	**58.53**	**100**	29.41	53.23	53.08	76.47	**27.58**
	100	44.81	**50.98**	**97.18**	39.28	48.28	48.09	59.15	**38.6**
	500	39.76	41.27	**93.4**	88.83	42.89	43.06	57.15	89.11
3	10	90.46	**97.79**	**100**	21.23	91.78	92.17	100	19
	50	71.14	**80.82**	**100**	34.95	74.51	74.64	94.12	36.27
	100	66.28	**73.9**	**100**	54.73	69.42	69.45	81.4	**54.62**
	500	60.29	62.84	**99.39**	138.64	62.88	63.01	78.65	**133.19**
4	10	**99.78**	**99.95**	**100**	21.51	99.78	99.62	**100**	18.81
	50	89.26	**96.09**	**100**	41.32	91.48	91.51	99.18	43.96
	100	85.08	**91.46**	**100**	68.07	87.44	87.27	97.8	65.69
	500	80.34	**82.96**	**100**	181.91	81.55	81.44	93.2	180.43

Angle Minimization. In this study, we provided an angle minimization algorithm to prevent unnecessary turns of the rotating device, preventing premature wear of the device. This algorithm aims to reach high coverage ratios with less rotation. In this context, the comparison of the algorithm versions (with and without angle minimization) we created is shared in Table 4.

When we examined the results, it was observed that there was a significant decrease in the average amount of rotation of the jammers (minimum 1°, maximum 45°). In addition, while this performance is achieved, it has been observed that coverage decreases by up to 5% on average for certain scenario types where coverage rates are very close in many scenarios. Moreover, although a new constraint has been added, the operating speed is slowed at most around 10–15 milliseconds.

Table 4. Comparison table in terms of coverage percentages of the case where angle minimization is used and not used in the calculations. Bold font indicates the better result.

| | | Angle Minimization | | | | | | | | | |
| | | Angle Minimization ON | | | | | Angle Minimization OFF | | | | |
# of Jammers	# of Threats	All Threats %	UAV %	FCA UAV %	Angle Difference	Duration (ms)	All threats %	UAV %	FCA UAV %	Angle Difference	Duration (ms)
1	10	37.96	46.09	**93.75**	**6.85**	23.93	38.84	**47.01**	**93.75**	12.67	**18.62**
	50	25.34	32.22	50.53	**17.31**	32.03	25.96	**33.79**	**56.91**	20.48	**24.43**
	100	21.97	24.19	45.54	**22.83**	36.98	22.96	**25.57**	**52.69**	27.53	**29.39**
	500	20.73	21.53	25.82	**23.4**	52.99	21.34	**22.34**	28.27	29.46	**50.73**
2	10	72.93	95.2	100	**3.55**	29.19	79.28	**97.8**	100	17.39	**18.66**
	50	48.75	49.13	89.68	**12.48**	44.84	49.89	**51.51**	**93.55**	16.27	**29.13**
	100	60.59	61.76	**99.5**	**4.72**	55.31	60.61	**61.77**	**99.5**	5.13	**39.76**
	500	48.58	48.64	**81.52**	**5.98**	112.31	48.6	**48.65**	81.48	6	**99.08**
3	10	90.3	99.67	100	**2.01**	35.67	93.65	**100**	100	43.68	**20.7**
	50	80.12	88.99	97.19	**8.78**	66.99	82.28	**91.91**	100	17.44	**37.96**
	100	69.3	73.27	96.67	**15.12**	86.43	71.44	**75.87**	**99.65**	21.25	**54.29**
	500	52.97	54.27	82.77	**21.33**	178.45	54.87	**56.49**	87.8	26.8	**153.07**
4	10	96.64	99.87	100	**1.34**	41.31	100	100	100	47.85	**21.59**
	50	96.56	99.64	99.63	**3.17**	67.77	97.71	100	100	15.93	**47.33**
	100	93.8	97.6	100	**5.44**	96.54	94.2	**98.06**	100	11.2	**77.45**
	500	88.63	90.51	99.33	**9.55**	296.05	90.65	**92.85**	100	13.54	**291.02**

Physical Constraint. In this set of experiments, we examine the effect of using physical constraints on performance. There is a random 20° physical constraint for each jammer. We used dataset 4 for the experiments in this section. The results are shared in Table 5.

When we examined the results, we saw that even when we added a physical constraint, the quality of the results decreased only slightly. This drop is acceptable because it can no longer rotate to the angles it needs to rotate to cover it. In addition, since it cannot turn to those angles, it tries to turn from the opposite side and cover the threat. This turn, causes an increase in the amount of angle difference.

Table 5. Comparison table in terms of coverage percentages of the case where a physical constraint exist and not exist in the calculations. Bold font indicates the better result.

| | | Physical Constraint | | | | | | | | | |
| | | Physical Constraints Exist | | | | | Physical Constraint NOT Exist | | | | |
# of Jammers	# of Threats	All Threats %	UAV %	FCA UAV %	Angle Difference	Duration (ms)	All threats %	UAV %	FCA UAV %	Angle Difference	Duration (ms)
1	10	35.28	44.37	90.63	8.33	22.16	**37.96**	**46.09**	**93.75**	**6.85**	**21.47**
	50	23.49	29.97	47.34	20.35	30.36	**25.34**	**32.22**	**50.53**	**17.31**	**28.48**
	100	**22.26**	**25.01**	**47.22**	23.33	35.48	21.97	24.19	45.54	**22.83**	**33.74**
	500	20.68	21.39	23.17	25.96	52.47	**20.73**	**21.53**	**25.82**	**23.4**	**54.24**
2	10	**74.15**	94.8	95.24	4.68	26.44	72.93	**95.2**	**100**	**3.55**	**26.4**
	50	**48.97**	**49.45**	**90.97**	12.34	42.26	48.75	49.13	89.68	**12.48**	**41.83**
	100	60.35	61.35	98.74	5.04	**50.79**	**60.59**	**61.76**	**99.5**	**4.72**	51.54
	500	48.52	48.59	81.49	**6.07**	**108.36**	**48.58**	**48.64**	**81.52**	5.98	112.1
3	10	**90.91**	99	97.78	2.24	34.91	90.3	**99.67**	**100**	**2.01**	**34.78**
	50	76.67	83.53	90.96	12.56	64.04	**80.12**	**88.99**	**97.19**	**8.78**	**59.98**
	100	66.74	70.67	91.58	17.14	83.41	**69.3**	**73.27**	**96.67**	**15.12**	**79.86**
	500	52.67	54	82.17	21.82	187.04	**52.97**	**54.27**	**82.77**	**21.33**	**178.6**
4	10	**99.48**	**99.93**	100	**0.56**	42.75	96.64	99.87	100	1.37	**42.95**
	50	91.74	97.8	97.79	5.59	77.72	**96.56**	**99.64**	**99.63**	**3.17**	**65.19**
	100	93.51	97.12	99.88	5.54	95.7	**93.8**	**97.6**	**100**	**5.44**	**93.62**
	500	**89.86**	**91.56**	**99.48**	**6.44**	261.57	88.63	90.51	99.33	9.55	**282.04**

6   Conclusion

In this study, we developed a solution specially designed for TEWA problems where the weapon is a directional jammer. Directional jammers, unlike regular weapons, work on angular coverage, so producing a particular solution for them will provide higher coverage performances in the problems using these devices. These devices are frequently used in anti-drone systems, and it is valuable to find effective solutions.

In this paper, we designed the threat parameters with more detail. We furthermore modeled the problem closer to real-life problems than in the previous studies. We took a greedy method and changed it to a genetic algorithm-based threat evaluation. We thought that we could achieve more realistic results with this approach. In genetic algorithm-based threat evaluation, We included parameters to cover all its different categorizations: their proximity (threat distance, threat speed), capabilities (threat type), and intents (threat direction). Then, for optimization of the numerical values of these parameters, an intuitive parameter setting was created in accordance with the problem structure as a starting point for individuals of the genetic algorithm. Previous studies are replicated and compared with the proposed pre-computation module, which uses a genetic algorithm. It was seen that the genetic algorithm-based threat evaluation provides better results than previous studies.

The proposed approach, unlike previous approaches, can more easily adapt to different datasets from different environments. Also, with a parameter setting given, the proposed mathematical model allows the algorithm to suggest angles that satisfy the physical constraints of the device, and it is ensured that the algorithm does not suggest angles where jammers are not desired to do jamming. It provides results that prevent devices from making unnecessary rotational movements. Thus, the solution produced by our algorithm does not cause additional wear for the rotational device that carries a directional jammer. Additionally, in the method we developed for physical constraints, although the problem has extra constraints, we did not encounter a severe decrease in coverage rates.

References

1. Andersen, A.C., Pavlikov, K., Toffolo, T.A.: Weapon-target assignment problem: exact and approximate solution algorithms. Annals of Operations Research, pp. 1–26 (2022)
2. Athans, M.: Command and control (c2) theory: a challenge to control science. IEEE Trans. Autom. Control **32**(4), 286–293 (1987)
3. Cai, H., Liu, J., Chen, Y., Wang, H.: Survey of the research on dynamic weapon-target assignment problem. J. Syst. Eng. Electron. **17**(3), 559–565 (2006)
4. Chang, T., Kong, D., Hao, N., Xu, K., Yang, G.: Solving the dynamic weapon target assignment problem by an improved artificial bee colony algorithm with heuristic factor initialization. Appl. Soft Comput. **70**, 845–863 (2018)
5. Davidson, P.P., Blum, C., Lozano, J.A.: The weighted independent domination problem: ILP model and algorithmic approaches. In: Hu, B., López-Ibáñez, M. (eds.) EvoCOP 2017. LNCS, vol. 10197, pp. 201–214. Springer, Cham (2017). https://doi.org/10.1007/978-3-319-55453-2_14
6. Gül, K.: Model and procedures for the jammer and target allocation problem. Master's thesis, Middle East Technical University (2018)
7. Johansson, F.: Evaluating the performance of TEWA systems. Ph.D. thesis, Örebro universitet (2010)
8. Johansson, F., Falkman, G.: A bayesian network approach to threat evaluation with application to an air defense scenario. In: 2008 11th International Conference on Information Fusion, pp. 1–7. IEEE (2008)
9. Kline, A., Ahner, D., Hill, R.: The weapon-target assignment problem. Comput. Oper. Res. **105**, 226–236 (2019)
10. Kline, A.G., Ahner, D.K., Lunday, B.J.: Real-time heuristic algorithms for the static weapon target assignment problem. J. Heuristics, 377–397 (2019)
11. Lee, Z.J., Su, S.F., Lee, C.Y.: Efficiently solving general weapon-target assignment problem by genetic algorithms with greedy eugenics. IEEE Trans. Syst. Man Cybern. Part B (Cybernetics) **33**(1), 113–121 (2003)
12. Lloyd, S.P., Witsenhausen, H.S.: Weapons allocation is np-complete. In: 1986 Summer Computer Simulation Conference, pp. 1054–1058 (1986)

13. Lötter, D., Van Vuuren, J.: Implementation challenges associated with a threat evaluation and weapon assignment system. In: Proceedings of the 2014 Operations Research Society of South Africa Annual Conference, pp. 27–35 (2014)
14. Lu, Y., Chen, D.Z.: A new exact algorithm for the weapon-target assignment problem. Omega **98**, 102138 (2021)
15. Ma, F., Ni, M., Gao, B., Yu, Z.: An efficient algorithm for the weapon target assignment problem. In: 2015 IEEE International Conference on Information and Automation, pp. 2093–2097. IEEE (2015)
16. Naeem, H., Masood, A.: An optimal dynamic threat evaluation and weapon scheduling technique. Knowl.-Based Syst. **23**(4), 337–342 (2010)
17. Park, S., Kim, H.T., Lee, S., Joo, H., Kim, H.: Survey on anti-drone systems: components, designs, and challenges. IEEE Access **9**, 42635–42659 (2021)
18. Shi, X., Yang, C., Xie, W., Liang, C., Shi, Z., Chen, J.: Anti-drone system with multiple surveillance technologies: architecture, implementation, and challenges. IEEE Commun. Mag. **56**(4), 68–74 (2018)
19. Shin, M.K., Lee, D., Choi, H.L.: Weapon-target assignment problem with interference constraints using mixed-integer linear programming. arXiv preprint arXiv:1911.12567 (2019)
20. Toet, A., de Waard, H.: The Weapon-Target Assignment Problem. TNO Human Factors Research Institute (1995)
21. Tyurin, V., Martyniuk, O., Mirnenko, V., Open'ko, P., Korenivska, I.: General approach to counter unmanned aerial vehicles. In: IEEE 5th International Conference Actual Problems of Unmanned Aerial Vehicles Developments, pp. 75–78 (2019)
22. Wu, L., Wang, H.Y., Lu, F.X., Jia, P.: An anytime algorithm based on modified ga for dynamic weapon-target allocation problem. In: 2008 IEEE Congress on Evolutionary Computation (IEEE World Congress on Computational Intelligence), pp. 2020–2025. IEEE (2008)
23. Xin, B., Wang, Y., Chen, J.: An efficient marginal-return-based constructive heuristic to solve the sensor-weapon-target assignment problem. IEEE Trans. Syst. Man Cybern. Syst. **49**(12), 2536–2547 (2018)
24. Yaghini, M., Karimi, M., Rahbar, M.: A set covering approach for multi-depot train driver scheduling. J. Comb. Optim. **29**(3), 636–654 (2015)
25. Zhao, Y., Chen, Y., Zhen, Z., Jiang, J.: Multi-weapon multi-target assignment based on hybrid genetic algorithm in uncertain environment. Int. J. Adv. Rob. Syst. **17**(2), 1729881420905922 (2020)

Improving the Size and Quality of MAP-Elites Containers via Multiple Emitters and Decoders for Urban Logistics

Neil Urquhart[✉] and Emma Hart

School of Computing, Engineering and the Built Environment,
Edinburgh Napier University, Edinburgh, UK
{n.urquhart,e.hart}@napier.ac.uk

Abstract. Quality-diversity (QD) methods such as MAP-Elites have been demonstrated to be useful in the domain of combinatorial optimisation due to their ability to generate a large set of solutions to a single-objective problem that are diverse with respect to user-defined features of interest. However, filling a MAP-Elites container with solutions can require careful design of operators to ensure complete exploration of the feature-space. Working in the domain of urban logistics, we propose two methods to increase exploration. Firstly, we exploit multiple decodings of the same genome which can generate different offspring from the same parent solution. Secondly, we make use of a multiple mutation operators to generate offspring from a parent, using a multi-armed bandit algorithm to adaptively select the best operator during the search. Our results on a set of 48 instances show that both the number of solutions within the container and the qd score of the container (indicating quality) can be significantly increased compared to the standard MAP-Elites approach.

Keywords: MAP-Elites · Urban Logistics · Multi-Arm Bandit

1 Introduction and Motivation

The use of quality-diversity methods such as the Map of Phenotypic Elites (MAP-Elites) [18,21] that generate multiple high-quality but diverse solutions to an optimisation problem is very well-established in the field of robotics. Increasingly the approach is finding applications in combinatorial optimisation domains such as routing [28], constrained optimisation [25] and instance-generation [2]. Specifically, the MAP-Elites algorithm generates a set of solutions to a single-objective optimisation problem that are diverse with respect to a set of features defined by a user. The algorithm attempts to fill a discretised container of cells with solutions, such that each cell contains the single best solution (*elite*) for the combination of features that map to the cell. The algorithm is often referred to as an *illumination algorithm* due its ability to highlight high-performing regions

© The Author(s), under exclusive license to Springer Nature Switzerland AG 2023
J. Correia et al. (Eds.): EvoApplications 2023, LNCS 13989, pp. 35–52, 2023.
https://doi.org/10.1007/978-3-031-30229-9_3

of the feature-space, but has also been shown to increase the quality of solutions found by maintaining stepping-stones in the population that lead to good solutions [12]. From a user perspective, the illumination aspect of the approach provides new insight into the characteristics of solutions that influence quality, as well as providing the user with *choice*, i.e. by returning a large set of solutions with differing characteristics, enabling a user to switch to an alternative solution in response to sudden changes. It is important to note that the concept behind MAP-Elites is distinct from multi-objective algorithms which return a set of solutions that are pareto-optimal with respect to two or more objectives. In contrast MAP-Elites returns solutions which are diverse with respect to a user-defined feature-space for a single objective problem, with the maximum number of potential solutions returned defined by the user when specifying the container size.

The canonical MAP-Elites algorithm is surprisingly simple, generating new solutions by applying a mutation operator to existing 'elite' solutions to generate new samples. However in many combinatorial optimisation applications, multiple mutation operators are applicable: for example including problem-specific operators in additional to more generic operators that modify or swap-values. The main contribution of this paper is therefore to propose a set of mutation operators for use in an urban-logistics problem to be used in conjunction with MAP-Elites to generate new solutions. In order to select the most appropriate operator at a given point in the search, we make use of a multi-armed bandit to select which operator to apply as the search progresses. In addition, we attempt to further increase solution diversity by introducing two mechanisms of decoding from genotype to phenotype. Our results demonstrate that the use of multiple decoders can increase the number of solutions found, showing some solutions can only be found by one of the decoders. The use of a Multi-Arm Bandit (MAB) algorithm to prioritise the choice of mutation operator when generating new solutions is demonstrated to improve overall performance in terms of the contents of the container. We assess performance using two metrics; quantity of solutions (s) and quality-diversity (qd) the degree to which high-quality solutions are found.

2 Related Work

Quality-diversity (QD) Algorithms [22] explore a feature space of possible solutions to a given problem, returning a diverse set of solutions to a problem, and facilitating location of optimal solutions by providing a series of stepping-stones through the search-space. One of the most common QD approaches is Map-Elites [18], which partitions the feature space into a series of cells, each of which stores the best found solution for feature combination associate with a cell. Its use is now prolific in the robotics and games communities[1]. However, it is increasingly finding application in the context of combinatorial optimization. For example, Urquhart *et al.* have used MAP-Elites to solve an urban logistics problem [28];

---

[1] A repository of QD papers is maintained at https://quality-diversity.github.io/papers which clearly illustrates this.

it has been used to obtain improved solutions to the Travelling Thief Problem [19], as well as to create new TSP instances [2]; Cardoso et al. [4,5] use it in neural architecture search.

Since its inception, a number of improvements have been proposed to the original algorithm with respect to the manner in which new solutions are generated from existing solutions stored in the grid of cells—a process determined by what is commonly referred to in QD literature as an 'emitter'. In the canonical MAP-Elites framework, a single emitter 'emits' solutions by uniformly selecting a single solution from the grid and returning a mutated copy of this solution. Fontaine et al. [11] extend this idea by proposing the concept of using multiple emitters to generate new solutions, where each emitter is based on running a CMA-ES algorithm [16], seeded by solutions from the grid. The key insight of this approach is leveraging the selection and adaptation rules of CMA-ES to optimise good solutions, while also efficiently exploring new areas of the search space via MAP-Elites. Cully [6] further refines this idea by using a multi-armed bandit to select the most appropriate emitter, leading to further improvements in both efficiency and quality. The use of CMA-ES as the underlying emitter restricts this line of research to the continuous optimisation domain. We therefore propose a variant adapted for combinatorial optimisation inspired by the work just discussed: specifically defining multiple mutation operators with different properties by which to generate new solutions from a single solution selected from a cell, combined with a multi-armed bandit to adaptively select the best operator. The idea of adaptive operator selection has previously been used in the context of evolutionary algorithms (e.g. [7]) and therefore could be expected to afford the same benefits to QD approaches. To be consistent with QD literature, in the remainder of this paper we use the term emitter to denote a method of generating offspring from an existing solution or solutions. However it should be clear that in this context, an emitter is in fact equivalent to the term *operator* used in the wider literature, and that applying each emitter only generates a single offspring.

3   Methodology

Vehicle routing problems (VRPs) [9,10] are a class of NP-hard combinatorial optimisation problems which require a set of routes to be constructed to allow a set of customers to be serviced. Many variants of the VRP have been devised taking into account many different real-world constraints including capacities, costs and environmental impact [13].

We consider a capacitated, mixed fleet vehicle routing problem. Each instance comprises a set of visits each with a demand (representing the quantity of goods to be delivered). Visits must be grouped into routes, and each route allocated to a vehicle type (e.g. van, pedestrian or cargo bicycle). Each vehicle type has a unique set of constraints including capacity, speed, emissions and costs. There is a single objective to minimise the distance travelled for any given solution. Each valid solution can be described by a set of characteristics which are of interest to the end user.

Two settings are considered, which are differentiated by the types of vehicles available to deliver goods. Each setting has a set of characteristics of user interest, given in Table 1. For each setting, multiple problem instances are used based on those utilised by Augerat [1]. The Augerat instances are enhanced by the addition of a model for alternative vehicles (e.g. bikes, pedestrians etc.) which contains time, cost and CO_2 values for each mode. The cost model and instances can be found in the GitHub repository[2] that accompanies this paper.

Table 1. A summary of the two settings used within this paper.

Setting A: 26 instances		Setting B: 22 instances	
Solution Characteristics	Vehicles	Solution Characteristics	Vehicles
CO2	Cargo Bike	CO2	Cargo Bike
Time Span	Large Van	Total Cost	Small Van
Fixed Cost	Pedestrian	Total Routes	Large Van
Running Cost		Cycles Used	
		Cycle Dels (%)	

3.1 MAP-Elites

MAP-Elites [18] uses a dynamically growing population to drive evolution. Each solution in this population is mapped to a cell in a discretised container defined by c axes. Each axis represents a characteristic of a solution that varies between a lower and upper bound specific to the characteristic and is discretised into cells. Each cell contains at most 1 solution, representing the solution with the best objective value found for the given combination of characteristics. In its simplest form, the container is first initialised with a batch of solutions. Following this, new solutions are generated at each iteration by selecting a solution at random from the container and applying an emitter to generate one or more new solutions. New solutions either map to an empty cell, replace an existing solution in a cell (if its objective value is better) otherwise they are discarded.

In this paper we propose two modified versions of MAP-Elites designed to increase the coverage within the containers produced by MAP-Elites:

- Round Robin MapElites (RR-ME): at each iteration, select a solution from the container at random and apply an emitter to create a new child. Repeat for each emitter in the emitter pool.
- Multi-Armed Bandit MapElites (MAB-ME): At each iteration, select a solution at random and apply the emitter determined by the output of a multi-armed bandit

Algorithm 1 provides a high-level overview of the algorithm using pseudo-code. Further details of each step are provided in the following sections.

3.2 Set up and Initialisation

In order to normalise solution characteristics to the scale used in the container it is necessary to define the *range* of each characteristic. For the specific application in question, the precise bounds of each characteristic are not known *a*

---

[2] https://github.com/NeilUrquhart/Improving-MAP-Elites-Coverage/.

Algorithm 1. MAP-Elites Algorithm, adapted from [18], with main modifications highlighted in red

```
procedure MAP-ELITES ALGORITHM
    (P ← ∅, X ← ∅)                    ▷ Empty N-dimensional map of elites: solutions X and their performances P
    E ← pool of emitters                                       ▷ set of emitters (variation operators) available
    for iter = 1 → I do
        if iter < G then
            x' ← add next solution from I                                        ▷ Initialise (see section 3.2)
        else                                             ▷ Subsequent solutions generated from elites in the map
            x' ← randomSelection(X)
            if RR-MapElites == TRUE then
                e ← nextEmitter(E)                                                  ▷ next emitter from pool
            else
                e ← applyMAB(E)                                               ▷ use MAB to choose emitter
            end if
            x' ← applyEmitter(X, e)                                          ▷ create modified version of x
        end if
        b' ← featureDescriptor(x')                                            ▷ Calculate characteristics
        p' ← performance(x')                                                  ▷ Calculate objective value
        if P(b') = ∅ or P(b') < p'  then   ▷ If the appropriate cell is empty or its occupants's performance is
≤ p
            P(b') ← p'     ▷ store the performance of x' in the map of elites according to its feature descriptor
            X(b') ← (x')          ▷ store the solution x' in the map of elites according to its feature descriptor b
        end if
    end for
    return feature-performance map(P and X)
end procedure
```

priori. Therefore we estimate them by running a single-objective EA on every instance using each characteristic as the objective function (100K evaluations). The process is repeated 10 times for each instance and characteristic. The lower and upper bounds of each characteristic are then determined from this data. All solutions generated in this phase are added to an initialisation set I and are used to initialise the MAP-Elites algorithm.

3.3 Representation

Solutions are represented using a *grand tour* representation [20]. An individual specifies a set of *visits* which are decoded in the order specified into a solution that consists of multiple *routes*, each with an allocated vehicle type. Each visit is specified by three values:

- **Del_ID** an identifier of a location according to the problem definition
- **New_Route (NR)** If set, this denotes that the visit marks the start of a new route
- **Preferred_Vehicle** Specifies the vehicle type to be used for the route

Algorithm 2. The Simple Decoder (SD)

```
routes = []
current = newRoute()
for gene in chromosome do
    if gene.newRoute || !current.hasCap(gene.visit) then
        routes.add(current)
        current = newRoute()
    end if
    current.add(gene.visit)
end for
```

In order to progress from a the grand-tour representation to a solution consisting of a set of routes a *decoder* function is required. Two different decoders are evaluated (Algorithms 2 and 3) within the RR-ME algorithm.

Algorithm 3. The Advanced Decoder (AD)

```
routes = []
for gene in chromosome do
    if gene.newRoute then
        r = newRoute
        r.add(gene.visit)
        routes.add(r)
    end if
    for route in routes do
        if (route.veh == gene.veh)and(route.hasCap(gene.visit)) then
            route.add(gene.visit)
            found = true
            break
        end if
    end for
    if !found then
        r = newRoute
        r.add(gene.vehicle)
        routes.add(r)
    end if
end for
```

- The simple decoder (*SD*) adds each visit to a current route in the order that they appear in the individual. A new route is started if the *NR* flag of the gene is set or the capacity constraint of the current route is broken.
- The advanced decoder (*AD*) allocates a visit to a route that is being serviced by the vehicle specified in the gene. It will also attempt to fill up existing routes to make maximum use of capacity, by checking each existing route for suitability, unless a gene has its *NR* flag set (see above).

3.4   Emitters

In order to generate new solutions from existing points in a container, we propose a set of emitters with varying properties:

- **rndMove()** Move a randomly selected gene to a new position
- **addRoute(*vehicle*)**[3] Select a gene at random, the preferred vehicle to *vehicle* and set the new NR to 1.
- **delRoute(*vehicle*)** Select a gene at random that has a preferred vehicle type of *vehicle* and its new route flag set to 1, then reset the route flag to 0.

Additionally, a *crossover* emitter is also utilised. This uses a variant of two point crossover to create a new solution (with both parents selected randomly): a section of tour from parent *A* defined between two randomly chosen points is copied directly to the child, with the missing genes then added in the order that they appear in parent *B*.

3.5   Performance Metrics

In all reported experiments, for each instance, we run MAP-Elites 10 times (with an evaluation budget of 1 million evaluations per run). The 10 containers returned are combined into a *grand container* retaining a single elite solution

---

[3] For *addRoute* and *delRoute*, emitters are created for each vehicle type used.

per filled cell. This approach generates one grand container for each problem instance.

We evaluate containers using two standard metrics from the QD literature. We define s as the number of solutions contained within the container. The qd score is defined as the sum of $\frac{1}{f}$ for each individual in the container where f is the fitness of that individual [21]. Higher values represent better results in both cases. After evaluating the qd and s metrics across the range of instances we measure the significance of any observed differences using a Friedman test [8]. Where the Friedman test indicates that a difference is significant we perform the Nemenyi post-hoc test [17] in order to understand where the significance occurs, to access these metrics we utilise the SciPy library (https://scipy.org/).

4 Decoding Strategies

We first conduct experiments to understand whether varying the decoder employed improves the s and/or qd scores. We test three versions of MAP-Elites: (1) using the SD decoder; (2) using the AD decoder; (3) decoding using both decoders and attempting to add both solutions to the container (denoted double-decoder DD). These experiments use the RR-Map-Elites method described above, i.e. at each iteration of the algorithm, the emitter is selected on a round-robin basis.

Table 2 shows the s and qd scores of the SD, AD and DD. The AD variant appears to perform poorly compared to the SD variant, while DD variant appears to perform marginally better than SD or AD. Table 3 shows the instances where each variant produces the best performing grand container according to metric (s, qd). It is clear that the AD never produces a result that outperforms the other variants on an instance. Table 3 suggests that the DD variant is most likely to produce a container that performs better according to our metrics.

In order to understand whether the choice of decoder has a *significant* impact on either metric, we apply a Friedman test for repeated measures analysis of variance by rank. The resulting p-values (Table 4) show that we are able to reject the null hypothesis that the choice of decoder has no impact. To better understand where the significant differences occur, we apply a Nemenyi post-hoc test. The resulting p-values are shown in Table 5. This highlights a significant difference between AD - SD, and AD - DD, with the latter of each pair providing better performance.

Recall that DD decodes solutions using both SD and AD methods. While it might be tempting therefore to attribute the performance of DD to the fact that its SD component is simply responsible for contributing solutions, Fig. 1 indicates that in fact, AD does contribute a small number of solutions to the grand container. Hence, it may be these solutions that form stepping stones for SD to find higher performing solutions. An alternative explanation is that AD is capable of reaching some areas of the search-space that SD cannot access.

Table 2. The ss and qd scores obtained on each of the grand fronts across both settings, the difference is the % of SD mean.

Instance	s SD	AD	DD	qd SD	AD	DD
A-n32-k5	354	324	356	0.014	0.012	0.014
A-n45-k6	345	306	337	0.008	0.007	0.008
A-n34-k5	319	314	320	0.012	0.011	0.012
A-n36-k5	339	328	338	0.012	0.011	0.012
A-n37-k5	344	311	354	0.014	0.011	0.014
A-n45-k7	329	301	324	0.009	0.007	0.009
A-n46-k7	343	304	337	0.009	0.007	0.009
A-n48-k7	308	270	317	0.008	0.006	0.008
A-n53-k7	340	316	337	0.007	0.006	0.008
A-n38-k5	329	314	332	0.012	0.010	0.012
A-n39-k5	334	328	346	0.011	0.009	0.011
A-n39-k6	323	301	335	0.012	0.010	0.012
A-n54-k7	333	290	338	0.008	0.006	0.008
A-n55-k9	326	287	328	0.007	0.005	0.007
A-n60-k9	313	295	322	0.006	0.005	0.006
A-n61-k9	322	275	320	0.006	0.004	0.006
A-n62-k8	338	289	332	0.005	0.004	0.005
A-n63-k9	316	287	320	0.004	0.003	0.004
A-n63-k10	325	281	326	0.005	0.004	0.005
A-n64-k9	319	289	321	0.004	0.004	0.004
A-n65-k9	326	280	337	0.005	0.004	0.005
A-n69-k9	325	277	320	0.005	0.003	0.005
A-n80-k10	318	257	317	0.004	0.002	0.004
A-n33-k5	340	314	337	0.017	0.015	0.017
A-n33-k6	337	312	331	0.013	0.011	0.013
A-n37-k6	311	304	316	0.008	0.008	0.008

Instance	s SD	AD	DD	qd SD	AD	DD
B-n31-k5	396	336	398	0.016	0.013	0.016
B-n34-k5	305	232	300	0.005	0.004	0.005
B-n35-k5	362	280	364	0.007	0.005	0.007
B-n38-k6	289	209	274	0.005	0.004	0.005
B-n39-k5	360	282	339	0.008	0.006	0.008
B-n41-k6	242	201	252	0.004	0.003	0.004
B-n43-k6	284	242	289	0.005	0.004	0.005
B-n44-k7	296	234	289	0.005	0.004	0.004
B-n45-k5	314	271	324	0.006	0.005	0.006
B-n50-k7	311	232	289	0.005	0.004	0.005
B-n50-k8	265	223	262	0.003	0.002	0.003
B-n51-k7	208	168	203	0.002	0.002	0.002
B-n52-k7	331	271	328	0.006	0.005	0.006
B-n56-k7	329	278	334	0.006	0.005	0.006
B-n57-k7	311	263	320	0.004	0.004	0.005
B-n57-k9	268	219	262	0.003	0.002	0.003
B-n63-k10	232	193	232	0.002	0.002	0.002
B-n64-k9	277	244	278	0.003	0.003	0.003
B-n66-k9	264	222	258	0.003	0.002	0.003
B-n67-k10	249	223	253	0.002	0.002	0.002
B-n68-k9	261	231	270	0.002	0.002	0.002
B-n78-k10	280	233	280	0.003	0.002	0.003

Table 3. Based on the raw results shown in Table 2. For each setting we count the number of instances where each variant finds the best result according to each of the metrics (s and qd).

		Setting A	Setting B
s	SD	11	10
	AD	0	0
	DD	15	12
qd	SD	12	14
	AD	0	0
	DD	14	8

Table 4. The p values generated by the Friedman test. By convention values of less than 0.05 are regarded as an indicator of significance.

Setting	Metric	p
A	s	2.9^{-9}
	qd	3.2^{-9}
B	s	4.6^{-8}
	qd	4.5^{-8}

Table 5. The values returned by the Nemenyi post-hoc test, a value of less than 0.05 is regarded as an indicator of significance (highlighted in italics). The arrows indicate whether the variant at the head of the column or row produced the best performance.

		qd			s		
		SD	AD	DD	SD	AD	DD
A	**SD**	1	←0.001	0.9	1	←0.001	0.8
	AD	↑0.001	1	↑0.001	↑0.001	1	↑0.001
	DD	0.9	←0.001	1	0.82	←0.001	1
B	**SD**	1	←0.001	0.6	1	← 0.001	0.9
	AD	↑0.001	1	↑0.001	↑0.001	1	↑0.001
	DD	0.6	←0.001	1	0.9	← 0.001	1

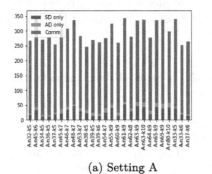

(a) Setting A

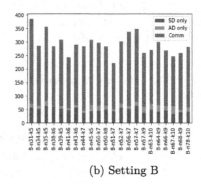

(b) Setting B

Fig. 1. Origin of solutions within the *grand* containers produced MAP-Elites incorporating the simple decoder (SD) and advanced decoder (AD). The "comm" solutions are those were found by both decoders. The x-axis shows each problem instance.

5   Managing a Collection of Emitters and Decoders

We now address the question of how to best utilise a diverse set of emitters and multiple decoders to optimise both qd and s. We first extend the set of emitters described in the previous section as follows:

- **NNSwap** - select a gene at random. Swap it to a position subsequent to the gene representing the closest visit
- **NNRoute** - Randomly select a route, and reorder using Nearest Neighbour heuristic.
- **SwapDelFrom(*vehicle*)** Randomly identify a gene that has *vehicle* as its preference and change to another vehicle
- **SwapDelTo(*vehicle*)** Randomly identify a gene that does **not** have *vehicle* as its preference and switch to *vehicle*.

The RR-MapElites algorithm described in Algorithm 1 is updated. Where an emitter takes a parameter, e.g. *vehicle* in *SwapDelTo (<vehicle>)*, an emitter is created for each possible value of *vehicle* available within the setting. Thus, for setting A (see Table 1) the emitters *SwapDelTo (CargoBike)*, *SwapDelTo (LargeVan)* and *SwapDelTo (Pedestrian)* would be included. As 4 out of the 7 emitters used are parameterised and 3 vehicle types are used in each setting, the size of the emitter pool is therefore 19 emitters. Solutions are created by applying the dual decoder to mutated genome, although the difference between SD and DD is not significant, DD does provide a small number of unique solutions via AD. The new variant with the larger emitter pool is named *Multi-Emitter (MuE)*. Each emitter in the pool is applied in turn on a round-robin basis as before.

Table 6 shows the differences in *qd* and *s* scores between the *DD* and *MuE* variants (also shown, but not yet discussed are the *MAB* variants). Table 7 shows the "winners" for the *DD* and *MuE* variants. For setting B *DD* produces a better performance in 21 out of 22 instances. For setting A, applying a paired-test shows no significant difference between the results w.r.t the *s* metric. On the other hand, there is statistical evidence that *MuE* outperforms DD with respect to the *qd* metric ($p \le 0.05$).

Table 6. The difference in mean values (as %s) between *DD* and the multi-emitter variants.

	s				qd			
Setting	MuE	MAB-ϵg	MAB-UCB	MAB-dϵg	MuE	MAB-ϵg	MAB-UCB	MAB-dϵg
A	99.6%	100.5%	100.6%	100.7%	100%	100%	100%	100%
B	95.3%	97.7%	96.5%	101.0%	100%	100%	100%	100%

Table 7. The performance of *DD* and *MuE* compared in terms of the number of instances where each variant performs best. In order to establish the significance of the differences noted, we apply a paired t-test. The values of *p* obtained are shown, as a *p* value of less than 0.03 is regarded as significant.

		Setting A	Setting B
s	DD	14	21
	MuE	12	1
	p	0.268	5.048^{-5}
qd	DD	20	21
	MuE	26	1
	p	0.0012	$5.8929^{-0.7}$

5.1 Emitter Selection via Multi-Arm Bandits

Section 5 highlights the need for a mechanism to prioritise the use of emitters/decoders that are producing solutions that improve the quantity of solutions and quality of the container rather than simply utilise each in turn. Hence, we follow the approach discussed in Sect. 2 proposed in the context of robotics [6] that uses a multi-arm bandit (MAB) algorithm to manage the choice of emitters at each iteration of the algorithm. We consider three versions of MAB[4] that use different strategies to optimise the reward. Specifically, we consider three different strategies commonly used in MAB literature:

- *MAB-ϵg* uses an Epsilon Greedy strategy [24]
- *MAB-ucb* uses the UCB (Upper Confidence Bound) strategy [3]
- *MAB-dϵg* uses a Diminishing Epsilon Greedy strategy [23]

Table 6 shows the results obtained with each *MAB* variant. As in previous sections, we further breakdown the results by examining the number of times each variant produces the best performing solution. Table 8 shows that of the three Multi Armed Bandit variants, *MAB-dϵg* appears most likely to produce a container which performs best according to the s and qd metrics. A Friedman test is utilised to calculate significance. The results are given in Table 10 which shows that the differences are significant with the exception of the qd score for setting A.

As the Friedman tests denote a significant difference in 3 cases, we carry out a Nemenyi post-hoc test to discover where that significance lies. The results of which may be seen in Table 9. Figures in italics note significance and the arrows point to the variant (row/column header) that produced the best results. The results can be summarised as follows: for the qd metric the improvements shown using *MAB-dϵg* are significant in the majority of cases, the exception being in setting A when *MAB-ucb* produces results that are not significantly different. When considering the size (s) of the archive, none of the variants exhibited a significant difference over *DD* in setting A, but for setting B we see that significant differences between variants are noted in favour of *DD*.

Table 8. The quantities of instances where each variant produces the best performing archive. Unlike the results shown in Tables 3 and 7 these results contain some instances where the best performing variant was tied (i.e. for setting A s there was one instance where both *DD* and *MAB-ϵg* produced the joint best performing solutions), in these occurrences each of the tied variants was awarded 0.5.

Metric	Variant	Setting A	Setting B
s	DD	5	8.5
	MAB-dϵg	7.5	12.5
	MAB-ϵg	7	1
	MAB-UCB	6.5	0
qd	DD	1	0
	MAB-dϵg	16.5	19
	MAB-ϵg	3	3
	MAB-UCB	5.5	0

---

[4] implemented using jxnl https://github.com/jxnl/bandits-java.

Table 9. The values returned by the Nemenyi post-hoc test, a value of less than 0.05 is regarded as an indicator of significance (highlighted in italics). The arrows indicate whether the variant at the head of the column or row produced the best performance.

		qd				s			
		DD	MAB-eg	MAB-ucb	MAB-deg	DD	MAB-eg	MAB-ucb	MAB-deg
A	DD	1	0.562	↑*0.008*	↑*0.001*	No significant difference noted.			
	MAB-eg	0.562	1	←*0.001*	↑*0.001*				
	MAB-ucb	←*0.008*	↑*0.001*	1	0.501				
	MAB-deg	←*0.001*	←*0.001*	0.501	1				
B	DD	1	↑*0.001*	0.152	↑*0.009*	1	← *0.001*	←*0.001*	0.9
	MAB-eg	←*0.001*	1	0.297	↑ *0.001*	↑*0.001*	1	0.63	↑*0.001*
	MAB-ucb	0.152	0.297	1	↑*0.001*	↑ *0.001*	0.63	1	↑*0.001*
	MAB-deg	←*0.009*	← *0.001*	←*0.001*	1	0.9	←*0.001*	←*0.001*	1

Table 10. Friedman tests for significance calculated over the results obtained for instances within both settings. Each p value is calculated using the results obtained from DD, MAB-*deg*, MAB-*eg* and MAB-*UCB*.

Setting	Metric	p
A	s	9.019^{-9}
	qd	0.232
B	s	1.146^{-9}
	qd	1.76^{-8}

6   Discussion

Recall that the goal of this research was to investigate whether the use of multiple decoders and multiple emitters can improve containers of solutions obtained on instances of two urban logistics settings, in terms of their qd and s scores. With

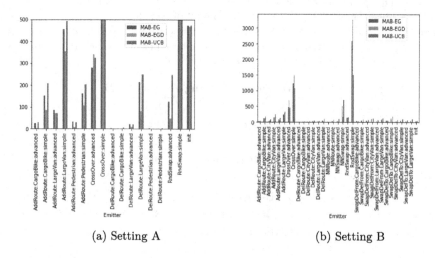

(a) Setting A   (b) Setting B

Fig. 2. The origin of solutions in the grand containers (emitter/decoder). Note that emitters that did not contribute to the final containers are not included.

respect to the use of multiple decoders, the results in Sect. 4 show that there are some solutions that are only produced by specific decoders. Therefore, using the dual-decoder algorithm can address this: however, note that this increases the number of solution evaluations performed at each iteration of the MAP-Elites algorithm.

Tables 11, 12, 13 and 14 summarise the qd and s scores obtained from all experiments conducted. When considering s, the $MAB\text{-}d\epsilon g$ and $MAB\text{-}ucb$ variants each produce the highest scoring containers in 30% of setting A instances and for setting B $MAB\text{-}d\epsilon g$ finds the highest scoring containers 95% of the time. In terms of qd, the $MAB\text{-}d\epsilon g$ variant performs best in both settings finding containers with the highest qd score in 65% and 95% for settings A and B. Figure 2 shows the emitter/decoder that were recorded as being used to create the solu-

Table 11. Setting A - s by instance by decoder and multi-arm bandit.

	Instance	SD	AD	MuE	MAB-ϵg	MAB-ucb	MAB-$d\epsilon g$
0	A-n32-k5	263	250	**266**	265	265	265
1	A-n45-k6	341	318	**350**	342	342	344
2	A-n34-k5	263	260	**275**	268	268	268
3	A-n36-k5	273	267	**280**	270	270	270
4	A-n37-k5	251	235	262	263	**268**	262
5	A-n45-k7	293	287	**316**	311	313	315
6	A-n46-k7	296	273	294	304	308	**311**
7	A-n48-k7	334	289	337	350	336	**346**
8	A-n53-k7	271	253	**285**	278	276	275
9	A-n38-k5	238	224	247	**252**	251	245
10	A-n39-k5	256	257	266	267	267	**271**
11	A-n39-k6	251	249	260	261	**265**	262
12	A-n54-k7	260	250	272	270	**276**	275
13	A-n55-k9	313	286	317	319	**325**	**325**
14	A-n60-k9	248	247	258	263	**265**	261
15	A-n61-k9	333	291	335	347	346	**347**
16	A-n62-k8	270	253	285	294	293	**296**
17	A-n63-k9	320	288	333	**339**	338	338
18	A-n63-k10	321	294	328	**347**	343	345
19	A-n64-k9	266	257	287	289	**290**	**290**
20	A-n65-k9	322	293	334	**348**	344	344
21	A-n69-k9	319	300	329	340	344	**345**
22	A-n80-k10	282	255	285	**301**	**301**	297
23	A-n33-k5	332	303	**335**	333	333	333
24	A-n33-k6	245	234	**250**	234	234	234
25	A-n37-k6	259	250	262	260	**266**	264
Tot.		0	0	8	5	8	8

tions that make up the final container. Those emitter/decoders which did not produce any solutions are not included in Fig. 2. Note that emitters based on the nearest neighbour heuristic have very little success in setting A, but far more success in setting B, justifying the use of a Multi-Arm-Bandit to select those emitters best suited to the problem. Additional conclusions provided by an in-depth analysis of the data include:

– Initialising MAP-Elites with optimised solutions from the single-objective EA is important: solutions created at this stage remain in the final container (see Fig. 2) and note the solutions that have the origin "init" which have remained unchanged from the initialisation phase.

Table 12. Setting A - qd by instance by decoder and multi-arm bandit.

	Instance	SD	AD	MuE	$MAB\text{-}\epsilon g$	$MAB\text{-}ucb$	$MAB\text{-}d\epsilon g$
0	A-n32-k5	0.008480	0.007575	0.008479	**0.008511**	**0.008511**	**0.008511**
1	A-n45-k6	0.008114	0.006716	0.008087	0.008205	0.008151	**0.008289**
2	A-n34-k5	0.009050	0.007954	**0.009181**	0.009141	0.009141	0.009141
3	A-n36-k5	0.008938	0.008065	**0.008988**	0.008930	0.008930	0.008930
4	A-n37-k5	0.009664	0.008072	0.009673	0.009803	**0.009906**	0.009835
5	A-n45-k7	0.007753	0.006270	0.007896	0.007989	0.007977	**0.008082**
6	A-n46-k7	0.007733	0.006113	0.007466	0.007756	0.007750	**0.008078**
7	A-n48-k7	0.007616	0.005650	0.007404	**0.008024**	0.007700	0.008021
8	A-n53-k7	0.005763	0.004489	0.005721	0.005804	0.005726	**0.005806**
9	A-n38-k5	0.007460	0.006392	0.007470	0.007626	0.007648	**0.007536**
10	A-n39-k5	0.007194	0.006521	0.007224	0.007319	0.007304	**0.007370**
11	A-n39-k6	0.008442	0.007251	0.008483	0.008593	**0.008665**	0.008611
12	A-n54-k7	0.006135	0.004787	0.006069	0.006222	0.006221	**0.006298**
13	A-n55-k9	0.006611	0.004618	0.006345	0.006639	**0.006688**	0.006843
14	A-n60-k9	0.004605	0.003847	0.004522	0.004693	**0.004756**	0.004733
15	A-n61-k9	0.005672	0.004171	0.005366	0.005816	0.005698	**0.005844**
16	A-n62-k8	0.003823	0.003139	0.003760	0.004016	0.003973	**0.004043**
17	A-n63-k9	0.004234	0.002989	0.004036	0.004322	0.004304	**0.004377**
18	A-n63-k10	0.004946	0.003531	0.004723	0.005088	0.005018	**0.005140**
19	A-n64-k9	0.003200	0.002637	0.003166	0.003295	0.003276	**0.003337**
20	A-n65-k9	0.004843	0.003511	0.004540	0.005022	0.004912	**0.005081**
21	A-n69-k9	0.004903	0.003735	0.004730	0.004990	0.005008	**0.005076**
22	A-n80-k10	0.003169	0.002183	0.002926	0.003262	0.003198	**0.003282**
23	A-n33-k5	**0.017040**	0.013398	0.016884	0.016969	0.016969	0.016969
24	A-n33-k6	**0.008077**	0.007177	0.008051	0.007925	0.007925	0.007925
25	A-n37-k6	0.006204	0.005541	0.006180	0.006208	**0.006279**	0.006279
Tot.		2	0	2	2	6	17

Table 13. Setting B - *s* by instance by decoder and multi-arm bandit.

	Instance	SD	AD	MuE	MAB-εg	MAB-ucb	MAB-dεg
0	B-n31-k5	371	330	396	**403**	386	**403**
1	B-n34-k5	274	234	281	295	285	**296**
2	B-n35-k5	336	295	324	349	330	**360**
3	B-n38-k6	281	225	286	292	275	**293**
4	B-n39-k5	300	255	296	310	**315**	**315**
5	B-n41-k6	235	195	234	234	235	**241**
6	B-n43-k6	279	233	285	295	294	**299**
7	B-n44-k7	276	246	266	279	276	**288**
8	B-n45-k5	283	265	291	318	306	**321**
9	B-n50-k7	277	251	272	287	276	**308**
10	B-n50-k8	268	236	262	272	274	**292**
11	B-n51-k7	216	179	212	226	211	**228**
12	B-n52-k7	287	246	295	300	**309**	305
13	B-n56-k7	329	286	335	319	**336**	**336**
14	B-n57-k7	331	286	338	325	337	**358**
15	B-n57-k9	239	222	255	257	259	**266**
16	B-n63-k10	248	221	246	257	249	**251**
17	B-n64-k9	276	259	289	292	279	**313**
18	B-n66-k9	255	222	255	263	255	**266**
19	B-n67-k10	218	215	238	238	242	**256**
20	B-n68-k9	252	220	258	254	264	**267**
21	B-n78-k10	264	240	280	282	277	**303**
Tot.		0	0	0	1	3	21

Table 14. Setting B - *qd* by instance by decoder and multi-arm bandit.

	Instance	SD	AD	MuE	MAB-εg	MAB-ucb	MAB-dεg
0	B-n31-k5	0.012838	0.010932	0.013114	**0.013394**	0.012987	0.013367
1	B-n34-k5	0.003933	0.003290	0.003954	0.004167	0.003983	**0.004193**
2	B-n35-k5	0.005829	0.004860	0.005563	0.005893	0.005557	**0.005952**
3	B-n38-k6	0.004741	0.003688	0.004716	0.004808	0.004583	**0.004824**
4	B-n39-k5	0.005557	0.004466	0.005307	0.005658	0.005635	**0.005725**
5	B-n41-k6	0.003405	0.002735	0.003318	0.003363	0.003332	**0.003426**
6	B-n43-k6	0.004260	0.003408	0.004216	0.004388	0.004296	**0.004446**
7	B-n44-k7	0.003874	0.003343	0.003688	0.003913	0.003791	**0.004009**
8	B-n45-k5	0.004848	0.004444	0.004754	0.005210	0.004939	**0.005272**
9	B-n50-k7	0.004453	0.003765	0.004270	0.004455	0.004323	**0.004726**
10	B-n50-k8	0.002588	0.002177	0.002456	0.002586	0.002528	**0.002721**
11	B-n51-k7	0.002162	0.001747	0.002094	0.002211	0.002095	**0.002253**
12	B-n52-k7	0.004661	0.003871	0.004625	0.004697	0.004790	**0.004804**
13	B-n56-k7	0.005297	0.004454	0.005066	0.005112	0.005142	**0.005312**
14	B-n57-k7	0.004322	0.003562	0.004146	0.004194	0.004186	**0.004417**
15	B-n57-k9	0.002166	0.001898	0.002186	0.002269	0.002213	**0.002322**
16	B-n63-k10	0.002020	0.001705	0.001986	0.002074	0.001997	**0.002055**
17	B-n64-k9	0.003116	0.002751	0.003110	0.003238	0.003076	**0.003402**
18	B-n66-k9	0.002558	0.002187	0.002486	0.002661	0.002486	**0.002645**
19	B-n67-k10	0.002029	0.001809	0.002060	0.002153	0.002109	**0.002255**
20	B-n68-k9	0.001945	0.001625	0.001907	0.001949	0.001950	**0.001981**
21	B-n78-k10	0.002264	0.001985	0.002233	0.002360	0.002272	**0.002499**
Tot.		0	0	0	1	0	21

- Simple random-swap based emitters are widely used, therefore should not be dismissed in favour of more complex operators
- The pattern of emitter usage differs between settings: e.g. in setting A, none of the nearest-neighbour heuristic emitters create a solution that is incorporated in the final container, yet they are widely used in setting B.

7 Conclusion

In previous work [27–29] we have demonstrated that MAP-Elites can be used provide a diverse set of optimised solutions in a combinatorial optimisation setting, specifically capacitated vehicle routing. While there have been significant developments to the MAP-Elites algorithm since its first introduction, the vast majority have been evaluated in the context of robotics and generative design [6,11,15], using continuous representations. Little attention has been paid to translating these to a combinatorial optimisation setting where there is significant potential for using the family of MAP-Elites algorithms to provide users with increased choice [26,27,29]. We proposed two methods for increasing the solutions returned (s) and diversity (qd) in a combinatorial setting of MAP-Elites: the first used multiple decodings of the same genome, while the second used a range of emitters with a multi-arm bandit to dynamically choose operators to generate new solutions. This direction is particularly promising, given that there is a rich history of developing domain-specific operators within vehicle-routing research that can be drawn on to further increase the set made available to the MAB. Furthermore, we note that our use of multiple decodings of the same genome has a conceptual parallel in some of the more recent work in MAP-Elites in the continuous domain that focuses on the data-driven, automatic generation of new representations of solutions during a run of MAP-Elites, to improve search [14].

References

1. Augerat, P., Belenger, J., Benavent, E., Corberan, A., Naddef, D., Rinald, G.: Computational results with a branch and cut code for the capacitated vehicle routing problem. Technical report Technical Report RR 949-M, University Joseph Fourier, Grenoble, France (1995)
2. Bossek, J., Neumann, F.: Exploring the feature space of tsp instances using quality diversity. In: Proceedings of the Genetic and Evolutionary Computation Conference, GECCO 2022, pp. 186–194. Association for Computing Machinery, New York (2022). https://doi.org/10.1145/3512290.3528851
3. Bouneffouf, D., Parthasarathy, S., Samulowitz, H., Wistuba, M.: Optimal exploitation of clustering and history information in multi-armed bandit. In: Proceedings of the 28th International Joint Conference on Artificial Intelligence, IJCAI 2019, pp. 2016–2022. AAAI Press (2019)
4. Cardoso, R.P., Hart, E., Kurka, D.B., Pitt, J.: WILDA: wide learning of diverse architectures for classification of large datasets. In: Castillo, P.A., Jiménez Laredo, J.L. (eds.) EvoApplications 2021. LNCS, vol. 12694, pp. 649–664. Springer, Cham (2021). https://doi.org/10.1007/978-3-030-72699-7_41

5. Cardoso, R.P., Hart, E., Pitt, J.V.: Diversity-driven wide learning for training distributed classification models. In: Proceedings of the 2020 Genetic and Evolutionary Computation Conference Companion, pp. 119–120 (2020)

6. Cully, A.: Multi-emitter MAP-elites. In: Proceedings of the Genetic and Evolutionary Computation Conference. ACM (jun 2021). https://doi.org/10.1145/3449639.3459326

7. DaCosta, L., Fialho, A., Schoenauer, M., Sebag, M.: Adaptive operator selection with dynamic multi-armed bandits. In: GECCO 2008, pp. 913–920. Association for Computing Machinery, New York (2008). https://doi.org/10.1145/1389095.1389272

8. Daniel, W.: Applied Nonparametric Statistics, 2nd edn. Brooks (1997)

9. Dantzig, G., Fulkerson, R., Johnson, S.: Solution of a large-scale traveling-salesman problem. J. Oper. Res. Soc. Am. **2**(4), 393–410 (1954). http://www.jstor.org/stable/166695, publisher: INFORMS

10. Dantzig, G.B., Ramser, J.H.: The truck dispatching problem. Manage. Sci. **6**(1), 80–91 (1959). https://doi.org/10.1287/mnsc.6.1.80, _eprint: https://doi.org/10.1287/mnsc.6.1.80

11. Fontaine, M.C., Togelius, J., Nikolaidis, S., Hoover, A.K.: Covariance matrix adaptation for the rapid illumination of behavior space. In: Proceedings of the 2020 Genetic and Evolutionary Computation Conference, pp. 94–102 (2020)

12. Gaier, A., Asteroth, A., Mouret, J.B.: Are quality diversity algorithms better at generating stepping stones than objective-based search? In: Proceedings of the Genetic and Evolutionary Computation Conference Companion, pp. 115–116 (2019)

13. Gilbert Laporte, P.T.: Vehicle routing: historical perspective and recent contributions. EURO J. Transp. Logistics **2**(1–2) (2013). https://doi.org/10.1007/s13676-013-0020-6

14. Hagg, A., Berns, S., Asteroth, A., Colton, S., Bäck, T.: Expressivity of parameterized and data-driven representations in quality diversity search. In: Proceedings of the Genetic and Evolutionary Computation Conference, pp. 678–686 (2021)

15. Hagg, A., Wilde, D., Asteroth, A., Bäck, T.: Designing air flow with surrogate-assisted phenotypic niching. In: Bäck, T., Preuss, M., Deutz, A., Wang, H., Doerr, C., Emmerich, M., Trautmann, H. (eds.) PPSN 2020. LNCS, vol. 12269, pp. 140–153. Springer, Cham (2020). https://doi.org/10.1007/978-3-030-58112-1_10

16. Hansen, N., Ostermeier, A.: Completely derandomized self-adaptation in evolution strategies. Evol. Comput. **9**(2), 159–195 (2001). https://doi.org/10.1162/106365601750190398

17. Hollander, M., Wolfe, A.D., Chicken, E.: Nonparametric Statistical Methods, 3rd edn. Brooks (2015)

18. Mouret, J.B., Clune, J.: Illuminating search spaces by mapping elites (2015). 10.48550/ARXIV.1504.04909. https://arxiv.org/abs/1504.04909

19. Nikfarjam, A., Neumann, A., Neumann, F.: On the use of quality diversity algorithms for the traveling thief problem. In: Proceedings of the Genetic and Evolutionary Computation Conference, pp. 260–268 (2022)

20. Potvin, J.Y.: State-of-the art review-evolutionary algorithms for vehicle routing. INFORMS J. Comput. **21**(4), 518–548 (2009)

21. Pugh, J.K., Soros, L., Szerlip, P.A., Stanley, K.O.: Confronting the challenge of quality diversity. In: Proceedings of the 2015 on Genetic and Evolutionary Computation Conference, pp. 967–974. ACM (2015)

22. Pugh, J.K., Soros, L.B., Stanley, K.O.: Quality diversity: a new frontier for evolutionary computation. Front. Robot. AI **3**, 40 (2016)

23. Shure, L.: Multi-armed bandit problem and exploration vs. exploitation trade-off (Oct 2016). https://blogs.mathworks.com/loren/2016/10/10/multi-armed-bandit-problem-and-exploration-vs-exploitation-trade-off/#f7f0010d-a0df-4bb9-9368-8d438cbf10d7
24. Sutton, R.S., Barto, A.G.: Multi-arm Bandits, chap. 2, pp. 31–47. MIT Press (2015)
25. Sánchez, M., Cruz-Duarte, J.M., Ortiz-Bayliss, J.C., Amaya, I.: Sequence-based selection hyper-heuristic model via map-elites. IEEE Access **9**, 116500–116527 (2021). https://doi.org/10.1109/ACCESS.2021.3106815
26. Urquhart, N., Guckert, M., Powers, S.: Increasing trust in meta-heuristics by using map-elites. In: Proceedings of the Genetic and Evolutionary Computation Conference Companion, pp. 1345–1348 (2019)
27. Urquhart, N., Hart, E., Hutcheson, W.: Using map-elites to support policy making around workforce scheduling and routing. at - Automatisierungstechnik **68**(2), 110–117 (2020). https://doi.org/10.1515/auto-2019-0107
28. Urquhart, N., Höhl, S., Hart, E.: An illumination algorithm approach to solving the micro-depot routing problem. In: Auger, A., Stützle, T. (eds.) Proceedings of the Genetic and Evolutionary Computation Conference, GECCO 2019, Prague, Czech Republic, July 13–17, 2019, pp. 1347–1355. ACM (2019). https://doi.org/10.1145/3321707.3321767
29. Urquhart, N., Höhl, S., Hart, E.: Automated, explainable rule extraction from MAP-elites archives. In: Castillo, P.A., Jiménez Laredo, J.L. (eds.) EvoApplications 2021. LNCS, vol. 12694, pp. 258–272. Springer, Cham (2021). https://doi.org/10.1007/978-3-030-72699-7_17

An Evolutionary Hyper-Heuristic
for Airport Slot Allocation

David Melder[(✉)] [iD], John Drake[iD], and Sha Wang[iD]

University of Leicester, Leicester, UK
{djjm1,john.drake,sw643}@leicester.ac.uk

Abstract. A large number of airports across Europe are resource constrained. With long-term growth in air transportation forecast to rise, more airports are expected to feel the imbalance between increased demand and limited resource capacity. This imbalance will lead to increasingly constrained scenarios, which require sophisticated solution methods to produce viable schedules. The use of 'slots' is one of the core mechanisms for managing access to constrained resources at airports. This paper presents a genetic algorithm-based hyper-heuristic approach to construct feasible solutions to the single airport slot allocation problem. To evaluate the proposed approach, we compare the solutions found by a number of previously developed and newly proposed constructive heuristics, over a range of real-world data sets. Our results show that the hyper-heuristic outperforms any individual constructive heuristic in all test instances, overcoming the drawbacks that a single heuristic faces when required to solve instances with different problem features.

Keywords: Hyper-Heuristic · Constructive Heuristic · Airport Slot Allocation · Genetic Algorithm · Scheduling

1 Introduction

Across Europe, air traffic is expected to grow by 1.9% per year, resulting in a 53% increase in flights by 2040 compared to 2017 levels [8]. Current capacity expansion plans fall short of the predicted demand, with an estimated 1.5 million flights (160 million passengers) set to be unaccommodated [8]. Furthermore, the imbalance between capacity and demand will result in a sevenfold increase in flight delays [9]. Runway expansion is not always a viable option to increase capacity due to a number of factors, such as political, environmental and other constraints [2,24]. Therefore, alternative measures to help alleviate the capacity-demand imbalance are needed. Economic based pricing is one such method where the price of a slot[1] is based on demand, or may be auctioned to airlines who value them the most [11,23].

---

[1] A slot is a period of time assigned to an airline in which they assume the right to use the full range of airport infrastructure [13].

J. Correia et al. (Eds.): EvoApplications 2023, LNCS 13989, pp. 53–68, 2023.
https://doi.org/10.1007/978-3-031-30229-9_4

An alternative approach is airport slot scheduling, or slot allocation, the focus of this study. Slot allocation is a combinatorial optimisation problem with the goal of optimising the allocation of slots to airlines, following a set of procedural rules, whilst adhering to a given set of constraints [23]. The slot allocation problem is complex due to the relatedness of airlines requesting the same slots across the scheduling period, and the specific capacity constraints that link requests together. In Europe, airports where demand exceeds capacity are known as coordinated airports. At these airports, a neutral schedule coordinator is used to allocate slots to arrival and departure movements (flights). Twice per year, before the summer and winter seasons begin, airlines submit requests to use slots to the coordinator(s) of the intended airports. The allocation of slots is governed by procedural rules that have been developed by the International Air Transportation Authority (IATA), known as the Worldwide Airport Slot Guidelines (WASG).

The focus of this paper is to create a feasible schedule for airlines at a given airport through the development of a novel hyper-heuristic framework. The hyper-heuristic adopts an evolutionary approach that adapts to the size of the given problem instance. The framework evolves candidate solutions, returning an ordered sequence of low-level constructive heuristics that are used to schedule requests. We will demonstrate that the proposed approach overcomes some of the drawbacks of the constructive heuristics currently used in the slot scheduling literature.

2 Related Work

Generating feasible solutions to slot allocation problems is itself not a trivial task, with both exact and heuristic methods adopted in the literature. Due to the size and complexity of solving such problems, heuristic approaches have been popular to solve larger instances when exact methods become intractable [17], or incorporated as warm-start solutions for matheuristic methods [18]. The focus of this paper lies within the category of heuristic approaches, for a review of exact methods we refer the interested reader to Zografos et al. [23].

The slot allocation problem has been modelled using a number of different objectives. Total displacement was first used to model the problem, where the time difference between requested and allocated slots is minimised [24]. Since then other metrics have been introduced, such as minimising the number of displaced slots [16,17], maximum displacement [19,22] and the number of rejected slots (where it is not possible to allocate a slot to a request) [16,19]. Fairness has also been incorporated into slot allocation models as a way to appropriately balance displacement across different airlines within the overall system [10,14]. Additionally, passenger-centric models have also been developed to improve passenger itineraries and transfer times [20].

A number of different constraints have been incorporated into slot allocation models. Declared capacity is used to ensure that the number of arrival and departure movements do not exceed airport limits. Capacity constraints are often set over rolling periods, where the number of arrival, departure and total movements are limited to declared thresholds over a period of consecutive slots [16,24]. Turnaround time constraints have also been considered, to ensure that an arrival movement is scheduled before a corresponding departure movement [14,22]. Terminal and apron constraints have been included to better represent the real-world problem, but increase the complexity and size of the model [18]. Some previous models have enforced that all movements are allocated a slot, not allowing any rejected requests [24]. However, as a constraint this can only work where capacity exceeds demand. When considering slot allocation as a network of multiple airports, rather than just a single airport, additional constraints also need to be considered. One example is the 'en-route' constraint, which ensures that the flight times between origin and destination airports are feasible [2].

Heuristic methods for solving the slot allocation problem typically involve a constructive and perturbative phase. A feasible solution is constructed, then attempts to improve the current solution are performed. Iterated Local Search and Variable Neighborhood Search were used by Pellegrini et al. [17] to solve the slot allocation problem, randomly selecting requests to allocate in both the constructive and perturbative stages. In the constructive phase, a Tabu back-tracking method is used to create a feasible solution, before a destroy and repair heuristic is used to improve the solution. A large-scale neighbourhood search was also used by Ribeiro et al. [18]. In the constructive phase, requests are split into groups depending on the number of requested operation days and sequentially assigned. A destroy and repair heuristic is then used to remove requests across sequential slots and reallocated. A two-stage heuristic approach was also used by Benlic [2] for slot allocation at the network level, where slot allocation problems for multiple airports are considered simultaneously. These previous studies have considered constructive heuristics that employ random ordering and allocate requests by most operation days. The use of alternative constructive heuristics to create feasible schedules for the slot allocation problem was investigated by Wang et al. [21]. Results showed that constructive heuristics perform differently across problem instances, therefore there is not one single global best to use for all cases.

Constructive heuristics performing differently across different problem instances lends itself to the idea that a more generalised approach, such as a hyper-heuristic, could be beneficial when developing solutions for the slot allocation problem, helping to overcome drawbacks that individual heuristics may encompass [3,6]. Hyper-heuristics are an approach to automate the design and selection of heuristics to solve optimisation problems [3]. Hyper-heuristics can be categorised into two groups: heuristic selection and heuristic generation. Selection hyper-heuristics pick a heuristic to use at a given point in the search, whereas generation hyper-heuristics generate new heuristics based on the components of other heuristics [3,7]. Furthermore, hyper-heuristics can be considered to be either *disposable*, where the framework is used to develop heuristics that can be applied to a single problem instance (or type of instance), or *reusable*, where the heuristics developed can be applied again to unseen instances [7].

Hyper-heuristics have been used for educational timetabling problems, with better results achieved using a larger number of low-level heuristics [4]. However, this increases the search space and computational time. Genetic algorithm based hyper-heuristics have previously been used in the context of scheduling for training courses to increase the generality of solving different problem instances [6]. The authors compared an adaptive layer on crossover and mutation rates against non-adaptive crossover and mutation rates. Results showed that a non-adaptive method typically outperformed the adaptive method. However, the requirement to tune parameters to each instance is costly. In the long run, using an adaptive method may increase generality and remove the need for parameter tuning. Adaptive length genes have also previously been used, allowing for the identification of good blocks of genes which can be expressed as a single gene to keep the desired genetic material [12]. Constraint satisfaction problems (CSP) have been solved using hyper-heuristics on real and generated instances [15]. The authors presented a messy-type genetic algorithm that uses variable length individuals to generate a new heuristic offline to be used on unseen CSP problems. Results suggested the hyper-heuristic can produce good solutions when compared against existing individual low-level heuristics. Boolean satisfiability problems (SAT) problems have been solved using a grammar based hyper-heuristic framework, where disposable heuristics are developed online, adapting to the specific problem instance as it is being solved [1]. In airport slot scheduling, requests can change every season and problem features vary depending on the airport. Thus a hyper-heuristic framework that can adapt to a variety of instances with different features could provide a more general solution approach, solving a wider range of problem instances effectively.

3   Solution Approach

This paper presents a hyper-heuristic genetic algorithm framework to generate solutions to the single airport slot allocation problem, across a complete scheduling season. The aim is to construct a high-quality feasible solution to the slot allocation problem, that can be used as a solution directly, or as a warm-start solution for further optimisation by alternative heuristics or exact methods. Given a set of requests, a feasible solution is constructed to maximise the number of accommodated requests whilst minimising total displacement. Displacement is measured in terms of the number of time periods between the requested and allocated slots. The mathematical model is based upon the one presented by Zografos et al. [24], adapted to allow for rejected requests, and is provided in detail in Appendix A.

3.1   Allocation Algorithm

The allocation algorithm used in this paper is based on the approach presented by Wang et al. [21]. However, paired arrivals and departures are treated as a single request and are scheduled or rejected simultaneously. Given a request, the algorithm first attempts to allocate a slot to the arrival and then the departure movements. A dynamic count of the capacity available for the airport is updated as slots are allocated. Starting with the arrival movements for a given request, if a requested slot is infeasible, the algorithm searches backwards and then forwards to find a feasible slot and if available selects the slot with the least displacement. If the displacement is equal when searching for a new slot to allocate, the earlier slot is selected. After an exhaustive search, if an arrival slot is not available, the corresponding departure slot is not allocated. Otherwise, the search then tries to find a corresponding departure slot to allocate. If a departure slot is not available, the corresponding arrival movement is not scheduled and is *rejected*. Otherwise the arrival and departure movement are scheduled and capacity is updated.

3.2   Request Ordering Heuristics

A solution can be constructed by iteratively adding requests (paired arrival and departure movements) to a partial solution in a given order until all requests have been allocated. The order in which they are added can be determined by a number of potential features of the given set of requests. For example, requests can be sorted by increasing number of operation days across the season. In this study, we use a set of seven static, dynamic and random low-level heuristics to order requests as described below. We introduce three novel low-level heuristics for this problem (MOW, LOW and LOD), in addition to a set of four previously proposed heuristics from the literature (RO, MOD, UC and SC) [21]. The seven low-level constructive heuristics are as follows:

- *Random Ordering (RO):* Randomly orders requests for assignment (Stochastic).
- *Most Operation Days (MOD):* Orders requests by decreasing order of operational days (Static).
- *Least Operation Days (LOD):* Orders requests by increasing order of operational days (Static).
- *Most Operation Weeks (MOW):* Orders requests by decreasing order of operational weeks (Static).
- *Least Operation Weeks (LOW):* Orders requests by increasing order of operation weeks requested (Static).
- *Unscheduled Conflicts (UC):* Orders requests that have the most conflicts, with regard to requested slots, for unallocated requests (Dynamic).
- *Scheduled Conflicts (SC):* Orders requests that have the most conflicts, with regard to requested slots, for allocated requests (Dynamic).

We note that for the final two heuristics (UC and SC), a conflict occurs when the same slot is requested. For example if two requests intend to operate the same arrival and departure slots, the total conflicts would be $2*Number\ of$ *Request Days*. Additionally, for these two heuristics the dynamic ordering of requests must be recalculated as each request is allocated.

3.3   Evolutionary Hyper-heuristic Approach

The search space and output of the hyper-heuristic approach is an ordered sequence of low-level heuristics that are used to select requests to be scheduled. Given a set of requests, a heuristic is used to select a request to schedule at each step until all requests have been allocated (or rejected). Once selected, the request is removed from the initial set. The hyper-heuristic uses a genetic algorithm representation to evolve candidate heuristic sequences, as demonstrated in Algorithm 1.

Algorithm 1. Hyper-Heuristic Framework

Input: Set of requests M, set of low-level heuristics H
1: generate initial population of heuristic sequences
2: **for** $i = 0$; $< iterations$; $i + +$ **do**
3:     select two parents randomly
4:     crossover parents to create two children
5:     mutate children
6:     replace lowest quality individuals in population if worse than generated children
7: **end for**
8: **return** best individual in the population

The rationale behind most constructive heuristics is to first schedule the most constrained requests. However, we also include heuristics that do not adhere to this principle. This is based on the notion that when used in the hyper-heuristic two possible scenarios may occur. Either the heuristics will not be included because they degrade the objective value and will not survive the evolutionary process, or including them in the search may improve both diversity and the final objective value and be beneficial to the search process.

For a given instance, the chromosome length is equal to the number of requests to be scheduled. Each gene is a low-level heuristic used to select the next request to allocate. The initial population is seeded with the seven heuristics introduced in the previous sub-section, with the remaining individuals generated randomly by setting each gene to be one of the heuristics with equal probability. Fitness is calculated by iterating through the chromosome and selecting a request to order depending on the gene. Each selected request is passed to the allocation algorithm and is either allocated or rejected. If the request is allocated, the displacement is equal to the absolute value of the difference between the requested slot period and the allocated period for each requested day. If the request is rejected, we set an arbitrarily large constant as its displacement value to enforce the model to maximise the number of requests allocated. At each iteration of the genetic algorithm, two parents are selected from the population. Single point crossover is applied to generate two child chromosomes. The child chromosomes are mutated such that a percentage of genes are swapped with each mutated gene being replaced at random with a new low-level heuristic. Both children are then compared to the population and if better replace weaker individuals, therefore implementing a steady state replacement schematic. We note that the stronger of the two child chromosomes will be retained if both are better than only a single individual in the population. To improve computational efficiency, the evaluation of the child solution will terminate early if the current fitness of a child solution is greater than the overall worst fitness in the population.

Figure 1 provides an illustration of four candidate sequences using the example requests provided in Table 1, along with the allocation of the best found solution using the allocation algorithm described in Sect. 3.1. In this example, we have seven slots available (T = {0-6}) and a maximum of one arrival and departure per slot. Three low-level heuristics are used, MOD, LOD and RO. As there are five requests, each candidate solution has five genes, each representing a heuristic used to select a request. The candidate sequences will select requests differently, producing different values for total displacement (TD) and total rejected requests (TRR).

Table 1. Slot allocation request examples.

Request	Arrival Time	Departure Time	Total Operation Days
R1	2	3	2
R2	1	4	3
R3	2	5	10
R4	3	4	5
R5	1	3	7

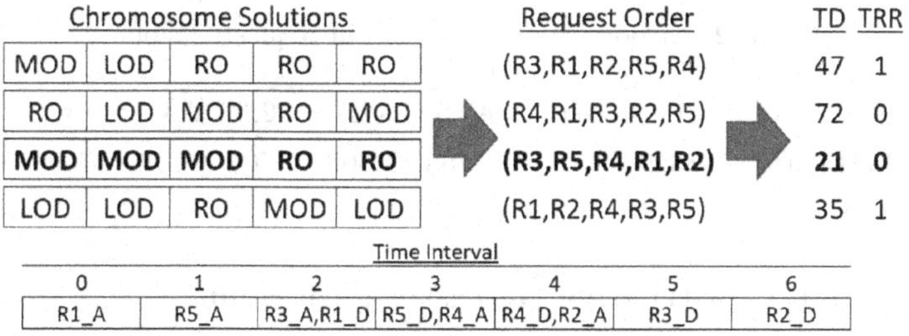

Fig. 1. Example of chromosomes using three request ordering heuristics to schedule five requests and the allocation of the best found solution.

4   Experimental Results

In this section we provide experimental analysis and results for the hyper-heuristic framework. We first show results from scheduling requests using each of the seven low-level heuristics introduced in Sect. 3.2 independently. Following this, we present some initial parameter tuning results to decide the values of

the main parameters for the remaining experiments. We then test and compare
the hyper-heuristic against each low-level heuristic, for a number of problem
instances with different request sizes and distributions to evaluate the effectiveness of the proposed approach.

The constructive and hyper-heuristic approaches are used to solve three real-
world and eight synthetic slot allocation problem instances (the synthetic data
sets are based on real-world instances). A summary of the total requested slots in
each instance can be seen in Table 2, with the number of requested slots increasing from type A1 airports to A3 airports. To assess the approaches tested, we
use metrics that have previously been used in airport slot scheduling literature,
total displacement (1), total unallocated requests (2) and total unallocated slots
(3) (mathematical notation can be found in Appendix A).

$$\sum_{m \in M} \sum_{t \in T} X_m^t \cdot |t - t_m| \cdot D_m \tag{1}$$

$$\sum_{m \in M} Y_m \tag{2}$$

$$\sum_{m \in M} Y_m \cdot D_m \tag{3}$$

Previous studies have allocated requests considering the priority levels from
the WASG guidelines [5,24]. In this study, requests are allocated across priority
levels together. However, one could allocate hierarchically and fix capacity levels
after the allocation of each priority level. We set the minimum turnaround time
to be 30 min for each request.

Table 2. Airport summary for total requested slots.

A1_1	A1_2	A1_3	A1_4	A1_5	A2_1	A2_2	A2_3	A3_1	A3_2	A3_3
15398	14214	15130	15992	14176	34768	34308	34704	47750	47590	43500

4.1 Individual Constructive Heuristic Performance

Figure 2 shows a bar plot of the performance of each of the seven low-level
heuristics for each airport. Scheduling the most constrained requests first tends
to result in a better objective value being obtained [21]. MOD and MOW are
the top performing heuristics across instances. However, there is not one single
heuristic that outperforms the rest in all cases. LOD and LOW typically perform
the worst with UC, SC and RO varying between the best and worst objective
values for each airport. Excluding RO, all requests and slots are allocated except
for A1_4 where SC results in 2 requests and 30 slots rejected.

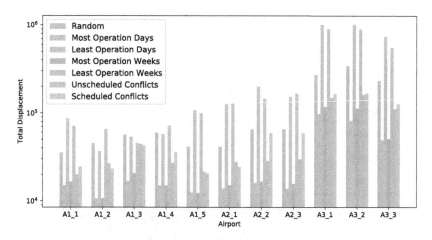

Fig. 2. Bar plot of constructive heuristic objective values on airport data sets.

4.2 Hyper-Heuristic Parameter Tuning

To tune the hyper-heuristic parameters, we use five repeated hyper-heuristic runs with varying levels of mutation rate (0.01, 0.05, 0.1, 0.2, 0.3) and population size (10, 20, 30, 100, 500, 1000) on data set A1_1. The stopping criteria for tuning the hyper-heuristic is set to 3000 iterations. For these preliminary tuning experiments, we use static and random heuristics (MOD, LOD, MOW, LOW, RO) because of the additional time required by the dynamic heuristics to update the order of requests after each allocation is made. We chose this subset as the goal of tuning is not necessarily to find the best objective value for a specific problem, but to see which combination of parameter values lead to reductions in displacement.

Figure 3 shows a boxplot of total displacement found over five runs of the hyper-heuristic for each parameter combination. We observe that as the size of the candidate population increases, in the case of 500 and 1000, there is higher total displacement. This is likely due to the larger population size, meaning that the search has not converged as well as a population with a smaller pool within the given stopping criteria. One could overcome this issue at the expense of increased iterations leading to a longer search time. A Kruskal-Wallis statistical test at the 95% confidence interval is carried out between the different combinations of parameter tuning groups. Results show a statistically significant difference between the groups.

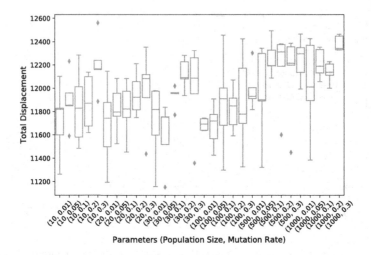

Fig. 3. Boxplot of hyper-heuristic run on tuning parameters.

The top five combinations of parameters, with regard to average displacement over the five runs, are further analysed. A Kruskal-Wallis statistical test at the 95% confidence interval shows no difference in this group. Within the top five, we observe smaller mutation rates and smaller population sizes. Table 3 shows the average total displacement and percentage improvement (over the repeated hyper-heuristic runs) across iterations compared with the best constructive heuristic objective value found (14952 using MOD). Interestingly, we find better objective values compared to MOD in the initial population where genes in the chromosome sequences are selected at random, before the evolutionary phase of the hyper-heuristic begins (i.e. at 0 iterations). The average displacement in the larger population sizes have better objective values in the initial set up stage before the algorithm runs, likely due to the fact that more solutions are being generated.

Table 3. Improvement in terms of total displacement and percentage improvement compared to the best heuristic value (14952) over iterations.

Parameter values (Pop. Size, Mut. rate)	Number of iterations						
	0	500	1000	1500	2000	2500	3000
(30, 0.05)	13971.2 (6.6%)	12083.2 (19.2%)	11916.4 (20.3%)	11844.8 (20.8%)	11790.4 (21.1%)	11790.4 (21.1%)	11602.8 (22.4%)
(100, 0.05)	13335.8 (10.8%)	12239.8 (18.1%)	12152.8 (18.7%)	11840.2 (20.8%)	11826.4 (20.9%)	11679.2 (21.9%)	11679.2 (21.9%)
(100, 0.01)	13535.6 (9.5%)	12158.2 (18.7%)	11932.2 (20.2%)	11839.0 (20.8%)	11783.2 (21.2%)	11760.8 (21.8%)	11690.2 (21.8%)
(20, 0.01)	14223.9 (4.9%)	12008.6 (19.7%)	11855.8 (20.7%)	11819.1 (21.0%)	11731.1 (21.5%)	11698.3 (21.8%)	11698.3 (21.8%)
(30, 0.01)	13918.4 (6.9%)	12132.2 (18.9%)	12049.8 (19.4%)	11985.8 (19.8%)	11944.6 (20.1%)	11705.2 (21.7%)	11705.2 (21.7%)

Considering all tuning combinations, the first 500 iterations yield average improvements of around 19% with the remaining iterations yielding on average an additional 3% improvement, indicating that the majority of improvement is occurring at the earlier stages of the algorithm. This improvement can be observed in Fig. 4 which shows the average improvement in displacement, across the top 5 parameter sets at every 10 iterations. We observe that the smaller population sizes have better objective values after 500 iterations once the selection pressure starts to lead to a more intensive search on these individuals. As our method only accepts better solutions into the population, the larger population groups should have more diversity which one may only see the benefit of at later iterations. However, we see similar improvements in displacement at the latest stage of the search and observe that the smaller populations have a slightly better displacement at this point. We therefore opt to use a candidate population size of 30 and a percentage mutation rate of 0.05 in our remaining experiments, as we see the best overall percentage improvement leading to the lowest objective value using these parameters. These parameters will be used to evaluate the hyper-heuristic approach against individual constructive heuristics on a wider variety of problem instances.

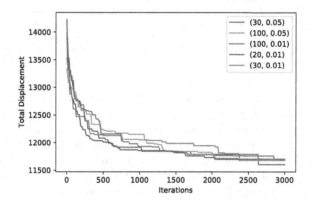

Fig. 4. Average improvement in displacement over iterations for different parameter settings (population size, mutation rate).

4.3 Hyper-Heuristic Search

In this section we compare the results of applying the hyper-heuristic against the performance of the best performing constructive heuristic for the instance. Based on the tuning in the previous subsection, we use a population size of 30 and mutation rate of 0.05. Stopping criteria is set to 3000 iterations and additionally a time limit of 4 and 8 h is imposed on the larger type A2 and A3 airports respectively, as increasing the number of requests significantly increases the run time of the hyper-heuristic framework. As the scheduling process itself occurs twice a year and coordinators have a large time window to develop a schedule, we do not deem the run times to be an issue. Each data set is run five times using the hyper-heuristic to assess average performance.

Table 4 shows the results of each hyper-heuristic run against the best individual constructive heuristic for each airport instance. We see improvements on every run of the hyper-heuristic, with improvements in the average total displacement in the range of 1.9% to 19.6%. Larger improvements are seen with the smaller type A1 airport instances. We find that using the hyper-heuristic leads to all requests being allocated.

Table 4. Total displacement and percentage improvement compared to best constructive heuristic used for each instance, across each hyper-heuristic run. Highlighted in bold is the best result for each airport instance.

Instance	Best Heuristic	Run_0	Run_1	Run_2	Run_3	Run_4	Average Improvement
A1_1	14952, MOD	11883 (20.5%)	**11872 (20.6%)**	12221 (18.3%)	12136 (18.8%)	11994 (19.8%)	19.6%
A1_2	10653, MOD	10162 (4.6%)	**9861 (7.4%)**	9864 (7.4%)	10025 (5.9%)	10207 (4.2%)	5.9%
A1_3	16847, MOD	**14901 (11.6%)**	14946 (11.3%)	15087 (10.4%)	15155 (10.0%)	15182 (9.9%)	10.6%
A1_4	14989, MOW	14549 (2.9%)	14247 (5.0%)	14404 (3.9%)	14453 (3.6%)	**14177 (5.4%)**	4.2%
A1_5	12365, MOW	**10875 (12.1%)**	10917 (11.7%)	10996 (11.1%)	10913 (11.7%)	11004 (11.0%)	11.5%
A2_1	13892, MOD	**12722 (8.4%)**	12751 (8.2%)	12864 (7.4%)	12926 (7.0%)	12739 (8.3%)	7.9%
A2_2	16049, MOD	14934 (6.9%)	**14911 (7.1%)**	15273 (4.8%)	15165 (5.5%)	15017 (6.4%)	6.1%
A2_3	13745, MOD	12799 (6.9%)	12996 (5.4%)	13079 (4.8%)	12847 (6.5%)	**12718 (7.5%)**	6.2%
A3_1	97693, MOD	91485 (6.4%)	92819 (5.0%)	91022 (6.8%)	92189 (5.6%)	**90200 (7.7%)**	6.3%
A3_2	81714, MOD	**79578 (2.6%)**	80245 (1.8%)	80827 (1.1%)	79932 (2.2%)	80428 (1.6%)	1.9%
A3_3	49320, MOD	**46059 (6.6%)**	46990 (4.7%)	46487 (5.7%)	46816 (5.1%)	46856 (5.0%)	5.4%

We next examine the low-level heuristics selected within the hyper-heuristic solution. For each instance and repeated run (55 total) we find the most selected heuristic across the solution chromosome. We observe that MOD, MOW and SC are the only dominating heuristics, selected most in 34.5%, 20% and 45.5% of runs respectively. Although SC by itself produced poorer quality schedules in comparison to MOD and MOW, it dominates in the hyper-heuristic solution more often than any other heuristic. Furthermore, we look at the final solution obtained to identify potential patterns in how the chromosomes are evolving across the runs. MOD, MOW and SC dominate the selection across solutions however all other heuristics are selected at least once, suggesting that although the aforementioned heuristics dominate the gene space, including the additional heuristics brings a level of diversity which helps to improve the overall displacement. We observe that when selected, the less dominating heuristics (RO, LOW, LOW and UC) tend to be at the later end of the chromosome, whereas the dominating heuristics appear throughout the sequence. An example of this can be seen in Fig. 5, which shows a cumulative count of the constructive heuristics selected from the beginning to the end of the gene. The solution is dominated by MOD, MOW and SC but we observe an increase in the use of the remaining heuristics towards the end of the chromosome.

5   Conclusions

In this paper, we have developed a genetic algorithm-based hyper-heuristic framework for solving the single airport slot allocation problem. We have shown

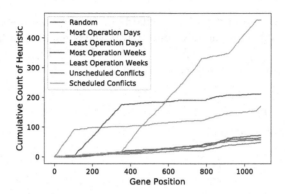

Fig. 5. Cumulative heuristic count over the run of solution A2_2 chromosome.

that the proposed method outperforms any single low-level constructive heuristic across all problem instances, reducing displacement in the range of 1.9% to 19.6%. Heuristics that have weaker individual performance are beneficial within the hyper-heuristic approach. Furthermore, such heuristics, when used within the hyper-heuristic, can dominate the chromosome space with regard to frequency of occurrence. We observe that constructive heuristics that schedule the most constrained requests are dominant in the early-mid stages of the chromosome, whilst the remaining heuristics occur more frequently towards the end. We also see in some instances that randomly selecting low-level heuristics for each gene produces better objective values than the best constructive heuristic, even before the evolutionary process starts.

Future work will look at increasing the generality of the model through the use of adaptive mutation and crossover rates to eliminate the need for model tuning. Additionally, the solutions yielded by this approach will be tested as warm start solutions, to assess whether providing a better initial solution for a destroy and repair type phase results in better final solutions. Finally, a natural extension of this would be to investigate network level slot allocation models, which consider the requests at multiple airports simultaneously.

Appendix A

Table 5 shows model parameters and inputs for the mathematical formulation presented below. Adapted from [24], we introduce a new decision variable Y_m which is equal to 1 if a request is rejected, 0 otherwise. The objective function (A.1) is used to maximise the number of requests allocated and to minimise displacement across all scheduling days requested. By setting constant Q to be arbitrarily large we enforce that requests are allocated. Furthermore, we introduce the addition of Equations A.2 and A.3. Equation A.2 enforces that a request is either allocated or rejected. Equation A.3 enforces that if an arrival (departure) request is allocated or rejected, the corresponding departure (arrival) request is allocated or rejected. Equation A.4 ensures that capacity constraints are met whilst Equation A.5 ensures a minimum feasible turnaround time for corresponding arrival and departure requests operating on the same day. If the departure is scheduled for the day after arrival, the equation can be replaced by $\sum_{t\in[0,k)} X^t_{dep} + \sum_{t\in[n-t^{dep}_{arr}+k,n)} X^t_{arr} \leq 1$, $k \in [1, t^{dep}_{arr})$ Equation A.6 and A.7 are binary constraints for the decision variables.

$$Minimise \sum_{m\in M}\sum_{t\in T} X^t_m \cdot |t - t_m| \cdot D_m + Q \cdot Y_m \tag{A.1}$$

$$Subject\ to \sum_{t\in T} X^t_m + Y_m = 1 \qquad\qquad \forall\, m \in M \tag{A.2}$$

$$\sum_{t\in T} X^t_{arr} - \sum_{t\in T} X^t_{dep} = 0 \qquad\qquad \forall\, \{arr, dep\} \in P \tag{A.3}$$

$$\sum_{m\in M}\sum_{t\in T^S_C} a^d_m \cdot b^c_m \cdot X^t_m \leq u^{dt}_c \qquad\qquad \forall\, c \in C, d \in D, s \in T_c \tag{A.4}$$

$$\sum_{t\in[0,k)} X^t_{dep} + \sum_{t\in[k-t^{dep}_{arr},n)} X^t_{arr} \leq 1 \quad \forall\, \{arr, dep\} \in P, k \in [t^{dep}_{arr}, n) \tag{A.5}$$

Table 5. Notation for slot allocation model.

Decision Variables	
X^t_m	Equal to 1 if request m is allocated to time t, 0 otherwise
Y_m	Equal to 1 if request m is rejected, 0 otherwise
Model Parameters	
$T = \{0, ..., n-1\}$	Set of coordination time intervals per day
D	Set of calendar days $d \in D$
M	Set of requests $m \in M$
$M_{arr}(M_{dep}) \in M$	Subset of arrival(departure) requests in M
P	Set of corresponding arrival and departure request pairs $\{M_{arr} \in M, M_{dep} \in M\}$
C	Set of airport capacity constraints $c \in C$
D_m	Number of operation days for request m
t_m	Requested slot time for request $m \in M$
a^d_m	1 if m requests to operate on day d, 0 otherwise
$u^{dt}_c \geq 0$	Capacity of constraint c on day d at time interval $t \in T_c$ with $T_c = \{t \in T | t < n - t_c + 1\}$
$s \in T_c$	For a specified start time s, let the set of consecutive intervals to check be $T^s_c = \{t \in T | s \leq t < s + t_c\}$
$t_c > 0$	Duration of capacity constraint c
b_{cm}	Unit of capacity consumed by request m for capacity constraint c
$t^{arr}_{dep} \geq 0$	Turnaround time for $p \in P$

$$X^t_m \in \{0,1\} \forall\, m \in M, t \in T \tag{A.6}$$

$$Y_m \in \{0,1\} \forall\, m \in M \tag{A.7}$$

References

1. Bader-El-Den, M., Poli, R.: Generating SAT local-search heuristics using a GP hyper-heuristic framework. In: Monmarché, N., Talbi, E.-G., Collet, P., Schoenauer, M., Lutton, E. (eds.) EA 2007. LNCS, vol. 4926, pp. 37–49. Springer, Heidelberg (2008). https://doi.org/10.1007/978-3-540-79305-2_4
2. Benlic, U.: Heuristic search for allocation of slots at network level. Transp. Res. Part C Emerging Technol. **86**, 488–509 (2018)
3. Burke, E., Gendreau, M., Hyde, M., Kendall, G., Ochoa, G., Özcan, E., Qu, R.: Hyper-heuristics: a survey of the state of the art. J. Oper. Res. Soc. **64**(12), 1695–1724 (2013)
4. Burke, E., McCollum, B., Meisels, A., Petrovic, S., Qu, R.: A graph-based hyper-heuristic for educational timetabling problems. Eur. J. Oper. Res. **176**(1), 177–192 (2007)
5. Castelli, L., Pellegrini, P., Pesenti, R.: Airport slot allocation in Europe: economic efficiency and fairness. Int. J. Revenue Manage. **6**(1/2), 28–44 (2012)
6. Cowling, P., Kendall, G., Han, L.: An investigation of a hyperheuristic genetic algorithm applied to a trainer scheduling problem. In: Proceedings of the 2002 Congress on Evolutionary Computation (CEC 2002), pp. 1185–1190. IEEE (2002)
7. Drake, J., Kheiri, A., Özcan, E., Burke, E.: Recent advances in selection hyper-heuristics. Eur. J. Oper. Res. **285**, 405–428 (2020)
8. Eurocontrol.: European Aviation in 2040 - Annex1: Flight Forecast to 2040. Technical report, Eurocontrol (2018). https://www.eurocontrol.int/publication/challenges-growth-2018
9. Eurocontrol.: European Aviation in 2040 - Annex4: Network Congestion. Technical report, Eurocontrol (2018). https://www.eurocontrol.int/publication/challenges-growth-2018
10. Fairbrother, J., Zografos, K., Glazebrook, K.: A slot-scheduling mechanism at congested airports that incorporates efficiency, fairness, and airline preferences. Transp. Sci. **54**(1), 115–138 (2020)
11. Gillen, D., Jacquillat, A., Odoni, A.: Airport demand management: the operations research and economics perspectives and potential synergies. Transp. Res. Part A Policy Practice **94**, 495–513 (2016)
12. Han, L., Kendall, G., Cowling, P.: An adaptive length chromosome hyper-heuristic genetic algorithm for a trainer scheduling problem. Recent Advances In Simulated Evolution And Learning, pp. 506–525 (2004)
13. IATA, Aci, WWACG: Worldwide Airport Slot Guidelines (WASG). Technical report (2020). https://www.iata.org/contentassets/4ede2aabfcc14a55919e468054d714fe/wasg-edition-1-english-version.pdf
14. Jiang, Y., Zografos, K.: A decision making framework for incorporating fairness in allocating slots at capacity-constrained airports. Transp. Res. Part C Emerg. Technol. **126**, 103039 (2021)
15. Ortiz-Bayliss, J.C., Terashima-Marín, H., Conant-Pablos, S.E.: Combine and conquer: an evolutionary hyper-heuristic approach for solving constraint satisfaction problems. Artif. Intell. Rev. **46**(3), 327–349 (2016). https://doi.org/10.1007/s10462-016-9466-x
16. Pellegrini, P., Bolić, T., Castelli, L., Pesenti, R.: SOSTA: an effective model for the simultaneous optimisation of airport SloT allocation. Transp. Res. Part E Logistics Transp. Rev. **99**, 34–53 (2017)

17. Pellegrini, P., Castelli, L., Pesenti, R.: Metaheuristic algorithms for the simultane-ous slot allocation problem. IET Intel. Transport Syst. **6**(4), 453–462 (2012)
18. Ribeiro, N., Jacquillat, A., Antunes, A.: A large-scale neighborhood search app-roach to airport slot allocation. Transp. Sci. **53**(6), 1772–1797 (2019)
19. Ribeiro, N., Jacquillat, A., Antunes, A., Odoni, A., Pita, J.: An optimization approach for airport slot allocation under IATA guidelines. Transp. Res. Part B Methodol. **112**, 132–156 (2018)
20. Ribeiro, N., Schmedeman, P., Birolini, S., Jacquillat, A.: Passenger-centric slot allocation at schedule-coordinated airports. SSRN Electron. J., 1–43 (2021)
21. Wang, S., Drake, J., Fairbrother, J., Woodward, J.: A constructive heuristic app-roach for single airport slot allocation problems. In: 2019 IEEE Symposium Series on Computational Intelligence (SSCI 2019), pp. 1171–1178 (2019)
22. Zografos, K., Androutsopoulos, K., Madas, M.: Minding the gap: optimizing airport schedule displacement and acceptability. Transp. Res. Part A Policy Practice **114**, 203–221 (2018)
23. Zografos, K., Madas, M., Androutsopoulos, K.: Increasing airport capacity utilisa-tion through optimum slot scheduling: review of current developments and identi-fication of future needs. J. Sched. **20**(1), 3–24 (2017)
24. Zografos, K., Salouras, Y., Madas, M.: Dealing with the efficient allocation of scarce resources at congested airports. Transp. Res. Part C Emerg. Technol. **21**(1), 244–256 (2012)

A Fitness Landscape Analysis Approach for Reinforcement Learning in the Control of the Coupled Inverted Pendulum Task

Ferrante Neri[1]([✉]) [iD] and Alexander Turner[2] [iD]

[1] NICE Group, School of Computer Science and Electronic Engineering,
University of Surrey, Guildford, UK
f.neri@surrey.ac.uk
[2] MRL, School of Computer Science, University of Nottingham, Nottingham, UK
alexander.turner@nottingham.ac.uk

Abstract. Fitness Landscape Analysis (FLA) for loss/gain functions for Machine Learning is an emerging research trend in Computational Intelligence that offered an alternative view on how learning algorithms work and should be designed. The vast majority of these recent studies investigate supervised learning whereas reinforcement learning remains so far nearly unaddressed. This paper performs a FLA on the reinforcement learning of a deep neural network for a simulated robot control task and focuses on ruggedness and neutrality. Two configurations of the physical system under investigation are considered and studied separately to highlight differences and similarities. Furthermore, the results of the performed FLA are put into relation with performance of the learning algorithm with the aim of achieving an understanding of most suitable parameter setting. Numerical results indicate a correlation between ruggedness and exploration to enable more optimal reinforcement learning. In the presence of high ruggedness the algorithm displays its best performance when the control parameters are set to enable a high degree of exploration. Conversely, when the landscape appears less rugged a less exploratory behaviour seems to contribute to the best performance of the learning algorithm.

Keywords: Fitness landscape analysis · Reinforcement learning · Evolutionary strategies

In computational intelligence, the term fitness landscape refers to the tuple composed of a fitness function and the domain where this function is defined [29]. Fitness Landscape Analysis (FLA) [18] is a group of techniques to analyse and characterise the landscape. More specifically, FLA aims at extracting some features of an optimisation problem such as the number of optima [7] and the correlation between pairs of variables [25]. This approach is well established in the discrete domain [21,22,28] and has been emerging also in the continuous domain [13,17,19].

In recent years, researchers investigated the landscape of the loss function of neural networks with various architectures. A comprehensive study on loss

J. Correia et al. (Eds.): EvoApplications 2023, LNCS 13989, pp. 69–85, 2023.
https://doi.org/10.1007/978-3-031-30229-9_5

landscapes of feed-forward neural networks was published in [2]. The landscapes of generative adversarial networks and weight-elimination neural networks are investigated in [24] and [3], respectively. Some work has been done to visualise the basin of attraction of loss functions [5] while some other works focus on techniques to sample data for analysing the landscape [4]. These are examples of several studies which are motivated by an observation reported in [9] which states that the fitness landscapes of loss functions are non-convex, complex and poorly understood.

Although [9] explicitly refers to supervised learning, the idea has been very recently introduced for the reinforcement learning domain [16]. Following this intuition, the present paper performs a first systematic investigation of FLA on a reinforcement learning. This study is carried out within the context of robot control in two configurations. Furthermore, this article makes use of the results of FLA to customise the learning algorithm used to train a neural network. More specifically, we use the results of ruggedness and neutrality to set the perturbation parameter of the $(1 + 1)$ Evolution Strategies (ES).

It must be remarked that in this article we intentionally employ a simplistic learning approach to perform the reinforcement learning: we chose a naive implementation of the ES as its algorithmic behaviour is clear and depends only on one parameter. The meaning of this parameter within the fitness landscape is intuitive and allows us to explain how linking this parameter to the parameters identified during FLA can enhance the performance of the algorithm. The remainder of this article is organised in the following way. Section 1 formulates the robotic task, i.e., the couple inverted pendulums task. Section 2 introduces the neural network and the fitness used in this article and briefly explains how the reinforcement learning is carried out. Section 3 describes the FLA carried out in this paper including how the data are collected through random walk and processed to estimate ruggedness and neutrality of the learning problem. Section 4 briefly outlines the ES functioning. Section 5 presents the FLA results and make some considerations on how FLA can inform the ES parameter setting. Then the results of the proposed approach against traditional ES setting are compared. Section 6 provides the conclusion of this study.

1   The Coupled Inverted Pendulums Task

The coupled inverted pendulums task is a control task designed to mimic complex real world control tasks such as the locomotion of multi-legged robots [12]. It provides a general purpose task for which many different control algorithms can be compared against to measure their ability to solve the task and to better understand their properties and functionality. This alleviates the time pressures associated with building bespoke tasks in which to evaluate a particular control algorithm.

The task consists of a 1-dimension track which contains between 1 - 5 carts. Each cart has a pendulum mounted at its center hanging below it. If there is more than one cart in the simulation, the carts are tethered together to restrict their

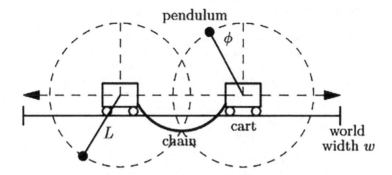

Fig. 1. Pendulum task·The coupled inverted pendulums task configured for two carts. The objective is to move the carts in such a way to move the pendulums to the upright position and maintain their balance [12].

movement. If this is the case, the movements of the carts must be coordinated so that the tether is not stretched or the carts make contact. If any cart hits another cart, or the boundary of the simulation, the simulation is halted and its fitness returned at the time point. The objective of the task is to move the cart(s) in such a way as to swing the pendulum into the upright position and maintain it there. The number of the carts within the simulation is directly related to the difficulty of the task. With a single cart, there is no interference on the track by other carts, so the cart can move with freedom along the track. With greater number of carts, the coordination and use of space becomes important, requiring additional criteria to be optimised to complete the task. There is a significant difference in complexity in the task between one and two pendulums.

Each of the cart(s) is controlled individually, and values specifying the speed at which the cart(s) is to move are provided at each time step. At each time step, each cart provides 10 sensor readings that describe its current state which can be seen in Table 1. The difference between the two actuator values specifies the speed at which the cart will move on the next time step. The simulation will run for a maximum of 3000 time steps. We use an aggregate fitness function which specifies the percentage of time steps that all pendulums spent in the upper equilibrium position

2   Robotic Control by Deep Reinforcement Learning

Reinforcement learning is a subsection of machine learning where an agent seeks to maximise a given reward over a number of timesteps [30]. Reinforcement learning differs from the more classical supervised learning as it does not require the use of labelled training data which can often be time consuming to acquire. There are a broad range of reinforcement leaning algorithms available, gaining a resurgence of popularity after deep neural networks were used in combination with reinforcement learning to play a range of computer games at better than human level [23]. This system utilised deep neural networks in a process called

Table 1. Parameters of the coupled inverted pendulums task.

ID	Sensor Name	System to sensor mapping
S_0	Pendulum Angle 0	$\phi \in [0, 0.5\pi] \rightarrow [127, 0]$, 0 else
S_1	Pendulum Angle 1	$\phi \in [1.5\pi, 2\pi] \rightarrow [0, 127]$, 0 else
S_2	Pendulum Angle 2	$\phi \in [0.5\pi, \pi] \rightarrow [127, 0]$, 0 else
S_3	Pendulum Angle 3	$\phi \in [\pi, 1.5\pi] \rightarrow [0, 127]$, 0 else
S_4	Proximity 0	Distance left $\rightarrow [0, 127]$
S_5	Proximity 1	Distance right $\rightarrow [0, 127]$
S_6	Cart Velocity 0	$v \in [-2, 0] \rightarrow [127, 0]$, 0 else
S_7	Cart Velocity 1	$v \in [0, 2] \rightarrow [0, 127]$, 0 else
S_8	Angular Velocity 0	$w \in [-5\pi, 0] \rightarrow [127, 0]$, 0 else
S_9	Angular Velocity 1	$w \in [0, 5\pi] \rightarrow [0, 127]$, 0 else
A_i	Actuators 0	$A_i \in [0, 127]$, for $i \in 0, 1$
u	Motor Control 0	$2(A_0/127 - A_1/127) \rightarrow [0, 1]$

deep Q learning [32]. Deep neural networks have multiple layers of neurons in between the input and output layer [15]. In this study, for each cart, we have used neural network with five layers containing 10, 10, 5, 5, 2 neurons in each of it and depicted in Fig. 2. The activation function used in this study is a sigmoid.

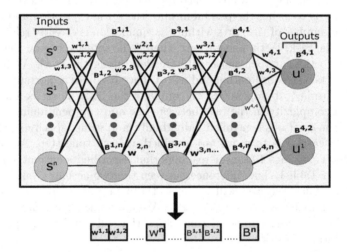

Fig. 2. Illustration of how the data from the neural network gets converted to a vector consisting of the weights and biases of the network. The sensor values and the actuator values are described in Table 1.

A candidate set of parameters of a neural network is here encoded as a vector whose elements are weights and biases of the network

$$\left(w^{1,1}, w^{1,2} \ldots, b^{1,1}, b^{1,2}, \ldots \right) \qquad (1)$$

where the first index refers to the layer and the second index to the position of the neuron within the layer. We may observe that the length of the vector in

Eq. 1 is 207 as it contains $10 \times 10 + 10 \times 5 + 5 \times 5 + 5 \times 2 = 185$ weights $w^{i,j}$ and $10 + 5 + 5 + 2 = 22$ biases. Each of these parameters is defined within the interval $[-1, 1]$.

In a nutshell, each neural network can be seen a mathematical model of each of the carts. The ten inputs of the network are the parameters $S_0, S_2, \ldots, S_9$ listed in Table 1 describing the kinematics of the cart and listed while the two outputs of the network are the acceleration values of each wheel (pair of wheels in the three-dimensional case) of the cart, A_1 and A_2 respectively.

Since, as shown in Fig. 1, in our problem multiple interconnected carts are considered, the candidate solution $\mathbf{x}$ is composed of weights and biases of multiple neural networks of the type in Eq. 1, each of them representing a cart. In order to assess the quality, i.e., the fitness f, associated with a candidate solution $\mathbf{x}$ a simulation is run. For each time step of the simulation, the section of $\mathbf{x}$ representing each cart is extracted and its outputs are calculated and its kinematics parameters $S_0, S_2, \ldots, S_9$ updated. During the simulation, the time during which each pendulum in the upright position is recorded. The fitness $f(\mathbf{x})$ to be maximised is the fraction of time during which each pendulum is in the upright position. Simulations are continued for a prearranged observation window and interrupted if a chain is broken or the carts have collided. In this study we consider two configurations, the first is with only one cart (*single pendulum task*) and three carts connected by two chains (*triple pendulum task*). Let us indicate with n_{pend} the number of carts/pendulums involved.

The pseudocode describing the calculation of the fitness function $f(\mathbf{x})$ for the coupled inverted pendulums studied in this article and depicted in Fig. 1 is shown in Algorithm 1.

3   Fitness Landscape Analysis

In order to analyse the learning problem formulated above, we have collected samples of the landscape by using a random walk structure, see [14], and then processed the samples to measure ruggedness and neutrality [2, 27]. The following subsections outline the chosen implementation of random walk employed to collect the data and the procedures used to analyse them and draw some conclusions about the problem.

3.1   Progressive Random Walk

Due to the continuous nature of the domain, we selected the progressive random walk as proposed in [19, 20]. The general idea of this method is that points are sampled one by one and, over time, each point is a neighbour of the previous sample. As a result, a sequence of points is stored in a data structure V.

More specifically, the initial point $\mathbf{x_0} = (x_1, x_2, \ldots, x_n)$ is sampled within the decision space identified by the weights, i.e., $[-1, 1]^n$. Then, for $l = 0, 1, 2, \ldots, L-1$ a point $\mathbf{x_{l+1}}$ is sampled as

$$\mathbf{x_{l+1}} = \mathbf{x_l} + \Delta \mathbf{x_l} \tag{2}$$

Algorithm 1. Fitness function f for the coupled inverted pendulums task using deep neural networks

---

INPUT The candidate solution $\mathbf{x}$ containing weights and biases of two neural networks (one for each cart)

$k = 1$

while Carts within the simulation are in bounds, proximity of each other and within 3000 step limit ($k \leq 3000$).

do

    for $j = 1 : n_{pend}$ carts **do**

        extract and normalise the sensor values $S_0, S_1, \ldots S_9$ for cart j from the simulation, see Table 1

        execute the network and collect the acceleration values for each wheel

        execute one step of the pendulum simulation and record the time $t_{up}^{k,j}$ during which the pendulum is upright

        update the sensor values $S_0, S_1, \ldots S_9$ representing the new state of the cart

    end for

    $k = k + 1$

end while

Calculate $t_{up} = \sum_{k=1}^{3000} \sum_{j=1}^{n_{pend}} t_{up}^{k,j}$ and normalise it $t_{up} = \frac{t_{up}}{\text{total time}}$

OUTPUT The fitness value $f(\mathbf{x}) = t_{up}$

---

where

$$\Delta\mathbf{x_l} = (b_{l1}\Delta x_{l1}, b_{l2}\Delta x_{l2}, \ldots, b_{ln}\Delta x_{ln})$$

is a vector whose elements are the product of two scalars. $\forall j = 1, 2, \ldots, n$:

- b_{lj} is a random value which can take either the value 1 or the value -1
- Δx_{lj} is a moving vector belonging to the interval $[0, \epsilon]$, where ϵ is the radius of the neighbourhood in the continuous space

The concept of neighbourhood $\mathcal{N}_\epsilon$ of radius ϵ and centre $\mathbf{x_l}$ is that set of points $\mathbf{x_{l+1}}$ such that

$$\forall j = 1, 2, \ldots n : \left| x_{lj} - x_{(l+1)j} \right| \leq \epsilon.$$

Thus, Eq. (2) means that a point (vector) $\mathbf{x_{l+1}}$ is generating by adding or subtracting small quantities to the elements of a point $\mathbf{x_l}$. Every time a point is sampled, its fitness value is calculated. Both the sampled vector $\mathbf{x_l}$ and the corresponding fitness value $f(\mathbf{x_l})$ are saved in V. At the end of the process the algorithm returns the data structure of the type

$$V = \begin{pmatrix} \mathbf{x_0} & f(\mathbf{x_0}) \\ \mathbf{x_1} & f(\mathbf{x_1}) \\ \cdots & \cdots \\ \mathbf{x_l} & f(\mathbf{x_l}) \\ \cdots & \cdots \\ \mathbf{x_L} & f(\mathbf{x_L}) \end{pmatrix}. \tag{3}$$

Algorithm 2 outlines the implementation details of the progressive random walk.

Algorithm 2. Progressive Random Walk

INPUT n, f, ϵ
sample a random vector n-dimensional vector $\mathbf{x_0}$ within the domain $[-1, 1]^n$
store $f(\mathbf{x_0})$ in V
initialise $l = 0$
while allocated budget to the random walk **do**
    generate a random vector

$$(b_{l1}, b_{l2}, \ldots, b_{ln}) \in \{-1, 1\}^n$$

    generate a random vector

$$(\Delta x_{l1}, \Delta x_{l2}, \ldots, \Delta x_{ln}) \in [0, \epsilon]^n$$

    build the vector $\Delta \mathbf{x_l} = (b_{l1}\Delta x_{l1}, b_{l2}\Delta x_{l2}, \ldots, b_{ln}\Delta x_{ln})$ by multiplying element-by-element the elements of the two generated vectors
    calculate $\mathbf{x_{l+1}} = \mathbf{x_l} + \Delta \mathbf{x_l}$
    calculate $f(\mathbf{x_{l+1}})$ and store it in V
    update l, i.e., $l = l + 1$
end while
RETURN V

3.2 Measure of Ruggedness

Ruggedness is a property of fitness landscape that estimates the number and distribution of optima within the decision space [17,31]. Qualitatively, we may say that a landscape is rugged if it contains many optima in small area of the decision space. The opposite of a rugged landscape is a smooth landscape. In this study we estimate the ruggedness of the landscape by using the data stored in the data structure V and by applying the First Entropic Measure of ruggedness (FEM) outlined in [17] and consisting of the following steps.

To illustrate the method to calculate the FEM let us consider the column of the data structure V in Equation (3) reporting the fitness values. For simplicity of notation let us indicate with f_l the value $f(\mathbf{x_l})$ $\forall l$. We may consider this column of fitness values as a time series $\{f_l\}_{l=0}^{L}$ where L indicates the largest value of L, that is the last sample index.

The calculation of FEM requires, besides the time series $\{f_l\}_{l=0}^{L}$, a scalar value namely neutrality threshold and here indicated with η is required. To measure the ruggedness the time series $\{f_l\}_{l=0}^{L}$ is scanned by means of a moving window to assess whether the landscape has a local increase in value, a local decrease in value or is approximately flat.

More specifically, a symbolic sequence of the type $S(\eta) = s_1, s_2, \ldots, s_L$ is generated by means of the following mapping

$$s_{l+1} = \Psi(l, \eta) = \begin{cases} -1, & \text{if } f_{l+1} - f_l < -\eta \\ 0, & \text{if } |f_{l+1} - f_l| < \eta \\ 1, & \text{if } f_{l+1} - f_l > \eta \end{cases}$$

Algorithm 3. Entropic Measure of Ruggedness $H(\eta)$

INPUT η, V
extract the last column of V to obtain the sequence $f_0, f_1, \ldots, f_L = \{f_l\}_{l=0}^{L}$
for l=0:L-1 **do**
   if $f_{l+1} - f_l < -\eta$ **then**
      $s_{l+1} = -1$
   else if $f_{l+1} - f_l > \eta$ **then**
      $s_{l+1} = 1$
   else if $|f_{l+1} - f_l| < \eta$ **then**
      $s_{l+1} = 0$
   end if
end for
with the calculated values, compose the sequence $S(\eta) = s_1, s_1, \ldots, s_L$
for l=1:L **do**
   check $s_l s_{l+1} = pq$ and increment the corresponding counter $n_{[pq]}$
end for
for each of the 6 pq types of sequence **do**
   calculate $P_{[pq]} = \frac{n_{[pq]}}{L-1}$
end for
$H(\eta) = 0$
for each of the 6 pq types of sequence **do**
   calculate $H(\eta) = H(\eta) - P_{[pq]} \log_6 P_{[pq]}$
end for
RETURN $H(\eta)$

Thus the sequence S records a symbol "-1" to indicate an abrupt decrease in the fitness value, a symbol "1" to indicate an abrupt increase in the fitness value, a symbol "0" to indicate a locally flat landscape.

The sequence $S(\eta)$ is then scanned and the variations in the landscape recorded in six counters. These counters record the transitions $-1 \to 0$, $-1 \to 1$, $0 \to -1$, $0 \to 1$, $1 \to -1$ and $1 \to 0$ respectively. Let us indicate these counters as $n_{[pq]}$ with $p, q \in \{-1, 0, 1\}$ and $p \neq q$. For a given pair of values p and q, we may then introduce

$$P_{[pq]} = \frac{n_{[pq]}}{L-1}$$

that is the quota of $p \to q$ transitions within the sequence $S(\eta)$. The $P_{[pq]}$ are the building blocks to calculate the entropic measure of ruggedness $H(\eta)$ that is [17,20]

$$H(\eta) = - \sum_{p,q \in \{-1,0,1\} \text{ and } p \neq q} P_{[pq]} \log_6 P_{[pq]} \tag{4}$$

Algorithm 3 shows the procedure to calculate, for a given η value the corresponding entropic measure of ruggedness.

Equation (4) shows that the entropic measure of ruggedness is a parametric entity which depends on an experimental choice, i.e., it depends on the selected neutrality threshold η. It must be observed that different conclusions can be drawn about the landscape for different values of η.

The studies reported in [17,20] show that if η is taken large enough then the ruggedness measurements report a flat landscape regardless of the fitness. Let us indicate with η^* the smallest value of η which causes a flat measurement of the landscape. Since ruggedness cannot be overestimated, the most reliable value of η is that corresponding to the maximum entropic measure of ruggedness. The latter is the FEM:

$$FEM = \max_{\eta \in [0,\eta^*]} H(\eta).$$

3.3 Measure of Neutrality

Neutrality is a property of fitness landscape that estimates distribution and frequency of areas with one objective function value. These areas can be visualised as plateaus. The study in [1] conceptualises neutrality and proposes two metrics to estimate the neutrality of a fitness landscape. To measure the neutrality the data structure V calculated by the progressive random walk is used. With reference to equation (3), the rows of V are scanned by using moving window that takes into consideration three rows at the time, that is for $l = 0, 1, \ldots, L - 2$ the sub-sequences $\{x_l, x_{l+1}, x_{l+2}\}$ and $\{f_l, f_{l+1}, f_{l+2}\}$ are considered. The three step sub-sequence $\{x_l, x_{l+1}, x_{l+2}\}$ is considered to be neutral if

$$|\max\{f_l, f_{l+1}, f_{l+2}\} - \min\{f_l, f_{l+1}, f_{l+2}\}| \leq \xi \tag{5}$$

Every time a three step sub-sequence is detected, a counter $s_{neutral}$ is incremented. The first neutral metric M_1 measures the quota of neutrality within the random walk:

$$M_1 = \frac{s_{neutral}}{L - 1} \tag{6}$$

where it may be observed that the random walk of $L + 1$ samples contains $L - 1$ three step sub-sequences.

The second neutral metric M_2 measures the width of the largest neutral structure

$$M_2 = \frac{\omega_{max}}{L - 1} \tag{7}$$

where ω_{max} is the maximum number of consecutive three point sub-structures within the random walk identified by the time series $\{f_l\}_{l=0}^{L}$.

Algorithm 4 displays the procedure to calculate the two neutrality metrics.

4 (1 + 1) Evolution Strategies for Reinforcement Learning

To perform the reinforcement learning, that is to maximise the function f of the vector $\mathbf{x}$ of the parameters of the neural network we used a naive $(1 + 1)$ Evolution Strategies (ES) [10]. The version of ES employed in this study sample one initial parent vector $\mathbf{x}^P = (x_1^p, x_2^p, \ldots x_n^p)$ within the decision space. Then, the algorithm generates an offspring vector x^o whose components, with a probability p_m, are calculated as

$$x_j^o = x_j^p + y_j$$

Algorithm 4. Neutrality Measures M_1 and M_2

INPUT ξ, V

extract the columns of V to obtain the sequences $\mathbf{x_0}, \mathbf{x_1}, \ldots, \mathbf{x_L} = \{\mathbf{x_l}\}_{l=0}^{L}$ and
$f_0, f_1, \ldots, f_L = \{f_l\}_{l=0}^{L}$

initialise $s_{neutral} = 0$

initialise the parameter $k = 1$ and $\omega_k = 0$

for l=0:L-2 **do**

    consider the three step sub-sequence $\{\mathbf{x_l}, \mathbf{x_{l+1}}, \mathbf{x_{l+2}}\}$ with the fitness values $\{f_l, f_{l+1}, f_{l+2}\}$

    if $|\max\{f_l, f_{l+1}, f_{l+2}\} - \min\{f_l, f_{l+1}, f_{l+2}\}| \leq \xi$ **then**

        increment $s_{neutral}$

        increment ω_k

    else

        increment k and pose $\omega_k = 0$

    end if

end for

$\omega_{max} = \max_k \omega_k$

$M_1 = \frac{s_{neutral}}{L-1}$

$M_2 = \frac{\omega_{max}}{L-1}$

RETURN M_1 and M_2

---

where y_j is sampled from a normal distribution of mean value zero and variance σ^2, that is

$$y_j \sim \mathcal{N}\left(0, \sigma^2\right), \forall j = 1, 2, \ldots, n$$

If one component x_j^o of the offspring candidate solution $\mathbf{x^o}$ falls outside the bounds, the toroidal correction is applied [10]. Let $[a, b]$ be the interval where the weight x_j^o is defined. If $x_j^o > b$ then x_j^o is corrected as

$$x_j^o = a + \left(|x_j^o - b| \mod |b - a|\right)$$

where mod is the modulo function. If $x_j^o < a$ then x_j^o is corrected as

$$x_j^o = b - \left(|x_j^o - a| \mod |b - a|\right).$$

The fitness of the offspring is compared against that of the parent. If the offspring outperforms the parent, it replaces it, otherwise the parent solution is retained. The training is stopped when a condition on the budget is exceeded. For the sake of clarity, the pseudocode of this naive $(1 + 1)$ ES is displayed in Algorithm 5 where $r \sim \mathcal{U}[0, 1]$ indicates that the number r is sampled from the interval $[0, 1]$ with a uniform probability.

5 Results of the Ruggedness and Neutrality

We have run the Progressive Random Walk described in Algorithm 2 for $50,000$ steps for each of the two configurations, i.e., single and triple pendulum tasks.

Algorithm 5. Naive (1+1) Evolution Strategies used for reinforcement learning

INPUT the fitness f, bounds of the weights $[a, b]$, σ, p_m
sample one parent candidate solution $\mathbf{x}^P$ within the decision space
while condition on the computational budget **do**
    for $j = 1 : n$ **do**
      sample $r \sim \mathcal{U}[0, 1]$
      if $r < p_m$ **then**
        sample $y_j \sim \mathcal{N}\left(0, \sigma^2\right)$
      else
        $y_j = 0$
      end if
      $x_j^o = x_j^p + y_j$
    end for
    save the offspring candidate solution $\mathbf{x}^o = (x_1^o, x_2^o, \ldots, x_n^o)$
    if $f\left(\mathbf{x}^o\right) \geq f\left(\mathbf{x}^P\right)$ **then**
      $\mathbf{x}^P = \mathbf{x}^o$
    end if
end while
RETURN the best solution $\mathbf{x}^P$

At the end of the random walk, a data structure V containing $50,000$ samples is generated. From these data the first entropic measures of ruggedness $H(\eta)$ are calculated. The data of ruggedness $H(\eta)$ are plotted with corresponding error bars in Fig. 3.

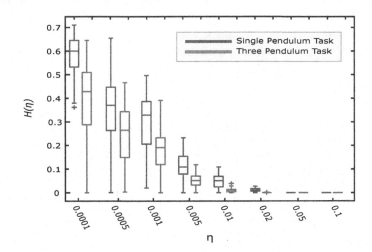

Fig. 3. First entropic measure of ruggedness with dependence on the threshold η for the single and triple pendulum tasks for different values of epsilon

Results in Fig. 3 clearly indicate that for both configurations, the maximum values of ruggedness achieved for the smallest measured values of η, set to 0.0001

in this case. Furthermore, we may observe that the single pendulum task is characterised by higher values of ruggedness than the triple pendulum task. We may visualise these results by thinking that ruggedness correspond to oscillations in the fitness values. Thus the results achieved report that the oscillations for the single pendulum task has a greater amplitude than those associated with the triple pendulum task. Furthermore, the frequency of these oscillations is very high for both configurations.

The results about neutrality displaying the values of M_1 and M_2 in Algorithm 4 for 400 independent runs are shown in Figs. 4 and 5, respectively.

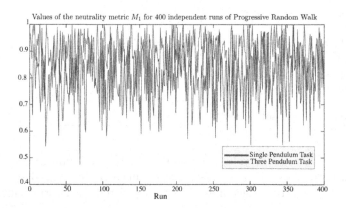

Fig. 4. Neutrality metric M_1 of the single and triple pendulum tasks over 400 runs (ξ=10e-4) size

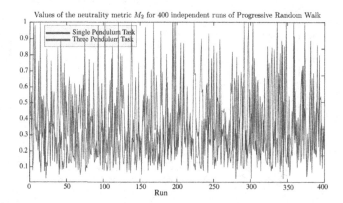

Fig. 5. Neutrality metric M_2 of the single and triple pendulum tasks over 400 runs (ξ=10e-4) size

Numerical results about neutrality in Figs. 4 and 5 broadly confirm and agree with the results about ruggedness. We may observe that the landscape associated with the triple pendulum task is more neutral and contains larger neutral regions that the landscape associated with the single pendulum task. Although single runs can lead to various results, we may observe that, on average, for the triple pendulum task $M_1 \approx 0.9$ and for for the single pendulum task $M_1 \approx 0.75$. Results about M_2 are more run dependant than those about M_1 but, on average, takes higher values for the triple pendulum task than on the single pendulum task.

5.1   Relation Between Control Parameters and FLA

In this work, we investigate ways to feed the results of FLA into the parameter setting of the training algorithm to customise the learning to the problem features under investigation. The naive $(1+1)$ ES used to train the neural network has two parameters to set, that is σ and p_m, see Algorithm 5. To investigate the impact of these parameters on the performance of the reinforcement learning, we have plotted the fitness values for each p_m value and for multiple σ values. For each parameter configuration, 40 independent runs have been executed; each run has been interrupted after $5,000$ generations. The results for single and three and pendulum tasks are displayed in Figs. 6 and 7 respectively. The plots show the mean values and the corresponding error bars. From both Figs. 6 and 7, we observe that σ has a higher impact on the performance than p_m. Nonetheless, a high value of p_m, i.e. a replacement on average of 50% of the parameters can yield an excessively exploratory behaviour and thus a worsening in the performance. Conversely, too small values of p_m do not allow the evolution to progress and the solutions to escape neutral areas of the fitness landscape.

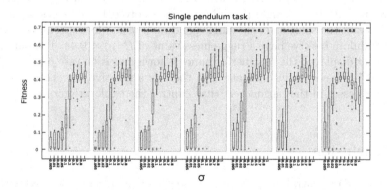

Fig. 6. Performance of the $(1+1)$ ES depending on multiple mutation rate p_m and standard deviation σ for the reinforcement learning problem applied to the single pendulum task.

The results on the single pendulum task in Fig. 6 show that the best results are obtained for a mutation rate $p_m = 0.05$ and for $\sigma = 1$. This result can provide context with the results about ruggedness. The landscape associated with the single pendulum task is highly rugged, on average $H(\eta) = 0.6$. The parameter

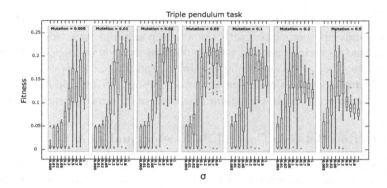

Fig. 7. Performance of the $(1 + 1)$ ES depending on multiple mutation rate p_m and standard deviation σ for the reinforcement learning problem applied to the triple pendulum task.

σ can be interpreted as a step size of ES exploration and $\sigma = 1$ corresponds to half of the width of the domain (that is $[-1, 1]$). According to our interpretation, a large value of σ is well-suited to handle high ruggedness since it allows to by-pass the local minima generated due the ruggedness in the neighbourhood of the candidate solution $\mathbf{x}^P$, see Algorithm 5. Ruggedness handling has an analogy with noise handling [6] where larger steps correspond to an implicit averaging of the noise [26].

This interpretation is confirmed by the results of the triple pendulum task depicted in Fig. 7. The latter task which is characterised by a lower ruggedness value than that of the single pendulum is best solved by σ values lower than those that suit the solution of the single pendulum task. More specifically, for the triple pendulum task we have a ruggedness value of $H(\eta) = 0.44$. The parameter setting that displayed the best performance is for a mutation rate $p_m = 0.02$ and for $\sigma = 0.5$. These results indicate that there is a positive correlation between the ruggedness and the parameters that control the exploration of the learning algorithm.

6   Conclusion

Ruggedness and neutrality of the fitness landscape analysis associated with the reinforcement learning problem for a neural network is investigated in this article. This neural network is used to control the coupled inverted pendulums task. Two configurations and thus two fitness landscapes have been analysed: single (one cart) and triple pendulum tasks respectively. The reinforcement learning has been carried out by Evolution Strategies.

Numerical results show that the landscape associated with the single pendulum task is more rugged than that of the triple pendulum task. Correlated to this result, the landscape associated with the single pendulum task is less neutral than that of the triple pendulum task. An analysis on the performance indicates that ruggedness metrics contain useful information for the setting of the control parameters of Evolution Strategies. More specifically, we observed a positive correlation between the ruggedness of the landscape and the step size that ensures the best performance of the learning algorithm.

Future work will focus on modelling this correlation and proposing an explicit rule to set these parameters. Furthermore, it is worth noting that the landscapes of single and triple pendulum problems, albeit different in several aspects, have common features such as a monotonic descending tend of the first entropic measure of ruggedness with respect to the random walk threshold. Thus, a future development of this study, in the fashion of multi-tasking optimisation [8], could attempt to perform simultaneous FLA [11] and multiple instances of reinforcement learning at the same time.

References

1. van Aardt, W.A., Bosman, A., Malan, K.M.: Characterising neutrality in neural network error landscapes. In: 2017 IEEE Congress on Evolutionary Computation (CEC), pp. 1374–1381 (2017)
2. Bosman, A.: Fitness Landscape Analysis of Feed-Forward Neural Networks. Ph.D. thesis, University of Pretoria, South Africa (09 2019)
3. Bosman, A.S., Engelbrecht, A.P., Helbig, M.: Fitness landscape analysis of weight-elimination neural networks. Neural Process. Lett. 48(1), 353–373 (2018)
4. Bosman, A.S., Engelbrecht, A.P., Helbig, M.: Progressive gradient walk for neural network fitness landscape analysis. In: GECCO (Companion), pp. 1473–1480. ACM (2018)
5. Bosman, A.S., Engelbrecht, A.P., Helbig, M.: Visualising basins of attraction for the cross-entropy and the squared error neural network loss functions. Neurocomputing 400, 113–136 (2020)
6. Branke, J.: Evolutionary optimization in dynamic environments. Ph.D. thesis, Universität Karlsruhe (2000)
7. Caamaño, P., Prieto, A., Becerra, J.A., Bellas, F., Duro, R.J.: Real-valued multi-modal fitness landscape characterization for evolution. In: Wong, K.W., Mendis, B.S.U., Bouzerdoum, A. (eds.) ICONIP 2010. LNCS, vol. 6443, pp. 567–574. Springer, Heidelberg (2010). https://doi.org/10.1007/978-3-642-17537-4_69
8. Choong, H.X., Ong, Y.S., Gupta, A., Lim, R.: Jack and masters of all trades: One-pass learning of a set of model sets from foundation ai models (2022). https://doi.org/10.48550/ARXIV.2205.00671. https://arxiv.org/abs/2205.00671
9. Choromanska, A., LeCun, Y., Ben Arous, G.: Open problem: the landscape of the loss surfaces of multilayer networks. In: Grünwald, P., Hazan, E., Kale, S. (eds.) Proceedings of The 28th Conference on Learning Theory. Proceedings of Machine Learning Research, vol. 40, pp. 1756–1760. PMLR, Paris, France (2015)
10. Eiben, A.E., Smith, J.E.: Introduction to Evolutionary Computing, Second Edition. Natural Computing Series. Springer (2015). https://doi.org/10.1007/978-3-662-44874-8

11. Gupta, A., Ong, Y.S., Da, B., Feng, L., Handoko, S.D.: Landscape synergy in evolutionary multitasking. In: 2016 IEEE Congress on Evolutionary Computation (CEC), pp. 3076–3083 (2016)
12. Hamann, H., Schmickl, T., Crailsheim, K.: Coupled inverted pendulums: a benchmark for evolving decentral controllers in modular robotics. In: Proceedings of the 13th Annual Conference on Genetic and Evolutionary Computation, pp. 195–202 (2011)
13. Jana, N.D., Sil, J., Das, S.: Continuous fitness landscape analysis using a chaos-based random walk algorithm. Soft. Comput. **22**, 921–948 (2018)
14. Lang, R., Engelbrecht, A.: On the robustness of random walks for fitness landscape analysis. In: 2019 IEEE Symposium Series on Computational Intelligence (SSCI), pp. 1898–1906 (2019)
15. LeCun, Y., Bengio, Y., Hinton, G.: Deep learning. Nature **521**(7553), 436–444 (2015)
16. Liu, F.Y., Qian, C.: Prediction guided meta-learning for multi-objective reinforcement learning. In: 2021 IEEE Congress on Evolutionary Computation (CEC), pp. 2171–2178 (2021)
17. Malan, K.M., Engelbrecht, A.P.: Quantifying ruggedness of continuous landscapes using entropy. In: 2009 IEEE Congress on Evolutionary Computation, pp. 1440–1447 (2009)
18. Malan, K.M., Engelbrecht, A.P.: A survey of techniques for characterising fitness landscapes and some possible ways forward. Inf. Sci. **241**, 148–163 (2013)
19. Malan, K.M., Engelbrecht, A.P.: A progressive random walk algorithm for sampling continuous fitness landscapes. In: 2014 IEEE Congress on Evolutionary Computation (CEC), pp. 2507–2514 (2014)
20. Malan, K.: Characterising continuous optimisation problems for particle swarm optimisation performance prediction. Ph.D. thesis, University of Pretoria (2014)
21. Merz, P., Freisleben, B.: Fitness landscape analysis and memetic algorithms for the quadratic assignment problem. IEEE Trans. Evol. Comput. 4(4), 337–352 (2000)
22. Merz, P.: Advanced fitness landscape analysis and the performance of memetic algorithms. Evol. Comput. **12**(3), 303–325 (2004)
23. Mnih, V., Kavukcuoglu, K., Silver, D., Graves, A., Antonoglou, I., Wierstra, D., Riedmiller, M.: Playing atari with deep reinforcement learning. arXiv preprint arXiv:1312.5602 (2013)
24. Moses, J., Malan, K.M., Bosman, A.S.: Analysing the loss landscape features of generative adversarial networks. In: Proceedings of the Genetic and Evolutionary Computation Conference Companion, pp. 1692–1699 (2021)
25. Neri, F.: Generalised pattern search with restarting fitness landscape analysis. SN Comput. Sci. **3**(2), 110 (2022)
26. Neri, F., del Toro Garcia, X., Cascella, G.L., Salvatore, N.: Surrogate assisted local search in PMSM drive design. COMPEL - Int. J. Comput. Math. Electr. Electron. Eng. **27**(3), 573–592 (2008)
27. Pimenta, C.G., de Sá, A.G.C., Ochoa, G., Pappa, G.L.: Fitness landscape analysis of automated machine learning search spaces. In: Paquete, L., Zarges, C. (eds.) EvoCOP 2020. LNCS, vol. 12102, pp. 114–130. Springer, Cham (2020). https://doi.org/10.1007/978-3-030-43680-3_8
28. Reeves, C., Rowe, J.E.: Genetic Algorithms: Principles and Perspectives. Springer (2002)
29. Smith, T., Husbands, P., O'Shea, M.: Fitness landscapes and evolvability. Evolutionary Comput. **10**(1), 1–34 (2002)

30. Sutton, R.S., Barto, A.G.: Reinforcement learning: An introduction. MIT Press (2018)
31. Vassilev, V.K., Fogarty, T.C., Miller, J.F.: Information characteristics and the structure of landscapes. Evol. Comput. **8**(1), 31–60 (mar 2000)
32. Watkins, C.J.C.H., Dayan, P.: Technical note q-learning. Mach. Learn. **8**, 279–292 (1992)

Local Optima Networks for Assisted Seismic History Matching Problems

Paul Mitchell[1,2](✉) ⓘ, Gabriela Ochoa[3] ⓘ, Yuri Lavinas[4] ⓘ,
and Romain Chassagne[1,5] ⓘ

[1] Heriot-Watt University, The Avenue, Edinburgh EH14 4AS, UK
pcm2@hw.ac.uk
[2] TAQA Bratani Limited, TAQA House, Prime Four Business Park, Kingswells,
Aberdeen AB15 8PU, UK
paul.mitchell@taqaglobal.com
[3] University of Stirling, Stirling FK9 4LA, UK
gabriela.ochoa@stir.ac.uk
[4] University of Tsukuba, 1-1-1 Tennodai, Tsukuba 305-8577, Japan
[5] Now at Bureau de Recherches Géologiques et Minières (BRGM), 3 avenue
Claude-Guillemin, BP 36009, 45060 Orléans, Cedex 02 Orléans, France
r.chassagne@brgm.fr

Abstract. Despite over twenty years of research and application, assisted seismic history matching (ASHM) remains a challenging problem for the energy industry. ASHM attempts to optimise the subsurface reservoir model parameters by matching simulated data to the observed production and time-lapse (4D) seismic data, leading to greater confidence in the assimilated models and their predictions. However, ASHM is a difficult and expensive task that has had mixed results in industry, and a new approach to the problem is required. In this work, we examine ASHM from a different perspective by exploring the topology of the optimisation fitness landscape. Many methods for fitness landscape analysis (FLA) have been developed over the past thirty years, but in this work, we extend the use of local optima networks (LONs) to the real-world and computationally expensive ASHM problem. We found that the LONs were different for objective functions based on both production data and time-lapse reservoir maps, and for each dimensionality. Objective functions based on well pressures and oil saturation maps had the highest success rate in finding the global optimum, but the number of suboptimal funnels increased with dimensionality for all objective functions. In contrast, the success rate and strength of the global optima decreased significantly with increasing dimensionality. Our work goes some way to explaining the mixed results of real ASHM problems in industry, and demonstrates the value of fitness landscape analysis for real-world, computationally expensive problems such as ASHM.

Keywords: Assisted seismic history matching · Local optima networks · Fitness landscape analysis

J. Correia et al. (Eds.): EvoApplications 2023, LNCS 13989, pp. 86–101, 2023.
https://doi.org/10.1007/978-3-031-30229-9_6

1   Introduction

In the energy industry, oil and gas production [25], CO_2 storage [11], geothermal energy [8], and more recently Hydrogen storage [26] have been exploited over many decades for numerous purposes. However, to simulate and optimise the production and injection of fluids from the reservoirs, detailed computer models of the subsurface geology, or *reservoir models*, are required [19]. These can be challenging to construct and often contain many errors and uncertainties. The main sources of data that are used to define the models are three-dimensional seismic images of the subsurface geology and the in-situ properties of the rocks measured in well bores. However, well-bore data is only measured at sparse locations in the reservoir and contains measurement errors, and seismic data, despite being recorded over the whole field, has weak signals, is noisy, and contains artefacts. The interpretation and integration of the data to construct reservoir models is a highly skilled task, but the resultant models are typically a simplified representation of the real subsurface geology, and have many possible alternatives [2]. Consequently, their predictions and forecasts often have a wide range of uncertainty.

During production, the reservoir model can be calibrated and constrained using reservoir *history matching* [16]. The model parameters are adjusted to achieve a good match between the predicted production data from the model and the measured, or *observed*, production data from the wells. In addition, time-lapse, or 4D, seismic data are sometimes recorded over the field, which provide three-dimensional images, or snapshots, of fluid movement and pressure changes within the reservoir over time [20]. This can be used in *assisted seismic history matching* (ASHM) [12], where the model parameters are automatically adjusted so that the model accurately predicts both the historical production data from the wells and the time-lapse seismic data [17]. The quality of the history-matched models is usually quantitatively assessed by measuring the misfit, or *fitness*, between the modelled and observed data using a predefined objective function [6] and metric [5]. However, since the data are sparse and contain errors [21], and since the initial reservoir models are so uncertain, there is often low confidence in the resulting assimilated models [17]. After more than twenty years of research and application, ASHM remains a challenging and expensive task with mixed success in industry, and a new approach to the problem is required.

In this work, we take a fresh look at ASHM and examine it from the perspective of the fitness landscape. This is the multidimensional surface that defines the misfit between the model and observed data over the entire parameter search-space [22]. We propose that characterising the topology of the fitness landscape before optimisation will inform the problem setup, guide the choice of optimisation strategy, and lead to better assimilated models. We believe that this new approach will result in a deeper understanding of the ASHM problem and will advance the technology in industry.

However, characterising the topology of fitness landscapes is an extremely challenging task. Fitness landscapes are complex, multidimensional surfaces that are difficult to compute and visualise. Many methods for fitness landscape analysis (FLA) have been developed during the past thirty years or more [9], but they have typically been applied to combinatorial problems or continuous problems with analytical solutions. Exploratory landscape analysis (ELA) is a popular approach to FLA [10], but recently, local optima networks (LONs) [13,14], which depict the global structure of fitness landscapes as graphs, has been extended to continuous optimisation problems [1]. LONs use a basin-hopping approach followed by local minimisation to identify funnels and local or global minima within the fitness landscape structure. They have been applied to benchmark problems and, more recently, to real-world problems with analytical functions [3], but they have not been applied to complex real-world problems such as ASHM.

This paper aims to extend the use of LONs to the computationally expensive real-world problem of ASHM, and to explore the structure of ASHM fitness landscapes for the first time. We have implemented the algorithm to run in parallel on a cluster computer, and calculated the LONs for several objective functions using a realistic reservoir model. These are based on both production data from wells and time-lapse reservoir maps extracted from the simulation model, which represent ideal 4D seismic data. Furthermore, we have explored the impact of the number of parameters by computing LONs in four, seven, and ten dimensions.

The paper is structured as follows. In Sect. 2.1, we define the local optima network model. Then, in Sect. 2.2, we describe the reservoir model used for the experiments and the method used to sample the fitness landscapes. In Sect. 3, we present the ASHM local optima networks for the reservoir model based on four objective functions and in three different dimensions, as well as their network and performance metrics. We discuss the results and their implications for ASHM in Sect. 4. Finally, we summarise our observations and conclusions in Sect. 5, as well as our thoughts on future work.

2   Method

2.1   Local Optima Networks

To analyse and visualise the structure of the studied landscapes, we consider the compressed monotonic local optima network (CMLON) model [3,15]. We formalise the notions of fitness landscapes and local optimum in continuous optimisation before defining the CMLON model. We also define the notions of monotonic sequence and funnel, which are relevant to our analysis. Thereafter, we describe the process of sampling and constructing the network models.

Definitions

Fitness Landscape. Is a triplet $(\mathbf{X}, N, f)$ where $\mathbf{X} \in \mathbb{R}^n$ is the set of all real-valued solutions of n dimensions, *i.e..*, the search space; N is a function that assigns to every solution $\mathbf{x} \in \mathbf{X}$ a set of neighbours $N(\mathbf{x})$; and $f : \mathbb{R}^n \to \mathbb{R}$ is the fitness function. A potential solution $\mathbf{x}$ is denoted as vector $\mathbf{x} = (x_1., x_2, \ldots, x_n)$, and the neighbourhood is based on hypercubes. Formally, the neighbourhood of a candidate solution $\mathbf{x}_k$ is defined as, $\mathbf{x}_j \in N(\mathbf{x}_k) \leftrightarrow |x_{ki} - x_{ji}| < s_i, i = \{1, \ldots, n\}$ where $\mathbf{s} = (s_1, s_2, \ldots, s_n)$ is a vector that represents the size of the neighbourhood in all dimensions.

Local Optimum. Is a solution $\mathbf{x}^* \in \mathbf{X}$ such that $\forall \mathbf{x} \in N(\mathbf{x}^*)$, $f(\mathbf{x}^*) \leq f(\mathbf{x})$.

Compressed Monotonic LON Model. In the standard LON model [13], nodes are local optima and edges represent any possible transition among optima with a given perturbation operator. The CMLON model [3,15] is a coarser model that compresses connected local optima at the same (or very similar) fitness values into single nodes, and has edges representing improving transitions only. The purpose of the CMLON model is to study landscapes with neutrality, that measures the complement of the proportion of compressed nodes to the total number of local optima, providing an indication of plateaus within the landscape, as well as to explore the landscape's funnel structure. To define the graph model, we first define the nodes and edges.

Compressed Local Optimum. A compressed local optimum is a single node that represents a set of connected nodes in the LON model with the same (or very similar) fitness value.

Monotonic Perturbation Edges. There is a monotonic perturbation edge from a local optimum l_1 to a local optimum l_2, if l_2 can be obtained from a random perturbation of l_1 followed by a local minimisation process, and $f(l_2) \leq f(l_1)$. The edge is called monotonic because the transition between two local optima is non-deteriorating. Edges are weighted with the number of times a transition between two local optima occurred.

Compressed Monotonic LON. It is the directed graph $CMLON = (CL, CE)$ where nodes are compressed local optima CL, and edges $CE \subset ME$ are aggregated from the monotonic edge set ME by summing up the edge weights.

Monotonic Sequence. A monotonic sequence is a path of connected local optima where their fitness values are always decreasing. Every monotonic sequence has a natural end, which represent a funnel bottom, also called sink in graph theory.

Funnel. We can characterise funnels in the CMLON as all the monotonic sequences ending at the same compressed local optimum (funnel bottom or sink).

Sampling and Constructing the Network Models. Our methodology for sampling and constructing the networks is based on the *basin-hopping* algorithm [23]. Basin-hopping is an iterative algorithm, where each iteration is composed of a random perturbation of a candidate solution, followed by a local minimisation process and an acceptance test. Specifically, the CMLON model construction uses a variant called monotonic basin-hopping (MBH) [7] where the acceptance criterion considers only improving solutions.

This is the first attempt to apply local optima networks to a computationally expensive real-world problem, such as ASHM. In order to construct the models, several basin-hopping runs are conducted. Each run produces a search trajectory, which is recorded and stored as a set of nodes (local minima) and edges (consecutive transitions). Note that different runs can in principle traverse the same nodes and edges, even if they start from different initialisation points. The models are constructed in a post-processing stage, where the trajectories generated by a fixed number of runs (100 in our implementation) are aggregated to contain only unique nodes and edges.

ASHM problems require simulation and evaluation of the objective function to compute each fitness sample. This is computationally expensive and may take several minutes or even hours to run. To ease this, the original LONs algorithm (https://github.com/gabro8a/LONs-Numerical.git) was re-implemented on a multi-node high-performance cluster computer. Each LON takes approximately four days to compute using fifty CPUs (AMD ^{TM}Opteron 6348 processors operating at, 1400 MHz) for each objective model. Consequently, some parameters have been estimated, and further evaluation is required. For example, the perturbation strength, which is an important parameter [1], was fixed at 2.5% of the normalised search space distance for each parameter, but this should be determined more rigorously in future work. The computational time also increases with dimensionality. Reservoir simulations were performed using the ^{TM}Eclipse 100, version 2018.2, black-oil reservoir simulator.

The CMLON network visualisations were generated using the original LONs post-processing software, which is based on the R implementation of the igraph package (https://igraph.org/r/). The parameters for the CMLON visualisations are listed in Table 1. The *Best* value is the minimum fitness of the objective function for each CMLON. The number of iterations was determined by examining the convergence of each run, and was chosen to be one hundred and fifty for all experiments. However, the increase in value of the *best* fitness with dimensionality suggests that more iterations may be needed to improve convergence in higher dimensions. The position threshold, ϵ, defines where two solutions represent the same local optimum, i.e., the difference between each of their components is less than the threshold value, ϵ.

2.2   Reservoir Model and Fitness Computation

The Brugge model, shown in Fig. 1, is a full-field synthetic reservoir model created for the Society of Petroleum Engineers (SPE) Applied Technology Workshop (ATW) on production optimisation in 2008 [18]. Twenty oil production wells and ten water injection wells are used to develop the field; the oil producers are placed near the crest of the structure in the oil zone, and the water injectors are sited down-flank in the water zone. Four geological formations were defined within the reservoir (Schelde, Mass, Waal, and Schie), which are populated with realistic petrophysical values for porosity, net-to-gross, and permeability. The model also contains a geological fault near the crest of the structure.

One of the supplied model realisations (FY-SF-KM-1-1) was selected to represent the true reservoir model, which was then simulated to generate the *observed* dataset used for the experiments. This included oil production rates (WOPR) and bottom-hole pressure (WBHP) measurements for each well, and time-lapse reservoir maps (between zero and ten years of production) of reservoir pressure (PRESSURE) and oil saturation (SOIL) extracted from the simulation model. The *observed* production data were used to control subsequent reservoir simulations by total liquid rate, but the proportions of produced oil and water, as

Table 1. CMLON parameters for the Brugge model SHM fitness landscapes.

Objective	Dimensions	Best	Step size	Iterations	Runs	ϵ
WOPR	4	1.032	2.5%	150	100	0.1
	7	2.277	2.5%	150	100	0.1
	10	20.325	2.5%	150	100	0.1
WBHP	4	1.597	2.5%	150	100	0.1
	7	1.616	2.5%	150	100	0.1
	10	1.678	2.5%	150	100	0.1
SOIL	4	7.410×10^{-10}	2.5%	150	100	1×10^{-6}
	7	9.030×10^{-7}	2.5%	150	100	1×10^{-6}
	10	8.000×10^{-6}	2.5%	150	100	1×10^{-6}
PRESSURE	4	2.390×10^{-3}	2.5%	150	100	0.1
	7	3.550×10^{-2}	2.5%	150	100	0.1
	10	2.490×10^{-1}	2.5%	150	100	0.1

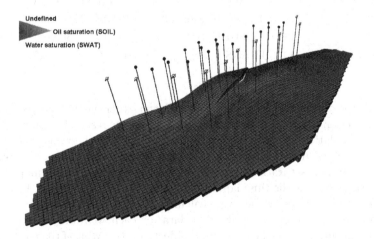

Fig. 1. Three-dimensional view of the Brugge reservoir model. The green zone is the oil bearing reservoir and the blue zone is the water bearing reservoir. The wells are shown by the black lines that penetrate the reservoir from above. (Color figure online)

well as bottom-hole pressure, were not constrained. These were dependent on the reservoir model and fluid properties.

The reservoir model properties were perturbed to create different realisations of the initial model, and their simulated data were compared with the observed data to calculate the model's fitness. Scalars were applied to the petrophysical property grid values of net-to-gross ratio (NTG) and permeabilities (PERMX and PERMY), measured in millidarcies (mD), within each formation. The parameters and their ranges are listed in Table 2. In the four-dimensional problems, the net-to-gross and permeability values were varied for the Waal formation, as well as the transmissibility of the fault. The same properties were also varied for the Maas formation in the seven-dimensional problems, and, in addition, the same properties were varied for the Schie formation in the ten-dimensional problems.

Table 2. Reservoir model parameters and their geological property ranges. The table shows the mean of the property values for each formation, as well as the minimum and maximum scalar values in parentheses.

Formation	Property	Mean	Minimum	Maximum
Maas	NTG_MAAS	0.88	0.62 (0.7)	0.97 (1.1)
	PERMX_MAAS (mD)	90	45 (0.5)	900 (10.0)
	PERMY_MAAS (mD)	90	45 (0.5)	900 (10.0)
Waal	NTG_WAAL	0.97	0.68 (0.7)	1.0 (1.1)
	PERMX_WAAL (mD)	814	407 (0.5)	1628 (2.0)
	PERMY_WAAL (mD)	814	407 (0.5)	1628 (2.0)
Schie	NTG_SCHIE	0.77	0.39 (0.7)	1.0 (1.3)
	PERMX_SCHIE (mD)	36	18 (0.5)	360 (10.0)
	PERMY_SCHIE (mD)	36	18 (0.5)	360 (10.0)
Fault	FLT_TRANS	1	0.1 (0.1)	2.0 (2.0)

Four objective functions were defined for the experiments, two based on the well production data, and two based on time-lapse reservoir maps extracted from the simulation models. The oil production rates of the wells (WOPR) and bottom-hole pressure measurements (WBHP) were used for the production based objectives, and oil saturation maps (SOIL) and reservoir pressure maps (PRESSURE) were used for the time-lapse reservoir map based objectives. For the well based objectives, fitness values were calculated as the mean square error (MSE) of the model and observed production data for each well and averaged for an overall model fitness. For the map based objectives, the MSE of the modelled and observed reservoir maps was calculated for each formation, and also averaged for an overall model fitness.

3   Results

The Brugge model CMLONs are shown as two-dimensional networks in Fig. 2 and as three-dimensional representations in Fig. 3. In the three-dimensional visualisations, fitness is shown on the vertical axis, where lower values are better. In all networks, the pink nodes are global funnels and the blue nodes are suboptimal funnels. The global optima are highlighted by bright red circles, and the suboptimal funnel bottoms are dark blue circles. The size of the nodes is related to the strength of the incoming connections, and the weight of the edges is related to the number of times a transition occurs between two nodes in the sampling process. The CMLONs are shown for four different objective functions; two are based on well production data (WOPR, WBHP), and two are based on time-lapse reservoir maps (SOIL, PRESSURE). The CMLONs dimensionality, n, increases from left to right for each objective function. The network and performance metrics, which are described in Table 3, are shown in Fig. 4 and their values are listed in Table 4.

Table 3. Network and performance metric descriptions.

Metric	Description
Success rate	Proportion of basin-hopping runs that reach the global minimum
Deviation	Mean deviation from the global minimum (unnormalised)
Nodes	Number of nodes in the CMLON (compressed local optima)
Funnels	Number of sinks (CMLON nodes without outgoing edges)
Neutral	Proportion of CMLON nodes to the number of local optima (complement)
Strength	Normalised incoming strength of the globally optimal funnels

The CMLONs in Fig. 2 and their three-dimensional representations in Fig. 3 show that the networks have different characteristics for each objective function and for each dimensionality. The four-dimensional CMLON for the WOPR objective has many suboptimal nodes that are clustered toward the centre of the network, and most search trajectories follow a path toward them. In higher dimensions, the funnels are more dispersed and some optimal nodes appear. It has a moderate success rate in four dimensions, which diminishes in higher dimensions. The CMLON for WBHP has high-strength optimal nodes in four-dimensions, which are also clustered toward the centre of the network. The global optimum is reached by many of the search trajectories. It has the highest success rate of all the experiments. In higher dimensions, the search trajectories disperse to form many suboptimal funnels, and the success rate decreases. The CMLONs for the SOIL objective are denser than the other objectives and have one hundred funnels in all dimensions. In four dimensions, there are many optimal and suboptimal funnels, and the success rate is high; however, the optima are dispersed. In higher dimensions, the trajectories spiral downwards toward suboptimal funnel bottoms, and the success rate diminishes markedly. The four-dimensional CMLON for the PRESSURE objective has a broad distribution of suboptimal nodes and a few global optima. The majority of the search trajectories descend rapidly towards low fitness values and then move towards the centre of the network. In higher dimensions, the suboptimal optimal funnels disperse, and there are only a few global funnels. The success rate is moderate in four dimensions, but also diminishes in higher dimensions.

In four dimensions, there is a distinct difference in character between the well production objectives (WOPR and WBHP) and the time-lapse reservoir map objectives (SOIL and PRESSURE). The optima of the production-based objectives are clustered toward the centre of the CMLON, whereas the optima of the time-lapse reservoir map based objectives are more dispersed. However, the differences are less apparent in higher dimensions. The SOIL objective is distinctive by its high density of funnels in all dimensions, which are dispersed across the network. They also have more evenly distributed fitness, as indicated by Figs. 3g to 3i. The WBHP and SOIL objectives have the highest success rates, which suggests they may be preferable for optimisation.

The network and performance metrics are shown as graphs in Fig. 4, and are tabulated in Table 4. They show that the total number of optima (optimal and suboptimal) is large for all objective functions, and that they increase with dimensionality for all objectives. There are noticeably more optima for the SOIL objective than the other objectives. The WOPR, SOIL, and PRESSURE objectives have only a few funnels in four dimensions, but increase markedly in higher dimensions. The SOIL objective is the exception and has one hundred funnels for all dimensions. Overall, the success rate is high for the objectives in four dimensions, but decreases considerably in seven and ten dimensions. The WBHP and SOIL objectives have the highest success rate and the highest strength in four dimensions, which is related to their larger proportion of global optima. The strength of the global optima nodes follows a similar trend to the success rate; where the objectives have high strength nodes in four dimensions, but it decreases in higher dimensions. The WOPR objective is an exception because despite its moderately high success rate, it has low strength. This may be explained by its high deviation, indicating the optima are dispersed. The neutral values are quite different for each objective function. WBHP and SOIL have high neutrality in four dimensions, which decreases in higher dimensions for SOIL, but increases slightly for WBHP in seven dimensions before decreasing in ten dimensions. PRESSURE has lower neutrality for all dimensions, which increases slightly in higher dimensions, and WOPR has negligible neutrality for all dimensions. Deviation depends on the measurement unit of the objective function, which varies widely in these experiments. It is lowest in four dimensions for all the objective functions, but increases with dimensionality. This follows because the funnels become more dispersed in higher dimensions.

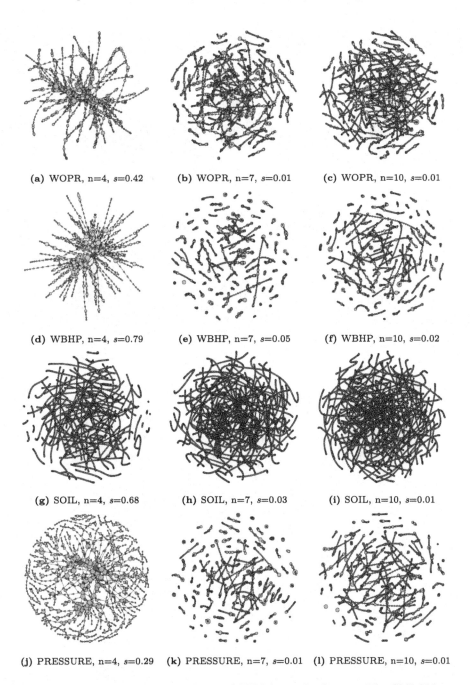

(a) WOPR, n=4, s=0.42 **(b)** WOPR, n=7, s=0.01 **(c)** WOPR, n=10, s=0.01

(d) WBHP, n=4, s=0.79 **(e)** WBHP, n=7, s=0.05 **(f)** WBHP, n=10, s=0.02

(g) SOIL, n=4, s=0.68 **(h)** SOIL, n=7, s=0.03 **(i)** SOIL, n=10, s=0.01

(j) PRESSURE, n=4, s=0.29 **(k)** PRESSURE, n=7, s=0.01 **(l)** PRESSURE, n=10, s=0.01

Fig. 2. CMLONs for the Brugge reservoir ASHM fitness landscapes. The CMLONs are shown for four objective functions, and in $n = \{4, 7, 10\}$ dimensions. The success rate, s, is shown for each CMLON.

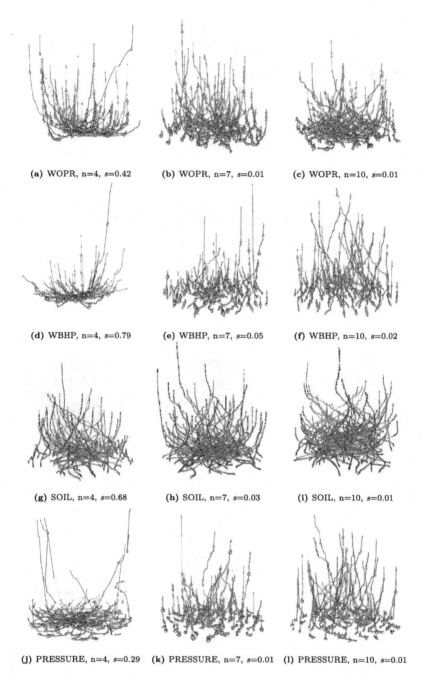

(a) WOPR, n=4, s=0.42   (b) WOPR, n=7, s=0.01   (c) WOPR, n=10, s=0.01

(d) WBHP, n=4, s=0.79   (e) WBHP, n=7, s=0.05   (f) WBHP, n=10, s=0.02

(g) SOIL, n=4, s=0.68   (h) SOIL, n=7, s=0.03   (i) SOIL, n=10, s=0.01

(j) PRESSURE, n=4, s=0.29   (k) PRESSURE, n=7, s=0.01   (l) PRESSURE, n=10, s=0.01

Fig. 3. Three-dimensional CMLONs for the Brugge reservoir ASHM fitness landscapes. The vertical axis represents fitness. The CMLONs are shown for four objective functions, and in $n = \{4, 7, 10\}$ dimensions. The success rate, s, is shown for each CMLON.

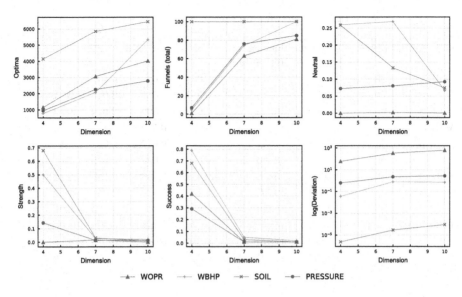

Fig. 4. CMLOM network and performance metrics for the ASHM fitness landscapes.

4   Discussion

The CMLONs presented in Sect. 3 provide some insight into the structure of the Brugge model ASHM fitness landscapes. The high proportion of suboptimal funnels suggests that the landscapes are complex surfaces with many local minima, even in low-dimensional spaces. This suggests that there may be many non-unique solutions, which is consistent with the results seen in industry [17], and may explain the low-confidence in the models. The number of suboptimal funnels increases in higher dimensions, and they spread out, indicating that there is a reduced likelihood of locating the global optimum. We have explored the problems for four, seven, and ten dimensions; however, ASHM problems for real reservoir models typically have hundreds or thousands of uncertain parameters, or dimensions, which makes them more challenging. The network and performance metrics show that problem difficulty and non-uniqueness increase with dimensionality, by increasing suboptimal funnels and decreasing success rates. It is possible that these trends would extend to higher dimensions, which would exacerbate the problem difficulty, but that requires further investigation.

Four objective functions have been investigated in this work, but which of them should be used and how they should be combined is an open question [5]. The production-based objectives have slightly higher success rates than the time-lapse reservoir map objectives, although this difference diminishes in higher dimensions. The WBHP and SOIL objectives have the highest success rates, which indicates that both production-based objectives and time-lapse reservoir map objectives are useful in seismic history matching. Combining these may lead to better objective functions with even higher success rates. CMLONs may

Table 4. Network and performance metrics.

Objective	Dimension	Optima	Funnels (opt,subopt)	Neutral	Strength	Success	Deviation
WOPR	4	1153	(0,1)	0.000	0.000	0.420	5.88×10^1
	7	3063	(1,62)	0.002	0.016	0.010	3.31×10^2
	10	4016	(0,81)	0.000	0.000	0.010	6.12×10^2
WBHP	4	778	(2,3)	0.260	0.500	0.790	3.60×10^{-2}
	7	2086	(2,72)	0.269	0.027	0.050	7.81×10^{-1}
	10	5328	(2,98)	0.066	0.020	0.020	7.07×10^{-1}
SOIL	4	4141	(68,32)	0.259	0.680	0.680	2.42×10^{-6}
	7	5851	(3,97)	0.133	0.030	0.030	2.92×10^{-5}
	10	6441	(1,99)	0.074	0.010	0.010	9.26×10^{-5}
PRESSURE	4	947	(1,6)	0.072	0.143	0.292	6.20×10^{-1}
	7	2260	(1,75)	0.080	0.013	0.010	2.28
	10	2775	(1,84)	0.092	0.012	0.010	2.82

provide a method for this, potentially leading to improved objective functions with less non-uniqueness and higher success rates.

This work is a first attempt to apply LONs to a computationally expensive real-world problem such as ASHM, and has some limitations. The perturbation step-size was chosen, based on previous experience, to be 2.5% of the normalised ($[0,1]$) parameter space distance, but it is an important parameter which requires further tuning [1]. The number of iterations was selected by examining the convergence of each run, and balancing it against the computation time. One hundred and fifty iterations was sufficient for the four dimensional problems, but more iterations may be required for higher dimensions. The increased value of the best fitness, the larger proportion of suboptimal funnels, and the reduced success rates seen in these experiments may be related to poorer convergence in higher dimensions; however, significantly increasing the number of iterations could be impractical.

The Brugge model was chosen for these experiments because it is a realistic full-field reservoir model, and its reservoir simulations are very fast. This makes it possible to calculate hundreds of thousands of fitness samples for the LONs. In contrast, reservoir simulations for real producing fields take much longer, and it will be much more challenging to compute LONs in a reasonable time. New methods, such as surrogate or proxy models instead of costly numerical simulations, may be required to significantly speed up simulations and fitness computations [24]. Time-lapse reservoir maps extracted from the simulation models were used to represent ideal time-lapse seismic data in these experiments. However, real time-lapse seismic data are recorded as acoustic signals, which must be inverted to pressure and saturation maps to compare with reservoir simulation models [4]. Since the signals are weak and the processed data contains noise and other errors, fitness calculations for real data will have some uncertainty. Local optima networks may provide the means to explore the impact of these issues on the fitness landscapes, and their implications for the assimilated models.

The results of this work support our initial hypothesis that characterising the ASHM fitness landscapes will help to understand the problem at a deeper level and guide its definition to achieve more reliable assimilated models. Our work demonstrates the value of analysing fitness landscapes for computationally expensive problems. Local optima networks provide a means to extract key features of the landscape's structure, and network graphs allow them to be visualised intuitively. The results of this work go some way to explaining the mixed results of ASHM seen in industry [17], and offer the potential to help design better objective functions for improved models and more confident predictions.

5 Conclusion

In this work, we have extended the application of local optima networks (LONs) to the computationally expensive, real-world optimisation problem of assisted seismic history matching (ASHM). We have compared the compressed monotonic LONs (CMLONs) of four different objective functions based on both well production data and time-lapse reservoir maps and in three different dimensions.

We found that the CMLONs of the ASHM landscapes have different characteristics for each of the objective functions and dimensionalities. They typically have a few global funnels in four dimensions, but are dominated by a larger number of suboptimal funnels in higher dimensions, where there are few if any global funnels. This implies that there is only a small likelihood of finding the global optimum in higher dimensional problems, which may help to explain the uncertainty in the results of real ASHM problems. Furthermore, it demonstrates the benefit of characterising fitness landscapes before optimisation, even for computationally expensive problems, and may provide a means to design more successful objective functions.

This work represents a first attempt to characterise the fitness landscapes of real-world ASHM problems, but there are many unresolved issues to address. At this stage of research, our goal is to identify methods that can characterise the main features of the fitness landscape and provide a more in-depth understanding of the ASHM optimisation problem. Local optima networks provide a unique view on the global structure of the landscape, but are computationally expensive. This work has investigated relatively low-dimensional problems, but real ASHM problems will have many more parameters.

The Brugge model was selected for these experiments because it is a realistic reservoir model, with a known solution, and fast reservoir simulations. This allowed for sufficiently well sampled LONs to be computed in a reasonable time. In the future, we intend to further extend the application of LONs to a real producing reservoir with real time-lapse seismic data However, real reservoir models have considerably longer simulations, and it may be more challenging to compute sufficiently sampled LONs. Faster methods for reservoir simulation, such as proxy models, may be required. Furthermore, real time-lapse seismic data contain many sources of data errors, and the impact of these on the fitness landscapes will be considered. Since the true model is unknown for real reservoirs, we will also investigate the impact of model uncertainty on the fitness landscapes and its implications for the assimilated models.

Acknowledgement. We thank the SPECIES society for funding a visiting scholarship for Yuri Lavinas to the University of Stirling, Scotland, UK.

References

1. Adair, J., Ochoa, G., Malan, K.M.: Local optima networks for continuous fitness landscapes. In: Proceedings of the Genetic and Evolutionary Computation Conference Companion, GECCO 2019, pp. 1407–1414. Association for Computing Machinery, New York (2019)
2. Arnold, D., Demyanov, V., Tatum, D., Christie, M., Rojas, T., Geiger, S., Corbett, P.: Hierarchical benchmark case study for history matching, uncertainty quantification and reservoir characterisation. Comput. Geosci. **50**, 4–15 (2013)
3. Contreras-Cruz, M.A., Ochoa, G., Ramirez-Paredes, J.P.: Synthetic vs. real-world continuous landscapes: a local optima networks view. In: Filipič, B., Minisci, E., Vasile, M. (eds.) BIOMA 2020. LNCS, vol. 12438, pp. 3–16. Springer, Cham (2020). https://doi.org/10.1007/978-3-030-63710-1_1
4. Corte, G., Dramsch, J., Amini, H., Macbeth, C.: Deep neural network application for 4d seismic inversion to changes in pressure and saturation: optimising the use of synthetic training datasets. Geophysical Prospecting (2020)
5. Hallam, A., Chassagne, R., Aranha, C., He, Y.: Comparison of map metrics as fitness input for assisted seismic history matching. J. Geophys. Eng. **19**(3), 457–474 (2022)
6. He, Y., Aranha, C., Hallam, A., Chassagne, R.: Optimization of subsurface models with multiple criteria using lexicase selection. Oper. Res. Perspectives **9**, 159–172 (2022)
7. Leary, R.H.: Global optimization on funneling landscapes. J. Global Optim. **18**(4), 367–383 (2000)
8. Lund, J.W., Toth, A.N.: Direct utilization of geothermal energy 2020 worldwide review. Geothermics **90**, 101915 (2021)
9. Malan, K.M.: A survey of advances in landscape analysis for optimisation. Algorithms **14**(2), 40 (2021)
10. Mersmann, O., Bischl, B., Trautmann, H., Preuss, M., Weihs, C., Rudolph, G.: Exploratory landscape analysis. In: Proceedings of the 13th Annual Conference on Genetic and Evolutionary Computation, pp. 829–836. Association for Computing Machinery (2011)
11. Michael, K., Golab, A., Shulakova, V., Ennis-King, J., Allinson, G., Sharma, S., Aiken, T.: Geological storage of co2 in saline aquifers-a review of the experience from existing storage operations. Int. J. Greenhouse Gas Control **4**(4), 659–667 (2010)
12. Mitchell, P., Chassagne, R.: 4d assisted seismic history matching using a differential evolution algorithm at the harding south field. In: 81st EAGE Conference and Exhibition 2019, vol. 2019, pp. 1–5. European Association of Geoscientists & Engineers (2019)

13. Ochoa, G., Tomassini, M., Vérel, S., Darabos, C.: A study of nk landscapes' basins and local optima networks. In: Proceedings of the 10th Annual Conference on Genetic and Evolutionary Computation, pp. 555–562 (2008)
14. Ochoa, G., Veerapen, N.: Mapping the global structure of tsp fitness landscapes. J. Heuristics **24**(3), 265–294 (2018)
15. Ochoa, G., Veerapen, N., Daolio, F., Tomassini, M.: Understanding phase transitions with local optima networks: number partitioning as a case study. In: Hu, B., López-Ibáñez, M. (eds.) EvoCOP 2017. LNCS, vol. 10197, pp. 233–248. Springer, Cham (2017). https://doi.org/10.1007/978-3-319-55453-2_16
16. Oliver, D.S., Chen, Y.: Recent progress on reservoir history matching: a review. Comput. Geosci. **15**(1), 185–221 (2011)
17. Oliver, D.S., Fossum, K., Bhakta, T., Sandø, I., Nævdal, G., Lorentzen, R.J.: 4d seismic history matching. J. Petroleum Sci. Eng. **207**, 109119 (2021)
18. Peters, L., et al.: Results of the brugge benchmark study for flooding optimization and history matching. SPE Reservoir Evaluation Eng. **13**(03), 391–405 (2010)
19. Ringrose, P., Bentley, M.: Reservoir Model Design, 2 edn. Springer (2021)
20. Sambo, C., Iferobia, C.C., Babasafari, A.A., Rezaei, S., Akanni, O.A.: The role of time lapse(4d) seismic technology as reservoir monitoring and surveillance tool: a comprehensive review. J. Natural Gas Sci. Eng. **80**, 103312 (2020)
21. Souza, R., Lumley, D., Shragge, J.: Estimation of reservoir fluid saturation from 4d seismic data: effects of noise on seismic amplitude and impedance attributes. J. Geophys. Eng. **14**(1), 51–68 (2017)
22. Stadler, P.F.: Fitness landscapes. Appl. Math. Comput. **117**, 187–207 (2002)
23. Wales, D.J., Doye, J.P.: Global optimization by basin-hopping and the lowest energy structures of Lennard-Jones clusters containing up to 110 atoms. J. Phys. Chem. A **101**(28), 5111–5116 (1997)
24. Werth, B., Pitzer, E., Affenzeller, M.: Surrogate-assisted fitness landscape analysis for computationally expensive optimization. In: Moreno-Díaz, R., Pichler, F., Quesada-Arencibia, A. (eds.) EUROCAST 2019. LNCS, vol. 12013, pp. 247–254. Springer, Cham (2020). https://doi.org/10.1007/978-3-030-45093-9_30
25. Zhang, G., Qu, H., Chen, G., Zhao, C., Zhang, F., Yang, H., Zhao, Z., Ma, M.: Giant discoveries of oil and gas fields in global deepwaters in the past 40 years and the prospect of exploration. J. Natural Gas Geosci. **4**(1), 1–28 (2019)
26. Zivar, D., Kumar, S., Foroozesh, J.: Underground hydrogen storage: a comprehensive review. Int. J. Hydrogen Energy **46**(45), 23436–23462 (2021)

Extending Boundary Updating Approach for Constrained Multi-objective Optimization Problems

Iman Rahimi[1] , Amir H. Gandomi[1,2(✉)] , Mohammad Reza Nikoo[3] , and Fang Chen[1]

[1] Faculty of Engineering and Information Technology, University of Technology Sydney, Sydney, Australia
{Gandomi,fang.chen}@uts.edu.au
[2] University Research and Innovation Center (EKIK), Óbuda University, 1034 Budapest, Hungary
[3] Department of Civil and Architectural Engineering, Sultan Qaboos University, Muscat, Oman
m.reza@squ.edu.om

Abstract. To date, several algorithms have been proposed to deal with constrained optimization problems, particularly multi-objective optimization problems (MOOPs), in real-world engineering. This work extends the 2020 study by Gandomi & Deb on boundary updating (BU) for the MOOPs. The proposed method is an implicit constraint handling technique (CHT) that aims to cut the infeasible search space, so the optimization algorithm focuses on feasible regions. Furthermore, the proposed method is coupled with an explicit CHT, namely, feasibility rules and then the search operator (here NSGA-II) is applied to the optimization problem. To illustrate the applicability of the proposed approach for MOOPs, a numerical example is presented in detail. Additionally, an evaluation of the BU method was conducted by comparing its performance to an approach without the BU method while the feasibility rules (as an explicit CHT) work alone. The results show that the proposed method can significantly boost the solutions of constrained multi-objective optimization.

Keywords: Constraint handling · Multi-objective Optimization · Evolutionary computation · NSGA-II

1 Introduction

Multi-objective optimization problems (MOOPs), are faced in several sectors of the engineering field. However, finding an optimal solution for these problems has been a challenge for scholars. Since evolutionary algorithms (EAs) do not need concavity or convexity and can yield several alternative solutions in a single run [1], they are usually used for MOOPs [2]. Additionally, EAs can be integrated with specific decomposition algorithms [1] and, generally, perform better in dealing with some MOOPs [3–5].

© The Author(s), under exclusive license to Springer Nature Switzerland AG 2023
J. Correia et al. (Eds.): EvoApplications 2023, LNCS 13989, pp. 102–117, 2023.
https://doi.org/10.1007/978-3-031-30229-9_7

Constraint handling technique (CHT) combined with an evolutionary algorithm is called constrained evolutionary algorithm optimization (CEAO) and has been used to tackle real-world constrained optimization problems [2].

In the literature, researchers have proposed numerous CHTs, for example, based on feasibility and infeasibility regions [3, 6, 7], priority assignment [4], tournament selection and selection operator [4, 5]. Some studies in the literature have provided surveys for CHTs [2, 5, 8].

However, most studies are classified as explicit CHTs. In the work [7], the authors tested a method known as boundary updating (BU), as an implicit CHT, which updates variable bounds by directly using constraints and applied it to several single-objective optimization problems. In the current study, the BU method is coupled with an explicit CHT (here feasibility rules), and for the strategy without BU, an explicit CHT (feasibility rules) is still applied to solve the problem. As a contribution, in this paper, the latter work [7] is extended by applying the BU method to some MOOPs using non-dominated sorting genetic algorithm-II (NSGA-II), and also the BU is coupled with feasibility rules as it is superior to the penalty function approach. The NSGA-II is one of the most popular and efficient EAs, as it employs an elitist principle, a diversity mechanism [6], and has been widely used to solve constrained MOOPs [2, 9]. This research is motivated by the lack of efficient constraint-handling techniques for multi-objective optimization.

The remainder of the study is organized as follows. Section 2 presents the proposed method. Section 3 illustrates the numerical study applied in this research. Finally, the last section gives the conclusion.

2 Proposed Method

Equations (1–3) present a simplified constrained optimization problem:

$$\text{maximize (minimize) } F(x) = (f_1(x), \dots, f_t(x)) \tag{1}$$

$$\text{s.t.} h_i(x) \leq 0 \quad \text{for } i \in \{1, \dots, n\} \tag{2}$$

$$g_j(x) = 0 \text{ for } j \in \{1, \dots, m\} \tag{3}$$

where $F(X)$ is the objective vector that consists of several objectives (t is the number of objective functions); N and M are the numbers of inequality and equality constraints, and $X = (x_1, \dots, Xn)$ is a vector of decision variables that are subjected to lower bound (LB) and upper bound (UB) vectors. Rather than producing a single solution, these equations yield several pareto optimal solutions instead [10].

Since optimization is an iterative process that the boundaries can be changed during the process, the i-th decision variable boundaries in each iteration own the dynamic nature; the proposed method uses the constraints to narrow down variable space and then forces the algorithm to focus its search in the feasible region by limiting the feasible search space for the variable (s). In the BU method, the boundaries are changing iteratively and are updated during the optimization procedure. Mathematically it could be written as follows [7]:

$$\exists i \in \{1, \dots, m\} : \left[\forall j \in \{1, \dots, n\} : x_i \geq l_{i,j}(x \neq i) \cup x_i \leq u_{i,j}(x \neq i) \right] \tag{4}$$

where li,j and ui,j are dynamic lower bound and upper bound for ith decision variable. Regarding updating the bounds, there are some scenarios as follows:

$$\text{If } lb_i = -\infty \text{ and } ub_i = +\infty: \tag{5}$$

$$lb^u = l_{i,j}(x_{\neq i}) \tag{6}$$

$$ub^u = u_{i,j}(x_{\neq i}) \tag{7}$$

Else:

$$1b^u = \min\left(\max\left(l_{i,j}(x_{\neq i}), lb_i\right), ub_i\right) \tag{8}$$

$$ubb^u = \max\left(\min\left(u_{i,j}(x_{\neq i}), ub_i\right), lb_i\right) \tag{9}$$

where (lb^u, ub^u) are updated boundaries. To start with the BU method, a repairing variable, which can handle the greatest number of constraints without overlapping with other repairing variables, should be selected; in this case, if there is more than one candidate, the one that handles the greatest number of constraints is selected. Also, if there is still another candidate for variable selection, one variable is selected randomly [7].

If a repairing variable handles the first ki constraints:

$$lb^u = \min\left(\max\left[l_{i,1}(x_{\neq i}), \dots, l_{i,k}(x_{\neq i}), lb_i\right], ub_i\right) \tag{10}$$

$$ub^u = \max\left(\min\left[u_{i,1}(x_{\neq i}), \dots, u_{i,k}(x_{\neq i}), ub_i\right], lb_i\right) \text{ (where } k_i \leq m) \tag{11}$$

If another repairing variable (xr) is defined:

$$lb^u = \min\left(\max\left[l_{i,1}(x_{\neq i,r}), \dots, l_{i,k}(x_{\neq i,r}), lb_i\right], ub_i\right) \tag{12}$$

$$ub^u = \max\left(\min\left[u_{i,1}(x_{\neq i,r}), \dots, u_{i,k}(x_{\neq i,r}), ub_i\right], lb_i\right) \tag{13}$$

$$lb_r^u = \min\left(\max\left[l_{i,k_{i+1}}(x_{\neq r}), \dots, l_{r,k_i+k_r}(x_{\neq r}), lb_r\right], ub_r\right) \tag{14}$$

$$ub_r^u = \max\left(\min\left[u_{i,k_{i+1}}(x_{\neq r}), \dots, u_{r,k_i+k_r}(x_{\neq r}), ub_r\right], lb_r\right) \tag{15}$$
$$r \in \{1, \dots, n\}, r \neq i$$

In the BU approach, a repairing variable could be substituted with a generalized semi-independent variable to avoid discontinuity among iterations [7] and then rewritten with a lower and upper bound:

$$x \in \{x_1, \dots, x_h, x_{h+1}, \dots, x_n\} \rightarrow mx \in \{p_1, \dots, p_h, x_{h+1}, \dots, x_n\} \tag{16}$$

where $p_1, \dots, ph$ are semi-independent variables.

$$x \in \{x_1, \dots, x_h, x_{h+1}, \dots, x_n\} \rightarrow mx \in \{p_1, \dots, p_h, x_{h+1}, \dots, x_n\} \tag{17}$$

$$x_i = lb^u + p_i(ub^u - lb^u) \text{ for } i = 1, \ldots, h \text{ where } 0 \leq p_i \leq 1 \qquad (18)$$

After selecting repairing variables, the search operator, here NSGA-II, will be applied to the problem, the boundaries of non-repairing variables will be checked, and mx- vector will be updated. During the solution procedure, the boundaries of repairing variables will be updated, and then the semi-independent variables $(pi, i = 1, \ldots, h)$ are remapped to the actual variables using updated boundaries. In the end, those constraints that were not involved in the repairing variable boundary along with fitness values will be evaluated using actual variables. The BU method is applied in the steps indicated in algorithm 1.

Algorithm 1. Implementation of the BU method in a MOOP

Initialization

While the criteria are not met

      applying the search algorithm (here NSGA-II)

      applying the BU method and updating the mx-vector

      Check the boundaries of non-repairing variables

      Update the boundaries of repairing variables

      Remap p variable to the original variables using updated boundaries.

      Evaluate the constraint violations

      Evaluating fitness function

End while

As an illustrative example, the following MOOP is considered to explain the BU.

$$\min f_1(x) = \left(x^2 + x^2\right) \qquad (19)$$

$$\min f_2(x) = -(x_1 - 1)^2 - x^2 \qquad (20)$$

$$g_1(x_1 + x_3) = x_1 + x_3 \leq 1000 \qquad (21)$$

$$-2 \leq x_1 \leq 2 \qquad (22)$$

$$-2 \leq x_2 \leq 2 \qquad (23)$$

$$-2 \leq x_3 \leq 2 \qquad (24)$$

Either x_1 or x_3 can be selected as a repairing variable of the single constraint. All variables are in the range of -2 to 2. From the variable selection strategy [7], as explained previously in this study, the repair variable, x_3, is selected as a repairing variable. With respect to x_3, the inequality constraint is solved, and the repairing variable, here x_3, is used as the lower and upper bound functions. The repairing variable is substituted with the mapped variable, p, which is in the range of 0 to 1. Therefore, the lower and upper bounds of the variables are $\{-2, -2, 0\}$ and $\{2, 2, 1\}$, respectively. When the search algorithm is applied to the problem, the repairing variable is substituted with the mapped variable, and the boundary of the repairing variable is mapped for the rest of the search cycle [7]. The original variable can be calculated by the following formula in order to compute the objectives and constraints:

$$x_3 = lb_3^u + p_i \times \left(ub_3^u - lb_3^u\right) \tag{25}$$

where ub_3^u and lb_3^u are the lower and upper bounds for repairing variable x_3, respectively. The updated bounds can be written as follows:

$$lb_3^u = \min(\max(lb_3, 1000 - x_1), ub_3) \tag{26}$$

$$ub_3^u = \max(\min(ub_3, 1000 - x_1), lb_3) \tag{27}$$

In the end, the repairing variable is remapped to the actual boundary when the fitness values are evaluated. Figure 1 compares Pareto solutions found by the BU method coupled with a CHT (here feasibility rules) and by the method without the BU method (only the feasibility rules approach is considered as an explicit constraint handling technique) for a population size of 100. As a termination criterion, the number of generations was set to 100. From Fig. 1, it is obvious that the BU method produced a greater number of non-dominated solutions than the approach without BU.

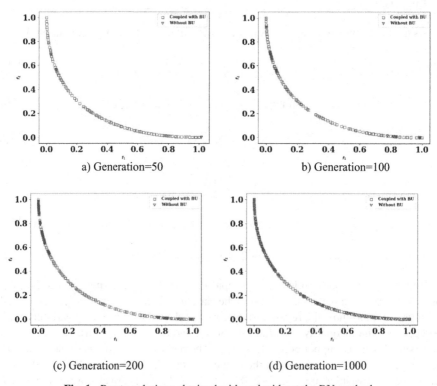

a) Generation=50

b) Generation=100

(c) Generation=200

(d) Generation=1000

Fig. 1. Pareto solutions obtained with and without the BU method

3 Numerical Study

This section presents two benchmark problems and two real-world problems to verify the performance of the proposed method, in which the BU method is implemented with NSGA-II. To evaluate the results, the following metrics were evaluated:

- The number of non-dominated solutions: The BU method was coupled with an explicit constraint handling technique and compared to the approach without BU (only feasibility rules were applied to the optimization problem). Since the BU method aims to reduce the search space, the number of non-dominated solutions was determined for comparison.
- Constraint violation (CV): Compared to unconstrained optimization problems, constrained MOOPs are more challenging since a large proportion of infeasibility regions appears in the search space (it means the hit ratio is low), which makes solving the constrained problem very challenging, especially for highly constrained and can lead to some difficulties such as convergence-, diversity-, and feasibility-hardness. Therefore, even finding one feasible solution is a major achievement. The BU method aims to reduce the search space by narrowing down variable space and cutting the infeasible search space. To compare the effectiveness of the BU method, the first feasible solution that could be found by the BU method was compared to the method without BU, and the whole population is tracked against generations as well.
- Performance indicators: Generational Distance (GD) [11], Generational Distance Plus (GD+) [12], Inverted Generational Distance (IGD) [13], Inverted Generational Distance Plus (IGD+) [12], and Hypervolume are five indicators that were analyzed.
- Running metric [14]: It is possible to trace the difference in the objective space followed by generations.
- The population and the maximum number of experiments were set to 100 and 11, respectively. All experiments in this study were performed in Python 3.10 and using the Pymoo library [15].

3.1 Osyczka and Kundu (OSY)

The OSY problem, proposed by [16], has two objective functions, six constraints, and six decision variables. For the OSY problem, $x1$, $x4$, and $x6$ were selected as repairing variables as they can handle all constraints. To evaluate the performance of the BU method, we ran the algorithm and traced it in every generation. The algorithm without the BU method found the first feasible solution after 1000 evaluations (only feasibility rules are applied to the optimization process), while the BU method coupled with the feasibility rules found the first feasible solution after only 250 evaluations (Fig. 2).

In addition, using the BU method, at generation 250, the whole population fell in the feasibility region, while without using the BU method after 1375 evaluations, the whole population converged in the feasibility area (Fig. 3). Figure 4 illustrates the number of non-dominated solutions found by both methods from generations 1 to 40, indicating that the BU method found more Pareto solutions in the first 40 generations. Figure 5 presents Pareto solutions obtained by the BU and without the BU in 20 generations.

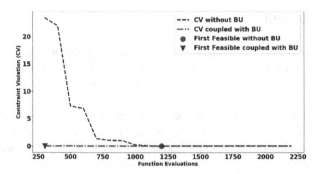

Fig. 2. First feasible solution found with and without the BU method (OSY problem)

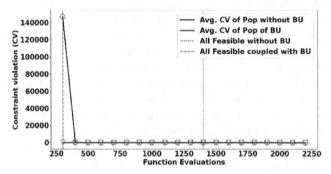

Fig. 3. Convergence of the whole population to the feasible solution with and without the BU method (OSY problem)

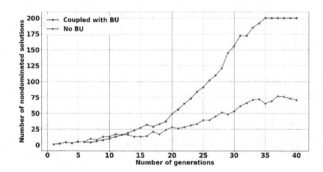

Fig. 4. Pareto solutions obtained by NSGA-II with BU and without BU for the OSY problem

Figures 6, 7, 8 and 9 display the performance indicator values obtained by the method with and without the BU method for the OSY problem. As mentioned, GD, GD+, IGD, and IGD+ were selected as the indicators. Sampling and crossover are two important factors for evolutionary algorithms, which strongly influence the convergence speed and are often used in noisy optimization, where reducing the negative impact of noise is a key matter; also, sampling is a popular strategy for dealing with noise. Therefore, experiments

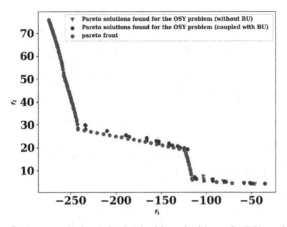

Fig. 5. Pareto solutions obtained with and without the BU method

were conducted with different operators, namely real random sampling, Latin hypercube sampling, and also two different real-SBX crossover and uniform crossover [17]. In these figures, the mean, median, standard deviation, and best, and worst values of the experiments using the BU method were much lower than the same method without BU. Table 1 presents the Summary of p-value of the Wilcoxon rank sum test ($\alpha = 0.05$) over all runs; as it can be seen, there is a significant difference overall run for all experiments excluding binary crossover.

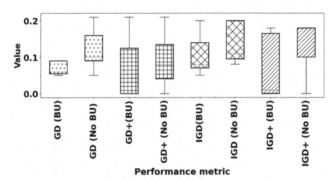

Fig. 6. Performance indicator values obtained with and without the BU method for the OSY problem (Sampling = Real-Random)

3.2 Bin and Korn Test Problem (BNH)

The BNH test problem, proposed by [3], has two objective functions, two constraints, and two decision variables. Figure 10 presents the results found by selecting the repairing variable, $x1$, for this problem. However, because a smaller number of Pareto solutions is not always a good indicator for this problem, two other indicators, namely IGD and HV,

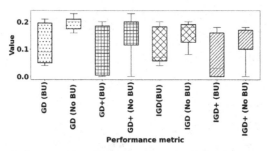

Fig. 7. Performance indicator values obtained with and without the BU method for the OSY problem (Sampling = Latin hypercube sampling)

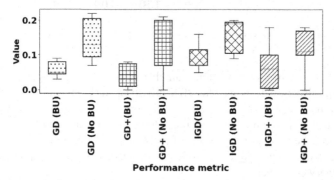

Fig. 8. Performance indicator values obtained with and without the BU method for the OSY problem (Crossover = real_sbx)

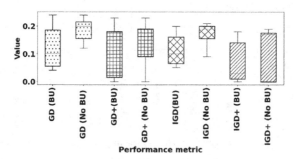

Fig. 9. Performance indicator values obtained with and without the BU method for the OSY problem (Crossover = bin_ux)

were considered. Figures 11–12 present the IGD and HV values found by the BU method in early generations compared with the original algorithm without BU (11 experiments have been conducted).

Table 1. *p*-value summary of the Wilcoxon rank sum test over all runs (BU vs. without BU).

Indicator	Sampling				Crossover			
	real-random		hypercube		real_sbx		bin_ux	
	p-value	h	*p*- value	h	*p*-value	h	*p*-value	h
GD	0.0068	+	0.0009	+	0.009	+	0.0050	+
IGD	0.009	+	0.009	+	0.004	+	0.0050	+
GD +	0.033	+	0.007	+	0.009	+	0.0656	~
IGD +	0.013	+	0.0356	+	0.12	+	0.0969	~

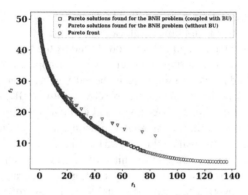

Fig. 10. Pareto solutions found with and without the BU method for the BNH problem (generations = 20)

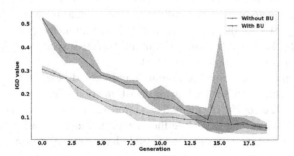

Fig. 11. IGD values for BNH problem

3.3 Welded Beam Design Problem

The welded beam design problem is a well-known engineering optimization benchmark problem [18, 19], which examines the tradeoff between the strength and cost of a beam. The two objective functions are minimizing the fabrication cost of the beam and minimizing the deflection of the end of the beam. This problem has four variables and four

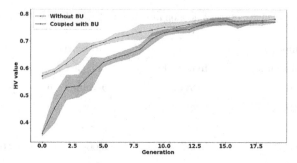

Fig. 12. HV values for BNH problem

constraints, where all constraints are handled by beam thickness. Herein, the NSGA-II was applied to this problem, and fitness values were evaluated.

Figure 13 shows the Pareto optimal solutions found by the two approaches, with and without the BU method, considering 20 generations. It is apparent that NSGA-II using the BU method can find the Pareto solutions in the early generations (38) and, thus, can find the global optimum solutions faster than NSGA-II without BU. The results of IGD, IGD+, GD, and GD+ values for NSGA-II with and without BU are presented in Fig. 14.

As it was mentioned earlier, compared to unconstrained optimization problems, constrained MOOPs are more challenging since a large proportion of infeasibility regions appears in the search space (it means the hit ratio is low), which makes solving the constrained problem very difficult, especially for highly constrained and can lead to some issues such as convergence-, diversity-, and feasibility-hardness.

The results for welded beam design problem show that the BU approach found the first feasible solution and the whole feasible population after 300 evaluations. Comparatively, the NSGA-II algorithm without the BU method found the first feasible solution after 1100 evaluations, and the whole population is feasible after 1500 evaluations (Figs. 15 and 16). Herein, the performance of the running metric proposed in [14] was evaluated, which illustrates the difference in the objective space and, specifically, the improvement from one generation to the next.

Figure 17 demonstrates how the running metric could be applied to visualize and compare the performance of NSGA-II using the BU method, where $t = 11$ to 13. Figure 17 shows that NSGA-II using the BU method was improved from 10e9 to 10e0 in generation 6, which is a significant change, while the method without BU was improved almost from 10e2 to 10e0 in generation 12. On the other hand, NSGA-II using the BU method converges to optimal solutions in the early generations, whereas the algorithm without BU needs to be further evaluated to find the optimal solutions.

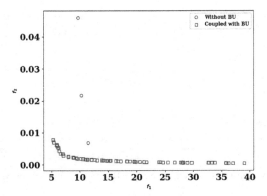

Fig. 13. Pareto solutions obtained for the welded beam design problem with and without the BU method

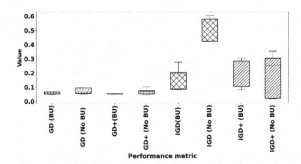

Fig. 14. Performance indicator values obtained with and without the BU method

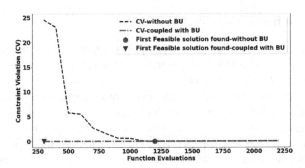

Fig. 15. Constraint violation evaluated by NSGA-II (first feasible solution)

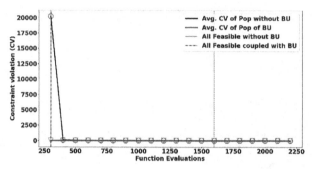

Fig. 16. Constraint violation evaluated by NSGA-II (convergence of the whole population to feasible solutions)

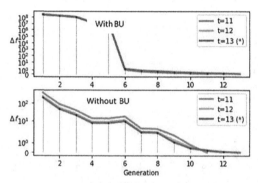

Fig. 17. Running metric for the multi-objective welded beam design problem with and without the BU method

3.4 Cantilevered Beam Design Problem

The Cantilevered design problem is an example of a large-scale size problem. In the cantilevered beam design problem, the stepped cantilever beam must be able to carry a prescribed end load [20]. This problem is originally a single-objective optimization problem considering minimizing the beam volume; however, this study has extended the original problem and added one more objective function, which minimizes end deflection subject to various engineering design variables. The beam supports the given load, P, at a fixed distance L from the support. Besides, designers of the beam can vary the width (bi) and height (hi) of each section. The problem is highly constrained and large-dimensional; however, for this example, five segments ($N = 5$), including 10 dimensions (10 constraints) are considered. In this example, $x2, x4, x6, x8$, and $x10$ are considered as the repairing variables. This example illustrates how the BU method can be used to solve a constrained nonlinear optimization problem. Figure 18 presents the optimal solutions found by NSGA-II with and without the BU method for the problems in generation 5.

Figure 19 presents the feasibility of the first solutions and the whole population of the cantilevered beam design problem using both methods. The NSGA-II with the BU approach found the first feasible solution and the whole feasible population after 100

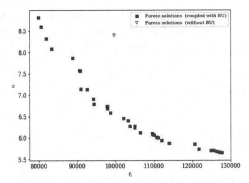

Fig. 18. Pareto solutions found for multi-objective cantilevered beam design problem using NSGA-II with and without the BU method (generations = 5)

evaluations. Comparatively, the method without BU found the first feasible solution 500 evaluations and the whole population is feasible after 1100 evaluations.

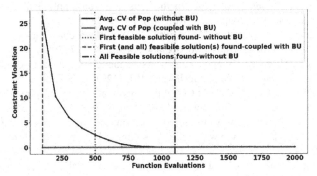

Fig. 19. Feasibility of first and whole population of cantilevered beam design problem using both methods

4 Conclusion

This study presents an extension of an implicit CHT called the BU method [7] for solving the MOOPs efficiently. Several performance indicators were obtained to evaluate the BU method. The results show that the NSGA-II algorithm using the BU method could find feasible solutions much faster than without the BU method (when only an explicit CHT is applied to an optimization process). Also, the performance indicators show significant improvement by implementing the optimization algorithm with the BU method in some cases. Moreover, the BU method found the first feasible solution and feasible solutions for the whole population in earlier generations.

As a future research direction, it is suggested to apply other population-based multi- and many-objective optimization algorithms to real-world optimization problems. Also, it is suggested to investigate the impact of binary crossover for other multi-objective evolutionary algorithms on other real-world problems. Also, it is suggested that the BU method is coupled with other constraint-handling techniques, which are available in the literature.

References

1. Poojari, C.A., Beasley, J.E.: Improving benders decomposition using a genetic algorithm. Eur. J. Oper. Res. **199**(1), 89–97 (2009)
2. Rahimi, I., Gandomi, A. H., Chen, F., Mezura-Montes, E.: A Review on Constraint Handling Techniques for Population-based Algorithms: from single-objective to multi-objective optimization. Archives (2022)
3. Binh, T.T., Korn, U.: MOBES: a multiobjective evolution strategy for costrained optimization problems. In: The Third International Conference on Genetic Algorithms (Mendel 97), 25, 27 (1997)
4. Jiménez, F., Verdegay, J. L., Gómez-Skarmeta, A.F.: Evolutionary techniques for constrained multiobjective optimization problems. In: Workshop on Multi-Criterion Optimization Using Evolutionary Methods GECCO-1999 (1999)
5. Salcedo-Sanz, S.: A survey of repair methods used as constraint handling techniques in evolutionary algorithms. Comput. Sci. Rev. **3**(3), 175–192 (2009)
6. Deb, K., Pratap, A., Agarwal, S., Meyarivan, T.: A fast and elitist multiobjective genetic algorithm: NSGA-II. IEEE Trans. Evol. Comput. **6**(2), 182–197 (2002)
7. Gandomi, A.H., Deb, K.: Implicit constraints handling for efficient search of feasible solutions. Comput. Methods Appl. Mech. Eng. **363**, 112917 (2020)
8. Mezura-Montes, E., Coello, C.A.C.: Constraint-handling in nature-inspired numerical optimization: past, present and future. Swarm Evol. Comput. **1**(4), 173–194 (2011)
9. Brownlee, A.E.I., Wright, J.A.: Constrained, mixed-integer and multi-objective optimisation of building designs by NSGA-II with fitness approximation. Appl. Soft Comput. **33**, 114–126 (2015)
10. Deb, K.: Multi-objective optimisation using evolutionary algorithms: an introduction. Springer (2011)
11. van Veldhuizen, D.A.: Multiobjective evolutionary algorithms: classifications, analyses, and new innovations. Air Force Institute of Technology (1999)
12. Ishibuchi, H., Masuda, H., Tanigaki, Y., Nojima, Y.: Modified distance calculation in generational distance and inverted generational distance. In: Evolutionary Multi-Criterion Optimization: 8th International Conference, EMO 2015, Guimarães, Portugal, March 29–April 1, 2015. Proceedings, Part II 8, pp. 110–125 (2015)
13. Coello Coello, C.A., Reyes Sierra, M.: A study of the parallelization of a coevolutionary multi-objective evolutionary algorithm. In: MICAI 2004: Advances in Artificial Intelligence: Third Mexican International Conference on Artificial Intelligence, Mexico City, Mexico, April 26-30, 2004. Proceedings 3, pp. 688–697 (2004)
14. Blank, J., Deb, K.: A running performance metric and termination criterion for evaluating evolutionary multi-and many-objective optimization algorithms. IEEE Congress on Evolutionary Computation (CEC) **2020**, 1–8 (2020)
15. Blank, J., Deb, K.: Pymoo: Multi-objective optimization in python. IEEE Access **8**, 89497–89509 (2020)
16. Osyczka, A., Kundu, S.: A new method to solve generalized multicriteria optimization problems using the simple genetic algorithm. Struct. Optim. **10**, 94–99 (1995)

17. Qian, C., Yu, Y., Tang, K., Jin, Y., Yao, X., Zhou, Z.-H.: On the effectiveness of sampling for evolutionary optimization in noisy environments. Evol. Comput. **26**(2), 237–267 (2018)
18. Deb, K., Sundar, J.: Reference point based multi-objective optimization using evolutionary algorithms. In: Proceedings of the 8th Annual Conference on Genetic and Evolutionary Computation, pp. 635–642 (2006)
19. Ray, T., Liew, K.M.: A swarm metaphor for multiobjective design optimization. Eng. Optim. **34**(2), 141–153 (2002)
20. Vanderplaats, G.: Very large scale optimization. In: 8th Symposium on Multidisciplinary Analysis and Optimization, 4809 (2002)

Multi-agent vs Classic System of an Electricity Mix Production Optimization

Solofohanitra Rahamefy Andriamalala(✉)📧, Toky Axel Andriamizakason📧,
and Andry Rasoanaivo

Université d'Antananarivo MISA, Antananarivo, Madagascar
soloforahamefy@gmail.com
http://misa-madagascar.com/

Abstract. Aiming to respond to a real-world complex problem, optimizing an electricity mix production, MixSimulator has been developed. This paper compares two approaches: the classic method and the multi-agent system based method (MAS). Each agent performs various functions such as producing electricity and updating the availability (power plants), predicting the oncoming demand, and handling all the information to provide optimized planning of the production. It takes into account technological, economic, and environmental constraints. Evolutionary algorithms (DE, (1+1)-ES and NgOpt) from the Nevergrad library are used to generate solutions. The results show that the multi-agent system based method outperforms the classic one thanks to its ability to react to events and provide dynamic schedule.

Keywords: Energy systems modeling · Agent-based modeling · Optimization · Heuristics · Real world problems

1 Introduction

As part of the simulation of the electric transmission grid, an electricity mix represents the distribution of different primary energy sources with the aim of producing electricity to supply a region or a country. The modeled system simulates the production of electricity in a given region over a period T. It must consider the technical characteristics of the power plants and the demand curve of the consumers in order to produce the right amount of electricity at each time-step while minimizing the electricity production cost under some constraints. To achieve this, the python-based MixSimulator is in development [1]. The most recent version of the application is a classical heuristic approach to solve the problem using derivative-free black-box optimizers from nevergrad [13]. This work presents the multi-agent feature which allows MixSimulator to react to a multitude of events, and thus, to readjust the production schedule over time.

As MixSimulator focuses on optimizing the production cost of the mix, it doesn't take into account the distribution part of the network but only the amount of demand at an outgoing substation.

© The Author(s), under exclusive license to Springer Nature Switzerland AG 2023
J. Correia et al. (Eds.): EvoApplications 2023, LNCS 13989, pp. 118–128, 2023.
https://doi.org/10.1007/978-3-031-30229-9_8

The paper is structured as follows. Section 2 gives the different methods used in the application, a classic approach, and a multi-agent-based approach. Experimentations comparing the performance of a group of optimizers are then presented in Sect. 3. Every aspect of the results is discussed in Sect. 4. Finally, conclusions are given in Sect. 5.

2   Methods

2.1   Previous Works

[4] is a related work which compares agent-based approaches and classical optimization techniques. The MAS architecture is more suitable for developing planning and control applications as developed in [18]. [12] shows that MAS is one of the solutions provided to solve scheduling problems in real-world environments. Besides, [3] is a MAS for solving dynamic scheduling problems using an evolutionary computation approach. The choice of black-box optimization is due to its efficiency through complex problems as shown in [9] and the paper [15] presents an overview of what derivative free algorithms are. A lot of works on using Multi-agent as an optimization technique in power engineering are presented in [11]. One of them is [5] which simulates an implementation of multi-agent system (MAS) in the distributed management of microgrids.

2.2   Classic Approach

The study of the state of a system composed of conversion facilities must go through the definition of variables in order to describe the behavior of this one [6]. An installation is characterized by its installed power α and its instantaneous power β. As the role of the system is to respond to any demand at any time t, there are so-called controlled power stations (thermal power stations) which can vary their diffused power to meet the demand. The coefficients of utilization (also known as duty cycle) are the simplest factors that allow this variation to be presented. On the other hand, intermittent power plants based on wind or solar power are affected by periods of involuntary interruptions.

The main function can be divided in two specific parts. Given x the vector of the $K_{t,i}$, the first part is the cost function based on [2]. Let d be the number of hours of operation of the energy mix, n_p the number of plants present in the mix, M_i the variable cost: it is the fuel expenditures in case of non-renewable stations and set to 0 otherwise. Let $K_{t,i}(\leq availability_{t,i})$ be the coefficients of utilization, α_i the installed power and $F_{t,i}$ the fixed cost (operations and maintenance expenditures).

$$cost(x) = \sum_{t=1}^{d} \left(\sum_{i=1}^{n_p} (M_i \times K_{t,i} \times \alpha_i + F_{t,i}) \right) \qquad (1)$$

The second part represents the constraint function which is the sum of the demand and the environment constraints. Let d be the number of hours of operation of the energy mix, k the penalization cost and U_t the unsatisfied demand (or overproduction with $demand_t$ the actual demand at time t and n_p the number of powerplants) at time t, C the penalization due to overproduction of carbon, C_{limit} the limitation of carbon production (g/MWh) and C_{avg} the actual amount of carbon produced by the mix (g/MWh).

$$constraints\ penalization = \sum_{t=1}^{d} (k \times U_t) + (C \times \max\{0, (C_{avg} - C_{limit})\}) \quad (2)$$

where

$$U_t = |demand_t - \sum_{i=1}^{n_p} (K_{t,i} * \alpha_i)|$$

The final objective function $f(x)$ (3) can be defined as the sum of (1) and (2):

$$f(x) = \sum_{t=1}^{d} \left(\sum_{i=1}^{n_p} (M_i \times K_{t,i} \times \alpha_i + F_{t,i}) \right) + (k \times U_t) + C \times \max\{0, (C_{avg} - C_{limit})\}$$
$$(3)$$

For example, the planning of 10 power plants over two days (48 h) represents an optimization of 480 variables. With a large time interval, we have to deal with a high-dimensional problem.

2.3   Multi-agent Based Approach

A multi-agent system can be defined as a collection of agents interacting and negotiating with each other in the same environment to solve an overall problem [11].

Our system is a collaborative multi-agent system based on cooperative interaction protocols, which means that all agents work in order to achieve only one common goal (reducing the production cost, always under the constraints stated above). Also, agents do not communicate with each other, but instead report all their information to the moderator. This architecture is also known as a mediator architecture where the moderator (the mediator in our case) proposes a dynamic schedule for power plants agents [7] to facilitate the coordination. As a result, the same objective function (3) used in the implementation of the classical optimization has been injected in the moderator and is therefore the objective function used in the MAS approach [4].

$$f(x) = \sum_{t=1}^{d} \left(\sum_{i=1}^{n_p} (M_i \times K_{t,i} \times \alpha_i + F_{t,i}) \right) + (k \times U_t) + C \times \max\{0, (C_{avg} - C_{limit})\}$$
$$(4)$$

By definition, the mediator (our moderator) is an agent that coordinates the other agents, in such a way that the production schedule is continuously improved. It also registers every agent and classifies them.

MixSimulator's MAS representation model is strongly inspired by the FIPA Agent Management reference Model [8] as seen in Fig. 1. In the Agent Platform (AP), reside all the agents (one per power plant and one for the demand estimation), but also the Agent Management System (the Moderator).

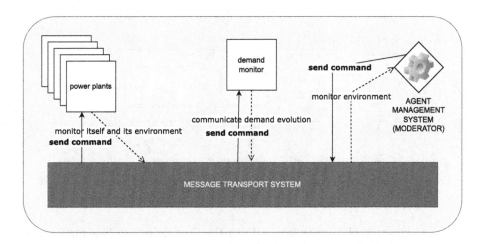

Fig. 1. MixSimulator as a MAS representation model. The Agent Platform is made of the power plants which monitor themselves and their environment autonomously, the demand monitor which analyzes previous consumption to predict future demand, and the agent management system (moderator). Each agent can communicate with the moderator thanks to the Message Transport System based on the observer/observable design pattern.

An initial optimization takes place in both classical and MAS approaches (at time $t = 0$, no unforseen events happens yet). The differences with the classic approach described above reside in the fact that agents can monitor their environment and thus inform the moderator of any change. As a simulation, the message is based on a predefined code sent between the agents and the moderator. Each agent has an implemented module to translate these codes and the value provided. This allows the moderator to react to these change immediately and readjust production planning from the moment t_i an event has occurred by re-optimizing the mix from t_i. This approach also offers fault tolerance and scalability to the system, which are essential features for an electricity mix simulator. The observation ability can be enhanced by using remote monitoring system (APIs), and predicted data (demand forecasting).

In addition, the application can switch between the two methods. The main architecture of MixSimulator (see Fig. 2) is actually build to take into account each method and bring with it an interface to the nevergrad platform and an integrate tool to evaluate different optimizer in one run.

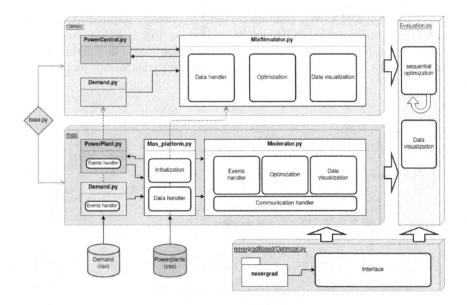

Fig. 2. Architecture of MixSimulator - From the class defined in base.py, you can choose which methods to use. The classic compartment (in orange) encompasses the direct optimization strategy. The MixSimulator class in MixSimulator.py has a Data handler method that collects data from different sources (Demand and Powerplants) to store it in respective classes that are defined by Demand.py and PowerCentral.py. These classes are then used by the optimization method of the MixSimulator class to calculate the optimal utilization coefficients according to the given parameters. We can see that the single MixSimulator class manages almost all of the process, unlike the MAS method (in green) which has the Mas_platform class and the event handler in addition. MixSimulator uses for both approach the same nevergrad interface for optimization and the Evaluation class (Evaluation.py) which sequentially evaluate multiple optimizers and visualize the results of multiple runs simultaneously. (Color figure online)

3 Experiments

The purpose of the following experiments is to compare the two implantations of MixSimulator. Each step of the budget gives us candidate vectors $K_{t,i}$, allowing us to evaluate the performance of the mix with low, mid and high budget.

3.1 Dataset

The experiment are based on data from Madagascar, and more precisely from the Interconnected Region of Toamasina.

The mix is composed of 10 generators and one of them is an hydroelectric plant located in Volobe. The others are thermal power stations located in Toamasina IV and Betainomby. The technical specifications of power plants available in this dataset have been taken from the "Office de regulation de l'électricité" website (Official Electricity Regulation organisation in Madagascar). Note that the most up to date information are from 2017. This is actually the main limitation of the experimentation.

The demand of Toamasina can be estimated by using the forecasting model provided by prophet [17] trained on the demand evolution from 2008 to 2018. However, in order to get a rigorous simulation, the current experiment has been done using the real data consumption of the year 2017. Since the environmental data of Toamasina is not yet publicly available at the time of writing, the only changes in the environment of our agents we could simulate are power plants going down (or up) at some time t. These scenarios are nevertheless relevant, since blackouts are recurring events in Madagascar, either planned by the authority or due to a lack of production or technical problems. There is no law on carbon limitation in Madagascar and this implies that the carbon cost C is set at 0\$. Regarding the demand gap penalization, we set k arbitrarily to 10 000\$ per MWh to make sure the demand constraints will be considered as an important constraint. We do not forget that the satisfaction of deliveries remains the primary purpose of an electricity mix. The mix does not take into account the costs related to transport and transformation to low voltage in substations.

All data is available in the repository of the project [1].

3.2   Scenarios

In order to compare the performance of the two approaches, we randomly generated an unique scenario consisting of repeated failures and failure recoveries of some of the 10 power plants (See Table 1) over 48 h (which represents an optimization of 480 variables). For each approach (MAS and the classic approach), and for each of the three best performing algorithms (Nevergrad optimizers) on the classic approach (Differential Evolution [16,19], the adaptative metaoptimizer NGOpt [10] and (1+1)-ES [14]), we executed 10 runs of 1000 budgets based on an instance of this same scenario. The budget is the maximum iteration value given to the Black-Box optimizer. In other words, it is the maximum iteration that is allocated to the algorithm in the search space to explore solutions. This gives us, for each optimization approach, the average result of each of the three optimizers on a low (100, 300) and semi-high (500, 1000) budget optimization. For all of these runs, we set the num_workers (number of thread) to only 1, so we don't apply any parallelism here.

We compare the value of the objective function $f(x)$ of the two implementations as following: let be $Avg\_cost_{classic}$ the average value of the cost function of the classical optimization at a given budget over the 10 runs, and Avg_cost_{MAS} the average value of the cost function of the MAS optimization at the same given budget over the 10 runs

$$Perf(\%) = \frac{(Avg\_cost_{classic} - Avg\_cost_{MAS})}{Avg\_cost_{classic}} \times 100$$

The metric presented above allows us to measure how well does a method perform compared to the other. This evaluation brings a qualitative information.

The higher the metric $(Perf(\%))$ is, the better the MAS approach performs compared to the classic approach. Another measure is what we call the demand gap which is the sum of U_t (see Sect. 2.2) along the time interval d and which gives the sum of the absolute difference between the actual demand and the mix production. The nearer to 0 the demand gap is, the better it is. We can see below how the demand gap is defined: let α_i be the installed power, $K_{t,i}$ the coefficients of utilization of each n_p power plant, $demand_t$ the actual demand at t and d the time interval

$$demand\_gap = \sum_{t=1}^{d}(U_t)$$

$$demand\_gap = \sum_{t=1}^{d}(|demand_t - \sum_{i=1}^{n_p}(K_{t,i} * \alpha_i)|)$$

4   Discussions

Based on the results from the experiments (See Table 1), we can see that the MAS implementation of MixSimulator performs significantly better than the Classic implementation.

Using the DE algorithm, the MAS based method consistently outperforms the classical method. The difference goes from 25.09/30.36% (from budget 100 to

Table 1. Performance comparison of the MAS and Classic approach.

Budget	Optimizer	Type	Objective function ($)	Perf(%)
100	DE	classic	3 190 553	+25.09%
		MAS	2 389 906	
	NGOpt	classic	3 366 063	+23.48%
		MAS	2 575 693	
	OnePlusOne	classic	2 646 677	+35.79%
		MAS	1 699 455	
300	DE	classic	2 616 888	+30.36%
		MAS	1 822 437	
	NGOpt	classic	3 021 924	+32.50%
		MAS	2 039 724	
	OnePlusOne	classic	1 790 570	+40.53%
		MAS	1 064 896	
500	DE	classic	2 212 279	+34.91%
		MAS	1 439 994	
	NGOpt	classic	2 912 171	+36.39%
		MAS	1 852 521	
	OnePlusOne	classic	1 353 154	+40.95%
		MAS	798 970	
1000	DE	classic	1 629 684	+41.56%
		MAS	952 310	
	NGOpt	classic	2 912 171	+36.39%
		MAS	1 852 521	
	OnePlusOne	classic	764 569	+24.80%
		MAS	574 949	

300) to 34.91/41.56% (from budget 500 to 1000). The Fig. 3 presents the average (over 10 runs) evolution of DE optimization and this specific case shows that the MAS method gives better result than the classic method. Note that the Demand gap is the difference between the demand provided by the demand monitor and the production proposed by the schedule.

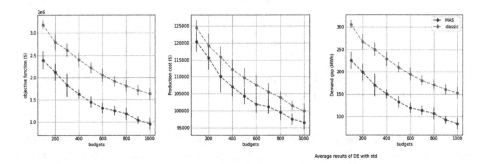

Fig. 3. The image on the left is the evolution of the objective function through the budget for the MAS based (in blue) and the classic (in red) implementation using the DE algorithm. The image in the middle shows how the production cost evolves through coefficients of utilization given by the optimizer. The image on the right is the evolution of the demand gap using the DE algorithm. The nearer to 0 the demand gap is, the better it is. The DE performs averagely better with the Agent-based method than with the classic one. Let's note that the error bar plotted on each image represents the standard deviation over the 10 runs. (Color figure online)

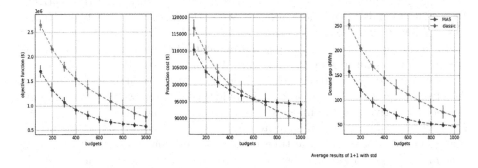

Fig. 4. The cost function converges almost at the same point at budget 1000 using the (1+1) algorithm. However, this convergence is faster at a lower budget for MAS. Regarding the demand gap, the classic optimization outperforms MAS.

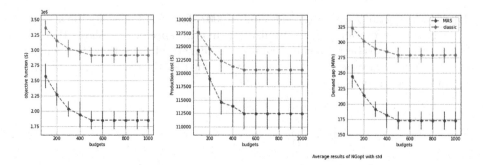

Fig. 5. MAS outperforms classic optimization (both on the demand gap and the cost function) using the NGOpt optimizer.

Using OnePlusOne, the MAS approach performs from +24.80% to +40.95% better than the classic approach. OnePlusOne provides also the best solution of this experiment. Indeed, at 1000 budgets, we get the lowest production cost value: 89 536.28\$ for the classic and 94 110.26\$ for the MAS. (See Fig. 4)

Using the meta-model NGOpt, the MAS approach performs from +23.48% to +36.39% better than the classic approach.

Note that the *Perf* is highly dependent on the number of events in the scenario. The more power plants go down at $t > t_0$, the more MAS system tends to outperform the classic one. This is the expected result, as MAS fine-tunes its initial planning for each new unexpected event. That is also why MAS is more time-consuming than a classic optimization: with DE (respectively NGOpt, OnePlusOne), the classical approach takes in average only 1030 (respectively 1945, 973) seconds to complete one run, while the MAS approach takes 10 469 (respectively 15 922, 7952) seconds.

In future work, simulating more environmental behavior like quantity variation in storage, solar irradiation and precipitation is a must. Regarding the other renewable energy like solar and wind energy, we plan to evaluate the model with data from different regions in order to include them. As renewable power stations are more sensitive to environmental change, including them is necessary to take full advantage of the MAS architecture. Also, as the probability of environmental change increases alongside the time interval, evaluating the model on a larger time interval (more than one week) and on a larger optimizer benchmark reveals those advantages even more.

5   Conclusion

This paper aims to compare the two optimization methods available in MixSimulator, and therefore show how the combination of MAS and evolutionary computation can be relevant to energy modeling. The two different approaches are the classical optimization technique for minimizing the cost of the mix production and the Agent-based approach of the optimization which brings as a plus dynamic scheduling. Our MAS implementation is a collaborative one, which means we have one Moderator which handles all the optimization jobs. The fact that dynamic schedule optimization has the ability to accommodate to various events and environment changes, allows the MAS based method to perform significantly better. We used nevergrad optimizers (DE, NGOpt, and 1+1) to generate solutions with both MAS and classic methods. Both methods converge as the budget increases (See Fig. 3, Fig. 4, Fig. 5). However, the experiments show that the MAS implementation of the simulator outperforms the classic implementation by 27% (respectively 33%, 36% and 36%) for a 100 (respectively 300, 500, 1000) budget optimisation in average over the three optimizers. By simulating scenarios and environmental behavior, mixsimulator can be used as a decision making tool, especially in our time when investments in different energy sources impact the economic and social face of the world in the near future.

References

1. Andriamalala, S.R., Andriamizakason, T.A., Rasoanaivo, A.: MixSimulator - an electricity mix simulator and optimization (2020). https://GitHub.com/Foloso/MixSimulator
2. Bonin, B., Safa, H., Laureau, A., Merle-Lucotte, E., Miss, J., Richet, Y.: MIXOPTIM: a tool for the evaluation and the optimization of the electricity mix in a territory. Eur. Phys. J. Plus **129**(9), 198 (2014). https://doi.org/10.1140/epjp/i2014-14198-7, https://hal-cea.archives-ouvertes.fr/cea-01089278
3. Chen, Y.M., Wang, S.C.: An agent-based evolutionary strategic negotiation for project dynamic scheduling. Int. J. Adv. Manuf. Technol. **35**(3), 333–348 (2007). https://doi.org/10.1007/s00170-006-0830-x
4. Davidsson, P., Persson, J.A., Holmgren, J.: On the integration of agent-based and mathematical optimization techniques. In: Nguyen, N.T., Grzech, A., Howlett, R.J., Jain, L.C. (eds.) KES-AMSTA 2007. LNCS (LNAI), vol. 4496, pp. 1–10. Springer, Heidelberg (2007). https://doi.org/10.1007/978-3-540-72830-6_1
5. Eddy, Y.F., Gooi, H.B., Chen, S.X.: Multi-agent system for distributed management of microgrids. IEEE Trans. Power Syst. **30**(1), 24–34 (2014)
6. Law, A.M., Kelton, W.D.: Simulation Modeling and Analysis, 2nd edn. McGraw-Hill Higher Education, New York (1997)
7. Maturana, F., Shen, W., Hong, M., Norrie, D.: Multi-agent architectures for concurrent design and manufacturing. Environments **1**(2), 3 (1997)
8. McArthur, S., et al.: Multi-agent systems for power engineering applications - part ii: technologies, standards and tools for building multi-agent systems. IEEE Trans. Power Syst. **22**, 1753–1759 (2007). https://doi.org/10.1109/TPWRS.2007.908472
9. Meunier, L., et al.: Black-box optimization revisited: improving algorithm selection wizards through massive benchmarking. CoRR abs/2010.04542 (2020). https://arxiv.org/abs/2010.04542
10. Meunier, L., et al.: Black-box optimization revisited: improving algorithm selection wizards through massive benchmarking. arxiv:2010.04542 (2020)
11. Moradi, M.H., Razini, S., Mahdi Hosseinian, S.: State of art of multiagent systems in power engineering: a review. Renew. Sustain. Energy Rev. **58**, 814–824 (2016). https://doi.org/10.1016/j.rser.2015.12.339, https://www.sciencedirect.com/science/article/pii/S1364032115017220

12. Ouelhadj, D., Petrovic, S.: A survey of dynamic scheduling in manufacturing systems. J. Sched. **12**(4), 417–431 (2009)
13. Rapin, J., Teytaud, O.: Nevergrad - a gradient-free optimization platform (2018). https://GitHub.com/FacebookResearch/Nevergrad
14. Rechenberg, I.: Evolutionstrategie: Optimierung Technischer Systeme nach Prinzipien des Biologischen Evolution. Fromman-Holzboog Verlag (1973)
15. Rios, L.M., Sahinidis, N.V.: Derivative-free optimization: a review of algorithms and comparison of software implementations. J. Glob. Optim. **56**(3), 1247–1293 (2013). https://doi.org/10.1007/s10898-012-9951-y
16. Storn, R., Price, K.: Differential evolution - a simple and efficient heuristic for global optimization over continuous spaces. J. Glob. Optim. **11**(4), 341–359 (1997)
17. Taylor, S., Letham, B.: Forecasting at scale. Peerj Preprints **5**, e3190v2 (2017)
18. Wörner, J., Wörn, H.: Benchmarking of multiagent systems in a production planning and control environment. In: Kirn, S., Herzog, O., Lockemann, P., Spaniol, O. (eds.) Multiagent Engineering. International Handbooks on Information Systems, pp. 115–133. Springer, Heidelberg (2006). https://doi.org/10.1007/3-540-32062-8_7
19. Yang, Z., Tang, K., Yao, X.: Differential evolution for high-dimensional function optimization. In: 2007 IEEE Congress on Evolutionary Computation, pp. 3523–3530 (2007). https://doi.org/10.1109/CEC.2007.4424929

A Multi-brain Approach for Multiple Tasks in Evolvable Robots

Ege de Bruin$^{(\boxtimes)}$ ⓘ, Julian Hatzky$^{(\boxtimes)}$ ⓘ, Babak Hosseinkhani Kargar$^{(\boxtimes)}$ ⓘ, and A. E. Eiben ⓘ

Department of Computer Science, Vrije Universiteit Amsterdam,
Amsterdam, The Netherlands
egedebruin@gmail.com, julianhatzky@gmail.com
{b.hosseinkhanikargar,a.e.eiben}@vu.nl

Abstract. We investigate the joint evolution of morphologies (bodies) and controllers (brains) of modular robots for multiple tasks. In particular, we want to validate an approach based on three premises. First, the controller is a combination of a user-defined decision tree and evolvable/learnable modules, one module for each given task. Second, morphologies and controllers are evolved jointly for each task simultaneously by a multi-objective evolutionary algorithm. Third, after terminating the evolutionary process, the brain of the users' favorite morphology is optimized by a learning algorithm applied to the task-specific controller modules independently.

Keywords: Evolutionary Robotics · Morphological evolution · Controller evolution · Robot learning · Multi-Objective Optimization · Locomotion

1 Introduction

Evolutionary robotics is a field where evolutionary algorithms are used to evolve robots. In the early years of the field, the focus was mainly on evolving the controllers of robots with a fixed morphology [11,21], but more recently the evolution of robot morphologies has been addressed as well [2,6,10,26].

Much of the existing work is based on evolving bodies and brains for one task, where acquiring an adequate gait is a popular problem. However, simply moving around without a goal is not really practical. Functional robots need to move with purpose, for example, move towards a target in sight. There has been work done for this targeted locomotion, on robots with fixed morphologies [19] and robots with evolvable morphologies [15,16]. These works however assume that the robot knows where the target is, and if the target is not in sight it assumes the target is at the last known position. When it is completely unaware of where the target is, it will not search for it.

In this work we, consider a practically more relevant case, where the robot has to 1) find the target and 2) move to the target, and be able to repeat this for additional targets.

J. Correia et al. (Eds.): EvoApplications 2023, LNCS 13989, pp. 129–144, 2023.
https://doi.org/10.1007/978-3-031-30229-9_9

To achieve this, we want to end up with a robot body, that is fully capable to perform those tasks, together with a controller architecture to move the robot and that is a combination of a user-defined decision tree and evolvable brains, one brain for each given task.

The decision tree is to capture the users' domain knowledge where the solution needs not to be evolved, only coded. For instance, IF `target-not-in-sight` THEN `search-for-it` or IF `target-in-sight` THEN `move-towards-it`. The evolvable modules represent (morphology-dependent) sub-controllers specifying how that given task can be executed by the given body. Thus, the first research question we address is:

Can a multi-objective evolutionary algorithm deliver good bodies for modular robots for handling two tasks?

Our approach is, in essence, a multi-brain system: one brain for each task, combined through a decision tree. However, the first research question is focused on delivering good bodies for robots to handle two tasks regardless of doing the tasks separately or simultaneously. Hence, the second question we address is:

How much can a secondary learning stage that separately optimizes each task enhance robot performance in a given morphology?

To this end, we use a two-phase approach, where first the creatures are evolved to do the tasks simultaneously, and then the creatures learn to do the two tasks separately. This results in one morphology and two different brains, one for each task, which can be used in the robot's controller. Naturally, this procedure can be extended to an arbitrary number of tasks by creating an additional number of brains.

2   Background

Locomotion for modular robots is a difficult task, but there has been promising work with robot locomotion based on Central Pattern Generators (CPGs) [14]. Moreover, a neural network-based approach, for example using HyperNEAT [12], can be used to evolve good controllers for robot locomotion. For targeted locomotion, there are studies focusing on robots with fixed shapes [19]. Lan et al. propose a method for targeted locomotion of generic shapes by morphologically evolving the robots [16]. A CPG-based approach was used with a HyperNEAT generative encoding technique. This study was later extended to follow a moving target [15]. In this work, when the target went out of sight of the robot, the robot assumed the last-known angle towards the target to be the angle towards the target. These studies showed that a robot can evolve and learn to walk towards and follow a target.

Considering multiple objectives, the task of the robot influences its morphology [4]. This is especially the case when the tasks are conflicting, and when the multiple objectives are put into a single function. It is not certain that a single solution exists for the problems [7], and in a single fitness function, it is difficult

to find the correct trade-off between multiple behavioral terms [25]. It is therefore preferred to use a multi-objective function, and the NSGA-II algorithm has
been shown to be a good and fast algorithm for multi-objective functions [20]
[8].

A paper by Lipson et al. shows that when the morphology and brain of
the robot are evolved together, we run into premature convergence, and this is
especially seen in the morphology of the creature [5]. Moreover, when morphology
and brain are evolved together to optimize multiple tasks, this will lead to worse
results than when the creatures are evolved for the tasks separately [3]. Nygaard
et al. try to overcome these convergences with a two-phase approach [22]. First,
the morphology and brain are evolved together. Then, after convergence, the
morphology is fixed and the brain evolves further. This paper shows that after
convergence of evolving the morphology and brain together, the creature can
still perform better when only the brain is evolved further. This is acknowledged
by Eiben and Hart [9], who state that after the evolution of brain and body
together, the brain needs to learn the optimal way to use the body. A paper
by Lessin et al. showed results for an approach to evolve creatures for multiple
objectives [18]. The robot was initially evolved for a locomotion task, after which
most of the morphology was fixed and other tasks were learned after this. This
was later extended by a method to make the morphology more flexible after the
initial evolution phase [17]. This differs from our approach by evolving robots
to more specific actions, like turning left and turning right, and by evolving the
robot's morphology per action instead of by all actions together.

3   Experimental Work

3.1   Phenotypes and Simulation

The robots are simulated on a Gazebo-based simulator *Revolve* [13][1] The robot
design is based on RoboGen [1]. There are three different robot modules which
can be used as building blocks for the robots. At first, every robot has one *core*
component which, when the robot is built in real life, contains the controller
board. This block has four possible connections to other components. Another
component is the *fixed brick*, which has four possible slots to attach other components. Lastly, there is an *active hinge* component and this is the component
responsible for the robot's movement. It is a joint that can be attached on two
lateral sides. In the simulator the orientation of the robot is a virtual sensor, if
the robot would be built in real life the camera would be located in the core component of the robot. In Fig. 1 an example of robot with its possible components
is shown.

3.2   Central Pattern Generators

As a controller for the robots, Central Pattern Generators (CPGs) have been
shown to perform well for a task like targeted locomotion. CPGs are neural

---

[1] (see Revolve https://github.com/ci-group/revolve).

Fig. 1. A possible robot from RoboGen. The middle white component is the core component, the green square components are the fixed brick components, and the red parts are the active hinge components. (Color figure online)

networks that are responsible for the rhythmic movement of animals, without any sensory information or rhythmic inputs [14]. Because they are independent of higher control centers, this reduces time in the motor control loop and reduces the dimensionality to control movements. This concept is applied to robots, where CPG models are being used to control the locomotion of robots. In this work, CPGs are used in the hinges of the robots, with differential oscillators, which are responsible for rhythmic movement, as its main components. The oscillators generate patterns by calculating activation levels of neurons x and y shown at the top of Fig. 2. Moreover, there is an output neuron that outputs the value for the CPG model. For these neurons, the equations for calculating the difference per time step can be seen in Eqs. 1 and 2.

$$\Delta x = w_{yx}y + bias_x \tag{1}$$

$$\Delta y = w_{xy}x + bias_y \tag{2}$$

The activation functions of the output neurons are *tanh* functions. The hinges also affect each other, therefore the CPG components are extended by taking into account the neurons of neighboring hinges. To be more precise, for each pair of neighboring hinges the x neurons in the CPG models also have an influence on each other. Therefore, the activation values for the x and y nodes can be calculated by the equations in Eqs. 3 and 4.

$$x_i(t) = x_i(t-1) + \Delta x_i(t) + \sum_{j \in N_j} x_j(t-1) * w_{ji} \tag{3}$$

$$y_i(t) = y_i(t-1) + \Delta y_i(t) \tag{4}$$

In these equations, N_j are all neighboring hinges of hinge i, and Δx_i and Δy_i are calculated from Eqs. 1 and 2. To also take into account the angle from

the robot's vision direction towards the target, the CPG model is extended with
sensory input as shown in Fig. 2. When the angle towards the target is lower
than α, the target is assumed to be on the left and otherwise on the right. This
has an influence on the actuators, where the left actuators are influenced when
the target is on the left and visa versa. Whether the actuator is on the left or
on the right depends on its lateral position and the direction of the target. An
overview of this approach can be seen at the bottom of Fig. 2.

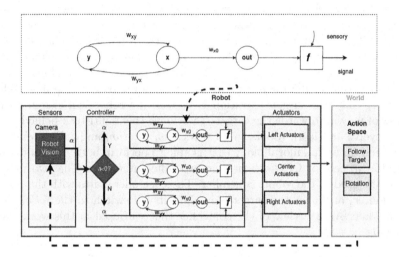

**Fig. 2. Overview of a CPG component with angle towards target taken into
account.** When the target is on the left, α is assumed to be lower than 0 and only the
left actuators are influenced. Otherwise the right actuators are influenced.

3.3   HyperNEAT and Compositional Pattern Producing Networks

We use HyperNEAT, which is a hypercube-based encoding to evolve large-scale
neural networks [23]. HyperNEAT is an effective choice when learning the weights
for the CPG controllers for a given task. The idea behind HyperNEAT is that
there is a substrate network, represented by for example a hyper-dimensional
cube, of which the coordinates of a node are used as input for a CPPN. The
CPPN is then evolved using NEAT [24] to find the optimal structure and activation functions of the CPPN for the problem at hand. For our work, to get the
correct connection weight values, the CPPN will have as input the coordinates
of the source and target CPG component, including a 1 if the target node is an
x-CPG-node, a -1 if it is a y-CPG-node, and 0 if it is an output-CPG-node. The
output of the CPPN is then the weight of the connection.

To evolve and generate the morphology of a robot a second CPPN is used.
This network takes as input the x, y and z coordinates of the possible location
for a module of the robot, and the length towards the core module as well as
input. The output is the module that is used, hinge, block or no module, and the

rotation of the module. The generation of the robots starts at the core module with coordinates (0,0,0), and from this module the robot is extended with new modules dependent on the outputs of the CPPN. This CPPN is also evolved using NEAT.

3.4 Two-Brain Approach

Other work related to targeted locomotion has mostly focused on only the locomotion part. For example, recent work by Lan et al. [15] showed progress in making evolvable robots move towards a moving target, but it was the robot's only task. When the target was out of sight of the robot, it assumed the latest known angle towards the target to be the current angle, so there was no specific task for searching for the target. In this work we aim to add an additional task to the robot, that enables it to explore its surrounding by rotating in search for the target. Once the target's position is known, the robot will switch back to the task of moving towards the target. To achieve this multi-task behaviour, we need to evolve and learn robots to do two tasks, *rotation* and *follow target*. We want to end up with a morphology that is able to perform both behaviours well, alongside two sets of CPG-weights, or two different *brains*. The two modules can then be used in the robots controller. The controller starts with the *rotation brain* module, and once the target is in sight it will switch to the *follow target brain* module. An overview of the controller that will used in this work can be seen in Fig. 3.

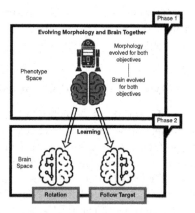

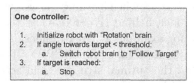

Fig. 3. The controller of the robot to end up with. There will be two sets of CPG-weights, or two brains: one for *rotation* and one for *follow target*. The robot will switch brains dependent on the angle towards the target.

Fig. 4. The two-phase approach to evolve and learn the robot to do two tasks. First a robot is evolved for both objectives, then it learns the two tasks separately.

3.5   Two-Phase Approach

The goal is to end up with one morphology alongside two different sets of CPG-weights for the two tasks, and we will do that using a two-phase approach [22]. First, the morphology of the robot is evolved to do both tasks at the same time. After that, the morphology of the robot is fixed and we will learn the robot twice. Once for the *rotation* task and once for the *follow target* task. An overview is given in Fig. 4.

Phase 1: Morphology Evolution. We want to have a morphology that can perform the two distinct tasks: *rotation* and *follow target*. To achieve this we use a multi-objective algorithm to simultaneously evolve the morphology and brain. Previous work has shown that the NSGA-II algorithm is a good and fast algorithm for multiple objectives [20]. The NSGA-II algorithm is an elitist genetic algorithm, which uses non-dominated sorting for selection. A non-dominated solution, in the context of multi-objective optimization, refers to a solution that cannot be surpassed by any other solution in terms of its performance on any objective, while maintaining equal or better performance on all other objectives. A front of such non-dominated solutions is called a Pareto-front. In the NSGA-II algorithm, there are multiple fronts, where the first front contains all non-dominated solutions, the second front contains all non-dominated solutions without the first front, and so on. A selection is made by choosing the solutions which are on the highest fronts, and ties are decided on crowding distance sorting to ensure diversity. This evolutionary process aims to arrive at a morphology that is good in doing both tasks, however not yet separately. This morphology can then be used to make the robot learn the tasks separately.

Phase 2: Task Learning. The multi-objective evolution of the morphology alongside the brain (phase 1) results in a brain that is optimized to perform both objectives simultaneously. However, since our goal is a clear separation between the desired behaviors, the idea of this phase 2 is to learn a brain for each task separately. We achieve this by fixing a chosen morphology after phase 1 and optimizing the weights of two randomly initialized copies of its CPG-brain (Fig. 4). These two sets of CPG-weights are then used by the controller (Fig. 3) to accomplish the execution of multiple separate tasks. Note that we randomly re-initialize the CPG-weights before the learning step, instead of keeping the already evolved weights intact. The reason is that by keeping the evolved CPG-weights, we would need to unlearn the opposing task instead of learning the desired task, and this would likely need an adjustment of the fitness function by penalizing the undesired behavior. Instead, the robot learns the task from scratch by randomly re-initializing the CPG-weights, so there is no unlearning involved and we can use the same fitness function for both phases.

3.6   Fitness Functions

There are two tasks for the robots, *rotation* and *follow target*, and for both tasks a separate fitness function is used. The robot should be able to rotate as quickly as possible to get a full overview of the environment, so for the objective value of rotation the forward orientation, which is the direction of the robot's vision, returned from the Revolve Simulator is used. At every time step the forward orientation of the robot is compared to the previous orientations, and this is summed up. This will then result in the total orientation of the robot, and the goal is to maximise this. The function for this can be seen in Eq. 5, where T are all time points during evaluation, and $o(t)$ is the forward orientation at time point t.

$$rotation = \sum_{t \in T} |o(t) - o(t-1)| \tag{5}$$

For the fitness function of the *follow target* task several factors are taken into account, and a visualisation can be seen in Fig. 5. There, $T0$ is the start position of the robot, $T1$ the end position and the red dotted line is the line towards the target. It is important that the robot is moving into the correct direction, so the distance travelled on the ideal trajectory line is taken into account, which is the distance between $p_0(x_0, y_0)$ and $p(x_p, y_p)$. Secondly, the distance between the end point of the robot and the ideal trajectory line is taken into account, which is the distance between $p_1(x_1, y_1)$ and $p(x_p, y_p)$ in Fig. 5. It is preferred that the robot moves in a straight line towards the target, so distance travelled is minimized simultaneously. Hence, the travelled path is also taken into account, shown in the figure as two different solid red trajectory lines. Finally, the angle between the optimal direction and travelled direction is also taken into account, as shown by the difference between β_0 and β_1. The combination of all this results in Eq. 6, where ϵ is an infinitesimal constant, $e1$ is the distance on ideal trajectory, $e2$ is the distance between the end point of the robot and the ideal trajectory, $e3$ is total distance travelled, and δ is the angle between the optimal direction and travelled direction. Finally, p1 is a penalty weight set to 0.01.

$$followTarget = \frac{e1}{e3 + \epsilon} \cdot \left(\frac{e1}{\delta + 1} - p1 \cdot e2 \right) \tag{6}$$

3.7   Experiment Parameters

For phase 1 we use a $(\mu + \lambda)$ selection mechanism with $\mu = 100$ and $\lambda = 50$ to update the population which is initially generated randomly. In each generation 50 offspring are produced by selecting 50 pairs of parents through tournament selection with replacement, creating one child per pair by crossover and mutation according to the MultiNeat implementation[2] Out of the $(\mu + \lambda)$ solutions μ solutions are selected through NSGA-II selection for the next generation. The evolutionary process is terminated after 300 generations and we do a total of 30 runs.

---

[2] (see MultiNeat https://github.com/MultiNEAT/).

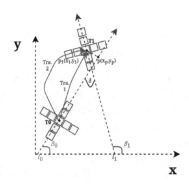

Fig. 5. Visualisation of calculating the *follow target* fitness value. $T0$ and $T1$ are the start and end point of the robot respectively. The red dotted line is the line towards the target, and the red solid lines are two possible trajectory lines of the robot. (Color figure online)

The morphologies for phase 2 are selected among the pareto front of generation 300 and then the specified objective is learned for additional 200 generations. For phase 2 the same set of parameters is used, but this time the morphology of the phenotype is fixed and only the brain weights are evolved for a single objective, again using MultiNeat (Table 1).

Table 1. Experiment parameters

Parameters	1, 2*	Description
Population	100, 100	Individuals per generation
Offspring	50, 50	Offspring per generation
Generations	300, 200	Termination condition
Mutation	NEAT, NEAT	Mutation operator
Crossover	NEAT, NEAT	Crossover operator
Parent selection	Tournament, Tournament	Parent selection operator
Survivor selection	NSGA-II, Tournament	Survivor selection operator
Tournament size	2, 2	Number of individuals used in the tournament
Evaluation time	50, 50	Duration of the test period per fitness evaluation in seconds
Runs	30, 30	Repetitions per experiment

*Values for Phase 1, Phase 2

4   Results

4.1   Phase 1: Morphology Evolution

With the multi objective evolutionary process using NSGA-II, many individuals perform well on the objectives at hand. Figure 7 shows that the dominating morphology of generation 300 is of type *snake* (Fig. 6c). However, after further investigation of the behaviour we decide to exclude this phenotype, since it not only rotates horizontally but also vertically, and a vertical rotation is not practical with the hardware we are using.

Hence, we decided to proceed by filtering out the snake morphologies. The other two morphologies we call *beyblade* (Fig. 6a) and *fancy* (Fig. 6b). Lastly, there is a category *other* that contains morphologies that are present in only a very small amount. The distribution shows that the *fancy* morphology is mostly good in rotating but not in following the target, whereas the *beyblade* scores well for both objectives, though not as high in *rotation* as *fancy*. The rotation score of the *beyblade* seems good enough for the task, and because following a target is the more difficult task it makes sense to continue to phase 2 with a *beyblade* morphology.

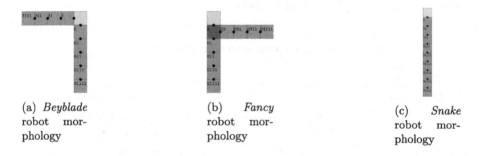

(a) *Beyblade* robot morphology

(b) *Fancy* robot morphology

(c) *Snake* robot morphology

Fig. 6. The three most occurring morphologies after 300 generations of evolution.

4.2   Phase 2: Task Learning

By following the argument of the foregoing Sect. 4.1, we choose the *beyblade* morphology that is given by the NSGA-II pareto-front of generation 300, for the second phase of learning. As explained in Sect. 3.5, the brain weights are reset to random values, and all morphologies of the *beyblade* robots are the same. Therefore, we can pick any *beyblade* morphology. To this end the morphology stays fixed and the brain weights are re-initialized and then the two objectives are learned separately, resulting in two separate brains. In Table 2 the fitness after 200 generations of learning - using the morphology of generation 300 - is compared to the fitness of the evolution-only approach, which is extended to 500 generations for a fair comparison. It can be observed that for the *rotation*

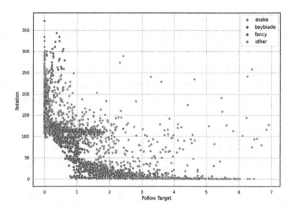

Fig. 7. Individuals in Generation 300 of 30 runs. It is observable that the snake morphology (orange, Fig. 6c) is dominating the population. The morphology is naturally good for the *follow target* objective, because of its straight line shape, and the exploitation of the rotation along its horizontal axis leads to high scores for rotation as well. However, as we state in Sect. 4.1 it comes with undesired behavioral trades. Besides the snake morphology, *fancy* (red, Fig. 6b) is very good in *rotation* although not so good in *follow target* and *other* (green) are potentially better in *follow target* than *rotation*. The *beyblade* (blue, Fig. 6a) morphology seems to lead to a more diverse distribution, getting decent scores for both objectives. *Rotation* is given in radians and *follow target* in meters. (Color figure online)

objective the learning results in a higher fitness, whereas for the learning of the *follow target* objective, the single objective learning performs slightly worse than NSGA-II.

Table 2. Phase 1 results versus phase 2 results. The results of the learning only approach with NSGA-II (phase 1) after 500 generations versus the separate learning of each behaviour for 200 generations (phase 2, after 300 generations of NSGA-II) averaged over 30 runs. It can be seen, that for rotation the learning leads to a higher overall fitness function while for the follow line objective, the NSGA-II approach performs better. The rotation performance is given in radians and the follow line performance in meters.

	evolution only	evolution + learning
rotation	66.38	**82**
follow line	**2.02**	0.72

4.3   Combining the Tasks

To achieve a more complex *search-and-chase behavior* we now use the controller to switch back and forth between the two learned sets of CPG-weights. We showcase this by using the beyblade morphology of phase 1, together with the two best-learned brains of phase 2. These two brains are used in the controller of the robot (Fig. 3). Figure 8 shows the trajectory of 30 runs for the robot, as well as the average trajectory. There is one target and its position is at coordinates (0,5). We estimate the performance of the robot by using the Mean Absolute Error (MAE) of its trajectory versus the optimal (shortest) trajectory. The MAE is calculated as MAE = $\frac{1}{N}\sum_{i=1}^{N} |Robot_i - Ideal_i|$, with N being the number of data points that are sampled, and *Robot* and *Ideal* being the coordinates of the robot and the optimal trajectory at time-step i. Compared to the optimal trajectory, the green dotted line, the average trajectory has an MAE of 0.6, with a standard deviation of 1.3. While none of the individual runs is able to transition to the target with an optimal, or close to optimal, trajectory all of them do reach the target. It seems that the robot keeps correcting itself when it is not going in the right direction, and once the target is reached it stays there.

Figure 8b shows the trajectory of 30 runs for two targets, target 1 is at coordinates (0,3) and target 2 is at coordinates (3,0). The robot will first search for the first target and move towards it. Once the first target is reached, it will search for the second target and move towards it. Compared to the ideal trajectory the average trajectory has an MAE of 0.9, and the standard deviation of the trajectories is 1.2. The trajectories again do not seem optimal, but in 27 cases the robot reached both targets. Overshooting the target does not seem to be a big problem and since more distance is traveled between the starting point and the first target, the plain average over the trajectories does not reach the target.

Figure 9 shows the distribution of the MAE of the 30 trajectories for both the one target scenario and the two target scenario. Both distributions look

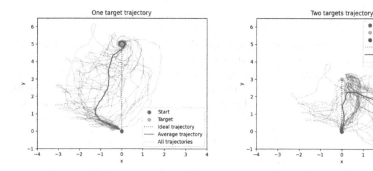

(a) The MAE of the average trajectory is 0.6 with a standard deviation of 1.3.   (b) The MAE of the average trajectory is 0.9 with a standard deviation of 1.2.

Fig. 8. Trajectories in meters of 30 runs of the resulting beyblade morphology with the learned controller. The trajectories are of a single beyblade robot, with its best-learned *rotation* and *follow target* brains, starting at coordinates (0,0). The dark blue line is the average trajectory of the 30 runs and the other blue lines are the individual runs. The left Figure is in a two target environment and the right Figure is in a one target environment. (Color figure online)

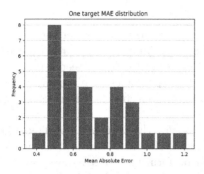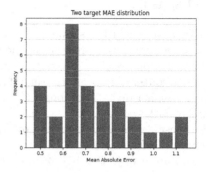

(a) **The distribution of the MAE of all 30 trajectories for the one target situation.** The average MAE is 0.6

(b) **The distribution of the MAE of all 30 trajectories for the two target situation.** The average MAE is 0.7

Fig. 9. Mean Absolute Error distribution of the 30 trajectories

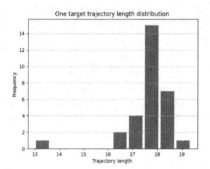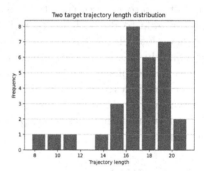

(a) **The distribution of the length of all 30 trajectories for the one target situation.** Most lengths are between 16 and 19, but there is one short outlier at 13. The ideal trajectory is 5

(b) **The distribution of the length of all 30 trajectories for the two target situation.** Most lengths are between 14 and 20, but there are some below 12. The ideal trajectory is 7.2

Fig. 10. Length of trajectory distribution of the 30 trajectories

similar, as most MAE are between 0.5 and 0.7. This is to be expected, as we work with the same robot in both scenarios. Figure 10 shows the distribution of the trajectory length of the 30 trajectories. The lengths are very similar to each other, except for some outliers. As mentioned before, for the two target scenario 27 were able to reach both targets, meaning 3 of them did not. These are the outliers between lengths 8 and 12.

5   Discussion

The NSGA-II evolution shows an expected and steady increase in the ability of the phenotypes to score well on the objectives. Another observation is that through this process, the diversity among the morphologies is decreasing over time, and converging to mostly three types of morphologies. A mechanism that ensures a higher diversity amongst the population could be a great improvement in this regard.

In phase 2, the learning of the rotation task for the *beyblade* morphology is clearly outperforming the NSGA-II performance. However, while the learning of the *rotation* objective is even able to outperform the NSGA-II we can not observe that for the learning of the *follow target* objective. We hypothesize that this is because the phenotype has a natural bias for rotation and the rotation behaviour does not drastically harm the robots abilities to follow the target. Moreover, another possibility is that the selection pressure is not high enough and thus the learning algorithm can not keep up with the elitist strategy of NSGA-II. However, our goal here is not only to outperform NSGA-II in first place, but also to *unlearn* the other task. The best morphologies out of the first phase are able to do both tasks well, but only simultaneously. The resulting trajectory plots of Fig. 8 show as well that on average the robots follow the target directly. Lastly, it is to point out that we use the same objective functions for phase 1 and phase 2 in order to have better comparable results. However, when the goal is a robot that is able to perform the desired tasks as accurately and separately as possible, a penalty term in phase 2, that penalizes the task that shall be unlearned has great potential to vastly improve on the resulting behaviour.

5.1   Future Work

As of now, only the learning for the *beyblade* morphology is analyzed. In the future it can be interesting to also apply the two phase approach to the *fancy* morphology as well as others. Furthermore, the state-machine controller could be exchanged by a reward-driven system like reinforcement learning. Another interesting idea would be to test the approach with different or more objectives and other types of environments. It is also of interest how the learning after generation 300 would behave without re-initialization of the brain weights, using more of a Lamarckian approach. Lastly, the forward orientation of the robot, which is also the potential view field, is not necessarily its direction of movement. The robot might be moving backwards, sideways or diagonally while facing forward. We would like to investigate how the enforcing of a forward orientation changes the phenotype space and behavioural space of the evolution.

6   Conclusion

The first question that we ask is: *Can a multi-objective evolutionary algorithm deliver good bodies for modular robots for handling two tasks?* We demonstrate that this is possible with the two-phase approach. Initially, we evolve a set of candidate morphologies using the multi-objective NSGA-II algorithm, and then we further separate the desired behaviors with a second phase of learning. To this end, we compare the effects of evolving the phenotypes for the two objectives (*follow target* and *rotation*) with only evolving the brain for the specified objectives separately. We show that using a two-phase approach leads to a better separation of desired behaviors and an increase in overall fitness performance for at least one objective. The second question we ask is: *How much can a secondary learning stage that separately optimizes each task enhance robot performance in a given morphology?* By using the evolved morphology of phase 1 and the two separately learned brains of phase 2, together with a controller that switches between the brains, we demonstrate that the robot is able to navigate through its environment while performing the tasks of searching and chasing one or multiple targets. Videos of the robot can be found on a video playlist[3].

References

1. Auerbach, J., et al.: Robogen: robot generation through artificial evolution. In: ALIFE 14: The Fourteenth International Conference on the Synthesis and Simulation of Living Systems, pp. 136–138 (2014)
2. Beer, R.D.: The dynamics of brain–body–environment systems. In: Handbook of Cognitive Science, pp. 99–120. Elsevier (2008)
3. Carlo, M.D., Ferrante, E., Ellers, J., Meynen, G., Eiben, A.E.: The impact of different tasks on evolved robot morphologies. In: Proceedings of the Genetic and Evolutionary Computation Conference Companion. ACM, July 2021
4. Carlo, M.D., Zeeuwe, D., Ferrante, E., Meynen, G., Ellers, J., Eiben, A.: Robotic task affects the resulting morphology and behaviour in evolutionary robotics. In: 2020 IEEE Symposium Series on Computational Intelligence (SSCI). IEEE, December 2020
5. Cheney, N., Bongard, J., Sunspiral, V., Lipson, H.: On the difficulty of co-optimizing morphology and control in evolved virtual creatures. IN: Proceedings of the Artificial Life Conference 2016, July 2016
6. Cheney, N., Bongard, J., SunSpiral, V., Lipson, H.: Scalable co-optimization of morphology and control in embodied machines. J. Roy. Soc. Interface **15** (2018)
7. Coello, C.C.: Evolutionary multi-objective optimization: a historical view of the field. IEEE Comput. Intell. Mag. **1**(1), 28–36 (2006)
8. Deb, K., Pratap, A., Agarwal, S., Meyarivan, T.: A fast and elitist multiobjective genetic algorithm: NSGA-II. IEEE Trans. Evol. Comput. **6**(2), 182–197 (2002)
9. Eiben, A.E., Hart, E.: If it evolves it needs to learn. In: Proceedings of the 2020 Genetic and Evolutionary Computation Conference Companion. ACM, July 2020
10. Eiben, A., et al.: The triangle of life: evolving robots in real-time and real-space. In: Lio, P., Miglino, O., Nicosia, G., Nolfi, S., Pavone, M. (eds.) Proceedings of the 12th European Conference on the Synthesis and Simulation of Living Systems (ECAL 2013), pp. 1056–1063. MIT Press (2013)
11. Floreano, D., Husbands, P., Nolfi, S.: Evolutionary robotics. In: Siciliano, B. and Khatib, O. (ed.) Handbook of Robotics, 1st edn, pp. 1423–1451. Springer, Heidelberg (2008). https://doi.org/10.1007/978-3-540-30301-5_62

---

[3] (see shorturl.me/JtySjtH).

12. Haasdijk, E., Rusu, A.A., Eiben, A.E.: HyperNEAT for locomotion control in modular robots. In: Tempesti, G., Tyrrell, A.M., Miller, J.F. (eds.) ICES 2010. LNCS, vol. 6274, pp. 169–180. Springer, Heidelberg (2010). https://doi.org/10.1007/978-3-642-15323-5_15

13. Hupkes, E., Jelisavcic, M., Eiben, A.E.: Revolve: a versatile simulator for online robot evolution. In: Sim, K., Kaufmann, P. (eds.) EvoApplications 2018. LNCS, vol. 10784, pp. 687–702. Springer, Cham (2018). https://doi.org/10.1007/978-3-319-77538-8_46

14. Ijspeert, A.J.: Central pattern generators for locomotion control in animals and robots: a review. Neural Netw. 21(4), 642–653 (2008)

15. Lan, G., van Hooft, M., Carlo, M.D., Tomczak, J.M., Eiben, A.: Learning locomotion skills in evolvable robots. Neurocomputing 452, 294–306 (2021)

16. Lan, G., Jelisavcic, M., Roijers, D.M., Haasdijk, E., Eiben, A.E.: Directed locomotion for modular robots with evolvable morphologies. In: Auger, A., Fonseca, C.M., Lourenço, N., Machado, P., Paquete, L., Whitley, D. (eds.) PPSN 2018. LNCS, vol. 11101, pp. 476–487. Springer, Cham (2018). https://doi.org/10.1007/978-3-319-99253-2_38

17. Lessin, D., Fussell, D., Miikkulainen, R.: Adopting morphology to multiple tasks in evolved virtual creatures. In: Artificial Life 14: Proceedings of the Fourteenth International Conference on the Synthesis and Simulation of Living Systems. The MIT Press, July 2014

18. Lessin, D., Fussell, D., Miikkulainen, R.: Open-ended behavioral complexity for evolved virtual creatures. In: Proceedings of the 15th annual conference on Genetic and evolutionary computation. ACM, July 2013

19. Matos, V., Santos, C.P.: Towards goal-directed biped locomotion: combining CPGs and motion primitives. Robot. Auton. Syst. 62(12), 1669–1690 (2014)

20. Moshaiov, A., Abramovich, O.: Is MO-CMA-ES superior to NSGA-II for the evolution of multi-objective neuro-controllers? In: 2014 IEEE Congress on Evolutionary Computation (CEC). IEEE, July 2014

21. Nolfi, S., Floreano, D.: Evolutionary Robotics: The Biology, Intelligence, and Technology of Self-organizing Machines. MIT Press, Cambridge (2000)

22. Nygaard, T.F., Samuelsen, E., Glette, K.: Overcoming initial convergence in multi-objective evolution of robot control and morphology using a two-phase approach. In: Squillero, G., Sim, K. (eds.) EvoApplications 2017. LNCS, vol. 10199, pp. 825–836. Springer, Cham (2017). https://doi.org/10.1007/978-3-319-55849-3_53

23. Stanley, K.O., D'Ambrosio, D.B., Gauci, J.: A hypercube-based encoding for evolving large-scale neural networks. Artif. Life 15(2), 185–212 (2009)

24. Stanley, K.O., Miikkulainen, R.: Evolving neural networks through augmenting topologies. Evol. Comput. 10(2), 99–127 (2002)

25. Trianni, V., López-Ibáñez, M.: Advantages of task-specific multi-objective optimisation in evolutionary robotics. PLoS ONE 10(8), e0136406 (2015)

26. Weel, B., Crosato, E., Heinerman, J., Haasdijk, E., Eiben, A.E.: A Robotic Ecosystem with Evolvable Minds and Bodies. In: 2014 IEEE International Conference on Evolvable Systems, pp. 165–172. IEEE Press, Piscataway (2014)

A Quality-Diversity Approach to Evolving a Repertoire of Diverse Behaviour-Trees in Robot Swarms

Kirsty Montague[1], Emma Hart[1](✉) (iD), Geoff Nitschke[2](iD), and Ben Paechter[1](iD)

[1] Edinburgh Napier University, Edinburgh, UK
{k.montague,e.hart,b.paechter}@napier.ac.uk
[2] University of Capetown, Cape Town, South Africa
geoff.nitschke@uct.ac.za

Abstract. Designing controllers for a swarm of robots such that collaborative behaviour emerges at the swarm level is known to be challenging. Evolutionary approaches have proved promising, with attention turning more recently to evolving repertoires of diverse behaviours that can be used to compose heterogeneous swarms or mitigate against faults. Here we extend existing work by combining a Quality-Diversity algorithm (MAP-Elites) with a Genetic-Programming (GP) algorithm to evolve repertoires of *behaviour-trees* that define the robot controllers. We compare this approach with two variants of GP, one of which uses an implicit diversity method. Our results show that the QD approach results in larger and more diverse repertoires than the other methods with no loss in quality with respect to the best solutions found. Given that behaviour-trees have the added advantage of being human-readable compared to neural controllers that are typically evolved, the results provide a solid platform for future work in composing heterogeneous swarms.

Keywords: Swarm-robotics · Quality-Diversity ·
Genetic-Programming

1 Introduction

Evolutionary techniques were first proposed to design controllers for swarms of robots as far back as 1992 [17]. A steadily increasing amount of work since then has shown that EAs are capable of designing controllers for a range of tasks, demonstrating that self-organising behaviours can emerge from a swarm in which each member executes the same controller. Recently, attention has begun to turn towards evolving repertoires of controllers which exhibit behavioural diversity. This is particularly prevalent in the literature relating to evolution of controllers for individual robots, for example, Cully et al. [7], but has also begun to permeate the swarm robotics field. Having access to a repertoire of behaviourally diverse controllers has multiple potential advantages. For example, it can facilitate fault-tolerance by providing alternative methods of control when there are faulty sensors or actuators; it raises the possibility of adapting to unseen

© The Author(s), under exclusive license to Springer Nature Switzerland AG 2023
J. Correia et al. (Eds.): EvoApplications 2023, LNCS 13989, pp. 145–160, 2023.
https://doi.org/10.1007/978-3-031-30229-9_10

situations in novel environments not known when the controllers were evolved; it enables composition of a *heterogeneous* swarm with respect to control. The latter is pertinent given the breadth of literature across disciplines suggesting that groups of diverse problem-solvers can outperform groups of high-ability problem-solvers [16]. Thus, there is considerable need for a library of diverse behavioural repertoires that can be used to improve swarm task-performance.

Existing work relating to evolution of diverse repertoires for swarms has tended to use neural network or parameterised controllers in conjunction with a *quality-diversity* algorithm. The latter are a family of methods first introduced in [27] that evolve multiple solutions that are diverse with respect to one or more user-defined characteristics. This study evaluates a method to evolve diverse behaviours on various swarm-robotic tasks, where behaviours result from controllers defined by *behaviour-trees* (BTs). BTs are structures that describe switching between a finite set of tasks [4,23], enabling complex behaviours to be described from a series of simple tasks. In addition, they have the added advantage of being human-readable: this allows them to be analysed, extended and verified [29]. A broad range of recent work has investigated ways in which Behaviour Trees can be leveraged for fully-autonomous robotic control [1,2,5,12,23] in both individual and swarm robotics.

The paper makes two contributions to the field of BTs and swarm-robotics. Firstly, we show that genetic-programming (GP) can be used to evolve diverse BTs representing *primitives* defining general skills required to realise foraging behaviours. Evolving primitives rather than a single BT that realises an overall foraging behaviour is advantageous in that primitives can be combined in multiple ways in a higher-level control strategy to achieve multiple tasks. We compare two GP methods for evolving primitives. The first requires separate runs of a GP algorithm to produce each primitive. The second simultaneously evolves all three primitives using an implicit population diversity mechanism. We then develop a third method that combines the quality-diversity algorithm MAP-Elites [25] with GP to explicitly evolve a repertoire of behaviours that are diverse with respect to three axes describing behavioural characteristics. We demonstrate our QD-GP algorithm significantly outperforms the two GP approaches in terms of metrics that quantify both the quality and diversity of the generated behaviours, although the GP method using implicit diversity is surprisingly competitive in terms of both quality and diversity metrics.

2 Background

In the context of swarm robotics, evolutionary algorithms have been used to evolve a range of controller *types:* most commonly, neural controllers are learned, with an EA learning the network topology, weights or both [30]. The parameters of parametric controllers have also been evolved [8]. However, another line of enquiry has focused on the evolution of *behaviour-tree* (BT) controllers. They are often presented as a viable architecture for robotics on the basis that they remove the trade-off between reactivity and modularity inherent in finite state machines [6], allowing complex behaviours to be encoded while still maintaining independence between components. In addition they have the added advantage of being human-readable; this allows them to be analysed, extended and verified [29]. Successful examples include evolution of behaviour tree controllers for swarms of kilobot robots foraging in a simulated arena, developed in a series of papers [18–20]. Behaviour trees for swarms of e-puck robots [24] were also evolved by [21] and compared to finite-state machine controllers generated by AutoMoDe Chocolate [10] and neural networks generated by an approach dubbed Evostick [11]. The authors found that both of the modular design approaches transferred to reality more successfully than the controllers evolved by Evostick. [14] introduce the Instinct Evolution Scheme to control a swarm of miniature autonomous agents with constrained resources and reconfigurable hardware. They propose a method where Grammatical Evolution is used to evolve BTs from a set of action nodes derived from a Pareto front of configurations for each hardware module which satisfy local objectives (for example precision versus power consumption).

Quality-diversity algorithms [27] are a relatively new family of evolutionary algorithms that aim to find a maximally diverse collection of individuals (often with respect to a user-defined measure of behaviour) in which each individual is as high performing as possible. While the field has rapidly expanded to include a large set of algorithm variants, most derive from two fundamental algorithms: MAP-Elites [25] and Novelty-Search with Local Competition (NSLC) [22]. Both methods return an archive of diverse, high-quality behaviors in a single run. Many successful applications of QD methods to evolve diverse behavioural repertoires for single robots exist[1]. However, applications to swarm-robotics to provide diverse behavioural repertoires are still under-represented in the literature, despite the potential benefits sketched in Sect. 1.

Engebråten et al. [8] evolve a repertoire of diverse controllers for a multi-function swarm using MAP-Elites. Robots were controlled using parametric or weighted controllers inspired by the use of artificial potential fields and evolved to conduct perimeter surveillance and network creation tasks. In Gomes et al. [13], evolution of a diverse set of task-agnostic behaviours is tackled. Robots have neural controllers, and novelty-search with local-competition (NSLC) [22] is used to evolve a set of diverse behaviours that can be utilised in multiple tasks. The authors demonstrated that repertoires of general swarm behaviours can be generated, yielding a wide diversity of high-quality behaviours. Steyven et al. [15] describe the first distributed model of the MAP-Elites algorithm for embodied evolution: each robot in a swarm runs its own MAP-Elites algorithm, with a robot exchanging or updating its individual container when coming into range of another robot. Their algorithm EDQD was demonstrated to produce a diverse range of controllers in a simple swarm following task.

Thus, despite a plethora of literature exploring QD methods to generate diverse behavioural repertoires for individual robots, such methods are under-explored in swarm-robotics. This study extends related work by combining GP

---

[1] For an up-to-date list of relevant papers, see https://quality-diversity.github.io.

with MAP-Elites to evolve a repertoire of BT controllers, representing generic *building-block* primitives, combined to realise swarm-robotic foraging tasks.

3   Methodology

The goal of the paper is twofold. Firstly, we investigate whether GP can be used to evolve BTs representing three primitive behaviours that are important components of swarm-foraging tasks (*go-to-food, go-to-nest, increase-neighbourhood-density*). Similarly to the work described by Gomes et al. [13], the motivation behind this approach is that general low-level behaviours can in future be composed into more complex behaviours in multiple different tasks. Secondly, we investigate the extent to which a diverse repertoire of swarm controllers can be evolved. In this work, we assume that swarms are homogeneous in that all robots run the same controller, however as noted in Sect. 1, the motivation behind this is that the evolved repertoire has broad swarm-robotic task applicability.

Two GP algorithms for generating BTs are compared. GP_1 evolves a BT to optimise the behaviour of a *single* primitive, hence needs to be run separately for each of three primitives *go_to_food, go_to_nest, increase_density*. The variants are thus labelled GP_f, GP_n, GP_d respectively. In contrast, $GP_{f,n,d}$ *simultaneously* evolves BTs for all three primitives: an *implicit* diversity maintenance mechanism is used in which the selection method compares solutions on a randomly chosen objective each time it as called. We then compare these methods to a quality-diversity approach (MAP-Elites) which evolves behaviours for each objective that are diverse with respect to three user-defined descriptors. Like GP_1, this also needs to be run once per objective (consequently labelled QD_n, QD_f, QD_d).

3.1   Environment, Robots and Simulator

The simulator and task implemented are inspired by the foraging kilobots experiment described in Jones et al. [18]. This requires robots to travel between nest and food regions in an arena as shown in Fig. 1. Our swarm consists of 9 robots. We consider evolution of three behavioural fragments which can be combined by a higher-level algorithm to achieve more complex swarm behaviours. The fragments considered attempt to maximise the following objectives:

- **O1: Increase neighbourhood density** maximises the difference between the density of neighbouring robots at the beginning and end of each trial by subtracting the initial density from the final density.
- **O2: Move towards the nest region** uses the distance estimated by each robot based on the shortest route by hops via neighbouring robots from its location at the start and end of each trial. The difference is maximised by subtracting the initial distance from the final distance.

- **O3: Move towards the food region** maximises the robots' estimated difference in distance to the food region in the same way as the nest region, by subtracting the initial estimated distance from food from the final estimated distance from food.

The robots used are *footbots*, shown in Fig. 2, simulated using ARGoS [26] and based on marXBots as described in Bonani et al. [3]. They move on a combination of tracks and wheels (treels). The sensing and actuation capabilities of the robots in this scenario are given in Table 1. If a robot is within 50 mm of the centre of the arena, it is considered to be within the nest region, while if it is located more than 50 mm plus the width of the gap between nest and food then it is considered to be within the food region. A *blackboard* provides the interface between the evolving behaviour-trees and the footbot control software. Information received from sensors and other robots in each update cycle is condensed into the blackboard entries listed below, each of which return true or false, with two separate entries as necessary for *left* and *right*. Note that 'left' means that a detection event occurred anywhere in the left hemisphere of the robot (assuming a virtual line drawn through the centre of the robot from front to back). 'Right' is interpreted in the same manner.

- Is this robot in the food region
- Is this robot in the nest region
- Is the shortest route to food to this robot's left/right
- Is the shortest route to the nest to this robot's left/right
- Is the closest neighbouring robot to this robot's left/right

Each update cycle, the robots transmit their ID and estimated distances from the food and nest regions to their neighbours. Distances are calculated using the shortest hop distance via neighbouring robots after adding the extra distance to the robot in question. Robots currently within the nest or food regions will send a default estimate of 0 mm and robots that are out of range default to 500 mm. Messages arriving from robots more than 100 mm away are ignored while messages within this range arrive with 95% probability (as per Jones et al. [18]). Also in keeping with Jones et al. [18]), behaviour trees are ticked (executed) 2 Hz while sensing and communication takes place 8 Hz, and the range and bearing data is averaged over seven samples to estimate neighbourhood density and the distance to the nest and food regions.

A single evaluation consists of ten trials, each lasting 20 s (40 ticks). In each trial, the robots start in a random position within 100 mm of the centre of a 500 mm × 500 mm arena. To encourage robust behaviours, the trials are equally split between two arena configurations which differ in the size of the gap between the nest and food regions (50 mm and 70 mm) while the radius of the nest region remains constant at 50 mm.

3.2 Behaviour Tree Representation

All nodes in a typical BT return one of three strings, *success*, *failure*, or *running* (the node's operation will continue to execute until at least the next tick).

Table 1. Footbot Reference Model

Ground sensor	Informs the robot if it is in the nest or food regions
Range	Distance to each neighbour
Bearing	Angle of each neighbour
Signal	Information payload from each robot in range containing the cumulative distance to nest and food regions derived from hops
Motors	Left and right wheels controlled independently, with $\pm$ 10% variation between robots

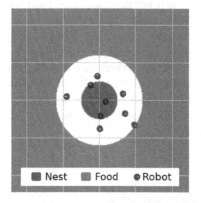

Fig. 1. The arena layout, with nine robots initialised in random starting positions.

Fig. 2. A screenshot of a footbot robot in the arena taken in ARGoS.

This results in a uniform interface where all nodes are compatible, removing any possibility of creating an invalid tree along with any requirement for type safety. The root node of a behaviour tree is executed (ticked) at regular intervals. If the root has child nodes, the tick propagates through the tree as each branch node which is executed ticks one or more of its children. Composition nodes form the branches and control the flow of execution through the tree by prioritising between several child nodes in different ways (Table 2). The composition nodes in these experiments have memory so if a child node returns *running* the parent does likewise. Upon resuming execution they continue by ticking the same child in their list the next time they themselves are ticked.

Leaf nodes can be either conditions or actions. Condition nodes control the flow of execution by returning *success* or *failure* depending whether their condition is satisfied, which determines whether their parent can continue to tick any subsequent children. The action nodes in these experiments all represent locomotion behaviours, returning *running* initially (in effect pausing the BT for one tick) and returning *success* on the next tick. In doing so, the action node terminates and allows execution to flow back to the parent. Condition or action nodes do not have children so they can be exchanged freely by the mutation and

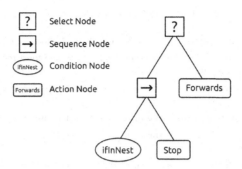

Fig. 3. An example behaviour tree.

crossover operators. An example behaviour tree is shown in Fig. 3, with a select node at its root and a sequence node as the first child. The tree checks if the robot is in the nest, in which case it stops, otherwise it moves forward.

Behaviour Tree Nodes: BTs nodes in these experiments can fulfil one of three roles as outlined below:

- **Composition nodes** control the flow of execution through the tree. There are three types of node (sequence, select and probabilistic), described in Table 2. The role of these nodes is to govern execution of child nodes, accounting for which was executed last.
- **Condition nodes** have no children and return *success* or *failure* depending whether their condition is satisfied. Nodes are listed in Table 3 and determine the direction of the nearest neighbour (left/right); direction of the shortest hop distance to food or nest (left/right); whether the robot is within the nest or food region. Where necessary, the robots' immediate vicinity within 100 mm is divided into left and right hemispheres and the node's return value indicates whether the given half contains the object of interest.
- **Action nodes** have no children. Their purpose is to send commands to the robots' motors to provide movement. These are listed in Table 4.

Table 2. Composition Nodes

Sequence node with memory and 2, 3 or 4 children	Executes each child until one fails
Select node with memory and 2, 3 or 4 children	Ticks each child until one succeeds
Probabilistic node with memory and 2, 3 or 4 children	Ticks one child chosen at random

Table 3. Condition Nodes

If on food	Returns success if the robot is within the food region
If food to left	Returns success if the shortest route to the food region is to the robot's left
If food to right	Returns success if the shortest route to the food region is to the robot's right
If in nest	Returns success if the robot is within the nest region
If nest to left	Returns success if the shortest route to the nest region is to the robot's left
If nest to right	Returns success if the shortest route to the nest region is to the robot's right
If robot to left	Returns success if the nearest robot is to this robot's left
If robot to right	Returns success if the nearest robot is to this robot's right

Table 4. Action Nodes

Stop	No movement for one tick
Forwards	Move forwards for one tick
Forwards left	Right wheel forwards for one tick, rotating the robot anti-clockwise
Forwards right	Left wheel forwards for one tick, rotating the robot clockwise
Reverse	Move backwards for one tick
Reverse left	Right wheel in reverse for one tick, rotating the robot clockwise
Reverse right	Left wheel in reverse for one tick, rotating the robot anti-clockwise

3.3   Algorithms

Controllers for three behaviour fragments are evolved using a benchmark evolutionary algorithm and two variations which employ different strategies for promoting behavioural diversity.

Benchmark Algorithms: We compare results obtained from running a classical genetic-programming algorithm with quality-diversity approach. The GP algorithm is implemented using DEAP [9] with the parameters specified in Table 5. We evaluate two versions of this algorithm:

1. GP_o: this optimises a single objective o (one of the three functions specified in Sect. 3.1), and therefore evolves a BT for one of three primitive behaviours. Experiments are conducted for each of three primitives in turn from each of the three variants (GP_f, GP_n, GP_d).

2. $GP_{f,n,d}$: in order to *implicitly* encourage diversity within the population, each time the selection method is called, one of three objectives is chosen with uniform probability, and the individuals in the tournament are compared according to the chosen objective. A list of the elite solution for each objective is maintained throughout the run. As the intention is to evolve three

separate trees representing each primitive, the population size is tripled and the maximum evaluation budget is set to three times the budget of GP_o.

All of the nodes used to define BTs return the same type (a string which is either "success", "failure" or "running"), so there is no requirement for the algorithm to be strongly typed. Pressure is applied towards smaller trees by selecting the shortest in the event that their fitness scores are exactly equal and at least one of them is more than three nodes deep.

MAP-Elites for Evolution of Diverse Behaviours: We use the QD algorithm MAP-Elites [25]. This is implemented using the Python Library QDPy; the reader is referred to the documentation for a full description of the algorithm which follows the original definition of MAP-Elites. MAP-Elites differs from classical evolutionary algorithms in that the population is stored in a container discretised into cells according to behaviour descriptors defined by a user. Each solution is mapped to a single cell according to the derived behavioural descriptors; the cell stores up to three solutions which have the best objective-values found for that cell during the run. We define three behavioural axes describing phenotypic aspects of the robots during execution of the BT controller that promote *phenotypic* (behavioural) diversity: (1) Ratio of forwards vs backwards movements; (2) Ratio of clockwise vs anti-clockwise rotations; (3) Ratio of condition nodes vs action nodes. Quality is calculated using the same three objective functions as the GP approaches described in the previous section. As with GP_o, the QD method must be run separately for each of three primitives. These methods are labelled (QD_n, QD_f & QD_d). All parameters required to run the GP algorithm defined in DEAP and the MAP-Elites algorithm defined in QDPy are listed in Table 5 and were derived from [18] where possible or by preliminary empirical testing otherwise.

4   Results

We present results from two sets of experiments: (1) A comparison of the quality of the best BT controller found from each of three methods for each primitive (GP_o, $GP_{f,n,d}$, QD_o); (2) A comparison of the diversity of behaviours found by each of the three algorithms per primitive. *Quality* is determined with respect to objective fitness for each of the three primitives. We measure diversity according to the *coverage* metric. That is, the proportion of bins filled, and finally also report the QD score which captures both quality and diversity. This is obtained by summing the highest fitness values found in each grid bin (Q_i) [28] $\sum_{i=1}^{t} Q_i$. All metrics are calculated over 10 runs with a maximum budget of 25,000 evaluations per objective[2]. To evaluate whether there is statistical significance between pairs of results, we first apply a Shapiro-Wilk test to test for normality. If the null hypothesis is rejected ($p \leq 0.05$) then a Mann-Whitney test is used for comparison, otherwise a t-test is applied (again with a confidence level of 0.05).

---

[2] $GP_{n,f,d}$ is allocated 75000 evaluations as it simultaneously solves 3 objectives.

Table 5. Genetic Programming & Quality-Diversity Parameters

GP Parameters	
Generations	1000
Population size	25 per objective
Tournament size	3 (GP_o) and 5 $(GP_{f,n,d})$
Elites	1 per objective
Probability of crossover	0.8
Probability of mutating by inserting a subtree	0.05
Probability of mutating by shrinking a subtree	0.1
Probability of mutating by replacing one node	0.5
Trials per evaluation	10
Arena configurations per evaluation	2
Trial length	20 s (40 ticks)
QD Parameters	
Bins	$8 \times 8 \times 8$
Max items per bin	3
Initial batch size	100
Batch size	50
Generations	500

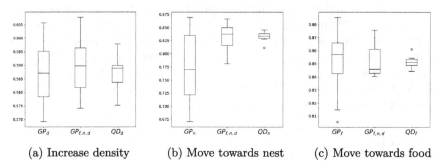

(a) Increase density   (b) Move towards nest   (c) Move towards food

Fig. 4. Box-plots of the best fitness obtained for all three algorithms and objectives with a design budget of 25,000 evaluations per objective and averaged over ten runs, or twenty runs where available

4.1   Comparison of Objective Fitness

While the main goal of the paper is to generate diverse repertoires of behaviours with respect to each of the three primitives, it is instructive to understand whether searching for diversity in addition to quality has an adverse impact on quality. Figure 4 shows box-plots over the 10 runs per objective value and algorithm. From a qualitative perspective, the QD approach appears more robust in that the variation across repeated runs is much smaller than the two GP methods. Statistical tests applied between each pair of algorithms per-objective show that the null-hypothesis cannot be rejected between any tests for the *go-to-food* and *density* objectives, whereas for the *go-to-nest* objective, a statistically significant difference is observed between GP_n and $GP_{f,n,d}$, and between GP_n and QD_n, with GP_n providing the poorer result in each case.

Our results clearly demonstrate that for two objectives (food, density), there is no evidence that the $GP_{f,n,d}$ approach, that aims to maximise all objectives simultaneously, results in any difference in quality when compared to results from the corresponding single objective GP variant. Furthermore, there is no evidence that evolving for both quality and diversity reduces the quality of the best solutions found for these two objectives.

4.2   Coverage and QD Scores

Figures 5 and 6 show boxplots for the three respective metrics for each of three objectives. Results from pairwise statistical testing are also given in Table 6. It is immediately clear that the QD method provides significantly better coverage

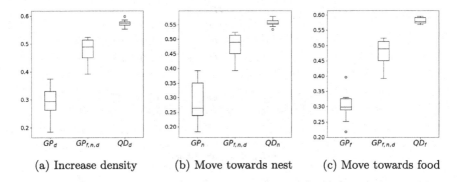

    (a) Increase density   (b) Move towards nest   (c) Move towards food

Fig. 5. Coverage for each algorithm and objective with a design budget of 25,000 evaluations per objective and averaged over ten runs

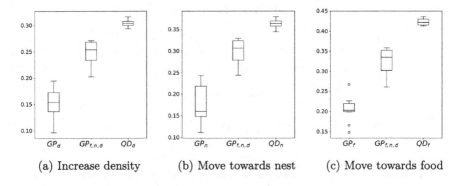

    (a) Increase density   (b) Move towards nest   (c) Move towards food

Fig. 6. Quality diversity scores for each algorithm and objective with a design budget of 25,000 evaluations per objective and averaged over ten runs

Table 6. Statistical testing results showing pairwise comparisons. Statistically significant results within a confidence interval of 0.05 are shown in bold.

Metric	Objective	Comparison	p-value	Type of test
Best fitness	Increase density	GP_d vs $GP_{n,f,d}$	0.3505	T-test
		GP_d vs QD_d	0.9695	T-test
	Go to nest	$\mathbf{GP_n}$ vs $\mathbf{GP_{n,f,d}}$	**0.0004**	T-test
		$\mathbf{GP_n}$ vs $\mathbf{QD_n}$	**0.0066**	T-test
	Go to food	GP_f vs $GP_{n,f,d}$	0.4406	Mann-Whitney
		GP_f vs QD_f	0.9422	T-test
Coverage	Increase density	$\mathbf{GP_d}$ vs $\mathbf{GP_{n,f,d}}$	< 0.0001	T-test
		$\mathbf{GP_d}$ vs $\mathbf{QD_d}$	< 0.0001	T-test
	Go to nest	$\mathbf{GP_n}$ vs $\mathbf{GP_{n,f,d}}$	< 0.0001	T-test
		$\mathbf{GP_n}$ vs $\mathbf{QD_n}$	< 0.0001	T-test
	Go to food	$\mathbf{GP_f}$ vs $\mathbf{GP_{n,f,d}}$	< 0.0001	T-test
		$\mathbf{GP_f}$ vs $\mathbf{QD_f}$	< 0.0001	T-test
QD Score	Increase density	$\mathbf{GP_d}$ vs $\mathbf{GP_{n,f,d}}$	< 0.0001	T-test
		$\mathbf{GP_d}$ vs $\mathbf{QD_d}$	< 0.0001	T-test
	Go to nest	$\mathbf{GP_n}$ vs $\mathbf{GP_{n,f,d}}$	< 0.0001	T-test
		$\mathbf{GP_n}$ vs $\mathbf{QD_n}$	< 0.0001	T-test
	Go to food	$\mathbf{GP_f}$ vs $\mathbf{GP_{n,f,d}}$	< 0.0001	T-test
		$\mathbf{GP_f}$ vs $\mathbf{QD_f}$	< 0.0001	T-test

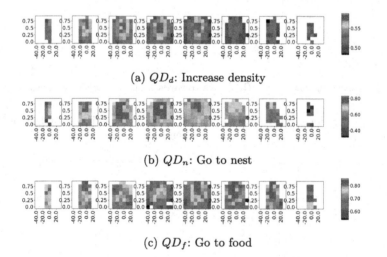

(a) QD_d: Increase density

(b) QD_n: Go to nest

(c) QD_f: Go to food

Fig. 7. Heatmaps for QD_o (three separate objectives)

and QD score than the EA approaches, indicating that the method is able to produce diversity in behaviour while also maintaining quality. In terms of the best solution found, the QD approach generally finds solutions that are at least

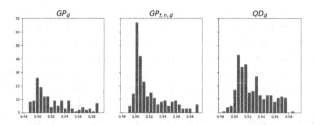

(a) Increase density GP_d 148 solutions, $GP_{f,n,d}$ 255 solutions, QD_d 290 solutions

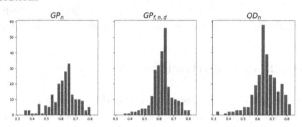

(b) Go to nest GP_n 201 solutions, $GP_{f,n,d}$ 255 solutions, QD_n 283 solutions

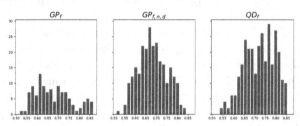

(c) Go to food GP_f 112 solutions, $GP_{f,n,d}$ 255 solutions, QD_f 305 solutions

Fig. 8. Histograms showing the distribution of fitness values per algorithm (one row per objective)

on a par with the other methods. That is, there is no statistical evidence that solution task performance is different (Sect. 4.1). Figure 7 provides a visualisation of the evolved QD container for each objective. Since the container has dimensions $8 \times 8 \times 8$ it is presented as 8 separate 8×8 containers[3].

Figure 8 plots the distribution of fitness values in the repertoires generated by each algorithm for each objective from a single run of each. The figure also indicates the size of each repertoire. Although all methods produce a spread of values, the QD approaches produce repertoires that encapsulate many different

---

[3] Given the obvious difference between the QD and EA methods, we only show the QD containers given page limit constraints.

behaviours that also have high quality (task performance). That is, the centre of mass of the distribution is shifted to the right in the QD case. Note that the $GP_{f,n,d}$ method finds much larger repertoires than the single objective GP method for all three objectives, and that the repertoires also contain solutions that tend to be of higher quality. One possible explanation for this is that by randomly selecting an evaluation function to score an individual during each round of tournament selection, some useful transfer of knowledge occurs which increases fitness. Similarly, Evolutionary Multi-Task Optimisation (EMTO) [31] methods have demonstrated that solving multiple tasks in combinatorial optimisation via a shared population can improve quality and is thus worthy of further investigation. Alternatively, the random choice of objective function per round of selection is likely to also encourage wider exploration of the search-space.

5   Conclusions and Further Work

The goal of this research was to compare methods for generating a repertoire of diverse behaviours for three primitive tasks that are useful components for synthesising foraging behaviours in swarms. To the best of our knowledge, this study was the first experimental comparison of various methods for generating diverse behaviours via *behaviour-tree* controllers, with respect to three different primitives. Our study compared a standard GP algorithm to a GP algorithm using an implicit diversity mechanism (to simultaneously evolve three behaviours), to a quality-diversity algorithm (MAP-Elites), using an explicit diversity mechanism.

Results showed, inline with existing literature that has evolved neural controllers, that QD is capable of evolving a diverse repertoire of high quality BT controllers. Furthermore, this study demonstrated that a GP algorithm using an implicit diversity mechanism is capable of evolving high (task-performance) quality controllers, although the repertoire size attained is smaller than that of the QD approach (Fig. 5). An advantage of using behaviour trees over neural networks is their readability. Using BT controllers in swarms therefore potentially offers a route to creating swarms with explainable behaviour in the future.

The evolved repertoires provide a platform for two avenues of future work. On the one hand, the diverse behaviours evolved for each primitive can be combined to deliver a broad range of complex foraging behaviours, for example, evolving a 'meta-BT' that uses the primitives themselves as nodes. Other high-level strategies might also be learned that utilise these primitives, for example, reinforcement learning. On the other hand, the existence of the evolved repertoires opens up the possibility of constructing swarms in which the robots have heterogeneous *behaviours*. This would require a search algorithm to discover useful compositions, but would provide new insights into the benefits of composing a swarm from a single high-performing behaviour or multiple behaviours that are diverse with respect to quality. Composing swarms where each entity has a diverse form of movement also potentially offers some mitigation against faults or breakages and thus ensures swarm-robotic resilience.

References

1. Banerjee, B.: Autonomous acquisition of behavior trees for robot control, pp. 3460–3467, October 2018. https://doi.org/10.1109/IROS.2018.8594083
2. Biggar, O., Zamani, M.: A framework for formal verification of behavior trees with linear temporal logic. IEEE Robot. Autom. Lett. 5(2), 2341–2348 (2020). https://doi.org/10.1109/LRA.2020.2970634
3. Bonani, M., et al.: The marxbot, a miniature mobile robot opening new perspectives for the collective-robotic research, pp. 4187–4193, January 2010. https://doi.org/10.1109/IROS.2010.5649153
4. Colledanchise, M., Marzinotto, A., ögren, P.: Performance analysis of stochastic behavior trees. In: 2014 IEEE International Conference on Robotics and Automation (ICRA), pp. 3265–3272 (2014). https://doi.org/10.1109/ICRA.2014.6907328
5. Colledanchise, M., Natale, L.: Improving the parallel execution of behavior trees, September 2018. https://doi.org/10.1109/IROS.2018.8593504
6. Colledanchise, M., ögren, P.: Behavior trees in robotics and AI: an introduction. CoRR abs/1709.00084 (2017). http://arxiv.org/abs/1709.00084
7. Cully, A., Clune, J., Tarapore, D., Mouret, J.B.: Robots that can adapt like animals. Nature 521(7553), 503–507 (2015)
8. Engebråten, S.A., Moen, J., Yakimenko, O., Glette, K.: Evolving a repertoire of controllers for a multi-function swarm. In: Sim, K., Kaufmann, P. (eds.) EvoApplications 2018. LNCS, vol. 10784, pp. 734–749. Springer, Cham (2018). https://doi.org/10.1007/978-3-319-77538-8_49
9. Fortin, F.A., De Rainville, F.M., Gardner, M., Parizeau, M., Gagné, C.: Deap: evolutionary algorithms made easy. J. Mach. Learn. Res. Mach. Learn. Open Source Softw. 13, 2171–2175 (2012)
10. Francesca, G., et al.: Automode-chocolate: automatic design of control software for robot swarms. Swarm Intell. 9, 125–152 (2015)
11. Francesca, G., Brambilla, M., Brutschy, A., Trianni, V., Birattari, M.: AutoMoDe: a novel approach to the automatic design of control software for robot swarms. Swarm Intell. 8(2), 89–112 (2014). https://doi.org/10.1007/s11721-014-0092-4
12. Giunchiglia, E., Colledanchise, M., Natale, L., Tacchella, A.: Conditional behavior trees: definition, executability, and applications. In: 2019 IEEE International Conference on Systems, Man and Cybernetics (SMC), pp. 1899–1906 (2019). https://doi.org/10.1109/SMC.2019.8914358
13. Gomes, J., Christensen, A.L.: Task-agnostic evolution of diverse repertoires of swarm behaviours. In: Dorigo, M., Birattari, M., Blum, C., Christensen, A.L., Reina, A., Trianni, V. (eds.) ANTS 2018. LNCS, vol. 11172, pp. 225–238. Springer, Cham (2018). https://doi.org/10.1007/978-3-030-00533-7_18
14. Hallawa, A., De Roose, J., Andraud, M., Verhelst, M., Ascheid, G.: Instinct-driven dynamic hardware reconfiguration: evolutionary algorithm optimized compression for autonomous sensory agents, pp. 1727–1734, July 2017. https://doi.org/10.1145/3067695.3084202
15. Hart, E., Steyven, A.S., Paechter, B.: Evolution of a functionally diverse swarm via a novel decentralised quality-diversity algorithm. In: Proceedings of the Genetic and Evolutionary Computation Conference, pp. 101–108 (2018)
16. Hong, L., Page, S.E.: Groups of diverse problem solvers can outperform groups of high-ability problem solvers. Proc. Natl. Acad. Sci. 101(46), 16385–16389 (2004)

17. Husbands, P., Harvey, I.: Evolution versus design: controlling autonomous robots. In: Proceedings of the Third Annual Conference of AI, Simulation, and Planning in High Autonomy Systems' Integrating Perception, Planning and Action', pp. 139–140. IEEE Computer Society (1992)
18. Jones, S., Studley, M., Hauert, S., Winfield, A.: Evolving behaviour trees for swarm robotics. In: Groß, R., et al. (eds.) Distributed Autonomous Robotic Systems. SPAR, vol. 6, pp. 487–501. Springer, Cham (2018). https://doi.org/10.1007/978-3-319-73008-0_34
19. Jones, S., Studley, M., Hauert, S., Winfield, A.: A two teraflop swarm. Front. Robot. AI **5**, 11 (2018). https://doi.org/10.3389/frobt.2018.00011
20. Jones, S., Winfield, A., Hauert, S., Studley, M.: Onboard evolution of understandable swarm behaviors. Adv. Intell. Syst. **1** (2019). https://doi.org/10.1002/aisy.201900031
21. Kuckling, J., Ligot, A., Bozhinoski, D., Birattari, M.: Behavior trees as a control architecture in the automatic modular design of robot swarms. In: Dorigo, M., Birattari, M., Blum, C., Christensen, A.L., Reina, A., Trianni, V. (eds.) ANTS 2018. LNCS, vol. 11172, pp. 30–43. Springer, Cham (2018). https://doi.org/10.1007/978-3-030-00533-7_3
22. Lehman, J., Stanley, K.O.: Evolving a diversity of virtual creatures through novelty search and local competition. In: Proceedings of the 13th Annual Conference on Genetic and Evolutionary Computation, pp. 211–218. ACM (2011)
23. Marzinotto, A., Colledanchise, M., Smith, C., Ogren, P.: Towards a unified behavior trees framework for robot control, pp. 5420–5427, May 2014). https://doi.org/10.1109/ICRA.2014.6907656
24. Mondada, F., et al.: The e-puck, a robot designed for education in engineering. In: Proceedings of the 9th Conference on Autonomous Robot Systems and Competitions, vol. 1, pp. 59–65. IPCB: Instituto Politécnico de Castelo Branco (2009)
25. Mouret, J.B., Clune, J.: Illuminating search spaces by mapping elites. arXiv preprint arXiv:1504.04909 (2015)
26. Pinciroli, C., et al.: Argos: a modular, parallel, multi-engine simulator for multi-robot systems. Swarm Intell. **6**, 271–295 (2012)
27. Pugh, J.K., Soros, L.B., Stanley, K.O.: Quality diversity: a new frontier for evolutionary computation. Front. Robot. AI **3**, 40 (2016)
28. Pugh, J.K., Soros, L.B., Szerlip, P.A., Stanley, K.O.: Confronting the challenge of quality diversity. In: Proceedings of the 2015 Annual Conference on Genetic and Evolutionary Computation, pp. 967–974 (2015)
29. Scheper, K., Tijmons, S., De Visser, C., Croon, G.: Behavior trees for evolutionary robotics. Artif. Life **22** (2016). https://doi.org/10.1162/ARTL_a_00192
30. Trianni, V.: Evolutionary Swarm Robotics: Evolving Self-Organising Behaviours in Groups of Autonomous Robots, vol. 108. Springer, Heidelberg (2008). https://doi.org/10.1007/978-3-540-77612-3
31. Wei, T., Wang, S., Zhong, J., Liu, D., Zhang, J.: A review on evolutionary multi-task optimization: trends and challenges. IEEE Trans. Evol. Comput. (2021)

Evolutionary Based Transfer Learning Approach to Improving Classification of Metamorphic Malware

Kehinde O. Babaagba[1]([⊠]) [iD] and Mayowa Ayodele[2] [iD]

[1] School of Computing, Edinburgh Napier University, Edinburgh, UK
k.babaagba@napier.ac.uk
[2] Fujitsu Research of Europe, Slough, UK
mayowa.ayodele@fujitsu.com

Abstract. The proliferation of metamorphic malware has recently gained a lot of research interest. This is because of their ability to transform their program codes stochastically. Several detectors are unable to detect this malware family because of how quickly they obfuscate their code. It has also been shown that Machine learning (ML) models are not robust to these attacks due to the insufficient data to train these models resulting from the constant code mutation of metamorphic malware. Although recent studies have shown how to generate samples of metamorphic malware to serve as training data, this process can be computationally expensive. One way to improve the performance of these ML models is to transfer learning from other fields which have robust models such as what has been done with the transfer of learning from computer vision and image processing to improve malware detection. In this work, we introduce an evolutionary-based transfer learning approach that uses evolved mutants of malware generated using a traditional Evolutionary Algorithm (EA) as well as models from Natural Language Processing (NLP) text classification to improve the classification of metamorphic malware. Our preliminary results demonstrate that using NLP models can improve the classification of metamorphic malware in some instances.

Keywords: Metamorphic Malware · Machine Learning · Evolutionary Algorithm · Transfer Learning · Natural Language Processing · Text Classification

1 Introduction

Detecting metamorphic malware have posed a severe challenge to antivirus engines among other detectors. This is because metamorphic malware comprises a group of complex malware that transform their codes between generations. They do this using various obfuscation techniques to evade detection by malware detectors. Some of the mutation techniques they employ include but are not limited to junk code insertion (this inserts garbage code to the malware's program

© The Author(s), under exclusive license to Springer Nature Switzerland AG 2023
J. Correia et al. (Eds.): EvoApplications 2023, LNCS 13989, pp. 161–176, 2023.
https://doi.org/10.1007/978-3-031-30229-9_11

code which do not affect the behavior of the malware), instruction replacement (this substitutes valid instruction code with its equivalent without distorting the functionality of the program code), and instruction re-ordering (which distorts the flow of control of program by reordering its instruction code).

A few techniques have been employed in detecting this class of malware. The authors in [2] summarises the detection approaches to include Opcode-Based Analysis (OBA), Control Flow Analysis (CFA) and Information Flow Analysis (IFA). The various detection approaches depend on the kind of information employed when carrying out the analysis. Several other detection techniques have been suggested that involve the use of ML methods e.g. Decision Trees (DT) in [10], Hidden Markov Models (HMM) in [37], Support Vector Machines (SVM) in [36] as well as a hybrid of both feature based and sequential ML models in [9].

However, ML techniques often involve using training data consisting of samples of known malware. Since metamorphic malware keep changing their code, there is usually insufficient training data, hence impeding ML model generality. Although previous work such as [6,7] and [8] have provided methods for generating mutant samples of malware, this task involves several iterations to generate sufficient executable samples which can be very computationally expensive.

A good approach to tackle this problem would be to use transfer learning, a machine learning technique wherein the knowledge generated from a task is stored and reused in another task, often a related task [30]. Specifically, evolutionary-based transfer learning is where the training data are generated using an Evolutionary Algorithm (EA) [16]. EAs are population-based metaheuristics that draw inspiration from processes occurring in biological evolution which guide a population to adapt towards a desired goal. Transfer learning is particularly useful in situations where there is a lack of sufficient training data. This technique has been used in several domains such as NLP [34] and computer vision [20]. It has also been used in malware analysis, particularly, in image processing applications [33] among others. However, this has not been used in text classification within the context of malware analysis. Furthermore, in metamorphic malware detection, it has not been used to the best of our knowledge.

In [7] and [8], two EAs were used to generate training data which are used for the generation of the evolved malware mutants used in this work. These methods led to improved metamorphic malware detection within the context of feature-based and sequential classifiers [9]. The problem of limited training data was however noted. In this work, we use evolutionary-based transfer learning designed specifically for cases with limited data as a way of improving the classification and detection of metamorphic malware. Also, we seek to compare the classification performance of NLP models.

In this work, we address the following research questions:

1. Can NLP language models be used in an evolutionary-based transfer learning context to improve the classification of metamorphic malware?
2. Which of these NLP models provides the best classification performance for metamorphic malware?

We answer these questions by carrying out experiments that first compare the performance of ML models without the use of language models from NLP. Then, we test for performance improvement of the ML models using the NLP language models. We then analyse the NLP models to determine which one leads to the best improvement in classification scores of the ML models.

Summarily, the major contributions of this paper are as follows: Our experiments show that the use of NLP language models in an evolutionary-based transfer learning context improves the classification of metamorphic malware in some instances. Also, we show that BERT language model has the best classification performance compared to the other language models tested.

The rest of the paper is structured as follows. Section 2 provides a review of related works. In Sect. 3, we describe the methodology of our research. Then we present and discuss our experimental settings in Sect. 4. Results are presented in Sect. 5. Section 6 summarises and concludes the paper, it also provides direction for future research.

2   Background

As earlier established, several approaches have been designed for metamorphic malware analysis and detection. These include statistical analysis of their binaries as in the work of [27] that used Linear Discriminant Analysis (LDA) and [39] that used Longest Common Subsequence (LCS). Other techniques include the use of control-flow graph matching [1], subroutine depermutation [19], code normalization [4], and similarity-based approaches like structural entropy [11] as well as compression based classification [28] among others.

Machine learning techniques have also been used in classifying and detecting metamorphic malware. As earlier mentioned techniques such as HMM have been used by the authors in [5]. They analysed the performance of the HMM using 4 distinct compilers as well as handwritten assembly code with results showing the effectiveness of HMMs in the detection of metamorphic malware. The work in [10] employed decision trees in metamorphic malware detection with classification results indicative of the reliability of decision trees in metamorphic malware classification. In [35], the authors used a single class SVM for detecting metamorphic malware with success. Furthermore, a combination of machine learning techniques has also been used for detecting metamorphic malware in [3]. Some authors [6,7], and [8] have also generated metamorphic malware to serve as training data to improve the classification of machine learning models.

Transfer learning as a technique for use in instances of small training sets or small labeled data has received a lot of research attention. There are applications in a number of domains such as in medical applications as seen in the work of [29] that modified the AlexNet [26] in order to detect Alzheimer's disease. It has also been used in bioinformatics such as in [32] that used it for the study and prediction of associations in genotype-phenotype using Label Propagation Algorithm (LPA) [21]. In the transportation domain, it was applied in [15] to process similar images derived during varying conditions. It has also gained attention in NLP as seen in the work of [34].

Transfer learning has increasingly been used in malware detection. It has been used in computer vision for instance as in [13] where a computer vision-based deep transfer learning was proposed for classifying static malware using knowledge from objects appearing in nature. Also, [33] used Deep Neural Network (DNN) built from the ResNet-50 architecture in classifying malicious software. The malware was converted to grayscale images. Then the DNN that had been trained previously on the ImageNet dataset is used in classifying the malicious samples. Similarly, the work of [12] built a deep learning model pre-trained on a large set of image data to improve the classification of malware. Transfer learning has also been employed in Generative Adversarial Network (GAN) settings as seen in the works of [23] and [24]. As in [24], their model comprised of a generator that created adversarial samples. The detector learns the characteristics of the malicious samples using a deep autoencoder (DAE). Prior to training the GAN, the DAE uses the learned features of malware to create data and transfers the learned information to improve the training of the GAN generator. Their method was shown to outperform other models designed for the same application. However, after an exhaustive literature review, a study that employed evolutionary-based transfer learning for metamorphic malware analysis and detection could not be found.

In this work, we use evolutionary-based transfer learning of language models in NLP to improve the classification of metamorphic malware. We use evasive and diverse mutant variants of malware previously created in [7] and [8]. The effects of using these samples to improve current ML detection models including both feature-based and sequential-based models were analysed [9]. In this paper, we study if using these mutants in a model that employs transfer learning from NLP can improve their classification accuracy.

3   Methodology

In this section, we explain briefly the two EAs employed in the creation of the training set (details can be found in [7,8]). thereafter, we explain how we collect and process the data collected. We then explain the transfer learning models employed in our experimentation.

3.1   Creation of the Evolved Malware Mutants

Algorithm 1 presents an EA originally proposed in [7]. The EA is a mutation only population-based algorithm used in creating the evolved malware mutants. In Line 1, an initial population P consisting of n randomly generated mutants is created. These mutants are optimised for either the behavioral similarity between a variant and the original malware; the structural similarity between a variant and the original malware; the detection rate with respect to 63 detection-engines. Furthermore, we use the EA in [8] which employs MAP-Elites, a Quality Diversity (QD) algorithm, to generate mutants that are structurally s and behaviorally b diverse to the original malware. Given each feature $<s, b>$, the algorithm seeks to find mutants associated with that feature that is as evasive as possible with results leading to the generation of more diverse mutants that retain their evasive ability. We ensure that the mutants created using both methods are still malicious by testing them against Droidbox[1]. This is a sandbox designed for the monitoring and dynamic analysis of mobile software. The sandbox works by executing the samples and then studying their behaviors by logging useful data relating to the sample such as its registry calls, process related operations among others.

To generate the final population of mutants, $max\_iterations$ generations of mutation steps are performed. During each generation, a new population R is created by randomly selecting k mutants from the initial population P (Line 4). In Line 6, one of three mutation types, which are, Garbage Code Insertion (GCI) (inserts a piece of junk code, e.g. a line number into the original program code), Instructional Reordering (IR) (adds a goto statement in the original program code that jumps to a label that does nothing) and Variable Renaming (VR) (renames a variable with another valid variable name in the original program code), is selected. In Line 7, a new mutant solution x_{new} is generated by performing mutation (using a randomly selected mutation type (mut_type) on the best solution in R (x_{best}). If the fitness of this new solution x_{new} is better than the worst solution in P, it replaces such solution, otherwise, x_{new} is discarded. At the end of $max\_iterations$ generations, the final population P is returned.

Algorithm 1. Evolutionary Algorithm [7]

1: initialize population P of size n.
2: assign fitness $f(x)$ to each mutant $x \in P$
3: **while** $max\_iterations$ not reached **do**
4:     $R \leftarrow$ randomly select k variants from P
5:     $x_{best} \leftarrow \text{argmin}\, \{f(x), x \in R\}$
6:     $mut\_type \leftarrow$ select a mutation operator at random with uniform probability
7:     $x_{new} \leftarrow mutate(mut\_type, x_{best})$
8:     $fit_{new} \leftarrow f(x_{new})$
9:     $x_{worst} \leftarrow \text{argmax}\, \{f(x), x \in P\}$,
10:     $fit_{worst} \leftarrow f(x_{worst})$
11:     **if** $fit_{new} < fit_{worst}$ **then**
12:         replace x_{worst} in P with x_{new}
13:     **end if**
14: **end while**
15: **return** P

---

[1] Droidbox - https://www.honeynet.org/taxonomy/term/191.

3.2   Data Collection and Processing

The samples used in this work comprise of Android malware which are archived as APK files. The main aim is to analyse metamorphic malware. However, due to the difficulty associated with the collection of these malware, we create mutant samples of existing popular malicious family as proxy which define prospective mutants. The samples comprise of both benign and malicious data.

The APK files comprise of 60 benign samples. These samples were collected from three categories namely; communication, entertainment and security. Equal number of samples were collected from each category resulting in 20 samples from each category. We chose these groups because they represent the behaviour of most Android clean files. The benign samples were collected from Google play store[2] (the samples were downloaded using Apkdownloader[3]) and Wondoujia play store[4].

The parent malware of the mutant variants of malware described in 3.1 was collected from Contagio Minidump[5] and Malgenome[6]. These comprise of three malware families and they are Dougalek[7], Droidkungfu[8] and GGtracker[9]. The three families were chosen based on their malicious payload and they belong to four groups described below:

1. Privilege Escalation: The complexity of the Android platform, owing to the fact that it comprises both Linux kernel and Android framework which have over 90 libraries, makes it prone to attacks in the form of privilege escalation. Droidkungfu [17], an Android malware family first discovered in May 2011, is an example of a malware family that uses privilege escalation. It is one of the families gotten from the MalGenome dump. It uses encryption to obfuscate its code in order to go undetected by detectors. It includes encrypted root exploits and malicious payloads that are in touch with C&C servers, from which they get instructions to be executed. This family of malware is considered in our analysis.
2. Remote control: This feature allows mobile malicious attackers to gain remote control of the phone. Malware families that have this functionality are in communication with remote C&C servers. Droidkungfu is also an example of a malware family that uses remote control malicious payload and is considered in our analysis.

---

[2] Google Play - https://play.google.com/store?hl=en.

[3] Apkdownloader -https://apps.evozi.com/apk-downloader/.

[4] Wondoujia Play - www.wandoujia.com.

[5] Contagio Minidump - http://contagiominidump.blogspot.com/2015/01/android-hideicon-malware-samples.html.

[6] Malgenome - http://www.malgenomeproject.org/.

[7] Dougalek - https://www.trendmicro.com/vinfo/us/threat-encyclopedia/malware/androidosdougalek.a.

[8] Droidkungfu - https://www.f-secure.com/v-descs/trojan_android_droidkungfu_c.shtml.

[9] GGtracker - https://www.f-secure.com/v-descs/trojan_android_ggtracker.shtml.

3. Financial charges: Some malicious attacks are launched to deliberately extort money from the users infected in form of financial charges. They subscribe users to premium services without proper authorisation, and in most cases, the infected parties are unaware of such services. GGtracker [18] is an example of such a family of malware. It is one of the families in the MalGenome dump and subscribes the infected users to various US premium services without their consent. It is also one of the families of malware studied.

4. Personal information stealing: There are also other malware families whose major goal is to collect information. This information could be on the infected user's account, contact list, text messages, among others. Malware families such as Dougalek [38] from the Contagio minidump and GGtracker, fall into this category and are analysed in our study.

The malicious samples also comprise malware gotten from the web that belongs to the aforementioned families collected from Contagio Minidump.

In order to get the features of the samples collected we carry out dynamic analysis of the samples using tools such as Strace[10] and MonkeyRunner[11]. Strace runs the samples in order to study its behavior and keeps track of each system call the samples make. It uses MonkeyRunner to execute the sample's main activity and MonkeyRunner is employed in simulating user interaction with the sample. This is then used to generate sequential features of the samples.

We use the log stored by Strace to derive each sample's sequential features. Thereafter, we generate a time-ordered system-calls list and this forms the feature vector of the samples.

3.3 NLP Language Models

In this section, we describe the language models employed in this work. The language models selected comprise of some of the most recent and commonly used models in NLP. They are briefly explained below:

BERT [14] is an acronym for Bidirectional Encoder Representations from Transformers. It was created for the pretraining of deep representations that are bidirectional, from the texts that are not labeled by taking into consideration the contextual information of the text that is, by working out both the left and right context of the token. Consequently, the pre-trained BERT models can be easily adjusted and tuned with only an extra output layer to produce advanced models for a large number of NLP tasks. This model is pre-trained on a massive unlabelled text corpus which includes the whole of Wikipedia (this has about 2.5 billion words) and Book Corpus (this comprises of about 800 million

---

[10] Strace - https://linux.die.net/man/1/strace.
[11] Monkeyrunner - https://developer.android.com/studio/test/monkey.

words). After being tested on about 11 NLP tasks, it produces novel state-of-the-art results such as improving the GLUE score by 7.7%, the MultiNLI accuracy by 4.6%, and the SQuAD v2.0 Test F1 by 5.1%, among others.

As illustrated in Fig. 1, the BERT model comprises both pre-training and fine-tuning steps. The pre-training task occurs with the training of the model on instances that are unlabelled for various distinct pre-training tasks. The fine-tuning process, on the other hand, begins with the initialization of the model with pre-trained parameters. Then, using the labeled data derived from the downstream tasks, the parameters are fine-tuned. BERT uses $O(n^2)$ time and space with regard to the length of the sequence.

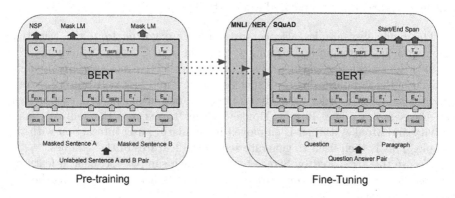

Fig. 1. A BERT model illustrating its pre-training and fine-tuning tasks [14]

GloVE [31] A number of models that use unsupervised techniques to understand word representations often rely on and use word occurrences statistics in a corpus to learn from word representations. However, a number of unanswered questions exist regarding how meaningful these statistics are as well as if the word vectors generated from them provide meaningful representations. The GloVe (Global Vectors) model was presented as an unsupervised learning algorithm used to represent words that directly capture the global corpus statistics in the model. It generates vector representations for words and trains on a composite of global word-word co-occurrence statistics from a corpus. It has been shown to produce representations with striking and meaningful linear substructures of the word vector space.

An example can be seen in making a quantitative distinction between man and woman as seen in Fig. 2. To do that, an association has to be built beyond one number to the pair of words by a model, for instance through the vector difference between their word vectors. In such an example, GloVe is well suited for computing such vector differences such that the meaning derived from the collocation of the two words is maximally represented.

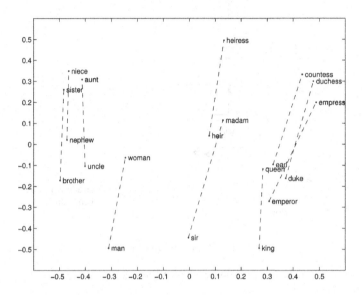

Fig. 2. Linear substructure for quantitatively distinguishing between man and woman using GloVe model [31]

FastText [22] This model was generated by researchers at Facebook AI Research (FAIR) lab to serve as a library for learning word representations as well as sentence classification more effectively. Unlike other word vectors that consider each word as the lowest unit in which we are seeking to find its representation, FastText considers each word as a n-grams of character, in which n can take values from 1 to the word length. It is beneficial in that it can discover the vector representation for uncommon words as these words can be split into character n-grams. It incorporates pre-trained language models learned in over 157 different languages and includes the whole of Wikipedia.

For complex and rare words that would have been difficult to represent, other than return a zero vector or a random vector with low magnitude, FastText will split those words into character n-grams and use the vectors of the generated character n-grams to produce the final word vector. This kind of embedding has been shown to outperform other embeddings, particularly on smaller data sets and its architecture is given in Fig. 3.

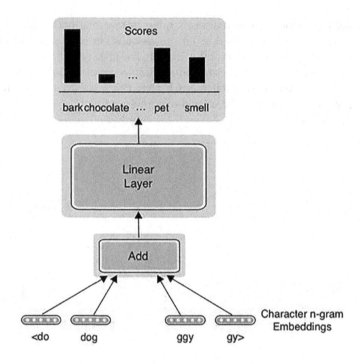

Fig. 3. FastText Architecture [22]

Table 1. Evolutionary based Parameter Settings

EA	Settings	
	EA	MAP-Elites
Bootstrap	NA	20
Selection	Tournament	Random
Population Size	20	NA
Iterations	120	120
Mutation Rate	1	1

4   Experimental Settings

The parameters used by the EA is the same as one in [7] and [8] and presented in Table 1.

Our experiments were implemented using Scikit-learn libraries for Python, including the use of the Keras library[12]. The models (explained in Sect. 3.3 of the paper) and their hyper-parameters were empirically tuned. As a result of its documented success in terms of its accuracy and computational power, "Adam"

---

[12] Keras - https://github.com/fchollet/keras.

optimiser [25] was employed. The binary cross entropy function was used as the loss function (this function was chosen as our classification is binary). Moreover, as our problem is a classification problem, we employ a Dense output layer comprising one neuron with a sigmoid activation function. We employed a batch size of 6 so as to space out the updates of weight. The model was fitted using just four epochs as it speedily over-fitted the problem.

The training set comprises 60 benign and 60 malicious samples. The 60 benign samples comprise 20 entertainment applications, 20 security applications, and 20 communication applications. The 60 malicious samples comprise 20 malware from the Dougalek family, 20 malware from the Droidkungfy family, and 20 malware from the GGtracker family. We will refer to the data combination as 6020combo from here on.

Also, we consider increasing the malicious samples for training by considering 60 benign samples and 157 malicious samples (50 from Dougalek family, 55 from the Droidkunfu family, and 52 from the GGTracker family). We will refer to this increased data combination as 6050combo from here on.

For testing, we use a dataset comprising of 27 benign samples, 23 malicious samples (10 dougalek family, 5 droidkunfu family and 8 ggtracker family) for the 6050combo. For the 6020combo, we use a dataset consisting of 27 benign samples, 16 malicious samples (10 dougalek family, 3 droidkunfu family and 3 ggtracker family).

The approach proposed in this paper provides a robust solution for detecting novel mutants of malware which represent the type of malware found in real environments.

5   Results

In this section, we analyse results based on the experimental settings described in the previous section. We particularly provide answers to our research questions in the subsections below.

5.1   Can NLP Language Models Be Used in an Evolutionary-Based Transfer Learning Context to Improve the Classification of Metamorphic Malware?

To answer our first research question, we conduct experiments with and without the use of the NLP language models and observe if there was an improvement in the classification accuracy and F1 score by reason of using the language models. This was done for the 6020combo and the 6050combo as shown in Tables 2 and 3.

We see that when we do not use a language model, we get an accuracy of 0.63 for the 6020combo and an accuracy of 0.54 for the 6050combo. Although the same results are obtained when we use the GloVe and FastText models, we see that for both the 6020combo and 6050combo data, we get an improved

Table 2. Comparing Accuracy obtained on the test sets for the 6020combo and 6050combo models using No language Model, BERT, FastText and GloVE language models

Test Sets	Accuracy			
	No Language Model	BERT	FastText	GloVE
6020combo	0.63	**0.93**	0.63	0.63
6050combo	0.54	**0.90**	0.54	0.54

classification score of 0.93 and 0.9 respectively using the BERT model. There is at least one model—BERT that results in improved classification accuracy.

Similarly, when F1 Score - which computes the harmonic mean of precision and recall is employed as an evaluation metric, we notice a similar trend showing that in two instances i.e., BERT (0.91 for the 6020combo and 0.9 for the 6050combo) and GloVE (0.8 for the 6020combo and 0.7 for the 6050combo) the F1 Score is higher when an NLP language model is employed than when no language model is used which results in an F1 Score of 0.5 for the 6020combo and 0.4 for the 6050combo. It is important to note that the BERT model has been shown to be significantly better than other language models when smaller data sets are involved [14].

Table 3. Comparing F1 Score obtained on the test sets for the 6020combo and 6050combo models using No language Model, BERT, FastText and GloVE language models

Test Sets	F1 Score			
	No Language Model	BERT	FastText	GloVE
6020combo	0.5	**0.91**	0.5	0.8
6050combo	0.4	**0.9**	0.4	0.7

5.2 Which of These NLP Models Provides the Best Classification Performance for Metamorphic Malware?

In this section, we compare the performance of the three language models to see which one produces the best classification accuracy and F1 Score. From Table 2, we see that the classification accuracy of both FastText and GloVe models are the same for both the 6020combo and 6050combo models. However, we see that the BERT model performs significantly better than the other two models producing an accuracy of 0.93 and 0.9 for the 6020combo and 6050combo data respectively.

Table 3 also shows that compared to the other models BERT has a better F1 Score of 0.91 for the 6020combo and 0.9 for the 6050combo. It is interesting to note that when we use the F1 Score, GloVE (0.8 for the 6020combo and 0.7 for the 6050combo) outperforms FastText (0.5 for the 6020combo and 0.4 for the 6050combo) for both the 6020combo and 6050combo.

6   Conclusion

We have established that metamorphic malware represents a difficult class of malware to detect due to the way they change their codes stochastically. Another problem with detecting these malware classes, particularly using ML models is that there is insufficient training data for ML models to learn from. Generating these data is very time-consuming and computationally expensive.

In this paper, we have presented an approach to address the aforementioned problem that employs an evolutionary-based transfer learning method to improve the classification of metamorphic malware. The results show that the use of BERT model leads to better classification accuracy and F1 Score compared to when a language model is not used. Furthermore, we demonstrate that the use of BERT model also yields the best accuracy and F1 Score on both data tested as compared to the other two language models employed.

Future work could compare more NLP models as well as use transfer learning from other application areas for improved classification and detection of metamorphic malware.

References

1. Alam, S., Traore, I., Sogukpinar, I.: Annotated control flow graph for metamorphic malware detection. Comput. J. **58**(10), 2608–2621 (2015). https://doi.org/10.1093/comjnl/bxu148
2. Alam, S., Traore, I., Sogukpinar, I.: Current trends and the future of metamorphic malware detection. In: Proceedings of the 7th International Conference on Security of Information and Networks. SIN 2014, pp. 411–416. ACM, New York (2014)
3. Alazab, M., Venkatraman, S., Watters, P., Alazab, M.: Zero-day malware detection based on supervised learning algorithms of API call signatures. In: Proceedings of the Ninth Australasian Data Mining Conference. AusDM 2011, vol. 121, pp. 171–182. Australian Computer Society Inc., Darlinghurst (2011). http://dl.acm.org/citation.cfm?id=2483628.2483648
4. Armoun, S.E., Hashemi, S.: A general paradigm for normalizing metamorphic malwares. In: 2012 10th International Conference on Frontiers of Information Technology. pp. 348–353 (Dec 2012). DOI: 10.1109/FIT.2012.69
5. Austin, T.H., Filiol, E., Josse, S., Stamp, M.: Exploring hidden Markov models for virus analysis: a semantic approach. In: 2013 46th Hawaii International Conference on System Sciences, pp. 5039–5048, January 2013. https://doi.org/10.1109/HICSS.2013.217

6. Aydogan, E., Sen, S.: Automatic generation of mobile malwares using genetic programming. In: Mora, A.M., Squillero, G. (eds.) EvoApplications 2015. LNCS, vol. 9028, pp. 745–756. Springer, Cham (2015). https://doi.org/10.1007/978-3-319-16549-3_60

7. Babaagba, K.O., Tan, Z., Hart, E.: Nowhere metamorphic malware can hide - a biological evolution inspired detection scheme. In: Wang, G., Bhuiyan, M.Z.A., De Capitani di Vimercati, S., Ren, Y. (eds.) DependSys 2019. CCIS, vol. 1123, pp. 369–382. Springer, Singapore (2019). https://doi.org/10.1007/978-981-15-1304-6_29

8. Babaagba, K.O., Tan, Z., Hart, E.: Automatic generation of adversarial metamorphic malware using MAP-elites. In: Castillo, P.A., Jiménez Laredo, J.L., Fernández de Vega, F. (eds.) EvoApplications 2020. LNCS, vol. 12104, pp. 117–132. Springer, Cham (2020). https://doi.org/10.1007/978-3-030-43722-0_8

9. Babaagba, K.O., Tan, Z., Hart, E.: Improving classification of metamorphic malware by augmenting training data with a diverse set of evolved mutant samples. In: 2020 IEEE Congress on Evolutionary Computation (CEC), pp. 1–7. IEEE (2020)

10. Bashari Rad, B., Masrom, M., Ibrahim, S., Ibrahim, S.: Morphed virus family classification based on opcodes statistical feature using decision tree. In: Abd Manaf, A., Zeki, A., Zamani, M., Chuprat, S., El-Qawasmeh, E. (eds.) ICIEIS 2011. CCIS, vol. 251, pp. 123–131. Springer, Heidelberg (2011). https://doi.org/10.1007/978-3-642-25327-0_11

11. Baysa, D., Low, R.M., Stamp, M.: Structural entropy and metamorphic malware. J. Comput. Virol. Hack. Tech. **9**(4), 179–192 (2013). https://doi.org/10.1007/s11416-013-0185-4

12. Bhodia, N., Prajapati, P., Troia, F.D., Stamp, M.: Transfer learning for image-based malware classification. CoRR abs/1903.11551 (2019). http://arxiv.org/abs/1903.11551

13. Chen, L.: Deep transfer learning for static malware classification. CoRR abs/1812.07606 (2018). http://arxiv.org/abs/1812.07606

14. Devlin, J., Chang, M., Lee, K., Toutanova, K.: BERT: pre-training of deep bidirectional transformers for language understanding. CoRR abs/1810.04805 (2018). http://arxiv.org/abs/1810.04805

15. Di, S., Zhang, H., Li, C., Mei, X., Prokhorov, D., Ling, H.: Cross-domain traffic scene understanding: a dense correspondence-based transfer learning approach. IEEE Trans. Intell. Transp. Syst. **19**(3), 745–757 (2018)

16. Eiben, A.E., Smith, J.E.: What is an evolutionary algorithm? In: Eiben, A.E., Smith, J.E. (eds.) Introduction to Evolutionary Computing. NCS, pp. 25–48. Springer, Heidelberg (2015). https://doi.org/10.1007/978-3-662-44874-8_3

17. F-Secure: Trojan:Android/DroidKungFu.C (2019). https://www.f-secure.com/v-descs/trojan_android_droidkungfu_c.shtml

18. F-Secure: Trojan:Android/GGTracker.A (2019). https://www.f-secure.com/v-descs/trojan_android_ggtracker.shtml

19. Fiñones, R.G., Fernandez, R.: Solving the metamorphic puzzle. Virus Bull. 14–19 (2006). https://www.virusbulletin.com/virusbulletin/2006/03/solving-metamorphic-puzzle/

20. Gao, J., Ling, H., Hu, W., Xing, J.: Transfer learning based visual tracking with gaussian processes regression. In: Fleet, D., Pajdla, T., Schiele, B., Tuytelaars, T. (eds.) ECCV 2014. LNCS, vol. 8691, pp. 188–203. Springer, Cham (2014). https://doi.org/10.1007/978-3-319-10578-9_13

21. Hwang, T., Kuang, R.: A heterogeneous label propagation algorithm for disease gene discovery. In: Proceedings of the 2010 SIAM International Conference on Data Mining, pp. 583–594. SIAM (2010)

22. Joulin, A., Grave, E., Bojanowski, P., Mikolov, T.: Bag of tricks for efficient text classification. In: Proceedings of the 15th Conference of the European Chapter of the Association for Computational Linguistics: Volume 2, Short Papers, pp. 427–431. Association for Computational Linguistics, Valencia, April 2017. https://aclanthology.org/E17-2068
23. Kim, J.Y., Bu, S.J., Cho, S.B.: Malware detection using deep transferred generative adversarial networks. In: Liu, D., Xie, S., Li, Y., Zhao, D., El-Alfy, E.S.M. (eds.) ICONIP 2017. LNCS, vol. 10634, pp. 556–564. Springer International Publishing, Cham (2017). https://doi.org/10.1007/978-3-319-70087-8_58
24. Kim, J.Y., Bu, S.J., Cho, S.B.: Zero-day malware detection using transferred generative adversarial networks based on deep autoencoders. Inf. Sci. **460–461**, 83–102 (2018). https://doi.org/10.1016/j.ins.2018.04.092
25. Kingma, D.P., Ba, J.: Adam: a method for stochastic optimization. CoRR abs/1412.6980 (2014)
26. Krizhevsky, A., Sutskever, I., Hinton, G.E.: Imagenet classification with deep convolutional neural networks. In: Pereira, F., Burges, C.J.C., Bottou, L., Weinberger, K.Q. (eds.) Advances in Neural Information Processing Systems, vol. 25, pp. 1097–1105. Curran Associates, Inc. (2012)
27. Kuriakose, J., Vinod, P.: Ranked linear discriminant analysis features for metamorphic malware detection. In: 2014 IEEE International Advance Computing Conference (IACC), pp. 112–117, February 2014. https://doi.org/10.1109/IAdCC.2014.6779304
28. Lee, J., Austin, T.H., Stamp, M.: Compression-based analysis of metamorphic malware. Int. J. Secur. Netw. **10**(2), 124–136 (2015). https://doi.org/10.1504/IJSN.2015.070426
29. Maqsood, M., et al.: Transfer learning assisted classification and detection of Alzheimer's disease stages using 3D MRI scans. Sensors (Basel, Switzerland) **19**(11), 2645 (2019). https://doi.org/10.3390/s19112645
30. Pan, S.J., Yang, Q.: A survey on transfer learning. IEEE Trans. Knowl. Data Eng. **22**(10), 1345–1359 (2010)
31. Pennington, J., Socher, R., Manning, C.D.: Glove: global vectors for word representation. In: Proceedings of the 2014 Conference on Empirical Methods in Natural Language Processing (EMNLP), pp. 1532–1543 (2014)
32. Petegrosso, R., Park, S., Hwang, T.H., Kuang, R.: Transfer learning across ontologies for phenome-genome association prediction. Bioinformatics **33**(4), 529–536 (2016). https://doi.org/10.1093/bioinformatics/btw649
33. Rezende, E., Ruppert, G., Carvalho, T., Ramos, F., de Geus, P.: Malicious software classification using transfer learning of resnet-50 deep neural network. In: 2017 16th IEEE International Conference on Machine Learning and Applications (ICMLA), pp. 1011–1014 (2017)
34. Ruder, S., Peters, M.E., Swayamdipta, S., Wolf, T.: Transfer learning in natural language processing. In: Proceedings of the 2019 Conference of the North American Chapter of the Association for Computational Linguistics: Tutorials, pp. 15–18. Association for Computational Linguistics, Minneapolis, June 2019. https://doi.org/10.18653/v1/N19-5004
35. Sahs, J., Khan, L.: A machine learning approach to android malware detection. In: 2012 European Intelligence and Security Informatics Conference, pp. 141–147, August 2012. https://doi.org/10.1109/EISIC.2012.34
36. Sahs, J., Khan, L.: A machine learning approach to android malware detection. In: 2012 European Intelligence and Security Informatics Conference (2012)

37. Toderici, A.H., Stamp, M.: Chi-squared distance and metamorphic virus detection. J. Comput. Virol. **9**(1), 1–14 (2013)
38. TRENDMICRO: ANDROIDOS_DOUGALEK.A (2012). https://www.trendmicro. com/vinfo/us/threat-encyclopedia/malware/androidos_dougalek.a
39. Vinod, P., Laxmi, V., Gaur, M.S., Kumar, G.V.S.S.P., Chundawat, Y.S.: Static CFG analyzer for metamorphic malware code. In: Proceedings of the 2Nd International Conference on Security of Information and Networks. SIN 2009, pp. 225–228. ACM, New York (2009). https://doi.org/10.1145/1626195.1626251

Evolving Lightweight Intrusion Detection Systems for RPL-Based Internet of Things

Ali Deveci[1]([✉]) [ID], Selim Yilmaz[1,2] [ID], and Sevil Sen[1] [ID]

[1] WISE Laboratory, Department of Computer Engineering,
Hacettepe University, Ankara, Turkey
alideveci1984@hotmail.com, ssen@cs.hacettepe.edu.tr
[2] Department of Software Engineering,
Muğla Sıtkı Koçman University, Muğla, Turkey
selimyilmaz@mu.edu.tr

Abstract. With the integration of efficient computation and communication technologies into sensory devices, the Internet of Things (IoT) applications have increased tremendously in recent decades. While these applications provide numerous benefits to our daily lives, they also pose a great potential risk in terms of security. One of the reasons for this is that devices in IoT-based networks are highly resource constrained and interconnected over lossy links that can be exposed by attackers. The Routing Protocol for Low-Power and Lossy Network (RPL) is the standard routing protocol for such lossy networks. Despite the efficient routing built by RPL, this protocol is susceptible to insider attacks. Therefore, researchers have been working on developing effective intrusion detection systems for RPL-based IoT. However, most of these studies consume excessive resources (e.g., energy, memory, communication, etc.) and do not consider the constrained characteristics of the network. Hence, they might not be suitable for some devices/networks. Therefore, in this study, we aim to develop an intrusion detection system (IDS) that is both effective and efficient in terms of the cost consumed by intrusion detection (ID) nodes. For this multiple-objective problem, we investigate the use of evolutionary computation-based algorithms and show the performance of evolved intrusion detection algorithms against various RPL-specific attacks.

Keywords: IoT · RPL attacks · intrusion detection · multi-objective optimization · genetic programming

1 Introduction

IoT, which enables a variety of devices to be connected to each other, is one of the most breakthrough advancements in our era. A great deal of IoT applications have found use in various domains including smart home, smart city, logistic monitoring, e-health, and the like. That's why the number of smart devices enabling such IoT applications has long been increasing. The total installed base of these devices is estimated to be 75 billion, a five-fold increase in ten years, and machine-to-machine (M2M) connections are estimated to constitute half of global connections by 2030 [6,22].

Low Power and Lossy Networks (LLN) are a type of IoT that provide lossy communication among IPv6-enabled resource-constrained devices. They are characterized by their constrained communication with high packet loss, low throughput, and limited frame size [1]. In a typical implementation of LLN, each of the resource-constrained nodes can communicate with each other, but also connects to a special node, called the LLN Border Router (LBR) in order to connect to the Internet. In order to build routes among nodes in such a constrained network, RPL was developed by IETF-ROLL in 2012 [1], and is adopted as a standard routing protocol for LLNs today.

Although RPL is good at building efficient routes between nodes in an LLN, it is still very susceptible to attacks, especially insider attacks. The results of such attacks can be vital considering the applications of LLNs in critical systems such as healthcare, smart home. Therefore, researchers have been working on developing effective intrusion detection systems for RPL-based IoT. However, most studies in the literature mainly focus on detecting attacks against RPL and overlook the suitability of developed IDS to such low-power and lossy networks. Therefore, the main aim of this study is to explore on developing intrusion detection systems that show both high accuracy and low cost.

Here, a distributed intrusion detection architecture in which a global ID node is placed at the root node and some other nodes participate in intrusion detection by sending their local information to the root node is explored. Although involving the monitoring nodes brings about an additional burden to the network and devices, they enable the global ID node to capture intrusions on a global scale, hence more effectively. Here, the size of the information collected and sent by the monitoring nodes becomes important in terms of resource consumption. In addition to increasing communication cost, large packets sent by monitoring nodes might also lead to fragmentation. Therefore, in this study, while developing intrusion detection systems, beside their accuracy, the information used and sent by the monitoring nodes is taken into account. For simplicity, this information regarding the cost of intrusion detection and communication is taken as the number of features used for training in this study. Therefore, the detection accuracy and the number of features extracted from both the ID node and the monitoring nodes must be tuned simultaneously to generate an effective and efficient IDS, which is the main motivation of this study.

In order to solve this multi-objective optimization problem, we employ genetic programming (GP) due to its ability to explore search space efficiently for complex environments such as LLNs and to also handle multiple objectives (i.e., accuracy and the number of features in this study) simultaneously. The main objective of GP is to evolve a detection program (or model) that finds a good trade-off between accuracy and the minimum number of features used. Only the evolved features are extracted and sent by the monitoring ID nodes to the central ID node, which then periodically runs the evolved program.

To handle multiple objectives by GP, we employ Non-dominated Sorting Genetic Algorithm II (NSGA-II) [8], one of the most popular Pareto-based evolutionary multi-objective algorithms. The following four attacks are covered in this study: worst parent, hello flood, increased version, and decreased rank attacks. Various network scenarios with these four attacks in which attackers are placed in different locations are evaluated and discussed. The experimental results show that the increase in the number of nodes and the number of data packets used in intrusion detection also increases the number of features used as expected, resulting in an increase in power consumption and a decrease in network performance. For WP, HF, and IV type attacks, GP can produce a satisfactory ID program with an average detection accuracy of 94%. On the other hand, limiting the number of features has an adverse impact on the detection of DR.

The paper is organized as follows. The background information on RPL and insider attacks against RPL as well as the methods used in this study, namely GP and NSGA-II are given in Sect. 2. The related studies in the field of intrusion detection in RPL are given in Sect. 3. The proposed approach is given in detail in Sect. 4. The experimental settings and results are provided in Sect. 5. The strengths, limitations, and possible future directions of this study are discussed in Sect. 6. Finally, Sect. 7 concludes this study.

2 Background

This section gives background information on both RPL, specific attacks against RPL and the methods used for detecting these attacks, namely evolutionary computation techniques.

2.1 RPL

RPL is a distance vector and source routing protocol and becomes standard for low-power and lossy networks [13]. RPL aims to create Destination Oriented Directed Acyclic Graphs (DODAG). A gateway node, known as the root node, is responsible for the formation of DODAG by broadcasting control messages in RPL called the DODAG Information Object (DIO). DIO messages are initially transmitted only by the root node to construct an upward route from the sensor nodes to itself. Nodes receiving multiple DIO packets from their neighbors select the most suitable candidate parent nodes considering the rank values in the DIO packets, determine their rank, and transmit the modified DIO packet. The Destination Advertisement Object (DAO) packets, however, are used to construct downward routes. The DAO packet is unicast by all nodes to their selected parents. The downward routes in DODAG are operated in two modes: *storing mode* and *non-storing mode*. In the non-storing mode, only the root node keeps the routing table, and hence the nodes rely on only the root node for forwarding their packets to the destination nodes. In the storing mode, a routing table is kept by each node in the network; and instead of sending every incoming packets to the root node, the nodes forward them to the next hop that is on the route to the destination address. Another control packet, called DODAG Information Solicitation (DIS), is broadcast by a new node to join DODAG. Upon receiving the DIS packet, a node returns with a DIO packet, thus sharing the DODAG configuration that is necessary for the requesting node. The objective function such as hop count specifies how a node computes its rank value that is used in parent selection. Although there are many OF types in the literature, Objective Function Zero (OF0) and Minimum Rank with Hysteresis Objective Function (MRHOF) are proposed as the default OF for RPL-based networks [9].

RPL Attacks: Although RPL has some security mechanisms specified in its RFC [1], it is susceptible to insider attacks. Attacks against RPL are classified according to what they primarily target in the literature [16]:

- **Attacks targeting resources:** Attackers aim to make legitimate nodes exhaust their resources by forcing them to perform unnecessary processing, causing also the available links to be down.
- **Attacks targeting topology:** Attackers dramatically affect the construction of the RPL topology in a non-optimal way or lead to the isolation of some nodes from the topology.
- **Attacks targeting traffic:** Attackers interfere with network traffic and try to change the traffic pattern.

This study focuses on the following attacks targeting resources and topology:

- Decreased Rank (DR): In this type of attack, the attacker node illegitimately advertises a lower rank value to other nodes in the network. As a result, benign nodes inevitably send their packet through the attacker node. Consequently, the entire network traffic may be controlled by this malicious node. This attack is considered a first step for the forthcoming attacks, increasing the severity of the attackers.
- Increased Version (IV): The version number is necessary for the global repair in RPL. It is propagated in DIO packets throughout the network and is increased only by the root node. In this type of attack, the attacker node illegitimately increases the value of the version field, resulting in unnecessarily rebuilding the networks.
- Hello Flood (HF): The main purpose here is to unnecessarily increase the size of network traffic by generating a large number of DIS packets, leading to a dramatic consumption of the network's resources.
- Worst Parent (WP): In this attack, the attacker node intentionally selects its worst parent to route incoming packets to degrade the routing performance of the network. The consequence of this attack is unoptimized routes between the nodes, which reduces the performance of the network.

2.2   Evolutionary Computation

Evolutionary computation is inspired by natural evolution and has been shown to be very effective in solving many problems in different domains, including intrusion detection [21,24]. Due to being very good at discovering complex characteristics of a system and being able to solve multi-objective optimization problems, it is explored in this study. While there are many evolutionary computation techniques in the literature, GP which represents candidate individuals as trees is used here. In the following, GP and one of the multi-objective evolutionary algorithms, namely NSGA-II, are introduced in detail.

Genetic Programming: GP [14] is one of the most popular evolutionary-based computation techniques that is inspired by the 'survival of the fittest' theory [11]. Because GP is a very simple yet effective learning approach, it has been used to solve a wide range of real-world problems in many research domains.

As a population-based learning algorithm, GP aims to find possibly the best solution across generations. A number of agents, called individuals, participate in the population, and each of them represents a candidate solution to the problem. Individuals are encoded with a tree structure, called GP tree, where terminal and non-terminal types of node take part. The terminals and non-terminals form the leaf and intermediate nodes of the GP tree, respectively.

Evolution in GP starts with an individual set that is initially generated randomly. The individuals are then evaluated and assigned a *fitness value* that indicates how well this candidate solution can solve the targeted problem. Afterward, individuals undergo three genetic operators in each generation that include *selection*, *crossover*, and *mutation* to breed their offspring. In selection, a pair of individuals is selected, and the fitness value of an individual plays a key role here to determine if it is reproduced in the next generation. The selection is made based on the selection operators such as tournament selection where a number of individuals is picked randomly first, then the fittest individual is selected from that subpopulation. They produce two new offspring individuals in crossover by replacing the subtrees rooted at the crossover point randomly determined. In mutation, however, subtrees of the offspring individuals are also replaced by, contrary to crossover, randomly generated new subtrees. Hence, better individuals are aimed to be evolved through generations. GP reaches the end of the generation once the termination condition is satisfied. There may be different conditions such as reaching the total number of generations, approximating well to the ideal or optimum solutions, and the like. The general steps of the GP algorithm are given in Algorithm 1.

Multi-objective Optimization: Multi-objective optimization is a task that aims to solve problems that involve two or more conflicting objectives. Most of the problems in real life today fall into this category. Until now, a great deal of effort has been put into developing evolutionary-based heuristic approaches that effectively solve such problems. Among them, the Pareto-based approaches are the most popular, and the majority of studies adopt the Pareto strategy where a set of solutions is achieved (called non-dominated solution or Pareto optimal solution) rather than a single solution. In the Pareto-based approaches, a solution x is said to be better than another solution y provided that x 'dominates' y for

Algorithm 1: Basic steps of GP.

1 Initialize population;
2 **repeat**
3 | Calculate the individuals' fitness;
4 | Sort and rank populations by fitness value;
5 | Reproduce/Regenerate the new population using GP operators (mutation, crossover etc.);
6 **until** *a termination criterion is satisfied*;
7 **return** best-of-run individual

---

every objective. So, the main goal of the Pareto-based approaches is to have a set of solutions, called a Pareto optimal set, that is not dominated at the end of the optimization process. The set of objective values corresponding to the Pareto optimal set is called the Pareto front.

As being one of the most popular evolutionary-based approaches targeting multi-objective problems, NSGA-II [8] relies on Pareto domination of solutions in the objective space. Here, the candidate solutions from the previous and current population are split into several fronts according to their Pareto dominance and crowding distances, and the solutions that belong to better fronts are allowed to survive to the next generations. By doing so, the non-dominated solutions survived at the best front are obtained for the problem. In this study, we adopt the selection and survival strategies of NSGA-II to handle multiple objectives.

3 Related Work

A considerable attention has been paid since the birth of the RPL protocol. These attempts are categorized mainly as studies that *i*) analyze the vulnerabilities of the RPL protocol under attack and *ii*) develop solutions to secure the protocol against different types of attacks. These studies are briefly discussed here.

In [2], the performance of RPL is investigated against version number attack with one to three attackers. It is shown that the number of attackers have a clear impact on packet delivery ratio. However, if the attacker is positioned closer to the root node, it increases end-to-end delay and power consumption. A comprehensive analysis is given in [9] to reveal how the performance of the RPL changes as a function of different objective functions when the network is subject to routing attacks with a varying number of attackers. It is shown that RPL is more vulnerable to these attacks when MRHOF is adopted.

A number of solutions have been proposed in the literature to secure RPL-based networks. For this purpose, the researchers have not only modified the protocol, but also developed IDSs that are integrated into the network. The first proposed IDS for RPL is SVELTE [20] which uses anomaly- and signature-based detection methods. Another anomaly-based study based on a game-theoretic model is proposed in [12]. It relies on two parts: *i*) a stochastic game for detection and *ii*) an evolutionary game for confirmation. The stochastic game model calculates the standard RPL rules as a zero-sum game, and the proposed scheme confirms the accuracy of the detection by applying evolutionary methods. In [18], a trust-based model is proposed for the detection of rank and black hole attacks. Here, trust- and mobility-based metrics are evaluated. The proposed model has two parts: i)trust formation including trust metrics, trust index computation, trust rating, and ii)attack detection, including isolation of malicious nodes. Here, a fuzzy threshold-based system is used to calculate the trust formation and attack detection metrics. Rank values are checked with the sequence value of DIO messages for detection of rank attacks, the trust index value of the preferred parent is checked whether it is lower than threshold value for the detection of black hole attacks. The trustworthiness of the nodes is considered for the selection of parent nodes. In [17], a solution is proposed not only to detect the version number attack, but also to locate the attacker nodes. Here, monitoring nodes periodically send the collected feature data towards the root, which then detects attacks and locates the attacker by inspecting incoming data.

Recently, machine learning-based IDS has also been proposed for intrusion detection in RPL. A multi-class classification based detection model is proposed in [27] against rank and wormhole attacks. Here, the light gradient boosting machine algorithm with one-sided sampling method is used in attack detection. In another ML-based IDS, a deep neural network approach is used in [25] to detect decreased rank, hello flooding, and version number attacks. Moreover, they introduce a dataset called IRAD. Another deep learning-based model is developed for the detection of hello flooding attack in [4]. Here, Gated Recurrent Unit (GRU)-based deep learning method with a Recurrent Neural Network (RNN) approach is used for classifying the nodes. Another neural network-based system is proposed in [15] to identify the normal behavior of the nodes. A recent neural network-based IDS is proposed in [5]. Contrary to other studies, they consider not only the routing layer, but also the link layer to extract features that are fed to the network to learn a model. The involvement of features related to the link layer has been reported to decrease the false positive rate. The use of evolutionary-based algorithms for the generation of the IDS model is investigated in [3].

Transfer learning-based approaches have also been proposed for intrusion detection in IoT. A deep transfer learning (DTL) approach called MultiMaximum Mean Discrepancy AE (MMD-AE), based on AutoEncoder (AE) and allows the transfer knowledge is proposed in [23]. Here, although no IoT protocol is targeted, general attacks such as TCP/UDP flooding attacks are targeted, and a labeled dataset is transferred to an unlabeled dataset in accordance with the proposed model. The results show that the proposed transfer learning approach has better experimental results (Area Under Curve (AUC) score) than the traditional approach. In [26], the knowledge is transferred to detect new types of attack and to evolve intrusion detection algorithms for new types of devices with different constraints. Here, while the energy usage of the devices is minimized, the detection accuracy is maximized.

While there are a few studies based on evolutionary computation in the literature [3,26], the current study differs from those by exploring different trade-offs between the intrusion detection accuracy and the cost of the evolved algorithm in terms of energy consumption and communication cost. Therefore, a different intrusion detection architecture is explored here, and communication cost is taken into account for the first time.

4 Evolving Intrusion Detection Algorithms

The main aim of this study is developing a suitable IDS for RPL-based IoT networks. Therefore, a central ID node is placed at the root node, which runs the evolved intrusion detection algorithm. In order to analyze the network traffic far from the central ID node, some monitoring nodes in which periodically collect the local data in their neighbourhood and sent it to the central ID node are participated in intrusion detection. Although these monitoring nodes enable the central ID node to detect attacks with a more satisfactory performance, it can have an adverse impact on the average lifetime of the network due to collecting their local information and on the communication cost due to their sending such information regularly to the central node. Therefore, the trade-offs between detection accuracy and cost need to be investigated. Hence, this study aims to evolve a lightweight ID model in terms of communication cost and energy consumption while effectively detecting malicious network traffic.

An ID program essentially contains a conditional statement represented by a GP tree. As stated earlier, there are terminal and non-terminal nodes in a GP tree. The terminal nodes are leaf nodes in the GP tree that represent the features collected by the nodes that are extracted from the RPL control and data packets in a flow. In this study, we used 35 different features that were proposed in [26]. Traffic flows are used for the construction of these features. To do that, the flows are first windowed within the specific time intervals by both the root and the monitoring nodes. The optimum interval time of the window is found to be 60 s in this study. The windowed feature data is then collected by the monitoring nodes and aggregated at the root node and provided to the GP tree as input data. In addition to the RPL-related features, randomly generated numbers are also assigned to leaf nodes to enable a more effective search by GP individuals. The non-terminal nodes represent arithmetic, comparison, and logical operators in the evolved model. It is worth stating that the root node in the GP tree is constrained to be either a comparison or a logical operator, so that it returns a Boolean value. An example of a GP tree that represents a candidate ID program is given in Fig. 1, and the program corresponding to this tree is given thereafter. The number of features employed in the ID model is of very high importance in determining to what extent the model leads to additional cost in terms of the memory and energy consumption of the monitoring nodes and communication load in the network. Moreover, the high number of features might result in fragmentation and the increase in the number of packets. As shown in the results of the preliminary experiment given in Fig. 2, the overall

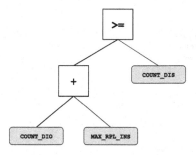

Fig. 1. An example GP tree.

power consumption of the nodes increases linearly with increasing data packets. Therefore, in addition to the detection accuracy, the number of distinct features employed in the intrusion detection algorithm is taken as the second objective which needs to be optimized simultaneously in order to ensure an efficient and high-performance ID model. The GP algorithm is used to learn an ID model (program) that is optimal with respect to these objectives. Hence, the GP tree corresponds to the detection algorithm. While detection accuracy is measured by running the evolved program in the network with and without attackers, the number of features is measured just after the candidate GP tree is constructed by counting the distinct features in the leaf nodes. In order to optimize these objectives simultaneously, the selection and survival strategies of NSGA-II [8] are integrated into the GP algorithm. Therefore, the set of Pareto dominant individuals found after each evolution step is taken into account to determine both the survival of the parent individuals and the breeding of new offspring for the next steps. In GP, each individual represents a candidate program that is evaluated according to Pareto dominance. The GP individuals that dominate others (i.e., give higher accuracy with fewer distinct features) are called *Pareto dominant ID programs*, which we aim to learn in this study.

---

1 **if** *((COUNT_DIO + MAX_RPL_INS) >= COUNT_DIS)* **then**
2 | alert(intrusion)
3 **end**

In this study, ECJ [10], a Java-based evolutionary computation toolkit, is used for the implementation of the GP and NSGA-II algorithms. The GP program is terminated at the 1000th generation. In order to point an ideal value for the number of generations, we have performed pre-experiments and found that the change in the performance of GP is not significant after the 1000th generation. The parameter settings of GP are listed in Table 1, and the other settings not listed in the table are the default parameters of the ECJ.

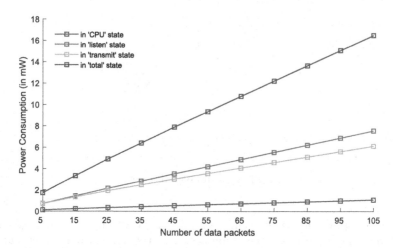

Fig. 2. The change in average power consumption under different states with varying number of packets.

5 Experimental Results

An overview of the simulation environment is given in this section in detail. In addition, the experimental results obtained from the standalone and collaborative IDS architectures are also discussed comparatively thereafter.

5.1 Simulation Environment

In the experiment, a grid topology, shown in Fig. 3, is used with 30 nodes, including the root node. Among them, three nodes (10% of the nodes) are set as attacker nodes, where their positions are randomly chosen. The nodes in the topology are positioned in the network such that each node is 20 m away from another node, and the transmission range between the nodes is limited to 25 m

Table 1. GP parameters and their values.

Parameters	Value
Non-terminals	$+$, $-$, $*$, $/$, sin, cos, log, ln, sqrt, abs, exp, ceil, floor, max, min, pow, mod, $<$, $\leq$, $>$, $\geq$, $==$, $! =$, and, or
Terminals	features in [26] and rnd(0,1)
Generation Size	1000
Population Size	100
Crossover and mutation probability	0.9 and 0.1
Max. depth of GP tree	20

so that the nodes can communicate with their neighbor nodes. The arrows in Fig. 3 represent the preferred parent of the child nodes in DODAG in a benign environment. The Cooja simulator [19] that emulates the LLN nodes is used in the experiments. Supported by Cooja for the emulated nodes, the Contiki operating system [7] (version 2.7) that also involves the implementation of RPL is used. Zolertia Z1 platform is chosen as the mote type for the nodes.

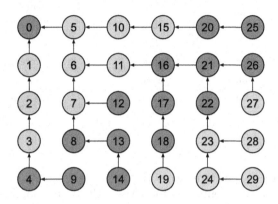

Fig. 3. Simulation network (Color figure online)

By adopting this experimental environment and the settings, we have thoroughly evaluated the performance of the proposed evolutionary-based IDS against four targeted routing attacks; WP, HF, IV, and DR. To do that, eight different scenarios are generated, and the learning is repeated 10 times for each scenario that is individually simulated for three hours. The main motivation of the scenario-based evaluation is to thoroughly discuss how well the evolved ID program can detect when monitoring or attacker nodes are re-positioned after the learning step. Therefore, these scenarios differ from each other in terms of the position of the monitoring and attacker nodes in the training and testing network environment. These scenarios are outlined in Table 2. It is seen from the table that the monitoring nodes in the network are grouped according to their proximity to the root node that is represented with blue in Fig. 3. The monitoring nodes are either placed closely to the root node (1–3 hops away from the root node, represented with yellow in Fig. 3) or far (4–6 hops away, represented with purple in Fig. 3) from the root node. It is worth pointing out here that each of the monitoring nodes is randomly chosen from a different hop level (that is, a single monitoring node is chosen per hop and the average of these nodes' data is used in the experiments). Hence, in each scenario 10% the nodes are responsible for monitoring. In addition, in order to see the effect of attackers' positions, in some scenarios, the attackers are placed randomly, but differently from its corresponding training setting.

Table 2. Position settings of the monitoring ID nodes and the attacker nodes used in the experiments.

Scenario	Location of monitoring ID nodes		Location of attackers
	In training	In testing	in training and testing
S1	close	close	same
S2	close	close	different
S3	far	far	same
S4	far	far	different
S5	close	far	same
S6	close	far	different
S7	far	close	same
S8	far	close	different

5.2   Results

In order to evaluate the performance of the proposed IDS, we have considered the Pareto front set that represents the objectives of the Pareto dominant individuals found for each simulated scenario. Because we have run the GP algorithm 10 times for each scenario, we have aggregated 10 Pareto front sets and extracted the extreme points from these sets to reveal how well the GP algorithm could achieve the best performance in terms of accuracy (ACCR) and number of features (NoF). These extreme points are shown in Table 3 separately for each of the two objectives. For example, '0.945 (8)' stated for the ACCR implies that the GP could reach 0.945 accuracy with a model that has eight distinct features. Similarly, '1 (0.850)' stated for NoF implies that the GP model has only one distinct feature and could reach 0.850 accuracy.

The results enable us to evaluate the performance of the proposed IDS from different points of view. The first is that DR is a harder-to-detect routing attack as compared to other attacks when considering the average of the accuracies from eight scenarios (it is 0.88 overall). The difference in the average accuracy performances obtained from the other attacks is not significant, and GP succeeds to evolve a satisfactory ID program for these attacks (it is 0.94 overall). As for the evaluation according to the average of NoF, it is seen that the ID program requires the more features for the DR attack (it is 10.5 in overall). This clearly suggests that GP is unable to evolve an ID program that gives higher detection accuracy even after a rigorous evaluation of massive feature data. Furthermore, the difference in the accuracy performances obtained from the extreme points with respect to ACCR and NoF ranges from 4.5% to 12.0%. They are the WP and DR attacks that give the highest and least difference in performance, respectively. This shows that limiting the number of features in ID program has an adverse impact on the detection capability to some degree as expected, and this varies according to the targeted attack. Note that these give the largest difference in the performances of GP, and even a few increases in the number of features in ID program yield much better accuracy.

When it comes to the change in IDS performance when only attackers are repositioned with the same configurations of the monitoring nodes (that is, the comparison of S1 with S2, S3 with S4, and so forth), a slight performance degradation is often observed, and the change here is no more than 2.5%. This is not surprising because the locations of attackers are positioned randomly from the entire network, and there are cases studied in these scenarios where the ID program evolved and tested when the attacker nodes are, respectively, in the vicinity and away from the monitoring or root node. As for the change in IDS performance as a function of different configurations of monitoring nodes by keeping the attacker's positions the same and different (that is, comparison of S1 with S3 and S2 with S4), it is seen that the performance of the evolved ID program slightly improves (up to 2.5%) when monitoring nodes are positioned within the first three hops.

In order to reveal how monitoring nodes are helpful in improving the attack detection capability of the ID program, we replicated the simulations by adopting standalone architecture where the root node is in charge of intrusion detection alone. The simulation here is run with two attacker configurations that are denoted 'same' and 'different' which again represent the cases where attackers are positioned at the same and different locations when ID program is evaluated in the testing environment, respectively. The results are shown in Table 4. Note that S1 through S4 in Table 3 should only be considered to ensure a fair comparison between the performances of collaborative and standalone architectures. The results suggest that the performance of the ID program increases with the

Table 3. The best performances with respect to accuracy (ACCR) and number of features (NoF) obtained in the Pareto front set.

Scenario	Objective	WP	HF	IV	DR
S1	ACCR	0.945 (8)	0.950 (13)	0.955 (12)	0.880 (11)
	NoF	1 (0.850)	1 (0.865)	1 (0.870)	1 (0.810)
S2	ACCR	0.940 (7)	0.945 (12)	0.950 (12)	0.870 (10)
	NoF	1 (0.840)	1 (0.855)	1 (0.870)	1 (0.800)
S3	ACCR	0.950 (11)	0.945 (10)	0.935 (8)	0.870 (10)
	NoF	1 (0.845)	1 (0.860)	1 (0.860)	1 (0.820)
S4	ACCR	0.940 (8)	0.940 (9)	0.925 (6)	0.865 (8)
	NoF	1 (0.840)	1 (0.855)	1 (0.850)	1 (0.820)
S5	ACCR	0.935 (8)	0.930 (9)	0.940 (10)	0.880 (11)
	NoF	1 (0.835)	1 (0.845)	1 (0.830)	1 (0.835)
S6	ACCR	0.935 (12)	0.925 (11)	0.945 (12)	0.875 (10)
	NoF	1 (0.830)	1 (0.840)	1 (0.825)	1 (0.810)
S7	ACCR	0.930 (9)	0.925 (12)	0.935 (10)	0.900 (12)
	NoF	1 (0.835)	1 (0.830)	1 (0.835)	1 (0.800)
S8	ACCR	0.925 (8)	0.920 (7)	0.910 (9)	0.890 (12)
	NoF	1 (0.825)	1 (0.830)	1 (0.810)	1 (0.805)

Table 4. The extreme points with respect to ACCR and NoF in the Pareto front sets obtained by the standalone architecture.

Position	Objective	WP	HF	IV	DR
Same	ACCR	0.910 (9)	0.920 (12)	0.935 (11)	0.865 (10)
	NoF	1 (0.835)	1 (0.840)	1 (0.830)	1 (0.790)
Different	ACCR	0.905 (8)	0.915 (11)	0.925 (12)	0.860 (9)
	NoF	1 (0.825)	1 (0.830)	1 (0.825)	1 (0.780)

collaborative architecture, and the difference reaches 4%. However, no significant difference is observed in terms of NoF.

6 Strengths, Limitations, and Future Directions

The primary target of our attempt is to reduce the operational cost of IDS considering the constrained resource of LLN nodes. Therefore, without sacrificing detection accuracy, we aim to minimize the number of distinct features that cause exhaustive resource consumption of nodes, as well as adversely affect the communication cost of monitoring nodes. The obtained results show the applicability of the proposed multi-objective approach to effectively and efficiently detect intrusions in LLNs. Because the depth of GP tree is highly correlated to the number of terminal nodes (i.e., feature nodes) in most cases, we implicitly control the size of the tree, and hence the length of the program. Therefore, the code bloating problem in GP is handled in our approach. However, even few, the LLN nodes also suffer from the frequency of the operations in the ID program that are to be executed. Although this execution cost is not included as an individual objective, it can be easily minimized by penalizing solutions that require intensive computation.

The existence of monitoring nodes may reduce the communication performance of LLNs, but they are undeniably important in detecting intruders effectively. We here perform the simulations by grouping the monitoring nodes as close (1–3 hops away from the root node) and far nodes (4–6 hops away from the node); however, it is of high importance to thoroughly investigate the number and the position of these nodes, which could also be studied in the future. In addition, the proposed approach is tested on a centralized architecture in which the root node is responsible for raising the alarm. That's why, the single point of failure, which occurs when the root node is down by the intruders, is the drawback of our detection system. To overcome this, one can rely on a fully decentralized architecture where multiple nodes are in charge of the detection task simultaneously. For this architecture, developing different local detection programs for each individual node is a must. To do that, it is worth studying the federated learning approach to have local programs that are informed globally and constructed collaboratively.

As the main objective of this study is to show applicability of multi-objective GP in developing IDS for resource-constrained LLNs, we only targeted four attack types, and it can be extended by involving other types of attacks. In addition, the positions of the LLN nodes are stationary in the experiment; however, if not all, in most real-world IoT applications, a portion of these nodes is mobile. Therefore, it is worth testing our proposed approach in a mobile environment, which could be another future direction of this study.

7 Conclusion

In this study, we explore the use of the Pareto-based multi-objective approach to efficiently and effectively detect four different attack types specific to RPL. To the best of the author's knowledge, this is the first study that aims to simultaneously optimize the detection accuracy of ID programs and their costs including communication cost of ID nodes. To do that, a massive number of simulations are generated, and the different ID programs are evolved by using these simulations. The evaluations are made on the basis of Pareto sets obtained from the evolved programs. Among all experiments, the average accuracy was 92.2%, while the variance in these accuracies was 0.08%, demonstrating that the proposed approach provides satisfactory results in detecting targeted attacks. It is also worth stating here that our IDS model converges to an accuracy level above 90% after reaching around 30% of the generations. In the future, we plan to explore the applicability of our approach when a portion of LLN nodes are mobile.

References

1. Alexander, R., et al.: RPL: IPv6 routing protocol for low-power and lossy networks. RFC 6550, March 2012. https://doi.org/10.17487/RFC6550, https://www.rfc-editor.org/info/rfc6550
2. Arış, A., Oktuğ, S.F.: Analysis of the RPL version number attack with multiple attackers. In: 2020 International Conference on Cyber Situational Awareness, Data Analytics and Assessment (CyberSA), pp. 1–8. IEEE (2020)
3. Aydogan, E., Yilmaz, S., Sen, S., Butun, I., Forsström, S., Gidlund, M.: A central intrusion detection system for RPL-based industrial internet of things. In: 2019 15th IEEE International Workshop on Factory Communication Systems (WFCS), pp. 1–5. IEEE (2019)
4. Cakir, S., Toklu, S., Yalcin, N.: RPL attack detection and prevention in the internet of things networks using a GRU based deep learning. IEEE Access 8, 183678–183689 (2020)
5. Canbalaban, E., Sen, S.: A cross-layer intrusion detection system for RPL-based internet of things. In: Grieco, L.A., Boggia, G., Piro, G., Jararweh, Y., Campolo, C. (eds.) ADHOC-NOW 2020. LNCS, vol. 12338, pp. 214–227. Springer, Cham (2020). https://doi.org/10.1007/978-3-030-61746-2_16
6. Cisco: Visual networking index: Forecast and trends, 2017–2022 White paper. https://www.cisco.com/c/en/us/solutions/collateral/service-provider/visual-networking-index-vni/white-paper-c11-741490.html. Accessed 04 Apr 2020

7. Contiki-Ng: contiki-ng/contiki-ng (2004). https://github.com/contiki-ng/contiki-ng/wiki. Accessed 13 July 2021
8. Deb, K., Pratap, A., Agarwal, S., Meyarivan, T.: A fast and elitist multiobjective genetic algorithm: NSGA-II. IEEE Trans. Evol. Comput. **6**(2), 182–197 (2002)
9. Dogan, C., Yilmaz, S., Sen, S.: Analysis of RPL objective functions with security perspective. In: SENSORNETS, pp. 71–80 (2022)
10. ECJ: A Java-based evolutionary computation research system (2017). https://cs.gmu.edu/eclab/projects/ecj. Accessed 04 Apr 2022
11. Eiben, A.E., Smith, J.E., et al.: Introduction to Evolutionary Computing. Natural Computing Series, vol. 53. Springer, Heidelberg (2003). https://doi.org/10.1007/978-3-662-44874-8
12. Gothawal, D.B., Nagaraj, S.: Anomaly-based intrusion detection system in RPL by applying stochastic and evolutionary game models over IoT environment. Wirel. Pers. Commun. **110**(3), 1323–1344 (2020)
13. Herberg, U., Clausen, T.: A comparative performance study of the routing protocols load and RPL with bi-directional traffic in low-power and lossy networks (LLN). In: Proceedings of the 8th ACM Symposium on Performance Evaluation of Wireless Ad Hoc, Sensor, and Ubiquitous Networks. PE-WASUN 2011, pp. 73–80. Association for Computing Machinery, New York (2011). https://doi.org/10.1145/2069063.2069076
14. Koza, J.R.: Genetic programming as a means for programming computers by natural selection. Stat. Comput. **4**(2), 87–112 (1994)
15. Li, F., Shinde, A., Shi, Y., Ye, J., Li, X.Y., Song, W.: System statistics learning-based IoT security: Feasibility and suitability. IEEE Internet Things J. **6**(4), 6396–6403 (2019)
16. Mayzaud, A., Badonnel, R., Chrisment, I.: A taxonomy of attacks in RPL-based internet of things. Int. J. Netw. Secur. **18**, 459–473 (2016)
17. Mayzaud, A., Badonnel, R., Chrisment, I.: A distributed monitoring strategy for detecting version number attacks in RPL-based networks. IEEE Trans. Netw. Serv. Manage. **14**(2), 472–486 (2017)
18. Muzammal, S.M., Murugesan, R.K., Jhanjhi, N.Z., Humayun, M., Ibrahim, A.O., Abdelmaboud, A.: A trust-based model for secure routing against RPL attacks in internet of things. Sensors **22**(18), 7052 (2022)
19. Osterlind, F., Dunkels, A., Eriksson, J., Finne, N., Voigt, T.: Cross-level sensor network simulation with Cooja. In: Proceedings. 2006 31st IEEE Conference on Local Computer Networks, pp. 641–648. IEEE (2006)
20. Raza, S., Wallgren, L., Voigt, T.: Svelte: real-time intrusion detection in the internet of things. Ad Hoc Netw. **11**(8), 2661–2674 (2013)
21. Sen, S.: A survey of intrusion detection systems using evolutionary computation. In: Bio-inspired Computation in Telecommunications, pp. 73–94. Elsevier (2015)
22. Statista: Internet of things (IoT) connected devices installed base worldwide from 2015 to 2025 (in billions). https://www.statista.com/statistics/471264/iot-number-of-connected-devices-worldwide/. Accessed 10 Apr 2022
23. Vu, L., Nguyen, Q.U., Nguyen, D.N., Hoang, D.T., Dutkiewicz, E.: Deep transfer learning for IoT attack detection. IEEE Access **8**, 107335–107344 (2020)
24. Wu, S.X., Banzhaf, W.: The use of computational intelligence in intrusion detection systems: a review. Appl. Soft Comput. **10**(1), 1–35 (2010)
25. Yavuz, F.Y., Devrim, Ü., Ensar, G.: Deep learning for detection of routing attacks in the internet of things. Int. J. Comput. Intell. Syst. **12**(1), 39 (2018)

26. Yılmaz, S., Aydogan, E., Sen, S.: A transfer learning approach for securing resource-constrained IoT devices. IEEE Trans. Inf. Forensics Secur. **16**, 4405–4418 (2021)

27. Zahra, F., Jhanjhi, N., Brohi, S.N., Khan, N.A., Masud, M., AlZain, M.A.: Rank and wormhole attack detection model for RPL-based internet of things using machine learning. Sensors **22**(18), 6765 (2022)

A New Prediction-Based Algorithm for Dynamic Multi-objective Optimization Problems

Kalthoum Karkazan[1], Haluk Rahmi Topcuoglu[1(✉)], and Shaaban Sahmoud[2]

[1] Computer Engineering Department, Faculty of Engineering, Marmara University,
Istanbul, Turkey
kalthoumkarkazan@marun.edu.tr, haluk@marmara.edu.tr
[2] Computer Engineering Department, Fatih Sultan Mehmet Vakif University,
Istanbul, Turkey
ssahmoud@fsm.edu.tr

Abstract. The mechanism for reacting to the changes in an environment when detected is the key issue that distinguishes various algorithms proposed for dynamic multi-objective optimization problems (DMOPs). The severity of change is a significant approach to identify the dynamic characteristics of DMOPs. In this paper, a prediction-based strategy based on utilizing the degree of the changes is presented to address environmental changes. In case of a change detection in the given DMOP, the severity of change is evaluated and an appropriate reaction mechanism is followed based on the degree of the observed change. To accelerate the convergence process, the algorithm may respond multiple times for the same change. The performance of our algorithm is evaluated by comparing it with dynamic multi-objective evolutionary algorithms using six benchmarks. The effectiveness of our algorithm is demonstrated in the experimental study where it outperforms other compared algorithms in most of the tested instances considered.

Keywords: dynamic multi-objective optimization · severity of change · change detection · prediction-based optimization

1 Introduction

Recently, Multi-Objective Optimization Problems (MOPs) attract researchers' interest due to their contribution as a key tool in many decision-making processes [20]. Evolutionary Algorithms (EAs) have proven their efficiency in solving many optimization problems from different domains. Therefore, researchers proposed many successful Multi-objective Evolutionary Algorithms (MOEAs) for solving static MOPs [4,25].

In real-world environments, many MOPs become dynamic in nature since objective function(s), some of their environmental settings and/or parameters may change over time [11], where they are referred to as Dynamic Multi-objective Optimization Problems (DMOPs). The primary motivation behind

© The Author(s), under exclusive license to Springer Nature Switzerland AG 2023
J. Correia et al. (Eds.): EvoApplications 2023, LNCS 13989, pp. 194–209, 2023.
https://doi.org/10.1007/978-3-031-30229-9_13

the techniques for solving DMOPs is to generate a diversified Pareto front of the problem and track its changes over time as quickly as possible [24]. Traditional MOEAs are not directly applicable to DMOPs because the various forms of dynamism dramatically decrease the optimizer's performance. This opens the window for further improvements and proposals of new algorithms that can deal with DMOPs more efficiently.

In this paper, we target to develop a novel change reaction strategy for DMOPs in order to respond to the environmental changes effectively. Different response mechanisms are developed within this strategy to deal with each change adaptively according to its degree of change. When an environmental change is detected, the algorithm calculates the severity of the change which will be used to decide which scheme should be followed. If the severity of the change is high enough, its value will be compared to other values from previous changes to decide whether the POF is following the same behavior in the last few changes or not. Accordingly, the initial position of the new POF is estimated using an adaptive change response strategy.

Moreover, our proposed algorithm injects a varying ratio of randomly generated individuals to maintain good diversity level in the population. This ratio is adjusted to be high in the first generations after each detected change. It will be decreased rationally as the generations evolves. Finally, after each change, the proposed algorithm may react multiple times to fasten the convergence process. The performance of our algorithm is evaluated as part of an empirical study using various DMOPs with different comparison metrics. Specifically, it outperforms related work for most of the test cases with respect to mIGD and mIGDB metrics, which validates the efficiency of our algorithm.

We organize the rest of the paper as follows. Section 2 gives a short summary of the previously proposed similar work. Section 3 describes the proposed algorithm. Section 4 presents the experimental settings for comparison. The results of the experimental study are provided in Sect. 5. We summarize the conclusions in Sect. 6.

2   Dynamic Multi-objective Evolutionary Algorithms

The multi-objective optimization problem is a problem with more than one conflicting objective function [9]. On the other hand, a dynamic multi-objective optimization problem (DMOP) has a dynamic nature based on a change in time at the objectives, parameters, and/or constraints of the problem. A DMOP is defined mathematically as follows:

minimize f(x,t) = $(f_1(x,t), f_2(x,t), ...f_{mt}(x,t))$
subject to

$$g_i(x,t) \leq 0, i = 1, 2, ...p$$
$$h_j(x,t) = 0, j = 1, 2, ...q \qquad (1)$$
$$x = (x_1, x_2, ...x_n) \; and \; x \in [x_{min}, x_{max}]$$

where m represents the number of objectives, t is the discrete-time instants. Functions g, h, and f represent the inequality, equality constraints, and objective function respectively. q and p represent the number of the equality and inequality constraints [14].

When comparing two solutions for a multi-objective optimization problem (MOP), one of them may have better values for some of the objective functions while having worse values for the rest. To address this problem, a solution for a MOP can be evaluated according to its dominant relationship with the other solutions. The solution is considered to be dominated by any other solution that is better than it in all objective functions. If a solution is not dominated by any other solution, it is called a non-dominated solution. The set of all non-dominated solutions in the objective space is referred as Pareto optimal front (POF), while the same corresponding set in the decision space is denoted by Pareto optimal set (POS). These two sets actually represent the optimal trade-off set for a given MOP in objective and decision spaces respectively [5].

When solving DMOPs using EAs, the main challenge appears due to the dynamism of problem to be optimized. In literature, different techniques have been developed for converging the new POFs as quickly as possible and/or tracking POFs as close as possible. Memory-based, diversity maintenance-based and prediction-based techniques are among the common approaches presented in the literature [16].

Population prediction strategy (PPS) [26] is one of the prediction-based algorithms where the authors target to predict the whole population when any change has been detected. They do that by dividing the Pareto set into a center point and manifolds. Then, the center point at each time step is stored and a univariate autoregression model is used to predict the new center position. The SGEA algorithm [10] is another prediction-based algorithm that predicts half of the population when a change is detected by calculating the direction and the movement step size of the last two consecutive environments. The other half is chosen from the current population with the goal of maintaining diversity as much as possible. In another algorithm, the directed search strategy (DSS) [23], they apply two approaches when an environmental change is detected. The first one is to reinitialize the population according to the predicted moving direction as well as a local search on an orthogonal direction to the moving direction. The second one introduces good individuals according to the moving direction of the Pareto set in the previous two consecutive generations.

Hybridization of a memory and a prediction-based technique is also proposed in the literature [14], which stores information related to each change in an archive. If a change is detected, it is compared to the historical information and responded to it using the memory information if any similar change has been determined. Additionally, an adaptive prediction-based DMOEA is proposed which reacts to the environmental changes according to the characteristics of the Pareto set [13]. If the Pareto set changes over time, the algorithm responds by using a classification prediction strategy. In addition, a dynamic mutation strategy will be activated to adapt to changes when the Pareto set remains static. The type of the DMOP can be utilized to predict or estimate the new POF [17].

Another study utilizes the Borda sorting concept to choose the worst individuals in the population to be re-initialized when a change is detected [15]. To solve DMOPs, a predictive method based on a grey optimization mechanism is proposed in [21]. It merges two different mechanisms which are clustering the solutions as well as using a grey prediction model of the clusters' centroids to generate the initial population.

3   The Proposed Algorithm

Our proposed algorithm utilizes the NSGA-II algorithm as a baseline and enhances its ability to deal with environmental changes. When an environmental change is detected, the newly developed mechanism is fired to optimize the population in the new conditions as soon as possible (see Algorithm 1). In the following subsections, the proposed algorithm's stages and mechanisms are explained in detail.

Algorithm 1. The general structure of a detection-based DMOEA

Initialize the population
$generation \leftarrow 0$
repeat
  Call change detection procedure
  if a change is detected in the environment **then**
    Call reaction procedure for the detected change /* Given Algorithm 2 */
  else
    Apply crossover operator
    Apply mutation operator
  end if
  Evaluate each individual in the population
  $generation \leftarrow generation + 1$
  $\vdots$

until *Termination condition is satisfied*

3.1   Change Detection Strategy

At each generation, a number of individuals are selected randomly and used as indicators or sensors for the environment's status. Those indicators will be saved in an archive along with their fitness values. In the next generation, the new fitness values for all of the indicators are recalculated and compared with the stored old values. If they are similar, the environment status is considered to be stable and no change is detected. On the other hand, in case of any difference in the fitness values of indicators, a flag that indicates encountering an environmental change is triggered and the severity of change is calculated [2,19]. The severity of change is calculated according to Eq. 2. It should be noted that calculating the severity of change based on this equation solves problems that are encountered when calculating the severity of change for each objective function separately and choosing the maximum value to be set as the overall severity (SC) [18].

$$SC_j = \sum_{i=1}^{S} \frac{F_{i,j}(t) - F_{i,j}(t-1)}{F_{i,j}(t) + s} \tag{2}$$
$$SC = \lambda \times \max(SC_1, SC_2, ..., SC_M)$$

In this equation, SC_j represents the severity of change for the objective function j, and S is the number of selected indicators to detect the environmental change. $F_{i,j}(t)$ and $F_{i,j}(t-1)$ are two terms to represent the value for objective function j for sensor i in the current and previous generations, respectively. The term s is a small constant used to avoid the undefined results when $F_{i,j}(t)$ is equal to zero; and λ is used to adapt the equation for various numbers of objective functions. It is set to be $M-1$ in this study as suggested by its original author where M is equal to the objectives of the given problem.

3.2   Change Response Strategy

When a change is detected, the first step is to calculate the severity of the change and to store the current estimated POF just before the change for further processing. After that, the current population is evolved over one more iteration. This step is important to help in estimating the direction and position of the new POF. In the following step, the current population is examined and the non-dominated solutions are stored as a primary estimation for the true POF in the new environment. Then, it compares the value of the severity of change for the current change with the average value for change severity. If it is greater than or equal to the average value, the response mechanism is fired. Otherwise, the algorithm will not react to the change as the change does not show significant severity of change.

When the situation requires a reaction, the reaction mechanism starts by calculating the centroid of the new POF and the old POF and subtracting them from each other. Equation 3 shows how to calculate the centroid for a given POF at time t. The algorithm subtracts this value from each individual in the new POF. This step creates a set of auxiliary solutions for the POF of the new environment. Those auxiliary solutions will have a centroid that is coinciding the old POF centroid. Having the same centroid helps in finding the correct corresponding solutions between the new and old POFs [1]. The algorithm then pairs the points in the new and old POFs. The Euclidean distance between every single solution in the auxiliary solutions and all individuals in the old POF is calculated in the variables space, and the one with the smallest value is chosen as its parent. The change response strategy is presented in Algorithm 2.

$$C_j t = \frac{1}{|POF_t|} \sum_{x_j \in POF_t} x_j \tag{3}$$

Algorithm 2. The Individual Generation Strategy

$POP_0 \leftarrow$ last population in the previous environment

$POF_0 \leftarrow$ approximated optimal solutions from POP_0

$F(POP_0) \leftarrow$ evaluate POP_0 /*according to the new environment*/

$POP_{new} \leftarrow$ Apply MOEA for one iteration /*to evolve the population*/

$POF_{new} \leftarrow$ non-dominated form POP_{new}

if $SeverityOfChange >= AvgSeverityOfChange$ **then**

    $Centroid_{diff} \leftarrow Centroid(POF_0) - Centroid(POP_{new})$

    if $|POP_{new}| < |POF_0|$ **then**

        $NumOfPairs \leftarrow |POP_{new}|$

    else

        $NumOfPairs \leftarrow |POF_0|$

    end if

    for $NumOfPairs$ **do**

        find the parent of each individual in POP_new from the POF_0

    end for

    if $SeverityOfChange >=$ the three previously calculated $SeverityOfChanges$ **then**

        for $|POP_{new}|$ **do**

            Calculate the new position of the individual by applying the step size algorithm

        end for

    else

        for $|POP_{new}|$ **do**

            Calculate the new position of the individual by applying the cluster representer algorithm

        end for

    end if

end if

---

where x_j is the objective value j for individual x in the population Pop at time t and $|POF_t|$ is the cardinality of the population Pop at time t. C_jt is the centroid of the pareto optimal front POF for the objective function j at time t.

3.3 Prediction Strategies for the Individuals

After having decided whether to react or not, the algorithm classifies the change and reacts to it according to one of the following two cases:

- *Mechanism for Low Severity of Change:* The severity value is compared to the previous three severity of change values. If it is less than all of them, the step size of the old and new POFs' centroids is calculated by subtracting them from each other and using it to find the initial position of the individuals in the new environment (see Eq. 4).

  This approach is developed because having a severity of change that is smaller than that of previous generations indicates that the old POF and the new POF are close to each other; and the step size can be used to estimate the new positions of individuals of the new environment.

$$X_{t+1} = X_t + (C_{t+1} - C_t) \tag{4}$$

- *Mechanism for High Severity of Change:* If the severity is greater than the average severity of the last three changes, the current population will be divided into clusters. The number of clusters will be determined according to the number of non-dominated solutions in the new environment and each non-dominated solution will be considered as its cluster's representative. Then, all other solutions will be distributed into those clusters according to each solution's smallest Euclidean Distance value from the clusters' representatives. After that, each individual's position will be updated according to its cluster's representative step size. The step size for each representative is calculated by subtracting its decision variables values from its parent's decision variables values.

$$X_{t+1} = X_t + |Rp_t - (Rp_t + (C_{t+1} - C_t) + 1)| \tag{5}$$

Here Rp_t is the closest element in the new POF to X_t, which is calculated using the Euclidean distance. The term $Rp_t + 1$ is the parent of Rp_t in the old POF.

3.4   Multiple Reactions

Finally, periodic reactions are applied after the new environment has been running for a specific number of generations. In our empirical study, the algorithm reacts periodically after every five generations. This step helps in accelerating the convergence process and prevents the solutions from sticking to the local optimum. The main reason that makes this step important is that after each generation more information is collected about the new environment and the prediction process will be more accurate, which can help in catching up any error in the previous prediction process. To maintain the diversity of the population, some randomly generated individuals will be injected to the population.

When a change is detected, individuals in the early generations are expected to be far away from the real POF. At this stage, increasing the diversity of the population is very important. To do that, a large number of randomly generated individuals are incorporated into the population in early generations. In this paper, 30% of the population size is injected randomly in the first react. As more generations are executed, the number of randomly generated individuals is reduced. In other words, the percentage of the randomly introduced individuals will be determined adaptively according to the number of generations that passed after the detection of a change. This is achieved by decreasing the percentage of the randomly generated individuals as shown in the following equation:

$$d = \lfloor c_{count}/g \rfloor + 1$$
$$p = p_{initial}/Rep \tag{6}$$

In this equation, c_{count} is the number of generations passed since the last change has been detected, g refers to the chosen number of generations to react, and $p_{initial}$ is the chosen initial percentage for the randomly generated individuals.

4   Experimental Design

4.1   Test Problems

A total of six test problems are considered to assess the performance of the algorithm, where four of them are from FDA test problems (FDA1, FDA3, FDA4, and FDA5) [7] and two of them are from dMOP test problems [8]. These test suites cover the widely used three types of DMOPs based on classification presented by Farina et al. [7].

Table 1. Information about the used benchmarks

Problem	Type	Objective number	Number of variables
FDA1	Type 1	2	11
FDA3	Type 2	2	10
FDA4	Type 1	3	12
FDA5	Type 2	3	12
dMOP1	Type 3	2	10
dMOP2	Type 2	2	10

The details of the selected test problems are summarized in Table 1, where the time used in these tests is defined as

$$t = \frac{1}{n_t} * \left\lfloor \frac{\tau}{fr} \right\rfloor \tag{7}$$

where fr is the change frequency, τ is the generations counter, and n_t is the severity of change.

4.2   Performance Metrics

Three most popular performance metrics are used for comparing the performance of the DMOEAs, which are summarized below.

- Mean Inverted Generational Distance (mIGD): It is the modified version of the inverted generational distance (IGD) metric [12]. In this method, the mean value of the IGD metric in all generations is calculated as shown below

$$mIGD = \frac{1}{|T|} \sum_{t \in T} IGD(PF_t, PF_t^*) \tag{8}$$

where IGD metric [3] is used to compare diversity and convergence of the algorithms for static multi-objective optimization problems.

$$IGD(PF_t, PF_t^*) = \frac{\sum\limits_{v \in PF_t} d(v, PF_t^*)}{|PF_t|} \quad (9)$$

Here, uniformly distributed points in the true PF is given in the set PF_t; and an approximation of the true PF generated by the algorithm at the $t-th$ time step is represented with the set PF_t^*. The term $d(v, PF_t^*)$ is the minimum Euclidean distance between a point v in PF_t and points in PF_t^*.

- *Mean Inverted Generational Distance just before the change (mIGDB):* The idea behind this metric is to calculate the mean value of the IGD value in time steps that are just before the next change rather than calculating the IGD for all generations [26].

$$mIGDB = \frac{1}{C_n} \sum_{t=1}^{C_n} IGD(PF_t, PF_t^*) \quad (10)$$

where C_n is the number of changes in the run and the IGD value is calculated between the true and the estimated POF immediately before the change.

- *Average Maximum Spread (aMS):* This metric calculates how well the estimated POF is spread on the actual POF. It is actually developed to be used in static multi-objective environments. To adapt it for dynamic environments, the average of its values in all generations is calculated as shown below [8]:

$$aMS = \sqrt{\frac{1}{M} \sum_{k=1}^{M} \left[\frac{min\left[\overline{POF_k}, \overline{POF_k^*}\right] - max\left[\underline{POF_k}, \underline{POF_k^*}\right]}{\overline{POF_k} - \underline{POF_k}} \right]} \quad (11)$$

In this equation, M is the number of fitness functions; and $\overline{POF_k}$ refers to the maximum of the kth objective of the true POF while $\underline{POF_k^*}$ refers to the minimum of the kth objective of the estimated POF.

4.3   Algorithms in the Empirical Study

Three DMOEAs are used in the comparison study: the dynamic version of the NSGA-II Algorithm (DNSGA-II [6] algorithm), the MOEA algorithm based on decomposition(MOEA/D [25]), and the multi-model prediction approach (the MMP algorithm [16]).

- *Multi-model prediction approach (MMP):* The main idea of this algorithm is to react upon the type of the Pareto optimal set (PS) change when detecting an environmental change. Based on this, a multi-model prediction approach that classifies the PS change into four different types is proposed [16]. It suggests different methods to detect the type of change as well as a different response method for each type. When a change is detected, the new individuals are predicted adaptively according to appropriate models based on the estimated type of PS.

– *Dynamic non-dominated sorting genetic algorithm II (DNSGA-II):* This algorithm is adapted for dynamic problems by introducing new individuals when a change in the environment is detected [6]. The newly introduced solutions can be generated either randomly or by mutating some of the existing individuals. In this paper, we used the mutating method as it performs better according to their results.

– *Multiobjective evolutionary algorithm based on decomposition (MOEA/D):* It is a commonly used algorithm for solving static multi-objective optimization problems. As its name indicates, the main problem is decomposed into multiple smaller and easier problems to solve [25]. Those problems will be solved in parallel and the individuals for each one of them will contribute to improving its neighboring problems.

Table 2. Patameter Settings

Parameter	Setting
Population size (N)	100
Cross-over probability	1
Mutation probability	$1/n$
Change frequency	5,10, and 20
Number of generations	$10 * \tau_t$
Number of executions	30
Change detection sensors	5% of the population size

5 Experimental Results

In this section, the effectiveness of our algorithm is evaluated and compared with other DMOEAs within the context of our experimental framework. The parameter settings of the algorithms used in our comparison study are from the related references. In experiments, the severity of change (n_t) is set to 10, and the frequency of change (τ_t) is set to 5, 10, and 20. The values of other parameters are shown in Table 2. Tables 3, 4 and 5 show the results of mIGD, mIGDB, and aMS metrics, respectively. The best results from each of the four algorithms are denoted in bold. In the experiments, 500 uniformly distributed points on the true POF are selected to compute the mIGD and mIGDB metrics' values. To indicate the significance between the results of the compared algorithm, the Wilcoxon rank-sum test has been carried out at 0.05 significance level [22]. At each row, the best result gained is highlighted in bold while the plus sign next to a result indicates that the bold result is statistically significant than this result.

As seen in Table 3, the proposed algorithm outperforms the other algorithms in most of the test instances (16 out of 18 cases). The mIGD metric evaluates both convergence and diversity when measuring an algorithm's performance, so getting competitive results in this metric means that the proposed algorithm is

Table 3. Mean values of mIGD metric obtained by the algorithms

Problem	(τ_t, n_t)	DNSGA-II	MMP	MOEA/D	Proposed Algorithm
FDA1	(5,10)	0.173+	0.172+	0.782+	**0.147**
	(10,10)	0.074+	0.077+	0.513+	**0.067**
	(20,10)	**0.041**	0.042	0.295+	0.047
FDA3	(5,10)	0.203+	0.214+	0.676+	**0.135**
	(10,10)	0.087	0.067	0.447+	**0.053**
	(20,10)	0.048+	**0.040**	0.311+	0.042+
FDA4	(5,10)	0.884+	0.371+	0.544+	**0.326**
	(10,10)	0.584	0.268+	0.389	**0.223**
	(20,10)	0.219+	0.184+	0.272+	**0.147**
FDA5	(5,10)	0.780+	0.355+	1.091+	**0.310**
	(10,10)	0.467+	0.237+	0.970+	**0.218**
	(20,10)	0.224+	0.175+	0.830+	**0.164**
dMOP1	(5,10)	0.129	0.134	0.609+	**0.127**
	(10,10)	**0.115**	**0.115**	0.394+	**0.115**
	(20,10)	0.105	0.107+	0.217+	**0.104**
dMOP2	(5,10)	0.639+	0.173	0.381+	**0.162**
	(10,10)	0.142+	0.101+	0.257+	**0.071**
	(20,10)	0.058+	0.055+	0.145+	**0.037**

Table 4. Mean values of mIGDB metric obtained by the algorithms

Problem	(τ_t, n_t)	DNSGA-II	MMP	MOEA/D	Proposed Algorithm
FDA1	(5,10)	0.128+	0.127+	0.553+	**0.108**
	(10,10)	0.035+	0.035+	0.229+	**0.034**
	(20,10)	**0.012**	**0.012**	0.045+	0.031+
FDA3	(5,10)	0.161+	0.165+	0.433+	**0.099**
	(10,10)	0.064	0.046	0.200+	**0.037**
	(20,10)	0.046+	0.042+	0.147+	**0.039**
FDA4	(5,10)	0.829+	0.310+	0.369+	**0.269**
	(10,10)	0.459+	0.200+	0.218+	**0.153**
	(20,10)	0.132+	0.121+	0.119	**0.094**
FDA5	(5,10)	0.715+	0.285+	0.934+	**0.244**
	(10,10)	0.354+	0.170+	0.790+	**0.158**
	(20,10)	0.151+	**0.133**	0.683+	0.140+
dMOP1	(5,10)	0.109	0.113	0.438+	**0.106**
	(10,10)	**0.098**	0.099	0.157+	**0.098**
	(20,10)	0.096	0.097	0.079+	**0.095**
dMOP2	(5,10)	0.565+	0.126	0.262+	**0.116**
	(10,10)	0.068+	0.045+	0.105+	**0.026**
	(20,10)	0.015	**0.013**	0.023+	0.015+

Table 5. Mean values of aMS metric obtained by the algorithms

Problem	(τ_t,n_t)	DNSGA-II	MMP	MOEA/D	Proposed Algorithm
FDA1	(5,10)	0.977+	**0.987**	0.881+	0.985
	(10,10)	0.991	**0.992**	0.909+	0.987+
	(20,10)	**0.995**	**0.995**	0.938+	0.984+
FDA3	(5,10)	0.916	0.917	0.805+	**0.937**
	(10,10)	0.939+	0.951	0.838+	**0.971**
	(20,10)	0.970	**0.981**	0.850+	**0.981**
FDA4	(5,10)	**1.000**	**1.000**	0.959+	**1.000**
	(10,10)	**1.000**	**1.000**	0.979+	**1.000**
	(20,10)	**1.000**	**1.000**	0.990+	**1.000**
FDA5	(5,10)	**1.000**	**1.000**	0.826+	**1.000**
	(10,10)	**1.000**	**1.000**	0.825+	0.999+
	(20,10)	0.998	**0.999**	0.830+	0.997+
dMOP1	(5,10)	**0.990**	0.985	0.890+	0.988
	(10,10)	**0.995**	0.993	0.928+	0.989+
	(20,10)	**0.998**	0.997	0.956+	0.991+
dMOP2	(5,10)	0.984	**0.990**	0.917+	0.987+
	(10,10)	0.993	**0.994**	0.945+	0.993+
	(20,10)	**0.997**	**0.997**	0.971+	0.994+

able to successfully converge to the true POF quickly and maintains the diversity, as well. On the other hand, the MMP algorithm provides the second-best results for all of the test instances.

The results of the mIGDB metric are shown in Table 4. The results of this metric confirm the results of the previous metric and demonstrate that the proposed algorithm obtains the best results on the majority of the tested instances (15 out of 18). These results ensure the effectiveness of our newly incorporated adaptive mechanisms to deal with environmental changes dynamically. The MPP algorithm again obtains the second good results in this metric.

Table 5 shows the results of the aMS metric which evaluates the diversity level of the obtained solutions. Although the proposed algorithm does not obtain the best results, its performance is competitive with other algorithms. The DNSGA-II and the MMP algorithms provide slightly better aMS results than the proposed algorithm. When those results are compared, it can be observed that there is no significant difference for most of the test instances.

To analyze the behavior of the compared DMOEAs during the evolving process, we conduct another experiment to view the performance of algorithms after each generation as shown in Fig. 1. As we see in the figure, the proposed algorithm shows a more stable performance compared to the other algorithms for most of the test problems. The proposed algorithm recovers its performance after each environmental change faster than other algorithms for most of the

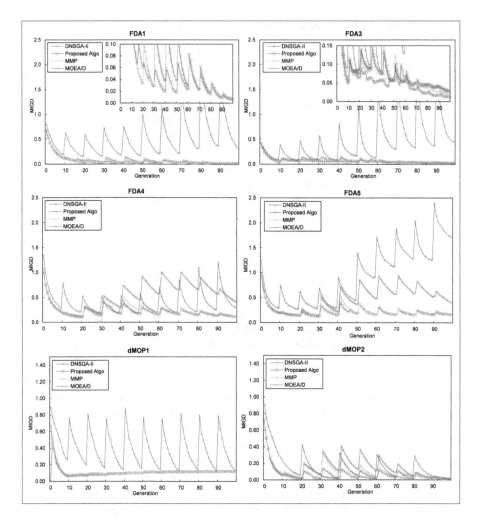

Fig. 1. Evolution curves of average mIGD values for the test instances used in the experimental study with $\tau_t = 10$ and $n_t = 10$

test problems. As discussed before, this is a direct result of the utilized adaptive prediction mechanism.

Figure 2 shows the final population distribution of the four algorithms for four different time steps. Because of the page limitations, only one of the used test instances (the FDA1 case) is includes. It is a type 1 DMOP where the POF is static in all time steps and only the POS is changing. The blue dots represent the positions of the population solutions produced by the considered algorithm. The more evenly distributed the points, and the closer to the POF line, the better estimation for the problem POF. The plots in this figure represent the obtained solutions by each algorithm with the best mIGD value among 30 independent

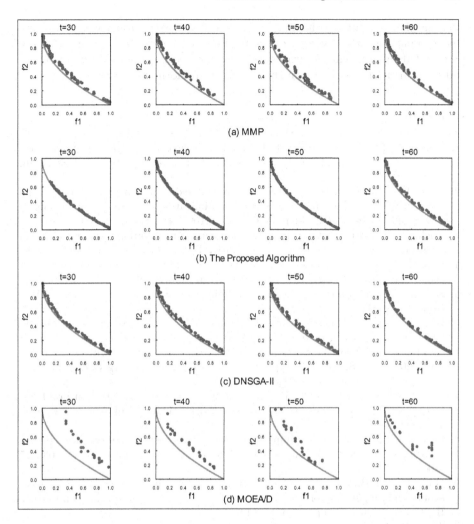

Fig. 2. Final population distribution of the four algorithms at six-time steps on FDA1 with $\tau_t = 10$ and $n_t = 10$

runs. This figure points out that the proposed algorithm can accelerate the process of convergence to the POF and produce more reliable solutions quickly.

6 Conclusions

The main aim of this paper is to propose a dynamic multi-objective optimization algorithm. The main idea of the proposed algorithm is to follow different response schemes according to the measured change severity. Those schemes are diverging from not responding at all to letting each individual follow its cluster's representative behaviour to locate its initial location in the new environment after the change. Our algorithm introduces an adaptive solution injection mechanism to maintain the diversity of the population. According to our results, the proposed algorithm achieved very good results in most of the used test instances and confirms its performance in three different performance metrics.

References

1. Ahrari, A., Elsayed, S., Sarker, R., Essam, D., Coello Coello, C.A.: Weighted pointwise prediction method for dynamic multiobjective optimization. Inf. Sci. **546**, 349–367 (2021)
2. Altin, L., Topcuoglu, H.R.: Impact of sensor-based change detection schemes on the performance of evolutionary dynamic optimization techniques. Soft. Comput. **22**(14), 4741–4762 (2017). https://doi.org/10.1007/s00500-017-2660-1
3. Coello, C.A., Cortes, N.C.: Solving multiobjective optimization problems using an artificial immune system. Genet. Program Evolvable Mach. **6**(2), 163–190 (2005)
4. Deb, K., Pratap, A., Agarwal, S., Meyarivan, T.: A fast and elitist multiobjective genetic algorithm: NSGA-II. IEEE Trans. Evol. Comput. **6**(2), 182–197 (2002)
5. Deb, K.: Multiobjective Optimization Using Evolutionary Algorithms. Wiley, Hoboken (2001)
6. Deb, K., Rao N., U.B., Karthik, S.: Dynamic multi-objective optimization and decision-making using modified NSGA-II: a case study on hydro-thermal power scheduling. In: Obayashi, S., Deb, K., Poloni, C., Hiroyasu, T., Murata, T. (eds.) EMO 2007. LNCS, vol. 4403, pp. 803–817. Springer, Heidelberg (2007). https://doi.org/10.1007/978-3-540-70928-2_60
7. Farina, M., Deb, K., Amato, P.: Dynamic multiobjective optimization problems: test cases, approximations, and applications. IEEE Trans. Evol. Comput. **8**, 425–442 (2004)
8. Goh, C.K., Tan, K.C.: A competitive-cooperative coevolutionary paradigm for dynamic multiobjective optimization. IEEE Trans. Evol. Comput. **13**(1), 103–127 (2009)
9. Gong, D., Xu, B., Zhang, Y., Guo, Y., Yang, S.: A similarity-based cooperative coevolutionary algorithm for dynamic interval multiobjective optimization problems. IEEE Trans. Evol. Comput. **24**(1), 142–156 (2020)
10. Jiang, S., Yang, S.: A steady-state and generational evolutionary algorithm for dynamic multiobjective optimization. IEEE Trans. Evol. Comput. **21**(1), 65–82 (2017)
11. Jorgen, B.: Evolutionary optimization in Dynamic Environments. Kluwer, Norwell (2001)
12. Li, H., Zhang, Q.: Multiobjective optimization problems with complicated pareto sets, MOEA/D and NSGA-II. IEEE Trans. Evol. Comput. **13**(2), 284–302 (2009)
13. Li, J., Liu, R., Wang, R.: Handling dynamic multiobjective optimization problems with variable environmental change via classification prediction and dynamic mutation. Inf. Sci. **608**, 970–995 (2022)
14. Liang, Z., Zheng, S., Zhu, Z., Yang, S.: Hybrid of memory and prediction strategies for dynamic multiobjective optimization. Inf. Sci. **485**, 200–218 (2019)
15. Orouskhani, M., Shi, D., Cheng, X.: A fuzzy adaptive dynamic NSGA-II with fuzzy-based Borda ranking method and its application to multimedia data analysis. IEEE Trans. Fuzzy Syst. **29**(1), 118–128 (2021)

16. Rong, M., Gong, D., Pedrycz, W., Wang, L.: A multimodel prediction method for dynamic multiobjective evolutionary optimization. IEEE Trans. Evol. Comput. **24**(2), 290–304 (2020)
17. Sahmoud, S., Topcuoglu, H.R.: A type detection based dynamic multi-objective evolutionary algorithm. In: Sim, K., Kaufmann, P. (eds.) EvoApplications 2018. LNCS, vol. 10784, pp. 879–893. Springer, Cham (2018). https://doi.org/10.1007/978-3-319-77538-8_58
18. Sahmoud, S., Topcuoglu, H.R.: Exploiting characterization of dynamism for enhancing dynamic multi-objective evolutionary algorithms. Appl. Soft Comput. **85**, 105783 (2019)
19. Sahmoud, S., Topcuoglu, H.R.: Hybrid techniques for detecting changes in less detectable dynamic multiobjective optimization problems. In: Proceedings of the Genetic and Evolutionary Computation Conference Companion, pp. 1449–1456 (2019)
20. Stewart, T., et al.: Real-world applications of multiobjective optimization. In: Branke, J., Deb, K., Miettinen, K., Słowiński, R. (eds.) Multiobjective Optimization. LNCS, vol. 5252, pp. 285–327. Springer, Heidelberg (2008). https://doi.org/10.1007/978-3-540-88908-3_11
21. Wang, C., Yen, G.G., Jiang, M.: A grey prediction-based evolutionary algorithm for dynamic multiobjective optimization. Swarm Evol. Comput. **56**, 100695 (2020)
22. Wilcoxon, F.: Individual comparisons by ranking methods. Biometrics Bull. **1**(6), 80–83 (1945)
23. Wu, Y., Jin, Y., Liu, X.: A directed search strategy for evolutionary dynamic multiobjective optimization. Soft. Comput. **19**(11), 3221–3235 (2014). https://doi.org/10.1007/s00500-014-1477-4
24. Yang, S., Yao, X.: Evolutionary Computation for Dynamic Optimization Problems. Springer, Heidelberg (2013). https://doi.org/10.1007/978-3-642-38416-5
25. Zhang, Q., Li, H.: MOEA/D: a multiobjective evolutionary algorithm based on decomposition. IEEE Trans. Evol. Comput. **11**(6), 712–731 (2007)
26. Zhou, A., Jin, Y., Zhang, Q.: A population prediction strategy for evolutionary dynamic multiobjective optimization. IEEE Trans. Cybern. **44** (2013)

Epoch-Based Application of Problem-Aware Operators in a Multiobjective Memetic Algorithm for Portfolio Optimization

Feijoo Colomine Durán[1] , Carlos Cotta[2,3(✉)] ,
and Antonio J. Fernández-Leiva[2,3]

[1] Laboratorio de Computación de Alto Rendimiento (LCAR), Universidad Nacional Experimental del Táchira (UNET), San Cristóbal, Venezuela
`fcolomin@unet.edu.ve`
[2] Dept. Lenguajes y Ciencias de la Computación, ETSI Informática, Universidad de Málaga, Campus de Teatinos, 29071 Málaga, Spain
`afdez@lcc.uma.es`
[3] ITIS Software, Universidad de Málaga, Málaga, Spain
`ccottap@lcc.uma.es`

Abstract. We consider the issue of intensification/diversification balance in the context of a memetic algorithm for the multiobjective optimization of investment portfolios with cardinality constraints. We approach this issue in this work by considering the selective application of knowledge-augmented operators (local search and a memory of elite solutions) based on the search epoch in which the algorithm finds itself, hence alternating between unbiased search (guided uniquely by the built-in search mechanics of the algorithm) and focused search (intensified by the use of the problem-aware operators). These operators exploit Sharpe index (a measure of the relationship between return and risk) as a source of problem knowledge. We have conducted a sensibility analysis to determine in which phases of the search the application of these operators leads to better results. Our findings indicate that the resulting algorithm is quite robust in terms of parameterization from the point of view of this problem-specific indicator. Furthermore, it is shown that not only can other non-memetic counterparts be outperformed, but that there is a range of parameters in which the MA is also competitive when not better in terms of standard multiobjective performance indicators.

Keywords: Memetic algorithms · multiobjective optimization ·
portfolio selection · sharpe index · intensification/diversification

1 Introduction

One of the crucial aspects for the successful application of metaheuristic optimization algorithms endowed with problem-aware search operators is the balance

This work is supported by Spanish Ministry of Economy under project Bio4Res (PID2021-125184NB-I00 – `http://bio4res.lcc.uma.es`) and by Universidad de Málaga, Campus de Excelencia Internacional Andalucía Tech.

J. Correia et al. (Eds.): EvoApplications 2023, LNCS 13989, pp. 210–222, 2023.
https://doi.org/10.1007/978-3-031-30229-9_14

between intensification (the use of this knowledge to focus the search in particular search directions/regions) and diversification (a more exploratory behavior aimed to find solutions not perfectly aligned with the preferences dictated by the knowledge being exploited). This is particularly true in the case of multiobjective optimization scenarios, in which the output is generally desired to broadly cover the set of efficient solutions. We approach this issue in this work in the context of portfolio selection, and more precisely in the optimization of portfolios on the basis of their performance (the returns of the investment) and their risk (the inherent variance of share values for those assets included in the portfolio). There is a whole series of theoretical elements and studies related to the relationship risk-return [11], considering that the potential return or loss of investments is not static, but always depends on the evolution of the market. In this sense, the Markowitz model [13] has become a fundamental theoretical reference for the selection of investment portfolios (see Sect. 2.1), although its mathematical complexity (let alone the presence of different types of constraints) has sometimes caused its application in practice not to be as extensive. Of course, this also makes this area ripe for the application of multiobjective evolutionary algorithms (MOEAs) – see [16].

We approach the issue of intensification/diversification balance in the context of this problem via a memetic approach that features problem-aware operators (local search and a memory of elite solutions – see Sect. 2.2). These operators exploit a problem-specific indicator, namely the Sharpe index [17], a measure of the relationship between return and risk in a given investment portfolio. Our main goal is to (i) establish the usefulness of these operators, and (ii) to study which regime of application results in better performance, either when considered from the problem perspective (that is, the values attained for the problem-specific indicator) or from the point of view of standard indicators of multiobjective performance. To this end, we consider a simple strategy in which intensification happens during a specific time-frame of the execution of the algorithm. We have conducted a sensitivity analysis of the parameters governing the application of this strategy (Sect. 3.2) as well as a comparison with other non-memetic MOEAs (Sect. 3.3). Our findings indicate that despite its simplicity this strategy is flexible enough to allow modulating the intensification of the search so as to not only outperform other algorithms on the problem-specific indicator, but also to obtain competitive if not better results in terms of quality of the Pareto front attained.

2   Materials and Methods

As mentioned in Sect. 1, we have considered the use of memetic algorithms for the optimization of investment portfolios with cardinality restrictions in the context of the Markowitz model. We shall now describe in greater detail how the target problem is defined in this context, and how our algorithmic approach is structured.

2.1   The Optimization Problem

The model developed by Markowitz [13] aims to obtain a portfolio of n assets that achieves the highest possible performance with the lowest risk. For this, average returns R_i for each asset and the different covariances σ_{ij} between the returns of assets (for $i = j$, this is σ_i^2, namely the variance of returns of the i-th asset) considered are calculated. Having performance and risk as the measures of interest, this model lends itself naturally to a multiobjective formulation as follows:

$$\min \sigma^2(\mathbf{w}) = \sum_{i=1}^{n} \sum_{j=1}^{n} w_i w_j \sigma_{ij} \tag{1}$$

$$\max E[R(\mathbf{w})] = \sum_{i=1}^{n} w_i E(R_i) \tag{2}$$

subject to

$$\sum_{i=1}^{n} w_i = 1 \tag{3}$$

where $\mathbf{w} = \langle w_1, \ldots, w_n \rangle$, with $w_i \geqslant 0$ being the proportion of the investor's budget allocated to the i-th financial asset ($\mathbf{w}$ thus comprises the decision variables subject to optimization), $\sigma^2(\mathbf{w})$ is the variance of the portfolio given weights $\mathbf{w}$, and $E[R(\mathbf{w})]$ is the expected return of the portfolio defined by $\mathbf{w}$.

Some other restrictions can be added to this model, such as (i) cardinality, i.e., a maximum of k weights is different from zero,

$$|\{w_i > 0 \mid 1 \leqslant i \leqslant n\}| \leqslant k \tag{4}$$

or allocation limits, i.e., any asset can take up to a maximum percentage ρ of the portfolio

$$\forall i \in \{1, \ldots, n\}: \ w_i \leqslant \rho. \tag{5}$$

We herein consider the case of cardinality constraints. The set of pairs $[\sigma^2(\mathbf{w}), E[R(\mathbf{w})]]$ or risk-return combinations of all efficient portfolios (that is, of those portfolios in which the return cannot be increased without increasing the risk and vice versa, the risk cannot be reduced without having the return reduced as well) is called the efficient (Pareto) front $\mathcal{P}$. Once known, the investor, according to his preferences and the level of risk he is willing to assume, will choose his optimal portfolio [10] from $\mathcal{P}$. while there is an obvious subjective component in the investor preferences, the financial literature also provides an objective performance measure that can be used by experts in order to quantify and compare how well a particular portfolio compares to another one. This measure is the Sharpe index [17], and it is defined as follows:

$$S(\mathbf{w}) = \frac{E[R(\mathbf{w})] - R_0}{\sigma(\mathbf{w})} \tag{6}$$

where R_0 is the assumed risk-free return. Thus, $E[R(\mathbf{w})] - R_0$ is the excess return obtained by taking some risk, and by dividing this quantity by $\sigma(\mathbf{w})$ (which is the standard deviation of the portfolio return, namely the square root of $\sigma^2(\mathbf{w}))$, we obtain a measure of the excess return per unit of risk. Thus, higher values of this index indicate the portfolio performs better in the presence of risk.

Quite interestingly, the Sharpe index also has an interesting geometric interpretation. The equation described in (6) is a straight line that is tangent to the curve determined by the optimal set of portfolios $\mathcal{P}$ in the plane (using risk and return as horizontal and vertical coordinates respectively) as a result of the solution of the Markowitz model. Furthermore, given two portfolios $\mathbf{w_1}$ and $\mathbf{w_2}$, if the latter dominates (in a Pareto sense) the former –i.e., $E[R(\mathbf{w_1})] \leqslant E[R(\mathbf{w_2})]$ and $\sigma^2(\mathbf{w_1}) \geqslant \sigma^2(\mathbf{w_2})$, with at least one of the inequalities being strict– then it must have a higher Sharpe index value (because $\mathbf{w_2}$ would have a larger numerator and a smaller denominator in Eq. (6)). This suggests that the Sharpe ratio can be used within a multi-objective optimizer to move towards the Pareto front.

2.2   A Memetic Approach

As anticipated in Sect. 1, one of the key issues in successfully applying metaheuristics to a given optimization task is to endow the former with problem-aware components [19]. This is precisely one of the central tenets of memetic algorithms (MA) [14]. From a very specific point of view, these techniques arise from the combination of population-based optimization algorithms (often responsible for providing exploration/diversification capabilities) and some form of local search and/or problem-aware operators (which is in turn responsible for exploitation/search intensification in promising regions of the search space). While there are many other possibilities in the framework of memetic algorithms (e.g., see [15]), this basic skeleton suffices to build highly effective optimization methods. More importantly, these methods can be considered as a complementary problem-solving strategy, aimed at benefiting from existing algorithmic ideas, combining them synergistically.

Given the multiobjective nature of the problem tackled, the population-based search component must be adapted to such an optimization scenario. To this end, we have considered the use of the IBEA method [21] as the underlying evolutionary search engine, due to its good performance in this problem domain [5]. Following the problem description in Sect. 2.1, we have picked $[\sigma(\mathbf{w}), E[R(\mathbf{w})]]$ as the two objective functions. Solutions (i.e., the collection of weights corresponding to each of the funds in the portfolio) are represented as real-valued vectors. After being subject to the variation operators –simulated binary crossover [8] (with parameter $\eta = 0$) and polynomial mutation [9] (with parameter $\eta_m = 20$) in this case– the cardinality constraint is enforced by picking the largest k nonzero weights, setting the remaining entries to zero, and normalizing values so as to have adding to 1.0. As for the intensifying components, our algorithm incorporates two such elements:

- *Local search* (LS): solutions are subject to local improvement via the application of a first-ascent hill-climbing method with probability P_{LS}. This procedure works by mutating a single non-zero weight in the portfolio, and accepting the

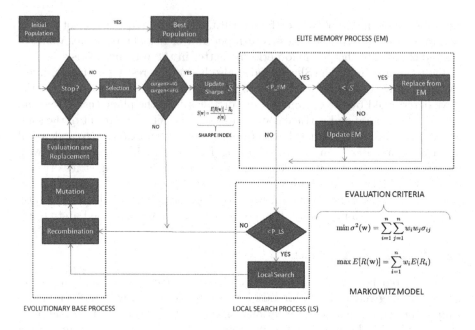

Fig. 1. Flowchart of the memetic approach considered

new solution if it has a better Sharpe index than the original solution (and this Sharpe index is also better than the population average $\bar{S}$). This is done until a certain fixed computational budget allocated to this component is exhausted.

– *Elite memory* (EM): throughout the execution of the algorithm a memory of elite solutions is kept. This is an ordered list of the best individuals (according to the Sharpe index) generated by the algorithm in any given moment. This list is initially empty and has some fixed maximum size Θ. Every time a solution is selected, it is incorporated to the elite memory if there is still available space in it; if the elite memory is at full capacity, the Sharpe index of the solution is checked against that of the worst solution currently in the elite memory, and the latter is substituted if worse. This elite memory is used as an intensification procedure as follows: if a selected solution is below the average Sharpe value of the population $\bar{S}$, a correction procedure is activated with probability P_{EM}, whereby the solution is substituted by a random member of the elite memory.

While the adequate parameterization of these two components is an interesting topic in itself [6], there is another relevant factor to be considered in light of the underlying multiobjective scenario, namely the extent to which the application of these operators could narrow down the search too much towards specific direction of the optimal Pareto front. Indeed, obtaining good convergence and broad coverage of this optimal Pareto front remains a major challenge for many multiobjective metaheuristic optimization methods. To tackle this issue, we have considered the application of these two components within a fixed predetermined time window during the execution of the algorithm. To be precise, we determine the initial generation IG and the final generation FG in which these operators are applied (see Fig. 1). The rationale behind this mechanism is to possibly allow the algorithm to expand its population broadly along the directions towards which the optimal Pareto front is located during an initial phase (which will be termed the *deployment* stage), followed by an intensification stage whose duration is modulated by the particular values chosen for parameters IG and FG), and in which the search prioritizes directions towards better values of the Sharpe index, and a final phase (which we shall term the *securing* stage) in which the algorithm, whose population would be ideally close to at least some regions of the optimal Pareto front, advances in a Sharpe-unaware way in order to extend along the broadest collection of non-dominated solutions. Certainly, this is a procedure that admits further refinements and more elaborated strategies, but it serves as a first baseline to assess the potential use of this approach.

2.3 Data Used in the Analysis

The data considered for the experimental validation comprises the monthly closing prices in the Colombian Stock Market (*Bolsa de Valores de Colombia* – BVC) of twenty capital funds for years 2010 to 2016, obtained from the BVC website[1] [4]. As a general rule, it is convenient to pick a time-frame for the data of at least five years, so as to –in an ideal scenario– capture a full market cycle with the corresponding fluctuation in share prices. On the other hand, a very large time-frame (say, ten years or longer) is not advisable either, since the real value of the fund might not be appropriately represented by such old performances. In this sense, the dataset considered conveniently fits these considerations.

3 Results

Having defined the algorithmic framework and data considered for the experimentation, we proceed to an experimental analysis of the MA in two phases: firstly, we shall conduct a sensitivity analysis to determine adequate time-frames for the intensification stage; then, we perform a comparative analysis of the best parameterizations of the MA with respect to other non-memetic MOEAs. This will be respectively done in Sect. 3.2 and Sect. 3.3 but prior to that, let us describe the computational setting for the experiments.

---

[1] The data is also available in our group's repository, at https://osf.io/wg7mn/.

3.1 Experimental Setup

In addition to the MA, we have considered four non-memetic MOEAs, namely, NSGA-II [7], SPEA2 [20], SHV [1] and HypE [2]. The PISA library (a platform-independent interface and programming language for search algorithms) proposed by Bleuler et al. [3] has been used for the base MOEAs. In all cases, the cardinality constraint is set to $k = 18$, the crossover rate is $P_X = 0.8$, the mutation rate is $P_M = 0.005$, and the population size is 400. With respect to the MA, elite memory has a size $\Theta = 30$. Algorithms were run for a maximum number of 20,000 fitness evaluations –i.e., 50 generations– (18,000 in the case of the MA to account for the overhead of local search), and 30 runs were performed for each of them.

To evaluate the results provided by each method, a double perspective has been considered, namely, the quality of the Pareto front obtained and the quality of the portfolio selected from this front. For the latter aspect, the Sharpe index is considered as mentioned in Sect. 2.1. More precisely, we pick a single portfolio from the Pareto front generated in each run using the Sharpe index as selection criterion. Regarding the evaluation of the Pareto fronts, two well-known performance indicators were used: the hypervolume indicator [22] and the generational distance indicator (GD) [18]. The former provides an indication of the region in the search space that is dominated by the front (thus, the bigger the better). This measure requires a reference point which in this case is defined by the maximum risk and minimum benefit observed in solutions from the best known Pareto front (the combined Pareto front obtained by all algorithms under comparison). As for the second indicator, it estimates the extent to which a certain front is close to another (the true Pareto optimal front if known, or a reference front otherwise). Again, we take the combined Pareto front (that is, the Pareto front obtained by combining the fronts provided by all algorithms under comparison) as the reference set. Being a measure of distance to the reference set, the lower the value of GD, the better. In all cases (Sharpe index, hypervolume, GD), the final outcome of each batch of runs is thus a collection of performance values which will be subsequently subject to statistical analysis. All experimental results are available in our public data repository[2].

3.2 Sensitivity Analysis

In order to analyze the behavior of our MA in response to a different parameterization of the epoch-based intensification, we have conducted experiments in three different scenarios, namely, using either local search or elite memory in isolation ($P_{LS} = 1, P_{EM} = 0$ and $P_{LS} = 0, P_{EM} = 1$ respectively), and using both components simultaneously ($P_{LS} = 1, P_{EM} = 1$). This provides a more detailed view of the effects that these parameters have on these components and any synergistic behavior that may show up.

The results are presented in the Table 1. Focusing firstly on the results for the configuration based on local search only, we can see that the best performing

---

[2] https://osf.io/qs8ae/.

Table 1. Results of the MA for different values of IG and FG. The best (x^*), median $(\tilde{x})$, quartile deviation (QD) and coefficient of quartile deviation (CQD) are shown for each performance indicator. The best configuration for each indicator is marked by $\star$; the remaining configurations are marked with symbols that indicate statistically significant difference at $\alpha = 0.1(\circ)$, $\alpha = 0.05(\bullet)$, and $\alpha = 0.01(\blacksquare)$ with this best value.

$P_{LS} = 1, P_{EM} = 0$

		Sharpe index				hypervolume				GD			
IG	FG	x^*	$\tilde{x}$	QD	CQD	x^*	$\tilde{x}$	QD	CQD	x^*	$\tilde{x}$	QD	CQD
0	10	0.408	0.406	0.001	0.139%	0.838	0.825	0.005	0.659% •	0.669	1.156	0.193	15.938% ■
0	20	0.408	0.406	0.000	0.097% •	0.839	0.830	0.005	0.560%	0.641	0.896	0.084	9.281% ★
0	30	0.408	0.407	0.000	0.118%	0.841	0.832	0.005	0.621% ★	0.644	0.970	0.188	18.709%
0	40	0.408	0.407	0.000	0.102%	0.839	0.831	0.003	0.362%	0.668	0.907	0.160	16.794%
0	50	0.408	0.407	0.000	0.071% ★	0.838	0.831	0.005	0.576%	0.746	1.031	0.131	12.658% ○
10	20	0.407	0.405	0.000	0.116% ■	0.834	0.824	0.004	0.480% •	0.933	1.460	0.201	14.452% ■
10	30	0.407	0.405	0.001	0.145% ■	0.837	0.825	0.002	0.286% ■	0.878	1.315	0.140	10.421% ■
10	40	0.407	0.406	0.001	0.134% ■	0.831	0.825	0.004	0.471% ■	0.870	1.286	0.223	16.750% ■
10	50	0.407	0.406	0.001	0.125% ■	0.837	0.825	0.004	0.446% •	0.829	1.335	0.212	15.758% ■
20	30	0.405	0.404	0.001	0.168% ■	0.832	0.818	0.004	0.445% ■	1.206	1.742	0.176	10.374% ■
20	40	0.406	0.404	0.001	0.153% ■	0.831	0.817	0.003	0.331% ■	1.003	1.682	0.206	12.107% ■
20	50	0.406	0.404	0.001	0.220% ■	0.827	0.815	0.004	0.510% ■	1.097	1.752	0.274	14.768% ■
30	40	0.405	0.403	0.001	0.164% ■	0.822	0.809	0.005	0.623% ■	1.489	2.093	0.299	14.522% ■
30	50	0.405	0.403	0.001	0.149% ■	0.824	0.809	0.006	0.684% ■	1.474	2.069	0.177	8.285% ■
40	50	0.405	0.402	0.001	0.171% ■	0.822	0.804	0.007	0.847% ■	1.378	2.209	0.314	14.724% ■

$P_{LS} = 0, P_{EM} = 1$

		Sharpe index				hypervolume				GD			
IG	FG	x^*	$\tilde{x}$	QD	CQD	x^*	$\tilde{x}$	QD	CQD	x^*	$\tilde{x}$	QD	CQD
0	10	0.405	0.403	0.002	0.384% ■	0.824	0.806	0.011	1.348% ■	1.178	1.987	0.373	18.412% ■
0	20	0.409	0.408	0.000	0.057% ★	0.839	0.818	0.029	3.673% ○	0.594	1.005	0.493	37.962% ★
0	30	0.409	0.408	0.000	0.089% •	0.823	0.799	0.032	4.144% ■	0.866	1.620	0.639	32.413% ■
0	40	0.408	0.407	0.000	0.115% ■	0.802	0.762	0.014	1.786% ■	1.986	2.662	0.278	10.570% ■
0	50	0.408	0.407	0.000	0.115% ■	0.742	0.694	0.013	1.817% ■	4.290	6.235	1.073	16.811% ■
10	20	0.409	0.408	0.000	0.084%	0.841	0.817	0.024	2.950%	0.536	1.273	0.350	30.365%
10	30	0.408	0.407	0.000	0.080% ■	0.829	0.811	0.015	1.888% ■	0.973	1.596	0.287	19.102% •
10	40	0.408	0.407	0.001	0.125% ■	0.799	0.755	0.009	1.232% ■	1.965	2.774	0.243	9.004% ■
10	50	0.408	0.407	0.001	0.147% ■	0.744	0.707	0.015	2.105% ■	3.692	5.745	0.614	10.858% ■
20	30	0.407	0.406	0.000	0.038% ■	0.834	0.826	0.003	0.409% ★	0.769	1.309	0.168	12.737%
20	40	0.407	0.406	0.000	0.107% ■	0.816	0.799	0.004	0.542% ■	1.651	2.287	0.260	11.143% ■
20	50	0.408	0.406	0.001	0.167% ■	0.752	0.730	0.010	1.349% ■	3.665	5.217	0.815	15.920% ■
30	40	0.407	0.406	0.001	0.130% ■	0.819	0.804	0.004	0.492% ■	1.507	2.503	0.301	12.340% ■
30	50	0.407	0.406	0.001	0.152% ■	0.765	0.746	0.012	1.658% ■	3.608	5.114	0.610	11.248% ■
40	50	0.407	0.405	0.001	0.184% ■	0.789	0.774	0.009	1.176% ■	2.421	3.321	0.802	21.041% ■

$P_{LS} = 1, P_{EM} = 1$

		Sharpe index				hypervolume				GD			
IG	FG	x^*	$\tilde{x}$	QD	CQD	x^*	$\tilde{x}$	QD	CQD	x^*	$\tilde{x}$	QD	CQD
0	10	0.409	0.409	0.000	0.037% ■	0.823	0.749	0.009	1.253% ■	0.613	2.548	0.372	15.919% ■
0	20	0.409	0.409	0.000	0.026% ■	0.767	0.746	0.003	0.372% ■	2.063	2.679	0.104	3.922% ■
0	30	0.409	0.409	0.000	0.028% ■	0.754	0.741	0.002	0.240% ■	2.375	3.077	0.110	3.611% ■
0	40	0.410	0.409	0.000	0.020% ★	0.721	0.710	0.003	0.431% ■	4.677	5.609	0.279	5.047% ■
0	50	0.409	0.409	0.000	0.022% ■	0.773	0.743	0.001	0.177% ■	1.797	2.810	0.062	2.198% ■
10	20	0.409	0.409	0.000	0.029% ■	0.820	0.749	0.006	0.807% ■	0.673	2.530	0.162	6.395% ■
10	30	0.409	0.409	0.000	0.023% ■	0.805	0.745	0.004	0.540% ■	1.271	2.759	0.242	8.830% ■
10	40	0.409	0.409	0.000	0.021% ■	0.720	0.708	0.005	0.645% ■	4.583	5.538	0.491	8.651% ■
10	50	0.409	0.409	0.000	0.022% ■	0.722	0.709	0.005	0.714% ■	4.813	5.644	0.612	10.497% ■
20	30	0.409	0.408	0.000	0.063% ■	0.840	0.830	0.004	0.490% ★	0.781	1.245	0.302	23.308% ★
20	40	0.409	0.408	0.000	0.047% ■	0.807	0.788	0.005	0.648% ■	1.883	2.715	0.332	12.190% ■
20	50	0.409	0.409	0.000	0.054% ■	0.776	0.745	0.011	1.417% ■	3.076	5.284	1.165	20.992% ■
30	40	0.408	0.407	0.000	0.081% ■	0.824	0.804	0.004	0.522% ■	1.709	2.711	0.302	11.064% ■
30	50	0.409	0.408	0.000	0.071% ■	0.775	0.757	0.005	0.638% ■	4.307	6.009	1.093	17.833% ■
40	50	0.408	0.407	0.001	0.135% ■	0.788	0.776	0.008	1.047% ■	2.417	3.241	0.657	17.950% ■

parameterization for all performance indicators are those that start with the intensification from the beginning ($IG = 0$). There is a difference in the duration of the intensification depending on the indicator though: for hypervolume and GD (that is, the pure multiobjective performance indicators), it seems that this intensification should not last until the latter stages of the run (there is a gentle degradation in the results of these indicators for larger values of FG, not enough in some cases to be statistically significant[3], although the trend seems to be present there); for the Sharpe index the best values are however obtained when local search is applied to the end of the algorithm. We can understand this behavior in terms of the pressure that local search exerts towards the region of the Pareto front containing high values of the Sharpe index. While deepening into this region can provide improvements in hypervolume and GD as well, it is clear that the latter also benefit from expanding the search toward Sharpe-suboptimal regions during the securing stage.

Let us now observe the results for the configuration based on the use of elite memory only. In this scenario all indicators seem to prefer a short intensification burst, albeit it must be noted that Sharpe index seems to benefit from an early application of the intensification, whereas hypervolume and GD also provide good results (the best ones for this configuration, or statistically undistinguishable form the best ones) when a short deployment phase is performed before the intensification stage. In this case, one has to take into account that the elite memory is a procedure whereby promising solutions generated in previous steps are retrieved and inserted back in the population. This has an obvious impact on the population diversity, which can be less detrimental during the initial stages in which diversity is still large. This is consistent with the degraded results obtained when this intensification only takes place at the end of the run: the population will be much closer to convergence and the benefits of reintroducing good known solutions are outweighed by the loss of diversity.

Finally moving to the joint use of local search and elite memory as intensification operators, the results are consistent with those of the previous configurations and seem to be not just more sharply defined, but also indicate both elements operate synergistically but without highly non-linear interactions. Thus, from the point of view of Sharpe index, the best configuration seems to be using a very low IG (as in the case of both local search and elite memory alone), and a slightly lower value of FG (larger than when elite memory is used on its own, but smaller than when using local search only). As to hypervolume and GD, these two indicators provide better results in the intermediate stages of the run (i.e., a deployment stage until $IG = 20$, allowing the population to spread in all directions, followed by a short intensification until $FG = 30$ that pushes forward the population towards the Sharpe-optimal region, and a final securing stage in which the population expands again from this bridgehead towards all directions of the Pareto front). From the point of view of absolute performance, it is clear that the combined used of both local search and elite memory is conducive to

---

[3] Statistical significance is here determined with the Mann-Whitney U test [12] at the significance level α indicated in each case.

Table 2. Results for the selected MA parameterization and non-memetic MOEAs. The best (x^*), median ($\tilde{x}$), quartile deviation (QD) and coefficient of quartile deviation (QCD) are shown for each performance indicator. The best configuration for each indicator is marked by $\star$; the remaining configurations are marked with symbols that indicate statistically significant difference at $\alpha = 0.1(\circ)$, $\alpha = 0.05(\bullet)$, and $\alpha = 0.01(\blacksquare)$ with this best value.

Algorithm	Sharpe index				hypervolume				GD			
	x^*	$\tilde{x}$	QD	CQD	x^*	$\tilde{x}$	QD	CQD	x^*	$\tilde{x}$	QD	CQD
$MA^{1,0}_{0,20}$	0.408	0.406	0.000	0.097% ■	0.839	0.830	0.005	0.560% ■	0.641	0.896	0.084	9.281% ★
$MA^{1,0}_{0,30}$	0.408	0.407	0.000	0.118% ■	0.841	0.832	0.005	0.621% ★	0.644	0.970	0.188	18.709%
$MA^{1,0}_{0,40}$	0.408	0.407	0.000	0.102% ■	0.839	0.831	0.003	0.362% ■	0.668	0.907	0.160	16.794%
$MA^{1,1}_{0,40}$	0.410	0.409	0.000	0.020% ★	0.721	0.710	0.003	0.431% ■	4.677	5.609	0.279	5.047% ■
$MA^{1,0}_{0,50}$	0.408	0.407	0.000	0.071% ■	0.838	0.831	0.005	0.576% ■	0.746	1.031	0.131	12.658% ○
$MA^{1,1}_{20,30}$	0.409	0.408	0.000	0.063% ■	0.840	0.830	0.004	0.490% ■	0.781	1.245	0.302	23.308% ■
NSGA-II	0.402	0.398	0.001	0.282% ■	0.808	0.782	0.006	0.772% ■	2.256	3.408	0.543	15.377% ■
SPEA2	0.401	0.398	0.002	0.407% ■	0.803	0.779	0.008	1.032% ■	2.190	3.339	0.336	9.706% ■
HypE	0.406	0.405	0.000	0.115% ■	0.830	0.820	0.004	0.503% ■	0.771	1.158	0.178	14.996% ■
SHV	0.392	0.380	0.004	1.115% ■	0.733	0.677	0.022	3.203% ■	4.840	7.566	1.064	14.006% ■

much improved results in terms of Sharpe index, at the cost of losing some coverage of other regions of the Pareto front. It is thus relevant to understand how this tradeoff is substantiated, and certainly whether the results can favorably compare to those of other MOEAs. This is done in next section.

3.3   Performance Comparison

In order to gauge the performance of the MA with non-memetic MOEAs, we have conducted a comparative analysis. We have selected those parameterizations that provided the best results in terms of either of the performance indicators, as well as those for which no statistically significant difference was observed (not even at the $\alpha = 0.1$ level). The notation $MA^{x,y}_{a,b}$ is used to denote the MA with $P_{LS} = x$, $P_{EM} = y$, $IG = a$, and $FG = b$. The results are presented in Table 2.

As it can be seen, most MA parameterizations outperform the non-memetic MOEAs. This is particularly true for the Sharpe index indicator, for which MAs very significantly outperform the remaining algorithms. This is not surprising in light of the Sharpe-based intensification that takes place in the MAs, even if this intensification is done in different ways and with different parameterizations. It is nevertheless remarkable that in most cases, this superiority is also reflected in the standard multiobjective performance indicators. This indicates that progressing towards the Sharpe-optimal region of the Pareto front also provides very important advantages in terms of obtaining good fronts. This is illustrated in Fig. 2, in which we depict the combined front (that is, the front attained by combining the Pareto fronts produced by the 30 runs of each algorithm) for $MA^{1,1}_{0,40}$ (the MA parameterization most oriented to Sharpe-index optimization) and HypE (the best performing MOEA in terms of hypervolume and GD). Notice firstly that in Fig. 2 (left), HypE seems to have a better coverage of extreme regions of the Pareto front (low risk/low performance and high risk/high performance),

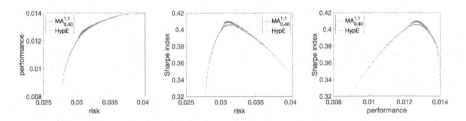

Fig. 2. Depiction of the combined Pareto fronts of $MA_{0,40}^{1,1}$ and HypE. The left plot shows the fronts achieved in the biobjective space. The middle and right plots show the Sharpe index for each solution in the front as a function of risk (middle) and performance (left).

although $MA_{0,40}^{1,1}$ also extends towards the high-risk end of the front. The MA has a clear edge in the Sharpe-optimal region though, into which it achieves a deeper advance. This bulge is more clearly depicted in Fig. 2 (middle)-(right), showing the clear gain attained the MA in this region, which noticeably contributes to improving the multiobjective performance.

While having a better support of the Sharpe-optimal region of the Pareto front is obviously interesting from the point of view of the decision-maker, it is also interesting to note that there some parameterizations (e.g., $MA_{0,(30|40|50)}^{1,0}$) that despite providing slightly worse results in terms of the values of the Sharpe index attained, still show almost no statistically significant difference with the best performing algorithms in terms of hypervolume and GD. It is thus possible to tune the intensification stage so as to achieve a wider coverage of the Pareto front.

4 Conclusions

Finding the right balance between intensification and diversification is of paramount importance in general heuristic optimization, and turns out to be of particular interest in multiobjective endeavors in which the very existence of a target collection of efficient solutions provides substantial richness to the issue. In this sense, and focusing in this latter context, one has to keep in mind that the determination of the near-optimal Pareto front is not the end of the optimization process, but must be regarded in connection to the ulterior decision-making stage, whereby the expert will select an appropriate solution from the front. This is something that becomes very pertinent in portfolio optimization problems: not only do they constitute a usual scenario for the use of multi-objective evolutionary algorithms but they are also endowed with a natural indicator of solution efficiency (Sharpe index in this case). Among other things, this provides a legitimate measure to indicate preferred search directions in the multiobjective space. Indeed, exploiting this indicator can provide a reasonable search heuristic. The question thus arises as to how to integrate this heuristic into the search process, and how to best exploit it. We have approached the first question via an MA that features local search and a memory of elite solutions, always on the basis of the Sharpe index. As to the second issue, we have initially proposed a simple scheme in which intensification happens during a pre-specified time-window of the algorithm execution. This simple approach has nevertheless been able to show that it is possible to modulate the performance of the MA with respect to both Sharpe-based measures as well as to multiobjective performance indicators. In fact, not only is it generally possible to outperform non-memetic MOEAs on the basis of the former problem-aware measure, but the MA can be made competitive with these MOEAs in terms of pure multiobjective performance (hypervolume and GD in our case).

The simplicity of the approach also paves the way for the consideration of more sophisticated strategies in the future. We believe that it would be of particular interest to determine an adaptive strategy, whereby the algorithm can adjust the use of these mechanisms during the run in response to population metrics and the state of the search. Needless to say, it would be convenient to extend these findings to other datasets or even to different domains.

References

1. Bader, J., Deb, K., Zitzler, E.: Faster hypervolume-based search using monte carlo sampling. In: Ehrgott, M., et al. (eds.) Multiple Criteria Decision Making for Sustainable Energy and Transportation Systems. Lecture Notes in Economics and Mathematical Systems, vol. 634, pp. 313–326. Springer, Berlin, Heidelberg (2010)
2. Bader, J., Zitzler, E.: HypE: an algorithm for fast hypervolume-based many-objective optimization. Evol. Comput. 19(1), 45–76 (2011)
3. Bleuler, S., Laumanns, M., Thiele, L., Zitzler, E.: PISA–a platform and programming language independent interface for search algorithms. In: Fonseca, C.M., et al. (eds.) Conference on Evolutionary Multi-Criterion Optimization (EMO 2003). Lecture Notes in Computer Science, vol. 2632, pp. 494–508. Springer, Berlin, Heidelberg (2003)
4. Colombian Stock Market, B.: Variable income market (2008). https://www.bvc.com.co/pps/tibco/portalbvc. Accessed Nov 2017
5. Colomine Duran, F.E., Cotta, C., Fernández-Leiva, A.J.: A comparative study of multi-objective evolutionary algorithms to optimize the selection of investment portfolios with cardinality constraints. In: Di Chio, C., et al. (eds.) EvoApplications 2012. LNCS, vol. 7248, pp. 165–173. Springer, Heidelberg (2012). https://doi.org/10.1007/978-3-642-29178-4_17
6. Colomine Durán, F., Cotta, C., Fernández-Leiva, A.J.: Sensitivity to partial lamarckism in a memetic algorithm for constrained portfolio optimization. In: Mora, A. (ed.) Evostar 2021 Late-Breaking Abstracts, arXiv:2106.11804. pp. 9–12 (2021)
7. Deb, K., Pratap, A., Agarwal, S., Meyarivan, T.: A fast and elitist multiobjective genetic algorithm: NSGA-II. IEEE Trans. Evol. Comput. 6(2), 182–197 (2002)
8. Deb, K., Beyer, H.G.: Self-adaptive genetic algorithms with simulated binary crossover. Evol. Comput. 9(2), 197–221 (2001)
9. Deb, K., Deb, D.: Analysing mutation schemes for real-parameter genetic algorithms. Int. J. Artif. Intell. Soft Comput. 4(1), 1–28 (2014)

10. Jin, M., Li, Z., Yuan, S.: Research and analysis on markowitz model and index model of portfolio selection. In: Proceedings of the 2021 3rd International Conference on Economic Management and Cultural Industry (ICEMCI 2021), pp. 1142–1150. Atlantis Press (2021)

11. Jorion, P.: Value at Risk: The New Benchmark for Managing Financial Risk. In: MacGraw-Hill International Editions: Finance series, McGraw-Hill (2001)

12. Mann, H.B., Whitney, D.R.: On a test of whether one of two random variables is stochastically larger than the other. Ann. Math. Stat. **18**(1), 50–60 (1947)

13. Markowitz, H.M.: Portfolio selection. J. Finan. **7**, 77–91 (1952)

14. Moscato, P., Cotta, C.: An accelerated introduction to memetic algorithms. In: Gendreau, M., Potvin, J.-Y. (eds.) Handbook of Metaheuristics. ISORMS, vol. 272, pp. 275–309. Springer, Cham (2019). https://doi.org/10.1007/978-3-319-91086-4_9

15. Neri, F., Cotta, C.: Memetic algorithms and memetic computing optimization: a literature review. Swarm Evol. Comput. **2**, 1–14 (2012)

16. Ponsich, A., Jaimes, A.L., Coello, C.A.C.: A survey on multiobjective evolutionary algorithms for the solution of the portfolio optimization problem and other finance and economics applications. IEEE Trans. Evol. Comput. **17**(3), 321–344 (2013)

17. Sharpe, W.F.: Mutual fund performance. J. Bus. **39**, 119–138 (1966)

18. Veldhuizen, D.A.V., Lamont, G.B.: Multiobjective evolutionary algorithms: analyzing the state-of-the-art. Evol. Comput. **8**(2), 125–147 (2000)

19. Wolpert, D., Macready, W.: No free lunch theorems for optimization. IEEE Trans. Evol. Comput. **1**(67), 67–82 (1997)

20. Zitzler, E., Laumanns, M., Thiele, L.: SPEA2: improving the strength pareto evolutionary algorithm for multiobjective optimization. In: Giannakoglou, K.C., et al. (eds.) Evolutionary Methods for Design Optimization and Control with Applications to Industrial Problems, pp. 95–100. International Center for Numerical Methods in Engineering (Cmine), Athens, Greece (2001)

21. Zitzler, E., Künzli, S.: Indicator-based selection in multiobjective search. In: Yao, X., et al. (eds.) Indicator-based selection in multiobjective search. LNCS, vol. 3242, pp. 832–842. Springer, Heidelberg (2004). https://doi.org/10.1007/978-3-540-30217-9_84

22. Zitzler, E., Thiele, L.: Multiobjective optimization using evolutionary algorithms—a comparative case study. In: Eiben, A.E., Bäck, T., Schoenauer, M., Schwefel, H.-P. (eds.) PPSN 1998. LNCS, vol. 1498, pp. 292–301. Springer, Heidelberg (1998). https://doi.org/10.1007/BFb0056872

Reducing the Price of Stable Cable Stayed Bridges with CMA-ES

Gabriel Fernandes[ID], Nuno Lourenço[ID], and João Correia[✉][ID]

CISUC, Department of Informatics Engineering, University of Coimbra,
3030 Coimbra, Portugal
gabrielf@student.dei.uc.pt, {naml,jncor}@dei.uc.pt

Abstract. The design of cable-stayed bridges requires the determination of several design variables' values. Civil engineers usually perform this task by hand as an iteration of steps that stops when the engineer is happy with both the cost and maintaining the structural constraints of the solution. The problem's difficulty arises from the fact that changing a variable may affect other variables, meaning that they are not independent, suggesting that we are facing a deceptive landscape.

In this work, we compare two approaches to a baseline solution: a Genetic Algorithm and a CMA-ES algorithm. There are two objectives when designing the bridges: minimizing the cost and maintaining the structural constraints in acceptable values to be considered safe. These are conflicting objectives, meaning that decreasing the cost often results in a bridge that is not structurally safe. The results suggest that CMA-ES is a better option for finding good solutions in the search space, beating the baseline with the same amount of evaluations, while the Genetic Algorithm could not. In concrete, the CMA-ES approach is able to design bridges that are cheaper and structurally safe.

Keywords: Genetic Algorithm · CMA-ES · Optimization · Cable Stayed Bridges

1 Introduction

Bridges are critical components of every transportation network infrastructure. They must be designed to be safe, robust and durable and simultaneously cost-effective and, sometimes, aesthetic pleasing, which are often competing objectives [3,17,18,27]. The restrictions on the structural design standards vary from country to country and specify the requirements that bridges must satisfy, such as safety versus heavy vehicle loads, high-velocity winds and earthquakes. The bridge must also be within serviceability requirements which specify the maximum deflections, stresses and oscillations when subject to dynamic actions, such as pedestrians' movements. Each bridge is planned by a structural design firm or a consortium of several companies. The process usually starts with a tender for the services, in which the choice of the company is based on a set of evaluation criteria where, usually, the best commercial proposal (the lowest price)

J. Correia et al. (Eds.): EvoApplications 2023, LNCS 13989, pp. 223–236, 2023.
https://doi.org/10.1007/978-3-031-30229-9_15

wins. Design firms must deliver a solution using the lowest possible resources (man-hours). For this reason, structural designers do not have the time to evaluate all possible solutions, even with simpler designs. As such, any technique or mechanism to help automate and optimize the design of bridges is, therefore very valuable as even a small percentage of optimization (without compromising the bridge safety and requirements) constitutes large sums of money saved for the public treasury.

Cable-stayed bridges (CSB) are one of the most complex type of bridges to design, due to the fact that they are highly static indeterminate. Thanks to the progress that we have seen in computational technologies, we can now build CSBs that are longer but, at the same time, safer.

In this work, we extend the study conducted in [4,5] which uses an Evolutionary Computation based approach to tackle the problem of designing CSBs. In concrete, the authors propose the use of a standard Genetic Algorithm (GA) [20] using a representation based on real numbers to represent each parameter that one most optimise to design a bridge. The results attained by the proposed approach were encouraging since the GA was able to optimize this type of bridge (with some variables fixed) in terms of the structural constraints. However, it was not able to reduce the costs when compared to a hand design without resorting to some tuning. For this work, we use the Covariance Matrix Adaptation Evolution Strategy algorithm (CMA-ES) [13] to see if it is able to surpass the results attained by the GA with the same amount of evaluations as well as the baseline solution.

In terms of contributions, we enumerate the following: (i) a study of a more complex problem, due to the number of cables being also evolved, instead of being static; (ii) a comparison between two optimization algorithms, GA and CMA-ES; (iii) the results suggest that a standard GA might not be enough to find efficient solutions.

Additionally, the CMA-ES algorithm was able to discover a bridge with a cost that is 4.656k € less than the one of the baseline solution. Taking into account that the solution was discovered automatically, without any human input, the result is impressive and opens for further application of evolutionary approaches in the automatic design of bridges.

The remainder of the paper is divided as follows. Section 2 briefly presents the related work. In Sect. 3, the problem is defined and in Sect. 4, it is explained how the GA and CMA-ES experiments were modelled. The obtained results are shown in Sect. 5, and our conclusions are listed in Sect. 6.

2 Related Work

The first works on the optimum design of CSBs focused on addressing the cable tensioning problem with fixed geometry and structural sections [1,23,26]. More recently, Genetic Algorithms (GA) have also been used to tackle this problem [14,16].

Including dynamic loads creates additional constraints for the design problem. Previous researches have focused on earthquakes [8,25], wind aerodynamics [2,19,22] and pedestrian induced action in cable stayed footbridges [9–11]. These dynamic loads may cause the bridge to vibrate, which is something that is detrimental. To mitigate this problem, there are some options, such as: (i) improve the sturdiness of the bridge by increasing its mass. This is something that is not desirable due to the potential increase in cost and the possibility of the resultant bridge not being as aesthetically pleasing as one might want; (ii) Including control devices like the ones used to retrofit the London Millennium Footbridge, [6,7], for example, viscous dampers or tuned mass dampers (TMDs).

Gradient-based optimization techniques have been used to optimize the bridge's geometry, sizing and cable tensioning [21,24]. GAs have also been used to optimize simultaneously these bridge properties [15], although with simpler models than the ones found in this and in the works on which this article is based [4,5]. These are based on the works of Ferreira and Simões [10,11], from where an already optimized solution (not necessarily a global optimum) was retrieved and then used as a baseline to help us understand if a given solution is in fact good. The literature tells us that gradient-based approaches are able to achieve more rapidly good results than the GA ones. However, compared to a GA, a gradient-based solution is far harder to parameterize, requiring more time and effort, while a GA is easier to get up and running.

As far as we know, there are no works applying CMA-ES to CSB design optimization. The CMA-ES ability to explore the search space via the exploitation of co-variance matrix properties of the genotype holds the potential to further optimize CSBs, beyond existing GA approaches.

3   Problem Definition

In this work, we are evolving configurations for cable-stayed footbridges, minimizing the overall cost of the structure while guaranteeing its structural safety. The structural safety of the bridge is accomplished when the values of the structural constraints are at most 1. For an in-depth explanation of this problem, the reader is advised to read [10].

Each individual is defined by an array composed of 22 variables, whose descriptions and domains can be seen in Table 2. These variables are then utilized to calculate the price of the bridge and the respective values for the structural constraints. Given that for a bridge to be considered secure, all of the constraints need to be less or equal to 1, we resort to using the maximum value of these constraints instead of the individual values. In addition to the variable parameters, the bridges also have fixed ones, which are presented in Table 1.

In our experiments, we compare our results to a baseline solution, a bridge configuration optimized by the same approach used in Ferreira's et al. work [10] with the same fixed parameters. This bridge has a cost of 91.354 k€ and a maximum of structural constraints of 0.9962.

Table 1. Values for the fixed parameters.

Bridge Length (LTotal)	220 m
Bridge Width	4 m
Tower Height below deck	10 m

Table 2. Cable-stayed bridges design variables description and domain values.

Variable Type	Description	Domain Values
Discrete		
DV0	Number of cables	3,4,5,6,7
Geometry		
DV1	Central span (tower to tower distance) of the structure	$[0.9, 1.2]$
DV2	Distance between the first and second cables anchorage in the lateral span of the deck	$[0.7, 1.3]$
DV3	Distance between the tower and the first cable in the central span	$[0.7, 1.3]$
DV4	Distance between the last cable anchorage and the bridge symmetry axis	$[0.7, 1.3]$
DV5	Height of the towers	$[0.1, 2.0]$
DV6	Distance where the cables are distributed in the top of the towers	$[0.1, 4.0]$
DV7	Distance between the top of each tower	$[0.1, 1.3]$
DV8	Distance between each tower at the base	$[0.1, 1.13]$
Control		
DV9	Transversal stiffness of the tower-deck connection	$[0.001, 1000]$
DV10	Vertical stiffness of the tower-deck connection	$[0.001, 1000]$
DV11	Transversal damping of the tower-deck connection	$[0.001, 1000]$
DV12	Vertical damping of the tower-deck connection	$[0.001, 1000]$
Sectional and tensioning		
DV13	Added mass of the concrete slab	$[0.1, 7.0]$
DV14	Deck section	$[0.1, 80.0]$
DV15	Deck section (triangular section)	$[0.5, 1.3]$
DV16	Tower section (rectangular hollow section)	$[0.4, 1.5]$
DV17	Tower section (rectangular hollow section)	$[0.1, 20.0]$
DV18	Tower section (rectangular hollow section)	$[0.3, 20.0]$
DV19	Tower section (rectangular hollow section)	$[0.3, 9.0]$
DV20	Cables pre-stress	$[0.7, 3.0]$
DV21	Cables cross section	$[0.5, 9.0]$

4   The Approach

To have a fair comparison with the previous works, we replicated the experiments conducted with the GA. The parameters used for the algorithm are almost the same as in [5], apart from the number of generations, and can be seen in Table 3.

The novelty that we add to this problem with this work is by experimenting with CMA-ES, implemented with Distributed Evolutionary Algorithms in Python (DEAP) framework [12]. The parameters used can be seen in Table 3. Since the sizes of the population are different, we ensure that the same number of evaluations is used, so that the results can be compared.

The initial individuals used to start the evolutionary process are created by uniformly sampling the domain intervals of each variable. This idea is also utilized in the mutation operator used in the Genetic Algorithm, meaning that when a gene is chosen to be mutated, the new value is also uniformly sampled from the domain of the specific variable. In order to deal with the unfeasible solutions generated by CMA-ES, we correct the specific variable to the minimum value of the domain if it is lower than it or to the maximum if it is greater than the maximum value. This process is not performed in the GA, because the variation operators ensure that the values of the variables are within the required domains, given that the crossover does not alter the values and the mutation operator, as previously stated, samples the new value from the domain. In CMA-ES, this correction is necessary because it is not guaranteed that the generated values are within the domain boundaries.

The fitness function used for both algorithms is presented in Eq. 1. $C(x)$ and $S(x)$ are the cost and the structural constraint value of individual x, respectively. The price returned by $C(x)$ is based on pre-determined pricing of the materials, and $S(x)$ returns the maximum value of the structural constraints of the individual [10]. In practice, we need to guarantee that the returned value of $S(x)$ is at most 1.0.

$$f(x) = \begin{cases} c_r/C(x), & \text{if } C(x) > c_r \\ 1 + 1/S(x), & \text{if } C(x) < c_r \wedge S(x) > 1.0 \\ 2 - (1.0 - S(x)) + c_r/C(x), & \text{if } C(x) < c_r \wedge S(x) \leq 1.0 \end{cases} \quad (1)$$

The fitness function aims to guide the population towards individuals that have a structural constraint of at most 1.0 and the lowest cost possible. First, by reducing the cost to a more acceptable value (c_r, it is fixed in our experiments, see Table 3, but can be changed), then search for individuals that are feasible structurally by rewarding individuals that have $S(x)$ values closer to 1.0, and finally find individuals that are both feasible and cost-effective (we want the lowest cost possible). Although we aim to minimize the cost of the structures, it is important to notice that we want to maximize the fitness value, thus defining this problem as a maximization problem. In the first branch, the fitness ranges from 0 to 1, in the second, from 1 to 2, and in the third, it is greater than 2.

Table 3. Algorithm's Parameters.

Parameter	Value
GA	
Generations	40 000
Population size	10
Tournament size	3
Crossover operator	Uniform Crossover
Crossover rate (per gene)	0.5
Mutation operator	per gene replacement
Mutation rate per gene	0.1
CMA-ES	Value
Generations	8 000
μ	25
λ	50
σ	0.5
Common	Value
Number of Runs	30
Elite size	1
c_r fitness constant	150
Number of evaluations	400 000

5   Experimental Results

The performance of the best individuals during the evolutionary process for the structural constraints $S(x)$ are presented in Fig. 1. Figure 3 presents the results regarding the cost $C(x)$, whilst Fig. 5 presents the results for the fitness function, $f(x)$. Results are averages of 30 independent runs.

Looking at Fig. 1, one can see that both approaches gradually improve the values of the structural constraints, by gradually getting close to the upper limit value of 1. Looking at the right panel, one can see that the GA has a steep descent in the value of the structural constraint, whilst CMA-ES performs a more slow descent. By the end of the optimization process, both approaches have reached approximately the same structural constraint values. To better understand the differences between the approaches we presented a boxplot of the results regarding the structural constraint for the best individuals. Looking at Fig. 2 one can see that CMA-ES is capable of optimizing this objective, but it has some runs where it fails by a relatively bigger margin, leading to a large mean value. The GA also fails (1 seed out of the 30), but by less. Both curves converge close to 1.0, which is the desired behavior, meaning that the structural safety of the bridge is being optimized.

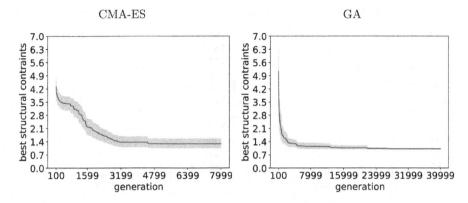

Fig. 1. Mean structural constraint values of the best individuals of 30 runs for CMA-ES (on the left) and for the GA (on the right) starting from generation 100 until generation 7 999 and 39 999 generations, respectively. We can see that CMA-ES stabilizes around the 4 800 generations mark and 28 000 for the GA. The first 100 generations ($[0 : 100[$) were not plotted due to the fact that the values of cost and structural constraints in these generations are extremely large, making the plots unreadable.

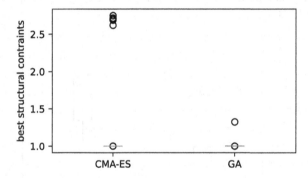

Fig. 2. Boxplot of the structural constraint of the best individual of each run (30 in total) for CMA-ES (on the left) and the GA (on the right).

In what concerns the cost, the results are depicted in Fig. 3. Looking at the results, one can see that during the first 1500 generations, the CMA-ES (left panel) does not seem to improve the values of the initial solutions. In fact, looking at the graph, one can see that it slightly increases cost. However, after generation 1500, the approach starts to rapidly improve the cost of the bridge. The curve stabilizes around generation 6000, which might be an indication that the CMA-ES reaches an optimum. The GA (right panel) exhibits roughly the same behavior in what concerns the optimization trend. These results might be explained by the fact that both approaches, in the first generations, focus on obtained bridges that have a good value in terms of structural constraints. After having such bridges the approaches start to reduce the cost, without compromising the integrity and safety of the bridge.

CMA-ES   GA

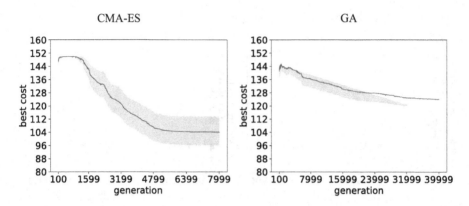

Fig. 3. Mean cost values of the best individuals of 30 runs for CMA-ES (on the left) and for the GA (on the right) starting from generation 100 until generation 7 999 and 39 999 generations, respectively. We can see that CMA-Es is able to achieve lower values of cost, however, it presents a higher variability between runs. It also appears to be stabilizing (we address this topic in the Experimental Results section). It appears that the GA is the opposite, continuing to optimize the cost even after the 40 000 generations. The explanation of why the first 100 generations are not plotted is presented in the caption of Fig. 1.

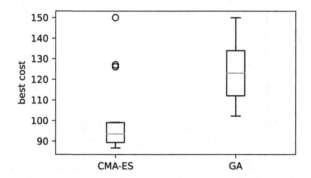

Fig. 4. Boxplot of the cost of the best individual of each run (30 in total) for CMA-ES (on the left) and the GA (on the right).

Another interesting result is that by the end of the evolutionary process, the best solutions obtained by the CMA-ES have a much lower cost than the ones discovered by the GA. To help with this analysis, we created a boxplot of the cost values for both approaches and show them in Fig. 4. Whilst CMA-ES is not as good as the GA at optimizing the structural constraints, it reaches brilliant results in terms of cost. In fact, one can see that the CMA-ES approach can not only find bridges with lower costs but also finds them consistently given the lower variance obtained when compared to the GA.

We also show the fitness plots, Figs. 5 and 6, to show the results for the $f(x)$ that combines both the cost $C(x)$ and the structural constraint $S(x)$.

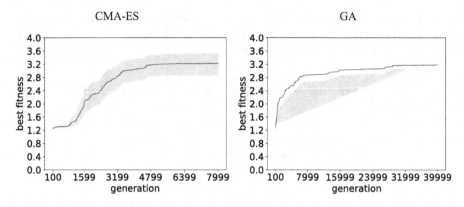

Fig. 5. Mean fitness values of the best individuals of 30 runs for CMA-ES (on the left) and for the GA (on the right) starting from generation 100 until generation 7 999 and 39 999 generations, respectively. It can be seen that CMA-ES is able to reach higher values of fitness, however, there is more variability between runs. The GA seems to be a more consistent algorithm, not showing as much variability. The explanation of why the first 100 generations are not plotted is presented in the caption of Fig. 1.

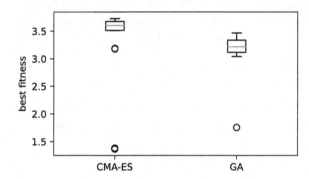

Fig. 6. Boxplot of the fitness of the best individual of each run (30 in total) for CMA-ES (on the left) and the GA (on the right).

Finally, Table 6 summarises the results. Looking at the values, it seems that CMA-ES, on average, does not differ that much from the GA. Even though the curves for the structural constraints are similar and the curves of the cost are very different, it appears that our fitness function is not very good at stretching the fitness values when the structural constraints are already satisfied, leading to similar values (but different) of fitness for candidate solutions that have a relatively different cost. For the optimization itself, it still classifies a better individual with higher fitness, however, when plotted, the difference is not very evident.

To understand if there are meaningful differences between the two approaches, we performed a statistical analysis. Since the samples do not follow a normal distribution, we used the Mann-Whitney non-parametric test with a significance level of $\alpha = 0.05$. The effect sizes are presented in Table 4, and it can be observed that there is a large effect size in all the metrics, meaning that the differences between both approaches are significant.

Table 4. Results of the statistical analysis using the Mann Whitney U test with a significance level $\alpha = 0.05$.

Feature	Effect Size
fitness	−0.510
$C(x)$	0.510
$S(x)$	−0.702

Table 5 presents the cost and the value of the structural constraint of the best solution for every approach. CMA-ES was able to achieve multiple solutions that beat the baseline (see Improvement Rate in Table 4) (with a good level of diversity, because the solutions have different numbers of cables), while the GA could not do it once (see Table 6). With the help of the differences, one can see that both the approaches optimized the structural constraints, however, CMA-ES was able to reduce the cost of the bridge by more than 4 k€, which is a substantial amount.

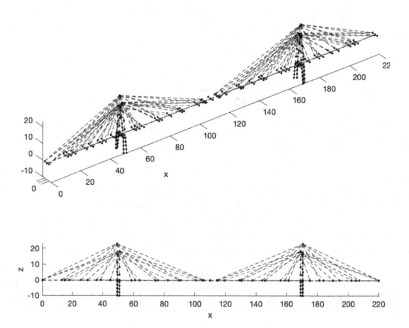

Fig. 7. Baseline bridge (black), versus the bridges optimized by the GA (pink) and by CMA-ES (blue). (Color figure online)

Figure 7 shows how the best bridges evolved by the two algorithms compare to the baseline and to each other. We can see that the CMA-ES is visually distinct from the baseline solution due to its height, but the one evolved by the GA is significantly different from the rest because it uses 3 cables, while the others use 4.

Table 5. Best solution of every approach and the difference between cost and structural constraints against the baseline approach. The best values for both the cost and the structural constraints are in bold.

	Baseline(B)	diff(B, GA)	best GA	diff(B, CMA-ES)	best CMA-ES
$C(x)$	91.354	−10.840	102.194	4.656	**86.698**
$S(x)$	**0.996**	−0.004	1.000	−0.004	1.000

CMA-ES finds really good solutions to this problem but appears to be a little extreme, meaning that when it is able to find a good solution it is really good, but when it is not able to, the result is not satisfactory, being expensive and not even structurally safe.

Table 6 further cements what was previously said, showing that, on average, CMA-ES is worse in optimizing $S(x)$ but is able to greatly reduce the cost of the structures when compared to the GA. However, the GA is more consistent, presenting less variability, seen by the standard deviation values.

Table 6. Stats from the experiments. Mean(...) is the average of the 30 runs and Improvement Rate is the rate of runs that were able to beat the baseline. The average values are presented along with the respective standard deviation.

	Mean(fitness)	Mean($C(x)$)	Mean($S(x)$)	Improvement Rate
CMA-ES	3.225 (± 0.862)	104.005 (± 23.042)	1.282 (± 0.652)	11/30
GA	3.181 (± 0.296)	123.987 (± 13.124)	1.01 (±0.059)	0/30

We decided to include Fig. 8 because we wanted to show how much CMA-ES improved the cost in relation to the one of the baseline. For this image, only the results of the 11 seeds that beat the baseline were included. We do not present a figure for the structural constraints, because, as previously stated, we are only using the results of the seeds that beat the baseline, meaning that the values of the structural constraint are all at most 1.0, given the inverse nature between the cost and the structural constraint.

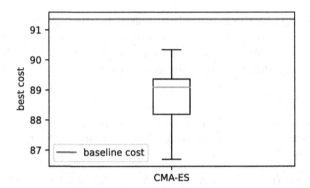

Fig. 8. Boxplot of the cost of the best 11 seeds of CMA-ES plotted with the baseline cost. This better highlights how much the CMA-ES was able to beat the baseline. All of the solutions that beat the baseline have a structural constraints value of at most 1.0, so these values are not plotted here because they would be basically all the same.

6   Conclusions

The design of Cable-stayed bridges (CSB) is one of the most complex designs in bridge engineering since they are highly static indeterminate and cannot be calculated by hand in a short amount of time. This task is mostly handled manually by Civil Engineers, where most of the research on CSB employs gradient-based optimization techniques which require programming the sensitivities of the problem. In this work, we perform a comparison of the performance of two evolutionary approaches, a standard GA and CMA-ES, in terms of cost and structural constraints. We further complement this analysis by comparing the results of both with a previously gradient-based optimized solution found in the literature.

In our results, CMA-ES was able to achieve a cost value of 86.698 k€, beating the baseline and GA costs, 91.354 k€ and 102.194 k€ respectively, while maintaining the structural constraints in acceptable values according to the safety codes. The behaviour of the GA and CMA-ES approach was analyzed in 30 different seeded runs. Under the same budget of evaluations, the GA was not able to beat the baseline solution not even once in terms of cost, despite being more consistent at optimizing the structural constraints when compared with the CMA-ES. The CMA-ES was able to beat the baseline 11 times by a significant margin. Statistical tests were performed with the results of the 30 runs of each algorithm, and the differences between the two were significantly different. The results suggest that CMA-ES performs better, under this setup, for this problem than a standard GA.

In future work, we intend to use quality-diversity algorithms, optimizing both the objective and exploring distinct solutions, to expand our knowledge of the search space as well as being able to retrieve multiple high-performing solutions from a single run and avoid local optimums that may exist.

Acknowledgements. This research was partially funded by the project grant BEIS (Bridge Engineering Information System), supported by Operational Programme for Competitiveness and Internationalisation (COMPETE 2020), under the POR-TUGAL 2020 Partnership Agreement, through the European Regional Development Fund (ERDF) and by the FCT - Foundation for Science and Technology, I.P./MCTES through national funds (PIDDAC), within the scope of CISUC R&D Unit - UIDB/00326/2020 or project code UIDP/00326/2020*.

References

1. Baldomir, A., Hernandez, S., Nieto, F., Jurado, J.: Cable optimization of a long span cable stayed bridge in La Coruña (Spain). Adv. Eng. Softw. **41**(7–8), 931–938 (2010). https://doi.org/10.1016/J.ADVENGSOFT.2010.05.001
2. Baldomir, A., Kusano, I., Hernandez, S., Jurado, J.: A reliability study for the Messina bridge with respect to flutter phenomena considering uncertainties in experimental and numerical data. Comput. Struct. **128**, 91–100 (2013). https://doi.org/10.1016/J.COMPSTRUC.2013.07.004
3. Chen, W.F., Duan, L. (eds.): Bridge Engineering Handbook. CRC Press (2014). https://doi.org/10.1201/b16467
4. Correia, J., Ferreira, F.: Designing cable-stayed bridges with genetic algorithms. In: Castillo, P.A., Jiménez Laredo, J.L., Fernández de Vega, F. (eds.) EvoApplications 2020. LNCS, vol. 12104, pp. 228–243. Springer, Cham (2020). https://doi.org/10.1007/978-3-030-43722-0_15
5. Correia, J., Ferreira, F., Maçãs, C.: Cable-stayed bridge optimization solution space exploration. In: Coello, C.A.C. (ed.) GECCO '20: Genetic and Evolutionary Computation Conference, Companion Volume, Cancún, Mexico, July 8–12, 2020, pp. 261–262. ACM (2020). https://doi.org/10.1145/3377929.3390033
6. Dallard, P.: The London millennium footbridge. Struct. Eng. **79**(22), 17–21 (2001)
7. Dallard, P., et al.: London Millennium bridge: pedestrian-induced lateral vibration. J. Bridge Eng. **6**(6), 412–417 (2001). https://doi.org/10.1061/(ASCE)1084-0702(2001)6:6(412)
8. Ferreira, F., Simoes, L.: Optimum design of a controlled cable stayed bridge subject to earthquakes. Structural Multidisciplinary Optimization **44**(4), 517–528 (2011). https://doi.org/10.1007/s00158-011-0628-9
9. Ferreira, F., Simões, L.: Optimum cost design of controlled cable stayed footbridges. Comput. Struct. **106–107**, 135–143 (2012). https://doi.org/10.1016/J.COMPSTRUC.2012.04.013
10. Ferreira, F., Simões, L.: Optimum design of a cable-stayed steel footbridge with three dimensional modelling and control devices. Eng. Struct. **180**, 510–523 (2019). https://doi.org/10.1016/j.engstruct.2018.11.038, https://linkinghub.elsevier.com/retrieve/pii/S0141029618314275
11. Ferreira, F., Simões, L.: Optimum design of a controlled cable-stayed footbridge subject to a running event using semiactive and passive mass dampers. J. Perform. Construct. Facilit. **33**(3), 04019025 (2019). https://doi.org/10.1061/(ASCE)CF.1943-5509.0001285
12. Fortin, F.A., De Rainville, F.M., Gardner, M.A., Parizeau, M., Gagné, C.: DEAP: evolutionary algorithms made easy. J. Mach. Learn. Res. **13**, 2171–2175 (2012)
13. Hansen, N., Ostermeier, A.: Completely derandomized self-adaptation in evolution strategies. Evol. Comput. **9**(2), 159–195 (2001). https://doi.org/10.1162/106365601750190398

14. Hassan, M.M.: Optimum design of cable-stayed bridges. Western Ontario University, Ph.d (2010)
15. Hassan, M.M., El Damatty, A.A., Nassef, A.O.: Database for the optimum design of semi-fan composite cable-stayed bridges based on genetic algorithms. Structure Infrastruct. Eng. **11**(8), 1054–1068 (2015). https://doi.org/10.1080/15732479.2014.931976
16. Hassan, M.: Optimization of stay cables in cable-stayed bridges using finite element, genetic algorithm, and B-spline combined technique. Eng. Struct. **49**, 643–654 (2013). https://doi.org/10.1016/J.ENGSTRUCT.2012.11.036
17. Hibbeler, R.C., Kiang, T.: Structural Analysis. Pearson Prentice Hall (2015)
18. Alan, H.: The art of structural engineering: the work of Jorg Schlaich and his team. Edition Axel Menges Stuttgart, London (1997)
19. Jurado, J.Á., Nieto, F., Hernández, S., Mosquera, A.: Efficient cable arrangement in cable stayed bridges based on sensitivity analysis of aeroelastic behaviour. Adv. Eng. Software **39**(9), 757–763 (2008). https://doi.org/10.1016/J.ADVENGSOFT.2007.10.004
20. Mitchell, M.: An Introduction to Genetic algorithms. MIT Press, London (1996)
21. Negrão, J., Simões, L.: Optimization of cable-stayed bridges with three-dimensional modelling. Comput. Struct. **64**(1–4), 741–758 (1997). https://doi.org/10.1016/S0045-7949(96)00166-6
22. Nieto, F., Hernández, S., Jurado, J.Á., Mosquera, A.: Analytical approach to sensitivity analysis of flutter speed in bridges considering variable deck mass. Adv. Eng. Software **42**(4), 117–129 (2011). https://doi.org/10.1016/J.ADVENGSOFT.2010.12.003
23. Qin, C.: Optimization of cable-stretching planning in the construction of cable-stayed bridges. Eng. Optim. **19**(1), 1–20 (1992). https://doi.org/10.1080/03052159208941217
24. Simões, L., Negrão, J.: Optimization of cable-stayed bridges with box-girder decks. Adv. Eng. Software **31**(6), 417–423 (2000). https://doi.org/10.1016/S0965-9978(00)00003-X
25. Simões, L.M.C., Negrão, J.H.J.O.: Optimization of cable-stayed bridges subjected to earthquakes with non-linear behaviour. Eng. Optim. **31**(4), 457–478 (1999). https://doi.org/10.1080/03052159908941382
26. Sung, Y.C., Chang, D.W., Teo, E.H.: Optimum post-tensioning cable forces of Mau-Lo Hsi cable-stayed bridge. Eng. Struct. **28**(10), 1407–1417 (2006). https://doi.org/10.1016/J.ENGSTRUCT.2006.01.009
27. Zienkiewicz, O., Taylor, R., Fox, D.: The finite element method for solid and structural mechanics. Elsevier (2014). https://doi.org/10.1016/C2009-0-26332-X

On the Evolution of Boomerang Uniformity in Cryptographic S-boxes

Marko Durasevic[1], Domagoj Jakobovic[1], Luca Mariot[2], Sihem Mesnager[3,4], and Stjepan Picek[5(✉)]

[1] Faculty of Electrical Engineering and Computing, University of Zagreb,
10000 Zagreb, Croatia
{marko.durasevic,domagoj.jakobovic}@fer.hr
[2] Semantics, Cybersecurity and Services Group, University of Twente,
Drienerlolaan 5, 7522 Enschede, NB, The Netherlands
l.mariot@utwente.nl
[3] Department of Mathematics, University of Paris VIII, 93526 Saint-Denis, Paris,
France
smesnager@univ-paris8.fr
[4] LAGA, CNRS, UMR 7539, Sorbonne Paris Cité, University of Paris XIII, 93430
Villetaneuse, France
[5] Digital Security Group, Radboud University, PO Box 9010, Nijmegen,
The Netherlands
stjepan.picek@ru.nl

Abstract. S-boxes are an important primitive that help cryptographic algorithms to be resilient against various attacks. The resilience against specific attacks can be connected with a certain property of an S-box, and the better the property value, the more secure the algorithm. One example of such a property is called boomerang uniformity, which helps to be resilient against boomerang attacks. How to construct S-boxes with good boomerang uniformity is not always clear. There are algebraic techniques that can result in good boomerang uniformity, but the results are still rare.

In this work, we explore the evolution of S-boxes with good values of boomerang uniformity. We consider three different encodings and five S-box sizes. For sizes 4×4 and 5×5, we manage to obtain optimal solutions. For 6×6, we obtain optimal boomerang uniformity for the non-APN function. For larger sizes, the results indicate the problem to be very difficult (even more difficult than evolving differential uniformity, which can be considered a well-researched problem).

Keywords: S-boxes · Permutations · Evolutionary Algorithms · Boomerang Uniformity

1 Introduction

S-boxes (Substitution boxes, (n,m) functions) are mathematical objects commonly used in block ciphers to provide resilience against various attacks [3]. To

© The Author(s), under exclusive license to Springer Nature Switzerland AG 2023
J. Correia et al. (Eds.): EvoApplications 2023, LNCS 13989, pp. 237–252, 2023.
https://doi.org/10.1007/978-3-031-30229-9_16

do so, S-boxes need to be carefully selected so that their cryptographic properties allow resilience against attacks. Today, the most common option is to use S-boxes with $n = m$. Since the search space of all S-boxes is huge (2^{n2^n}), an exhaustive search is not easy already for S-boxes of size 4×4, and for larger sizes, it is impossible. Thus, constructing S-boxes with good cryptographic properties has been an active research domain for more than 30 years on. The common options are to use 1) algebraic constructions, 2) random search, and 3) metaheuristics. Among these, algebraic constructions are the most accepted option as it guarantees the quality of the attainable cryptographic properties.

Nevertheless, not all constructions will achieve optimal values for all relevant cryptographic properties. Luckily, this is not a problem for well-known properties like nonlinearity and differential uniformity, as we know a number of constructions that will give excellent (if not optimal) properties. Still, as community knowledge progresses, new properties are developed, and for some of those, it becomes less clear what algebraic construction to use (if any).

One example of such a property is the boomerang uniformity [2], which has been proposed only recently. This property is not preserved under all notions of equivalence (discussed in Sect. 2). This means that we cannot "count" on every algebraic construction to provide an S-box with the desired boomerang uniformity. Consequently, it makes sense to investigate different ways how to construct S-boxes with good boomerang uniformity. In this paper, we use evolutionary algorithms (EAs) for this goal. Unfortunately, it is far from trivial to construct S-boxes with good boomerang uniformity. We already discussed one reason: the search space size. The other difficulty lies in the computational cost of evaluating the boomerang uniformity, which is $\mathcal{O}(2^{3n})$.

At the same time, it is relevant to investigate how well evolutionary algorithms can perform for this problem. Since we know the best possible values for boomerang uniformity for different S-box sizes, we can easily determine how well EAs perform. Thus, we can consider it a benchmark problem for EAs. Additionally, if we manage to find (many) solutions with optimal (or excellent) boomerang uniformity, some of those solutions could lead to new insights into which algebraic constructions should be used when requiring S-boxes with good boomerang uniformity.

To the best of our knowledge, there are no related works that consider how metaheuristics perform for the problem of constructing S-boxes with good boomerang uniformity. Still, related works (discussed in Sect. 3) indicate that EAs can evolve S-boxes with good cryptographic properties, especially for smaller sizes (e.g., 4×4 should be easy while 8×8 is very difficult).

In this work, we consider three different encodings to evolve S-boxes: integer, permutation, and CA-based encoding. Our main contributions are:

1. We are the first to consider the evolutionary algorithm approach to construct S-boxes with good boomerang uniformity.
2. We employ random search as the baseline technique and observe that for the smallest S-box size (4×4), the problem is easy.

3. For several sizes, we obtain optimal values for boomerang uniformity. More precisely, for 4×4 and 5×5, we reach optimal boomerang uniformity, and for 6×6, optimal boomerang uniformity for the non-APN function.
4. We optimize differential uniformity and boomerang uniformity together with a multi-objective optimization approach, where the results indicate that boomerang uniformity could be an even more difficult property to evolve than differential uniformity.
5. Based on the obtained results, we conclude that the CA-based encoding is the most effective as it gives the best results for smaller sizes, while none of the encodings work particularly well for larger S-box sizes.

2   Background

2.1   Notation

Let n, m be positive integers, i.e., $n, m \in \mathbb{N}^+$, and let $\mathbb{F}_2$ be the Galois field (GF) with two elements. By $\mathbb{F}_2^n$ and $\mathbb{F}_{2^n}$ we denote respectively the n-dimensional vector space over $\mathbb{F}_2$ and the field extension of $\mathbb{F}_2$ with 2^n elements. The addition of elements in $\mathbb{F}_2^n$ and $\mathbb{F}_{2^n}$ are denoted respectively with "+" and "$\oplus$".

2.2   S-boxes

An S-box (substitution box, or (n, m)-function) is a mapping F from n bits into m bits. An (n, m)-function F can be defined as a vector $F = (f_1, \cdots, f_m)$, where the Boolean functions $f_i : \mathbb{F}_2^n \to \mathbb{F}_2$ for $i \in \{1, \cdots, m\}$ are called the coordinate functions of F. As for every n, there exists a field $\mathbb{F}_{2^n}$ of order 2^n, and we can endow the vector space $\mathbb{F}_2^n$ with the structure of that field when convenient.

Equivalence Relations. A function $F : \mathbb{F}_2^n \to \mathbb{F}_2^n$ is linear if:

$$F(x) = \sum_{0 \leq i < n} a_i x^{2^i}, a_i \in \mathbb{F}_{2^n}. \tag{1}$$

A function F is affine if it is a sum of a linear function (Eq. (1)) and a constant term.

Two functions $F : \mathbb{F}_{2^n} \to \mathbb{F}_{2^m}$ and $G : \mathbb{F}_{2^n} \to \mathbb{F}_{2^m}$ are called:

1. Affine equivalent if $G = A \circ F \circ B$, where the mappings A and B are affine permutations on $\mathbb{F}_{2^m}$ and $\mathbb{F}_{2^n}$, respectively.
2. Extended affine equivalent (EA-equivalent) if $G = A \circ F \circ B + C$, where A and B are affine permutations on $\mathbb{F}_{2^m}$ and $\mathbb{F}_{2^n}$, respectively, and where C is an affine function from $\mathbb{F}_{2^n}$ to $\mathbb{F}_{2^m}$.
3. Carlet-Charpin-Zinoviev equivalent (CCZ-equivalent) if for an affine permutation $\mathcal{A}$ of $\mathbb{F}_{2^n} \times \mathbb{F}_{2^m}$, the image of the graph of F is the graph of G, i.e.: $\mathcal{A}(\{(x, F(x)), x \in \mathbb{F}_{2^n}\}) = \{(x, G(x)), x \in \mathbb{F}_{2^n}\}$ [4].

A cryptographic property is called invariant if it is preserved by a certain equivalence notion.

Balancedness. An (n, m)-function F is balanced if it takes every value of $\mathbb{F}_2^m$ the same number 2^{n-m} of times. For an S-box to be balanced, it needs to be a permutation. In the rest of this work, we will concentrate on bijective S-boxes, i.e., those where the input and output dimensions are the same ($n = m$).

Differential Uniformity and APN Functions. Let F be a function from $\mathbb{F}_2^n$ into $\mathbb{F}_2^m$ with $a \in \mathbb{F}_2^n$ and $b \in \mathbb{F}_2^m$, then:

$$D_F(a, b) = \{x \in \mathbb{F}_2^n : F(x) + F(x + a) = b\}. \tag{2}$$

The entry at the position (a, b) corresponds to the cardinality of the delta difference table $D_F(a, b)$ and is denoted as $\delta(a, b)$. The differential uniformity δ is defined as [12]:

$$\delta = \max_{a \neq 0, b} \delta(a, b). \tag{3}$$

The differential uniformity must be even since the solutions of Eq. (2) go in pairs. Indeed, if x is a solution of $F(x) + F(x + a) = b$, then $x + a$ is also a solution. The lower the differential uniformity, the better the S-box contribution to withstand the differential attack [3]. When the number of output bits is the same as the number of input bits, differential uniformity is equal to or greater than 2. Functions with differential uniformity equal to 2 are called Almost Perfect Nonlinear (APN) functions. APN functions exist for both odd and even numbers of variables. When discussing the differential uniformity parameter for permutations, the best possible (and known) value is 2 for any odd n and $n = 6$. For n even and larger than 6, this is an open question. For $n = 4$, the best possible differential uniformity equals 4, and for $n = 8$, the best-known value equals 4. Differential uniformity is invariant for all previously discussed equivalence relations.

Boomerang Uniformity. In 1999, Wagner introduced the boomerang attack, a cryptanalysis technique against block ciphers involving S-boxes [24]. Then, in 2018, Cid et al. introduced the concept of the Boomerang Connectivity Table (BCT) of a permutation F [5]. The same year, Boura and Canteaut introduced a property of cryptographic S-boxes called boomerang uniformity which is defined as the maximum value in the BCT [2].

Let F be a permutation over $\mathbb{F}_{2^n}$, and $a, b \in \mathbb{F}_{2^n}$. Then, the Boomerang Connectivity Table (BCT) of F is given by a $2^n \times 2^n$ table T_f:

$$T_F(a, b) = |\{x \in \mathbb{F}_{2^n} : F^{-1}(F(x) + a) + F^{-1}(F(x + b) + a) = b\}|. \tag{4}$$

The entry at the position (a, b) corresponds to the cardinality of the boomerang uniformity table $T_F(a, b)$ and is denoted as $\beta(a, b)$. The boomerang uniformity β is defined as [2]:

$$\beta = \max_{a, b \neq 0} \beta(a, b). \tag{5}$$

The boomerang uniformity is invariant for affine equivalence but not for extended affine and CCZ-equivalence [2]. It has been proved that $\delta \leq \beta$ for any function F [5]. Additionally, $\delta = 2$ if and only if $\beta = 2$. Moreover, for $n = 4$, the lowest boomerang uniformity that can be achieved is 6. For the mapping $F(x) = x^{2^n-2}$, the boomerang uniformity equals 6 if $n \equiv 0\ mod\ 4$ or 4 if $n \equiv 2\ mod\ 4$. Almost all permutations with optimal boomerang uniformity known today are extended-affine equivalent to the Gold function $(F(x) = x^d$, where $d = 2^i + 1$ and $gcd(i, n) = 1)$ [22]. More results about boomerang uniformity, especially for quadratic permutations, can be found in [11].

3   Related Work

As discussed in Sect. 1, metaheuristics are one of the options for designing S-boxes with good cryptographic properties. In 2004, Clark et al. were the first to use metaheuristics for the design of S-boxes [6]. There, the authors used the principles from the evolutionary design of Boolean functions to evolve S-boxes with sizes up to 8×8 and good nonlinearity. P. Tesar used a genetic algorithm and a total tree search to evolve 8×8 S-boxes with nonlinearity up to 104 [21]. Kazymyrov et al. used gradient descent to evolve S-boxes with good cryptographic properties [9]. The approach used here differs from previous works as the authors started with S-boxes with low differential uniformity, and they conducted several steps until they found an S-box with good nonlinearity. Picek et al. investigated the performance of CGP and GP to evolve 3×3 and 4×4 S-boxes [18]. The authors considered the nonlinearity and differential uniformity properties.

Picek, Rotim, and Cupic developed a cost function capable of reaching high nonlinearity values for several S-box sizes [13]. While optimizing nonlinearity only, the authors also reported differential uniformity. Picek and Jakobovic used genetic programming to evolve constructions resulting in S-boxes with good cryptographic properties [15]. The authors used a single-objective approach and considered differential uniformity. Picek et al. used genetic programming to evolve cellular automata rules that can be used to generate S-boxes [16]. The obtained results outperformed other metaheuristic techniques for sizes 5×5 up to 7×7. These results still represent state-of-the-art results obtained with metaheuristics.

These works show that metaheuristics can be used to construct small S-boxes with good cryptographic properties. Still, as the properties considered are CCZ-invariant, a simpler approach would be to use an algebraic construction (known to give optimal cryptographic properties) and then transform the S-box into an equivalent one (if required).

Regarding the cryptographic properties that are not invariant, some works consider the S-box side-channel resilience or their implementation properties. For instance, Ege et al. considered confusion coefficient [8] and Picek et al. transparency order [14]. Both properties are relevant to make the S-box more resilient against side-channel attacks. Picek et al. used evolutionary algorithms to evolve S-boxes that are power- or area-efficient [19]. Similarly, Picek et al. used genetic programming to find cellular automata rules that result in S-boxes with good implementation properties like latency, area, and power [17].

4   Experimental Setup

4.1   Encodings

S-boxes can be represented in several ways that differ substantially from one another. However, the previous related work (as discussed in Sect. 3) shows that no single representation is dominant; often, different representations offer the best results depending on both the size of the S-box and the optimized criteria. In this work, we employ three encodings to represent S-boxes: integer, permutation, and cellular automata-based encoding.

Integer Encoding. The simplest encoding represents an $n \times n$ S-box as a vector of integer values of size 2^n; each element of the vector encodes the S-box output for a corresponding S-box input. Since an $n \times n$ S-box has n output Boolean values, every element in the vector assumes values in $[0, \ldots, 2^n - 1]$.

For integer encoding, we define the following operators; there is a single mutation operator that selects a random gene and modifies its value in the defined range $[0, \ldots, 2^n - 1]$ with uniform distribution. Crossover is performed using one of three operators: single-point and two-point crossover function by combining a child genotype from two parents (individuals) using either a single or two break points when copying genes to the child. Finally, average crossover creates the child individual where each gene is a (rounded) mean value between corresponding parent genes.

Note that this representation and operators do not preserve balancedness since every possible output is obtainable for every input; in this case, we first enforce the balancedness property with a penalty term in the fitness function (next subsection) and then optimize for boomerang uniformity.

Permutation Encoding. Probably the most natural way to represent a *balanced* S-box is the permutation encoding. In this case, the individual is encoded with a permutation of size 2^n with elements in the range $[0, \ldots, 2^n - 1]$. This representation preserves the balancedness property.

Genetic operators used for this encoding were as follows; for mutation, three operators are used: insert mutation, inversion mutation, and swap mutation [23]. As for the crossover operators, we use partially mapped crossover (PMX), position-based crossover (PBX), order crossover (OX), uniform-like crossover (ULX), and cyclic crossover [23].

Cellular Automata with Genetic Programming. The third representation uses the fact that an S-box could be represented as a cellular automaton (CA), with transitions from the input bits as the current state to the output bits as the following state. The transitions of the cellular automaton can be defined by using a *local update rule*, which is simply a Boolean function of at most n bits with a single output bit. The CA local rule defines the next state of a given bit $c_i(t+1)$, based on the current state of the same bit and its adjacent bits: $c_i(t)$,

$c_{i+1}(t)$, $c_{i+2}(t)$, etc. The same approach is used in the design of some existing S-boxes, e.g., in Keccak [1].

In this case, the S-box is represented with a Boolean function that embodies a CA local rule. To evolve a suitable local rule, we employ genetic programming and encode the Boolean function as a tree. The n input Boolean variables of the S-box are used as GP terminals. The GP uses the function set consisting of several Boolean primitives: NOT (inverting its argument), XOR, AND, OR, NAND, and XNOR, each of which takes two input arguments. Finally, we use the function IF, which takes three arguments and returns the second one if the first one evaluates to *true*, and the third one otherwise. An individual obtained with GP is evaluated in the following manner: all the possible 2^n input states are considered, and for each state, the same rule (the evolved Boolean function) is applied in parallel to each of the bits to determine the next state (S-box output). As in the integer encoding, this representation also does not preserve balancedness, which has to be enforced with a suitable fitness function.

The genetic operators are simple tree crossover, uniform crossover, size fair, one-point, and context preserving crossover [20] (selected randomly), and subtree mutation. Previous experiments using GP suggest that having a maximum tree depth equal to the size of the S-box is sufficient (i.e., maximum tree depth equals n, which is also the number of terminals).

4.2   Fitness Functions

Since in all experiments we optimize boomerang uniformity, the first fitness function minimizes the property value:

$$fitness_1 = \beta. \tag{6}$$

The above fitness function is used with permutation encoding; for the other two representations, we add a penalty term that measures the distance from a balanced S-box, which is expressed simply as the number of missing output values, denoted as BAL. Only if this value equals zero the boomerang uniformity property is calculated and used in the minimization of the following fitness function:

$$fitness_2 = \begin{cases} 2^n + BAL & \text{if } BAL > 0 \\ \beta, & \text{otherwise.} \end{cases} \tag{7}$$

4.3   Algorithms and Parameters

In the single-objective optimization, we optimize the boomerang property of the S-box only. The search algorithm, used with all encodings, is a steady-state evolutionary algorithm with tournament selection. In each iteration, three individuals from the population are selected at random, and the worst of the three selected individuals is eliminated. A new individual is formed using crossover on the remaining two from the tournament, and the new individual is mutated with a given individual mutation probability. When either crossover or mutation is applied, only one of the available operators for that encoding is used, selected at random. Furthermore, we use a random search algorithm as a baseline.

Besides the single-objective, we also employ the multi-objective optimization in which both the boomerang and delta uniformity are optimized; both properties are minimized. In this part of the experiment, we use the well-known NSGA-II multi-objective evolutionary algorithm [7].

All encodings and algorithms are applied with the same population size of 500 individuals. The individual mutation probability is 0.7. All algorithms use the same stopping criterion, which is 500 000 function evaluations. These values were selected based on preliminary tuning results.

5 Experimental Results

This section outlines the experimental results, first by examining the performance of each encoding independently and then by comparing the best solutions of all encodings. Finally, we also investigate the possibility of simultaneously optimizing the boomerang uniformity and differential uniformity using a multi-objective evolutionary algorithm. Our experiments consider sizes from 4×4 to 8×8, as those sizes have the most practical relevance (being used in most of the modern block ciphers that are substitution-permutation networks, where bijectivity of S-boxes is mandatory).

5.1 Integer Encoding

Table 1 outlines the results obtained using the integer encoding. The results show that using a random search, it was impossible to obtain solutions with a boomerang uniformity for any S-box size except the smallest one since the obtained solutions were not balanced. The results are denoted with '-' in the table for all such cases. On the other hand, when using the evolutionary algorithm, the results show that quite stable results are obtained, usually with only a small dispersion among the individual executions. Even as the sizes of the S-boxes increase, the distribution of the solutions is still compact.

Table 1. Results obtained for the integer encoding

S-box size	EA			RS		
	Min.	Avg.	Std.	Min.	Avg.	Std.
4×4	6	7.2	1.35	8	14.8	3.35
5×5	10	10.9	1.01	-	-	-
6×6	12	13.9	1.11	-	-	-
7×7	16	17.2	1.35	-	-	-
8×8	18	20	2.29	-	-	-

5.2 Permutation Encoding

Table 2 outlines the results obtained for the permutation encoding. It is interesting to observe that for this encoding, both random search and the evolutionary algorithm perform quite similarly across all S-box sizes. For example, for smaller S-box values, both achieve the same performance since, in every run, they obtained solutions of equal quality. On the other hand, for larger sizes, the evolutionary algorithm obtains better solutions than a random search in a few runs and usually achieves better minimum values. Nevertheless, the average values of both methods are similar, especially for the largest S-box size, thus indicating a similar performance between both. For the two largest S-box sizes, we see that random search always gets stuck in solutions with the same fitness, whereas the evolutionary algorithm is sometimes able to obtain a solution of slightly better quality.

Table 2. Results obtained for the permutation encoding

S-box size	EA			RS		
	Min.	Avg.	Std.	Min.	Avg.	Std.
4×4	6	6	0	6	6	0
5×5	8	8	0	8	8	0
6×6	10	10.5	0.86	10	11.93	0.37
7×7	12	13.6	0.81	14	14	0
8×8	14	15.9	0.37	16	16	0

5.3 CA-Based Encoding

Table 3 shows the results obtained by the CA-based representation. For this encoding, we again see that for the smaller S-box sizes, both random search and the evolutionary algorithm perform equally well, with random search even achieving a better result on average for S-box of sizes 5×5, since random search always found the best-obtained value. However, as the size of the S-boxes increases, the performance of random search significantly deteriorates, and it constantly achieves worse results than the evolutionary algorithm. Regardless, the results obtained by the evolutionary algorithm can also be seen to deteriorate quite swiftly, which does suggest that this encoding might not be suitable for larger S-box sizes. This is aligned with the results in related works that consider, for instance, differential uniformity [10].

5.4 Representation Comparison

Table 4 outlines the best result achieved by each of the encodings for the tested S-box sizes. The best value obtained for each size is denoted in bold. The results

Table 3. Results obtained for the CA-based encoding

S-box size	EA			RS		
	Min.	Avg.	Std.	Min.	Avg.	Std.
4×4	6	6	0	6	6	0
5×5	2	2.93	2.15	2	2	0
6×6	4	15.5	3.39	12	23.2	4.72
7×7	16	26.1	6.06	28	50.2	7.97
8×8	42	73.9	16.5	112	132	16.76

show that for the smallest S-box size, all three encodings are equally successful since they all obtain a value of 6 for boomerang uniformity. For S-box sizes of 5×5 and 6×6, the CA-based encoding clearly achieves the best results, whereas the remaining two encodings achieve similar results, with the permutation encoding performing slightly better. However, for the two largest S-box sizes that were tested, we observe that the performance of the CA-based encoding is not as good, especially for the S-box size of 8×8, for which it achieves poor results. For these two sizes, the best results are obtained by permutation encoding, followed by integer encoding.

Table 4. Best results obtained by tested encodings for various S-box sizes

S-box size	Int	Perm	CA
4×4	**6**	**6**	**6**
5×5	10	8	**2**
6×6	12	10	**4**
7×7	16	**12**	16
8×8	18	**14**	42

Figure 1 shows the violin plots of the results obtained with the evolutionary algorithm for each encoding and S-box size. The distributions of the violin plots were cut off to outline better the minimum and maximum ranges that were obtained using each encoding. We can see that using the permutation encoding results in the lowest distribution of the results across all S-box sizes. On the other hand, the evolutionary algorithm seems to be the least stable when using the CA-based representation since even in cases when it obtained the best overall result, the solutions from the 30 runs were quite dissipated. As the size of the S-boxes increases, the dissipation of the CA-based results increases even further. Finally, the integer encoding also obtains quite dispersed results, more dispersed than the permutation encoding, but the dispersion does not seem to increase as significantly with the increase of S-box sizes.

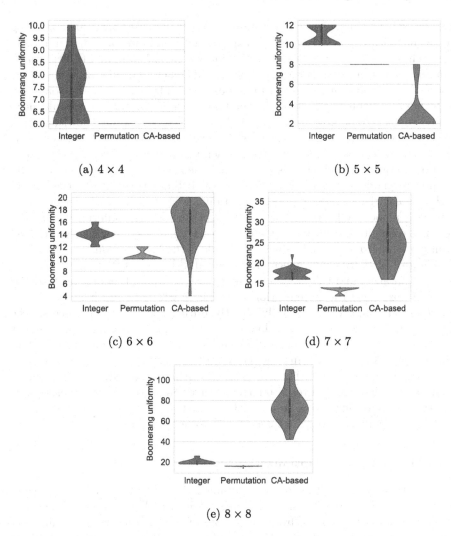

Fig. 1. Violin plots outlining the distributions of the results obtained by each encoding.

Based on the summary of results in this section, we see that no single encoding leads to the best results for all S-box sizes. For smaller S-box sizes, the CA-based encoding is most appropriate, whereas for larger S-box sizes, the permutation encoding achieves the best results. Since the integer encoding always achieves results worse than those of the permutation encoding, there is little benefit in using this encoding, especially as it does not provide any guarantee that it will obtain balanced S-boxes.

In comparison with theoretical results (algebraic constructions), we can also provide some observations. For 4×4, all encodings with EA reach a boomerang uniformity of 6, which is the optimal value. For 5×5, we reach optimal boomerang uniformity with CA-based encoding, using both EA and random search. For 6×6, we reach optimal boomerang uniformity for a non-APN function (differentially 4-uniform function). Further analysis shows this function is quadratic (algebraic degree 2). It is known that such functions have optimal values of boomerang uniformity. Considering that for better boomerang uniformity, we would need to find smaller differential uniformity (which, to the best of our knowledge, was never achieved with metaheuristics), we believe our result is excellent. For 7×7, we cannot reach the optimal value for boomerang uniformity. The best value is 12 with permutation encoding. It is known that the maximal value for differentially 4-uniform quadratic permutation is at most 4. Since there are known results with metaheuristics reaching APN functions [10], we can consider our result here rather poor, indicating that evolving for boomerang uniformity may well be more difficult than evolving for differential uniformity. For 8×8, the best results are far from optimal values achievable with algebraic constructions.

Figure 2 provides convergence graphs for all S-box sizes. Note how for 4×4 size, permutation and CA-based encoding reach optimal fitness rather fast. For larger S-box sizes, the permutation encoding starts with much better initial solutions but improves them only slightly during the evolution process. The integer encoding demonstrates a slower convergence, and after a certain number of evaluations, it can be observed that the solutions are not improved further. Finally, the CA-based encoding has the slowest convergence for large S-box sizes. However, the figures show that fitness is still being improved, meaning that it would be possible to achieve better results given more time. Nevertheless, it is questionable how long and, if at all, the CA-based encoding could reach solutions of the same quality as those obtained by the other two encodings.

5.5   Multi-objective Optimization

This section outlines the results obtained by simultaneously optimizing boomerang and differential uniformity using NSGA-II. Figure 3 shows the union of the Pareto fronts obtained from the 30 runs of NSGA-II for each encoding and S-box size. Since the figure denotes the union of all Pareto fronts, the outlined points do not necessarily form a single Pareto front. This was done to illustrate better the distribution of the obtained solutions for each encoding.

Even when considering multi-objective optimization, we see that the performance of the individual encodings follows a similar pattern as the one observed for the single-objective case. For the S-boxes of size 4×4, all encodings obtained the best possible result. The permutation and CA-based encodings are more stable, as each run obtains the best solution, whereas the integer encoding sometimes obtains much worse solutions, which can be seen from the larger dissipation of solutions in the objective space. For sizes 5×5 and 6×6, the overall best result is obtained by the CA-based encoding. For these sizes, the permutation encoding

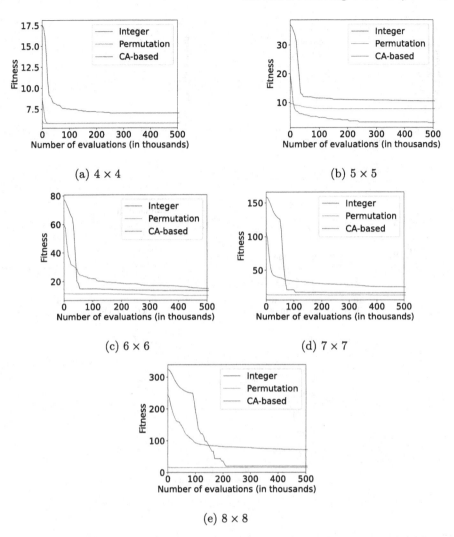

Fig. 2. Convergence patterns obtained for the tested encodings and S-box sizes.

clearly demonstrates to be superior to the integer encoding as it obtains better solutions for both criteria. For the S-box size 6 × 6, we also observe that the performance of the CA-based encoding started to deteriorate significantly, as in many cases, it obtained quite poor solutions. Thus, one could say that the best solution it obtained was more due to luck. Finally, for the remaining two sizes, we clearly see that the permutation-based encoding achieves the best result, which dominates the results obtained by the other two.

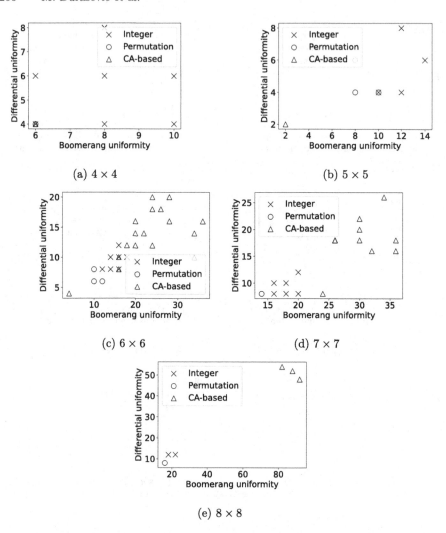

Fig. 3. Pareto fronts obtained for simultaneous optimization of boomerang and differential uniformity.

6   Conclusions and Future Work

This work is the first paper that considers the evolutionary algorithms' perspective in designing S-boxes with good values of boomerang uniformity. We run experiments for the relevant S-box sizes and three different solution encodings, and we obtain rather interesting results. For sizes 4 × 4 up to 5 × 5, we obtain optimal values for the boomerang uniformity. For 6×6, we obtain optimal boomerang uniformity for non-APN functions. Sizes larger than 6 × 6 exhibit rather poor results, and Pareto fronts indicate it is even somewhat easier to obtain good differential uniformity than boomerang uniformity.

In future work, it would be interesting to consider optimizing for nonlinearity and boomerang uniformity, as the link between those two properties is even less understood [11]. Moreover, it would be interesting to see how we can add additional information to the evolutionary search to improve the performance. For instance, it is known that differentially 4-uniform quadratic permutations have boomerang uniformity at most 12. Adding that to the evolutionary search could, on one side, make the search more efficient, but it would also make the fitness function more complex. Finally, since no single encoding is a clear winner, a promising search algorithm could use a distributed population with different encodings (e.g. permutation and CA-based) in different sub-populations to promote diversity.

References

1. Bertoni, G., Daemen, J., Peeters, M., Assche, G.V.: The Keccak reference, January 2011. http://keccak.noekeon.org/
2. Boura, C., Canteaut, A.: On the boomerang uniformity of cryptographic sboxes. IACR Transactions on Symmetric Cryptology 2018(3), 290–310 (2018). https://doi.org/10.13154/tosc.v2018.i3.290-310, https://tosc.iacr.org/index.php/ToSC/article/view/7304
3. Carlet, C.: Boolean Functions for Cryptography and Coding Theory. Cambridge University Press (2021). https://doi.org/10.1017/9781108606806
4. Carlet, C., Charpin, P., Zinoviev, V.: Codes, bent functions and permutations suitable for des-like cryptosystems. Des. Codes Cryptogr. **15**(2), 125–156 (1998). https://doi.org/10.1023/A:1008344232130
5. Cid, C., Huang, T., Peyrin, T., Sasaki, Y., Song, L.: Boomerang connectivity table: a new cryptanalysis tool. In: Nielsen, J.B., Rijmen, V. (eds.) EUROCRYPT 2018. LNCS, vol. 10821, pp. 683–714. Springer, Cham (2018). https://doi.org/10.1007/978-3-319-78375-8_22
6. Clark, J.A., Jacob, J.L., Stepney, S.: The design of s-boxes by simulated annealing. In: Proceedings of the 2004 Congress on Evolutionary Computation (IEEE Cat. No. 04TH8753), vol. 2, pp. 1533–1537 (2004)
7. Deb, K., Agrawal, S., Pratap, A., Meyarivan, T.: A fast and elitist multiobjective genetic algorithm: NSGA-II. IEEE Trans. Evol. Comput. **6**(2), 182–197 (2002)
8. Ege, B., Papagiannopoulos, K., Batina, L., Picek, S.: Improving DPA resistance of s-boxes: how far can we go? In: 2015 IEEE International Symposium on Circuits and Systems (ISCAS), pp. 2013–2016 (2015)
9. Kazymyrov, O., Kazymyrova, V., Oliynykov, R.: A method for generation of high-nonlinear s-boxes based on gradient descent. Cryptology ePrint Archive, Report 2013/578 (2013). https://eprint.iacr.org/2013/578
10. Mariot, L., Picek, S., Leporati, A., Jakobovic, D.: Cellular automata based S-boxes. Cryptogr. Commun. **11**(1), 41–62 (2018). https://doi.org/10.1007/s12095-018-0311-8
11. Mesnager, S., Tang, C., Xiong, M.: On the boomerang uniformity of quadratic permutations. Designs, Codes and Cryptography **88**(10), 2233–2246 (2020). https://doi.org/10.1007/s10623-020-00775-2
12. Nyberg, K.: Perfect nonlinear S-boxes. In: Davies, D.W. (ed.) EUROCRYPT 1991. LNCS, vol. 547, pp. 378–386. Springer, Heidelberg (1991). https://doi.org/10.1007/3-540-46416-6_32

13. Picek, S., Cupic, M., Rotim, L.: A new cost function for evolution of S-boxes. Evol. Comput. **24**(4), 695–718 (2016)
14. Picek, S., Ege, B., Batina, L., Jakobovic, D., Chmielewski, U., Golub, M.: On using genetic algorithms for intrinsic side-channel resistance: The case of AES s-box. In: Proceedings of the First Workshop on Cryptography and Security in Computing Systems. CS2 2014, pp. 13–18. Association for Computing Machinery, New York, NY, USA (2014). https://doi.org/10.1145/2556315.2556319
15. Picek, S., Jakobovic, D.: On the design of s-box constructions with genetic programming. In: Proceedings of the Genetic and Evolutionary Computation Conference Companion. GECCO 2019, New York, NY, USA, pp. 395–396. Association for Computing Machinery (2019). https://doi.org/10.1145/3319619.3322040
16. Picek, S., Mariot, L., Leporati, A., Jakobovic, D.: Evolving s-boxes based on cellular automata with genetic programming. In: Proceedings of the Genetic and Evolutionary Computation Conference Companion. GECCO 2017, New York, NY, USA, pp. 251–252. Association for Computing Machinery (2017). https://doi.org/10.1145/3067695.3076084
17. Picek, S., Mariot, L., Yang, B., Jakobovic, D., Mentens, N.: Design of s-boxes defined with cellular automata rules. In: Proceedings of the Computing Frontiers Conference. CF 2017, New York, NY, USA, pp. 409–414. Association for Computing Machinery (2017). https://doi.org/10.1145/3075564.3079069
18. Picek, S., Miller, J.F., Jakobovic, D., Batina, L.: Cartesian genetic programming approach for generating substitution boxes of different sizes. In: Proceedings of the Companion Publication of the 2015 Annual Conference on Genetic and Evolutionary Computation. GECCO Companion 2015, New York, NY, USA, pp. 1457–1458. Association for Computing Machinery (2015). https://doi.org/10.1145/2739482.2764698
19. Picek, S., Yang, B., Rozic, V., Mentens, N.: On the construction of hardware-friendly 4×4 and 5×5 s-boxes. In: Avanzi, R., Heys, H. (eds.) SAC 2016. LNCS, vol. 10532, pp. 161–179. Springer, Cham (2017). https://doi.org/10.1007/978-3-319-69453-5_9
20. Poli, R., Langdon, W.B., McPhee, N.F.: A field guide to genetic programming. Published via http://lulu.com and freely available at http://www.gp-field-guide.org.uk (2008). (With contributions by J. R. Koza)
21. Tesař, P.: A new method for generating high non-linearity S-boxes. Radioengineering **19**(1), 23–26 (2010)
22. Tian, S., Boura, C., Perrin, L.: Boomerang uniformity of popular S-box constructions. Des. Codes Crypt. **88**(9), 1959–1989 (2020). https://doi.org/10.1007/s10623-020-00785-0
23. Vlašić, I., Durasević, M., Jakobović, D.: Improving genetic algorithm performance by population initialisation with dispatching rules. Comput. Ind. Eng. **137**, 106030 (2019). https://doi.org/10.1016/j.cie.2019.106030, https://www.sciencedirect.com/science/article/pii/S0360835219304899
24. Wagner, D.: The boomerang attack. In: Knudsen, L. (ed.) FSE 1999. LNCS, vol. 1636, pp. 156–170. Springer, Heidelberg (1999). https://doi.org/10.1007/3-540-48519-8_12

Using Knowledge Graphs for Performance Prediction of Modular Optimization Algorithms

Ana Kostovska[1,2]([✉]) [ID], Diederick Vermetten[4] [ID], Sašo Džeroski[1,2] [ID],
Panče Panov[1,2] [ID], Tome Eftimov[1] [ID], and Carola Doerr[3] [ID]

[1] Jožef Stefan Institute, Ljubljana, Slovenia
{ana.kostovska,saso.dzeroski,pance.panov}@ijs.si
[2] Jožef Stefan International Postgraduate School, Ljubljana, Slovenia
[3] Sorbonne Université, CNRS, LIP6, Paris, France
Carola.Doerr@lip6.fr
[4] LIACS, Leiden University, Leiden, The Netherlands
d.l.vermetten@liacs.leidenuniv.nl

Abstract. Empirical data plays an important role in evolutionary computation research. To make better use of the available data, ontologies have been proposed in the literature to organize their storage in a structured way. However, the full potential of these formal methods to capture our domain knowledge has yet to be demonstrated. In this work, we evaluate a performance prediction model built on top of the extension of the recently proposed OPTION ontology. More specifically, we first extend the OPTION ontology with the vocabulary needed to represent modular black-box optimization algorithms. Then, we use the extended OPTION ontology, to create knowledge graphs with fixed-budget performance data for two modular algorithm frameworks, modCMA, and modDE, for the 24 noiseless BBOB benchmark functions. We build the performance prediction model using a knowledge graph embedding-based methodology. Using a number of different evaluation scenarios, we show that a triple classification approach, a fairly standard predictive modeling task in the context of knowledge graphs, can correctly predict whether a given algorithm instance will be able to achieve a certain target precision for a given problem instance. This approach requires feature representation of algorithms and problems. While the latter is already well developed, we hope that our work will motivate the community to collaborate on appropriate algorithm representations.

Keywords: Algorithm Performance Prediction · KG Completion · Evolutionary Computation · Black-box Optimization

1 Introduction

Reproducibility is slowly becoming the norm in many areas of computer science [11]. In the domain of black-box optimization, this means that many researchers are making available not just the code, but also large amounts

© The Author(s), under exclusive license to Springer Nature Switzerland AG 2023
J. Correia et al. (Eds.): EvoApplications 2023, LNCS 13989, pp. 253–268, 2023.
https://doi.org/10.1007/978-3-031-30229-9_17

of benchmark data. While this increasing availability of data is beneficial to the entire community, tools to structure and interpret data are not yet widely adopted. While performance data is becoming easier to use thanks to increasing interoperability between benchmarking environments, information about the algorithms that collected that data is not as readily available. This is in part due to the complexities inherent in describing optimization heuristics. Even within a single family of algorithms, differences in operator choices, parameter adjustment strategies, and hyper-parameter settings can result in very different algorithm behavior. If these design decisions can be stored in combination with the corresponding performance data, this would open the door to extracting knowledge from the vast amount of data generated every day. One way that this could be achieved is through the use of ontologies. The data structured with the help of ontologies can then be used in various predictive studies, such as algorithm performance prediction.

In the context of computer science, **ontologies** are "explicit formal specifications of the concepts and relations among them that can exist in a given domain" [5]. Ontologies are generalized data models, i.e., they model only general types of things that share certain properties, but do not contain information about specific individuals in the domain. On the other hand, data about specific individuals stored in a directed labeled graph in which the labels have a well-defined meaning that comes from an ontology is commonly referred to as a **Knowledge Graph (KG)**. **Knowledge Graph Embeddings (KGEs)** are low-dimensional feature-based representations of the entities and relations in a knowledge graph. They provide a generalizable context over the entire KG, which can be used for tasks such as KG completion, triple classification, link prediction, and node classification [2,19]. The flexibility of the KG model and the explicit storage of data relationships facilitate not only the management of data from different sources, but also the search and exploration of these data to discover new insights that would be very difficult to discover using classical ML approaches.

Several efforts have been made to conceptualize various aspects of domain knowledge about black-box optimization, such as Evolutionary Computation Ontology [20], Diversity-Oriented Optimization Ontology [1], Preference-based Multi-Objective Ontology [10], and Semantic Multi-Criteria Decision Making Ontology [12]. The above ontologies are strongly focused on specifics, leading to classifications of algorithms that allow users to query only for high-level relations. For example, searching for algorithms that can solve problems from a particular class, and searching for algorithms that have been applied to a specific engineering problem. The OPTImization Algorithm Benchmarking ONtology (OPTION) [9] formalizes knowledge about benchmarking optimization algorithms, focusing on the formal representation of data from the performance and problem landscape space, but currently lacks descriptors for optimization algorithms.

Despite various efforts to conceptualize knowledge in black-box optimization, the main goal of these studies has been limited to the representation and organization of domain knowledge. In other words, as far as we know, the obtained representations have never been used to test their performance in predictive studies. In this work, we propose a novel approach that leverages and evaluates black-box optimization knowledge (i.e., optimization algorithms, performance data, optimization problems, and problem landscape data) represented by a formal semantic representation in the form of an ontology and KGs for predicting algorithm performance. The predictive model is developed by utilizing KG embedding-based algorithm performance classifiers.

Our Contributions: We test the utility of using a formal semantic representation for black-box optimization data in the form of KGs to predict algorithm performance. The KGs contain information about the problem landscape, algorithm performance, and algorithm descriptor data. To capture this knowledge, we first extended the OPTION ontology, which already provides a vocabulary for representing problem landscape and algorithm performance data, to include representations (i.e., descriptors) that describe optimization algorithms. Proof-of-concept was performed by extending the OPTION ontology with algorithm descriptors representing two modular frameworks related to the CMA-ES (mod-CMA) and differential evolution (modDE) algorithms. The same ontology can be used to represent other modular algorithms. However, we note that describing all existing algorithms proposed outside of modular frameworks is a time-consuming and challenging process that requires the participation of the entire community to reach a consensus on the standard unified representation of black-box optimization algorithms.

Next, we converted the knowledge base of the extended OPTION ontology into a KG and used it to learn a binary classifier that can predict whether or not an algorithm can solve a given problem (represented as 'solved' and 'not-solved' relations in the KG) within a predefined target precision in a fixed-budget scenario. In the context of KGs, this task is referred to as *triple classification*. We performed an experimental evaluation of the proposed approach by predicting the performance of 324 modCMA and 576 modDE algorithm configurations on the 24 noiseless problem classes from the BBOB benchmark suite in 5 and 30 dimensions, respectively.

We explore different evaluation scenarios to assess the predictive power of our KG-based performance classifier. The results show that our classifier correctly predicts whether an algorithm achieves a certain target precision for a given instance in the case of balanced classification. However, in the case of imbalanced classification, the baseline (the classifier that predicts the majority class) is superior. We succeeded in improving the performance of the KG embedding-based classifier in the case of imbalanced classification by modifying the pipeline and training an additional predictive model built on the learned embeddings.

Paper Outline. In Sect. 2, we present our extension of the OPTION ontology for the formal representation of modular optimization algorithms. Section 3 describes the construction of the KGs as well as the proposed methodology for performance prediction of modular optimization algorithms. The experimental results are discussed in Sect. 4. Section 5 proposes a modification of the pipeline to address the case of imbalanced classification performance prediction. Finally, we conclude the paper with a summary of contributions and plans for future work in Sect. 6.

2 Formal Representation of Modular Optimization Algorithms

In this paper we consider two different families of evolutionary algorithms: Differential Evolution (DE) [17] and Covariance Matrix Adaptation Evolution Strategies (CMA-ES) [7]. Since these two algorithms have been well-researched for over a decade, many variations and modifications have been proposed. Some of these modifications may be relatively minor, such as proposing an alternative initialization of the population. Larger changes may affect the structure of the algorithm by introducing restart mechanisms or new adaptation schemes for internal parameters. Since most of these changes are proposed in isolation, it is often difficult to understand how these variations interact. All of this has led to the development of modular algorithms. These frameworks combine large sets of variations into a single code base, where arbitrary combinations of variations can be combined into a variety of possible algorithm configurations. This not only allows a fair comparison between two different variations of the algorithm but also a more robust analysis of the potential interplay between algorithm components.

For the CMA-ES, we use the modCMA framework [13], which contains many variants of the core algorithm. This ranges from modifications of the sampling distributions (including mirrored or orthogonal sampling) to different weighting schemes for recombination to different restart strategies.

For DE, we use the modDE package available at https://github.com/Dvermetten/ModDE, v0.0.1-beta. This framework provides a wide range of mutation mechanisms, with different modules for selecting the base component, the number of differences included, and the use of an archive for some of the difference components. In addition, the usual crossover mechanisms can be enabled, as well as update mechanisms for internal parameters based on several state-of-the-art DE versions. For the formal representation of modular optimization algorithms, we extend the OPTION ontology by creating a new ontology module that is fully compatible with OPTION. We adhere to the same ontology design principles as in OPTION. For example, we align the new classes with the same upper- and middle-level ontologies, we follow the specification-implementation-execution ontology design pattern, and we use relations from the Relations Ontology [16].

Our extension allows us to specify the different steps in the optimization process and link them to the corresponding module parameters (see Fig. 1). For this purpose, we introduced the *modular optimization algorithm* class[1] as a subclass

---

[1] In the rest of this paper, we will refer to the ontology classes in *italic*, while the relations between the classes will be written in `typewriter`.

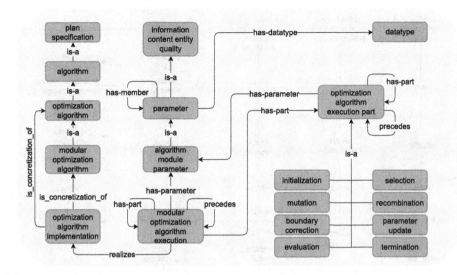

Fig. 1. The entities and relations included in the extension of the OPTION ontology for the representation of modular optimization algorithms.

of the *optimization algorithm* class, which is already defined in OPTION. For modular algorithms, we have also defined a specialized class *modular optimization algorithm execution*. Optimization algorithm execution can be a composition of several subprocesses (e.g., initialization, mutation, and recombination). To model this in the ontology, we have defined the *modular optimization algorithm execution part* class and linked it to the *modular optimization algorithm execution* class via the **has-part** relation. The algorithm execution flow is represented with the **precedes** relation. *Algorithm module parameters* are linked to both *modular optimization algorithm execution* and *modular algorithm execution part* through the **has-parameter** relation.

In Fig. 2 we illustrate the ontological representation of the modDE algorithm. In the ontology, we create specialized subclasses of the general classes corresponding to the modDE versions. For example, the *modDE execution* class is a subclass of *modular optimization algorithm execution*. It inherits all the properties of its superclass but also contains definitions that are unique to the modDE algorithm, such as the different execution parts, their execution order, and links to the modDE module parameters. We note here that in Fig. 2 only the execution parts such as initialization, mutation, and recombination are shown, while the others (i.e., boundary correction, evaluation, selection, parameter update, and termination check) have been omitted due to space constraints. Finally, in Fig. 2, we present two modDE configurations (as instances of the modDE class) that differ by the crossover type, which is a parameter that affects the recombination part of the optimization process. The modeling of the modCMA algorithm is done in a similar way.

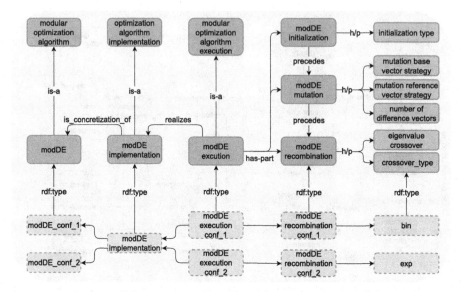

Fig. 2. An illustration of the representation of the modDE algorithm in the ontology and two examples of annotation of modDE configurations. Rectangular boxes correspond to the ontology classes. Dashed rectangular boxes correspond to the class instances.

3   Performance Prediction via KG Triple Classification

Representing optimization algorithms, benchmark problems, performance, and problem landscape data in a unified ontological framework facilitates the construction of KGs that can be used as data resources for a variety of predictive modeling tasks. In this paper, we investigate whether KG embeddings can be used to predict algorithm performance by performing the task of KG completion. More specifically, we are interested in predicting unseen performance relations between problem instances and algorithm configurations. This corresponds to the task of triple classification (i.e., whether an algorithm configuration solves a problem instance with a given target precision). This task can be translated to a binary classification task and easily addressed by using standard machine learning algorithms.

In this section, we first focus on the construction of the KG. Then we describe the details of the KG embedding-based pipeline for automated algorithm performance prediction.

3.1   Construction of the KG

The two main node types in the KGs are problem instances and algorithm instances/configurations (see Fig. 3). We collected data for the first five instances of each of the 24 noiseless BBOB problems [6] in dimensions $D = 5$ and

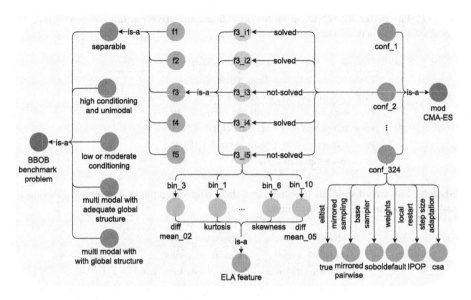

Fig. 3. A snippet of our KG constructed from the original OPTION ontology and the new algorithm representation module depicting its general structure.

$D = 30$, resulting in two problem sets (one for each dimension) with 120 problem instances each. Each problem instance is described with high-level and low-level landscape features. As high-level features, we used the five problem classes (i.e., separable, low or moderate conditioning, high conditioning, and unimodal multi-modal with adequate global structure and multi-modal with weak global structure) introduced in the BBOB test suite that group benchmark problems with similar properties. The low-level landscape features consist of 46 exploratory landscape analysis (ELA) features implemented in the R package `flacco` [8]. The ELA features are numerical representations of the problem instances that capture the characteristics of optimization problems. We use a publicly available dataset [15] containing the 46 ELA features computed for the first five instances of the 24 BBOB functions using the Sobol sampling strategy and a sample size of $100D$ with a total of 100 independent repetitions. For a more robust analysis, we use the median of the 100 calculated feature values. Finally, each ELA feature is discretized into 10 bins using the uniform binning strategy. These data are available through the OPTION ontology [9] knowledge base and we used their API to extract them.

The data described below were generated as part of this study and matched with data extracted from OPTION to create the KGs. The algorithm configurations are from two different modular algorithms, modular CMA-ES, and modular DE. Since it is computationally infeasible to collect data based on a complete enumeration of all possible combinations of modular CMA-ES and modular DE modules, we use a set of 324 and 576 configurations, respectively. Table 1 and Table 2 contain the details of the modules and parameter spaces used, which we

Table 1. The list of modCMA modules and their respective parameter space yielding a total of 324 algorithm configurations.

Module	Parameter space
Elitist	True, False
Mirrored_sampling	None, mirrored, mirrored pairwise
base_sampler	gaussian, sobol, halton
weights_option	default, equal, 1/2^lambda
local_restart	None, IPOP, BIPOP
step_size_adaptation	csa, psr

Table 2. The list of modDE modules and their respective parameter space yielding a total of 576 algorithm configurations.

Module	Parameter space
mutation_base	rand, best, target
mutation_reference	None, pbest, best, rand
mutation_n_comps	1, 2
use_archive	True, False
crossover	bin, exp
adaptation_method	None, shade, jDE
lpsr	True, False

then use to create a Cartesian product to obtain the different algorithm configurations. In KG, each algorithm configuration is represented as a node and is connected to the different modules via labeled links/edges.

To obtain performance data for each of these configurations, we perform 10 independent algorithm runs for each problem instance and calculate the median value. As a performance measure, we use the target precision achieved by the algorithm in the context of a fixed- budget (i.e., after a fixed number of function evaluations), and use the best precision achieved after $B = \{2\,000, 5\,000, 10\,000, 50\,000\}$ function evaluations.

Finally, the problem instances and the algorithm configurations are associated with a *solved* or *not-solved* edge, depending on the performance of the algorithm considering three different target precision thresholds, $T = \{1, 0.1, 0.001\}$ in the case of the $5D$ benchmark problems and $T = \{10, 1, 0.1\}$ for the $30D$ benchmark problems. More precisely, if an algorithm configuration achieves a target precision equal to or lower than the specified threshold for a given problem instance, we associate the algorithm configuration and the problem instance with a *solved* edge; otherwise, we associate the algorithm configuration and the problem instance with a *not-solved* edge.

3.2   KG Embedding-Based Pipeline for Automated Algorithm Performance Prediction

Our knowledge graph G can be represented as a collection of triples $\{(h, r, t)\} \subseteq E \times R \times E$, where E and R are the entity and relation set. One of the tasks

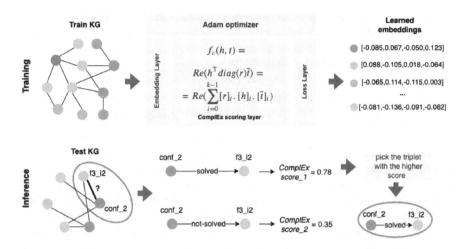

Fig. 4. KG embedding-based training and inference pipeline for triple classification.

in KG completion is to predict unseen relations r between two existing entities $(h, ?, t)$. In this paper, we focus on the $\{(a, s, p)\} \subseteq A \times S \times P$ triples, where $A \subset E$ and $P \subset E$ are the algorithm configuration and the problem instance set, respectively, and $S = \{solved, not\text{-}solved\} \subset R$ is the performance relation. To predict the unseen performance relation between algorithm configurations and problem instances $(a, ?, p)$, we perform triple classification.

Our proposed pipeline for predicting algorithm performance is shown in Fig. 4. For training the KG embeddings, we use the Ampligraph library [3]. In the training phase, we initialize the KG embeddings with the Xhavier initializer [4] and update them throughout several training epochs. During training, we minimize a loss function using Adam Optimizer that includes a ComplEx scoring function [18] – a model-specific function that assigns a score to a triple. Scoring functions for knowledge graph embeddings measure how far away two entities are relative to the relation in the embedding space. In general, the goal is to maximize the ComplEx model score for the positive triples and minimize it for the negative ones.

In the inference phase, we iterate over the $(a, ?, p)$ triples with a missing performance relation and calculate the ComplEx model score for the $(a, solved, p)$ and $(a, not\text{-}solved, p)$ triples by using the learned embeddings. We select the triple with the larger ComplEx score.

To find the best hyperparameters for the triple classifier, we used the grid search methodology, which performs an exhaustive search over the selected hyperparameters and their corresponding search spaces. Three different hyperparameters were selected for tuning: (1) k - the dimensionality of the embedding space; (2) optimizer_lr - the learning rate of the optimizer; and (3) loss - the type of loss function to be used during training, such as pairwise margin-based

Table 3. Hyperparameters of the KG embedding model and their corresponding values considered in the grid search.

Hyperparameter	Search space
k	$[50, 100, 150, 200]$
optimizer_lr	$[1e-3, 1e-4]$
loss	[PAIRWISE, NLL, SELF_ADVERSARIAL]

loss, negative loss probability, and adversarial sampling loss. The search spaces of the hyperparameters used in our study are shown in Table 3.

The optimal set of hyperparameters is estimated using a separate validation set. We initially set the number of training epochs to 500, but we activate a mechanism that terminates training early if 10 consecutive validation checks/epochs do not improve performance.

For the triple classifiers, we report the F1 score, which is defined as the harmonic mean of precision and recall: $F1 = \frac{2*Precision*Recall}{Precision+Recall} = \frac{2*TP}{2*TP+FP+FN}$. As a baseline, we use the classifier that predicts the majority class (solved/not-solved class). The same evaluation metric is used as a heuristic in the grid search step, where based on the F1 score we identify the best-performing model.

4   Evaluation Results

In this section, we report the results of two different evaluation scenarios based on how we select the data for training and testing. The first one uses leave random performance triplets out validation, while the second one uses leave problem instances/algorithm configurations out validation.

4.1   Leave Random Performance Triplets Out Validation

For our first set of experiments, we perform algorithm performance prediction using the method described in Sect. 3. Since we consider two dimensionalities of problems, four budgets, and three target precision thresholds, we have a total of 24 different KGs for each of the two algorithms (modCMA and modDE).

For each of the KGs, we split the performance triples in the ratio 60:20:20. That is, 60% of the triples are assigned to the training set, 20% to the validation set, and the remaining 20% to the test set. We do this in a stratified fashion, keeping the distribution of performance links as in the original KG. Since the split is based on a stratified sample of the performance links, performance links related to a particular problem instance or algorithm configuration can be split between the training/validation set and the test set. This approach can be used when the performance of algorithm configurations is known for the majority of problem instances in the selected problem portfolio but is unknown for some problem instances. Note that the training KG contains not only the performance triples but also other types of entities and relations, such as the high-level and low-level landscape features and the description of the algorithm configuration in terms of the modules and their parameters, while the validation and test sets contain only the links/triples of interest - 'solved' and 'not-solved' performance links.

The percentage of the 'solved' performance relations with respect to 'not-solved' ones for the modCMA and modDE KGs for the KG composite problem in 5 and 30 dimensions are shown in Table 4. We can notice that in some of the scenarios, we are dealing with imbalanced classification, especially in the case of $30D$ problems.

Table 4. The percentage of *solved* links for the modCMA and modDE algorithms in the KGs composed of a) $5D$ and b) $30D$ problems across the different fixed-budget scenarios and target precision thresholds.

	modCMA				modDE			
	2000	5000	10000	50000	2000	5000	10000	50000
1	62.9	68.2	71.3	78.9	27.7	42.2	58.1	81.4
0.1	46.8	54.1	57.1	63.7	13.2	23.3	33.1	62.8
0.001	36.9	47.8	50.7	55.9	9.4	14.4	21.7	56.2

(a) $5D$ problems

	modCMA				modDE			
	2000	5000	10000	50000	2000	5000	10000	50000
10	35.1	46.2	49.9	68.1	13.0	21.6	29.1	46.1
1	10.7	16.0	21.0	40.9	1.6	4.4	8.0	17.5
0.1	6.1	08.8	12.5	31.5	1.2	3.2	6.2	12.7

(b) $30D$ problems

Table 5 presents the F1 scores of the triple classifier and the percentage of improvement of the classifier compared to the baseline across the different fixed-budget scenarios and target precision thresholds for both $5D$ and $30D$ problems. Results show that the triple classifier improves the performance, in the case when we do not have imbalanced classification. In the case of imbalanced classification, we have a performance drop, meaning that in this case our proposed pipeline should be adjusted.

4.2 Leave Problem Instances/algorithm Configurations Out Validation

Our second set of experiments evaluates a practically relevant scenario when there is no performance data for a given problem instance/algorithm configuration. To evaluate the performance of the triple classifier in this setup, we have investigated two additional evaluation scenarios:

– **Leave problem instances out validation:** In this scenario, we use all performance triples of one problem instance from each of the 24 BBOB problems

Table 5. The F1 score and the percentage of improvement compared to the baseline of the modCMA and modDE algorithm performance triple classifier obtained using the ComplEx scoring model for the KGs composed of $5D$ and $30D$ problems across the different fixed-budget scenarios and target precision thresholds.

	2000	5000	10000	50000
1	0.922/19.43%	0.942/16.15%	0.944/13.33%	0.953/8.05%
0.1	0.905/30.22%	0.933/32.91%	0.937/29.06%	0.942/21.08%
0.001	0.893/15.37%	0.944/37.61%	0.944/40.48%	0.946/31.75%

(a) $5D$ problems - modCMA

	2000	5000	10000	50000
1	0.848/1.07%	0.876/19.67%	0.901/22.59%	0.946/5.46%
0.1	0.788/-15.18%	0.82/-5.53%	0.858/6.98%	0.922/19.43%
0.001	0.831/-12.62%	0.745/-19.20%	0.803/-8.65%	0.919/27.64%

(b) $5D$ problems - modDE

	2000	5000	10000	50000
10	0.937/19.06%	0.927/32.62%	0.939/40.78%	0.953/17.51%
1	0.902/-4.45%	0.808/-11.50%	0.855/-3.17%	0.929/25.03%
0.1	0.935/-3.41%	0.89/-6.71%	0.852/-8.68%	0.921/13.28%

(c) $30D$ problems - modCMA

	2000	5000	10000	50000
10	0.9/-3.23%	0.931/5.92%	0.947/14.10%	0.948/35.24%
1	0.504/-49.19%	0.792/-19.02%	0.846/-11.69%	0.87/-3.76%
0.1	0.695/-30.08%	0.735/-25.30%	0.835/-13.74%%	0.885/-5.04%

(d) $30D$ problems - modDE

for testing, select the performance triples from another problem instance for validation, and use the remaining three for training. For example, we use the first three instances of each of the 24 BBOB problems for training, the fourth instance for validation, and the fifth instance for testing. We repeat this five times so that each of the five instances appears once in the test set.

- **Leave algorithm configurations out validation**: in this scenario, the algorithm configurations are split with a 60:20:20 ratio and their performance triples are selected for training, validation, and testing, respectively. In order to assess the robustness of the results, we repeat this procedure five times independently.

We have applied these evaluation scenarios to the KGs comprised of $5D$ benchmark problems and modCMA algorithm configurations across the four different budgets with a target precision threshold of 0.1. The average F1 scores of the triple classifier (averaged over the five runs), their standard deviations, as well as the percentage of improvement, are displayed in Table 6. Similarly as in Sect. 4.1, our approach improves compared to the baseline when we have a balanced classification. Table 7 presents the evaluation results for the modDE performance classifier, where similar patterns can be observed as in the previous case.

Table 6. The F1 score and the percentage of improvement compared to the baseline of the modCMA algorithm performance triple classifier for the KGs where all performance links are removed for a subset of problems and algorithm configurations composed of 5D problems across the different budgets and a target precision threshold of 0.1.

	Leave-problems-out	Leave-algorithms-out
2000	0.728 (0.006)/4.90	0.893 (0.009)/28.67
5000	0.761 (0.018)/8.40	0.915 (0.011)/30.34
10000	0.766(0.008)/5.36	0.91 (0.011)/25.17
50000	0.797(0.014)/2.44	0.913 (0.002)/17.35

Table 7. The F1 score and the percentage of improvement compared to the baseline of the modDE algorithm performance triple classifier for the KGs where all performance links are removed for a subset of problems and algorithm configurations composed of 5D problems across the different budgets and a target precision threshold of 0.1.

	Leave-problems-out	Leave-algorithms-out
2000	0.854(0.061)/-8.07%	0.79(0.035)/-14.96%
5000	0.837(0.022)/-3.57%	0.85(0.021)/-2.07%
10000	0.796(0.024)/-0.75%	0.825(0.010)/2.87%
50000	0.83(0.010)/7.51%	0.822(0.013)/6.48%

5 Addressing the Problem of Imbalanced Classification

To solve the problem that arises in the case of imbalanced classification, we modify the pipeline described in Sect. 3.2. More specifically, after the KG training phase, we add an additional training layer, where we train a Random Forest (RF) classifier based on the learned embeddings. Our data instances are the performance triples. In order to generate the data for the RF classifier, we represent each (a, s, p) triple as a concatenation of the embedding vectors of the a and p entities. We perform inference by using the RF classifier instead of the ComplEx model scores.

We evaluate this approach using the most imbalanced scenario from the experiments in Sect. 4.1, i.e., the setup where we predict modDE performance on $30D$ problem instances with a target precision threshold of 0.1. We train RF classifier with 10 estimators, implemented in the scikit-learn library [14]. The rest of the hyperparameters are used with their default values. In Table 8, we compare the F1 scores of the classifiers trained using the pipeline presented in Sect. 3.2 with the scores of the RF classifiers described in this section. The results

Table 8. Comparison of the two proposed pipelines for modDE performance prediction on the $30D$ problem instances with a target precision of 0.1. Results are reported in the format: F1-score of the classifier/F1-score of the baseline/Percentage of improvement compared to the baseline.

	KG - ComplEx scoring	RF classifier
2000	0.695/0.994/-30.08%	0.999/0.994/0.52%
5000	0.735/0.984/-25.30%	0.998/0.984/1.43%
10000	0.835/0.968/-13.74%	0.996/0.968/2.91%
50000	0.885/0.932/-5.04%	0.991/0.932/6.27%

Table 9. Performance of the RF classifier for modDE performance prediction on the $30D$ problem instances with a target precision of 0.1. Results are reported in the format: Performance of the classifier/Performance of the baseline.

	RF classifier		
	AUC ROC	Average precision	G-mean
2000	0.994/0.5	0.962/0.012	0.963/0.0
5000	0.998/0.5	0.975/0.032	0.962/0.0
10000	0.998/0.5	0.977/0.062	0.954/0.0
50000	0.987/0.5	0.964/0.127	0.947/0.0

show that training a RF classifier on the learned embeddings improves the performance in terms of F1 score. As we are dealing with imbalanced classification, the choice of the evaluation measure is essential. In Table 9, we additionally report the AUC ROC, average precision, and geometric mean scores. We reach the same conclusion that the embedding-based RF classifier improves the performance prediction method. We believe that the results improve because there is a separability in the embeddings space that the RF models manage to capture when predicting the algorithm's performance. However, this assumption requires further investigation.

6   Conclusions and Future Work

In this paper, we investigate the predictive power of a formal semantic representation of black-box optimization for automated prediction of algorithm performance. To this end, we evaluate the feasibility of using KGs to predict the performance of the modCMA and modDE optimization algorithms on the noiseless BBOB functions. More specifically, our goal was to investigate whether we can train KG embeddings that can be used to predict performance links/triplets (*solved* or *not-solved* links) in the KG between algorithm configurations and problem instances with a given target precision in a fixed- budget scenario. The KGs combine the problem landscape and algorithm performance data with the data related to the modular algorithm configuration.

The results show that when we randomly select performance triples for the test set (a classic KG completion scenario), our proposed triple classifier outperforms the baseline in the cases where we have balanced classification. In the case of imbalanced classification, the performance of the classifier decreases and it is worse than the baseline. In our second set of experiments, we have a "more rigorous" evaluation scenario where we try to predict all the performance links belonging to the problem instances and algorithm configurations that appear in the test set (no performance links appear in the training set). We observe similar patterns as in the previous case. To solve the performance degradation problem in the case of imbalanced classification, we modify the proposed pipeline and train a Random Forest classifier on top of the learned embeddings.

For our future work, we plan investigate different class-balancing techniques in the case of imbalanced classification. Additionally, we plan to test different methods for training KG embeddings and improve their explainability. It would also be interesting to compare our results with other approaches such as relational learning and graph frequent pattern mining.

We have shown that KGs of experimental data about the modCMA and modDE modular optimization algorithms created from ontology knowledge bases can be used in predictive studies. It would be interesting to test the applicability of this approach to other modular frameworks. However, the more challenging task would be to extend it outside of modular frameworks, as we would need to develop a formal, standard vocabulary that can be used to represent algorithm operators, their hyperparameters, and interactions. We hope that our work will inspire the community to collaborate on the development of appropriate algorithm representations.

Acknowledgments. The authors acknowledge the support of the Slovenian Research Agency through program grant No. P2-0103 and P2-0098, project grants N2-0239 and J2-4460, a young researcher grant to AK, and a bilateral project between Slovenia and France grant No. BI-FR/23-24-PROTEUS-001 (PR-12040), as well as the EC through grant No. 952215 (TAILOR). Our work is also supported by ANR-22-ERCS-0003-01 project VARIATION, and via a SPECIES scholarship for Ana Kostovska.

Data and Code Availability.. Our source code, data, the OPTION ontology extension, the generated KGs, and figures are available at: https://github.com/KostovskaAna/KG4AlgorithmPerformancePrediction.git.

References

1. Basto-Fernandes, V., Yevseyeva, I., Deutz, A., Emmerich, M.: A survey of diversity oriented optimization: problems, indicators, and algorithms. In: Emmerich, M., Deutz, A., Schütze, O., Legrand, P., Tantar, E., Tantar, A.-A. (eds.) EVOLVE – A Bridge between Probability, Set Oriented Numerics and Evolutionary Computation VII. SCI, vol. 662, pp. 3–23. Springer, Cham (2017). https://doi.org/10.1007/978-3-319-49325-1_1
2. Chen, X., Jia, S., Xiang, Y.: A review: Knowledge reasoning over knowledge graph. Expert Syst. Appl. **141**, 112948 (2020)

3. Costabello, L., Pai, S., Le Van, C., McGrath, R., McCarthy, N., Tabacof, P.: Ampligraph: a library for representation learning on knowledge graphs (2019). Accessed 10 Oct 2019

4. Glorot, X., Bengio, Y.: Understanding the difficulty of training deep feedforward neural networks. In: Proceedings of the AISTATS. JMLR Workshop and Conference Proceedings, pp. 249–256 (2010)

5. Gruber, T.: Toward principles for the design of ontologies used for knowledge sharing? Int. J. Hum Comput Stud. **43**(5–6), 907–928 (1995)

6. Hansen, N., Finck, S., Ros, R., Auger, A.: Real-Parameter Black-Box Optimization Benchmarking 2009: Noiseless Functions Definitions. Research Report RR-6829, INRIA (2009). https://hal.inria.fr/inria-00362633

7. Hansen, N., Ostermeier, A.: Adapting arbitrary normal mutation distributions in evolution strategies: the covariance matrix adaptation. In: Proceedings of the CEC, pp. 312–317. IEEE (1996)

8. Kerschke, P., Trautmann, H.: The R-Package FLACCO for exploratory landscape analysis with applications to multi-objective optimization problems. In: CEC, pp. 5262–5269, July 2016. https://doi.org/10.1109/CEC.2016.7748359

9. Kostovska, A., Vermetten, D., Doerr, C., Dzeroski, S., Panov, P., Eftimov, T.: OPTION: OPTImization Algorithm Benchmarking ONtology. In: Proceedings of the GECCO, Companion Material (2021)

10. Li, L., Yevseyeva, I., Basto-Fernandes, V., Trautmann, H., Jing, N., Emmerich, M.: Building and using an ontology of preference-based multiobjective evolutionary algorithms. In: Trautmann, H., et al. (eds.) EMO 2017. LNCS, vol. 10173, pp. 406–421. Springer, Cham (2017). https://doi.org/10.1007/978-3-319-54157-0_28

11. López-Ibáñez, M., Branke, J., Paquete, L.: Reproducibility in evolutionary computation. ACM Trans. Evol. Learn. Optim. **1**(4), 1–21 (2021)

12. Mahmoudi, G., Muller-Schloer, C.: Semantic multi-criteria decision making semcdm. In: 2009 IEEE Symposium on Computational Intelligence in Multi-Criteria Decision-Making (MCDM), pp. 149–156. IEEE (2009)

13. de Nobel, J., Vermetten, D., Wang, H., Doerr, C., Bäck, T.: Tuning as a means of assessing the benefits of new ideas in interplay with existing algorithmic modules. In: Proceedings of the GECCO, Companion Material, pp. 1375–1384. ACM (2021)

14. Pedregosa, F., et al.: Scikit-learn: machine learning in Python. J. Mach. Learn. Res. **12**, 2825–2830 (2011)

15. Renau, Q., Doerr, C., Dreo, J., Doerr, B.: Experimental data set for the study "Exploratory Landscape Analysis is Strongly Sensitive to the Sampling Strategy", June 2020. https://doi.org/10.5281/zenodo.3886816

16. Smith, B., et al.: Relations in biomedical ontologies. Genome Biol. **6**(5), R46 (2005)

17. Storn, R., Price, K.: Differential evolution-a simple and efficient heuristic for global optimization over continuous spaces. J. Global Optim. **11**(4), 341–359 (1997)

18. Trouillon, T., Welbl, J., Riedel, S., Gaussier, É., Bouchard, G.: Complex embeddings for simple link prediction. In: International Conference on Machine Learning, pp. 2071–2080. PMLR (2016)

19. Wang, Q., Mao, Z., Wang, B., Guo, L.: Knowledge graph embedding: a survey of approaches and applications. IEEE Trans. Knowl. Data Eng. **29**(12), 2724–2743 (2017)

20. Yaman, A., Hallawa, A., Coler, M., Iacca, G.: Presenting the ECO: evolutionary computation ontology. In: Squillero, G., Sim, K. (eds.) EvoApplications 2017. LNCS, vol. 10199, pp. 603–619. Springer, Cham (2017). https://doi.org/10.1007/978-3-319-55849-3_39

Automatic Design of Telecom Networks with Genetic Algorithms

João Correia[1]([✉]), Gustavo Gama[1], João Tiago Guerrinha[2], Ricardo Cadime[2], Pedro Antero Carvalhido[2], Tiago Vieira[2], and Nuno Lourenço[1]

[1] University of Coimbra, CISUC, DEI, Coimbra, Portugal
{jncor,naml}@dei.uc.pt, gustavogama@student.dei.uc.pt
[2] Altice Labs, Aveiro, Portugal
{joao-t-guerrinha,ricardo-v-cadime,pedro-a-carvalhido,
tiago-s-vieira}@alticelabs.com

Abstract. With the increasing demand for high-quality internet services, deploying GPON/Fiber-to-the-Home networks is one of the biggest challenges that internet providers have to deal with due to the significant investments involved. Automated network design usage becomes more critical to aid with planning the network by minimising the costs of planning and deployment. The main objective is to tackle this problem of optimisation of networks that requires taking into account multiple factors such as the equipment placement and their configuration, the optimisation of the cable routes, the optimisation of the clients' allocation and other constraints involved in the minimisation problem. An AI-based solution is proposed to automate network design, which is a task typically done manually by teams of engineers. It is a difficult task requiring significant time to complete manually. To alleviate this tiresome task, we proposed a Genetic Algorithm using a two-level representation to design the networks automatically. To validate the approach, we compare the quality of the generated solutions with the handmade design ones that are deployed in the real world. The results show that our method can save costs and time in finding suitable and better solutions than existing ones, indicating its potential as a support design tool of solutions for GPON/Fiber-to-the-Home networks. In concrete, in the two scenarios where we validate our proposal, our approach can cut costs by 31% and by 52.2%, respectively, when compared with existing handmade ones, showcasing and validating the potential of the proposed approach.

Keywords: Genetic Algorithms · Automatic Design · Design of Networks

1 Introduction

In the last few years, the demand for high-quality internet services increased rapidly due to the growth in the usage of bandwidth-intensive applications such as Cloud-based Services, Video and Audio Streaming, and gaming. This pushes Internet Service Provider (ISP) to consider faster technologies to answer the

© The Author(s), under exclusive license to Springer Nature Switzerland AG 2023
J. Correia et al. (Eds.): EvoApplications 2023, LNCS 13989, pp. 269–284, 2023.
https://doi.org/10.1007/978-3-031-30229-9_18

increasing demand, such as optical fiber, i.e. Fiber-to-the-X (FTTX). FTTX refers to a wide range of deployment configurations for different scenarios in the last mile stage of telecommunications networks. The most common scenarios are the Fiber-to-the-Home (FTTH) connecting directly to the client's residence, and Fiber-to-the-Building (FTTB) reaching buildings directly. One way of providing this type of service is through a Gigabyte Passive Optical Network (GPON) based solution. GPON is a point-to-multipoint (P2MP) network which enables a single fiber system to serve multiple customer premises with a splitting process. This technology has many advantages compared to other types of networks (e.g., copper-based networks (DSL)), which have been the standard solution for many years. First, it allowed for faster data transmission and increased range, with optical fiber cables able to cover distances up to 20km. Secondly, it allows for boosted security. Thirdly, if there is the need to upgrade the network, it is possible to change only the endpoints and leave the fiber infrastructure intact. Furthermore, lastly, it is cost effective.

To deploy a GPON/FTTH network, there are many challenges that the ISP has to consider, involving significant investments. Deployment of the GPON/FFTH networks involves building new infrastructures (e.g., duct digging, laying new cables to connect to the homes) or updating and maintaining an existing one. As such, it is crucial to optimise the design of the network to reduce high costs. Considering that the source of difficulties while deploying a network is engineering or technical aspects of the design, proper design of the solution can lead to savings in the range of 30% [4].

However, to create a cost-effective GPON/FTTH, engineers have to consider several factors, such as the splitter position and its maximum ratio, the headend position, the optical distribution point position, the routes to consider, the maximum distance, the number of homes ones has to serve and the optical budget. Considering all these factors, it is a complex task requiring considerable time and effort. Therefore, automation of the process can minimise the effort required for the design process from days to a few hours or minutes whilst optimising the design in what concerns costs. Typically this process design process has several phases, from the physical infrastructure, passing from the design planning of the GPON/FFTH networks to the network itself. This goes without saying that there are multiple stages where automation can help save time and effort and be valuable for the telecom network company responsible for its deployment. Although multiple phases of the design process occur in the context of this work, we are concerned with the design of a network at the equipment level, assuming that the physical part is already handled.

Thus, this paper proposes a bio-inspired approach to the automatic design of GPON/FTTH networks. The automatic design of the network is framed as a combinatorial optimisation problem that we aim to solve using Evolutionary Computation. We designed a Genetic Algorithm (GA) where each individual has a two-level representation that maps into an existing physical infrastructure. The first level corresponds to a binary list defining the points of optimal distribution (PDO), and the second list indicates the connections between a PDO and the clients. We validate the approach by employing the proposed approach to optimise real-world networks and compare the automatically designed solutions with the existing ones, manually designed by engineering teams and currently deployed and functioning. The results show that for each benchmarked network, the GA is able to design networks with inferior costs to the existing handmade ones. In concrete, when compared to the handmade solutions, our approach can discover solutions that reduce the costs by 31% and 52.2% for the two scenarios considered.

The remainder of the article is as follows. In Sect. 2, we cover the background and definitions of the problem and the related work. Section 3 presents the approach for the automatic design of networks. Afterwards, in Sect. 4, we present the experimental setup and in Sect. 5 the results. In Sect. 6, overall conclusions are drawn, and future work is presented.

2 Background

A simple GPON network is composed of three components: the Optical Line Terminals (OLT) that represents the shared network equipment of the ISP, the different Optical Network Units (ONU), which represents the equipment on the client side (the modem) and the splitter.

Splitter Level and Ratio There is a single feeder optical fibre from the OLT that connects the splitter and splits the signal on multiple fibres with a $1 : N$ ratio that will link to the ONUs, with N being the number of output ports. To optimally design a GPON/FTTH, this splitter ratio and the level problem must be solved. Multiple configurations can be used. For example, in a *centralised approach*, the designer can consider a splitting ratio of $1 : 32$, which means that this splitter design can serve up to 32 homes (ONUs). In a *cascading splitting approach*, we can have different splitters connecting each other before reaching the user side [10]. This approach with multiple levels may use different ratios for the splitters. For example, a $1 : 4$ ratio for the first splitter level is followed by a $1 : 8$ for the second. The choice of approach depends on the region to serve. A centralised approach might adapt more to urban environments and crowded areas as it allows more flexibility and lowers costs. On the other hand, a cascaded approach is often used in scenarios where it is required to cover an extended area, for example, in Long-Reach Passive Optical Network [7].

The locations of the network equipment are crucial in the design of networks due to the high costs of the materials and the civil work involved. It is essential to consider the different equipment factors like the splitter's location, the location of the different ONUs (the clients), the headend position, the optical distribution point position (i.e. a street cabinet along the route used to manage interconnections, integrate fibre splicing, can also contain splitters and other components). When planning a cost-effective network design, all the equipment must be placed optimally to minimise the amount of cable used in the network and the overall cost.

If we are dealing with buildings as homes (an FTTB scenario), we might have to consider the demand of that location. It might require, for example, a demand of 50 ports for all the homes in the building, while in a simple house, we require only one. Thus, the cabinet containing the splitters and other components should be placed directly on the location of the building.

Another important aspect to consider is that the optical power budget has to be respected regardless of the choice of topology and design considerations. The optical power budget refers to the light required to transmit signals over a fibre connection within a certain distance. There are multiple sources of budget loss (in dB). The splitter can generate energy loss when dividing the signal, called the splitter loss (Table 1). It can also happen with other components of the network, like connectors. However, most of the loss is made up by the cable length, around 0.35 dB per km. Thus, it emphasises the importance of reducing the cable required to run the GPON. By minimising the cable's distance, we improve the network's efficiency (reduced loss).

Considering the aspects mentioned above, the problem of automatically designing a GPON network can be summarised as finding the most effective locations for the different types of equipment, the split ratios of the splitters for the different levels, and the optical power budget; while reducing the financial costs.

2.1 Related Work

The problem of designing fibre networks is often tackled with meta-heuristics techniques. In [11], an application of a Genetic Algorithm combined with Graph theory to design PONs for different network topologies (bus, tree and ring) is proposed for a multiple-level network (i.e., with multiple splitters levels). The main objective was to minimise the cost of implementation of the network with a focus on optimising the material needed. Like the number of splitters, their geographical location, their split ratio and the cable length. For that purpose, the authors' solution for the GA is based on the paths between the OLT (the shared equipment in the central office) to each ONU (the homes) obtained with the help of graph theory algorithms. Another similar work is performed in [6], where a GA is applied to solve the PON optimisation problem with multiple constraints. Once again, a multiple-level splitter design was considered. A single individual is represented by a double string (where the order of the elements matter), one string for the splitters equipment location (primary and secondary level) and ratio and another for the ONUs locations. The size of those strings depends on the number of ONUs to satisfy and the available material/splitters. This kind of representation allows the authors to satisfy certain constraints automatically. The author also did a comparison work between the GA approach and an ILP approach. Both seem to find exact solutions for smaller networks. However, as the instance of the problem increases, the ILP solution needs to catch up in terms of execution time for an acceptable solution where the GA prevails, even though it is not certain that the optimal global solution is found.

A more recent work proposed by [5], also detailed an application of the GA in combination with Graph theory to solve the problem of the design of PONs. Four different topologies were considered, using different configurations of splitters, including both centralised and cascaded approaches. The representation choice for the GA consisted of a binary representation to find the locations to place the equipment amongst a set of potential candidate spots. Their system was tested in real networks from Brazil, where the authors compared the application of the different topologies for different maps. Overall it depends on the type of map. Dense and non-dense maps were considered. For example, a cascaded approach might be more adapted in scenarios with fewer clients to reduce cable usage. The optical budget is also considered in the solutions to check the validity of the networks.

Other meta-heuristic approaches have also been used to address this network design problem. For example, in [2], a solution using Ant Colony Optimization (ACO) is used to design GPON/FTTH for a greenfield (i.e., deployment of a network to a fresh new site) with the input requirements of the potential locations for the materials (splitters and cable distribution locations). Their method allows them to approach near-optimal solutions. The authors compared with exact solutions solvers (simplex algorithms), and the difference in solution quality is less than 1%, with the advantage of a much better execution time for more significant instances ($50-90\%$ faster). Additional steps were required to achieve this result, like softening the constraints and post-optimisation. In [3], the same authors propose a similar approach with ACO with extra meta-heuristics, but this time for a problem evolving additional equipment (aggregating equipment/drop closures).

In addition, in the work of Ali et al. [1], a Local Search approach was introduced. A Local Search procedure modifies the candidate solutions locally until no improvement is detected (or other stop criteria). In this work, those modifications can be simply the assignment of a client to new network equipment or swapping the connection (link) between two clients. As for the guided aspect, a penalty function is used when the algorithm is stuck in a local optimum. The objective function will be changed (i.e., the weights of the features present in the current solution will be modified). This approach is compared to Simulated-annealing one and seems to outperform the latter significantly.

Finally, there are also applications of AI in other subproblems of telecommunications networks. The work of Javier Mata [8] proposes a survey of the vast applications of AI for telecommunication problems. Some of the referenced works are worth mentioning, for example, Morais et al. [9], proposes a GA algorithm to deal with topology design considering network survivability. Network survivability refers to the capacity of the network to operate under the presence of equipment failures. Once more, ILP approaches are also used to benchmark their approach.

3   Proposed Approach

We developed a system based on Genetic Algorithms to tackle the problem of the design of networks. As seen in Sect. 2, some design considerations must be made to obtain a robust solution for the problem. This section will overview the system developed and outline the principal algorithmic design choices considered. During the evolutionary process, the GA has to deal with the following constraints and challenges: (i) find the placement of the network equipment, precisely the optical distribution point positions (PDOs), to connect the client's homes to the network. The PDOs are capacitated, i.e. they have limited ports available; (ii) the routes to consider to minimize the resources needed to cover the totality of clients' homes, thus, the solution must consider that the routes that are "buried" (more expensive) should be avoided whenever it is possible; (iii) the demand of each client must be considered (1 unit of demand for fibre corresponds to 1 port allocated to the client); (iv) if we have big buildings such as apartments, the placement of the PDO is expected to be on-site, i.e. closest network equipment node, for "on-field" practicality, reducing the amount of drop cable in this case despite the demand of the location becoming higher; (v) the solution must find the optimal drop cable links based on the PDOs existing ports (vi) there are constraints related to maximum distance for the drop cables (e.g. limit the amount of cable that can be used to link one client to a PDO) (vii) constraints for the distribution cables, i.e., to restrict the network size and respect the optical budget; (viii) the design is impacted by weights, in the form of required equipment costs of a given solution; (ix) the constraints related to intersections of drop cables, i.e. a drop cable cannot intersect with a distribution cable. In the next sections we detailed aspects related to the characteristics of the employed GA-based solution.

3.1 Representation

In our approach, each individual in the population is composed of two arrays: (i) a binary representation, where each index corresponds to a network equipment candidate to the placement of a PDO, where if the value is one, the PDO is placed on that equipment; (ii) an integer based representation used to manage the drop cables links for the capacitated sub-problem related to the limited ports of the PDOs. Each index corresponds to a different client, and the values within correspond to the associated PDO (node id) for that specific client's home. If a house is not assigned to a PDO, the value is -1. This array depends on the first binary representation, where modifying the first might impact the second.

3.2 Variation Operators

Our variation operators are applied to the binary representation. The mutation modifies the population of individuals randomly by changing one gene. For this work we employed bit flip mutation which consists of simply modifying a bit from the first representation array, i.e., a 1 can become a 0 with a fixed probability and vice-versa.

Similarly, the crossover operator allows for variation in the population at each generation by combining the genetic material of two distinct individuals (the parents) and obtaining a new solution. The function of the crossover operator is the uniform crossover.

In the evolutionary system proposed, the variation operators modify "representation 1" which can require an update of "representation 2" since we are potentially modifying the placement of a PDO, and some homes might need to be reinspected to check their links. Both those operators are applied for a selected pool of individuals (the parents) based on fitness scores.

3.3   Local Search

Concerning the sub-problem related to the capacitated problem of the PDOs mentioned earlier, there are two options for the step of connecting each home to a PDO: (i) closest to the home method; (ii) perform local search with Hill Climbing to optimise the distribution. For our specific case, we end up mixing both. The algorithm starts by performing option (i), and if we end up with homes not allocated to their closest PDO, we store them for option (ii). A computation of the possible set of PDO nodes for each home allocated to their second (or more) closest PDO is required and a cache can be used to avoid repetitive computations. Then, the swap procedure can start by swapping drop cables with clients' homes from that set of PDOs. If the cost is reduced, the modified solution is kept.

It is important to note that every swap has to respect the constraints in limit drop cable range and ports for the PDOs (for both the current home inspected and the other client home swapped).

Another point worth mentioning is that we only enter this Local Search sometimes, which would be computationally costly. The algorithm only considers the Local Search step if homes are not allocated to their closest PDO, which in theory, only happens in more dense zones of the graph.

3.4   Fitness Function

Another critical aspect of the GA is the fitness function used to evaluate each solution's performance. The evolutionary system proposed evaluates the population initially and at every generation cycle. It represents the financial cost of building the network, where the objective is to minimise this cost.

The fitness function requires the graph, the individual to be evaluated, the precalculated costs for the drop cable and the different input weights (business rules) provided by the user. Three main materials greatly impact the cost of the solution returned: (i) Drop cable cost C_{Drop}; (ii)Distribution cable cost C_{Dist}; (iii) PDO costs C_{Pdo} which should depend on the maximum ports available.

The first step consists in computing the amount of drop cable needed, N_{Drop}, for the current solution (this can be done by accessing the cache containing the drop cable distance costs computed in the initial phase of the system). A simple iteration through the second representation (integer one) is sufficient to perform this step. It is important to note that a factor 10 is applied to the financial cost of the drop cable in the case of the buildings (MDUs) to guide the GA to place the PDOs near the buildings (if there is any on the map).

The second step focuses on the solution's needed distribution cable, N_{Dist}. This is the network's cable between the starting point and the different PDOs (total of N_{Pdo}). This step needs more computational power to perform since it is required to compute the shortest path between the PDO to the starting point. Dijkstra's algorithm is applied in this scenario, and the results are stored in a dynamic cache so other calls to this fitness function can eventually reuse them. This storage step is essential when dealing with more extensive networks and dense graphs to speed up the process.

The third step multiplies the input weights by the resources needed (cables and PDOs) with the following formula for the cost of the materials C_{Mat} :

$$C_{Mat} = C_{Drop} * N_{Drop} + C_{Dist} * N_{Dist} + C_{Pdo} * N_{Pdo} \qquad (1)$$

Finally, extra penalties can be applied to this cost C_{Mat} in case of broken constraints. In this case, the system may cover all the clients. So, every solution that does not have all the clients connected to the network suffers a constant penalty P; the constant P is fixed to the cost of a PDO (chosen by trial and error). This penalty is based on the number of homes missing to be allocated $H_{Missing}$. This value will be zero if all houses are covered and no penalty is added (minimisation). The fitness (f) result returned is the following:

$$f = C_{Mat} + H_{Missing} * P \qquad (2)$$

4   Experimental Setup

This section defines the experimental setup used to validate the proposed approach to the automatic design of networks. Our experiments aim to automatically find a solution for each dataset and compare it to the existing real-world handmade solution. We start by describing the details of the maps used, followed by the business rules and restrictions that the telecommunications partner imposed. Lastly, we present the parameters used in the Genetic Algorithm.

We evaluate our approach in two different scenarios from New York and New Jersey. Table 1 details the available maps for which we have a corresponding handmade solution. In this table, the number of location nodes (column 2) is sometimes superior to the sum of the number of valid network equipment nodes (column 4) and the total client nodes (column 5). This occurs because some of the equipment nodes got filtered since they were not part of the main connected component containing the starting point, i.e., isolated nodes.

Table 1. Aspects of the graphs for the two maps that compose the experiment.

Data set (map)	Number of locations (nodes)	Number of routes (edges)	Number of valid network equipment nodes	Total clients: Houses(SDUs) & Buildings (MDUs)	Sum of the distance of all routes
Map 1	188	81	80	103 & 5	2.11 km
Map 2	222	166	133	76 & 11	4.76 km

The business rules comprise the materials costs and the constraints related to the problem. They have a direct impact on the result returned by the algorithm. In this set of experiments, we used a configuration aimed at fewer PDOs, less distribution cable, and more drop cable. The configuration is pre-set with costs for several items that act on weights in the evaluation phase. Ultimately encourage the algorithm to converge to a solution with the aspects mentioned earlier. Each of those scenarios depends on a fixed set of weights representing the material costs (input provided by the user). Table 2 displays the experiments' weights. Note that the parameters can be adjusted depending on the restrictions. It is worth noting that these are the weights plugged into the Genetic Algorithm's fitness function.

Table 2. Sets of experimental weights for the materials.

Material	Cost
Cost PDO	300$ unit
Cost drop cable per meter	2$
Cost distribution cable per meter	5$

As for the constraints, they refer to everything that might limit the search space to find a solution. Each scenario is also limited by the constraints defined in Table 3. There are two types of constraints to consider. The ones concerning the PDOs: maximum number of ports per PDO and also the margin to leave for each PDO for future expansion of the network. For example, if a PDO with a limit of 12 ports is chosen with a margin of 10% to leave open, only 11 ports are operational in practice. We also have restrictions on the maximum distance allowed for the cables, like the maximum overall distance of the network (from the OLT to a client) and the limits imposed on the drop cables. For the drop cables, the designs are proposed for an 85 m limit, but it can vary depending on the data set. Our approach is able to deal with all the mentioned inputs, i.e. costs and constraints, and are adjustable and chosen by the user that interacts with our approach. From an experimental point of view, they are the inputs for our approach.

Table 3. Experimental constraints.

Constraint parameter	Value
Limit ports per PDO	12
Margin ports open per PDO	10%
Limit drop cable	85 m (range: 50-100)
Maximum range of the network	20 Km

Table 4. Genetic Algorithm parameters.

Parameter	Value
Population size	100
Number of generations	100
Mutation	Bitflip
Mutation rate	$2/len(genotype)$
Crossover	Uniform (0.5 per gene)
Crossover rate	0.85
Tournament selection	5
Survivors selection (elitism percentage)	10

Table 4 presents the configuration used in the GA for each evolutionary run. The experimental results that we are going to present are averages of 10 evolutionary runs. All the experiments were conducted in a computer with the following hardware specifications: Intel(R) Core(TM) i7-8700K CPU @ 3.70GHz; RAM: 16GB.

5   Experimental Results

In this section, we present the results obtained by our approach when designing networks for two maps. We analyse and compare the obtained networks with the handmade ones, considering the following metrics: number of PDOs, drop and distribution cable used. All the results for our approach are means of 30 independent runs.

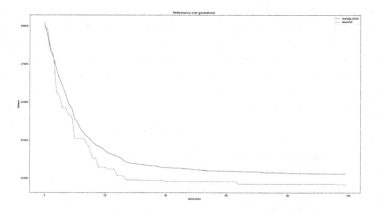

Fig. 1. Fitness over the generations for the dataset Map 1.

Figure 1 presents the average and best across the 100 generations over 10 runs. Looking at the results depicted in the plot, one can see that fitness rapidly decreases in the first 25 generations. After this point, the GA continues to improve the quality of the solutions but at a slower pace. These results show that one does not need to wait long to obtain good-quality results.

Table 5 summarises the results for the metrics. The second column shows the values for the handmade solution metrics, and the next column shows the results for the best, worst, median and means solutions obtained by the GA. When comparing the best GA solutions, we can see that the GA solutions reduce the number of PDOs (16). In contrast, it increases the distances for the drop cables.

We can also notice that the fitness score for the worst solutions increases considerably compared to the best solution (the best has a score of 23908 and the worst solution has a score of 204545). This high score is due to some penalties in the fitness function of the GA, more specifically in the case where some homes are not allocated to the network. A high constant penalty based on the homes left to be allocated is applied to guide the evolutionary algorithm towards complete coverage of the homes (which corresponds to a penetration rate of 100%).

The second column of the table (Handmade) contains the statistics for the handmade real-world deployed solution. When comparing it to the solutions generated by the GA, in terms of fitness score, the GA outperforms the handmade design when considering the best solution (and the median). This is because way too many PDOs are installed in the handmade solution. Every network equipment (like a pole or pedestal) has an installation point (PDO), which could be more optimal.

Table 5. Comparison between the handmade and the GA solutions for the Map 1.

Metrics	Handmade	GA			
		Best	Worst	Median	Mean
Number of PDOs	62	17	22	20	18.80 (±0.92)
Drop cable used (km)	2.75	4.18	3.76	4.07	4.15 (±0.12)
Distribution cable used (km)	2.08	1.74	1.77	1.74	1.70 (±0.03)
Fitness Score	34505	23908	204545	24561	38893(±5832)
Fitness Difference (GA-Handmade)		−10597	170040	−9944	−4388

One of the requirements for the PDOs was the number of ports available and the margin of ports to leave open. Table 6 shows the results for the PDO nodes with a building link. Those buildings (MDUs) can have a higher demand (i.e., more ports required) than a simple home (SDU) which is usually 1. We can see that even when multiple homes and buildings are connected to the same PDO, there is always a margin-left of ports as it never reaches the maximum limit. This allows for future network expansion in case new homes are built in those areas, for example, or another existing home changes its demand.

As for the types of splitters and cables, we needed the following extra equipment in Map 1: 2 splitters; 2 cables 16FO and; 3 cables 32FO. The demand for this network is 121 fibres. To serve that, we need at least 2 splitters of type 1 : 64. As for the cables, different types are needed based on the branches that come out of the starting point node. For example, a branch that requires 47 fibres requires a distribution cable of 32FO and another of 16FO (which results in 48 fibres, with 1 additional that is unused but still can be helpful for future expansion).

Regarding execution time, it took 30.5 seconds to do a single run to design this first network. This is advantageous compared to the handmade solution, which in theory, could take several hours for a network of this size. Overall, the total cost of the best-optimised version is 31% cheaper than the handmade counterpart.

Table 6. Amount of ports used in the PDOs that contain a connection to a building (MDU) for Map 1.

PDO node id (that contains a MDU link)	Ports used (out of 12 max.)	Amount of unique clients served (distinct SDUs/MDUs nodes)
0	6	5
9	9	4
24	9	8
65	7	5
76	11	5

Table 7. Comparison between the handmade and the GA solutions for the Map 2

		GA			
Metrics	Handmade	Best	Worst	Median	Mean
Number of PDOs	27	14	14	15	16 (±0.5)
Drop cable used (km)	3.53	3.09	3.29	2.90	3.01 (±0.01)
Distribution cable used (km)	2.76	1.65	1.57	1.76	1.71(±0.05)
Fitness Score	50216	24828	115013	25274	26635(±1247)
Fitness Difference (GA-Handmade)	–	−25388	64797	−24942	−23581

The second map studied is a map from New Jersey (Map 2), slightly bigger than Map 1, with more nodes and edges. This map provides a more challenging scenario since it has multiple possible paths. On Map 1, the number of possible paths was limited to reach a node on a particular branch, i.e., there was only one single path. Due to space constraints, we focus only on comparing the handmade

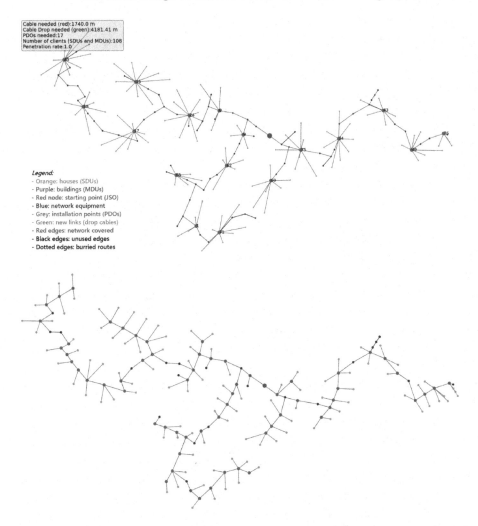

Cable needed (red):1740.0 m
Cable Drop needed (green):4181.41 m
PDOs needed:17
Number of clients (SDUs and MDUs):108
Penetration rate:1.0

Legend:
- Orange: houses (SDUs)
- Purple: buildings (MDUs)
- Red node: starting point (JSO)
- Blue: network equipment
- Grey: installation points (PDOs)
- Green: new links (drop cables)
- Red edges: network covered
- **Black edges: unused edges**
- Dotted edges: burried routes

Fig. 2. Visualisation of the solutions for Map 1. On the top we have the best solution of the proposed approach, on the bottom we have the existing real-world handmade and deployed one.

and the GA solutions in terms of the metrics. Looking at the results presented in Table 7, one can see that they are in line with the previous ones. The GA creates solutions with a small number of PDOs at the expense of increasing the distances of the drop cable. Regarding the computational time, despite being a larger map, it took, on average, 37s to design a whole network. Compared to the days it took to design the handmade solution, the computational overhead is negligible. The best solution found by the GA represents savings of 52.2% when comparing with the cost of to the handmade.

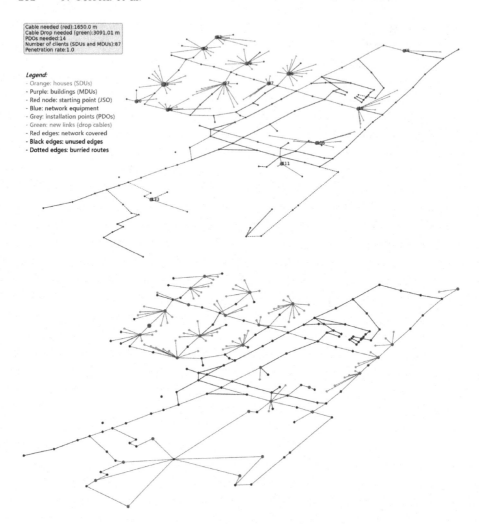

Fig. 3. Visualisation of the solutions for Map 2. On the top we have the best solution of the proposed approach, on the bottom we have the existing real-world handmade and deployed one.

In Figs. 2 and 3 we can observe the visual design solutions. In terms of solutions, they are distinct and present alternative solutions for engineers which can later be updated and enhanced. The tendency of the GA solutions is to explore and save on equipment and cut the costs of new equipment. Due to the nature of the fitness function, the outcome makes sense. We can observe that most PDOs have multiple collections when contrasting with the manually found ones. In general, the solutions found do not compromise the parameters set by the telecom company and end up giving a startup optimised and cost-saving solutions, minimising cable length, and number of PDOs and making use of the max drop cable constraint to save on multiple and short connections between clients and PDO.

6 Conclusion

Over the years, the demand for high-quality internet services increased rapidly due to the growth in the usage of bandwidth-intensive applications such as Cloud-based Services, Video and Audio Streaming, and gaming. This forces telecommunications companies to improve their network services to satisfy their clients. However, to create effective networks, the companies must consider several factors such as equipment cost, distance, the number of homes one has to serve and the optical budget. Considering all these factors, it is a complex task that requires time and human resources to do it. Therefore, by automating the design of these networks, we can minimise the effort required for the design process from days to a few hours or minutes whilst optimising the design in what concerns costs.

In this work, we propose an approach based on Genetic Algorithms to design telecommunications networks automatically. In concrete, we propose a representation based on two levels. The first level corresponds to a binary list defining the points of optimal distribution, and the second list indicates to which point a client is connected. The approach is validated by comparing the results obtained against two handmade real-world deployed solutions, using maps from New York and New Jersey. For each map, a comparison with a handmade solution was performed in order to assess the quality of the automatic design networks.

Overall, the GA can discover better solutions than real-world existing ones that were planned and handmade by teams of engineers. For the scenarios considered, our proposed solution can reduce the costs by 31% and 52.2%. Another important aspect of our approach is obtaining good-quality solutions in a relatively short computational time. Usually, the design of a new handmade solution can take up to multiple hours or even days of man-hours, depending on the complexity of the network. In contrast, the proposed system generates solutions in seconds, even for larger maps.

Acknowledgement. This work is partially funded by the project POWER (grant number POCI-01-0247-FEDER-070365), co-financed by the European Regional Development Fund (FEDER), through Portugal 2020 (PT2020), and by the Competitiveness and Internationalization Operational Programme (COMPETE 2020) and by the FCT - Foundation for Science and Technology, I.P./MCTES through national funds (PID-DAC), within the scope of CISUC R&D Unit - UIDB/00326/2020 or project code UIDP/00326/2020*.

References

1. Ali, R., Glover, T., Kampouridis, M., Tsang, E.: Guided local search for optimal GPON/FTTP network design. In: Computer Networks & Communications (Net-Com). UK

2. Andrej, C., Kin, P., Anis, O.: An enhanced ant colony optimization for FTTH access network design. In: 2012 17th European Conference on Networks and Optical Communications. Abu Dhabi, UAE (2012)
3. Andrej, C., Kin, P., Anis, O.: Using ant colony optimization to design GPON-FTTH networks with aggregating equipment. In: 2013 IEEE Symposium on Computational Intelligence for Communication Systems and Networks (CIComms) (2013)
4. Council, F.: Fiber-to-the-home handbook edition 7. Europe (2016)
5. Dias, L., Santos, A., Pereira, H., Almeida, R., Giozza, W., Sousa, R., Assis, K.: Evolutionary strategy for practical design of passive optical networks. Photonics **2022**(9), 278 (2022)
6. Kokungal, A.: Optimization of passive optical network planning. Appl. Math. Model. **35**(7), 3345–3354 (2011). Turkey (2011)
7. Lin, B.: Cascaded splitter topology optimization in LRPONs (2012)
8. Mata, J.: Artificial intelligence (AI) methods in optical networks: a comprehensive survey. UK (2018)
9. Morais, R., Pavan, C., Pinto, A., Requejo, C.: Genetic algorithm for the topological design of survivable optical transport networks. J. Opt. Commun. Netw. **3**, 17–26 (2011). Portugal (2011)
10. Papaefthimiou, K.: Algorithmic PON/P2P FTTH access network design for capex minimization. Serbia, Belgrade (2013)
11. Tany, V.: Design of passive optical networks using genetic algorithm. In: SBMO/IEEE MTT-S International Microwave and Optoelectronics Conference (IMOC). Brazil (2009)

RF+clust for Leave-One-Problem-Out Performance Prediction

Ana Nikolikj[1,2,3]($\boxtimes$) (iD), Carola Doerr[3] (iD), and Tome Eftimov[1] (iD)

[1] Computer Systems Department, Jožef Stefan Institute, 1000 Ljubljana, Slovenia
{ana.nikolikj,tome.eftimov}@ijs.si
[2] Jožef Stefan International Postgraduate School, 1000 Ljubljana, Slovenia
[3] Sorbonne Université, CNRS, LIP6, Paris, France
Carola.Doerr@lip6.fr

Abstract. Per-instance automated algorithm configuration and selection are gaining significant moments in evolutionary computation in recent years. Two crucial, sometimes implicit, ingredients for these automated machine learning (AutoML) methods are 1) feature-based representations of the problem instances and 2) performance prediction methods that take the features as input to estimate how well a specific algorithm instance will perform on a given problem instance. Non-surprisingly, common machine learning models fail to make predictions for instances whose feature-based representation is underrepresented or not covered in the training data, resulting in poor generalization ability of the models for problems not seen during training. In this work, we study leave-one-problem-out (LOPO) performance prediction. We analyze whether standard random forest (RF) model predictions can be improved by calibrating them with a weighted average of performance values obtained by the algorithm on problem instances that are sufficiently similar to the problem for which a performance prediction is sought, measured by cosine similarity in feature space. While our RF+clust approach obtains more accurate performance prediction for several problems, its predictive power crucially depends on the chosen similarity threshold as well as on the feature portfolio for which the cosine similarity is measured, thereby opening a new angle for feature selection in a zero-shot learning setting, as LOPO is termed in machine learning.

Keywords: Algorithm Performance Prediction · AutoML · Zero-Shot Learning · Single-Objective Black-Box Optimization

1 Introduction

Various algorithms for continuous single-objective optimization (SOO) have already been developed and their performance investigated through statistical analyses, in most cases reporting the average performance across a selected set of benchmark problem instances [26]. However, the algorithm instance behavior varies substantially depending on the problem instance that is being solved.

J. Correia et al. (Eds.): EvoApplications 2023, LNCS 13989, pp. 285–301, 2023.
https://doi.org/10.1007/978-3-031-30229-9_19

For this purpose, there is a predictive task known as automated algorithm selection, where the main goal is selecting the best-performing algorithm from a set of algorithm instances for a given problem instance [1,5,11,17]. To achieve it, automated algorithm performance prediction is a crucial step that should be done. The automated algorithm performance prediction is tackled as a supervised machine learning problem, such as classification (i.e., predicts whether or not the algorithm solves the instance given some precision) or a regression (i.e., predicts the performance of the algorithm as a real value). To train a supervised machine learning algorithm a set of examples, problem instances described by their landscape features used as input data (i.e., benchmark suite) and algorithm performance achieved on them used as target (i.e., solution precision) are required. Nowadays, if we want to generalize a supervised prediction model to another benchmark suite, whose problems were not involved in the training data, the predictive model performance decreases greatly. This occurs because the training data does not cover some regions of the landscape space involved in the test data, or the model is biased toward some over-represented landscape regions that are present in the training data. This is the reason why most of the studies focus on the Black-box Optimization Benchmarking (BBOB) benchmark suite [9] since it involves several instances from the same problem class involved in the training data which makes it a suitable resource for training an ML-supervised model. However, if we remove the instances from the same problem class from the training data (leave all instances from a single problem out for testing), the performance of the model for automated algorithm performance prediction decreases significantly [25]. This issue makes all approaches for automated algorithm performance prediction that are developed based on the BBOB benchmark suite difficult to generalize on other benchmark suites such as CEC [16], Nevergrad [23], etc. since in their definition there is only a single instance per each problem class. In practice, an application such as algorithm performance prediction requires making predictions for problem instances whose problem class (i.e., landscape properties) has not been seen previously by the underlying model or we have a leave-one-problem-out (LOPO) learning scenario.

Our Contribution. In this study, we propose an approach for LOPO automated algorithm performance prediction. Our RF+clust approach calibrates a classic random forest (RF) predictive model with a prediction obtained by a similarity relationship method that aggregates the ground algorithm instance performance for the most similar problem instances from the training data. In contrast to a classic KNN approach, we use a similarity threshold to decide which problems are taken into account for the calibration. The number of considered 'neighbors' can therefore differ between problems. We evaluate our approach on performance data of three differential evolution (DE) variants on the CEC 2014 and the BBOB benchmark suite of the COmparing Continuous Optimizers (COCO) environment [10]. We observe better results for RF+clust than for stand-alone RF performance prediction in a number of cases. However, there are also cases when similar landscape representations can lead to different performances of the algorithm, which can affect the prediction of the RF+clust approach. This further points out that in the future we need to focus on finding problem feature representations with sufficient discriminate power that will be also able to capture the performance of the algorithm.

Outline. The remainder of the paper is organized as follows: Sect. 2 surveys past work on automated algorithm performance prediction. The proposed LOPO regression method is introduced in Sect. 3. Section 4 details the benchmark problem suites and algorithms used for the validation of the proposed approach, the problem landscape features, as well as the machine learning algorithm tuning and evaluation. The results and discussion are provided in Sect. 5. Finally, the conclusions are drawn in Sect. 6.

2   Related Work

Next, we point out some of the works which are addressing the critical issue of generalization over new problem classes.

Bischl et al. [3] consider automated algorithm selection as a cost-sensitive classification task using one-sided Support Vector Machines. Problem instance-specific miss-classification costs are defined, unlike standard classification where all the errors in classification are penalized the same by the algorithm. The prede-fined miss-classification costs represent external information to aid the learning process. The approach was tested on problem instance feature representations consisting of "cheap" and "expensive" ELA features [18], with respect to the sample size required for their calculation. It was shown that the model is able to generalize over new instances, however, the prediction error gets worse for new problem classes, when the prediction is based only on the "cheap" ELA features representation. To discover the source of the larger model errors, analysis of the feature space is performed based on the euclidean distances between the problem instances representations. They conclude that the degree of classification performance tends to correlate with the proximity in feature space for the case of using the entire feature set, however, this was not that straightforward for the "cheap" features.

The work [7] brings to attention the possibility to *personalize* regression models (Decision Tree, Random Forest, and Bagging Tree Regression) to specific optimization problem classes. Instead of aiming for a single model that works well across a whole set of possibly diverse problems, the personalized regression approach acknowledges that different models may suit different problem types.

In [25] a classification-based algorithm selection approach is evaluated on the COCO benchmark suite [9] and artificially generated problems [6]. The results show that such a model has low generalization power between datasets and in the leave-one problem-out cross-validation procedure where each problem class was removed one at a time from the same dataset. However, a model trained and tested in a leave-instance-out scenario achieves much higher accuracy. A correlation analysis using the Pearson correlation coefficient was performed for the problem representations based on the "cheap" ELA features, showing that a large number of both, the COCO and the artificial problems are highly correlated within their own set of problems. i.e., the poor generalization is due to the differences between the two data sets in feature space.

The feasibility of a "per-run" algorithm selection scheme is investigated in [14], based on ELA features that are calculated from the observed trajectory of the algorithm (i.e., the samples the algorithm visits during the optimization procedure). This avoids the usually required additional evaluation of (quasi-)random samples implemented by classic per-instance algorithm selection schemes. Results for the COCO benchmark suite show performance comparable to the per-function virtual best solver. However, these results did not directly generalize to the other benchmark suites used in the experiments, namely the YABBOB suite from the Nevergrad platform [23].

3   LOPO Algorithm Performance Prediction

Let us assume a set of benchmark problem instances P_t^i, $i = 1, \ldots, n$, which are grouped into training problem classes P_t, $t = 1, \ldots, m$, and performance data for an algorithm instance A on the selected set of benchmark problem instances. To predict the performance y_q^i of the algorithm instance A on a problem instance P_q^i from a new problem class P_q that is not involved in the training data, we have proposed the following LOPO approach:

1) Represent the selected benchmark problem instances from the m problem classes by calculating the ELA features and linking them to the performance of the algorithm instance after a certain number of function evaluations.
2) Train a supervised regression model that uses the ELA features as input data and predicts the algorithm instance performance.
3) For a new problem instance P_q^i from a new problem class P_q that is not involved in the training data, use its ELA features as input data into the learned model to make the prediction y_q^i.
4) Select the k-nearest problem instances from the training set that are the most similar to the new problem instance based on their landscape features representation. The similarity is measured by a similarity metric s, and the selection is done by defining a prior similarity threshold. We selected cosine similarity as a similarity measure. Finally, all problem instances from the training data that have a similarity greater or equal to the predefined threshold are selected, from which the ground truth algorithm performance is retrieved $p_1, p_2, \ldots, p_k$. We need to point out here that the number of the nearest problem instances k, differs for different problem instances, so it is found by the selection rule and the predefined threshold.
5) The final prediction of the algorithm instance performance on the new problem instance is made by calibrating the prediction obtained by the learned model y_q^i with the ground truth algorithm instance performance retrieved for the selected nearest problem instances from the training data. This is performed as an aggregation procedure as follows: $\widehat{y}_q^i = \left(y_q^i + F(p_1, p_2, \ldots, p_k) \right) / 2$, where $F(p_1, p_2, \ldots, p_k)$ is an aggregation function, which can be for example weighted mean.

6) If there are no problem instances in the training data to which the new problem instance is similar above the threshold, only the prediction of the model is considered, y_q.

4   Experimental Design

Here, we are going to present all experimental details starting from the data that is involved and the techniques used for the ML learning process.

Problem Benchmark Suites. We evaluate the proposed method by using two of the most currently used benchmark problem suites in the field of numeric single-objective optimization. The first benchmark suite involved in the experiments is the 2014 CEC Special Sessions & Competitions (CEC 2014) suite. The suite consists of 30 problems where only one instance per problem is available. The problems are provided in dimension 10. The full problem list and descriptions of all the problems are available at [16]. The second problem set is the 24 noiseless single-objective optimization problems from the BBOB collection of the COCO benchmarking platform [10]. Different problem instances can be derived by transforming the base problem with predefined transformations to both its domain and its objective space, resulting in a set of different instances for each base problem class, that have the same global characteristics. We consider the first five instances of each BBOB problem, resulting in a dataset of 120 problem instances. In coherence with the CEC problem suite, the problem dimension D was set to 10.

Algorithm Performance Data. Performance data is collected for three different randomly selected Differential Evolution (DE) [27] configurations, on both the CEC 2014 and BBOB benchmark suites. The DE hyper-parameters are as presented in Table 1. We indexed the algorithm configurations starting from DE1 to DE3 for easier notation of the results. DE is an iterative population-based meta-heuristic. The population size of DE is set to equal the problem dimension D ($D = 10$ in our study). The three DE configurations were run 30 times on each problem instance, and we extracted the precision after a budget of $500D = 5000$ function evaluations. In our study, we consider the median target precision achieved in these 30 runs. Following the approach suggested in [11], we also consider the logarithm (log10) of the median solution precision. This algorithm performance measure estimates the order of magnitude of the distance of the reached solution to the optimum. Figure 1 presents DE1 performance (log-scale) obtained per benchmark problem on the CEC 2014 and Fig. 2 on the BBOB benchmark suite (aggregated for all problem instances).

Table 1. Differential Evolution (DE) configurations.

index	*strategy*	F	Cr
DE$_1$	Best/3/Bin	0.533	0.809
DE$_2$	Best/1/Bin	0.617	0.514
DE$_3$	Rand/Rand/Bin	0.516	0.686

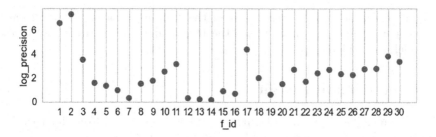

Fig. 1. DE1 solution precision (log-scale) per problem instance on the CEC 2014 suite.

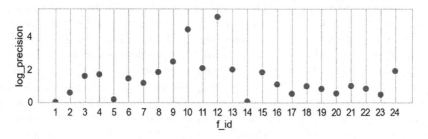

Fig. 2. DE1 solution precision (log-scale) on the first instance of each problem in the BBOB suite.

Exploratory Landscape Analysis (ELA). To create a feature representation that encodes problem properties, the static Exploratory Landscape Analysis (ELA) [18] features are used. The features are calculated by the evaluation of a sample of candidate solutions generated by systematic sampling of the decision space of the problem. The corresponding fitness values are then fed to different statistical and mathematical methods to calculate the feature values. The Improved Latin Hypercube Sampling (ILHS) [28] was used as a sampling technique, with a sample size of $800D$ (8000). In reality, this is a really big sample size, however, we are interested in whether the approach works, so we want to reduce the randomness from the feature computation [15]. For each benchmark problem instance, the calculation of the features was repeated 30 times, as it is a stochastic process and the median value was taken as the final feature value that numerically quantifies some property of the problem. The R package "flacco" [13] was utilized for their calculation. We selected all the ELA features which are cheap to calculate with regard to sample size, and do not require additional sampling. This way, a total of 64 features were calculated. The selected features are coming from the following groups: classical ELA (y-distribution measures, level-set, meta-model), Dispersion, Information Content, Nearest Better Clustering, and Principal Component Analysis.

Table 2. RF hyper-parameters and the considered search spaces.

hyper-parameter	search space
n estimators	{10, 20, 50, 70}
max features	{all, sqrt, log2}
max depth	{3, 5, 7, 10}
min samples split	{2, 5, 7, 10}

Regression Models for Algorithm Performance Prediction. For the learning process, we considered random forest (RF) regression [2], as it provides promising results for algorithm performance prediction [12] and is one of the most commonly used algorithms for algorithm performance prediction studies in evolutionary computation. The RF algorithm was used as implemented by the scikit-learn package [21] in Python. We have trained single-target regression (STR) models. That is, we have a separate model for predicting the performance of each of the three DE algorithms.

ML Model Evaluation. When splitting the data as described in Sect. 3 the evaluation of the automated algorithm performance prediction results in leave-one-problem-out fold validation. At each fold, a model was trained using one problem class (including all of its instances) left out for testing, while the remaining are used for training. In order to assess the accuracy of the models, we compute the Mean Absolute Error (MAE). The prediction errors are the absolute distances of the prediction to the true algorithm precision value on the new problem class.

Hyper-parameter Tuning for the Regression Models. The best hyper-parameters are selected for each RF model from the training portion of the fold. The hyper-parameters selected for tuning are *n estimators* - the number of trees in the random forest; *max features* - the number of features used for making the best split; *max depth* - the maximum depth of the trees, and *min samples split* - the minimum number of samples required for splitting an internal node in the tree. The ranges of the hyper-parameters have been selected concerning the data set size and the guidelines available in ML to avoid over-fitting. The best hyper-parameters for each problem class are presented in our repository [20] (Table 2).

Feature Selection. Taking into consideration the size of the datasets, in a scenario where 30 data instances (CEC 2014 benchmark suite) are available, and 64 features to describe them, we run the risk of overfitting our model. Therefore, we have performed feature selection. Since we have a LOPO scenario (i.e., leave all instances for a single problem out), in our case we ended with 30 ML predictive models. To select the top most important features for each model, the SHAP method [19] was utilized. Finally, the importance of the features was summarized across all the models, and the 10 and 30 most important features were used to train the models. These two sizes of feature portfolios were tested

Table 3. Mean Absolute Error (MAE) obtained by the RF models for predicting the performance of DE1 with different feature portfolios on the CEC 2014 benchmark problems

top features	aggregation	mae_train	mae_test
10	mean	0.504464	1.279261
10	median	0.536681	0.991274
30	mean	0.458142	1.318326
30	median	0.439434	0.770265
64	mean	0.498807	1.465239
64	median	0.480457	0.931685

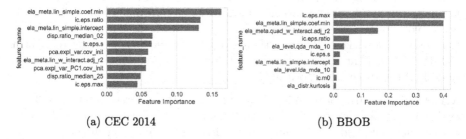

(a) CEC 2014   (b) BBOB

Fig. 3. The ten most important ELA features for predicting the performance of DE1 on the CEC 2014 and BBOB benchmark suites.

in order to compare the decrease in ML model performance (if any) when using different feature portfolios (Table 3). Also, the feature portfolio influences the proposed approach as it is based on the pairwise similarity of the features. So different feature portfolios can result in different instances retrieved as similar. Figure 3 shows the feature portfolio of the 10 most important features for CEC 2014 and BBOB accordingly.

5   Results

We apply the approach to three random DE configurations and two benchmark suites (CEC 2014 and COCO). Due to space limitations, however, we present here some selected results for algorithm DE1 and the CEC 2014 benchmark suite, while other results where similar findings are noticed, are available at [20].

In Fig. 4 we compare a classical supervised RF model and an RF+clust model in a leave-one-problem-out scenario. Figure 4 shows the prediction errors obtained by a standard RF model trained in the LOPO (corresponding to the "RF" denoted row on the heatmap) and errors of the proposed RF+clust approach (for similarity thresholds of 0.5, 0.7, and 0.9, also with corresponding rows on the heatmap). The predictions were obtained by using a feature portfolio of the ten most important features. Each cell of the heatmap represents the absolute error obtained by the models. The columns represent each problem instance

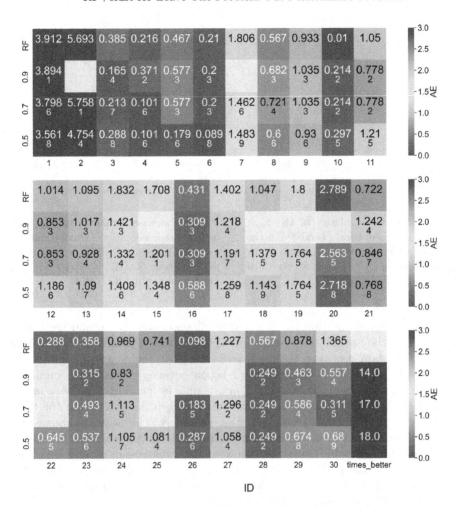

Fig. 4. The mean absolute error of the RF model and the RF+clust approach using a similarity threshold of .5, .7, and .9, for each problem in the CEC 2014 benchmark suite, for DE1 and the feature portfolio of ten most important ELA features.

separately, while the last column indicates on how many instances the approach showed lower prediction error). The numbers under the model error indicate the number of similar instances set above the corresponding threshold that were present in the training of the model. The blank cells in the heatmap are places where the RF+clust approach provides the equal result as the RF model because for those problems we could not find similar problem instances from the training data using the predefined threshold.

The figure shows that there are problems (1, 3, 6, 11, 12, 13, 14, 16, 17, 23, 24, 28, 29, 30) for which the RF+clust approach shows better predictive results than using a classically trained RF model. We can see that for the high similarity

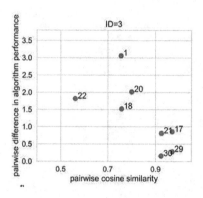

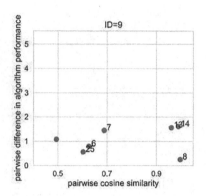

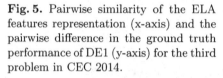

Fig. 5. Pairwise similarity of the ELA features representation (x-axis) and the pairwise difference in the ground truth performance of DE1 (y-axis) for the third problem in CEC 2014.

Fig. 6. Pairwise similarity of the ELA features representation (x-axis) and the pairwise difference in the ground truth performance of DE1 (y-axis) for the ninth problem in CEC 2014.

threshold (0.9), calibrating the classical prediction with the ground truth performance of the optimization algorithm of the retrieved similar problems from the training data decreases the predictive error. To provide an explanation for why this happens, Fig. 5 presents the relation between the pairwise similarity of the ELA features representation (x-axis) and the pairwise difference in the ground truth performance of the optimization algorithm (y-axis) for the third problem in the CEC 2014, with the other problems. The heatmap shows that the third problem has four similar problem instances over 0.9 (17, 21, 29, 30), as visible in Fig. 5. In addition, we can see that the difference in ground algorithm performance of the problem and the similar instances is low, so the algorithm has similar behavior on these problems (see also Fig. 1), and using them for the calibration helps to obtain better predictive errors.

There are also problems when the predictions are affected by the RF+clust approach, slightly worse than the prediction obtained from the classical RF model. Figure 6 presents the relation between the pairwise similarity of the ELA features representation (x-axis) and the pairwise difference in the ground truth performance of the optimization algorithm (y-axis) for the ninth problem in the CEC 2014. The heatmap shows that the ninth problem has three similar problem instances (8, 13, 14) according to Fig. 6. Here, we can see that one out of three problems is similarly based on the ELA representation and the algorithm has similar behavior on it. However, on the remaining two problems, we can see that even with high similarity in the landscape space, the difference in algorithm performance is larger in reality (see Fig. 1), so using the performance to calibrate the prediction yields a larger error. A similar scenario happens for the 21st problems, where the similarity is greater or equal to 0.9 but the difference in performance between them is very large (see Fig. 7). This indicates that there are problems for which the ELA features representations are not expressive enough

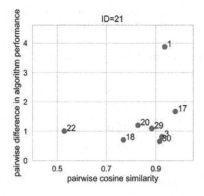

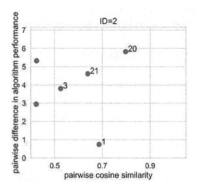

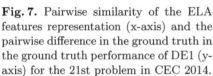

Fig. 7. Pairwise similarity of the ELA features representation (x-axis) and the pairwise difference in the ground truth in the ground truth performance of DE1 (y-axis) for the 21st problem in CEC 2014.

Fig. 8. Pairwise similarity of the ELA features representation (x-axis) and the pairwise difference in the ground truth in the ground truth performance of DE1 (y-axis) for the second problem in CEC 2014.

and they could not well describe the problems in such a scenario (i.e., similar ELA landscape representation may not lead to similar algorithm behavior in the performance space).

There are also problems such as the first and the second that are difficult to be solved by the optimization algorithm (see Fig. 1 for ground truth performance). Figure 8 presents the relation between the pairwise similarity of the ELA features representation (x-axis) and the pairwise difference in the ground truth performance of the optimization algorithm (y-axis) for the second problem in the CEC 2014. It is visible that this problem has very few similar instances, does not have similar instances over 0.9 at all, and also the difference in algorithm performance with similar instances over 0.5 is very large. This is the case where the test problem class is not covered enough by the train, however, even in such scenarios, we can be slightly better in the prediction results.

Looking back at the heatmap (Fig. 4), we can see that when the similarity threshold decreases (i.e., going to 0.5), some of the predictions are slightly worse. This was expected since having a lower threshold in the landscape space does not guarantee to capture similar performance in the performance space (see Figs. 6, 7, and 8). The heatmaps presenting the results for the other two DE algorithms lead to similar results and explanations.

In Table 4 we summarize how many out of the 30 problems the RF+clust approach provides better, worse, or equal predictions (when similar instances are not found, the prediction is not calibrated) than the classical (stand-alone) RF model on the CEC 2014 benchmark suite.

Table 4. Number of times the RF+clust is better, equal, or worse than stand-alone RF for predicting the performance of DE variants on the CEC 2014 benchmark suite.

algorithm name	model	# better	# equal	# worse
DE1	0.9	14.0	10.0	6.0
DE1	0.7	17.0	2.0	11.0
DE1	0.5	18.0	0.0	12.0
DE2	0.9	13.0	5.0	12.0
DE2	0.7	18.0	0.0	12.0
DE2	0.5	21.0	0.0	9.0
DE3	0.9	8.0	15.0	7.0
DE3	0.7	14.0	2.0	14.0
DE3	0.5	16.0	0.0	14.0

To investigate the sensitivity with a different feature portfolio, we repeated the experiments by selecting the top 30 most important features for predicting the performance of DE1 on the CEC 2014 benchmark suite. Figure 9 presents the absolute errors of the RF and RF+clust approaches obtained for each problem of the CEC 2014. Focusing on the similarity threshold of 0.9, the results show that the RF+clust approach provides better predictions (i.e., improvements) than the classical RF model for nine problems, worse predictions for five problems, and equal for 16 problems. We need to point out here that increasing the number of the top most important features from 10 to 30, also affects the similarity of the problem instances. From the heatmap is visible that now we are not able to detect similar problems for some of the problems (e.g., 1, 21, 23, 24, 28, 29, 30) for which we were able to detect similar problems above 0.9 when the feature portfolio of the 10 most important features is used. This further opens a new angle for feature selection that will have discriminate power to capture also differences that happen in the performance space.

In addition to the results obtained on the CEC 2014 benchmark suite, Fig. 10 presents the prediction results obtained for the DE1 algorithm for the first instance of each COCO problem. We selected only one instance here, for visualization purposes (our overall setting remains LOPO, i.e., we omit *all* instances of the left-out problem, and we use data from the five instances of the other problems for training. This also explains why the number of similar problem instances is larger in Fig. 10 compared to those for the CEC benchmark presented in Fig. 4). The results are in line with those obtained for the CEC 2014 benchmark. We point out that the top 10 most important features to train the prediction model differ from those selected on the CEC 2014 benchmark; see Fig. 3 for details. An indirect outcome of this study is that these two benchmark suites are different, which also supports previously published findings [24].

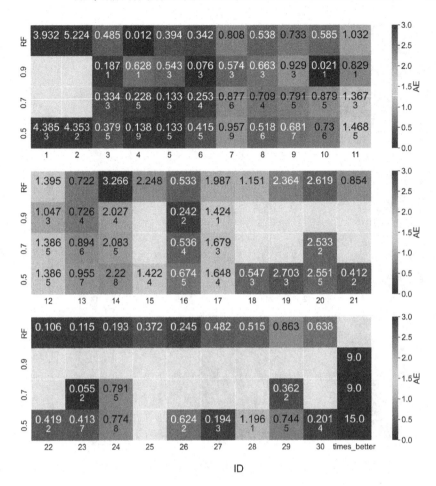

Fig. 9. The mean absolute error of the RF model and the RF+clust approach using a similarity threshold of .5, .7, and .9, for each problem in the CEC 2014 benchmark suite, for DE1 and the feature portfolio of 30 most important ELA features.

298   A. Nikolikj et al.

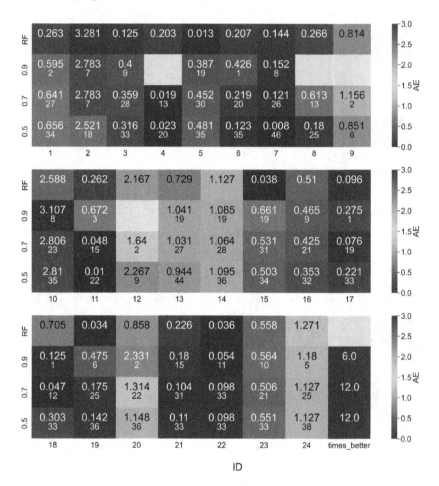

Fig. 10. The mean absolute error of the RF model and the RF+clust approach using a similarity threshold of .5, .7, and .9, for the first instance of each problem in the BBOB benchmark suite, for DE1 and the feature portfolio of ten most important ELA features.

6   Conclusions

In this study, we have proposed leave-one-problem-out (LOPO) for algorithm performance prediction. The idea behind the approach is to predict the performance of an optimization algorithm by using a supervised ML model on a problem that is not presented in the training data. First, a model is learned from a feature landscape representation of the problem instances from the train problem classes, and a prediction is made for an instance from a new problem class in a supervised manner. Second, based on the similarity relationship between the problem classes based on their feature landscape representation, the prediction for the new problem instance is calibrated by applying an aggregation procedure over the algorithm performance of the k-nearest problem instances from the training data.

The results performed on the CEC 2014 benchmark suite showed promising results and explanations about the strengths and weaknesses of the proposed approach. Better results are achieved for problems for which their landscape feature representation is similar to other problems from the training data and the algorithm behaves similarly on those problems. However, there are also problems for which the proposed approach can lead to slightly worse prediction results. This happens for problems for which the landscape feature representation leads to finding similar problems from the training data, however, the performance of the algorithm significantly differs. Such a result indicates that we need to find an expressive enough landscape feature representation that correlates with the algorithm's performance. Also, there are problems for which there are no similar problems in the training data, which further indicates that we need to enrich the data that is used in ML setup with new problems (e.g., merging different benchmark suites or using artificially problem generators [6]) by taking care that all landscape spaces approximate a uniform distribution in the problem space.

In the future, we are planning to test the approach on a more comprehensive algorithm portfolio. Next, instead of exploratory landscape features calculated by a global sampling, we are planning to calculate them using the trajectory data that was observed by the algorithm during the run (i.e., to capture also information about the algorithm behavior) [14]. We are also going to try different problem feature representations such as topological data analysis [22]. Last but not least, we are planning to merge different benchmark suites to select representative problem instances [4,8] that will allow us to represent all possible landscape spaces from the problem space with the same number of problems that will further help the LOPO approach to have better prediction results.

Acknowledgments. The authors acknowledge the support of the Slovenian Research Agency through program grant P2-0098, project grants N2-0239 and J2-4460, and a bilateral project between Slovenia and France grant No. BI-FR/23-24-PROTEUS-001 (PR-12040). Our work is also supported by ANR-22-ERCS-0003-01 project VARIATION.

References

1. Belkhir, N., Dréo, J., Savéant, P., Schoenauer, M.: Per instance algorithm configuration of CMA-ES with limited budget. In: GECCO, pp. 681–688 (2017)
2. Biau, G., Scornet, E.: A random forest guided tour. TEST **25**(2), 197–227 (2016). https://doi.org/10.1007/s11749-016-0481-7
3. Bischl, B., Mersmann, O., Trautmann, H., Preuß, M.: Algorithm selection based on exploratory landscape analysis and cost-sensitive learning. In: Proceedings of the 14th Annual Conference on Genetic and Evolutionary Computation, pp. 313–320 (2012)
4. Cenikj, G., Lang, R.D., Engelbrecht, A.P., Doerr, C., Korošec, P., Eftimov, T.: Selector: selecting a representative benchmark suite for reproducible statistical comparison. arXiv preprint arXiv:2204.11527 (2022)

5. Derbel, B., Liefooghe, A., Vérel, S., Aguirre, H., Tanaka, K.: New features for continuous exploratory landscape analysis based on the soo tree. In: FOGA, pp. 72–86 (2019)
6. Dietrich, K., Mersmann, O.: Increasing the diversity of benchmark function sets through affine recombination. In: Rudolph, G., Kononova, A.V., Aguirre, H., Kerschke, P., Ochoa, G., Tušar, T. (eds.) International Conference on Parallel Problem Solving from Nature, pp. 590–602. Springer, Cham (2022). https://doi.org/10.1007/978-3-031-14714-2_41
7. Eftimov, T., Jankovic, A., Popovski, G., Doerr, C., Korošec, P.: Personalizing performance regression models to black-box optimization problems. In: GECCO, pp. 669–677 (2021)
8. Eftimov, T., et al.: Less is more: Selecting the right benchmarking set of data for time series classification. Expert Syst. Appl. **198**, 116871 (2022)
9. Hansen, N., Auger, A., Finck, S., Ros, R.: Real-parameter black-box optimization benchmarking 2010: Experimental setup. Ph.D. thesis, INRIA (2010)
10. Hansen, N., Auger, A., Ros, R., Mersmann, O., Tušar, T., Brockhoff, D.: COCO: a platform for comparing continuous optimizers in a black-box setting. Optim. Meth. Software **36**(1), 114–144 (2021)
11. Jankovic, A., Doerr, C.: Landscape-aware fixed-budget performance regression and algorithm selection for modular CMA-ES variants. In: GECCO, pp. 841–849. ACM (2020)
12. Jankovic, A., Popovski, G., Eftimov, T., Doerr, C.: The impact of hyper-parameter tuning for landscape-aware performance regression and algorithm selection. In: Proceedings of the Genetic and Evolutionary Computation Conference, pp. 687–696 (2021)
13. Kerschke, P., Trautmann, H.: The r-package flacco for exploratory landscape analysis with applications to multi-objective optimization problems. In: 2016 IEEE Congress on Evolutionary Computation (CEC), pp. 5262–5269. IEEE (2016)
14. Kostovska, A., et al.: Per-run algorithm selection with warm-starting using trajectory-based features. arXiv preprint arXiv:2204.09483 (2022)
15. Lang, R.D., Engelbrecht, A.P.: An exploratory landscape analysis-based benchmark suite. Algorithms **14**(3), 78 (2021)
16. Liang, J.J., Qu, B.Y., Suganthan, P.N.: Problem definitions and evaluation criteria for the cec 2014 special session and competition on single objective real-parameter numerical optimization. Technical report Zhengzhou, China **635**, 490 (2013)
17. Malan, K.M., Engelbrecht, A.P.: Fitness landscape analysis for metaheuristic performance prediction. In: Richter, H., Engelbrecht, A. (eds.) Recent Advances in the Theory and Application of Fitness Landscapes. ECC, vol. 6, pp. 103–132. Springer, Heidelberg (2014). https://doi.org/10.1007/978-3-642-41888-4_4
18. Mersmann, O., Bischl, B., Trautmann, H., Preuss, M., Weihs, C., Rudolph, G.: Exploratory landscape analysis. In: GECCO, pp. 829–836 (2011)
19. Molnar, C.: Interpretable machine learning. Lulu. com (2020)
20. Nikolikj, A.: Rfclustgit (2023). https://github.com/anikolik/RF-clust
21. Pedregosa, F., et al.: Scikit-learn: machine learning in python. J. Mach. Learn. Res. **12**, 2825–2830 (2011)
22. Petelin, G., Cenikj, G., Eftimov, T.: Tla: Topological landscape analysis for single objective continuous optimization problem instances. In: 2022 IEEE Symposium Series on Computational Intelligence (SSCI). p. In Press. IEEE (2022)
23. Rapin, J., Teytaud, O.: Nevergrad - A gradient-free optimization platform. https://GitHub.com/FacebookResearch/Nevergrad (2018)

24. Škvorc, U., Eftimov, T., Korošec, P.: Understanding the problem space in single-objective numerical optimization using exploratory landscape analysis. Appl. Soft Comput. **90**, 106138 (2020)

25. Škvorc, U., Eftimov, T., Korošec, P.: Transfer learning analysis of multi-class classification for landscape-aware algorithm selection. Mathematics **10**(3), 432 (2022)

26. Stork, J., Eiben, A.E., Bartz-Beielstein, T.: A new taxonomy of global optimization algorithms. Natural Comput. **21**, 219–242 (2020). https://doi.org/10.1007/s11047-020-09820-4

27. Storn, R., Price, K.: Differential evolution-a simple and efficient heuristic for global optimization over continuous spaces. J. Global Optim. **11**(4), 341–359 (1997)

28. Xu, Q., Yang, Y., Liu, Y., Wang, X.: An improved Latin hypercube sampling method to enhance numerical stability considering the correlation of input variables. IEEE Access **5**, 15197–15205 (2017)

Evolving Non-cryptographic Hash Functions Using Genetic Programming for High-speed Lookups in Network Security Applications

Mujtaba Hassan[1]([✉]) [iD], Arish Sateesan[1] [iD], Jo Vliegen[1] [iD], Stjepan Picek[3] [iD], and Nele Mentens[1,2] [iD]

[1] imec-COSIC/ES&S, ESAT, KU Leuven, Leuven, Belgium
{mujtaba.hassan,arish.sateesan,jo.vliegen,nele.mentens}@kuleuven.be
[2] LIACS, Leiden University, Leiden, The Netherlands
[3] Digital Security Group, Radboud University, Nijmegen, The Netherlands
stjepan.picek@ru.nl

Abstract. Non-cryptographic (NC) hash functions are the core part of many networking and security applications such as traffic flow monitoring and deep packet inspection. For these applications, speed is more important than strong cryptographic properties. In Terabit Ethernet networks, the speed of the hash functions can have a significant impact on the overall performance of the system when it is required to process the packets at a line rate. Hence, improving the speed of hash functions can have a significant impact on the overall performance of such architectures. Designing a good hash function is a challenging task because of the highly non-linear and complex relationship between input and output variables. Techniques based on Evolutionary Computation (EC) excel in addressing such challenges. In this paper, we propose novel fast non-cryptographic hash functions using genetic programming, and we call the resulting hash functions the GPNCH (Genetic Programming-based Non-Cryptographic Hash) family. We choose to employ avalanche metrics as a fitness function because the networking and security applications we consider require hash functions to be uniform and independent. We evaluate the performance of GPNCH functions on FPGA and compare the delay, throughput, and resource occupation with the state-of-the-art NC hash functions that satisfy the avalanche criteria. We show that GPNCH functions outperform the other algorithms in terms of latency, operating frequency, and throughput at the modest cost of hardware resources.

Keywords: Evolutionary Computation · Genetic Programming · Non-cryptographic Hash Functions · FPGA · Avalanche Metrics

1 Introduction

All the hash functions (cryptographic and non-cryptographic) map messages of arbitrary length into fixed-length codes called hashes or digests. The output of any good hash function for a large set of inputs must be uniformly distributed and exhibit good randomness. Moreover, the hash code should be changed significantly by altering any single bit in the input, which is known as the avalanche effect. Cryptographic hash functions offer better security guarantees by holding properties such as pre-image, second pre-image, and collision resistance. On the other hand, for NC hash functions, such properties are relatively relaxed in exchange for fast execution time. Name any application where there is a need to do a fast lookup, NC hash functions can be used, e.g., hash tables, checksums and correction codes, finding duplicate records, caches, data compression, Bloom filters, transposition tables, etc. Probabilistic data structures such as Bloom filters [22], Hyperloglog [12], and Count-Min Sketches [7] depend upon NC hash functions to do probabilistic membership queries or counting. In network security applications where a few false positives are acceptable, and the address of stored data is not required, Bloom filters are the preferred choice because they provide constant time queries along with better space utilization [23].

Network traffic monitoring and filtering allow network managers to keep the network safe from malicious traffic and ensure that network bandwidth is utilized effectively. These tools often check for the source and destination IP addresses, sender and receiver ports, and protocols by extracting information from the packet header. A combination of these fields makes up a network flow ID. These flow IDs can be used as an index in a table to look for potential matches to redirect, count or block the packets. On average, hash-based search algorithms offer constant time complexity of $O(1)$, independent of input data size. In the worst case, where every hash value matches with the input key, they can behave like linear search $O(n)$, but this will only happen if they are poorly designed. This time complexity, together with low memory requirements, makes them an ideal choice for high-speed lookups in databases.

For high-speed networks like Terabit Ethernet networks (i.e., networks having a bandwidth of more than 100 Gbps), high-speed flow ID lookups need to be accomplished at a line rate. Therefore, implementing probabilistic architectures such as Bloom filters and Sketches on FPGA can lead to low latency, high throughput, and high operating frequencies [1]. Many large data centers that form the backbone of cloud computing are also using FPGAs to achieve high efficiency at a low-cost [4]. To meet the present and future bandwidth requirements, one way is to increase the computing capabilities of such architectures by improving the design of NC hash functions. This has led Sateesan et al. [29] to develop a hardware-oriented NC hash function called Xoodoo-NC for Bloom filter to process 96-bit network flow IDs. The authors significantly improved the performance of the Bloom filter compared to the FNV-1a [14] based implementation, which is known to have the smallest computational delay [29]. They partitioned a single hash output of 96 bits into several sets instead of using many hash functions. Claesen et al. [6], inspired by this work [29], proposed five more NC hash functions for FPGA-based high-speed lookups. One of their proposed algorithms, GIFT-NC, managed to slightly improve the latency and throughput of Xoodoo-NC at the cost of more hardware resources.

The main motivation behind our work is to build upon the work of [6] and [29] by improving the latency and throughput of hash functions on FPGAs since networks having a speed of 46 Terabits have become a reality, and in the future bandwidths, will continue to increase [5]. Therefore, network monitoring and security applications that process data at a line rate should be ready for these bandwidths. Designing hash functions is extremely challenging as the interconnection between input and output variables is highly nonlinear, convoluted, and abstruse, which makes them an ideal candidate for evolutionary computation (EC) techniques [11]. EC approaches [10] such as Cartesian Genetic Programming (CGP) and Genetic Programming (GP), which are inspired by Darwin's theory of natural selection [3], make an excellent choice to design and optimize digital circuits because both can easily be modeled in the hardware. They use the same logical operations that are the main building blocks in the modeling of digital circuits [26]. Surprisingly, there has been very little research done to use EC in the design of NC hash functions targeting efficient FPGA implementations. This drives our motivation to use GP to evolve digital circuits to design high-speed NC hash functions. GP is an example of the EC algorithms [10] in which the solution to a given problem is represented as a computer program with the help of syntax trees, as is shown in Fig. 1. All the solutions start from the embryonic stage and undergo the evolution process of selection, crossover, and mutation in search of finding the optimized individual representing the final solution.

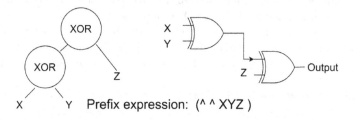

Fig. 1. Individual represented with GP (left), equivalent 3-bit parity checker (right), and prefix expression (bottom)

In this research, we show that our proposed family of NC hash functions outperforms the compared NC hash functions in terms of latency, operating frequency, and throughput on FPGA. Our contributions to the research domain are as follows:

- We are the first to design NC hash functions using EC for a 96-bit input/output block while fully satisfying all three avalanche metrics (entropy, bit dependence, and avalanche weight).
- Our proposed method outperforms the current state-of-the-art NC hash functions (which satisfy the avalanche criteria) on FPGA in terms of delay, operating frequency, and throughput.
- The proposed solution generates a family of independent NC hash functions. This could cater to different applications that require multiple, independent hash algorithms, like the Bloom filter.

The rest of the paper is organized as follows: Sect. 2 presents related work on the design of NC hash functions using various techniques. Section 3 discusses the design criteria of NC hash functions. Section 4 introduces the design and implementation of the GPNCH family. Results are discussed and compared with the state of the art in Sect. 5. Conclusions are drawn in Sect. 6.

2   Related Work

NC hash functions provide fast speed at the cost of less stringent security properties. As discussed in Sect. 1, this enables their use in many network security applications that require fast searching. On the other hand, hardware platforms such as FPGAs are increasingly being used to implement probabilistic architectures like Bloom filters, sketch data structures, and other network security applications because they offer high performance, low cost, reconfigurability, scalability, parallelism, and fast market availability [1].

There is not much work in literature where the performance of NC hash functions is evaluated on FPGA. As already discussed in Sect. 1, Sateesan et al. [29] proposed a novel NC hash function for Bloom filter called Xoodoo-NC that satisfies the avalanche criteria. The resulting architecture performed better than the other state-of-the-art Bloom filter implementations in terms of hardware operating frequency, delay, and occupied resources. Still, the authors did not report specifically on the performance and occupied resources of their used NC hash function called Xoodoo-NC. Extending the work of [29], Claesen et al. [6] chose five famous ciphers from literature (AES, GIFT, SKINNY, SPECK, and Pyjamask), reduced their internal states and rounds to derive NC hash functions while maintaining the same avalanche properties. They compared the performance on FPGA with state-of-the-art NC hash functions such as FNV1a, Xoodoo-NC, Murmur3, etc. One of their proposed algorithms, GIFT-NC, performed better than all in terms of computational delay though it consumed more Lookup Tables (LUTs) on FPGA.

Table 1. NC hash Functions using EC approaches

Author	EC Method	Block Size	Fitness Function	Implementation
Estebanez,2014 [11]	GP	32/32	Avalanche Score	Software
Dobai,2015 [9]	EA	32/16	Collision rate	Software
Grochol,2016 [15]	LGP	96/16	Collision rate	Software
Kidon,2017 [21]	GP	32/16	Collision rate	software
Grochol,2017 [16]	LGP,NSGA-II	96/16	Collision rate, execution time	software
Grochol,2018 [17]	LGP,NSGA-II	96/16	Collision rate, execution time	FPGA
Saez,2019 [28]	GP	96/16	Collision rate	Software
Grochol,2020 [18]	CGP	96/16	Collision rate, execution time	Software
Hu,2020 [20]	GP	96/16	Active flow estimation, Uniformity	software

The last decade has seen reasonable work on using EC techniques for designing NC hash functions. Table 1 summarizes some of the prominent work in EC to evolve NC hash functions using various fitness functions. Estebanez et al. [11] used strict avalanche criteria as a fitness function to evolve 32-bit NC hash functions using GP. They produced remarkable results on different key sets with only 25 nodes (tree size), but their work has certain limitations. First, they used a multiplication operator to produce a good entropy in the output, but using this operator in the design is not recommended as it takes up a lot of FPGA resources. Second, the hash output size is small. Finally, the authors used Monte Carlo simulation to calculate the avalanche probability matrix to save computation time that considers a small subset of a large set. This is necessary as it is not feasible to check for all $32*2^{32}$ combinations. But at the same time, it is also paramount that the avalanche probability matrix should be based on the worst-case values for all possible input/output differences. This is important as the resulting hash function will provide assurance against millions of different kinds of keys.

In literature, many people use collision rate as a fitness function to evolve NC hash functions, although it has been proven that this only works for specific datasets [11]. In [21], Kidon et al. have used GP to evolve 32-bit domain-specific NC hash functions for hash tables. Their used fitness function measured the successful hashes until the first collision is detected. The resulting hash function performed 2.7 to 7 times faster and inserted 1.6% more keys than the state-of-the-art for specific datasets. Saez et al. [28] used the same GP algorithm as in their earlier work [11] but only changed the fitness function to collision rate to evolve a NC hash function called GP-Hash-AdHoc. They compared their results to the state-of-the-art, and for some datasets, they managed to improve upon collision rates. Another very interesting work is done by Dobai et al. [9], they have used an evolutionary algorithm to evolve a pair of NC hash functions for cuckoo filtering to be implemented on FPGA. The use of LFSRs and NLFSRs in the design of 32-bit hash functions is expected to be efficient on FPGA but they did not evaluate the performance on FPGA. In [20], Hu et al. used multi-objective GP to evolve the NC hash function to process IPv6 flow IDs of 288-bit length. Their main objective is to achieve better packet flow estimation while having a smaller number of collisions. The proposed method used three rounds to produce 16-bit hash outputs, and the results were comparable to the state-of-the-art on specific datasets in terms of active flow estimation and uniformity.

Grochol et al. [15] proposed the use of Linear Genetic Programming (LGP) to evolve NC hash functions to process 96-bit Network flow IDs. Further improving their own work, Grochol et al. [16] proposed a family of 16-bit hash functions using LGP and NSGA-II. They have used multi-objective fitness functions (collision rate and execution time). Additionally, they implemented their algorithms on FPGA in [17]. One of their designs clocked 739 MHz but this work is not comparable to our work because they have optimized the design with a very different fitness function that does not produce uniform hash functions. We verified this by testing their NC hash functions on the avalanche criteria explained in Sect. 3.

On changing the most significant bit 200k times, the entropy is close to zero, which clearly indicates it has poor randomness and is highly vulnerable to network attacks. Grochol et al. also experimented with CGP [18] to evolve NC hash functions to process network flow IDs of IPv4 and IPv6 packets. They reported similar performance for IPv4 flow ID processing as their previous work [15] and achieved good collision results for 320-bit IPv6 flow IDs as well.

All the above-described EC approaches except [11] are designed for specific datasets, hence they have mostly used collision resistance as the fitness function. They can only work on specific datasets and fail to generalize. They also produce either 16-bit or 32-bit hash outputs. The majority of the techniques use multiplication operators as well, which is a costly operator in terms of resources and computation time in a hardware implementation. Probabilistic architectures like Bloom filters require hash functions to be independent and have uniform output distribution, otherwise, the theoretical and simulation results for false positive rate (FPR) will have large discrepancy [27]. To ensure the above condition, one way is to evolve NC hash functions using avalanche metrics as a fitness measure to have a uniform distribution.

3 Design Criteria

The quality of NC hash functions can be analyzed based on the collision resistance, output distribution, avalanche effect, and speed of execution [11].

Collision Resistance: When two different input values produce the same hash output, a hash collision occurs. Since collisions are inevitable in hash functions as the range of input values is larger than the output, it is important to reduce the collision rate for many applications. This property is highly data dependent [11] and works well with specific datasets only. For this reason, many EC techniques rely on this criterion for designing hash functions but are not suitable for Bloom filters for the above reasons.

Output Distribution: Uneven output distribution will lead to clustering issues in the hash outputs and will have poor randomness. So a good NC hash function should have uniform output distribution [19].

Speed: The NC hash functions are mostly used in search operations, therefore they should be fast. This criterion highly depends upon the type of architecture used [11]. In the case of FPGAs, the choice and number of used operators matter the most in terms of hardware resources.

Avalanche Effect: The Avalanche effect illustrates how significantly the output change for a small change in the input of the hash function. Satisfying the avalanche criteria ensures that there is sufficient randomness in the output, thus all the input statistical patterns are properly diffused. Avalanche criteria are not data-dependent. To measure the avalanche effect, we have used the three avalanche metrics that are typically used to evaluate hash functions and defined in the work of Daemen et al. [8]: avalanche dependence(D_{av}), weight(W_{av}), and entropy(E_{av}).

Avalanche dependence: This tells us how many bits in the output may flip in response to a single input bit flip. The property is defined as:

$$D_{av} = t - \sum_{1}^{t} h(p[i]).$$

(1)

Here, p is the probability vector, where $p[i]$ represents the probability of each bit i of output and t is the total number of bits in the output. Further, $h(p) = 1$ if $p = 0$, and otherwise it is zero. This property generalizes full diffusion and is only satisfied when $D_{av} \simeq t$ for all outputs subject to a single-bit change in the input.

Avalanche weight: This is the expected Hamming weight of the difference in output for a single-bit change in the input (Δ) and is defined as:

$$W_{av} = \sum_{1}^{t} p[i].$$

(2)

To satisfy the condition, 50% of the output bits should be changed for Δ, i.e., $W_{av} \simeq t/2$ for Hamming weight $= 1$. The avalanche weight generalizes the avalanche criteria.

Avalanche entropy: This metric can be used to measure the randomness of the distribution, and it is defined as:

$$E_{av} = \sum_{1}^{t} (-p[i] * log_2 p[i] - (1 - p[i]) * log_2(1 - p[i])).$$

(3)

Full entropy means that the output of the hash function is completely random and there are no useful patterns left in the data. This uncertainty makes it very difficult for the intruder to guess the sequence. Assuming an output bit size of t, an entropy of value t has the same degree of uncertainty as a uniformly distributed t-bit random value [25]. This metric generalizes the Strict Avalanche Criteria (SAC) [31], which state that the probability of each output bit should be $1/2$ on a single bit change in the input. This only happens when $E_{av} \simeq t$ is satisfied.

4 Design and Implementation

The primary objective of this research is to design NC hash functions for probabilistic data structures such as Bloom filters and Sketch data structures. Thus it is very important that the output of the hash function should be uniformly distributed [27]. Therefore, we seek to achieve the avalanche criterion as described in Sect. 3. This section delineates all the choices and decisions made during the design of the 96-bit GPNCH function.

4.1   Fitness Function

The following two scenarios are considered for a fitness function.

Scenario 1: In this scenario, we use the sum of Bit dependence, Avalanche weight, and Entropy as a fitness function.

$$fitness function = \sum (D_{av}, W_{av}, E_{av}). \tag{4}$$

The proposed GPNCH function input/output block size is 96-bit, so the perfect values are 96 for D_{av} and E_{av} and 48 for W_{av}. The sum of all these three metrics is 240, therefore, we have used this value as a fitness objective and evolved the individuals seeking to maximize this sum and get as close as possible to the perfect value of 240. That being said, it is not practical to calculate the hash values for all $96 * 2^{96}$ possible input bit strings for all the individuals in each generation, hence we decided to calculate the output difference one by one while keeping the rest of the input constant. This results in $96 * 25$ bit flips (M=25) in the first stage. We keep only the worst-case values for the three avalanche metrics(D_{av}, W_{av}, and E_{av}). In the second stage, we extract the best individual and again calculate the scores for $96 * 2.5$ million bit flips (M=2.5 million) to ensure many input patterns are covered. The reported results are of the second stage.

Scenario 2: In the second scenario, we use Entropy as a fitness function.

$$fitness function = E_{\mathrm{av}}. \tag{5}$$

We hypothesized (later confirmed by experiments) that if the candidate solution achieves the maximum possible entropy (96 bits) then it should also have perfect values for D_{av} and W_{av} consequently.

Table 2. Elementary operators used in state-of-the-art NC hash functions

Hash functions	Rotation	Shift	XOR	AND	OR	NOT	+	-	/	*
Xoodo NC	✓		✓	✓						
MurmurHash2		✓	✓							✓
MurmurHash3	✓		✓				✓			✓
FNV			✓							✓
SuperFastHash		✓	✓				✓			
Lookup3	✓	✓	✓				✓	✓		
FarmHash	✓	✓	✓	✓			✓			✓
CityHash	✓	✓	✓				✓			✓
SipHash	✓		✓							
DEK			✓							✓
GP-Hash-Adhoc/Generic	✓		✓							✓
NSGAHash6	✓		✓				✓			

4.2   Terminal and Non-terminal Set

For the non-terminal set, we gathered the elementary operators used by state-of-the-art NC hash functions as depicted in Table 2. The majority of the techniques use the multiplication operator as it produces significant entropy in the output, but it also consumes a lot of hardware resources on FPGA as is shown in [30], thus we did not use this operator. The use of a division operator is also avoided for the same reasons [2,24]. [11] has provided useful insights regarding the selection of operators for GP but our extensive experiments show that excluding any of the operators (Rotation, Swaps, XOR, AND, OR, NOT) degrades the avalanche results for 96-bit hash outputs. We have used 20 different rotation functions (10 left and 10 right) with circular shifts from 1 to 10 bits. The left and right circular rotations effectually produce the same results ($m \lll n = m \lll 96 - n$).

The swap operator also produces significant disorder in the output. Hence, we have used three swap operators of 48, 24, and 12 bits respectively. For the terminal nodes, we have used 96-bit inputs and ephemeral random constants (ERC). The ERC has the unique property that once randomly generated, it keeps its value throughout the evolution process.

4.3   Parameters Setting

The values of the hyperparameters control the overall run and help the algorithm fine-tune the results. All these values are tuned by doing extensive empirical analysis of various sets of combinations. Table 3 summarizes the settings under which the best results are obtained. Our experiments show that for a 96-bit hash output, the number of nodes should be large, otherwise avalanche metrics are not satisfied. This trial and error process suggested a required value of 550 for the number of nodes for the given primitive set. These seemingly large numbers

Table 3. Basic configuration of GPNCH functions

Parameter	Value
Input/output block size	96 bits
Basic function set	{XOR, AND, OR, NOT, ROT, Swaps, Constant}
Max generations	1500
Max nodes	550
Population size	300
Crossover Rate and type	0.9 & one point
Mutation Rate	0.3
Mutation operator	Uniform
Tree height(Min & Max)	5 &10
Tree height limit	75
Mutated tree height(Min & Max)	1 & 3
Mutated Tree type	Gen grow
Selection	Tournament
Tournament size	3
Elitism	10
Initialization	Gen full

of nodes are not the cause of concern since the final design contains a large proportion of nodes comprising rotations and swaps. These operations add negligible cost to FPGA resources. The selection of parents for the next generation of individuals has been performed through tournament selection. The size of the tournament is kept small to avoid premature convergence. Sometimes, the best solutions can be lost during evolution, hence elitism is used to preserve the best ones.

4.4 Stopping Criteria

To stop the evolution process, either of the following two conditions should be met:

1. The perfect fitness score, i.e., 240, is achieved.
2. The total number of generations is completed.

4.5 Software and Hardware Implementation

The Python DEAP framework [13], which is a fast prototyping evolutionary computing platform to design and test ideas, is used to code the GPNCH functions. It also supports multiprocessing to speed up the evaluations. The simulations are performed on a 32-core Intel Xeon 4208 CPU with a clock speed of 2.1 GHz.

The best individuals representing GPNCH functions are later implemented on the FPGA targeting Virtex Ultrascale+ xcvu7p-flvb2104-2-i chip using the Vivado 2022.1 to evaluate the operating speed, delay, throughput, and hardware resources. The general block diagram (Fig. 2) shows that only a single round is needed to generate hash outputs.

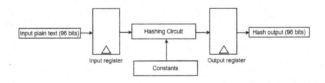

Fig. 2. GPNCH hash function hardware implementation

5 Results and Discussion

5.1 Software Evaluation

The first objective of this research work is to design the NC hash function that satisfies the avalanche metrics as described in Sect. 3 for a 96-bit block size. To achieve the said objective, around 450000 evaluations are performed across 1500 generations and a population size of 300 in each independent run. The fitness values are measured for both scenarios detailed in Sect. 4. They are plotted in Fig. 3. The plots show the typical GP trait of the rapid increase in earlier generations. Once the basic structure of the design is set, the algorithm fine-tunes the values and gradually moves toward the objective.

For scenario 1 (fitness = $\sum D_{av}, W_{av}, E_{av}$), the objective is to achieve the fitness value of 240. As can be seen in Fig. 3, the best fitness value is achieved in around 600 generations in one of the runs, and it does not improve further in later generations. On average, the individuals achieve the mean fitness value of approx. 220 in around 500 generations. In the second scenario (fitness = E_{av}), the goal is to get to the value of 96 to achieve the maximum possible entropy while also observing values for D_{av} and W_{av}. In this case, the best fitness is achieved in approximately 1400 generations. This shows the stochastic nature of the GP algorithm.

The complete results of avalanche metrics for both scenarios are compiled and compared against the selected state-of-the-art NC hash functions in Table 4. As can be seen from the avalanche scores, both variants produced almost perfect fitness values. In comparison with other algorithms, the GPNCH achieved better avalanche scores, and that too in only a single round. Scenario 2 also confirmed our hypothesis that if the perfect entropy of 96 bits is achieved then D_{av} and W_{av} are also satisfied. Perfect entropy is only possible when all the bits in the probability vector, as described by Equation (3) have a probability of 0.5. This means for each input bit change; every output bit will have a 50% chance to change. Consequently, every output bit could be affected by a single bit change in each input which automatically satisfies the D_{av}. Moreover, SAC criteria, as

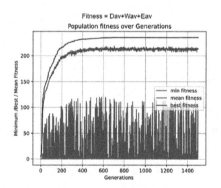 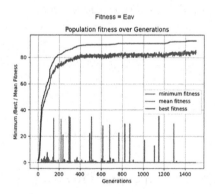

Fig. 3. Fitness function scores for both scenarios over the generations

Table 4. Comparison with the avalanche metrics of selected NC hash functions

NC hash	Hash block size	D_{av}	W_{av}	E_{av}	No.of rounds
Xoodo-NC	96/96	96	47.324	95.864	3
SPECK-NC	96/96	96	47.45	95.68	7
Pyjamask-NC	96/96	96	47.73	95.96	3
GIFT-NC	96/96	128	63.39	127.95	7
AES-NC	96/96	128	63.0	127.43	3
SKINNY-NC	96/96	128	63.72	127.95	7
GPNCH (scenario 1)	96/96	96	48	95.99	1
GPNCH (scenario 2)	96/96	96	48	95.97	1

described in Sect. 3 is also fully satisfied due to this perfect entropy. Further, the summation of all the bits in the probability vector leads to half of the bits being changed on average, which satisfies the W_{av} as well. We have also tried to experiment with using D_{av} and W_{av} as fitness functions separately but the E_{av} values were not up to the mark in all the runs. Due to the probabilistic nature of the GP algorithm, it is necessary to run the experiment enough times to ensure that results are consistent over several runs. As both fitness functions lead to equally good results, we chose the fitness function from scenario 1 to complete multiple runs. During the 10 independent runs, the average of the best fitness is approx. 237 when M = 250000. The complete results are shown in Fig. 4(a). This figure shows that the results are very consistent, and the algorithm is extremely robust. Moreover, for all the 10 runs, the best individual is always found between 600–1500 generations, so there is no need to go beyond that.

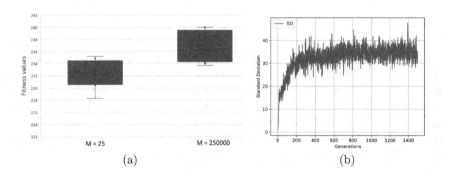

Fig. 4. (a) Box plot of fitness values for all the 10 independent runs; (b) Standard deviation of the population across generations for a single run

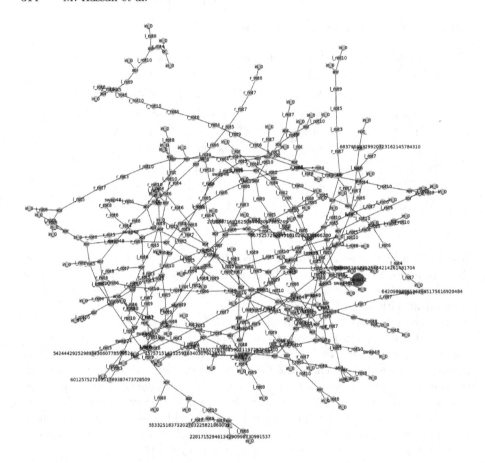

Fig. 5. Best individual representing the GPNCH function

The best fitness value, i.e., 239.99 is found during the 7th run of an algorithm. Figure 4(b) shows the standard deviation (SD) of the population for this run. The SD varies between 20 to 40, which is expected as the population size is high (300) and the mutation rate is 0.3. It indicates that the population has good variance at any given time, which is a good indicator of population diversity. Figure 5 shows the complete tree representing the one GPNCH function. As we said earlier, it can be observed that the majority of nodes are rotations and swaps, which barely add any resources when implemented on FPGA.

5.2   Hardware Evaluation

Obtaining high performance in hardware is the main focus of this research work. To carry out the evaluation, we selected the best run (7th) among the 10 completely independent runs. We extracted the top 10 individuals representing a family of 10 different GPNCH functions. All these NC functions are implemented on FPGA and the operating frequency, latency, and occupied hardware resources are measured and shown in Table 5. All the individuals achieved the fitness values of 239.99 which means they are equally good candidates to represent the NC hash function. This is a very important result as some networking search applications use separate independent uniform hash functions. The best individual is GPNCH-V, which has an operating frequency of 595.23 MHz, a throughput of 57.142 Gbps, and a latency of 1.68 ns. The complete design of all hash functions takes only 1 clock cycle to generate the result. When it comes to hardware resources, the design itself does not use any flip-flops and all individuals consume between 1446–1673 LUTs.

Table 5. Hardware evaluation of the top 10 best individuals

Best individuals	Fitness	Max. operating frequency(MHz)	Throughput(Gbps)	No. of LUTs	Flip-flops	Latency(ns)
GPNCH-I	239.99	595.23	57.142	1490	0	1.68
GPNCH-II	239.99	595.23	57.142	1471	0	1.68
GPNCH-III	239.99	595.23	57.142	1614	0	1.68
GPNCH-IV	239.99	591.71	56.804	1673	0	1.69
GPNCH-V	**239.99**	**595.23**	**57.142**	**1446**	**0**	**1.68**
GPNCH-VI	239.99	595.23	57.142	1475	0	1.68
GPNCH-VII	239.99	595.23	57.142	1462	0	1.68
GPNCH-VIII	239.99	595.23	57.142	1463	0	1.68
GPNCH-IX	239.99	595.23	57.142	1495	0	1.68
GPNCH-X	239.99	591.71	56.804	1483	0	1.69

Comparison with Related Work. As we mentioned in Sect. 1, the NC hash functions designed by [6] and [29] are by far the best among all the known state-of-the-art NC hash functions when it comes to performance on FPGA for a 96-bit block size while satisfying the avalanche properties. Therefore, to have a fair comparison, we first implemented Xoodoo-NC on the same FPGA (Virtex Ultrascale+ xcvu7p-flvb2104-2-i) used in our work and by the authors of [6]. We chose the number of rounds for Xoodoo-NC to be 2.5 as the avalanche properties are satisfied at this stage [30]. We compared the latency, throughput, maximal operating frequency, and hardware resources. The complete results are summarised in Table 6. It should be noted that in the implementation of GIFT-NC, AES-NC, and SKINNY-NC, the authors have kept the last 32 bits of the output block to zero, so effectively they have measured the hardware performance on the 96-bit block. As evident from Table 6, GPNCH-V outperforms Xoodoo-NC by 19% in throughput, frequency, and latency while an improvement of 62% is achieved compared to the GIFT-NC [6]. The speed-up we achieve comes at the cost of increased FPGA resources. In high-speed network monitoring and security applications, this increase in resources is negligible compared to the overall resources of the considered FPGA. We can therefore conclude that GPNCH offers a better alternative than state-of-the-art NC hash functions.

Finally, it is pointed out that the proposed solution, using Genetic Programming, is capable of generating multiple NC hash functions. Some approximate membership query data structures require multiple, independent NC hash functions, for example, Bloom filters, Cuckoo filters, and sketches. In contrast to the single NC hash functions that are reported in the literature, our solution could supply these data structures with the number of NC hash functions they require.

Table 6. Comparison with the related work

NC Hash function	Block Size	Max. operating frequency(MHz)	Throughput (Gbps)	No. of LUTs	Latency (ns)
Xoodoo-NC (2.5 rounds)[1]	96/96	500	48	480	2.00
SPECK-NC	96/96	171.6	16.47	432	5.82
Pyjamask-NC	96/96	279.4	26.82	832	3.58
GIFT-NC	96/128	369.7	35.49	546	2.70
AES-NC	96/128	255.9	24.56	2225	3.91
SKINNY-NC	96/128	202.4	19.44	2176	4.93
GPNCH-V	**96/96**	**595.23**	**59.95**	**1446**	**1.68**

[1]: these are the results of our implementation

6 Conclusion

This paper presents a novel non-cryptographic hash function family (GPNCH) through the use of Genetic Programming. We evolve non-cryptographic hash functions by satisfying the avalanche metrics for a 96-bit input/output block size. During the design phase, we propose two different fitness functions that can be used to satisfy the avalanche properties. We have simulated approx. 4.5 million evaluations to generate a family of the best possible uniform and independent NC hash functions. In order to evaluate the performance on the hardware setup, we have used an FPGA platform on which we compared the speed, the delay, the throughput, and the hardware resources with the state-of-art non-cryptographic hash functions. We show that our proposed method outperforms all the other non-cryptographic hash functions in terms of execution delay and throughput by at least 19%. This improvement is very important as the speed of probabilistic lookup and counting architectures depends heavily on the execution speed of hash functions. We next plan to use this GPNCH function family in the implementation of Bloom filters and report on the efficiency.

Acknowledgement. This work is supported by the ESCALATE project, funded by FWO (G0E0719N) and SNSF (200021L_182005). This work is also supported by Cybersecurity Research Flanders (VR20192203) and the Higher Education Commission, Pakistan.

References

1. The advantages of using FPGAs. https://www.af-inventions.de/en/services/fpga-development/fpga-advantages.html. Accessed 15 Sep 2022
2. FPGA optimization guide for intel®oneapi toolkits:developer guide. https://www.intel.com/content/www/us/en/develop/documentation/oneapi-fpga-optimization-guide/top/optimize-your-design/resource-use/data-types-and-operations/avoid-expensive-functions.html. Accessed 20 Jan 2023
3. Natural selection. https://education.nationalgeographic.org/resource/natural-selection/. Accessed 15 Oct 2022
4. Shifting to an FPGA data center future: How are FPGAs a potential solution? https://www.allaboutcircuits.com/news/shifting-to-a-field-programable-gate-array-data-center-future/. Accessed 15 Sept 2022
5. World's fastest internet network upgraded to staggering 46 terabit/s. https://newatlas.com/telecommunications/esnet6-worlds-fastest-internet-46-terabit-second/. Accessed 01 Nov 2022
6. Claesen, T., Sateesan, A., Vliegen, J., Mentens, N.: Novel non-cryptographic hash functions for networking and security applications on FPGA. In: 2021 24th Euromicro conference on digital system design (DSD), pp. 347–354 IEEE (2021)
7. Cormode, G., Muthukrishnan, S.: An improved data stream summary: the count-min sketch and its applications. J. Algorithms **55**(1), 58–75 (2005)
8. Daemen, J., Hoffert, S., Van Assche, G., Van Keer, R.: The design of xoodoo and xoofff. IACR Transactions on Symmetric Cryptology, p. 1–38 (2018)
9. Dobai, R., Korenek, J.: Evolution of non-cryptographic hash function pairs for FPGA-based network applications. In: 2015 IEEE Symposium Series on Computational Intelligence, pp. 1214–1219 IEEE (2015)
10. Eiben, A. E., Smith, J. E., et al.: Introduction to evolutionary computing, vol. 53 Springer, 2003. https://doi.org/10.1007/978-3-662-05094-1
11. Estébanez, C., Saez, Y., Recio, G., Isasi, P.: Automatic design of noncryptographic hash functions using genetic programming. Comput. Intell. **30**(4), 798–831 (2014)
12. Flajolet, P., Fusy, É., Gandouet, O., Meunier, F.: Hyperloglog: the analysis of a near-optimal cardinality estimation algorithm. Discrete Math. Theor. Comput. Sci. (2007)
13. Fortin, F.-A., De Rainville, F.-M., Gardner, M.-A.G., Parizeau, M., Gagné, C.: Deap: Evolutionary algorithms made easy. J. Mach. Learn. Res. **13**(1), 2171–2175 (2012)
14. Fowler, G., Vo, K.-P., Eastlake, D., Hansen, T.: The FNV non-cryptographic hash algorithm. IETF-draft. (2012)
15. Grochol, D., Sekanina, L.: Evolutionary design of fast high-quality hash functions for network applications. Proc. Genet. Evol. Comput. Conf. **2016**, 901–908 (2016)
16. Grochol, D., Sekanina, L.: Multi-objective evolution of hash functions for high speed networks. In: 2017 IEEE Congress on Evolutionary Computation (CEC), pp. 1533–1540. IEEE (2017)
17. Grochol, D., Sekanina, L.: Fast reconfigurable hash functions for network flow hashing in FPGAs. In: 2018 NASA/ESA Conference on Adaptive Hardware and Systems (AHS), pp. 257–263 IEEE (2018)
18. Grochol, D., Sekanina, L.: Evolutionary design of hash functions for IPv6 network flow hashing. In: 2020 IEEE Congress on Evolutionary Computation (CEC), pp. 1–8. IEEE, (2020)

19. Henke, C., Schmoll, C., Zseby, T.: Empirical evaluation of hash functions for mul-
    tipoint measurements. ACM SIGCOMM Comput. Commun. Rev. **38**(3), 39–50
    (2008)
20. Hu, Y., Cheng, G., Tang, Y., Wang, F.: A practical design of hash functions for
    ipv6 using multi-objective genetic programming. Comput. Commun. **162**, 160–168
    (2020)
21. Kidoň, M., Dobai, R.: Evolutionary design of hash functions for IP address hashing
    using genetic programming. In: 2017 IEEE Congress on Evolutionary Computation
    (CEC), pp. 1720–1727 IEEE (2017)
22. Kirsch, A., Mitzenmacher, M.: Less Hashing, Same Performance: Building a Better
    Bloom Filter. In: Azar, Y., Erlebach, T. (eds.) ESA 2006. LNCS, vol. 4168, pp.
    456–467. Springer, Heidelberg (2006). https://doi.org/10.1007/11841036_42
23. Luo, L., Guo, D., Ma, R.T., Rottenstreich, O., Luo, X.: Optimizing Bloom Fil-
    ter: challenges, solutions, and comparisons. IEEE Commun. Surv. Tutorials **21**(2),
    1912–1949 (2018)
24. Mannatunga, K., Perera, M.: Performance evaluation of division algorithms in
    FPGA. (2016)
25. NIST. Digital identity guidelines. Technical Report NIST Special Publication 800–
    63-3, INCLUDES UPDATES AS OF 03 Feb 2020, U.S. Department of Commerce,
    Washington, D.C. (2017)
26. Picek, S., Yang, B., Rozic, V., Vliegen, J., Winderickx, J., De Cnudde, T., Mentens,
    N.: PRNGs for Masking Applications and Their Mapping to Evolvable Hardware.
    In: Lemke-Rust, K., Tunstall, M. (eds.) CARDIS 2016. LNCS, vol. 10146, pp.
    209–227. Springer, Cham (2017). https://doi.org/10.1007/978-3-319-54669-8_13
27. Ramakrishna, M.: Practical performance of bloom filters and parallel free-text
    searching. Commun. ACM **32**(10), 1237–1239 (1989)
28. Saez, Y., Estebanez, C., Quintana, D., Isasi, P.: Evolutionary hash functions for
    specific domains. Appl. Soft Comput. **78**, 58–69 (2019)
29. Sateesan, A., Vliegen, J., Daemen, J., Mentens, N.: Novel Bloom filter algorithms
    and architectures for ultra-high-speed network security applications. In: 2020 23rd
    Euromicro Conference on Digital System Design (DSD), pp. 262–269. IEEE, (2020)
30. Sateesan, A., Vliegen, J., Daemen, J., Mentens, N.: Hardware-oriented optimization
    of Bloom filter algorithms and architectures for ultra-high-speed lookups in network
    applications. Microprocess. Microsyst. **93**, 104619 (2022)
31. Webster, A.. F.., Tavares, S.. E..: On the Design of S-Boxes. In: Williams, H.C. (ed.)
    Advances in Cryptology — CRYPTO '85 Proceedings, pp. 523–534. Springer Berlin
    Heidelberg, Berlin, Heidelberg (2000). https://doi.org/10.1007/3-540-39799-X_41

Use of a Genetic Algorithm to Evolve the Parameters of an Iterated Function System in Order to Create Adapted Phenotypic Structures

Habiba Akter$^{(\boxtimes)}$ ⓘ, Rupert Young ⓘ, Phil Birch ⓘ, and Chris Chatwin ⓘ

Department of Engineering and Design, School of Engineering and Informatics,
University of Sussex, Brighton, UK
{h.akter,r.c.d.young,p.m.birch,c.r.chatwin}@sussex.ac.uk

Abstract. This work investigates the use of Evolutionary Computation to generate fractal pattern structures representing the phenotype of an organism, using the Barnsley fern as an example. Genetic Algorithm is implemented as the search and optimisation tool to generate the fractal structure of the leaf pattern. The Genetic Algorithm evolves the parameters of the Iterated Function System and selects the resulting fractal structures, representing a generated phenotype, using box-counting dimension as a fitness metric. In this way, realistic self-similar fern structures are evolved over a few tens of generations. The algorithm is further extended to test its potential to generate other natural fractals.

Keywords: Fractal · Genetic Algorithm · Iterated Function Systems · Box-counting Dimension

1 Introduction

Barnsley [3,4] explored the generation of fractal phenotypic structures in nature. In particular, he showed Iterated Function Systems (IFS) can produce fractal patterns strikingly similar to ferns. Fractal structures observed in nature also include that of lungs, the vascular system and the fine structure of bone [2,7,12, 22]. The similarity of IFS-generated patterns and real-life biological organisms has inspired Barnsley to consider collaborating with biologists to investigate developmental gene pathways.

Indeed, some work in plant developmental biology has investigated in detail the generation of phenotypic structures by the plant hormone auxin. Jönsson et al. [19] proposed an auxin-driven polarised transport model for phyllotaxis leading to the generation of regular patterns. Also, Smith et al. [24] presented a model of phyllotaxis to create a regular arrangement of lateral organs around a central axis of the plant. More recent work by Bhatia et al. [5] suggests that an auxin response factor is involved to orientate the polarity of an auxin efflux carrier, forming a positive feedback loop to create a periodicity of organ formation in the plant.

© The Author(s), under exclusive license to Springer Nature Switzerland AG 2023
J. Correia et al. (Eds.): EvoApplications 2023, LNCS 13989, pp. 319–331, 2023.
https://doi.org/10.1007/978-3-031-30229-9_21

More generally, the recent developments in the field of Evolutionary Developmental biology (Evo-Devo) have emphasised that an organism's body is built by a sequential developmental process under genetic control. A particularly clear exposition of this modern view of evolutionary biology has been given by Carroll [8]. The emphasis is that genes build a body during the sequential developmental process under the control of genetic switches. This process can be directed at a particular stage in the development of the organism by evolutionary changes to produce variations in the final phenotype. Thus, transcription factors are selectively activated as part of developmental genetics to sequentially control phenotypic development. For example, in plants, Monopteros is a transcription factor that triggers the gene expression required for flower formation that is activated by auxin [19].

2   Motivation for the Research

The initial motivation for the paper is to focus on how Genetic Algorithms (GA), currently applied in engineering optimisation problems, can link to a developmental process leading to the generation of the phenotypic structure of an organism. GAs essentially model the process of meiosis in sexual reproduction i.e. gene swapping between chromosomes and chromosome shuffling. Thus an initial bit string is recombined in different combinations by this process and mapped to a particular output configuration whose fitness can then be assessed [13,14,17,18,20]. This has been proved successful in many applications [6], nevertheless, does not directly model the inherent developmental processes that occur to build the phenotype from the action of the genes that occurs in biology. This requires an essential developmental process between the value of a gene parameter and its final phenotypic expression [11].

The authors are interested in producing a more closely biologically analogous model and the IFS process employed by Barnsley seems ideal for this as fractal structures are so commonly observed in the morphology of biological organisms. Thus the IFS matrix parameters are identified with genes controlling development through an iterative process applied to the development of the organism to produce its final phenotypic appearance.

To do this, the box-counting dimension is employed as the fitness metric. The box-counting dimension of a fractal can plausibly be related to a physical metric for fitness by quantifying how efficient the structure would be in presenting an effective surface area to sunlight for a given tissue mass. Most importantly, the selection pressure is applied to the phenotype, but this is first generated by a developmental process under the control of the "genes" which are subjected to the reproduction process. The GA proceeds (by cross-over and limited mutation rate) and produces modified phenotypes which are again selected using the box-counting dimension (i.e. environmental fitness measure).

In the initial experiments conducted, after only 50 generations a realistic fern appearance is generated. Thus the parameters are modified iteratively by the algorithm mimicking the meiotic process occurring during sexual reproduction to evolve an effective structural phenotype. It should be noted that the fractal iteration is halted at a given resolution. Analogously, in biological systems, this will occur when the fractal structures near the minimal volumes required to construct them from cellular tissues.

Thus the analogy is made to the process by which genes build a body through the mechanism of embryonic development and can be directed at a particular stage in the development of the organism by evolutionary changes to produce variations in the final phenotype [8]. The fact that an evolutionary mechanism builds an organism's body under genetic control has motivated us to apply a GA to generate and evolve the parameters which successfully produce the fractal pattern representing the phenotypic structure of the Barnsley fern.

3   Design of the Algorithm

An algorithm has been designed combining the following three components:

- Genetic Algorithm (GA)
- Iterated Function System (IFS)
- Fractal Dimension Analyser (FDA)

The GA is used as the search and optimisation tool. It generates and evolves the parameters to be fed into the IFS which then generates images that are complex in nature. This section explains the steps involved in designing the algorithm.

- **Step 1:** A population of 80 chromosomes is generated. Each chromosome has 12 parameters which represent the coefficients of the affine transformation matrix describing the IFS for the Barnsley fern (as shown below, with Eq. 5). Each chromosome can thus be represented as follows:

$$P_i = [w_0, w_1, w_2, w_3, w_4, w_5, w_6, w_7, w_8, w_9, w_{10}, w_{11}] \qquad (1)$$

The values of the variables, w_i are generated within a range from -0.50 to 0.50, many simulations having been run to decide this range.

- **Step 2:** Each chromosome is then evaluated using its fractal dimension as a cost function. After a comparison of the Hausdorff, Perimeter and box-counting dimension methods, the box-counting dimension method has been employed [10,26]. Equation 2 calculates the box-counting dimension of an object.

$$D = \lim_{\epsilon \to 0} \frac{\log n}{\log \frac{1}{\epsilon}} \qquad (2)$$

Here, n is the number of boxes covering the points w_i, generated by the GA, if plotted in graphical form and ϵ is the size of boxes. The size of the boxes also needs to be set carefully after test and evaluation as it will directly impact the value of the box-counting dimension.

The higher fractal dimension represents a more complex fractal pattern and so can be used to optimises the fitness of the phenotypic structure. Thus, this becomes a "maximisation" problem and so the chromosomes are evaluated and ranked in the descending order of the measurements of their box-counting dimension. The fittest chromosome with the highest box-counting dimension, P', is then saved to ensure that the best member from each iteration is never lost.

- **Step 3:** A pre-selected number of the top-ranked population is sent to the mating pool for reproduction. A larger selection confirms more diversity in the offspring [21,23].

- **Step 4:** This step involves the reproduction operators, crossover and mutation, each with a certain probability, to generate offspring from the fitter parents selected in Step 3. The crossover probability is usually within the range of 50% to 80% and the mutation probability is much lower, within a range from 30%, reaching as low as 1% [9,15,16]. Previous research has shown that a higher probability of crossover and lower probability of mutation improves the efficiency of a GA [1,25].

Within this step, the following reproduction operations are performed:

- The first step of reproduction is the crossover, where for a certain probability, ρ_c, two parent chromosomes are selected and crossed over to produce offspring. A single-point crossover is implemented here. Let us assume P_1 and P_2 (as shown in Fig. 1) are selected for this procedure.

  At first, a random value, α, is generated between 0 and 1. After multiplying α by each element of P_1 and $(1 - \alpha)$ by each element of P_2 we add them together to produce the first chromosome. Similarly, multiplying α by each element of P_2 and $(1 - \alpha)$ by each element of P_1 and adding them together, the 2nd chromosome is produced.

Fig. 1. Parent population for crossover

The crossover procedure is shown in Eq. 3:

$$C_1 = \alpha \times P_1 + (1 - \alpha) \times P_2$$
$$C_2 = \alpha \times P_2 + (1 - \alpha) \times P_1$$
(3)

- For a certain mutation probability, ρ_m, a parent chromosome undergoes mutation. The aim here is to change the value of a randomly selected gene and create a mutated offspring. Equation 4 describes the mutated offspring:

$$M_1 = P_1 + \rho_m \times \sigma$$
(4)

Here, the standard deviation of a normal distribution, σ, is used to find a random number around the mean.

- **Step 5:** In the next step, the elite population, P', crossed over offspring, C_i, and mutated offspring, M_i, are merged together making the final population set, P_f.

P_f then undergoes the process of "environmental selection". To do so, it is evaluated again and the best chromosome is selected based on the highest value of the fractal dimension which replaces the P' calculated earlier. P' is in the same format shown in Eq. 1, i.e. there is no change in the length of the chromosome. The genes are then used as the parameters in the IFS to generate the fractal pattern.

Barnsley has described the IFS that generates the fractal structure of the spleenwort fern [3]. The four affine transformations used to generate the self similar fractal pattern can be represented as follows:

$$f(x, y) = \begin{bmatrix} a & b \\ c & d \end{bmatrix} \begin{bmatrix} x_n \\ y_n \end{bmatrix} + \begin{bmatrix} e \\ f \end{bmatrix}$$
(5)

For a certain probability, p, a function is selected and generates a specific portion of the fern:

- f_1: *Stem*
  - a. Fractal Width   b. Skew right   c. Skewed height
  - d. Height×10 decimal number   e. x- origin   f. y- origin

- f_2: *Form/ Smaller leaflet*
  - a. Leaf spread   b. Right spiral   c. Left spiral
  - d. Separation   e. Stem part shift   f. Flip left/ right
- f_3: *Left leaf*
  - a. Leaf curvature   b. Leaf length   c. Leaf width
  - d. Leaf skew   e. Leaf stem position   f. Steam stem height
- f_4: *Right leaf*
  - a. Leaf curvature   b. Leaf length   c. Leaf width
  - d. Leaf skew   e. Leaf stem position   f. Steam stem height

Equations 6, 7, 8 and 9 represent the transformation equations of the IFS which are chosen for $p = 1\%, 85\%, 7\%$ and 7%:

$$x_{n+1} = 0$$
$$y_{n+1} = 0.16y_n \tag{6}$$

$$x_{n+1} = w0 \times x_n + w1 \times y_n$$
$$y_{n+1} = w2 \times x_n + w3 \times y_n + 1.6 \tag{7}$$

$$x_{n+1} = w4 \times x_n + w5 \times y_n$$
$$y_{n+1} = w6 \times x_n + w7 \times y_n + 1.6 \tag{8}$$

$$x_{n+1} = w8 \times x_n + w9 \times y_n$$
$$y_{n+1} = w10 \times x_n + w11 \times y_n + 0.44 \tag{9}$$

It can be observed from the above equations that the value of the coefficient for Eq. 6 is not being generated by the GA. The reason is that this equation simply generates the stem.

The fern generated using these functions is the output image from the current iteration. At this stage, the final population set, P_f, of the current iteration, i, is then fed as the initial population P for the next iteration, $(i + 1)$.

- **Step 6:** The terminating condition of the algorithm is set to be the maximum number of iterations, i.e. these steps are followed until the maximum number of iterations is reached to produce the final image.

Algorithm 1 shows the pseudo-code.

Algorithm 1. GA-IFS for Fern

INPUT:

Current iteration, itr
Iteration limit, i_{max}
Crossover probability, ρ_c
Mutation probability, ρ_m
Number of IFS Iterations, it
Iteration limit for IFS, it_{max}
$p_{rng} \leftarrow$ range of p, (1-100)

OUTPUT: Barnsley fern, *img*

1: $P \leftarrow$ Randomly generated initial population
2: $N \leftarrow$ Size of P
3: $F \leftarrow$ Fitness score
4: $P' \leftarrow P$ with maximum F
5: $MP \leftarrow$ Mating pool for reproduction
6: $N_{mp} \leftarrow$ Size of mating pool
7: **while** $itr \le i_{max}$ **do**
8:      Sort P in ascending order of F
9:      Save P' as Elite
10:     Send fitter Chromosomes to MP
11:     Do Crossover on $\frac{\rho_m \times N}{100}$ of P
12:     $C \leftarrow$ Children after Crossover
13:     Do Mutation on $\frac{\rho_m \times N}{100}$ of P
14:     $M \leftarrow$ Children after Mutation
15:     $P_f \leftarrow P' \cup C \cup M$, Final set of Population
16:     Generate fern image, *img* from P'
17:     **function** IFS(P')
18:         **for** $it \le it_{max}$ **do**
19:             Generate p (from p_{rng})
20:             **if** $p = 1$ **then**
21:                 $f1$
22:             **else if** $p >= 2$ AND $p <= 86$ **then**
23:                 $f2$
24:             **else if** $p >= 87$ AND $p <= 93$ **then**
25:                 $f3$
26:             **else if** $p >= 94$ AND $p <= 100$ **then**
27:                 $f4$
28:             **end if**
29:         **end for**
30:         Plot *img*
31:     **end function**
32:     $itr \leftarrow (itr + 1)$
33: **end while**
        return *img*

4   Results

Thousands of simulations have been run using High-Performance Computing to select a set of GA parameters in order to generate the Barnsley fern. Figure 2 shows some initial attempts.

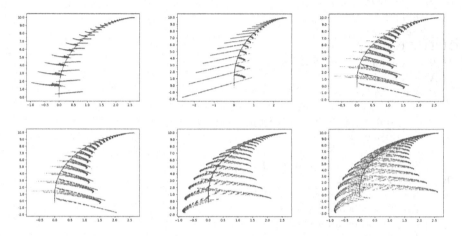

Fig. 2. Initial attempts of generating the Barnsley fern

It can be readily observed that none of these images is an accurate representation of the spleenwort fern. However, with the increment of the number of iterations of the algorithm, the generated random numbers are evolved and the phenotype of the Fern becomes more realistic. Table 1 includes the final selection of the parameters of the GA.

Table 1. Selected parameters of the GA

GA-Parameters	Values
Population size , N	80
Upper limit of the variables, w_{max}	0.50
Lower limit of the variables, w_{min}	−0.50
Fitness, F	1.87
Crossover probability, ρ_c	70%
Mutation probability ρ_m	20%
Crossover rate, r_c	0.06
Mutation rate, r_m	0.002
Terminating condition, i_{max}	40

The algorithm has been run for different values of i_{max} and proved that the convergence is clearly visible over the first 40 iterations with these selected parameters.

To verify the stages of the algorithm, results from each step of the 1st iteration are thoroughly evaluated. At this iteration, the worst box-counting dimension was calculated as 1.1983 and the best value was 1.4341. Figure 3 shows the two images of the fern with these costs.

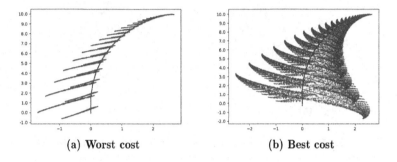

(a) Worst cost   (b) Best cost

Fig. 3. Barnsley fern generated after iteration 1

The final set of the population from this iteration is then passed to the 2nd iteration and so on.

For the final run, only the points with the best cost value are used in the IFS to generate the Barnsley fern. The same steps are followed in each iteration up to the 40th. Figure 4 shows the images generated using the best cost points of the 10th and 20th iterations.

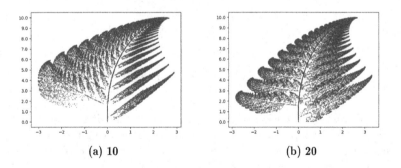

(a) 10   (b) 20

Fig. 4. Barnsley fern generated after iteration 10 and 20

It is clear that at each 10th step of the iteration loop, the Barnsley fern generated starts taking on a phenotypic structure that corresponds more closely to a spleenwort fern. The box-counting dimension of Fig. 4a is 1.7966 and that of Fig. 4b is 1.8313.

With increase in the number of iterations, the images gradually represent more natural fern. The output images from the 30th and 40th iterations are shown in Fig. 5.

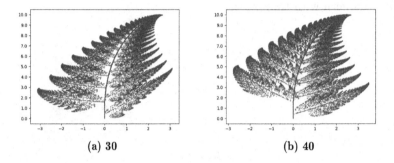

(a) 30   (b) 40

Fig. 5. Barnsley fern generated after iteration 30 and 40

The final output from the algorithm when it reaches the 50th iteration is in Fig. 6

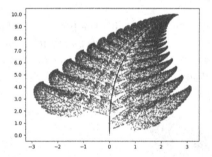

Fig. 6. Barnsley fern generated after the 50th iteration

The final fractal dimension of the Fern is 1.832. Thus it is apparent that 40 iterations of the developed algorithm is capable of generating the natural fractal structure of the Barnsley fern. The final optimal parameters evolved by the GA to generate the image shown in Fig. 5b are given in Table 2.

To get the precise value of the box-counting dimension, the algorithm is designed to ignore the same points while plotting the images to confirm that the points are not overlapped and crowded. However, the positive results of the algorithm that have been located are not numerous and widely distributed within the total population. Hence, the GA has been used to search the solution space thoroughly to obtain the optimal set of parameters.

Table 2. Values of the elite population

Coefficient	Value
w_0	0.50
w_1	0.48
w_2	−0.4
w_3	0.498
w_4	0.1293
w_5	0.378
w_6	0.21326
w_7	0.22668
w_8	−0.15528
w_9	−0.35
w_10	0.25963
w_11	0.29807

5 Further Test and Evaluation

The algorithm has also been tested to generate the fractal structure of a Maple leaf which also has the same stochastic IFS as the Barnsley fern, as shown in Eq. 5. A difference is that the probability of selecting each of the four transformation functions needs to be altered to the approximate values 10%, 35%, 35% and 20%. The complexity of the transformation functions of the Maple leaf is greater than that of the Barnsley fern. Hence a careful selection of the range of the values of each gene (w_i) is crucial to make the attempt successful. The finally evolved maple leaf is shown in Fig. 7.

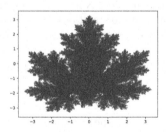

Fig. 7. Computer-generated maple leaf

6 Conclusions

This paper has presented the implementation of a Genetic Algorithm as an effective tool to search and evolve the parameters of an Iterated function System, the

Barnsley fern being used as an example. The GA has been used to measure and determine fitness by implementing the box-counting dimension as a cost function. Once suitable parameters have been selected by multiple runs of the GA, they are employed as the coefficients of the affine transformations to generate the fern patterns of rapidly improving fidelity in just tens of generations. Future work will investigate the optimisation of other Iterated Function Systems and explore their resemblance with the fractal patterns observed in nature.

Acknowledgement. This work was funded by the Leverhulme Trust Research Project Grant RPG- 2019-269 which the authors gratefully acknowledge.

References

1. Akter, H.: AS domain tunnelling for user-selectable loose source routing, Ph. D. thesis, Queen Mary University of London (2020)
2. Ball, P.: Pattern formation in nature: Physical constraints and self-organising characteristics. Archit. Des. **82**(2), 22–27 (2012)
3. Barnsley, M.F.: Fractals everywhere. Academic press, Boston (2014)
4. Barnsley, M.F., Demko, S.: Iterated function systems and the global construction of fractals. Proceed. Royal Soc. London. A. Math. Phys. Sci. **399**(1817), 243–275 (1985)
5. Bhatia, N., Bozorg, B., Larsson, A., Ohno, C., Jönsson, H., Heisler, M.G.: Auxin acts through monopteros to regulate plant cell polarity and pattern phyllotaxis. Curr. Biol. **26**(23), 3202–3208 (2016)
6. Birch, P., Young, R., Farsari, M., Chatwin, C., Budgett, D.: A comparison of the iterative Fourier Transform method and evolutionary algorithms for the design of diffractive optical elements. Opt. Lasers Eng. **33**(6), 439–448 (2000)
7. Bunde, A., Havlin, S. (eds.): Fractals in science. Springer, Heidelberg (1994). https://doi.org/10.1007/978-3-642-77953-4
8. Carroll, S.B.: Endless forms most beautiful: The new science of evo devo and the making of the animal kingdom. WW Norton & Company, USA (2005)
9. Chiu, C.S.: A genetic algorithm for multiobjective path optimisation problem. In: 2010 Sixth International Conference on Natural Computation, vol. 5, pp. 2217–2222. IEEE (2010)
10. Chu, H.T., Chen, C.C.: On bounding boxes of iterated function system attractors. Comput. Graph. **27**(3), 407–414 (2003)
11. Davidson, E.H.: Evolutionary bioscience as regulatory systems biology. Dev. Biol. **357**(1), 35–40 (2011)
12. Dullemeijer, P.: The self-made tapestry: pattern formation in nature. Science **284**(5420), 1627 (1999)
13. Gen, M., Cheng, R.: Genetic algorithms and engineering optimization, vol. 7. John Wiley & Sons (1999)
14. Goldberg, D.E.: Optimization, and machine learning. Genetic algorithms in Search (1989)
15. Goldberg, D.E., Deb, K.: A comparative analysis of selection schemes used in genetic algorithms. In: Foundations of genetic algorithms, vol. 1, pp. 69–93. Elsevier (1991)

16. Hassanat, A., Almohammadi, K., Alkafaween, E., Abunawas, E., Hammouri, A., Prasath, V.: Choosing mutation and crossover ratios for genetic algorithms-a review with a new dynamic approach. Information 10(12), 390 (2019)
17. Haupt, R.L., Haupt, S.E.: Practical genetic algorithms. John Wiley & Sons (2004)
18. Holland, J.H.: Adaptation in natural and artificial systems: an introductory analysis with applications to biology, control, and artificial intelligence. MIT press (1992)
19. Jönsson, H., Heisler, M.G., Shapiro, B.E., Meyerowitz, E.M., Mjolsness, E.: An auxin-driven polarized transport model for phyllotaxis. Proc. Natl. Acad. Sci. 103(5), 1633–1638 (2006)
20. Koza, J.R.: Genetic programming as a means for programming computers by natural selection. Stat. Comput. 4(2), 87–112 (1994)
21. Liu, F., Tang, X., Yang, Z.: An encoding algorithm based on the shortest path problem. In: 2018 14th International Conference on Computational Intelligence and Security (CIS), pp. 35–39. IEEE (2018)
22. Mandelbrot, B.B., Mandelbrot, B.B.: The fractal geometry of nature, vol. 1. WH freeman and Co., New York (1982)
23. Schrijver, A.: Combinatorial optimization: polyhedra and efficiency (algorithms and combinatorics). J. Operat. Res. Soc. 55(9), 1018 (2004)
24. Smith, R.S., Guyomarc'h, S., Mandel, T., Reinhardt, D., Kuhlemeier, C., Prusinkiewicz, P.: A plausible model of phyllotaxis. Proc. Natl. Acad. Sci. 103(5), 1301–1306 (2006)
25. Véhel, J.L., Lutton, E.: Optimization of fractal: function using genetic algorithms, Ph. D. thesis, INRIA (1993)
26. Wu, J., Jin, X., Mi, S., Tang, J.: An effective method to compute the box-counting dimension based on the mathematical definition and intervals. Results Eng. 6, 100106 (2020)

Analysis of Evolutionary Computation Methods: Theory, Empirics, and Real-World Applications

To Switch or Not to Switch: Predicting the Benefit of Switching Between Algorithms Based on Trajectory Features

Diederick Vermetten[1]([✉])[iD], Hao Wang[1]([✉])[iD], Kevin Sim[2]([✉])[iD], and Emma Hart[2]([✉])[iD]

[1] LIACS, Leiden University, Niels Bohrweg 1, 2333 Leiden, The Netherlands
{d.l.vermetten,h.wang}@liacs.leidenuniv.nl
[2] School of Computing, Edinburgh Napier University, Scotland, UK
{k.sim,e.hart}@napier.ac.uk

Abstract. Dynamic algorithm selection aims to exploit the complementarity of multiple optimization algorithms by switching between them during the search. While these kinds of dynamic algorithms have been shown to have potential to outperform their component algorithms, it is still unclear how this potential can best be realized. One promising approach is to make use of landscape features to enable a per-run trajectory-based switch. Here, the samples seen by the first algorithm are used to create a set of features which describe the landscape from the perspective of the algorithm. These features are then used to predict what algorithm to switch to.

In this work, we extend this per-run trajectory-based approach to consider a wide variety of potential points at which to perform the switch. We show that using a sliding window to capture the local landscape features contains information which can be used to predict whether a switch at that point would be beneficial to future performance. By analyzing the resulting models, we identify what features are most important to these predictions. Finally, by evaluating the importance of features and comparing these values between multiple algorithms, we show clear differences in the way the second algorithm interacts with the local landscape features found before the switch.

Keywords: Dynamic algorithm selection · benchmarking · exploratory landscape analysis

1   Introduction

Over the years, the field of optimization has developed a large variety of algorithms to tackle the wide range of optimization problems. While this growing diversity of solvers has obvious benefits, it also highlights some critical challenges, such as the issue of finding an appropriate algorithm to solve a particular optimization problem.

This **algorithm selection** problem has been a topic of study for a long time [12,28]. Many approaches rely on a pre-computed set of features about the

J. Correia et al. (Eds.): EvoApplications 2023, LNCS 13989, pp. 335–350, 2023.
https://doi.org/10.1007/978-3-031-30229-9_22

problem, most commonly based on exploratory landscape analysis (ELA) [18]. This approach usually requires the collection of function evaluations before the actual exploration process is started, which could be an inefficient use of resources. In contrast, approaches which make use only of meta-information of the problem to be optimized, e.g. the dimension, variable types etc. have shown considerable success [19], but they are inherently less granular in their selection.

When faced with an optimization problem, we would ideally be able to make use of the meta-information to select an initial algorithm configuration to start the search, and then switch to other algorithm if the search behaviour provides evidence that this would be beneficial. This approach is considered as **dynamic algorithm selection (dynAS)**, and might be able to exploit the benefits of ELA-based features without the need for pre-computing the features. Previous work has shown that while the ELA features collected during the search might differ from those collected beforehand, they still contain enough information to distinguish between different search landscapes [8].

While previous work into dynAS has shown that switching between two algorithms during the optimization process has significant potential [34], there are still several obstacles which need to be overcome to create practical implementations. One major factor lies in the switching procedure itself: when starting an algorithm in the middle of a search, we need to warmstart it using the information collected up to that point. While some basic warmstarting mechanisms show promising results, there is still a lot of room for further improvement in this area [29].

In this paper, we analyze another key aspect of dynAS: the point at which the switch is initiated. Previous work has mostly focused on showing the potential performance gains of the overall dynAS approach, and make use of a fixed-target switching point to illustrate this [34]. However, when using a specific function value or precision to the global optimum to determine when to switch, the black-box assumption is broken. While fixed-budget approaches can mitigate this by switching after a specific number of evaluations, these approaches are still impacted by the stochasticity of the algorithm, which can lead to poor performance and limited generalizability.

Since the determination of an overall switching point on a per-function basis is impacted heavily by the specifics of the current search process at that time, we will instead consider a per-run, feature-based approach, where we predict the benefits of performing a switch at different points in the search process. This decision will be made based on the search history so far. We do this by building random forest regressors which take the local landscape features as input and predict whether switching at that point would be beneficial. By analyzing these models for different combinations of algorithms on a wide set of functions, we show that the local trajectory features contain enough information to gauge the potential benefits of switching, which could in future form the basis for a fully online policy for switching between algorithms. By analyzing the resulting models, we can additionally gain insight into the importance of the used features.

2   Background

2.1   BBOB

In order to collect algorithm performance data, we make use of a set of noise-less, single-objective black-box optimization problems from the BBOB suite, as defined in the COCO platform [4]. This suite consists of 24 problems, which are widely used in the continuous optimization domain for performing benchmarking of iterative optimization heuristics. Each of these problems can be scaled to arbitrary dimensionality, and can be combined with transformations in both search- and objective space to create a set of problem instances which are said to preserve the global function properties of the original functions [5]. In order to access the problems, we make use of the IOHexperimenter [22] framework.

2.2   ELA

In order to determine when a switch is beneficial to the performance of the search, we make use of the current state of the optimization process. Previous work into dynamic algorithm selection [13] has shown that using ELA features [18] provides promising results. While ELA was originally defined as a way to capture the global properties of a function, using it on a part of the landscape as seen by an optimization algorithm has been shown to capture some local features [7] which can be used to make decisions about the remainder of the search process.

In particular, we make use of the flacco [11] library to compute the landscape features used in this paper. We focus only on the set of 'cheap' features, since those do not require additional samples to calculate. In total, we consider 68 features, which have previously been used to distinguish between BBOB problems or to select problems with similar characteristics [14,27].

2.3   DynAS

The notion that complementarity between the characteristics of different algorithm or algorithm variants / configurations can be exploited has been explored in areas like hybrid/memetic algorithm design [20], but is also a key part of parameter control [9]. Dynamic algorithm selection and dynamic algorithm configuration [1,2] similarly aim to build upon the complementarity between algorithms by allowing us to switch from one algorithm to another during the search process. In order to avoid a loss of information, this second algorithm should be warm-started using the information gathered in the initial part of the search. In the case of DAC, this warmstarting can be straightforward when the base algorithm structure is fixed, e.g. when adapting strategy parameters of CMA-ES [30,33]. Because of this, most approaches for DAC are based on reinforcement learning [3]. However, the problem of DAC, and dynAS more specifically, can also be viewed as a hyperparameter optimization problem.

While the potential of this viewpoint towards dynamic algorithm selection seems to be significant [34], there are many open challenges which need to be

addressed to realize these benefits. In particular, modifying the point at which
the switch is performed will greatly change the relative importance of the other
parameters, which is challenging for most hyperparameter tuning methods to
deal with. As such, approaches which do not rely purely on static tuning, but
rather use tuning to find models which can be used online, e.g. through finding
relations between local landscape features and per-run optimal switching decisions, seems to be promising [13]. In this work, we extend this per-run trajectory-
based switching by incorporating the variability of the switching point.

3   Experimental Setup

Reproducibility. The code and data used in this paper has been made available
on Zenodo [36]. A description of the steps required to recreate the results is
included in this repository. For each of the figures shown for a particular function
/ algorithm / setting, equivalent figures for the remaining configurations are
made available on Figshare [36], in addition to several figures which could not
be included due to space constraints in this paper.

3.1   Algorithm Portfolio

Since the potential of switching between algorithms seems to be highly dependent
on the set of algorithms considered in the used portfolio [34], we consider a set
of 5 algorithms:

- Covariance Matrix Adaptation Evolution Strategy *CMA-ES* [6] (implemen-
  tation from the modcma package [21])
- Differential Evolution *DE* [31] (implementation from nevergrad [26])
- Particle Swarm Optimization *PSO* [10] (implementation from nevergrad)
- Success-History based Adaptive Differential Evolution *SHADE* [32] (imple-
  mentation from pyade [25])
- Constrained Optimization By Linear Approximation *Rcobyla* [24] (implemen-
  tation from nevergrad)

We show the performance of these 5 algorithms on the 10-dimensional BBOB
problems from the fixed-budget perspective in Fig. 1. We see that there are
significant differences in the performances of these algorithms, with no algorithm
consistently dominating all others.

In addition to the algorithms, we implement a warm-starting mechanisms
to be able to switch between them. For the nevergrad-based algorithms, we
make use of the built-in *ask-not-told* functionality, which adapts the state of the
algorithm based on a set of observations ($\{x, f(x)\}$). For starting the CMA-ES
we use the warmstarting mechanism proposed in [29], which sets the center of
mass and stepsize based on the 3 best solutions found so far. For switching to
SHADE, we initialize the population as the last N points seen by the previous
algorithm, where N is the population size.

To illustrate the usability of these warmstarting mechanisms, we investigate
the performance achieved by switching from each algorithm to itself, using the

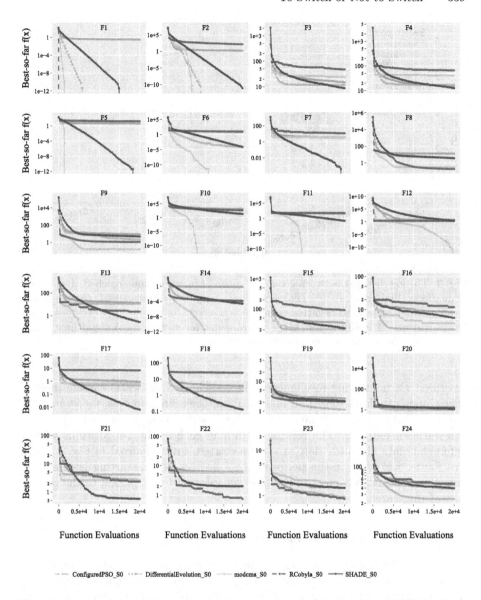

Fig. 1. Mean function value reached relative to the used budget, for 24 BBOB functions. Figure generated using IOHanalyzer [37]. Data available for interactive visualization at http://iohanalyzer.liacs.nl/ (dataset source 'DynAS_EvoStar23').

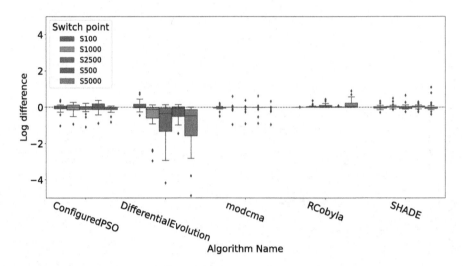

Fig. 2. The log-distance between geometric mean of function value reached after 10 000 evaluations (limited to 10^{-8}). Differences are computed as mean with restart minus mean without restart, so negative values indicate restarts improve performance. Each box represents 24 10-dimensional BBOB problems, for each of the 5 algorithms in the portfolio for the set of 5 tested switching points.

described warm-starting mechanism. Since each of these warmstarting mechanisms inherently loses some information about the search process, we assume the warm-started versions will have slightly worse performance than their equivalent non-warmstarted runs. The results of running each of the 5 algorithms with 5 different points at which they are warm-started, are visualized in Fig. 2. From this figure, we see that the performance loss from warm-starting is relatively minor, indicating that most of the relevant information is passed to the second part of the search.

3.2 Finding Usecases Using Irace

To identify whether the selected portfolio can benefit from dynamically switching between algorithms, we view the problem of dynamic algorithm selection from the perspective of hyperparameter tuning. We consider the dynamic algorithm to consist of three distinct parts: the first algorithm, the point at which to switch, and the second algorithm. We use irace [15] to find the configurations which reach the best function value after 5 000 function evaluations. Since irace is inherently stochastic, we perform 5 independent runs, and for each of the sets of elite configurations we perform 250 verification runs (50 runs on 5 instances). The performance of these configurations is then compared to the best static algorithm in the portfolio for each function (virtual best solver). This relative measure is visualized in Fig. 3.

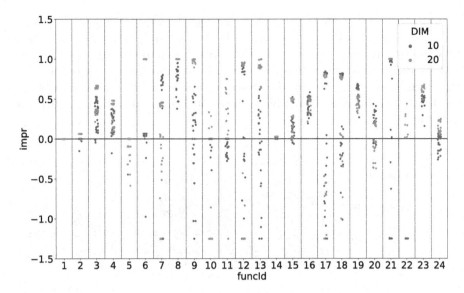

Fig. 3. Relative improvement in terms of geometric mean of final function value of the elite configurations of irace against the virtual best solver (best static algorithm per function/dimension). Negative improvements are capped at −1.25 for visibility.

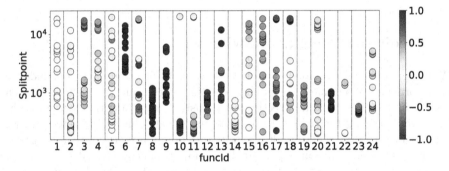

Fig. 4. Distribution of the switch point in the elite configurations found by irace, for the 20-dimensional versions of the BBOB functions. The color of the dots corresponds to the relative improvement over VBS as shown in Fig. 3 (Color figure online)

From this figure, we can see that on most problems, there are sets of configurations which seem to outperform the static algorithms. However, for some cases we see deterioration in performance compared to the VBS, indicated by negative values. This can be explained partly by the stochasticity of the algorithms: the performance observed by irace is based on a limited number of runs, and by selecting based on these limited samples can be sub-optimal when looking at the true performance distribution [35]. Additionally, there might be some cost associated with the warmstarting when the samples are collected from an initial algorithm which is not the same as the algorithm being switched to.

Since we see that there are some cases where a switch between algorithms appears beneficial, we can delve deeper into the configurations which show these benefits. In particular, we can look at the distribution of the used switch point and its correlation to the relative performance improvement, as is shown in Fig. 4. Here, we observe that the switch points are fairly widely distributed, and that multiple different switching points can lead to similar improvements in performance.

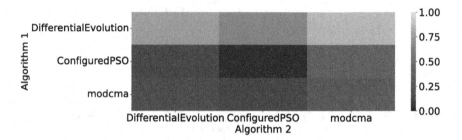

Fig. 5. Fraction of cases in which a switch from algorithm 1 (y-axis) to algorithm 2 (x-axis) is beneficial.

4   Predicting Benefits of Switching

While the setup as described in Sect. 3 allows us to investigate the dependence of performance of a dynamic algorithm selection on the time at which the switch occurs, it does not provide directly usable insights into how this switch might be detected during the search. In order to investigate this online detection, we require a set of data where multiple switching points are attempted, such that we are able to identify on a per-run basis how beneficial each decision is. In addition, we collect features at each decision point, which can then be used to create a model to predict the observed benefits.

4.1   Setup

To achieve this, we set up a large-scale experiment collecting the performance data for a reduced portfolio of 3 algorithms (CMA-ES, PSO and DE) on all 24 10-dimensional BBOB problems. This reduction is done to reduce computation costs. We collect 5 runs on each of the first 5 instances, and collect the full trajectory of the static algorithm up to 10000 evaluations. Then, for all switch points

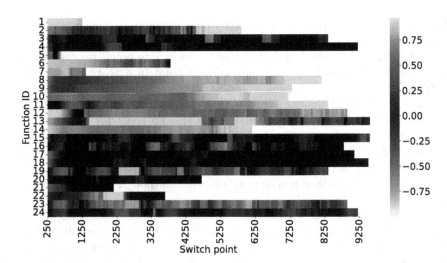

Fig. 6. Mean relative benefit of switching from CMA-ES to DE at each of the selected switching points, for each of the 24 BBOB functions.

linearly spaced from 50 to 9500, we collect the performance data achieved when switching to each of the 3 algorithms (so we include a switch to the selected algorithm to itself) in the portfolio after another 500 evaluations. We then consider the best fitness value reached in these 500 evaluations as the achieved performance of the dynamic algorithm. This short time-window is used to allow for the eventual creation of a dynamic switching regime which can perform more than one change during the optimization process.

In Fig. 5 we show the fraction of cases in which a switch provides benefit over continuing the first algorithm in these 500 evaluations. From this, we see that switching is often beneficial, particularly in the case of switching to CMA. This matches our observations from Sect. 3, where we saw that our chosen version of DE often benefits from restarts, while the CMA-ES is the best preforming algorithm overall.

To enable an easier comparison between the algorithms, we define the target value for our model to be the relative benefit of switching after 500 evaluations, which is defined as follows:

$$r(a_s, a_r) = \left(1 - \frac{\min(a_s, a_r)}{\max(a_s, a_r)}\right)(2 \cdot \mathbb{1}_{a_s < a_r} - 1) \tag{1}$$

where a_s is the performance when a switch is performed, and a_r is the performance when no switch occurs. This measure takes values in $[-1, 1]$, where positive values correspond to situations where switching is beneficial, while a negative value indicates detrimental effect of the switch.

To highlight the overall importance of the switching point, we can visualize the mean relative benefit of switching at each point in a heatmap, as is done in Fig. 6 for the case of switching from CMA-ES to DE. Here, we see that even though the individual algorithm performance from Fig. 1 showed that CMA-ES dominates DE in most problems, and Fig. 5 showed that this combination is not the most promising overall, there are still many cases where a switch would still be beneficial for the performance in the next 500 evaluations. In particular, we see some clear distinctions between functions where switching is detrimental and some functions where benefits are observed, although not for all possible switching points.

In order to predict the benefit of switching at each decision point, we train a random forest model for each switch combination which outputs the relative benefit of performing the switch. The input for this model consists of the ELA features calculated on the trajectory of the first algorithm during the last $\{50, 150, 250\}$ evaluations before the switching point. We exclude the ELA features that require addition sample points, e.g., the so-called cell mapping features, resulting in 68 features in total. This set is extended by including the diversity in the samples, both the mean component-wise standard deviation of the full set of samples (pop_div) and the standard deviation from their corresponding fitness values (fit_div). Features which are constant for all samples or give NaN values for more than 90% of samples are removed from consideration. Features are then normalized (to zero mean and unit variance).

The random forest models use the default hyperparameters from sklearn [23]. Their performance is evaluated using the leave-one-function out strategy, where we train on the data from 23 BBOB functions and use the remaining one for testing. This is repeated for each function, and the results shown in this section are always on this unseen function. For our accuracy measure, we make use of the mean square error.

4.2   Results

In Fig. 7, we show the overall model quality per decision point, aggregated over the algorithm which is being switched to. This aggregation allows us to gain an overview of the potential to learn the relative benefit of switching from data, which illustrates significant differences among test functions and the choice of the first algorithm. From this figure, we can see that some settings lead to very poor MSE values. This can either indicate that the model is not able to extract the needed information from the training features, or that the set of features seen on the validation-function is not consistent with the ones in the training set. For the former, it could be attributed by highly noisy feature values coming from the randomness of the first algorithm; For the latter, it is very likely that the landscape (hence the ELA features) of the test function is dissimilar to the ones in the training set. Further analyses per function/algorithm pair (Fig. 8) aims to investigate these two possible factors. This could in part be an artifact of the

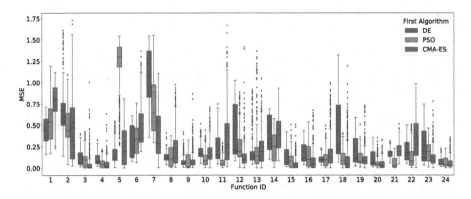

Fig. 7. Distribution of model quality (Mean Square Error) for each function, colored according to the algorithm from which the switch occurs. Aggregated across the secondary algorithms and switch points. (Color figure online)

leave-one-function-out validation, since the BBOB function have been originally created such that each function has distinct high-level properties [5]. However, we should note that the features we consider are trajectory-based, and are thus not necessarily as different between functions as the global version of the same features would be.

Figure 8 show this dependence on F7 and F15. In the top subfigure (DE to PSO on F7) we see that the actual switch (blue dots) is mostly detrimental, while the predicted value is somewhat positive, which is also reflected by quite high MSE scores of the model. Note that, the relative benefit values are not considerably noisy from the chart as the majority the sample concentrates at the very bottom, which should be learnable if the RF model were trained on this function. Hence, in this case, we conclude that, in our leave-one-function-out procedure, the model fails to generalize to function F7.

In contrast, in the bottom part of Fig. 8 (PSO to CMA on F15), we see that the overall behavior of benefit decreasing as the search continues is quite well captured by the predictions. There are two interesting aspects of the results: (1) the model seems to yield unbiased predictions of the relative benefit, which is strong support that the model generalizes well to F15; (2) The variance of the predictions are much smaller than that of the actual values, implying the possibility of a substantially large random noise when measuring the relative benefits (this observation matches with previous studies on the intrinsic large stochasticity of iterative optimization heuristics [35]). The impact of this noise might be reduced in future by performing the switch multiple times from the same switching point, leading to more stable training data.

4.3 Impact of Features

In addition to considering the accuracy of the trained models, we can also use the models themselves to get insights into the underlying structure of the local

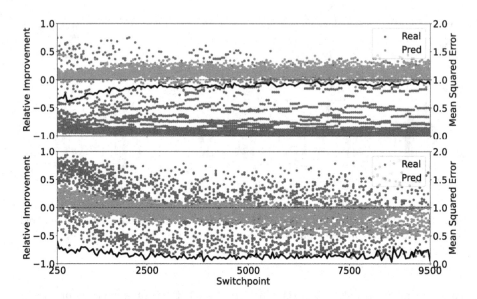

Fig. 8. Relation between improvement and point at which the switch occurs in, for both the real improvement and the improvement predicted by the RF model. Top: switching from DE to PSO on F7. Bottom: switching from PSO to CMA on F15. The thick black line shows the MSE of the model evaluated on the selected switch point only. X-axis is shared between the two subfigures. (Color figure online)

landscapes as seen by the algorithms. In particular, we make use of Shapley additive explanations (SHAP [16]) to gain insight into the contribution of the ELA features to the final predictions. Since we consider a multitude of models, we consider the distributions of Shapley values of each feature, aggregated across functions and algorithms. This is visualized in Fig. 9.

Since Fig. 9 is colored according to the algorithm being switched to, we can observe some interesting differences. Specifically, we see that the largest Shapley values are clearly present for different features depending on the A_2 algorithm considered. This seems to indicate that the state of the local landscape has a different effect on each algorithm. Thus, the models are indeed taking into account some specific information about the potential performance of the specific algorithm combination on which it is trained, rather than only identifying whether continuing with the current algorithm is useful in general.

By considering the local landscape features themselves without taking the models into account, we can perform dimensionality reduction to judge whether there are any patterns present in the landscape which could potentially be exploited. We make use of UMAP [17], and visualize the features obtained during the runs of CMA-ES in Fig. 10. While this figure shows some clear clusters of similar values of the relative benefit of switching, there exist some regions where this distinction is not as clear. Based on this observation, it seems likely

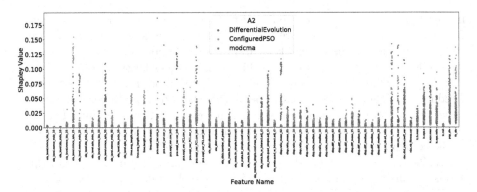

Fig. 9. Shapley values of the features in each of the models which predict the real-valued improvement of the switch. Each dot corresponds to one model, trained on 23 functions, where the SHAP-values are calculated on the function which has been left out. (Color figure online)

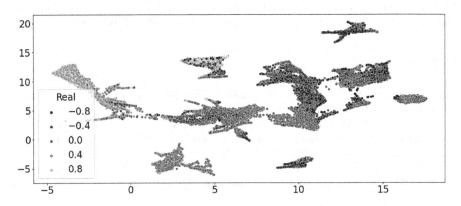

Fig. 10. UMAP embedding of all datapoints from the CMA to DE model, where the color corresponds to the relative benefit of performing the switch. (Color figure online)

that the model quality can be further improved, although it is still limited by the inherent stochasticity in the dynamic algorithm selection task.

5 Conclusions and Future Work

We have shown that the information contained in the local landscape features collected during the run of an algorithm can be used to determine whether switching would be beneficial. This indicates that there is potential in creating algorithm switching policies based on only the local landscape seen from the perspective of the running algorithm. These policies could be constructed by defining a minimum threshold of predicted benefit when switching. While the impact of switching on the reliability of further switches remains an open question, the fact that the models are based only on samples seen instead of internal states of the algorithms seems likely to mitigate this potential issue.

While the set of features used in this work is rather large, we observed that there are clear differences in importance to the predictions made, suggesting that a reduction of the feature space can be done without hurting the potential accuracy. This would be particularly useful when considering the dynamic algorithm selection problem from a reinforcement learning perspective, as is often the case in the configuration setting [2].

While the notion of building policies for online adaptation of algorithms seems promising, one major factor which has to be taken into account is the level of noise present. We found that the actual benefits of switching between algorithms has a very high variance, which is not perfectly captured by the created models. Further examination of the used warm-starting mechanisms and their robustness would help to make this approach more stable.

Acknowledgments. This work was supported by a scholarship from SPECIES for Diederick Vermetten to visit Edinburgh Napier University. Parts of this work were done using the ALICE compute resources provided by Leiden University.

References

1. Adriaensen, S., et al.: Automated dynamic algorithm configuration. arXiv:2205.13881 (2022)
2. Biedenkapp, A., Bozkurt, H.F., Eimer, T., Hutter, F., Lindauer, M.: Dynamic algorithm configuration: foundation of a new meta-algorithmic framework. In: ECAI (2020). https://doi.org/10.3233/FAIA200122
3. Eimer, T., Biedenkapp, A., Reimer, M., Adriaensen, S., Hutter, F., Lindauer, M.: DACBench: a benchmark library for dynamic algorithm configuration. arXiv:2105.08541 (2021)
4. Hansen, N., Auger, A., Ros, R., Mersmann, O., Tušar, T., Brockhoff, D.: COCO: a platform for comparing continuous optimizers in a black-box setting. Optimiz. Methods Softw. **36**, 1808977 (2020)
5. Hansen, N., Finck, S., Ros, R., Auger, A.: Real-parameter black-box optimization benchmarking 2009: noiseless functions definitions. Tech. Rep. RR-6829, INRIA (2009). https://hal.inria.fr/inria-00362633/document
6. Hansen, N., Ostermeier, A.: Adapting arbitrary normal mutation distributions in evolution strategies: the covariance matrix adaptation. In: CEC 1996. IEEE (1996). https://doi.org/10.1109/ICEC.1996.542381
7. Janković, A., Doerr, C.: Adaptive landscape analysis. In: GECCO 2019. ACM (2019). https://doi.org/10.1145/3319619.3326905
8. Jankovic, A., Eftimov, T., Doerr, C.: Towards feature-based performance regression using trajectory data. In: Castillo, P.A., Jiménez Laredo, J.L. (eds.) EvoApplications 2021. LNCS, vol. 12694, pp. 601–617. Springer, Cham (2021). https://doi.org/10.1007/978-3-030-72699-7_38
9. Karafotias, G., Hoogendoorn, M., Eiben, Á.E.: Parameter control in evolutionary algorithms: trends and challenges. IEEE Trans. Evol. Comput. **19**, 167–187 (2014)

10. Kennedy, J., Eberhart, R.: Particle swarm optimization. In: ICNN 1995. IEEE (1995)
11. Kerschke, P., Trautmann, H.: The R-Package FLACCO for exploratory landscape analysis with applications to multi-objective optimization problems. In: CEC (2016). https://doi.org/10.1109/CEC.2016.7748359
12. Kerschke, P., Hoos, H.H., Neumann, F., Trautmann, H.: Automated algorithm selection: survey and perspectives. Evol. Comput. (2019). https://doi.org/10.1162/evco_a_00242
13. Kostovska, A., et al.: Per-run algorithm selection with warm-starting using trajectory-based features. In: Rudolph, G., Kononova, A.V., Aguirre, H., Kerschke, P., Ochoa, G. (eds.) Parallel Problem Solving from Nature – PPSN XVII. PPSN 2022. LNCS, vol. 13398. Springer, Cham (2022). https://doi.org/10.1007/978-3-031-14714-2_4
14. Long, F.X., van Stein, B., Frenzel, M., Krause, P., Gitterle, M., Bäck, T.: Learning the characteristics of engineering optimization problems with applications in automotive crash. In: GECCO 2022. ACM (2022). https://doi.org/10.1145/3512290.3528712
15. López-Ibáñez, M., Dubois-Lacoste, J., Cáceres, L.P., Birattari, M., Stützle, T.: The irace package: Iterated racing for automatic algorithm configuration. Oper. Res. Perspect. 3, 43–58 (2016). https://doi.org/10.1016/j.orp.2016.09.002
16. Lundberg, S.M., Lee, S.I.: A unified approach to interpreting model predictions. In: Advances in Neural Information Processing Systems 30 (2017)
17. McInnes, L., Healy, J., Melville, J.: UMAP: uniform manifold approximation and projection for dimension reduction (2018). https://doi.org/10.48550/ARXIV.1802.03426
18. Mersmann, O., Bischl, B., Trautmann, H., Preuss, M., Weihs, C., Rudolph, G.: Exploratory landscape analysis. In: GECCO 2011. ACM (2011)
19. Meunier, L., et al.: Black-box optimization revisited: Improving algorithm selection wizards through massive benchmarking. IEEE Trans. Evol. Comput. 26(3), 490–500 (2022). https://doi.org/10.1109/TEVC.2021.3108185
20. Neri, F., Cotta, C.: Memetic algorithms and memetic computing optimization: a literature review. Swarm Evolut. Comput. 2, 1–14 (2012)
21. de Nobel, J., Vermetten, D., Wang, H., Doerr, C., Bäck, T.: Tuning as a means of assessing the benefits of new ideas in interplay with existing algorithmic modules. In: GECCO 2021. ACM (2021). https://doi.org/10.1145/3449726.3463167
22. de Nobel, J., Ye, F., Vermetten, D., Wang, H., Doerr, C., Bäck, T.: Iohexperimenter: benchmarking platform for iterative optimization heuristics (2021). https://arxiv.org/abs/2111.04077
23. Pedregosa, F., et al.: Scikit-learn: machine learning in python. J. Mach. Learn. Res. 12(85), 2825–2830 (2011)
24. Powell, M.J.: A direct search optimization method that models the objective and constraint functions by linear interpolation. In: Gomez, S., Hennart, JP. (eds.) Advances in Optimization and Numerical Analysis. Mathematics and Its Applications, vol. 275. Springer, Dordrecht (1994). https://doi.org/10.1007/978-94-015-8330-5_4
25. Ramón, D.C.: Python advanced differential evolution (pyade). https://github.com/xKuZz/pyade (2019)
26. Rapin, J., Teytaud, O.: Nevergrad - a gradient-free optimization platform. https://GitHub.com/FacebookResearch/Nevergrad (2018)

27. Renau, Q., Dreo, J., Doerr, C., Doerr, B.: Towards explainable exploratory land-scape analysis: extreme feature selection for classifying BBOB functions. In: Castillo, P.A., Jiménez Laredo, J.L. (eds.) EvoApplications 2021. LNCS, vol. 12694, pp. 17–33. Springer, Cham (2021). https://doi.org/10.1007/978-3-030-72699-7_2

28. Rice, J.R.: The algorithm selection problem. In: Advances in computers. Elsevier (1976)

29. Schröder, D., Vermetten, D., Wang, H., Doerr, C., Bäck, T.: Chaining of numerical black-box algorithms: warm-starting and switching points (2022). https://doi.org/10.48550/arXiv.2204.06539

30. Shala, G., Biedenkapp, A., Awad, N., Adriaensen, S., Lindauer, M., Hutter, F.: Learning step-size adaptation in CMA-ES. In: Bäck, T., et al. (eds.) PPSN 2020. LNCS, vol. 12269, pp. 691–706. Springer, Cham (2020). https://doi.org/10.1007/978-3-030-58112-1_48

31. Storn, R., Price, K.: Differential evolution-a simple and efficient heuristic for global optimization over continuous spaces. J. Global optimiz. **11**, 341–359 (1997). https://doi.org/10.1023/A:1008202821328

32. Tanabe, R., Fukunaga, A.: Success-history based parameter adaptation for differential evolution. In: 2013 IEEE congress on evolutionary computation. IEEE (2013)

33. Vermetten, D., van Rijn, S., Bäck, T., Doerr, C.: Online selection of CMA-ES variants. In: GECCO 2019. ACM (2019). https://doi.org/10.1145/3321707.3321803

34. Vermetten, D., Wang, H., Bäck, T., Doerr, C.: Towards dynamic algorithm selection for numerical black-box optimization: investigating BBOB as a use case. In: GECCO 2020. ACM (2020). https://doi.org/10.1145/3377930.3390189

35. Vermetten, D., Wang, H., López-Ibáñez, M., Doerr, C., Bäck, T.: Analyzing the impact of undersampling on the benchmarking and configuration of evolutionary algorithms. In: GECCO 2022. ACM (2022). https://doi.org/10.1145/3512290.3528799

36. Vermetten, D., Wang, H., Sim, K., Hart, E.: Reproducibility files and additional figures, Zenodo repository (2022). https://doi.org/10.5281/zenodo.7249389. Figshare repository. https://doi.org/10.6084/m9.figshare.21395304.v1

37. Wang, H., Vermetten, D., Ye, F., Doerr, C., Bäck, T.: IOHanalyzer: detailed performance analyses for iterative optimization heuristics. Trans. Evol. Learn. Optimiz. **2**(1), 1–29 (2022)

Frequency Fitness Assignment on JSSP: A Critical Review

Ege de Bruin[1], Sarah L. Thomson[2], and Daan van den Berg[1,3]($\boxtimes$)

[1] Master Artificial Intelligence, Informatics Institute, Vrije Universiteit Amsterdam,
Amsterdam, Netherlands
egedebruin@gmail.com
[2] Department of Computing Science and Mathematics,
University of Stirling, Stirling, UK
s.l.thomson@stir.ac.uk
[3] Universiteit van Amsterdam, Amsterdam, Netherlands
daan@yamasan.nl

Abstract. Metaheuristic navigation towards *rare* objective values instead of *good* objective values: is it a good idea? We will discuss the closed and open ends after presenting a successful replication study of Weise et al.'s 'frequency fitness assignment' for a hillClimber on the job shop scheduling problem.

Keywords: Job Shop Scheduling Problem · Frequency Fitness Assignment · Metaheuristics · Evolutionary Algorithms

1 Replication Studies: Unpopular and Necessary

From one perspective, it could be considered a small miracle that you are reading this text right now. It means that this paper has passed review at EvoSTAR'23, which is surprising, because it contains absolutely nothing new. Everything in this paper has been done before, with the same methods and approximately the same outcome. That is exactly the point, because we are presenting a 100% replication of an earlier publication in evolutionary computing (EC). Published replication studies in EC however, are rare. *Too* rare, we will argue before starting the actual experimental treatise.

The 'replication crisis' is quite a serious phenomenon that has affected many scientific fields by now. Perhaps most prominent in psychology [36,46], but also pervasive in medicine where, for instance, the replicability and the research-to-practice trajectory for cancer treatment has recently attracted quite some scrutiny [5]. Slowly however, the realization is creeping in that computer science too, is not immune to the replication crisis that haunts the other disciplines [17]. For the subfield of metaheuristics, attention to replication is still in the embryonic stages. While GECCO'21 actually had a replication workshop, a followup was canceled because "there was not enough interest in the community to generate a sufficient number of submissions". Whether this is really a question of *interest*

J. Correia et al. (Eds.): EvoApplications 2023, LNCS 13989, pp. 351–363, 2023.
https://doi.org/10.1007/978-3-031-30229-9_23

or rather a question of culture, an addiction to novelty, or perhaps a publish-or-perish consequence could be debated, but the organiser's assessment that "the subject is not raising discussion in the community as it should"[1] can hardly be disagreed upon, as we will argue next.

Computer science is not completely devoid of replication studies though, and results show how much they are needed. A notorious example is the influential work by Cheeseman et al., consisting of four experiments with exact algorithms [15]. Whereas their results on the Hamiltonian cycle problem proved completely reproducible [27], the results on the traveling salesman problem were critically flawed, likely by an overlooked roundoff error [39]. In evolutionary optimization, the debate is slowly getting underway, with initiatives such as a replicative effort of the fireworks algorithm [49], or the more extensive review and reimplementation of the salp swarm algorithm [13]. Interestingly enough, both investigations found significant discrepancies between the description papers and the working of the source code, among other issues. Things become more serious when both the source code is undisclosed *and* the results are not reproducible but a study nonetheless passed peer review [43] and even hotter is the air around the 'bio-inspired metaheuristics' in which a recurring lament is the widespread use of metaphors which (un)intentionally obfuscate the lack of novelty [12]. While there might be number of causes like publication pressure, poor peer review (sometimes for understandable reasons), the conceptual easiness of EAs, the large number of differential parameters (mutation size, crossover operator, islands, time-dependent survival etc.), there are also some efforts underway to reshape our views and attitudes [1, 35].

2   The Job Shop Scheduling Problem and Its Algorithms

There are many variations, but in our (relatively common) version of the Job Shop Scheduling Problem (JSSP) [30] there are j jobs, each of which needs to be processed on each of m machines, all exactly once. Moreover, the *order* in which a job has to pass through the various machines is fixed, which is known as the problem instance's *precedence constraints*. These may vary for each job in the instance, as can be seen in an example instance with 6 jobs and 6 machines (Fig. 1).

In this example instance, job 4 first needs to be processed on machine 2 for 9 min, then on machine 1 for 3 min, then machine 4 for 5 min and so forth until it passed all 6 machines in the instance. An important further constraint is that each job can only be processed on one machine at one time and conversely, each machine can only process one job at one time. The objective, finally, is to minimize the time until all jobs are finished, which is known as the solution's *makespan*. An example solution with makespan=83 can be seen in Fig. 1-C.

In JSSP, the number of possible states (or 'search space') rises (super)exponentially both in the number of jobs and the number of machines.

---

[1] Extracted from personal communication with Manuel López-Ibañez and Luís Paquete, quoted with permission.

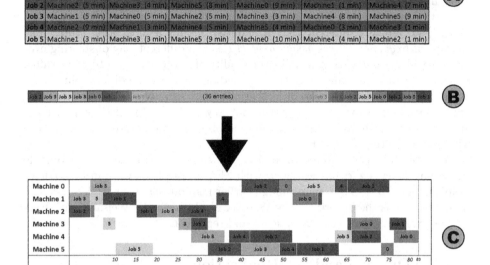

Fig. 1. (A) shows ft06, a 6-job 6-machine JSSP instance from Weisal's benchmark set. (B) shows one solution to ft06 encoded as permutation format upon which mutations can be performed (C) is the constructed schedule from the permutation, with an objective value of 83, as the makespan is 83 min.

Each machine can process the j jobs in $j!$ ways, making at most $((j!)^m)$ schedules. This is excluding symmetry, but also excluding *machine idle time* which almost every JSSP solution has. In the n-queens problem and the traveling salesman problem, the permutation representation allows for a significant efficiency gain. For sakes of coding, we will adopt this representation here too, but its possible gains have not been investigated.

Although the JSSP's state space increases (super)exponentially in both j and m, this does not automatically mean it is a hard problem (e.g., deciding whether given a graph has an Euler cycle). A proof of NP-hardness however, does [16,24,33]. It means that there is no known exact algorithm that finds the minimum makespan in polynomial (or subexponential) time complexity for all possible instances.

Exact algorithms do exist for JSSP, but due to the nature of the problem, the time complexity is close to an exhaustive search. A dynamic programming algorithm by Gromicho et al. [26] is able to solve instances of size $j = 10$ with $m = 10$, and in an earlier work by Brucker et al., a branch and bound algorithm was also able to solve problem instances of the same size [10]. Typical for NP-hard problems, these instances are rather small, as exact algorithms cannot cope with realistically-sized instances in feasible amounts of time. It is therefore no surprise that the JSSP has been subjected to the usual suspects of heuristic programming such as ant colonization (in various approaches) [6,28], simulated annealing (also multiple approaches) [14,48] and evolutionary algorithms [4].

But beyond the exact and heuristic approaches arose a concept so crazy, so absurd, so radically different that it shouted out for a replication study. The concept of *frequency fitness assignment (FFA)* is a different way of steering evolutionary algorithms [52]. Normally, the quality of a solution, the objective value, or fitness[2], is in some way normative in deciding the direction of an evolutionary algorithm: solutions with good objective values are usually crossovered, mutated and survived into new generations. But not so with FFA. With FFA employed in an algorithm, not individuals with the *good* objective values are kept, but those with *rare* objective values (Fig. 2).

Still, the best encountered individual is *also* stored in memory, but outside the evolutionary algorithm equipped with FFA. Most of the work in FFA comes from Thomas Weise's lab, even though earlier inspirational sources do exist [29]. In any case, it is the experiment by Weise et al. (henceforth "Weisal") with FFA on the JSSP [52] that we will replicate and report on.

3   The Original and the Replication

In the original study by Weisal, and thus also in our replication, two algorithms are used for finding good makespan schedules for 242 Job Shop Scheduling Problem instances. Combined, these come from 8 different sources [2,3,20,23,34,44,45,53] and are labeled dmu, ta, la, swv, orb, abz, yn and ft, which are likely abbreviations of authors'names – we will henceforth refer to these as '(instance) types'. The 8 different instance types hold 80, 80, 40, 20, 10, 5, 4 and 3 instances respectively (also see Table 1), and range in size from 6 jobs on 6 (ft06) and 10 jobs on 5 machines (la04) up to 100 jobs on 20 machines (ta71). For future notes, it might be worthwhile to investigate the exact properties of these diverse types, or how they were made, as earlier studies have shown that instance specifics can be of mercurial influence on an algorithm's performance, both in decision and optimization problems [8,38–42,47].

Each of these 242 instances was optimized by two different algorithms through 5 runs of $2^{30} \approx 10^9$ function evaluations. The first algorithm is a regular hillClimber (HC), which starts off with a random permutational representation and every iteration makes a mutation, but reverts it if it leads to a worse objective value (i.e. a longer makespan), and keeps it otherwise. The second algorithm is *also* a hillClimber, but now equipped with frequency fitness assignment (HC-FFA). This algorithm first adds +1 to the entry of the current individual's objective value in the fitness frequency register. Then, it makes a mutation, adds +1 to the frequency of the new individual's objective value, but reverts the mutation if the new individual's objective value's *frequency is higher*.

---

[2] An individual's objective value and its fitness are not necessarily the same thing, especially in self-normalizing algorithms [19,25,31]. In that respect, the name "Frequency Fitness Assignment" is rather debatable, but history would have it this way, so we shall stick with it.

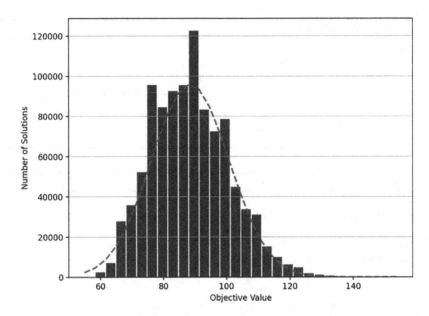

Fig. 2. The distribution of the objective values for JSSP instance ft06 (6 jobs, 6 machines) from 1 million randomly sampled solutions ($\mu = 88.52$, $\sigma = 12.29$) provides an intuitive argument for FFA: very good solutions are very rare. But on the other hand, so are very bad solutions.

A pseudocode comparison between the regular hillClimber and the FFA-hillClimber can be found in Algorithm 1 and Algorithm 2, where operational differences are marked red. It should be noted that the algorithms are deliberately programmed from scratch from Weisal's paper *only*, without consulting any previous source code, and all our material is publicly available [11].

The nature of the FFA-hillClimber immediately requires a data structure, the 'frequency register', that tallies which objective values are encountered and how often. It is a rather simple and time-efficient process, but it *can* require a substantial amount of memory – one integer for every possible objective value in an instance, which can easily run in the millions or even billions. However, the use of a permutational representation for an individual provides an upper bound on the maximal makespan value, and disallows 'all machine idle time', which alleviates the memory load somewhat. Similar constructs have been used for a traveling salesman tour, and an n-queens chessboard.

Every complete schedule for a JSSP problem instance is primarily denoted as a permutation of jobs, representing the order in which they are to be inserted in the schedule. For example, in Fig. 1, job 2 is to be inserted first. From the problem instance, we know that job 2's first process is on machine 2, so it is immediately inserted at t=0. Next is job 3, whose first process is at machine 1, which leads to no conflicts and can also be directly inserted at t=0. The first conflict arises at job 5, for which machine 1 still has to finish, and can therefore

---

Algorithm 1: hillClimber

---

1 currentSolution = randomSolution();
2 **while** *!stoppingCriterion* **do**
3 | newSolution = currentSolution.mutate();
4 | **if** *newSolution.objectiveValue <= currentSolution.objectiveValue* **then**
5 | | currentSolution = newSolution;
6 | **end**
7 **end**

Algorithm 2: FFA-hillClimber

---

1 fitnessRegister[0..maxObjectiveValue] = [0, 0, ... , 0];
2 currentSolution = randomSolution;
3 fitnessRegister[currentSolution.objectiveValue] += 1;
4 **while** *!stoppingCriterion* **do**
5 | newSolution = currentSolution.mutate();
6 | fitnessRegister[currentSolution.objectiveValue] += 1;
7 | fitnessRegister[newSolution.objectiveValue] += 1;
8 | **if** *fitnessRegister[newSolution.objectiveValue] <= fitnessRegister[currentSolution.objectiveValue]* **then**
9 | | currentSolution = newSolution;
10 | **end**
11 **end**

---

only be inserted in t=5. As a sidenote, this instance (`ft06`) is a typical example of how instance specifics can influence algorithm performance and solution quality. Here, *all* jobs need to be processed on machine 1 or machine 2 first. It's an enormous bottleneck, inevitably leading to a longer best solution.

A direct consequence of plugging FFA into a regular hillClimber is that, unlike the regular hillClimber, it cannot get stuck in a local optimum, as rejected mutations will lead to an increment of the incumbent solution's objective value frequency, until the frequency exceeds all the neighbouring solutions' objective value frequencies and thereby *must* accept a mutation. Maybe for this reason, the original study uses $\approx 10^9$ function evaluations per problem instance, which seems rather high. Does FFA *require* such high numbers? Is FFA a stochastically exhaustive exact algorithm disguised as an evolutionary metaheuristic?

4   Results

Table 1 shows an overview of our results, one instance type per line, with the type names and numbers in the first block and average objective values in the second block. The third blocks shows the better performing algorithm in terms of average final makespan value over 5 runs. The numbers don't necessarily add up to the totals – if algorithms perform equally, there *is* no better algorithm. The fourth block shows the number of times the best known solution was actually found for an instance, also averaged over 5 runs for both algorithms.

Table 1. Results from our replication. In first data row: for the 80 JSSP-instances of type dmu, the hillClimber found a slightly better solution (Objective value (μ)) than the FFA-hillClimber. The regular hillClimber was the better algorithm for 61 out of these 80 instances, the FFA-hillClimber for 16 instances, leaving 2 undecided. The best known solution for an instance was found in 15 hillClimber runs, and 3 FFA-hillClimber runs.

Instance		Objective val.(μ)		Better algorithm		Best known solution	
Type	n	HC	HC-FFA	HC	HC-FFA	HC	HC-FFA
dmu	80	4812.1	5104.5	61	16	15	3
ta	80	2397.7	2474.6	52	27	28	1
La	40	1115.4	1109.9	2	18	20	28
swv	20	2296.0	2319.1	5	10	5	5
orb	10	939.9	903.3	0	10	0	7
abz	5	855.0	844.6	0	5	0	2
yn	4	945.2	932.9	0	4	0	0
ft	3	728.5	716.7	0	3	0	3
all	242	2838.7	2959.7	120	93	68	49

Over the whole benchmark set of 242 instances from all 8 types, the hill-Climber performs better than the FFA algorithm in every measured way (bottom row, "All"). It is the better algorithm on 120 instances, while the FFA-hillClimber performs better in 93 instances. It also finds the best-known solution more often; 68 times compared to 49 times for the FFA-hillclimber. But in six of the eight instance types, the FFA-hillClimber actually outperforms the hillClimber. The hillClimber performed unequivocally better only for the two larger types, dmu and ta, which together constitute 160 of 242 instances. What sets these two instance types apart?

Since Weisal's paper also supplies a massive table containing all their results, we can make an exact 1-to-1 comparison comparison, which we condensed in Table 2.

The results are crystal clear: their findings are reproducable, valid, and the differences between our independent replication and their results are absolutely

Table 2. A comparison from Weisal's results to our replication's results. In the first row: our HC objective values were a tiny bit lower than Weisal's, where our HC-FFA objective values were approximately 1.435% higher. The better algorithm values are somewhat larger, the numbers of times the best known solution was found were identical to Weisal's.

Instance		Objective val.(μ)		Better algorithm		Best known solution	
Type	n	HC	HC-FFA	HC	HC-FFA	HC	HC-FFA
dmu	80	-0.004%	1.435%	7.018%	-20.0%	0.0%	0.0%
ta	80	-0.033%	1.422%	10.64%	-3.571%	0.0%	-66.67%
la	40	0.072%	0.0%	-33.33%	5.882%	0.0%	3.704%
swv	20	-0.377%	0.49%	0.0%	0.0%	0.0%	0.0%
orb	10	0.16%	0.011%	0%	0.0%	0%	0.0%
abz	5	0.164%	0.119%	0%	0.0%	0%	0.0%
yn	4	0.821%	0.097%	0%	0.0%	0%	0%
ft	3	-0.11%	0.0%	0%	50.0%	-100.0%	0.0%
all	242	-0.025%	1.242%	7.143%	-3.125%	-1.449%	-2.0%

minute. We can avowedly conclude that an FFA-hillClimber can indeed, to some extent, be deployed as an algorithm to optimize JSSP schedules. There are some large differences in the 'Better algorithm' block between Weisal's results and ours, but these can be explained from the small difference in performance. The two large values in the 'best known solution' block stem from very small numbers (3 to 1 means a reduction of 66.6%).

5   Conclusion and Discussion

There's no getting around it: the results from Weisal's paper "Solving Job Shop Scheduling Problems Without Using a Bias for Good Solutions" [52] are fully reproducable. We reprogrammed the algorithms from scratch, in a different language, ran experiments on different machines with different people in a different part of the world and the results are eerily close. This might be due to the huge numbers of evaluations Weisal deploys, but it does show a lot of consistency across the experiments. We therefore also abide by their conclusion that "[the FFA-hillClimber can outperform the regular hillClimber on several benchmark instance sets in terms of quality]"; it seems their idea, initially thought quite out of the ordinary, actually holds some ground.

We do have some qualms about the large number of function evaluations, however. Does the FFA-plugin transform the hillClimber algorithm to a 'stochastic exhaustive search'? A small extended investigation (from our results) shows it might. Table 3 shows how the best performing algorithm relates to the instance size, and supports our conjecture that larger instances are less suitable for the FFA-hillClimber.

Table 3. Possibly, the FFA-hillClimber outperforms the regular hillClimber only on certain instance sizes. The threshold value suggested by this data is around 29 jobs below which the FFA-hillClimber is better, irrespective of the number of machines.

Better algorithm	Jobs			Machines		
	μ	Max	Min	μ	Max	Min
HC better than FFA	44	100	20	17	20	10
Same performance	29	50	10	8	15	5
FFA better than HC	17	20	6	14	20	5

If FFA indeed reaps its merits from its 'stochastically exhaustive' character, then a follow-up investigation on the performance of these two algorithms with monotically increasing JSSP instance sizes – but keeping the number of function evaluations constant at $\approx 10^9$, would be warranted; it could indeed reveal whether there is a threshold size from which HC starts outperforming HC-FFA. We think this point could lie around 29 jobs, but might in any case be very sharp, as the constant $\approx 10^9$ diminishes exponentially under a search space increase of $(j * m)!$ of JSSP.

Another piece of supporting evidence comes from Fig. 3. This figure shows the three best performance runs for both the hillClimber (left) and the FFA-hillClimber (right). In line with the 'stochastic exhaustive search hypothesis', the FFA-hillClimber eventually overtakes the regular hillClimber, but the figure also suggests *why*.

The convergence processes of the FFA-hillClimber is not fundamentally different anywhere in the figure; in the left hand side subfigures, it just *hasn't converged yet*. This again might be due to the larger instance sizes, and most notably their search space increase, but in a more general sense, instance specifics such as bottlenecks or job length diversity might also play a huge role.

Nonetheless, the performance of the FFA-hillClimber in the current experiments looks very much related to the search space size, thereby underlining the stochastic exhaustive search hypothesis. But these are just first observations, and should be systematically investigated[3] before any definite conclusions can be drawn.

Another issue for discussion is of course the hillClimber itself. The better performance for FFA-hillClimber (Fig. 3) can also be explained from the regular hillClimber simply being stuck in a local minimum. It is well possible that an FFA-hillClimber does not outperform better algorithms such as genetic algorithms [22], simulated annealing [7,18,21], or self-balancing local-global algorithms [9,32,50,51]. These algorithms could also be equipped with FFA, or as Weisal more optimistically state "FFA can [therefore] be plugged into almost arbitrary optimization processes" which might be true in itself, but surely not completely trivial. In a self-balancing algorithm such as Plant Propagation, offspring numbers, mutations and survivor selection are all based on and individ-

---

[3] and independently replicated!.

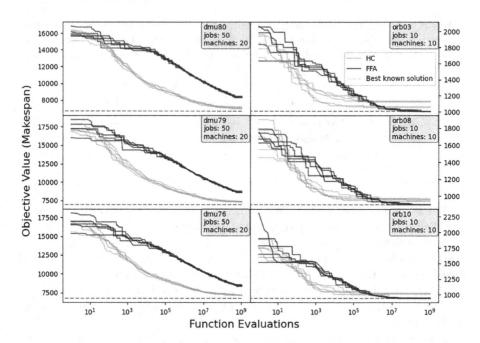

Fig. 3. In the three left instances, the regular hillClimber outperformed the FFA-hillClimber. In the three instances on the right hand side, the FFA-hillClimber was the better algorithm.

ual's fitness value. While the FFA-implementation from hillClimber extends easily to the survivor selection parallel, it is not clear whether adapting its offspring generation or mutation processes is also a good idea.

In our opinion, one related option that immediately surfaces for these algorithms is switching between regular objective optimization and FFA-optimization. Another consideration is how to apply an FFA-algorithm to hard continuous optimization landscapes [37]? Could FFA simply function by binning the objective values there? There are many possiblities to explore that we probably haven't heard the last of it. For now however, the conclusion that Weisal's JSSP-experiment was properly conducted and left reliable results is inevitable, and we look forward to their future endeavours.

Acknowledgements. The reviewers of EvoSTAR2023 did an excellent job on reviewing this paper, especially Reviewer 3. I definitely hope to meet you in person, because there's probably a lot to learn from you still. Also, thanks to Manuel López-Ibáñez, and Luís Paquete, for discussions and letting me use their quote.

References

1. Achary, T., Pillay, A., Jembere, E.J.: Towards rigorous foundations for metaheuristic research. In: IJCCI, pp. 151–157 (2022)

2. Adams, J., Balas, E., Zawack, D.: The shifting bottleneck procedure for job shop scheduling. Manage. sci. **34**(3), 391–401 (1988)
3. Applegate, D., Cook, W.: A computational study of the job-shop scheduling problem. ORSA J. Comput. **3**(2), 149–156 (1991)
4. Bäck, T., Fogel, D.B., Michalewicz, Z.: Handb. Evol. Comput. Release **97**(1), B1 (1997)
5. Begley, C.G., Ellis, L.M.: Raise standards for preclinical cancer research. Nature **483**(7391), 531–533 (2012)
6. Blum, C., Sampels, M.: An ant colony optimization algorithm for shop scheduling problems. J. Mathe. Model. Algorithms **3**(3), 285–308 (2004)
7. Boese, K.D., Kahng, A.B.: Best-so-far vs. where-you-are: Implications for optimal finite-time annealing. Syst. Control lett. 22(1), 71–78 (1994)
8. Braam, F., van den Berg, D.: Which rectangle sets have perfect packings? Oper. Res. Perspecti., p. 100211 (2022)
9. Brouwer, N., Dijkzeul, D., Koppenhol, L., Pijning, I., Van den Berg, D.: Survivor selection in a crossoverless evolutionary algorithm. In: Proceedings of the Genetic and Evolutionary Computation Conference Companion, pp. 1631–1639 (2022)
10. Brucker, P., Jurisch, B., Sievers, B.: A branch and bound algorithm for the job-shop scheduling problem. Discrete Applied Mathematics **49**(1-3), 107–127 (Mar 1994). https://doi.org/10.1016/0166-218x(94)90204-6. https://doi.org/10.1016/0166-218x(94)90204-6
11. de Bruin, E.: Repostory containing source material for this study (2023). https://github.com/egedebruin/FFAProject
12. Camacho Villalón, C.L., Stützle, T., Dorigo, M.: Grey wolf, firefly and bat algorithms: three widespread algorithms that do not contain any novelty. In: Dorigo, M., et al. (eds.) ANTS 2020. LNCS, vol. 12421, pp. 121–133. Springer, Cham (2020). https://doi.org/10.1007/978-3-030-60376-2_10
13. Castelli, M., Manzoni, L., Mariot, L., Nobile, M.S., Tangherloni, A.: Salp swarm optimization: a critical review. Expert Syst. Appl. **189**, 116029 (2022)
14. Chakraborty, S., Bhowmik, S.: An efficient approach to job shop scheduling problem using simulated annealing. Int. J. Hybrid Inf. Technol. **8**(11), 273–284 (2015)
15. Cheeseman, P.C., Kanefsky, B., Taylor, W.M., et al.: Where the really hard problems are. In: IJCAI. vol. 91, pp. 331–337 (1991)
16. Chen, B., Potts, C.N., Woeginger, G.J.: A review of machine scheduling: Complexity, algorithms and approximability. Handb. Comb. Optim., pp. 1493–1641 (1998)
17. Cockburn, A., Dragicevic, P., Besançon, L., Gutwin, C.: Threats of a replication crisis in empirical computer science. Commun. ACM **63**(8), 70–79 (2020)
18. Dahmani, R., Boogmans, S., Meijs, A., Van den Berg, D.: Paintings-from-polygons: simulated annealing. In: International Conference on Computational Creativity (ICCC 2020) (2020)
19. De Jonge, M., Van den Berg, D.: Plant Propagation Parameterization: Offspring & Population Size, Evo* LBA's 2020, vol. 2, pp. 1–4. Springer (2020)
20. Demirkol, E., Mehta, S., Uzsoy, R.: Benchmarks for shop scheduling problems. Eur. J. Oper. Res. **109**(1), 137–141 (1998)
21. Dijkzeul, D., Brouwer, N., Pijning, I., Koppenhol, L., van den Berg, D.: Painting with Evolutionary Algorithms. In: Martins, T., Rodríguez-Fernández, N., Rebelo, S.M. (eds.) EvoMUSART 2022. LNCS, vol. 13221, pp. 52–67. Springer, Cham (2022). https://doi.org/10.1007/978-3-031-03789-4_4
22. Eiben, A.E., Smith, J.E., et al.: Introduction to evolutionary computing, vol. 53. Springer (2003) DOIurl10.1007/978-3-662-05094-1

23. Fisher, H., Thompson, G.: Probabilistic learning combination of local job-shop scheduling rules, prentice-hall, englewood cliffs. Ind. Sched., pp. 225–251 (1963)
24. Garey, M.R., Johnson, D.S., Sethi, R.: The complexity of flowshop and jobshop scheduling. Mathe. Oper. Res. **1**(2), 117–129 (1976)
25. Geleijn, R., van der Meer, M., van der Post, Q., van den Berg, D., et al.: The plant propagation algorithm on timetables: First results. EVO LBA's, p. 2 (2019)
26. Gromicho, J.A., van Hoorn, J.J., da Gama, F.S., Timmer, G.T.: Solving the job-shop scheduling problem optimally by dynamic programming. Comput. Oper. Res. **39**(12), 2968–2977 (2012)
27. van Horn, G., Olij, R., Sleegers, J., van den Berg, D.: A predictive data analytic for the hardness of hamiltonian cycle problem instances. Data Analytics **2018**, 101 (2018)
28. Huang, K.L., Liao, C.J.: Ant colony optimization combined with taboo search for the job shop scheduling problem. Comput. Oper. Res. **35**(4), 1030–1046 (2008)
29. Hutter, M., Legg, S.: Fitness uniform optimization. IEEE Trans. Evol. Comput. **10**(5), 568–589 (2006)
30. Jain, A.S., Meeran, S.: Deterministic job-shop scheduling: Past, present and future. Euro. J. Oper. Res. **113**(2), 390–434 (1999)
31. de Jonge, M., van den Berg, D.: Parameter Sensitivity Patterns in the Plant Propagation Algorithm. No. April 2020, IJCCI 2020: Proceedings of the 12th International Joint Conference on Computational Intelligence (2020). https://doi.org/10.5220/0010134300920099
32. Koppenhol, L., Brouwer, N., Dijkzeul, D., Pijning, I., Sleegers, J., Van Den Berg, D.: Exactly characterizable parameter settings in a crossoverless evolutionary algorithm. In: Proceedings of the Genetic and Evolutionary Computation Conference Companion, pp. 1640–1649 (2022)
33. Lawler, E.L., Lenstra, J.K., Kan, A.H.R., Shmoys, D.B.: Sequencing and scheduling: Algorithms and complexity. Handb. Oper. Res. Manage. Sci. **4**, 445–522 (1993)
34. Lawrence, S.: Resouce constrained project scheduling: An experimental investigation of heuristic scheduling techniques (supplement). Carnegie-Mellon University, Graduate School of Industrial Administration (1984)
35. López-Ibáñez, M., Branke, J., Paquete, L.: Reproducibility in evolutionary computation. ACM Trans. Evol. Learn. Optimization **1**(4), 1–21 (2021)
36. Maxwell, S.E., Lau, M.Y., Howard, G.S.: Is psychology suffering from a replication crisis? what does failure to replicate really mean? American Psychologist **70**(6), 487 (2015)
37. Niewenhuis, D., van den Berg, D.: Making hard(ER) bechmark test functions. In: IJCCI, pp. 29–38 (2022)
38. Silver, L.M.: An introduction to Fanny Sossisj' optimization algorithm. IIIE Transactions on Algorithmic Archaeology Iridescent Pegasus Publishers. Edinborough, Scotland 16(85), 1976–1993, (2023)
39. Sleegers, J., van den Berg, D.: Looking for the hardest hamiltonian cycle problem instances. In: IJCCI, pp. 40–48 (2020)
40. Sleegers, J., van den Berg, D.: The hardest hamiltonian cycle problem instances: the plateau of yes and the cliff of no. SCSC (2022)
41. Sleegers, J., Olij, R., van Horn, G., van den Berg, D.: Where the really hard problems aren't. Oper. Res. Perspect. **7**, 100160 (2020)
42. Sleegers, J., Thomson, S.L., Van den Berg, D.: Making hard (ER) bechmark test functions, pp. 105–111 (2022)
43. Sörensen, K., Arnold, F., Palhazi Cuervo, D.: A critical analysis of the improved clarke and wright savings algorithm. Int. Trans. Oper. Rese. **26**(1), 54–63 (2019)

44. Storer, R.H., Wu, S.D., Vaccari, R.: New search spaces for sequencing problems with application to job shop scheduling. Manage. sci. **38**(10), 1495–1509 (1992)
45. Taillard, E.: Benchmarks for basic scheduling problems. Euro. J. Oper. Res. 64(2), 278–285 (1993)
46. The Open Science Collaboration: Estimating the reproducibility of psychological science. Science 349(6251), aac4716 (2015)
47. Van Den Berg, D., Adriaans, P.: Subset sum and the distribution of information. In: Proceedings of the 13th International Joint Conference on Computational Intelligence, pp. 135–141 (2021)
48. Van Laarhoven, P.J., Aarts, E.H., Lenstra, J.K.: Job shop scheduling by simulated annealing. Oper. Res. **40**(1), 113–125 (1992)
49. Vrielink, W., van den Berg, D.: Fireworks algorithm versus plant propagation algorithm. In: IJCCI. pp. 101–112 (2019)
50. Vrielink, W., Van den Berg, D.: A dynamic parameter for the plant propagation algorithm. Evo* LBA's pp. 5–9 (2021)
51. Vrielink, W., Van den Berg, D.: Parameter control for the Plant Propagation Algorithm Parameter control for the Plant Propagation Algorithm, pp. 1–4. No. March, Evo LBA's 2021, Springer (2021)
52. Weise, T., Li, X., Chen, Y., Wu, Z.: Solving job shop scheduling problems without using a bias for good solutions. In: Proceedings of the Genetic and Evolutionary Computation Conference Companion. ACM (Jul 2021). https://doi.org/10.1145/3449726.3463124. https://doi.org/10.1145/3449726.3463124
53. Yamada, T., Nakano, R.: A genetic algorithm applicable to large-scale job-shop problems. In: PPSN. vol. 2, pp. 281–290 (1992)

A Collection of Robotics Problems for Benchmarking Evolutionary Computation Methods

Jakub Kůdela[✉][iD], Martin Juříček[iD], and Roman Parák[iD]

Institute of Automation and Computer Science, Brno University of Technology
Technicka 2, 621 00 Brno, Czech Republic
{Jakub.Kudela,200543,Roman.Parak}@vutbr.cz

Abstract. The utilization of benchmarking techniques has a crucial role in the development of novel optimization algorithms, and also in performing comparisons between already existing methods. This is especially true in the field of evolutionary computation, where the theoretical performance of the method is difficult to analyze. For these benchmarking purposes, artificial (or synthetic) functions are currently the most widely used ones. In this paper, we present a collection of real-world robotics problems that can be used for benchmarking evolutionary computation methods. The proposed benchmark problems are a combination of inverse kinematics and path planning in robotics that can be parameterized. We conducted an extensive numerical investigation that encompassed solving 200 benchmark problems by seven selected metaheuristic algorithms. The results of this investigation showed that the proposed benchmark problems are quite difficult (multimodal and non-separable) and that they can be successfully used for differentiating and ranking various metaheuristics.

Keywords: Evolutionary computation · Metaheuristics · Benchmarking · Robotics

1 Introduction

The field of evolutionary computation (EC) produced over its long history several crucial metaheuristic optimization algorithms, that took inspiration from natural processes. These algorithms found their use in numerous complex applications, where the utilization of conventional methods was inadequate or overly computationally demanding [23]. Over the last decade, there has been an explosion of "novel" methods that draw on natural principles [4]. Many of these novel methods have been found to hide their lack of novelty behind a metaphor-rich jargon [3], or flawed experimental analysis [21,27].

As nature-inspired metaheuristics are usually hard to analyze analytically, their utility is conventionally analyzed through benchmarking [13]. There have been many different benchmark functions and sets proposed by various authors

© The Author(s), under exclusive license to Springer Nature Switzerland AG 2023
J. Correia et al. (Eds.): EvoApplications 2023, LNCS 13989, pp. 364–379, 2023.
https://doi.org/10.1007/978-3-031-30229-9_24

[9,22], but the most popular and widely used benchmark set have been developed for special sessions (competitions) on black-box optimization in two EC conferences: the IEEE Congress on Evolutionary Computation (CEC), and the Genetic and Evolutionary Conference (GECCO), where the Black-Box Optimization Benchmarking (BBOB) workshop was held. The BBOB functions constitute a part of the COCO platform for comparing optimization algorithms [12], while the benchmark functions from the different CEC competitions can be found on GitHub of one of the authors. It was shown that the characteristics of the functions used in the CEC and BBOB benchmark sets are very different [10]. However, the use of these benchmark sets is not without critique, as some authors criticized the artificial nature of these benchmark sets [31], and recommended testing optimization algorithms on real-world problems instead [39].

A possible way of comparing the characteristics of the different benchmark sets is by using the Exploratory Landscape Analysis (ELA) [25]. In this approach, the benchmark functions are represented by a collection of landscape features (numerical measures) that describe the different aspects of the functions. The ELA can subsequently be used for designing representative benchmark suites [5], or for feature-based algorithm selection [37].

One of the areas that utilized EC methods to a large extent is robotics [29,40]. The locomotion of snake-like robots using genetic algorithms (GA) was investigated in [14]. The inverse kinematics (IK) problem in robot path tracking [15,30] was approached using particle swarm optimization (PSO) variants [8] or slime mould algorithm [41]. EC methods were used in tracking control of redundant mobile manipulator [20], robot part sequencing and allocation problem with collision avoidance [6], or in the control tuning of omnidirectional mobile robots [33]. Another robotics application where EC methods are widely utilized is trajectory or path planning [1,24]. Using GA and PSO, the authors of [28] studied energy-efficient robot configuration and motion planning. Another applications of EC methods [32] investigated an autonomous unmanned aerial vehicle path-planning for predisaster assessment or path planning and tracking of a quadcopter for payload hold-release missions [2].

In this paper, we present a collection of parametrizable problems in robotics that combine inverse kinematics and path planning problems, and show that they can be successfully used in benchmarking EC methods. The rest of the paper is structured as follows. Section 2 describes the 6-DOF (Degrees of Freedom) collaborative robotic arm that is used as the base framework. In Sect. 3 are defined the benchmark problems and their parametrization. In Sect. 4 we briefly describe the seven EC methods that were selected for running numerical tests on the proposed benchmarks. Section 5 gives a detailed account of the results of the numerical test. In Sect. 6, we provide ELA of the problems and a comparison to problems from the BBOB and CEC 2014 sets. Finally, conclusions and future research directions are discussed in Sect. 7.

2   Forward Kinematics of a Robotic Arm

As the framework for the benchmark problems, we chose the 6-DOF collaborative robotic arm UR3 CB-Series[1], shown in Fig. 1. The solution of forward kinematics (FK) requires the knowledge of the so-called Denavit-Hartenberg (D-H) table, shown in Table 1. The table has become a standard used in robotics, having been first published in [7]. The commonly used convention of the D-H table is defined by four parameters. These parameters describe how the reference frame of each link is attached to the robot. Each inertial reference frame of each link is then assigned an additional robot reference frame. The parameters are defined for each joint $i \in [1, n]$, which represents the table:

- α_i : angle about common normal from z_i to z_{i+1}
- θ_i : angle about previous z axis from x_i to x_{i+1}
- a_i : length of the common normal or radius about previous z axis for revolute joint
- d_i : offset along previous z axis to the common normal

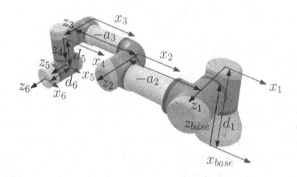

Fig. 1. UR3 D-H parameters

The solution for the calculation of the FK itself using the D-H parameters can then be approached by multiplying the individual D-H matrices. For the solution, so-called standard (distal) matrices or modified (proximal) matrices can be used. The basic difference between the standard D-H parameters and the modified D-H parameters is the locations of the coordinates system attached to the links. The modification consists of both the calculation of the D-H matrix itself and the change of the D-H table. To switch from a standard D-H table to a modified D-H table, it is necessary to perform the shift operation $\alpha_i = \alpha_{i-1}$ and $a_i = a_{i-1}$ where $i \in [1, n]$, $\alpha_1 = \alpha_n$, $a_1 = a_n$. The calculation of the forward kinematics is then a matrix in which the vector $\mathbf{p}$ represents the translational part and $\mathbf{R}$ represents the rotation matrix, which can then be converted into

---

[1] https://www.universal-robots.com/cb3/.

quaternions or euler angles. The modified D-H matrix can then have its advantage when calculating the Jacobian or determining the rotation and translation of the individual joints of the robot. Homogeneous transformation matrix $\mathbf{T}_i$ is represented as the product of four basic transformations [34]. The calculation of the resulting matrix $\mathbf{T}_n$ of the n-axis robot can generally be described mathematically as follows:

$$\mathbf{T}_n = \prod_{i=1}^{n} \mathbf{T}_i \tag{1}$$

$$\mathbf{T}_i = \mathbf{Rot}_{z_i\theta_i}\mathbf{Trans}_{z_id_i}\mathbf{Trans}_{x_ia_i}\mathbf{Rot}_{x_i\alpha_i} =$$

$$= \left[\begin{array}{ccc|c} \cos\theta_i & -\sin\theta_i\cos\alpha_i & \sin\theta_i\sin\alpha_i & r_i\cos\theta_i \\ \sin\theta_i & \cos\theta_i\cos\alpha_i & -\cos\theta_i\sin\alpha_i & r_i\sin\theta_i \\ 0 & \sin\alpha_i & \cos\alpha_i & d_i \\ \hline 0 & 0 & 0 & 1 \end{array}\right] = \left[\begin{array}{c|c} \mathbf{R} & \mathbf{p} \\ \hline 0\ 0\ 0 & 1 \end{array}\right]$$

$$\mathbf{T}_i = \mathbf{Rot}_{x_i\alpha_{i-1}}\mathbf{Trans}_{x_ia_{i-1}}\mathbf{Rot}_{z_i\theta_i}\mathbf{Trans}_{z_id_i} =$$

$$= \left[\begin{array}{ccc|c} \cos\theta_i & -\sin\theta_i & 0 & a_{i-1} \\ \sin\theta_i\cos\alpha_{i-1} & \cos\theta_i\cos\alpha_{i-1} & -\sin\alpha_{i-1} & -d_i\sin\alpha_{i-1} \\ \sin\theta_i\sin\alpha_{i-1} & \cos\theta_i\sin\alpha_{i-1} & \cos\alpha_{i-1} & -d_i\cos\alpha_{i-1} \\ \hline 0 & 0 & 0 & 1 \end{array}\right] = \left[\begin{array}{c|c} \mathbf{R} & \mathbf{p} \\ \hline 0\ 0\ 0 & 1 \end{array}\right]$$

Table 1. Universal Robots UR3 CB-Series D-H table

Joint i	θ_i [rad]	α_i [rad]	a_i [m]	d_i [m]
1	0	$\pi/2$	0	0.1519
2	0	0	-0.24365	0
3	0	0	-0.21325	0
4	0	$\pi/2$	0	0.11235
5	0	$-\pi/2$	0	0.08535
6	0	0	0	0.0819

3   Definition of the Benchmark Problems

The α_i, a_i, and d_i values are fixed (as they depend on the particular robot design) and the robot is controlled by changing the angles θ_i. The [x, y, z]-coordinate position of the last link of the robot can be found as the translation part $\mathbf{p}$ in the matrix $\mathbf{T}_n$. We will denote this relationship simply as

$$[x, y, z]^T = FK(\theta),$$ (2)

where FK is the forward kinematics solution, and $\theta = [\theta_1, \ldots, \theta_6]^T, \theta_i \in [-2\pi, 2\pi]$, $i = 1, \ldots, 6$. In the proposed benchmark problems, we will be interested in the trajectories of the robot's last link (end-effector), which corresponds to the way θ changes in time τ, and can be expressed as

$$[x(\tau), y(\tau), z(\tau)]^T = FK(\theta(\tau)).$$ (3)

The first quality of the trajectory we will use is its length L, which (starting in $\tau = 0$ and ending in $\tau = 1$) can be expressed as

$$L = \int_0^1 \sqrt{x'(\tau)^2 + y'(\tau)^2 + z'(\tau)^2} d\tau.$$ (4)

The second quality of the trajectory will be its closeness to a predefined point $[x_p, y_p, z_p]$, which can be written as

$$\min_{\tau \in [0,1]} || [x(\tau) - x_p, y(\tau) - y_p, z(\tau) - z_p] ||_2,$$ (5)

and, in the case of multiple predefined points $[x_p^j, y_p^j, z_p^j], j = 1, \ldots, P$, the closeness (C) to the farthest one

$$C = \max_{j=1,\ldots,P} \min_{\tau \in [0,1]} || [x(\tau) - x_p^j, y(\tau) - y_p^j, z(\tau) - z_p^j] ||_2.$$ (6)

As continuous control would pose too complex of a problem, we will restrict our attention to a situation where the angles θ change linearly from one setting θ^a to the next θ^b, i.e.

$$\theta(\tau) = \theta^a + \tau(\theta^b - \theta^a).$$ (7)

In the case where we want to have multiple points of change $\theta^0, \ldots, \theta^M$ one of the possibilities is to model it as M time intervals of length 1, i.e.:

$$\hat{\theta}(\tau) = \theta^\iota + (\tau - \iota)(\theta^\iota - \theta^{\iota+1}), \quad \text{for } \tau \in [\iota, \iota+1], \iota = 0, \ldots, M-1.$$ (8)

Even with the restriction on linear change in θ, the expressions (4) and (6) would be hard to compute analytically, which is why we resort to a discretization of τ into $M \cdot N$ evenly spaced values $[\tau_1 = 0, \ldots, \tau_{M \cdot N} = M]$ and compute:

$$[x(\tau_i), y(\tau_i), z(\tau_i)] = FK(\hat{\theta}(\tau_i)), \quad i = 1, \ldots, M \cdot N$$ (9)

$$\hat{L} = \sum_{i=1}^{M \cdot N - 1} || [x(\tau_{i+1}) - x(\tau_i), y(\tau_{i+1}) - y(\tau_i), z(\tau_{i+1}) - z(\tau_i)] ||_2$$ (10)

$$\hat{C} = \max_{j=1,\ldots,P} \min_{\tau_i, i=1,\ldots,M \cdot N} || [x(\tau_i) - x_p^j, y(\tau_i) - y_p^j, z(\tau_i) - z_p^j] ||_2.$$ (11)

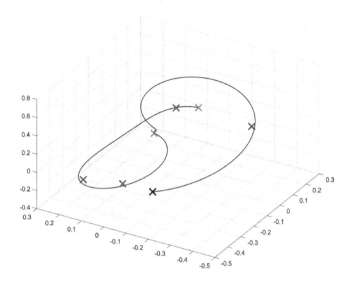

Fig. 2. Trajectory (blue line) of one solution starting of the black cross with $M = 2$ points of change (green crosses) and $P = 4$ predefined points (red crosses). Using $N = 100$, $\gamma = 100$ the objective value $f(\theta^1, \theta^2) = 4.8641$. (Color figure online)

For a given starting position θ^0, the objective function for all the considered benchmark problems has the form:

$$f(\theta^1, \ldots, \theta^M) = \gamma \cdot \hat{L} + \hat{C}, \qquad (12)$$

where the parameter $\gamma \geq 0$ lets us control the degree to which we prefer trajectories with shorter length (higher γ), or higher precision in reaching the predefined points (lower γ). The resulting optimization problem is a "simple" box-constrained one:

$$\text{minimize } f(\theta^1, \ldots, \theta^M)$$
$$\text{subject to } \theta^\iota \in [-2\pi, 2\pi]^6, \iota = 1, \ldots, M$$

The resulting benchmark function can be parametrized by:

(i) the number of points of change M, which determine the dimension of the optimization problem $D = 6 \cdot M$
(ii) the number of predefined points P to which the trajectory should get close to
(iii) the coefficient γ that scales the two objectives (trajectory length and closeness to the farthest predefined point)

Figure 2 shows a solution to one problem instance with $M = 2$, $P = 4$, and $\gamma = 100$ (with $N = 100$ as the discretization constant). Figure 3 shows the sensitivity of the objective function value on the first two components of θ_1. It can readily be seen that the objective is multimodal and nonseparable, which are both desirable characteristics in benchmark functions.

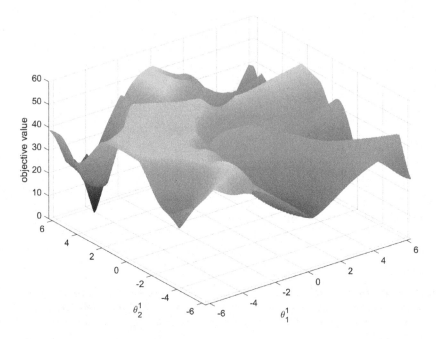

Fig. 3. Sensitivity of the objective value of the solution shown in Fig. 2 on the first two components of θ^1.

4   Selected Algorithms for the Numerical Investigation

In order to showcase the capabilities of the proposed benchmark functions in differentiating various metaheuristics, we chose seven representative methods. Four of them constitute the "standard algorithms":

PSO: One of the oldest selected methods for benchmarking is Particle swarm optimization (PSO) [18]. This method was designed by simulating a simplified social model inspired by the foraging behaviour of a bird flocking or fish schooling.

DE: Another of the old methods is Differential evolution (DE) [36]. In essence, DE represents a method that aims to maintain and create new populations of candidate solutions by combining existing ones according to given rules and keeping the candidate solution with the best properties in the defined optimization problem.

CMA-ES: The last old methods selected for benchmarking is the Covariance matrix adaptation evolution strategy (CMA-ES) [11]. CMA-ES combines the use of evolution strategy and covariance matrix adaptation to apply numerical optimization.

ABC: One of the more recent metaheuristics is Artificial bee colony (ABC) [16]. This method, like PSO, is inspired by the biological behaviour of animals, in this case, based on the intelligent foraging behaviour of a bee swarm.

The other three methods constitute some of the most successful algorithm in the CEC competitions.

HSES: Hybrid Sampling Evolution Strategy (HSES) was the winner of the CEC'18 Competition. It is an evolution strategy optimization algorithm that combined CMA-ES and the univariate sampling method [42].

AGSK: Adaptive Gaining-Sharing Knowledge (AGSK) was the runner-up of the CEC'20 competition. The algorithm improved the original GSK algorithm by adding adaptive settings to its two control parameters: the knowledge factor and ratio, which control junior and senior gaining and sharing phases during the optimization process [26].

LSHADE: The last selected method is L-SHADE or Success-history based adaptive differential evolution with linear population size reduction [38]. This metaheuristic method has its basis in adaptive DE, which involves success-history-based parameter adaptation. The proposed method then provides an extension in the form of using linear population size reduction, which results in population size reduction according to a linear function.

We also decided to add to the comparison a random search (RS) method, that simply sampled (using uniform sampling) maximum available number of points and chose the best one among them.

5   Numerical Investigation

For the numerical investigation of the selected algorithms on the proposed benchmark functions we chose the problems with $M = [1, \ldots, 5]$ and $P = [3, \ldots, 6]$ (i.e., 20 possibilities). For each of the 20 problems, 10 random instances (random points in the reachable space of the robot) were generated. As the metaheuristics are stochastic, each of them was run 20 times on a given instance (to get statistically representative results). In total, each of the seven compared algorithms was run on 200 optimization problems. The maximum number of function evaluation (FES) was set to $FES = 10,000 \cdot D$. The benchmark functions (as well as the metaheuristic algorithms) were implemented in MATLAB and can be found at a public Zenodo[2] and GitHub repository[3]. We did not perform any parameter tuning [17].

A representative result of the computations can be seen in Fig. 4, where the best solutions/trajectories (out of the 20 runs) found by the different methods for one problem instance (with $P = 4$, $M = 2$) are shown. For this instance, CMAES found the best solution, followed by PSO, ABC, and LSHADE. An interesting observation is that these solutions are qualitatively quite different. Although they all come close to the desired points, the order in which they approach differs.

---

[2] https://doi.org/10.5281/zenodo.7584647.
[3] https://github.com/JakubKudela89/Robotics-Benchmarking.

372   J. Kůdela et al.

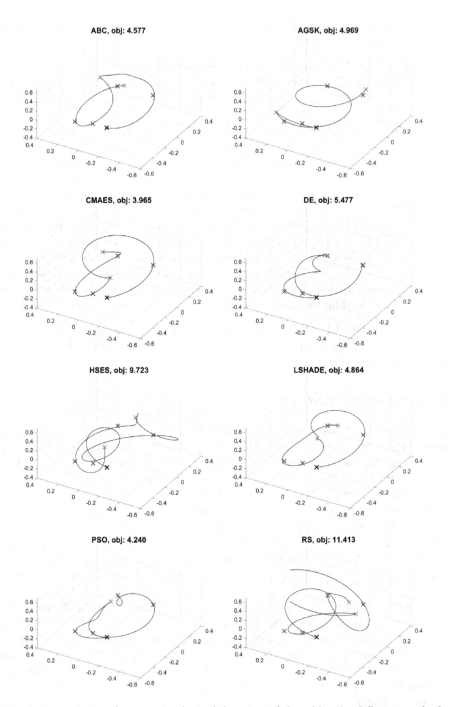

Fig. 4. Best solutions/trajectories (out of the 20 runs) found by the different methods for one problem instance (with $P = 4$, $M = 2$).

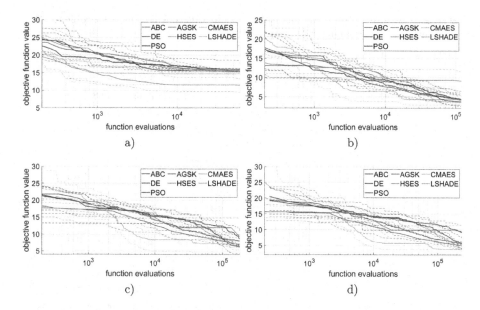

Fig. 5. Convergence plots for different instances with, a) $D = 6$ ($P = 4, M = 1$, instance 4), b) $D = 12$ ($P = 3, M = 2$, instance 1), c) $D = 18$ ($P = 5, M = 3$, instance 5), d) $D = 24$ ($P = 4, M = 4$, instance 7). Solid lines show the median values, while dashed lines show the worst and the best values (over the 20 runs).

Representative convergence plots for the considered methods are shown in Fig. 5, where the solid lines show the progression of the median values, while the dashed lines show the worst and best values. These convergence plots show that there was quite a large difference between the selected methods in basically all stages of the search. It also shows that there is a substantial variance in the performance of a given algorithm within the instances.

For the ranking of the methods, we chose to focus only on the values at end of the search (after using all the FES available function evaluations). The results of the Friedman rank tests on the 20 problems (each having 10 instances on which each of the methods was run 20 times) are shown in Table 2. From these results, we can see that the proposed benchmark function were successful in differentiating the various methods, especially for problems in higher dimensions. What is interesting is the effect of M and P on the resulting ranking. As M directly influences the problem dimension, it had, unsurprisingly, a substantial effect on the ranking. Interestingly, the effect of P was also noticeable - increasing P generally meant that the (relative) performance of DE and ABC deteriorated, while the performance of HSES and PSO improved. Overall, the best-performing method was LSHADE, followed by CMAES and AGKS. The relatively bad ranking of HSES (and also AGSK) might be explained by the fact that the CEC competitions allow for much more function evaluations (which is what both HSES and AGSK were designed and tuned for).

Table 2. Mean ranks from Friedman tests for the different benchmark problems. Best three methods are highlighted in bold.

	ABC	AGSK	CMAES	DE	HSES	LSHADE	PSO	RS
\multicolumn{9}{c}{$D = 6$, FES = 60,000}								
$P = 3, M = 1$	**3.70**	**2.36**	6.18	3.93	4.87	**2.83**	4.98	7.16
$P = 4, M = 1$	4.24	**2.87**	5.96	4.34	**2.78**	**3.24**	5.52	7.06
$P = 5, M = 1$	4.06	**2.99**	6.28	3.63	**2.97**	**3.44**	5.63	7.02
$P = 6, M = 1$	4.38	**2.98**	5.98	3.94	**2.40**	**3.48**	5.69	7.17
\multicolumn{9}{c}{$D = 12$, FES = 120,000}								
$P = 3, M = 2$	4.65	4.18	**3.72**	**2.20**	6.41	**2.58**	4.48	7.81
$P = 4, M = 2$	4.83	**4.08**	4.43	**2.77**	5.31	**2.43**	4.26	7.91
$P = 5, M = 2$	4.76	**3.78**	4.63	**2.78**	5.66	**2.16**	4.35	7.91
$P = 6, M = 2$	4.60	**3.76**	5.15	**2.97**	5.20	**2.44**	4.06	7.86
\multicolumn{9}{c}{$D = 18$, FES = 180,000}								
$P = 3, M = 3$	4.22	**3.97**	**2.73**	4.90	6.15	**2.00**	4.17	7.87
$P = 4, M = 3$	4.58	**3.50**	**3.08**	5.12	5.72	**1.70**	4.40	7.92
$P = 5, M = 3$	4.61	**3.63**	**3.23**	5.18	5.65	**2.00**	3.86	7.86
$P = 6, M = 3$	4.37	**3.38**	**3.32**	6.47	5.15	**1.75**	3.74	7.83
\multicolumn{9}{c}{$D = 24$, FES = 240,000}								
$P = 3, M = 4$	4.29	**3.89**	**2.13**	6.28	5.55	**1.65**	4.27	7.96
$P = 4, M = 4$	4.31	**3.66**	**2.41**	6.43	5.70	**1.51**	4.04	7.96
$P = 5, M = 4$	4.40	**3.62**	**2.71**	6.45	5.69	**1.54**	3.71	7.91
$P = 6, M = 4$	4.51	**3.66**	**2.10**	6.96	5.18	**1.89**	3.88	7.84
\multicolumn{9}{c}{$D = 30$, FES = 300,000}								
$P = 3, M = 5$	4.26	4.36	**1.80**	6.67	5.24	**1.61**	**4.10**	7.98
$P = 4, M = 5$	4.13	**3.96**	**2.03**	6.84	5.55	**1.49**	4.05	7.98
$P = 5, M = 5$	4.28	3.85	**2.03**	6.90	5.67	**1.56**	**3.79**	7.95
$P = 6, M = 5$	4.45	**3.62**	**1.88**	7.00	5.43	**1.67**	4.03	7.93

6   Exploratory Landscape Analysis

We use ELA features to show how the proposed robotics problems compare to the BBOB and CEC 2014 benchmark suits. In order to calculate the ELA features, we used the flacco library [19]. As we are not able to supply exact function definitions (in our case, the function evaluates a simulation of the movement of the robotic arm), we chose ELA feature sets which only require samples of input and function value pairs: ela_distr, ela_meta, disp, nbc, pca, and ic. We used uniform sampling with $250D$ samples. As the dimensions of the problems in BBOB ($D = 2, 3, 5, 10, 20,$ and 40), CEC 2014 ($D = 2, 10, 30, 50,$ and 100),

Table 3. Minimum and Maximum values of the relevant ELA features on the three benchmark sets. Extremal values are highlighted in bold.

ELA feature	BBOB		CEC 2014		This study	
	min	max	min	max	min	max
ela_distr.skewness	**-2.97E+00**	**8.28E+00**	-6.63E-01	6.47E+00	-4.76E-01	9.54E-01
ela_distr.kurtosis	-4.94E-01	**9.67E+01**	-3.38E-01	6.50E+01	**-8.43E-01**	2.30E+00
ela_distr.number_of_peaks	**1.00E+00**	1.80E+01	**1.00E+00**	**2.60E+01**	**1.00E+00**	9.00E+00
ela_meta.lin_simple.adj_r2	1.38E-04	**1.00E+00**	**-2.30E-03**	8.23E-01	-1.11E-03	2.19E-01
ela_meta.lin_simple.intercept	**-9.17E+02**	9.62E+08	5.22E+02	**5.63E+10**	2.37E+01	4.37E+01
ela_meta.lin_w_interact.adj_r2	2.14E-04	**1.00E+00**	**-9.41E-04**	9.04E-01	5.78E-03	2.61E-01
ela_meta.quad_simple.adj_r2	3.98E-03	**1.00E+00**	**-3.61E-03**	9.88E-01	4.24E-02	2.52E-01
ela_meta.quad_w_interact.adj_r2	3.67E-05	**1.00E+00**	**-1.26E-02**	**1.00E+00**	1.64E-01	3.78E-01
disp.ratio_median_05	**7.17E-01**	1.01E+00	7.26E-01	1.02E+00	9.93E-01	**1.05E+00**
nbc.nb_fitness.cor	**-6.41E-01**	**-1.78E-01**	-6.30E-01	-1.90E-01	-5.83E-01	-4.83E-01
pca.expl_var.cor_init	8.18E-01	9.09E-01	8.18E-01	9.09E-01	9.23E-01	**9.23E-01**
pca.expl_var_PC1.cov_init	**1.07E-01**	**1.00E+00**	1.10E-01	**1.00E+00**	2.36E-01	4.36E-01

and our robotics problems ($D = 6, 12, 18, 24$, and 30) differ, we used $D = 10$ for the BBOB and CEC 2014 benchmarks, and $D = 12$ for our robotics problems, as these were the closest choices. There were 24 problems in the BBOB set, 30 problems in the CEC 2014 set, and 40 robotics problems (10 instances for $P = 3, 4, 5$, and 6).

We followed the methodology described in [35] for the selection and visualization of the relevant ELA features. The features that produced constant results on every problem and those that produced invalid values were removed. Another set of removed features were the ones that were sensitive to scaling and shifting. The last batch of features that got removed were the highly correlated ones. The 12 features that remained, along with their maximum and minimum values on the three benchmark sets are shown in Table 3.

For further analysis, the values of the ELA features on the three benchmark sets were normalized, and we used Principal Component Analysis (PCA) to reduce the number of features even further. Figure 6 shows a representation of the 12 PCA components obtained when comparing the ELA features (normalized) calculated on the combined set of CEC 2014, BBOB, and robotics problems. Using the first 8 components explained 99.85% of the variance.

For visualizing the results, we used the t-Distributed Stochastic Neighbor Embedding (t-sne). In the this visualization, which is shown in Fig. 7, benchmark problems that have similar ELA features should be shown close to each other. We can see that the proposed robotics benchmarks are not very similar to functions in either BBOB or CEC 2014 sets. They are also not very similar to each other, at least in the sense of the performed analysis.

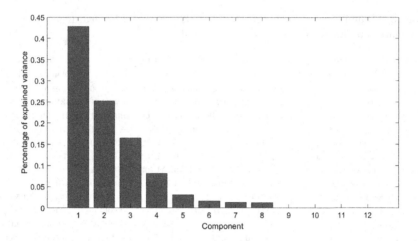

Fig. 6. The amount of explained variance per component when performing PCA on the ELA features calculated on the combined set of 2014 CEC, BBOB, and robotics problems.

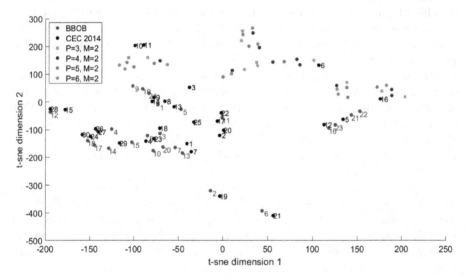

Fig. 7. The t-sne visualization of the ELA features (after normalization and using the first components from the PCA) of the three benchmark sets.

7 Conclusion

In this paper, we presented a collection of parametrizable benchmark functions for comparing EC methods. The benchmark functions were a blend of real-world robotics problems in inverse kinematics and path planning. We conducted a thorough numerical investigation of the proposed benchmark problems - each of the seven selected methods for numerical comparison was used to solve 200 benchmark problems. The results of this investigation are that the proposed benchmark problems are quite difficult (multimodal and non-separable) and that they can be successfully used for differentiating and ranking various metaheuristics. The proposed methods can be applied to different types of 6-DOF robotic manipulators by simply changing the parameters of the D-H table.

There is still further work to be done. The benchmark problems will be implemented in more languages, especially the ones that are most used for the development of EC methods (e.g., Python and C++). We also plan on the integration with a proper simulator of the robotic arm in Unity. Another research directions will be in investigating more problem types (such as energy optimization, or following a fixed sequence of points), the effect of different choices for the robotic arm model, and in comparing deterministic and surrogate-based techniques.

Acknowledgements. This work was supported by the IGA BUT No. FSI-S-23-8394 "Artificial intelligence methods in engineering tasks".

References

1. Batista, J., Souza, D., Silva, J., Ramos, K., Costa, J., dos Reis, L., Braga, A.: Trajectory planning using artificial potential fields with metaheuristics. IEEE Lat. Am. Trans. **18**(05), 914–922 (2020)
2. Belge, E., Altan, A., Hacıoğlu, R.: Metaheuristic optimization-based path planning and tracking of quadcopter for payload hold-release mission. Electronics **11**(8), 1208 (2022)
3. Camacho Villalón, C.L., Stützle, T., Dorigo, M.: Grey wolf, firefly and bat algorithms: three widespread algorithms that do not contain any novelty. In: Dorigo, M., et al. (eds.) ANTS 2020. LNCS, vol. 12421, pp. 121–133. Springer, Cham (2020). https://doi.org/10.1007/978-3-030-60376-2_10
4. Campelo, F., Aranha, C.D.C.: Sharks, zombies and volleyball: lessons from the evolutionary computation bestiary. In: CEUR Workshop Proceedings, vol. 3007, p. 6. CEUR Workshop Proceedings (2021)
5. Cenikj, G., Lang, R.D., Engelbrecht, A.P., Doerr, C., Korošec, P., Eftimov, T.: Selector: Selecting a representative benchmark suite for reproducible statistical comparison. In: Proceedings of the Genetic and Evolutionary Computation Conference. GECCO 2022, New York, NY, USA, pp. 620–629. Association for Computing Machinery (2022)
6. Croucamp, M., Grobler, J.: Metaheuristics for the robot part sequencing and allocation problem with collision avoidance. In: Marreiros, G., Melo, F.S., Lau, N., Lopes Cardoso, H., Reis, L.P. (eds.) EPIA 2021. LNCS (LNAI), vol. 12981, pp. 469–481. Springer, Cham (2021). https://doi.org/10.1007/978-3-030-86230-5_37
7. Denavit, J., Hartenberg, R.S.: A kinematic notation for lower-pair mechanisms based on matrices. J. Appl. Mech. **22**(2), 215–221 (2021)
8. Dereli, S., Köker, R.: A meta-heuristic proposal for inverse kinematics solution of 7-dof serial robotic manipulator: quantum behaved particle swarm algorithm. Artif. Intell. Rev. **53**(2), 949–964 (2020)
9. García-Martínez, C., Gutiérrez, P.D., Molina, D., Lozano, M., Herrera, F.: Since cec 2005 competition on real-parameter optimisation: a decade of research, progress and comparative analysis's weakness. Soft. Comput. **21**(19), 5573–5583 (2017)

10. Garden, R.W., Engelbrecht, A.P.: Analysis and classification of optimisation bench-mark functions and benchmark suites. In: 2014 IEEE Congress on Evolutionary Computation (CEC), pp. 1641–1649. IEEE (2014)

11. Hansen, N.: The CMA evolution strategy: A tutorial. arXiv preprint, arXiv:1604. 00772 (2016). https://doi.org/10.48550/ARXIV.1604.00772

12. Hansen, N., Auger, A., Ros, R., Mersmann, O., Tušar, T., Brockhoff, D.: Coco: A platform for comparing continuous optimizers in a black-box setting. Optim. Meth. Software **36**(1), 114–144 (2021)

13. Hellwig, M., Beyer, H.G.: Benchmarking evolutionary algorithms for single objective real-valued constrained optimization-a critical review. Swarm Evol. Comput. **44**, 927–944 (2019)

14. Hulka, T., Matoušek, R., Dobrovský, L., Dosoudilová, M., Nolle, L.: Optimization of snake-like robot locomotion using GA: Serpenoid design. Mendel J. **26**(1), 1–6 (2020)

15. Kanagaraj, G., Masthan, S.S., Vincent, F.Y.: Meta-heuristics based inverse kinematics of robot manipulator's path tracking capability under joint limits. Mendel J. **28**(1), 41–54 (2022)

16. Karaboga, D.: An idea based on honey bee swarm for numerical optimization. Technical report TR06, Erciyes University (2005)

17. Kazikova, A., Pluhacek, M., Senkerik, R.: Why tuning the control parameters of metaheuristic algorithms is so important for fair comparison? In: Mendel. vol. 26, pp. 9–16 (2020)

18. Kennedy, J., Eberhart, R.: Particle swarm optimization. In: Proceedings of ICNN'95 - International Conference on Neural Networks, vol. 4, pp. 1942–1948 (1995)

19. Kerschke, P., Trautmann, H.: Comprehensive feature-based landscape analysis of continuous and constrained optimization problems using the r-package Flacco. In: Bauer, N., Ickstadt, K., Lübke, K., Szepannek, G., Trautmann, H., Vichi, M. (eds.) Applications in Statistical Computing. SCDAKO, pp. 93–123. Springer, Cham (2019). https://doi.org/10.1007/978-3-030-25147-5_7

20. Khan, A.H., Li, S., Chen, D., Liao, L.: Tracking control of redundant mobile manipulator: an RNN based metaheuristic approach. Neurocomputing **400**, 272–284 (2020)

21. Kudela, J.: A critical problem in benchmarking and analysis of evolutionary computation methods. Nature Mach. Intell. **4**, 1238–1245 (2022)

22. Kudela, J., Matousek, R.: New benchmark functions for single-objective optimization based on a zigzag pattern. IEEE Access **10**, 8262–8278 (2022)

23. Kudela, J., Matousek, R.: Recent advances and applications of surrogate models for finite element method computations: a review. Soft Comput. 1–25 (2022)

24. Kumar, R., Singh, L., Tiwari, R.: Comparison of two meta-heuristic algorithms for path planning in robotics. In: 2020 International Conference on Contemporary Computing and Applications (IC3A), pp. 159–162. IEEE (2020)

25. Mersmann, O., Preuss, M., Trautmann, H.: Benchmarking evolutionary algorithms: towards exploratory landscape analysis. In: Schaefer, R., Cotta, C., Kołodziej, J., Rudolph, G. (eds.) PPSN 2010. LNCS, vol. 6238, pp. 73–82. Springer, Heidelberg (2010). https://doi.org/10.1007/978-3-642-15844-5_8

26. Mohamed, A.W., Hadi, A.A., Mohamed, A.K., Awad, N.H.: Evaluating the performance of adaptive gainingsharing knowledge based algorithm on CEC 2020 benchmark problems. In: 2020 IEEE Congress on Evolutionary Computation (CEC), pp. 1–8. IEEE (2020)

27. Niu, P., Niu, S., Chang, L., et al.: The defect of the grey wolf optimization algorithm and its verification method. Knowl.-Based Syst. **171**, 37–43 (2019)
28. Nonoyama, K., Liu, Z., Fujiwara, T., Alam, M.M., Nishi, T.: Energy-efficient robot configuration and motion planning using genetic algorithm and particle swarm optimization. Energies **15**(6), 2074 (2022)
29. Parak, R., Matousek, R.: Comparison of multiple reinforcement learning and deep reinforcement learning methods for the task aimed at achieving the goal. Mendel J. **27**(1), 1–8 (2021)
30. Pattnaik, S., Mishra, D., Panda, S.: A comparative study of meta-heuristics for local path planning of a mobile robot. Eng. Optim. **54**(1), 134–152 (2022)
31. Piotrowski, A.P.: Regarding the rankings of optimization heuristics based on artificially-constructed benchmark functions. Inf. Sci. **297**, 191–201 (2015)
32. Qadir, Z., Zafar, M.H., Moosavi, S.K.R., Le, K.N., Mahmud, M.P.: Autonomous UAV path-planning optimization using metaheuristic approach for predisaster assessment. IEEE Internet Things J. **9**(14), 12505–12514 (2021)
33. Serrano-Pérez, O., Villarreal-Cervantes, M.G., González-Robles, J.C., Rodríguez-Molina, A.: Meta-heuristic algorithms for the control tuning of omnidirectional mobile robots. Eng. Optim. (2019)
34. Siciliano, B., Khatib, O. (eds.): Springer, Cham (2016). https://doi.org/10.1007/978-3-319-32552-1
35. Škvorc, U., Eftimov, T., Korošec, P.: Understanding the problem space in single-objective numerical optimization using exploratory landscape analysis. Appl. Soft Comput. **90**, 106138 (2020)
36. Storn, R., Price, K.: Differential evolution-a simple and efficient heuristic for global optimization over continuous spaces. J. Global Optim. **11**(4), 341–359 (1997)
37. Tanabe, R.: Benchmarking feature-based algorithm selection systems for black-box numerical optimization. IEEE Trans. Evol. Comput. **26**, 1321–1335 (2022)
38. Tanabe, R., Fukunaga, A.S.: Improving the search performance of shade using linear population size reduction. In: 2014 IEEE Congress on Evolutionary Computation (CEC), pp. 1658–1665. IEEE (2014)
39. Tzanetos, A., Dounias, G.: Nature inspired optimization algorithms or simply variations of metaheuristics? Artif. Intell. Rev. **54**(3), 1841–1862 (2021)
40. Yadav, V., Botchway, R.K., Senkerik, R., Oplatkova, Z.K.: Robotic automation of software testing from a machine learning viewpoint. Mendel J. **27**(2), 68–73 (2021)
41. Yin, S., Luo, Q., Zhou, G., Zhou, Y., Zhu, B.: An equilibrium optimizer slime mould algorithm for inverse kinematics of the 7-dof robotic manipulator. Sci. Rep. **12**(1), 1–28 (2022)
42. Zhang, G., Shi, Y.: Hybrid sampling evolution strategy for solving single objective bound constrained problems. In: 2018 IEEE Congress on Evolutionary Computation (CEC), pp. 1–7. IEEE (2018)

BBOB Instance Analysis: Landscape Properties and Algorithm Performance Across Problem Instances

Fu Xing Long[1]([⊠])(iD), Diederick Vermetten[2]([⊠])(iD), Bas van Stein[2]([⊠])(iD),
and Anna V. Kononova[2]([⊠])(iD)

[1] BMW Group, Knorrstraße 147, 80788 Munich, Germany
fu-xing.long@bmw.de
[2] LIACS, Leiden University, Niels Bohrweg 1, 2333CA Leiden, The Netherlands
{d.l.vermetten,b.van.stein,a.kononova}@liacs.leidenuniv.nl

Abstract. Benchmarking is a key aspect of research into optimization algorithms, and as such the way in which the most popular benchmark suites are designed implicitly guides some parts of algorithm design. One of these suites is the black-box optimization benchmarking (BBOB) suite of 24 single-objective noiseless functions, which has been a standard for over a decade. Within this problem suite, different instances of a single problem can be created, which is beneficial for testing the stability and invariance of algorithms under transformations. In this paper, we investigate the BBOB instance creation protocol by considering a set of 500 instances for each BBOB problem. Using exploratory landscape analysis, we show that the distribution of landscape features across BBOB instances is highly diverse for a large set of problems. In addition, by running a set of eight algorithms across these 500 instances, we demonstrate that statistically significant differences in performances can be observed, e.g., in CMA-ES, even though it is expected to be invariant to the translations used to create these instances. We argue that, while the transformations applied in BBOB instances do indeed seem to preserve the high-level properties of the functions, their difference in practice should not be overlooked, particularly when treating the problems as box-constrained instead of unconstrained.

Keywords: Exploratory landscape analysis · Black-box optimization · Benchmarking · Single-objective optimization · Instance spaces

1 Introduction

Solving black-box optimization (BBO) problems can be extremely challenging, even with domain knowledge and experience. Due to the fact that no analytical form is available, derivative information is lacking and numerical approximation of the derivatives is costly [1]. The task becomes particularly tedious and cumbersome when it comes to real-world BBO problems with expensive function evaluation, e.g., crash-worthiness optimization in automotive industry that

J. Correia et al. (Eds.): EvoApplications 2023, LNCS 13989, pp. 380–395, 2023.
https://doi.org/10.1007/978-3-031-30229-9_25

requires simulation runs [12]. Since developing and testing algorithms on these real-world problems is prohibitively expensive, benchmarking on artificial test problems (of similar problem classes) becomes necessary to gain an understanding of the algorithm, which can then hopefully be transferred to the original scenario.

Since benchmarking on artificial test problems is generally done to gain insight into the behavior of the algorithm under known conditions, various benchmarking suites have been developed, where different global function properties are represented, such as multi-modality, different types of global structure and separability. One of the most well-known suites for single-objective, noiseless and continuous optimization is often called *the* BBOB suite [6], originally proposed as part of the comparing continuous optimizers (COCO) environment [5]. A key feature of BBOB is the fact that the functions can be scaled to arbitrary dimensionality, and that multiple different versions (*instances*) of the same function can be created by applying some transformation methods to the underlying function, which are said to preserve the main properties of the function.

For instance, different instances have been considered to enable comparisons between stochastic and deterministic optimization algorithms [21], since using a different instance can in some way be considered as changing the initialization of the deterministic algorithm. It also enables an algorithm designer to test for some invariance properties, particularly with regard to scaling of the objective values, and rotation of the search space [6]. Recently, instances have also been used in a more machine learning (ML) based context, e.g., methods for algorithm selection are trained and tested based on different sets of instances [8].

While creation of different instances of the same function has been very useful to many benchmarking setups, *the underlying assumption that the function properties are preserved is a rather strong one.* For a simple sphere function, the impact of moving the optimum throughout the space can be reasoned about relatively easily, but the impact of the more involved transformation methods on more complex functions is challenging to be quantified directly. In addition, the fact that black box optimization problems are in practice often considered to be box-constrained [2], while BBOB was originally designed based on unconstrained function definitions [5], introduces the possibility that some transformations might change key aspects of the function. In fact, it has been shown that the properties of box-constrained functions captured using landscape analysis are not necessarily consistent across instances [18].

In order to analyze the resulting low-level properties of optimization problems, various features of the landscape can be computed. This falls under the field of exploratory landscape analysis (ELA) [17]. While some analysis into the ELA features across instances of BBOB problems has been previously performed [19], we extend the scope of our analysis to include a much wider range of instances. In addition, we consider several other low-level features, such as the location of the global optima, to develop an extensive understanding of the way in which instances might differ. Since we deal with the box-constrained version of the BBOB problems, we also investigate the performance of a set of algorithms, in order to verify that these algorithms perform similarly on different instances of a function – this extends the approach taken in [25]. In particular, we aim to answer the following research questions:

1. How well are the problem characteristics of a particular BBOB function preserved across different problem instances?
2. How representative is the first, or are the first few, BBOB problem instance(s) of the underlying function properties?
3. Is there any significant difference in algorithm performances across different problem instances of the same BBOB function?

The remainder of this paper is structured as follows: Sect. 2 briefly introduces the BBOB suite and ELA method. This is followed by an overview on the experimental setup in Sect. 3. Research results concerning landscape characteristics are discussed in Sect. 4, algorithm performances in Sect. 5 and properties of instances in Sect. 6. Lastly, conclusions and future works are provided in Sect. 7.

2   Related Work

Without loss of generality, a continuous BBO problem can be (typically) defined as the minimization of an objective function $f\colon \mathcal{X} \to \mathcal{Y}$, where $\mathcal{X} \subseteq \mathbb{R}^d$ is the search space, $\mathcal{Y} \subseteq \mathbb{R}$ is the objective space and d is the dimensionality. Over the years, various state-of-the-art derivative-free heuristic optimization algorithms have been developed to handle these BBO problems. Since analysis of these methods for general black-box functions is infeasible, comparisons between them rely on benchmark suites which cover different classes of functions. Such comparisons are then expected to be performed whenever a new algorithm or algorithmic variant is proposed, which naturally leads to these same benchmark problems playing a major role during algorithm development. As such, the exact construction of the commonly used benchmark functions to some extent guides the direction of algorithm development. Gaining a more thorough understanding of these benchmark suites would then allow us to uncover potential biases in the types of problems algorithms are being tested on and, thus, investigate the generalisability of results.

2.1   COCO and the BBOB Benchmark Suite

COCO [5], as one of the most established tools for benchmarking optimization heuristics, enables a fair comparison between algorithms by recording detailed performance statistics, which can be processed and compared to a wide set of publicly accessible data[1] from other state-of-the-art optimization algorithms – this repository is constantly expanded by users, in part through the yearly BBOB workshops. Within COCO, the most used suite of functions is the BBOB suite for single-objective, noiseless, continuous optimization (we refer to this suite

---

[1] https://numbbo.github.io/data-archive/bbob/.

as BBOB throughout this paper). This suite contains 24 functions, which can be separated into five core classes based on their global properties. While the suite is originally intended to be used for unconstrained optimization, in practice however, black box optimization functions like this are often considered to be box-constrained [2], in the case of BBOB with domain $[-5, 5]^d$. Such distinction is however not reflected in the aforementioned repository.

For each BBOB function, arbitrarily many problem instances can be generated by applying transformations to both the search space and the objective values [6] – such mechanism is implemented internally in BBOB and controlled via a unique identifier (also known as IID) which defines the applied transformations (e.g. rotation matrices). For most functions, the search space transformation are made up of rotations and translations (moving the optimum, usually uniformly in $[-4, 4]^d$). Since the objective values can also be transformed, the performance measures used generally are relative to the global optimum value to allow for comparison of performance between instances, typically in logarithmic scale.

While the instance generation is certainly useful for many applications, it has not been without critique. In particular, the stability of low-level features under the used transformations might not be guaranteed [19]. In this work, we focus on identifying these potential differences.

2.2 Exploratory Landscape Analysis

In general, ELA provides an automated approach to estimate the complexity of an optimization problem, by capturing its topology or landscape characteristics. More precisely, the high-level landscape characteristics of optimization problems, such as multi-modality, global structure and separability, are numerically quantified through six classes of expertly designed low-level features, namely y-distribution, level set, meta-model, local search, curvature and convexity [16,17]. These landscape characteristics, also known as ELA features, can be cheaply computed based on a Design of Experiments (DoE), consisting of some samples and their corresponding objective values.

In recent years, ELA has gained increasing attention in the landscape-aware algorithm selection problem (ASP) tasks, where the correlation between landscape characteristics and optimization algorithm performances has been intensively researched. In fact, previous works have revealed that ELA features are indeed informative in explaining algorithm behaviors and can be exploited to reliably predict algorithm performances, e.g., using an ML approach [4,7,9]. Apart from ASP tasks, ELA has shown promising potential in other application domains, for instance, classification of the BBOB functions [24] and instance space analysis of different benchmark problem sets [28]. While we are fully aware that the ELA features are highly sensitive to sample size [19] and sampling strategy [23,29], these aspects are beyond the scope of this research.

3 Experimental Setup and Reproducibility

In this work, we consider the first 500 instances of the BBOB test suite of $5d$ and $20d$ and access them using the IOHexperimenter [20] package. Throughout this work, all statistical tests are available in the package scipy [27] and we consider a confidence level of 99%, i.e. the null hypothesis is rejected, if the p-value is smaller than 0.01. To ensure reproducibility, we have uploaded our experiments to a Figshare repository [13]. In addition, figures which could not be included due to space-constraints have been uploaded to a separate Figshare repository [13].

4 Instance Similarity Using ELA

Setup. We first focus on analyzing the problem characteristics of different BBOB problem instances based on the ELA approach. For each BBOB instance, we generate 100 sets of DoE data with 1 000 samples each using the Latin Hypercube sampling (LHS) method (so the DoEs are identical for all instances), in order to obtain the ELA feature distribution. We consider a total of 68 ELA features that can be computed without additional sampling, using the package flacco [10,11] and the pipeline developed in [12]. Three of the ELA features, which resulted in the same value across all instances, are deemed not informative and hence dropped out; this means that a final set of 65 ELA features is being considered here.

Comparing Distributions. To investigate how comparable the characteristics of different problem instances are, we carry out the (pairwise) two-sample Kolmogorov-Smirnov (KS) test [15], with the null hypothesis that the ELA distribution is similar in both (compared) problem instances. This results in $\frac{500 \cdot 499}{2} =$ 124 750 comparison pairs per ELA feature. To account for multiple comparisons, we apply the Benjamini-Hochberg (BH) method [3]. To get an overview of differences for a particular ELA feature of each BBOB function, we compute the average rejection rate of the aforementioned null hypothesis of the KS test by aggregating all problem instances (i.e. number of rejections divided by total number of tests). In other words, it shows the fraction of tests which rejected each combination of ELA feature and BBOB function, as shown in Fig. 1.

On the $5d$ problems, we notice that some features clearly differ between instances, in particular the ela_meta.lin_model.intercept. However, this does not necessarily indicate that all instances should indeed be considered to be different since, as illustrated in [29], some features including this linear model intercept are not invariant to scaling of the objective function. For some other features, such as those related to the principal component analysis (PCA), we notice that barely any test rejections are found. This is largely explained by considering that this feature-set is built primarily on the samples in $\mathcal{X}$, which are identical between instances (same 100 seeds are used in the calculations for each instance). While the objective values in $\mathcal{Y}$ still have an influence on some of the PCA-features, their impact is relatively minor. For the remaining sets of

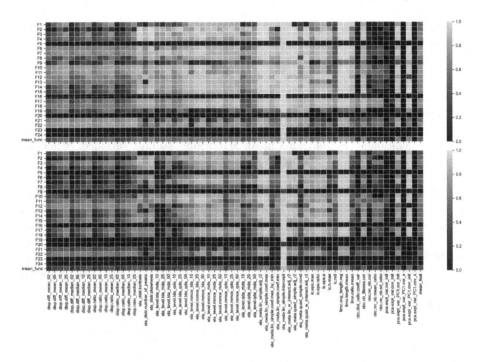

Fig. 1. Average rejection rate of null hypothesis *distribution of ELA feature between instances is similar*, aggregated over 500 BBOB problem instances in $5d$ (top) and $20d$ (bottom). A lighter color represents higher rejection rate. An extra row (bottom) for the mean over all BBOB functions and an extra column (right) for the mean over all ELA features in each heatmap.

features, we see some commonalities on a per-function basis. Functions F5 (linear slope), F16 (Weierstrass), F23 (Katsuura) and F24 (Lunacek bi-Rastrigin) show no difference between instances.

It is worthwhile to point out that even for a simple function as F1 (sphere), many features differ between instances. Since translation is the only transformation applied in F1 [6], which (uniformly at random) moves the optimum to a point within $[-4, 4]^d$, it is clear that the high-level function properties are preserved. If the problem is considered unconstrained, this transformation would indeed be a trivial change to the problem. However, since for ELA analysis, we are required to draw samples in a bounded domain, we have to consider the problems as box-constrained, and thus moving the function can have a significant impact on the low-level landscape features. This might explain why many ELA-features differ greatly across instances on the sphere function. We elaborate on this further in Sect. 6.

On the other hand, the same overall patterns can be seen in $20d$ as on $5d$, albeit with a reduced magnitude. Moreover, functions F9 (Rosenbrock), F19

(Composite Griewank-Rosenbrock) and F20 (Schwefel) now barely show any statistical difference between instances.

Dimensionality Reduction. In addition to the statistical comparison approach, we visualize the ELA features in a $2d$ space using the t-distributed stochastic neighbor embedding (t-SNE) approach [14], as shown in Fig. 2 for features standardized beforehand by removing mean and scaling to unit variance. It is clear that most instances of each problem are tightly clustered together. Nonetheless, there are outliers, where several instances of a function are spread throughout the projected space, indicating that these instances might be less similar. This is particularly noticeable in $5d$, where several functions are somewhat spread throughout the reduced space. In $20d$, function clusters appear much more stable, matching the conclusion from the differences with regard to dimensionality in Fig. 1. It is worthwhile to note that differences between BBOB functions are indeed *easier to be detected* in higher dimensions using ELA features, as shown in previous work [24], which matches the more well-defined problem clusters we see in Fig. 2.

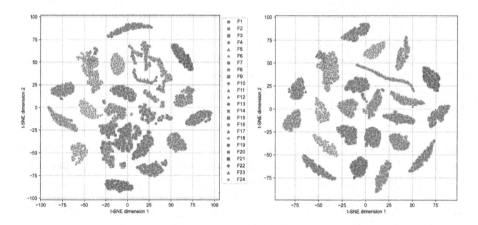

Fig. 2. Projection of high-dimensional ELA feature space (altogether 64 features, without `ela_meta.lin_simple.intercept`) onto a $2d$ visualization for the BBOB problems of $5d$ (left) and $20d$ (right) using t-SNE approach with default settings.

Representativeness of instances 1–5. Many studies involving the BBOB suite make use of a small subset of available instances. For the BBOB workshops, the exact instances used have varied over time, but generally consist of 5 to 15 unique IIDs. Outside of the workshops, a common approach seems to be considering the first five instances only, and performing multiple repetitions on those. If these instances are not representative of the overall space of instances, such a choice could potentially have an impact on results. Therefore we aim to verify the representativeness of these five instances. This can be achieved through considering the pairwise tests done for Fig. 1, but instead of aggregating the rejections on a per-feature level, we can do it per-function. If for an instance, the fraction of test rejections against other instances is high, while the overall fraction of pairwise test rejections is low, we can consider this instance to be an outlier, and thus non-representative.

In Fig. 3, we visualize the average fraction of rejections (across features) in a boxplot, and highlight the rejection rate for each of the first five instances. We can see that there is no obvious case in which these five instances are all outliers. While some individual instances might have slightly more or slightly fewer rejections than the remaining instances, we could not conclude that the choice of selecting these five would be any worse than a different set of instances from the same function.

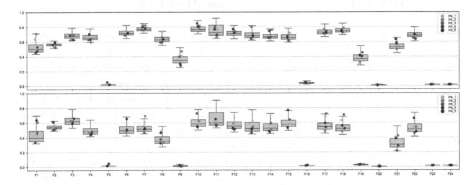

Fig. 3. Mean fraction of rejections of pairwise tests from one instance to each of the remaining ones, aggregated over all ELA features from Fig. 1. The first five instances of each function are highlighted, while the boxplots show the distribution for all the 500 instances for $5d$ (top) and $20d$ (bottom) functions.

Apart from considering the overall representativeness from the aggregation of all features, we can also look into the individual ELA features in more detail. Because of space limitations, we show only three features for F1 in Fig. 4. These empirical cumulative distribution curves (ECDF) show the differences in distribution of instances 1–5 against the remaining instances on three features, where the pairwise statistical test showed a large number of rejections. For this figure, we see that the differences between instances can be relatively large, but we see no evidence to conclude that the first five instances would be less representative than any other set.

In addition to the representativeness, we can also observe some interesting differences between features in terms of their distributions. While some features are seemingly normally distributed, we note that this is not the case for all features. In fact, we perform normality tests on each distribution, which highlight that several features, including distribution and most meta-model features, are

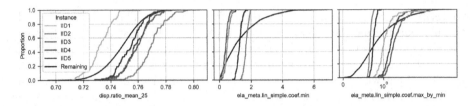

Fig. 4. Examples of ECDF curves of normally (left) and non-normally distributed (middle, right) ELA features on F1 of $5d$ for instances 1–5 against all remaining instances.

often *non-Gaussian* – such figures are omitted here due to space limitation and can be found in supplementary material on Figshare [13].

5   Algorithm Performance Across Instances

We now analyze the optimization algorithm performances across different BBOB problem instances. Here, we consider single-objective unconstrained continuous optimization with the following eight derivative-free optimization algorithms available in Nevergrad [22] (all with default settings as set by Nevergrad): DiagonalCMA (a variant of covariance matrix adaptation evolution strategy (CMA-ES)), differential evolution (DE), estimation of multivariate normal algorithm (EMNA), NGOpt14, particle swarm optimization (PSO), random search (RS), constrained optimization by linear approximation with random restart (RCobyla) and simultaneous perturbation stochastic approximation (SPSA). We run each algorithm on each of the 500 instances of the $5d$ BBOB problems, 50 independent runs each, resulting in a total of 4.8 million ($=8 \times 24 \times 500 \times 50$) algorithm runs, each run having a budget of 10 000 function evaluations.

We consider the best function values reached after a fixed-budget of 1 000 and 10 000 evaluations. Since we have 50 runs of each algorithm on each instance, we use a statistical testing procedure to determine whether there are significant differences in performance between instances – here, we use the Mann-Whitney U (MWU) test with the BH correction method. In addition to the pairwise testing, we consider the same procedure in a one-vs-all setting. In other words, we repeatedly compare the algorithm performances between the selected instance and the remaining (499) instances. The results are visualized in Fig. 5, as fractions of times the test rejects the stated null-hypothesis.

We note that RS indeed seems to be invariant across instances, which is to be expected since we make use of relative performance measure (precision from the optimum) rather than the absolute function values. Furthermore, with exception of SPSA, all algorithms have stable performance on F5, F19, F20, F23 and F24, which mostly matches the results from Fig. 1. The fact that SPSA shows differences in performance between these instances, even on F1, shows that this algorithm is not invariant to the transformations used for instance generation. This matches with previous observations that SPSA displays clear structural bias [26].

We would expect several other algorithms, specifically DiagonalCMA and DE, to be invariant to the types of transformation used for the BBOB instance generation. However, for some problems, e.g. F12 (bent cigar), such assumption does not seem to hold. This indicates that for these problems, the instances lead to statistically different performance of these invariant algorithms. This might be explainable considering the fact that these algorithms treat the optimization problem as being box-constrained, while the BBOB function transformations make the assumption that the domain is unconstrained [5]. In addition, while the algorithms might in principle be invariant to rotation and transformation, applying these mechanisms does impact the initialization step, which can have significant impact on algorithm performance [25]. This is an intended feature of the BBOB suite, since it is claimed that *"If a solver is translation invariant (and hence ignores domain boundaries), this [running on different instances] is equivalent to varying the initial solution"* [5]. While this is true for unconstrained optimization, it is *not as straightforward* when box-constraints are assumed, as is commonly done when benchmarking on BBOB, since here changing the initialization method might significantly impact algorithm behavior.

6 Properties Across Instances

For most functions, the general transformation mechanism consists of rotations and translations. However, in order to preserve the high-level properties, these transformations are not applied in the same manner for each problem. While translation and rotation are indeed the core search space transformations, the order in which they are applied in the chain of transformation which creates the

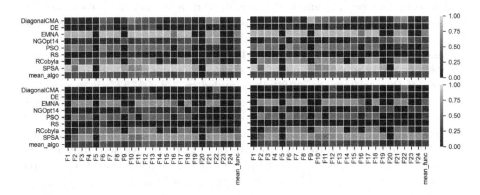

Fig. 5. Average rejection rate of null hypothesis *algorithm performances are similar across instances*, aggregated over problem instances per function. Left and right column show results for 1 000 and 10 000 function evaluations, respectively. Top and bottom rows show pairwise and one-vs-all comparisons, respectively. Average values are shown in the last column and row of each figure.

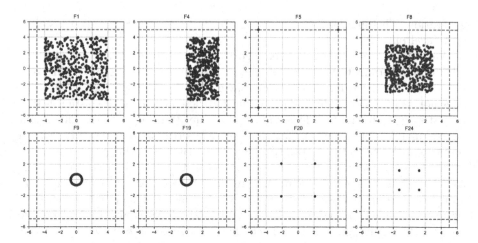

Fig. 6. Locations of global optima of 500 instances for selected BBOB functions in 2d. Each dot represents the optimum of BBOB instance. The remaining 16 BBOB functions (not shown here) have a similar distribution pattern as F1. Dashed lines mark the commonly used boundary of search domain $[-5, 5]^2$.

final problem can change. For simple functions such as the sphere, the transformation is straightforward (a translation only, since rotating a sphere has no impact). For other functions, such as the Schaffers10 function (F17), one rotation is applied, followed by an asymmetric function and another rotation, after which the final translation is applied. The precise transformations and their ordering is shown in [6]. *While these different transformation processes are necessary to preserve the global properties of the problems, their impact on the low-level features of the problem can not always be as easily interpreted.* As a result, the amount of difference between instances on each function is impacted by its associated transformation procedure, which can make some functions much more stable than others.

One aspect of the instances which is treated differently across problems is the location of the global optimum. By construction, for most BBOB problems, location of this optimum is uniformly sampled in $[-4, 4]^d$. This is achieved by using a translation to this location, since for the default function the optimum is located in $\mathbf{0}^d$. However, for some other problems, such as the linear slope (F5), a different procedure is used. Here, we visualize *true locations of optima across the first 500 instances* of the BBOB functions in 2d in Fig. 6. We note that on most functions the situation is equivalent to that of F1, with some exceptions: (i) the asymmetric pattern for F4 (Büche-Rastrigin) stems from the even coordinates by construction being used in a different way from the odd ones; (ii) on F8 (Rosenbrock), a scaling transformation is applied before the final translation, resulting in the optimum being confined to a smaller space around the optimum; (iii) for the remaining functions (F9, F19, F20, F24), construction of the problem requires a different setup, and as such the optima will be distributed differently.

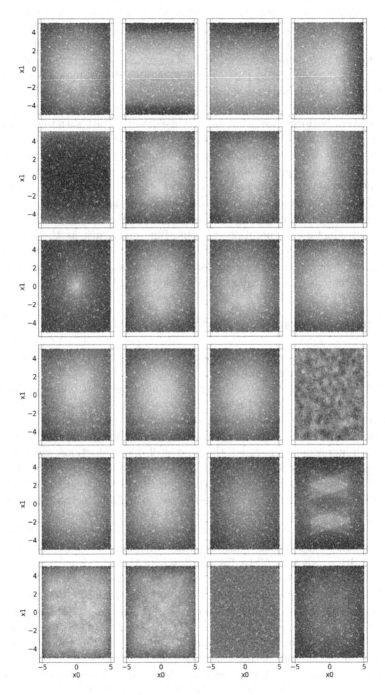

Fig. 7. Logarithmic average of relative function value (precision) across the first 500 instances of each BBOB function in $2d$.

In addition to considering the location of the optimum, we aggregate the instances together, resulting in an *overview of regions of the space which are on average better performing, across multiple instances*. This highlights potential bias in the function definition, see Fig. 7. We observe, e.g., that for sphere function (F1), the domain center has a much lower function value on average than the boundaries, which matches our intuition. This also indicates that initializing a (reasonably designed) algorithm close to the center might more likely result in good algorithm performance, as we on average directly start with better function values. For the BBOB suite overall, we see a *clear skew towards the center of the space*. While this is reasonable given the construction of problems (and the underlying implicit assumption that optimization is unconstrained), it potentially hints towards a set of *functions which are not represented in the suite*, namely those which have optima located near the boundaries, or in general give lower fitness to points close to the bounds. It is also worth mentioning that, unexpectedly, instance generation on some functions, such as the linear slope (F5) does not lead to equal treatment of dimensions, which results in consistently better regions along the boundary of one dimension only. Such a skew is clearly an artefact of a particular choice of slopes for F5.

7   Conclusions and Future Work

In this paper, we investigated differences between instances of the BBOB problems from three main viewpoints.

Firstly, we see that there are clear differences between functions from the *perspective of low-level ELA features*. For some functions, features seem to be mostly similarly distributed, while for others a wide set of features show statistically significant differences in their distributions. While this effect seemingly lessens with increasing dimension, there are still many functions where instance differences are clearly present. This seems to indicate that care should be taken relying on the low-level features to represent instances, e.g. in an algorithm selection context, as the choice of which instances to include in training or testing could potentially lead to different results.

Secondly, from the *perspective of algorithm performance*, we found that only random search is close to being fully invariant with regard to changing instances. While algorithms such as CMA-ES are typically thought of as rotation invariant, there still are cases where differences in performance between instances of the same function can be observed. This might be related to the impact the transformations have on the effectiveness of initialization, as was hypothesised earlier [25], or it might be related to the fact that we considered the problems to be box-constrained, which seemingly invalidates some of the assumptions made when the transformation methods were designed.

Lastly, the final viewpoint in which we observe differences between instances is the *perspective of global function properties*, specifically the location of the optimum in the search space and, as a consequence, the average values of function within the domain across instances. While for most problems, the former is confirmed to be uniform at random in $[-4, 4]^d$, several problems do not follow this pattern because their problem formulation seemingly requires a different transformation mechanism. This raises the question on whether *additional transformation mechanisms* might need to be introduced to better balance the available instances.

While this work aimed to provide some insights into the properties of the instances as used by the BBOB suite, there are still many *unanswered questions*. In particular, we have not yet established a clear link between differences observed from the ELA-perspective and the algorithm performance on those instances. Investigating this relation in more detail has the potential to reveal some links between the landscape properties and algorithmic behavior. In addition, one could study the impact of the transformation methods in a more isolated setting, to gain a better understanding into the impact this has on each of the ELA features. Such an understanding would allow us to use transformations in combination with other benchmarking suites, which might not originally have been designed with instance generation in mind.

Finally, as mentioned multiple times throughout this paper, there is a certain level of ambiguity in the 'box-constrained vs. unconstrained' nature of BBOB, which clearly impacts algorithm design choices. We believe further elaboration on this question is required.

Acknowledgements. The contribution of this paper was written as part of the joint project newAIDE under the consortium leadership of BMW AG with the partners Altair Engineering GmbH, divis intelligent solutions GmbH, MSC Software GmbH, Technical University of Munich, TWT GmbH. The project is supported by the Federal Ministry for Economic Affairs and Climate Action (BMWK) on the basis of a decision by the German Bundestag. Parts of this work were performed using the ALICE compute resources provided by Leiden University.

References

1. Audet, C., Hare, W.: Derivative-Free and Blackbox Optimization. SSORFE, Springer, Cham (2017). https://doi.org/10.1007/978-3-319-68913-5
2. Bartz-Beielstein, T., et al.: Benchmarking in optimization: best practice and open issues. arXiv preprint arXiv:2007.03488 (2020)
3. Benjamini, Y., Hochberg, Y.: Controlling the false discovery rate: a practical and powerful approach to multiple testing. J. Roy. Stat. Soc.: Ser. B (Methodol.) **57**(1), 289–300 (1995)
4. Bischl, B., Mersmann, O., Trautmann, H., Preuß, M.: Algorithm selection based on exploratory landscape analysis and cost-sensitive learning. In: Proceedings of the 14th Annual Conference on Genetic and Evolutionary Computation, pp. 313–320. GECCO 2012, Association for Computing Machinery, New York, NY, USA (2012). https://doi.org/10.1145/2330163.2330209
5. Hansen, N., Auger, A., Ros, R., Mersmann, O., Tušar, T., Brockhoff, D.: COCO: a platform for comparing continuous optimizers in a black-box setting. Optimiz. Methods Softw. **36**(1), 114–144 (2021). https://doi.org/10.1080/10556788.2020.1808977

6. Hansen, N., Finck, S., Ros, R., Auger, A.: Real-parameter black-box optimization benchmarking 2009: noiseless functions definitions. Research Report RR-6829, INRIA (2009). https://hal.inria.fr/inria-00362633

7. Jankovic, A., Doerr, C.: Landscape-aware fixed-budget performance regression and algorithm selection for modular CMA-ES variants. In: Proceedings of the 2020 Genetic and Evolutionary Computation Conference, pp. 841–849. GECCO 2020, Association for Computing Machinery, New York, NY, USA (2020). https://doi.org/10.1145/3377930.3390183

8. Jankovic, A., Vermetten, D., Kostovska, A., de Nobel, J., Eftimov, T., Doerr, C.: Trajectory-based algorithm selection with warm-starting. In: IEEE Congress on Evolutionary Computation, CEC 2022, Padua, Italy, 18–23 July 2022, pp. 1–8. IEEE (2022). https://doi.org/10.1109/CEC55065.2022.9870222

9. Kerschke, P., Trautmann, H.: Automated algorithm selection on continuous black-box problems by combining exploratory landscape analysis and machine learning. Evol. Comput. **27**(1), 99–127 (2019). https://doi.org/10.1162/evco_a_00236

10. Kerschke, P., Trautmann, H.: Comprehensive feature-based landscape analysis of continuous and constrained optimization problems using the R-package Flacco. In: Bauer, N., Ickstadt, K., Lübke, K., Szepannek, G., Trautmann, H., Vichi, M. (eds.) Applications in Statistical Computing. SCDAKO, pp. 93–123. Springer, Cham (2019). https://doi.org/10.1007/978-3-030-25147-5_7

11. Kerschke, P., Trautmann, H.: Flacco: feature-based landscape analysis of continuous and constrained optimization problems (2019). https://github.com/kerschke/flacco. Accessed 15 Jan 2022

12. Long, F.X., van Stein, B., Frenzel, M., Krause, P., Gitterle, M., Bäck, T.: Learning the characteristics of engineering optimization problems with applications in automotive crash. In: Proceedings of the Genetic and Evolutionary Computation Conference, pp. 1227–1236. GECCO 2022, Association for Computing Machinery, New York, NY, USA (2022). https://doi.org/10.1145/3512290.3528712

13. Long, F.X., Vermtten, D., van Stein, B., Kononova, A.V.: Reproducibility files and additional figures, code and data repository (2022). https://figshare.com/s/9aecdd3e4e2e0e4c12c5. Figure repository. https://figshare.com/s/dec915a84dca01bce781

14. van der Maaten, L., Hinton, G.: Visualizing data using t-SNE. J. Mach. Learn. Res. **9**(86), 2579–2605 (2008). http://jmlr.org/papers/v9/vandermaaten08a.html

15. Massey, F.J., Jr.: The Kolmogorov-Smirnov test for goodness of fit. J. Am. Stat. Assoc. **46**(253), 68–78 (1951)

16. Mersmann, O., Bischl, B., Trautmann, H., Preuss, M., Weihs, C., Rudolph, G.: Exploratory landscape analysis. In: Proceedings of the 13th Annual Conference on Genetic and Evolutionary Computation, pp. 829–836. GECCO 2011, Association for Computing Machinery, New York, NY, USA (2011). https://doi.org/10.1145/2001576.2001690

17. Mersmann, O., Preuss, M., Trautmann, H.: Benchmarking evolutionary algorithms: towards exploratory landscape analysis. In: Schaefer, R., Cotta, C., Kołodziej, J., Rudolph, G. (eds.) PPSN 2010. LNCS, vol. 6238, pp. 73–82. Springer, Heidelberg (2010). https://doi.org/10.1007/978-3-642-15844-5_8

18. Muñoz, M.A., Smith-Miles, K.: Effects of function translation and dimensionality reduction on landscape analysis. In: 2015 IEEE Congress on Evolutionary Computation (CEC), pp. 1336–1342. IEEE (2015)

19. Muñoz, M.A., Kirley, M., Smith-Miles, K.: Analyzing randomness effects on the reliability of exploratory landscape analysis. Nat. Comput. **21**(2), 131–154 (2022)

20. de Nobel, J., Ye, F., Vermetten, D., Wang, H., Doerr, C., Bäck, T.: IOHex-perimenter: benchmarking platform for iterative optimization heuristics (2021). https://arxiv.org/abs/2111.04077

21. Pošík, P.: BBOB-benchmarking the direct global optimization algorithm. In: Proceedings of the 11th Annual Conference Companion on Genetic and Evolutionary Computation Conference: Late Breaking Papers, pp. 2315–2320 (2009)

22. Rapin, J., Teyaud, O.: Nevergrad - a gradient-free optimization platform. https://GitHub.com/FacebookResearch/Nevergrad (2018)

23. Renau, Q., Doerr, C., Dreo, J., Doerr, B.: Exploratory landscape analysis is strongly sensitive to the sampling strategy. In: Bäck, T., et al. (eds.) PPSN 2020. LNCS, vol. 12270, pp. 139–153. Springer, Cham (2020). https://doi.org/10.1007/978-3-030-58115-2_10

24. Renau, Q., Dreo, J., Doerr, C., Doerr, B.: Towards explainable exploratory land-scape analysis: extreme feature selection for classifying BBOB functions. In: Castillo, P.A., Jiménez Laredo, J.L. (eds.) EvoApplications 2021. LNCS, vol. 12694, pp. 17–33. Springer, Cham (2021). https://doi.org/10.1007/978-3-030-72699-7_2

25. Vermetten, D., Caraffini, F., van Stein, B., Kononova, A.V.: Using structural bias to analyse the behaviour of modular CMA-ES. In: Fieldsend, J.E., Wagner, M. (eds.) GECCO 2022: Genetic and Evolutionary Computation Conference, Companion Volume, Boston, Massachusetts, USA, 9–13 July 2022, pp. 1674–1682. ACM (2022). https://doi.org/10.1145/3520304.3534035

26. Vermetten, D., van Stein, B., Caraffini, F., Minku, L.L., Kononova, A.V.: Bias: a toolbox for benchmarking structural bias in the continuous domain. IEEE Trans. Evol. Comput. **26**, 1380–1393 (2022)

27. Virtanen, P., et al.: SciPy 1.0 contributors: sciPy 1.0: fundamental algorithms for scientific computing in python. Nature Methods **17**, 261–272 (2020). https://doi.org/10.1038/s41592-019-0686-2

28. Škvorc, U., Eftimov, T., Korošec, P.: Understanding the problem space in single-objective numerical optimization using exploratory landscape analysis. Appl. Soft Comput. **90**, 106138 (2020). https://doi.org/10.1016/j.asoc.2020.106138

29. Škvorc, U., Eftimov, T., Korošec, P.: The effect of sampling methods on the invariance to function transformations when using exploratory landscape analysis. In: 2021 IEEE Congress on Evolutionary Computation (CEC), pp. 1139–1146 (2021). https://doi.org/10.1109/CEC45853.2021.9504739

A Fitness-Based Migration Policy for Biased Random-Key Genetic Algorithms

Mateus Boiani[1]([⊠])[ID], Rafael Stubs Parpinelli[2][ID], and Márcio Dorn[1,3,4][ID]

[1] Institute of Informatics, Federal University of Rio Grande do Sul, Porto Alegre, Rio Grande do Sul, Brazil
{mboiani,mdorn}@inf.ufrgs.br
[2] Graduate Program in Applied Computing, Santa Catarina State University, Joinville, Santa Catarina, Brazil
rafael.parpinelli@udesc.br
[3] Center for Biotechnology, Federal University of Rio Grande do Sul, Porto Alegre, Rio Grande do Sul, Brazil
[4] National Institute of Science and Technology - Forensic Science, Porto Alegre, Rio Grande do Sul, Brazil

Abstract. Population diversity management is crucial for the quality of solutions in Evolutionary Algorithms. Many techniques require assistance to handle diverse problem characteristics and may prematurely converge in local optima. Maintaining diversity enables the algorithm to search the space and produce better results effectively. Parallel models are a common approach to preserving diversity; however, design decisions affect optimization process characteristics. For example, the Island model's migration policy affects convergence speed. This study proposes and evaluates a fitness-based migration policy for the Biased Random-Key Genetic Algorithm (BRKGA) and compares it to two traditional strategies. The results in continuous search spaces demonstrate that the proposed policy can enhance BRKGA optimization with appropriate parameters.

Keywords: Parallel Metaheuristics · Island Model · Genetic Algorithms · Optimization

1 Introduction

Optimization problems frequently arise across diverse domains and involve maximizing resource utilization to meet objectives. Over recent years, metaheuristics have gained popularity as an alternative to traditional optimization methods, providing reasonable solutions within a reasonable time [6]. Examples of metaheuristics include Genetic Algorithm (GA), Differential Evolution (DE), Particle Swarm Optimization (PSO), and Ant Colony Optimization (ACO) [7,8,15,22]. However, these techniques often yield acceptable solutions relatively quickly, with no optimality guarantees. GAs are a well-known population metaheuristic

© The Author(s), under exclusive license to Springer Nature Switzerland AG 2023
J. Correia et al. (Eds.): EvoApplications 2023, LNCS 13989, pp. 396–410, 2023.
https://doi.org/10.1007/978-3-031-30229-9_26

inspired by Charles Darwin's theory of evolution. Essentially, population members represent potential solutions and evolve via recombination operators seeking convergence for the best solutions.

Dynamic landscapes and multiple local optima can hinder algorithm performance by trapping it in specific regions and leading to premature convergence. In GAs, population diversity is a crucial factor for success. Lack of diversity leads to the rapid diffusion of genetic material from elite solutions throughout the population, causing premature convergence. Maintaining diversity allows the population to explore new regions of the search space, resulting in improved solutions. Several strategies have been proposed to achieve and maintain population diversity [13,24,25]. One strategy is the Biased Random-Key Genetic Algorithm (BRKGA) proposed by Gonçalves and Resende [12], where the population is organized into groups, and genetic operators are applied to individuals from all groups, resulting in improved efficiency and diversity. Another popular strategy is the Island Model in Distributed Genetic Algorithms (DGAs), where the population is divided into smaller sub-populations (also called islands) [13]. Each sub-population evolves independently and periodically exchanges individuals through migration, guided by a migration policy. This dynamic can prevent premature convergence and promotes diversity through controlled migration frequency, number of migrated individuals, selection and replacement methods, and sub-population communication topology [2,13,23].

This paper proposes and explores a new migration policy for IM-DGAs based on the BRKGA. The Fitness-based Migration Policy (FBMP) proposed benefits from BRKGA's population structures to promote and maintain diversity using a dynamic that combines groups of individuals to alternate between exploration and exploitation. Moreover, since metaheuristics (including GAs) suffer from parameter sensibility, which affects the algorithm's behavior during the search, we adopt a problem-dependent offline parameter search via an iterated racing procedure (provided by irace package [19]). Thus, the proposal is assessed on the CEC'17 continuous optimization problems, and the results are compared against two traditional migration policies. Lastly, an analysis of convergence and diversity is present.

The remainder of the paper is organized as follows. Section 2 presents conceptual backgrounds to understand the work better. Section 3 presents the proposal for migration policy, followed by Sect. 4, presenting the experiments, results, and analysis. Lastly, the final considerations and future works are shown in Sect. 5.

2   Background and Literature Review

2.1   Biased Random-Key Genetic Algorithms

Genetic Algorithms (GAs) are population-based metaheuristics that iterate through multiple generations. Individuals in the population are combined via a crossover operator at each generation to form the next. In GA, solutions are represented as individuals, and the populations consist of multiple solutions. The objective function (also known as the fitness function) is used to evaluate each population member.

GAs aims to improve population quality from generation to generation by selecting the fittest individuals to crossover [14]. Therefore, pairs of individuals are chosen to participate in recombination (crossover) and mutation. For instance, in the Roulette Selection scheme, each individual has a probability of selection proportional to their fitness value. In this way, the fittest individuals are more likely to transfer genetic material to the next generation [11].

There is a common variation on GAs that consists in promoting a set of best individuals (called the elite group) to the next population, ensuring that the best solutions found so far are passed through generations. GAs equipped with this mechanism is called Elitist GAs. For a complete description of Evolutionary Computation and Genetic Algorithms, please refer to [18] and [9].

On the Biased Random-Key Genetic Algorithm (BRKGA), solutions are encoded as vectors of randomly generated numbers in the continuous interval [0, 1). The decoder function maps these vectors to solutions in the problem domain and computes their costs. The decoder-encoder mechanism makes the algorithm problem independent, and whenever necessary, the found solution can be decoded back into the problem domain [5,12].

A BRKGA specificity is the population structure, where individuals are arranged according to their fitness value, which aims to preserve the diversity of the population. Figure 1 illustrates the BRKGA schema, where the initial population sorted is separated into two groups: Elite and Non-elite. In the next generation, the Elite is preserved and wholly copied to the new population (Algorithm 1, line 4). Then, P_m mutant new individuals are randomly generated using a uniform distribution in the interval [0, 1). The mutant group aims to surpass local optima, introducing diversity (Algorithm 1, line 5). Lastly, the crossover between Elite and Non-elite is performed (Algorithm 1, lines 6–11). The crossover ensures the exploration of population diversity by combining individuals from different groups. A parent is randomly selected from the Elite group, while the second parent is randomly chosen from Non-elite (which includes mutants). Unlike traditional GAs, each crossover yields only one offspring. A biased coin flip decides which parent transfers genetic material to the offspring for each gene. The bias favors the Elite parent, but the algorithm designer can adjust the inheritance probability through the parameter ρ_e. Since Elite parents tend to have better fitness, the authors suggest a value between 50% and 70%.

2.2   Distributed Genetic Algorithms

Distributed Genetic Algorithm (DGA) is a popular technique that enhances the search space exploration and diversity in GAs. By dividing the global population into sub-populations, called islands, each island executes a GA that evolves independently. In the related literature, this technique is referred to as Island Model (IM) and is widespread among Evolutionary Algorithms (EAs) [2,13].

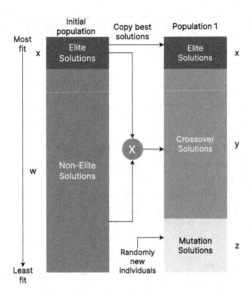

Fig. 1. BRKGA's population structure, transition from initial population to first generation. Adapted from [12].

In this model, a migration procedure occurs periodically to interchange information (selected individuals) between the islands, coordinated by a directed graph connecting sub-populations [13,23]. As a result, islands that were previously trapped on strong local attractors may be influenced by successful migrants. However, it is important to note that various design choices, such as the emigration and immigrant policies, migration frequency, number of migrants, migration topology, and homogeneity or heterogeneity of the algorithms employed by each island, may impact the behavior of the system. Furthermore, the IM benefits the system; it coordinates the search, improves exploration to converge to different regions of the search space, improves system resources utilization, and incorporates a robust mechanism for preserving diversity [2,13,23].

The IM-DGA operates by having each island execute an independent Genetic Algorithm (GA). The most common migration topologies are Ring, Star, and Fully Connected [1]. For in-depth details on Distributed Evolutionary Algorithms, see [13]. Since IM design choices affect the optimization process, in this work, we use the Ring and Fully-Connected topologies as baseline approaches to validate and compare the proposal. The best/replace worst fashion is adopted in the Ring, sending the elite individual to the right neighborhood island. Furthermore, the Fully-Connected topology consists of each island broadcasting η elite individuals; we call this policy (ηBest).

Algorithm 1. BRKGA – A population P with P_e *Elite*, $\bar{P}_e$ *Non-Elite*, and P_m
Mutant groups.

Require: $P, P_e, \bar{P}_e, P_m$
Ensure: $\bar{P}_e = P \setminus P_e$
 1: Create initial population P
 2: **while** stopping criteria not met **do**
 3:     Sort population
 4:     Copy P_e from population k to $k + 1$
 5:     Add P_m mutants individuals to population $k + 1$
 6:     **while** $(k + 1)$-th population $< P$ **do**
 7:         select a random individual from P_e
 8:         select a random individual from $\bar{P}_e$
 9:         produce offspring with a probability ρ_e
10:         add new offspring to population $k + 1$
11:     **end while**
12: **end while**
13: **return** fittest individual

2.3 Literature Review

The following presents a literature review regarding using BRKGA with multi-population systems and migration policies.

In [10], a stratified migration policy called distributed BRKGA (D-BRKGA) is proposed. In D-BRKGA, the migrant solutions are chosen from the three groups of individuals, i.e., Elite, Non-Elite, and Mutants. A parameter called migration rate defines the percentage of randomly selected individuals from each group to migrate - copied to replace worse individuals on the destination island in a Ring topology.

In [21], the unequal area facility layout problem is addressed by utilizing the BRKGA incorporating the IM with a Ring topology. This approach adopts the IM to enhance population diversity and exchange information about different regions of the search space. The proposed method yielded superior layout solutions for 14 out of 26 problems through empirical parameter tuning. Additionally, the satisfactory solutions for two previously unaddressed instances illustrate the benefits of DGAs for optimization.

More recently, to improve the variability of individuals and speed up convergence, Andrade *et al.* [3] explored a robust multi-parent version of BRKGA with the Ring IM on real-world problems. From the results, the authors emphasized that the standard BRKGA performance has significantly improved.

3 Proposed Method

This section presents a novel migration policy, called Fitness-based Migration Policy (FBMP), that aims to maintain population diversity and enhance search space exploration by exchanging information. The FBMP approach leverages the structured population of BRKGA and takes advantage of the existing diversity to, in a fully-connected topology, rearrange individuals from different islands. The migration policy is determined based on the fitness values, which serve as a similarity measure during the optimization process, providing sufficient information to guide the migration policy.

With individuals sorted based on their adaptability to the target optimization problem, the migration policy starts dividing the population into batches of size ω. The batch size is defined by Eq. 1, where NP is the population size, and ι is the number of islands. From this, the batch size defines the new arrangement of the individuals. Each island contains individuals from the i-th portion (batch) of all islands, such that the first island consists of the first ω individuals from each island, the second island consists of the next group of individuals, and so on.

$$\omega = \frac{NP}{\iota} \tag{1}$$

The proposed migration policy establishes a fully-connected communication structure to exchange individuals. Thus, it enables access to genetic material from all islands, achieving combinations that may not otherwise be accessible due to local convergence. However, to avoid a significant disparity after migration, the new population is formed by individuals in the same fitness range. Nonetheless, it does not necessarily imply similarity in fitness values, as each island may be at a different stage of evolution. As a result, the optimization process oscillates between exploitation and exploration. Figure 2 illustrates the proposal; it depicts a scenario with three islands, each containing six individuals. The $\lambda_{i,j}$ value represents the difference in fitness between the best and worst individuals in population i in migration step j.

In the proposal dynamics, groups (batches) of individuals compose populations based on their adaptability. Thus, the first batch is formed by the fittest individuals, followed by the second batch, which contains individuals with average fitness, and the third compiles the worst-performing individuals. This process leads to a decrease in λ values, indicating an exploitation process from step i to step $i + 1$. The transition between steps $i + 1$ and $i + 2$ represents an exploration process where individuals from each island are separated and sent to other islands, resulting in a more diversified new population. Figure 2 illustrates the migration policy behavior in an advanced-convergence scenario, where no further improvements have been achieved. Therefore, the proposed policy leads to populations that are sometimes diverse and sometimes composed of similar individuals. Moreover, despite the stochastic nature of genetic operators, the proposal primarily distinguishes itself by its ability to switch between exploration and exploitation according to the convergence level.

4   Experiments, Results and Analysis

The experiments were carried out on a computing node with 2 Intel Xeon Silver 4216 processors at 2.1 GHz, featuring 32 cores and 64 threads. The development

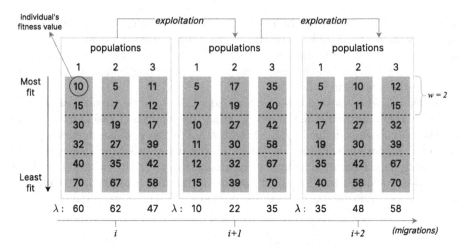

Fig. 2. FBMP's behavior over time with six individuals (NP), three islands (ι), and batch size equal to two (ω). From i to $i+1$, an exploitation behavior is observed, i.e., similar individuals are gathered in the same island (indicated by λ values). Then, a new migration results in an exploration behavior (from $i+1$ to $i+2$), i.e., similar individuals are spread over the islands.

environment is made of Ubuntu 20.04 operating system, and we use C and Python programming languages.

In the experiments, some parameters were established through empirical determination, while others passed through an automated algorithm configuration package that varies values to estimate the most suitable set. Each scenario/configuration run 31 times, starting from different random seeds with a fixed population size (NP) of 100. BRKGA parameters follows author's recommendation [12]. The population (P) comprises an elite group (P_e) of 10%; the crossover solutions contain 70%, and 20% are reserved for mutation solutions. The elite inheritance probability (ρ_e) was fixed at 0.70.

The CEC17's single-objective real parameter numeric optimization benchmark was used to analyze algorithms' effectiveness [4]. The benchmark contains a set of 29 functions structured into four groups: unimodal, simple multimodal, hybrid, and composition. Regarding dimensionality, we evaluate the proposal on 10, 30, 50, and 100 dimensions. More details and functions' characteristics are available in [4]. The results report the error metric. Equation 2 describes the metric, where F_i^* is the optimal value and F_i is the optimization result. Both terms are relative to the i-th function.

$$Error(F_i) = F_i - F_i^* \tag{2}$$

The irace[1] automatic algorithm configuration package [19] was used to evaluate the performance of each migration policy fairly, based on the parameters listed in Table 1. The table includes descriptions of each parameter, including its type and value range.

Table 1. Description and ranges of parameters for automatic parameter tuning with irace [19].

Parameter	Description	Type	Value range
τ	Migration frequency	Categorical	$\{32, 64, 128, 256, 512, 1024\}$
ι	Number of islands	Categorical	$\{2, 4, 5, 10\}$
η	Number of migrant individuals on ηBest policy	Integer	$[1, 10]$

Regarding offline tuning, seeking a reasonable proportion between experimentation time and results quality, instead of searching for the most suitable set of parameter values for all 29 functions in the test suite, we focus on finding the most appropriate set of parameters for each function group: unimodal, simple multimodal, hybrid, and composition.

The tuning budget (number of runs) for irace varies according to the number of instances (functions) variations; we define a budget of 500 runs per function. Thus, the first group with two functions has a budget of 1000, for example. Table 2 summarizes the results of the irace tuning. In total, 48 experiments were performed to determine suitable parameters for each group of problems varying the dimensionality and migration policy. Furthermore, for the whole experimentation, each execution has $D \cdot 10^4$ function evaluations with D referring to the problem dimensionality.

The results were evaluated using a statistical assessment method outlined in LaTorre, Muelas, and Peña [16]. We report the best, median, worst, mean and standard deviation of error for each benchmark function for all experiments. For brevity, this information can be found in the supplementary material[2]. Additionally, we used the Friedman test with a confidence level of 95% for multiple comparisons to determine if there are significant differences among the assessed algorithms [16,17]. Moreover, whenever Friedman's detects differences, the following metrics are computed for each algorithm:

– **Ranking**: relatively to the Friedman test, it computes the relative ranking of each algorithm according to its mean performance on each function and reports the average ranking computed through all the functions. For instance, given the following mean performance in a benchmark of three functions for algorithms A and B: A = (0.21, 3.45, 1.20), B = (2.25, 1.33, 0.80); their

---

[1] Available at https://cran.r-project.org/web/packages/irace version 3.4.1.
[2] Available at https://github.com/sbcblab/EVO23-FBMP.

Table 2. Parameters found by irace tuning: migration frequency (τ), number of islands (ι), and ηBest (η).

	D_{10}			D_{30}			D_{50}			D_{100}		
	τ	ι	η	τ	ι	η	τ	ι	η	τ	ι	η
Unimodal Functions $F_{1,3}$												
FBMP	64	10	–	32	10	–	32	10	–	128	10	–
Ring	32	10	–	32	10	–	64	10	–	32	10	–
ηBest	32	10	6	32	10	4	64	10	2	32	10	6
Simple Multimodal Functions F_{4-10}												
FBMP	256	10	–	64	10	–	64	10	–	128	10	–
Ring	32	10	–	64	10	–	128	10	–	32	10	–
ηBest	256	10	5	128	10	1	32	10	4	32	10	4
Hybrid Functions F_{11-20}												
FBMP	512	10	–	128	10	–	256	10	–	64	10	–
Ring	32	10	–	128	10	–	32	10	–	256	10	–
ηBest	256	10	8	256	4	2	128	10	1	32	10	1
Composition Functions F_{21-30}												
FBMP	64	10	–	256	10	–	512	10	–	1024	10	–
Ring	32	10	–	32	10	–	256	10	–	32	10	–
ηBest	64	10	4	256	10	1	256	10	2	64	10	10

relative ranking would be: Rank(A) = (1, 2, 2), Rank(B) = (2, 1, 1); and thus their corresponding average rankings are 1.67 and 1.33, respectively.
- **Best**: refers to the number of functions in which each algorithm obtains the best results compared to other algorithms.
- **Wins**: computes the difference between the number of times each algorithm is statistically better and worse according to the Wilcoxon Signed Rank Test in a pair-wise comparison (with a degree of confidence of 95%).

Table 3 presents the results obtained by varying the migration policy and applying the suitable parameters found on the offline tuning with irace. Also, it presents three different scenarios that depict a slight modification based on the first scenario's performance. The modification is described along with the per-scenario analysis.

The first scenario comprehends the algorithm proposed as stated. From the results, it is possible to observe that for problems of up to 50 dimensions, FBMP proved to be the best choice migration strategy in comparison with Ring and ηBest. Furthermore, *best* and *wins* metrics reveal the superiority of the results obtained by the proposed migration policy. On the other hand, for 100 dimensions, the result deteriorated; it indicates some points of attention related to the proposal. For instance, the islands still need to converge to local or global attractors, causing a premature migration that compromises the optimization's

Table 3. Average ranking (lower better), number of functions for which the algorithm obtains the *best* results and number of *wins* in pair-wise comparisons on the CEC'17 benchmark.

D	MP	1st scenario			2nd scenario			3rd scenario		
		Ranking	Best	Wins	Ranking	Best	Wins	Ranking	Best	Wins
	FBMP	1.80	13	3	1.44	18	10	1.82	14	3
10	Ring	2.00	11	5	2.33	8	−1	1.82	10	7
	ηBest	2.20	5	−8	2.22	3	−9	2.36	5	−10
	FBMP	1.62	16	12	1.50	17	19	1.52	17	17
30	Ring	1.81	11	13	1.88	10	11	1.83	10	13
	ηBest	2.57	2	−25	2.62	2	−30	2.65	2	−30
	FBMP	1.44	17	18	1.45	18	20	1.46	19	27
50	Ring	2.17	8	2	2.10	7	−1	2.08	6	−5
	ηBest	2.39	4	−20	2.45	4	−19	2.46	4	−22
	FBMP	2.35	7	−11	2.19	8	−11	1.65	15	16
100	Ring	1.55	15	10	1.62	15	10	2.00	8	−5
	ηBest	2.10	7	1	2.19	6	1	2.35	6	−11

progress, stopping the search abruptly. In addition, the parameters found by the offline tuning could be better; in this case, a more extended experimentation period and a higher budget for irace are necessary.

The hypothesis of premature migration emphasizes the possibility of losing valuable information after a migration procedure. The BRKGA's mutant group replaces part of the population with new randomly generated individuals to increase diversity. Consequently, some of the elite solutions that compose the fittest islands are thrown away. From this, two new test scenarios were proposed to evaluate the impact of the mutant group after migration. The second scenario assesses the impact of turn-off the mutant group for k generations with k equals 25 (empirically defined) after each migration. The third scenario assesses the same impact; however, it attaches the duration of the turn-off to the migration frequency. For instance, considering a migration frequency of 100 and a turn-off of 10%, the mutant group will be turned off by ten generations. We consider a turn-off of 20% of the migration frequency in the third scenario experimentation. Also, the hypothesis and motivation for the parameter attachment are that the irace tuning may suggest fewer migrations in cases with slow convergence; consequently, the turn-off for long periods enables the algorithm to exploit the set of solutions better.

From the second scenario results, the rank values for the 10 and 30 dimensions decreased. As an outcome, the shutdown of the mutant group for 25 generations has significantly impacted the optimization result. The FBMP could achieve new best values and increase the wins on the pair-wise comparison. However, for 50 and 100 dimensions, the impact is less expressive. For 50 dimensions, the ranking slightly increased with one new best value and two new wins. For 100 dimensions, the scenario almost presents no change, and the better ranking remains for the Ring topology.

Regarding the third scenario, for ten dimensions, we have a tie in the ranking between the Ring and the FBMP migration policies, with the FBMP presenting better absolute means and Ring with better pair-wise wins. This result is related to the difference in the number of positive cases where the Friendman test rejects the null hypothesis. The second scenario presents 9 cases, while the third presents 11. For 30 dimensions, the results are almost equivalent. However, the improvement is more expressive for the 50 and 100 dimensions. For 50 dimensions, the wins jump to 27, an increase of 35% compared to the second scenario and 50% to the first. Lastly, for 100 dimensions, the results highlight the FBMP as the best choice among the alternatives, with considerable increases in all metrics.

The results highlight that the FBMP is a promising migration policy with the potential to increase the BRKGA's performance. The temporary shutdown of the mutant group is an alternative solution to mitigate the problem of information loss after a migration. However, there is still room for improvement. For instance, it is necessary to perform new parameter tuning to address this new parameter and better understand the relationship between the shutdown period and the migration frequency.

In order to visually analyze the proposed migration policies' impact on the optimization process, charts depicting the algorithm convergence and diversity were generated. The charts play a crucial role in understanding the working process of the algorithm, as they provide valuable insights. The convergence chart shows the distance of the solution from the optimal one in each iteration. On the other hand, the diversity chart demonstrates the distribution of solutions in the search space, thus allowing us to determine whether the algorithm is engaged in local or global search at a given iteration. We compute the diversity based on the inertia momentum proposed by Morrison and De Jong [20].

Figure 3 depicts an example of the FBMP (3rd scenario) convergence and diversity behavior during optimization. From the figure, the characteristics of the proposal are evident. On the Island #1 convergence (Fig. 3a), the migration promotes intensification and enhances convergence since it is an elite population. Regarding diversity (Fig. 3b), the mutant group shutdown effects are noticeable, presenting momentaneous exploitation behavior that makes the diversity reach zero. When the mutant group is turned on, the BRKGA dynamics raise diversity in the population.

Figure 3c presents the Island #10 behaviors and depicts the elite individuals' migration effects. Here, we can observe that losing the elite group reference is similar to a population reset, increasing diversity (Figure 3d) and restarting the search for promising regions of the search space. It is interesting to highlight that the FBMP's fully-connected migration spread the information of promising regions to reach all islands, especially elite ones. According to convergence charts,

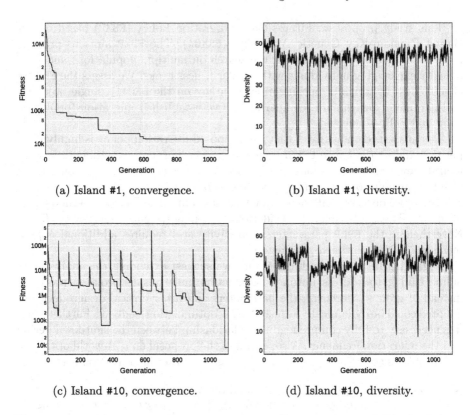

(a) Island #1, convergence.

(b) Island #1, diversity.

(c) Island #10, convergence.

(d) Island #10, diversity.

Fig. 3. FBMP' with temporary mutant-group turn-off: convergence and diversity behavior for Island #1 and #10.

we can observe the dynamic of population diversity. When the FBMP is applied, the diversity of Island #10 undergoes a sudden increase (exploration). Conversely, on Island #1, we can observe that the constant diversity level is interrupted in the migration movement giving rise to an exploitation behavior. Again, the mechanisms of BRKGA control back the diversity.

5   Conclusion and Future Works

Metaheuristics have proven to solve complex continuous optimization problems in real-world applications effectively. Premature convergence is a significant challenge metaheuristics faces, which affects their accuracy. Moreover, an essential aspect of metaheuristics success is maintaining a proper balance between exploration and exploitation during the search process. A technique to promote effective balance is utilizing parallel models, such as Distributed Genetic Algorithms (DGAs). In DGAs, multiple solutions populations simultaneously search the space and coordinate their search by exchanging information periodically, leading to the identification of promising regions and, consequently, better solutions.

This study proposes a Fitness-based Migration Policy (FBMP) based on the Biased Random-Key Genetic Algorithm (BRKGA). The FBMP aims to preserve population diversity, coordinating the search on multiple populations and periodically exchanging individuals with similar fitness ranks. We assess the FBMP's effectiveness in terms of optimization accuracy on the CEC'17 single objective real-parameter benchmark, comparing to two established migration topologies, Ring and Fully-connected.

The analysis demonstrates that the proposed migration policy is highly competitive and can enhance BRKGA's performance. Through experimentation, we identify and propose solutions to mitigate the proposal's issues. We could reach results that point out the FBMP as the best choice among the tested alternatives. Moreover, we observe that the parameters substantially affect the methods' performance. Thus, experiments consider offline tuning to reduce sensitivity effects. Nevertheless, the FBMP parameters' sensitivity still requires additional experiments to address untuned parameters.

Regarding future research directions, we suggest extending the analysis and offline tuning to address untuned parameters and consider isolated problems instead of groups. Moreover, the parameters and other critical design decisions adopted could be dynamically adjusted using optimization feedback. Furthermore, it is necessary to fully comprehend the relationship between the shutdown period and the migration frequency. Also, evaluate the proposal on real-world optimization problems and extend the comparison against other migration policies.

Acknowledgment. This work was supported by grants from the Fundação de Amparo á Pesquisa do Estado do Rio Grande do Sul [19/2551-0001906-8], Conselho Nacional de Desenvolvimento Científico e Tecnológico [408154/2022-5; 314082/2021-2; 440279/2022-4], the Coordenação de Aperfeiçoamento de Pessoal de Nível Superior [STICAMSUD 88881.522073/2020-01; PROBRAL 88881.198766/2018-01] - Brazil. This study was financed in part by the Coordenação de Aperfeiçoamento de Pessoal de Nível Superior - Brazil (CAPES) - Finance Code 001.

References

1. Alba, E.: Parallel metaheuristics: a new class of algorithms, vol. 47. Wiley-Interscience, New York, USA, 1st edn. (September 2005). https://doi.org/10.1002/0471739383
2. Alba, E., Luque, G., Nesmachnow, S.: Parallel metaheuristics: recent advances and new trends. Int. Trans. Oper. Res. **20**(1), 1–48 (2013). https://doi.org/10.1111/j.1475-3995.2012.00862.x
3. Andrade, C.E., Toso, R.F., Gonçalves, J.F., Resende, M.G.: The multi-parent biased random-key genetic algorithm with implicit path-relinking and its real-world applications. Eur. J. Oper. Res. **289**(1), 17–30 (2021). https://doi.org/10.1016/j.ejor.2019.11.037
4. Awad, N.H., Ali, M.Z., Suganthan, P.N., Liang, J.J., Qu, B.Y.: Problem definitions and evaluation criteria for the CEC 2017 special session and competition on single objective real-parameter numerical optimization. Nanyang Technological University, Singapore (October, Tech. Rep. (2016)

5. Bean, J.C.: Genetic algorithms and random keys for sequencing and optimization. ORSA J. Comput. **6**(2), 154–160 (1994). https://doi.org/10.1287/ijoc.6.2.154
6. Chopard, B., Tomassini, M.: An Introduction to Metaheuristics for Optimization. Springer Nature Switzerland AG, Gewerbestrasse, Springer Cham (2018). https://doi.org/10.1007/978-3-319-93073-2
7. Dorigo, M., Di Caro, G.: Ant colony optimization: a new meta-heuristic. In: Proceedings of the 1999 Congress on Evolutionary Computation - CEC99 (Cat. No. 99TH8406), vol. 2, pp. 1470–1477. IEEE, Washington, DC, USA (1999)
8. Eberhart, R., Kennedy, J.: A new optimizer using particle swarm theory. In: Proceedings of the Sixth International Symposium on Micro Machine and Human Science, 1995 (MHS'95). pp. 39–43. IEEE, Nagoya, Japan (1995)
9. Eiben, A.E., Smith, J.E.: Introduction to Evolutionary Computing. NCS, 2nd edn. Springer, Heidelberg (2015). https://doi.org/10.1007/978-3-662-44874-8
10. Ferreira De Faria Alixandre, B., Dorn, M.: D-BRKGA: A distributed biased random-key genetic algorithm. In: 2017 IEEE Congress on Evolutionary Computation (CEC), pp. 1398–1405 (2017). https://doi.org/10.1109/CEC.2017.7969467
11. Goldberg, D.E.: Genetic Algorithms in Search Optimization and Machine Learning, 13th edn. Addison-Wesley Professional, USA (1989)
12. Gonçalves, J.F., Resende, M.G.: Biased random-key genetic algorithms for combinatorial optimization. J. Heuristics **17**(5), 487–525 (2011). https://doi.org/10.1007/s10732-010-9143-1
13. Gong, Y.J., et al.: Distributed evolutionary algorithms and their models: a survey of the state-of-the-art. Appl. Soft Comput. **34**, 286–300 (2015). https://doi.org/10.1016/J.ASOC.2015.04.061
14. Holland, J.H.: Adaptation in Natural and Artificial Systems: An Introductory Analysis with Applications to Biology, Control, and Artificial Intelligence, 1st edn. MIT press, USA (1992)
15. Holland, J.H.: Genetic algorithms. Sci. Am. **267**(1), 66–73 (1992)
16. LaTorre, A., Muelas, S., Peña, J.M.: A comprehensive comparison of large scale global optimizers. Inf. Sci. **316**, 517–549 (2015). https://doi.org/10.1016/j.ins.2014.09.031
17. Lilja, D.J.: Measuring computer performance: a practitioner's guide. Cambridge University Press, Minneapolis, Minnesota, USA (2000). https://doi.org/10.1017/CBO9780511612398
18. Luke, S.: Essentials of Metaheuristics. Lulu, second edn. (2013). http://cs.gmu.edu/~sean/book/metaheuristics/
19. López-Ibáñez, M., Dubois-Lacoste, J., Pérez Cáceres, L., Birattari, M., Stützle, T.: The irace package: Iterated racing for automatic algorithm configuration. Oper. Res. Perspect. **3**, 43–58 (2016). https://doi.org/10.1016/j.orp.2016.09.002
20. Morrison, R.W., De Jong, K.A.: Measurement of population diversity. In: Collet, P., Fonlupt, C., Hao, J.-K., Lutton, E., Schoenauer, M. (eds.) EA 2001. LNCS, vol. 2310, pp. 31–41. Springer, Heidelberg (2002). https://doi.org/10.1007/3-540-46033-0_3
21. Palomo-Romero, J.M., Salas-Morera, L., García-Hernández, L.: An island model genetic algorithm for unequal area facility layout problems. Expert Syst. Appl. **68**, 151–162 (2017). https://doi.org/10.1016/j.eswa.2016.10.004
22. Storn, R., Price, K.: Differential evolution - a simple and efficient heuristic for global optimization over continuous spaces. J. Global Optim. **11**(4), 341–359 (1997)
23. Sudholt, D.: Parallel evolutionary algorithms. In: Kacprzyk, J., Pedrycz, W. (eds.) Springer Handbook of Computational Intelligence, pp. 929–959. Springer, Heidelberg (2015). https://doi.org/10.1007/978-3-662-43505-2_46

24. Črepinšek, M., Liu, S.H., Mernik, M.: Exploration and exploitation in evolutionary algorithms: a survey. ACM Comput. Surv. 45(3) (Jul 2013). https://doi.org/10.1145/2480741.2480752
25. Xu, J., Zhang, J.: Exploration-exploitation tradeoffs in metaheuristics: Survey and analysis. In: Proceedings of the 33rd Chinese Control Conference. pp. 8633–8638. IEEE, Nanjing, China (July 2014). https://doi.org/10.1109/ChiCC.2014.6896450

Nullifying the Inherent Bias of Non-invariant Exploratory Landscape Analysis Features

Raphael Patrick Prager[1]([✉]) [iD] and Heike Trautmann[1,2] [iD]

[1] Data Science: Statistics and Optimization, University of Münster, Münster,
Germany
{raphael.prager,heike.trautmann}@uni-muenster.de
[2] Data Management and Biometrics, University of Twente, Enschede,
The Netherlands

Abstract. Exploratory landscape analysis (ELA) in single-objective black-box optimization relies on a comprehensive and large set of numerical features characterizing problem instances. Those foster problem understanding and serve as basis for constructing automated algorithm selection models choosing the best suited algorithm for a problem at hand based on the aforementioned features computed prior to optimization. This work specifically points to the sensitivity of a substantial proportion of these features to absolute objective values, i.e., we observe a lack of shift and scale invariance. We show that this unfortunately induces bias within automated algorithm selection models, an overfitting to specific benchmark problem sets used for training and thereby hinders generalization capabilities to unseen problems. We tackle these issues by presenting an appropriate objective normalization to be used prior to ELA feature computation and empirically illustrate the respective effectiveness focusing on the BBOB benchmark set.

Keywords: Exploratory Landscape Analysis · Invariance · Automated Algorithm Selection

1 Introduction

Exploratory landscape analysis (ELA) [11] is a powerful tool for numerically characterizing landscape properties of single-objective black-box optimization problems $f : \mathbb{R}^D \to \mathbb{R}$ of unknown structure with exclusively real-valued parameters $x \in \mathbb{R}^D$ and a single, also real-valued objective value $f(x) \in \mathbb{R}$. Apart from fostering problem understanding, leading to improvements in analyzing problem hardness and algorithm design, the respective numerical features [9,15] are crucial input to automated algorithm selection (AAS) [6] models. AAS relies on machine learning techniques to model the interaction of algorithm performances and problem features, with the purpose to recommend the best suited algorithm for a problem instance at hand out of a pre-defined algorithm portfolio.

ELA research has flourished in recent years, ranging from introducing innovative feature sets (e.g. [7,10,12]), over AAS studies [8,13,16], the influence of

© The Author(s), under exclusive license to Springer Nature Switzerland AG 2023
J. Correia et al. (Eds.): EvoApplications 2023, LNCS 13989, pp. 411–425, 2023.
https://doi.org/10.1007/978-3-031-30229-9_27

the underlying sampling strategy of objective values based on which ELA features are computed prior to optimization [1,17] up to feature importance studies investigating expressiveness [18]. The latter also brought attention to the lack of invariance of a subset of ELA features, i.e., sensitivity to absolute objective values and thereby a lack of robustness regarding shift and scaling. Interestingly and precariously, some of these features which are not invariant turned out to be most expressive in terms of characterizing the underlying problem landscape. Renau et al. [19] build on this work, confirm respective findings and investigate explainability of models classifying specific BBOB benchmark problems [4], thereby acknowledging that oftentimes a carefully selected very small subset of expressive features is sufficient to yield a very high model accuracy.

In this work, we classify state-of-the art ELA features w.r.t. invariance to shifts and scale of the objective function which might arise due to different scales of measurement in real-world applications or in artificial benchmark sets via creating slightly varying instances of a problem class (as e.g. in BBOB). We point to some misconceptions of invariance of specific BBOB problems in previous works [19,20] and suggest an appropriate normalization procedure to be applied to the objective values of the initial sample used for ELA feature computation to overcome a potential lack of invariance. Via an empirical study on BBOB, strong effectiveness is demonstrated in terms of enabling problem discrimination via ELA features for the correct reasons, i.e., landscape properties in contrast to objective value ranges. Also, we illustrate a substantially positive effect on AAS studies by increasing generalization capability while simultaneously keeping the model accuracy at a similar level.

Section 2 provides relevant background on ELA, followed by a detailed discussion of invariance properties of ELA features in Sect. 3 and insights into the BBOB benchmark set in Sect. 4. Our suggested transformation function is provided in Sect. 5 and the respective positive effects on BBOB problem prediction and AAS models are presented in Sects. 6 and 7. Overall conclusions together with further research potential are discussed in Sect. 8.

2   Exploratory Landscape Analysis

Fitness landscape analysis is a popular tool to get insight into an optimization problem and its inherent properties. In general, this is done by different means of visualization as well as the calculation of certain numeric values. These values can act as information proxies about certain high-level properties such as the degree of multi-modality. These values or features have been developed and are used in many different optimization domains. Prevalent in the continuous single-objective area are the so called exploratory landscape analysis features. ELA serves as an umbrella term to cover various proposed feature sets over the last decade. The R package `flacco` [9] and the Python package `pflacco` [15] implement them respectively.

Mandatory for any given ELA feature is a small sample of the search space $\{x^{(1)}, ..., x^{(n)}\} \in X^{n \times D}$, where n describes the sample size and D the problem dimension. This sample is evaluated on an objective function $f : X \to \mathbb{R}$ such that each individual observation $x^{(1)}$ has its corresponding objective value $f(x^{(1)})$.

Influencing factors for any subsequent ELA feature calculation is the size n of the initial sample as well as the mechanism by which it is generated. This has been the subject of [1, 7] for sample sizes, and [17] for sampling techniques. Within this paper, we will use the a sample size of $50D$ which has been widely accepted in AAS settings [7, 8]. For the sampling technique, we use latin hypercube sampling (LHS), even while recent works have pointed out that a sample generated by Sobol sampling might yield better results [17]. We do this mainly to compare with older research directly, without the compounding effect which different sampling strategies might introduce.

The following listing gives a short introduction into the used feature sets. Note that we chose only ELA features which can be computed on the small sample of size $50D$ and do not require additional function evaluations. For a more comprehensive discussion we refer to their original and cited papers.

- **Classical ELA** (12 features): This feature set is a subset of the most prominent collection of landscape features in the area of single-objective, continuous optimization. The original proposition of this collection divides its numerical features into six categories of which two are used in this work: the meta model and y-distribution feature sets. The remaining four feature sets are ill-suited for the purpose of this work because they require additional function evaluations and are thereby not cheap to compute [11].
- **Dispersion** (16 features): The dispersion features divide the initial sample into different subsets of points of diverging quality, e.g., the points whose objective values lie below the 25%-quantile. The homogeneity of these groups is measured in contrast to the homogeneity of the entire sample [10].
- **Information Content** (5 features): A sequence of random walks over the initial sample serves as basis for the computation. Each step is compared with its consecutive step and metrics are derived to describe a landscape in terms of smoothness, ruggedness, and neutrality [12].
- **Nearest Better Clustering** (5 features): Based on the initial sample, first the distances of all observations to their nearest neighbour are computed. A second set of distances is derived by calculating the distances of all observations to their nearest better neighbour, where 'better' refers to an improvement in fitness. Several metrics and ratios are then computed based upon these two sets in isolation and conjunction [7].
- **Miscellaneous** (9 features): The last feature set comprises features which did not warrant their own publications or are simple in nature, such as features based on principal component analysis and the dimensionality of the optimization problem in question [9].

3   Invariance in Exploratory Landscape Analysis

As observable in [19, 20], the term 'invariance' in landscape analysis seems to have different connotations and is used ambiguously. An example is the feature `ela_meta.lin_int`, which is the intercept of a linear model and is classified in

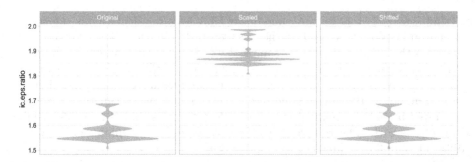

Fig. 1. Distribution of the feature `ic.eps.ratio` computed a 1000 times on the same samples $(X, f(X))$, its scaled version $(X, g(f(X)))$, where $g(f(X)) = 2 \cdot f(X)$, and its shifted version $(X, h(f(X)))$, where $h(f(X)) = f(X) + 100$. The used BBOB function is the first instance of the separable Rastrigin function of BBOB in dimension 2.

[20] as an invariant feature, whereas [19] deems it as a **non**-invariant feature. Both refer to invariance to shifts and scales. Without providing substantial proof, it is clear that the intercept b of a simple linear model $y = a \cdot x + b$ is clearly not invariant to shifts and scales which in turn makes the feature `ela_meta.lin_int` non-invariant. The second example we want to discuss pertains to the feature `ic.eps.ratio`. Here, [20] declares it as non-invariant whereas [19] does. The case for this feature is a little more difficult because the computation is based on sequence of random walks over the given sample. Essentially meaning that it is not deterministic. However, we can evaluate this experimentally by comparing the distribution of this feature's values. These are calculated first on a sample X and $f(X)$ and second on linear transformed values of $\hat{f}(X)$. Figure 1 displays this result. We can see that this feature is invariant to shifts but not scale invariant.

The notion of invariance, which we will use throughout this paper, is the same as in [19,20]. Meaning, we focus on the shift and scale invariance which ultimately means an invariance against linear transformations. We further limit this to the objective space and exclude the decision space from our consideration. This facilitates an understanding of the effects we want to measure but is also a limitation of our chosen benchmark which will be the topic in Sect. 4. This benchmark is box-constrained in the interval $[-5, 5]^D$ for any of its consisting optimization problems and therefore does not offer any potential for analyzing differently scaled decision spaces.

Given an initial sample X and a linear transformation function $g : f(X) \to \mathbb{R}$, a function which produces an arbitrary ELA feature $h : (X, f(X)) \to \mathbb{R}$, then this ELA feature is invariant to shifts and scaling of the objective function f, if and only if the following holds true:

$$h(X, f(X)) = h(X, g(f(X)))$$

When an ELA feature is not shift nor scale invariant, it means it uses the objective values $f(X)$ as an absolute value in its internal computation. Therefore, an arbitrary problem p and a version of p called $\hat{p}$, which is shifted by a constant b

Table 1. Non-invariant ELA features out of the 47 considered ones (c.f. Section 2). From left to right, the following information is displayed: (1) the ELA feature name, (2) and (3) the classification whether this particular ELA feature is shift **and** scale invariant of [19,20] respectively, (4) and (5) our classification in regards to shift invariance and scale invariance separately. The ■ indicates that a feature is **not** invariant and ○ the opposite.

Feature name	Class. in [20]	Class. in [19]	Invariant to	
			Shift	Scale
ic.eps.s	-	○	○	■
ic.eps.max	■	-	○	■
ic.eps.ratio	-	○	○	■
ela_meta.lin_simple.intercept	○	■	■	■
ela_meta.lin_simple.coef.min	○	-	○	■
ela_meta.lin_simple.coef.max	■	■	○	■
pca.expl_var.cov_init	○	-	○	■
pca.expl_var_PC1.cov_init	○	○	○	■

and scaled by a factor a, produce objective values in form of $f(X)$ and $a \cdot f(X+b)$. The difference in value ranges between both is subject to a and b but can be substantial. This in turn might yield very different values for the ELA feature in question. The properties of the problems p and $\hat{p}$ are expected to be the same while their respective ELA features are not. This creates ELA features which correlate in some fashion with the value range of f and its values cannot be interpreted independent from the problem as would be desirable.

At this point, we like to emphasize that the majority of ELA features are inherently shift and scale invariant. Yet, the few which are not invariant are often used in different works. A full enumeration of these non-invariant features as well as a juxtaposition of different classification of invariance can be found in Table 1.

4   Black-Box Optimization Benchmark

The Black-Box Optimization Benchmark (BBOB) [4] is widely spread in different works about ELA features and AAS. In general, these works validate a certain hypothesis or theorem based on the results they achieve on the BBOB function set. This makes the BBOB a foundational pillar of these works. To understand the deeply rooted influence of non-invariant ELA features, we will discuss some aspects of the BBOB functions in more detail.

The BBOB comprises 24 unique optimization problems termed as function where each function is referred to by its respective function identifier (FID). These functions can be used in arbitrary dimension D. Also, functions provide a set of different instances where each instance IID is a transformation of the base function induced by shifting, scaling and/or rotation.

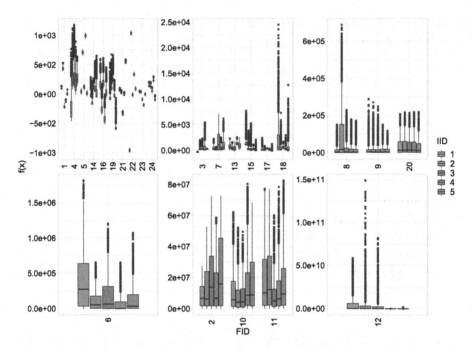

Fig. 2. Range of the objective values $f(x)$ of all 24 BBOB functions (FID) visualized as a set of boxplots. For each FID, the first five instances (IID) are shown over the dimensions 2, 3, 5, and 10. The size of the sample is $1\,000D^2$. The FIDs are compartmentalized into smaller subplots to illustrate the increasing values and value range of $f(x)$.

Each FID has unique properties such as the degree of multimodality, its global structure and so forth. One property, which is understandably often cast aside, is the range of the objective values. These can differ significantly between different FIDs whereas they remain somewhat stable between instances of a given function, i.e., the order of magnitude rarely changes.

This can be seen in Fig. 2, which displays all 24 functions and their first five instances. The functions are grouped in ascending order dependent on the order of magnitude of their objective value distribution. Note that this grouping is not based on a hard quantitative criterion and other groupings are also possible. Nevertheless, this does not invalidate the fact that we can group functions by their range of objective values across the first five instances (many researchers do not use more than that in their work). This in turn is troubling on its own. It is apparent that these groups of functions differ significantly compared to others w.r.t their value ranges. This in turn makes them distinguishable simply by this fact alone and ELA features which are not invariant to the absolute values of $f(x)$ incorporate this information. That, however, does not hold any value beyond this specific function, actively biases non-invariant ELA features, and cannot be translated to unseen functions.

5   Transformation Function

To negate the effect these different scales have on subsequent results, we consider the following transformation function for the objective values $f(X)$:

$$\hat{x}_i = \frac{x_i - x_{\min}}{x_{\max} - x_{\min}}, \tag{1}$$

which is also known as min-max-normalization. There exist a variety of other transformation functions in literature and practice such as standardization $\hat{x}_i = \frac{x_i - \bar{x}}{\sigma_x}$, robust scaling $\hat{x}_i = \frac{x_i - x_{\text{med}}}{\text{IQR}}$ where IQR is the interquartile range, and several more. Yet, min-max-normalization exhibits the desired properties for our case. These are the shift and scale invariance which can be translated to invariance against linear transformations in form of $\hat{x} = a \cdot x + b$ in general.

To prove this, let us assume that we normalize the objective values $f(X)$ for each problem independently. To improve readability we will refer in the following to $f(X)$ as Y. If we have a normalized sample Y then a linear transformation of it $\hat{Y} = a \cdot Y + b$ is identical to Y for $a > 0$ as shown in the following:

$$
\begin{aligned}
& \frac{(a \cdot y_i + b) - (a \cdot y_{\min} + b)}{(a \cdot y_{\max} + b) - (a \cdot y_{\min} + b)} \\
&= \frac{a \cdot y_i + b - a \cdot y_{\min} - b}{a \cdot y_{\max} + b - a \cdot y_{\min} - b} \\
&= \frac{a}{a} \cdot \frac{y_i - y_{\min}}{y_{\max} - y_{\min}}
\end{aligned}
\tag{2}
$$

While the case $a = 0$ is trivial, the case for $a < 0$ is not. Equation 2 does not hold because the respective minimum and maximum change when multiplied with a negative constant, i.e., $(-1)y_{\min} = \hat{y}_{\max}$ and vice versa. If we incorporate this negative constant already into the sample $\hat{Y}$ however:

$$
\begin{aligned}
& \frac{|a| \cdot \hat{y}_i - |a| \cdot \hat{y}_{\min}}{|a| \cdot \hat{y}_{\max} - |a| \cdot \hat{y}_{\min}} \\
&= \frac{|a|}{|a|} \cdot \frac{\hat{y}_i - \hat{y}_{\min}}{\hat{y}_{\max} - \hat{y}_{\min}} \\
&= \frac{(-1)y_i - (-1)y_{\max}}{(-1)y_{\min} - (-1)y_{\max}} \\
&= \frac{y_i - y_{\max}}{y_{\min} - y_{\max}},
\end{aligned}
\tag{3}
$$

we get values which are still scaled to $[0, 1]$ but the distribution of the values is mirrored, i.e., $\frac{y_i - y_{\min}}{y_{\max} - y_{\min}} = 1 - \frac{y_i - y_{\max}}{y_{\min} - y_{\max}}$. In reality, scaling an optimization problem with a constant $a < 0$, does not happen as a means to generate new instances of a given problem. Multiplying with a negative value transforms a minimization into a maximization problem and conversely a maximization into a minimization problem. Hence, normalization is not strictly scale invariant but we argue that for the purpose of this work this does not matter. We deem it as sufficient that normalization is invariant to an arbitrary linear transformation with $a > 0$.

6   BBOB Function Prediction

In this section, we illustrate the inherent bias of non-invariant features with a modelled classification scenario. In this, we predict an FID based on a set of ELA features as listed in Sect. 2. These ELA features are computed on an initial sample X of size $50D$ using latin hypercube sampling and its respective objective values $f(X)$. The transformation function, defined in Eq. 1, is used to create the normalized objective values $\hat{f}(X)$. Note that the minimal and maximal values of the objective space (required for the normalization) are extracted from the sampled points. Our procedure does not require the true minimum and maximum. In this particular experiment, we consider all 24 BBOB functions in $D = \{2, 3, 5, 10\}$ and their respective first five instances. For each optimization problem, we create 20 samples to mitigate any stochastic interference of our chosen sampling size and strategy. Some of the ELA features occasionally produce NA values. Instead of discarding these features as a whole, we choose a type of classifier which is able to deal with missing values. A classifier with that capability is the HistGradientBoostingClassifier of the Python package scikit-learn [14] and based on that we train two distinct models. The first model relies on ELA features computed on the sample $(X, f(X))$ whereas we use ELA features which are computed on the normalized sample $(X, \hat{f}(X))$ in the second model. For both, we use a 5-fold cross-validation strategy, where each fold consists of only a single IID, e.g., the first fold includes only the first instance of all 24 BBOB functions in all considered dimensions D. Efforts into hyperparameter tuning and feature selection are intentionally omitted to avoid a qualitative difference between these two models which can be traced back to a better found hyperparameter configuration of one model.

Figure 3 depicts the results of this undertaking. A cursory assessment reveals a clear disparity between the quality of the two models. The predictions which are based on the non-transformed case have an F1 score of 0.87 and therefore exhibit less faulty predictions. The case based on the normalized ELA features only has an F1 score of 0.779. Furthermore, the misclassified FIDs of the normalized case are spread out over a variety of different FIDs.

The higher accuracy in predictions can be traced back to the training setting of our model. As previously mentioned, we are forced to use a cross-validation variant where each IID (independent of the function) is a separate fold.

First, we focus on FIDs where the difference in accurate predicted FIDs between the two approaches is equal to or larger than 50. This involves functions 4, 6, 7, 9, 12, 14, 15, and 17. For FID 4, we can see for the non-transformed case that there are a number of misclassifications. These congregate mainly into FID 3 and 18. When we consult Fig. 2 again, we can see that these two functions share similar objective values $f(x)$ as FID 4. In fact, every other misclassification, FIDs

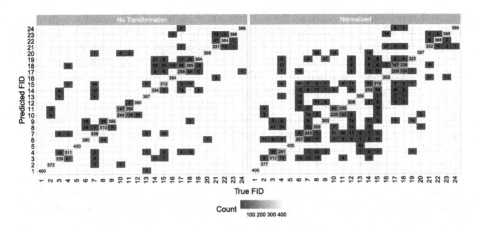

Fig. 3. Confusion heat map for the same classification scenario. The left hand side uses the ELA features which are computed on the raw sample. The right hand side uses ELA features which are computed on a normalized sample.

7 and 15, can be attributed to this proximity in value ranges, i.e., misclassified FIDs belong to the same sub-plot or an adjacent one.

This logic applies to any of the previous singled out functions 4, 6, ..., 17. Yet, this is only true for the non-transformed case. An entirely different picture is drawn when we shift our focus on the normalized case. Previously, we have seen a pattern in the classification errors, which the normalized case lacks. Misclassified functions do not exhibit any clear structure and often do not even overlap with the misclassification of the non-transformation case.

Noteworthy is that function 5 is not misclassified in either case. This is especially peculiar since this FID can be mistaken for a variety of other functions from an objective value range perspective. Another interesting fact is that the ELA features of functions belonging to the last BBOB function group seem to be expressive enough to somewhat counteract this phenomenon which we have seen before, i.e., the normalized case still has a high accuracy for these.

7   Automated Algorithm Selection

Up until now, we have highlighted the downsides of non-normalized objective values. Within this section, we show experimentally the positive effects this normalization process might have on new unseen optimization problems. For that, we fall back on an AAS scenario used in [8].

A schematic illustration of a generic AAS model is displayed in Fig. 4. Here, we benchmark a set of optimization algorithms on a set of problem instances and receive a performance data set. At the same time, we compute the ELA features of interest on the same set of problem instances which yields an ELA feature data set. For each problem instance, we now determine the best algorithm by consulting the performance data set. This serves as a so called 'classification

Table 2. Performance results of the six differently evaluated AAS models. The first row indicates which cross-validation strategy was chosen whereas the second row states whether objective values are not transformed or normalized prior to the computation of ELA features. The results are displayed for each combination of problem dimension (D) and BBOB function group (FGrp). The values in each individual cell represent the relative performance in contrast to the non-transformed or normalized counterpart of that cross-validation grouping. For instance, the third row and third column shows the relative performance of the non-transformed case with a cross-validation based on IIDs for BBOB function group 1 in dimension 2 with a value of 1. The horizontally adjacent cell displays the same results for the normalized case. Furthermore, the normalized AAS requires on average 1.19 more function evaluations to solve the optimization problems belonging to BBOB function group 1 in dimension 2 compared to its non-transformed counterpart. Different background colors indicate the better performing option for a given combination of problem dimension, function group and cross-validation strategy.

		IID		FID		Custom	
D	FGrp	No Trans.	Norm.	No Trans.	Norm.	No Trans.	Norm.
2	1	1.00	1.19	1.42	1.00	1.00	1.11
	2	10.25	1.00	77.13	1.00	1.00	1.39
	3	1.00	1.01	1.03	1.00	39.35	1.00
	4	1.00	2.42	1.00	3.58	1.00	1.01
	5	103.33	1.00	1.00	1.00	1.01	1.00
	all	1.00	1.15	1.00	1.20	1.05	1.00
3	1	1.09	1.00	3.83	1.00	4.16	1.00
	2	1.00	25.46	1.00	40.19	1.00	29.79
	3	10.65	1.00	3.38	1.00	1.00	1.58
	4	1.00	2.25	1.00	1.21	1.14	1.00
	5	1.00	1.99	1.00	1.00	1.77	1.00
	all	1.00	1.72	1.04	1.00	1.47	1.00
5	1	2.47	1.00	1.00	2.61	1.00	1.17
	2	1.00	1817.66	1.00	1.45	1.00	1.06
	3	1.00	984.04	1.00	18.42	2.52	1.00
	4	1.00	2.48	1.00	1.29	1.24	1.00
	5	1.00	1.02	1.51	1.00	1.33	1.00
	all	1.00	2.39	1.00	1.16	1.22	1.00
10	1	1.00	1.71	1.00	2.94	1.00	1.07
	2	1.00	5.09	7.53	1.00	89.74	1.00
	3	1.00	1.04	26.75	1.00	77.29	1.00
	4	1.00	2.76	1.00	1.49	1.09	1.00
	5	1.00	1.00	2.44	1.00	1.50	1.00
	all	1.00	2.46	1.19	1.00	1.25	1.00
all	1	1.07	1.00	1.00	1.29	1.29	1.00
	2	1.00	2.02	1.14	1.00	1.21	1.00
	3	1.00	1.66	1.00	1.52	2.94	1.00
	4	1.00	2.50	1.00	1.44	1.14	1.00
	5	1.48	1.00	1.36	1.00	1.41	1.00
	all	1.00	1.97	1.00	1.02	1.25	1.00

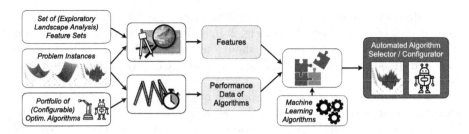

Fig. 4. Schematic illustration of a feature-based automated algorithm selection framework. Adapted from [8].

label' for our subsequent machine learning model. The inputs to this model are the respective entries of our ELA feature data set. Both, input features and labels, are used to train our AAS model. Note that an AAS model is not limited to a classification scenario but can also be modelled as regression or rank-based scenario [6].

In our particular case, that means we compute the 47 considered ELA features (c.f. Sect. 4) on the BBOB functions once on the sample $(X, f(X))$ and once on the normalized one $(X, \hat{f}(X))$. The exact experimental details such as the sampling technique and size for ELA, the considered BBOB problems, their dimensionality etc. is identical to the experiment described in the previous section. The algorithm portfolio consists of a variety of meta-heuristics such as IPOP-CMA-ES as well as local search methods. For a full account of the algorithms and their situational merits, we refer to [8]. We fetched the publicly available performance data of our algorithm portfolio from the COCO platform [2]. As a machine learning model, we choose once again the `HistGradientBoostingClassifier` of the Python package `scikit-learn` [14]. Similar to before, we omit the search for adequate hyperparameters or feature selection. The reasoning is akin to the one provided in Sect. 6. We are primarily interested in the intrinsic effect normalization has without obfuscating our results with random beneficial events for one model. This can take form in a slightly better found subset of features or a better hyperparameter configuration for one specific model.

Previously, we could only leave out one IID in our cross-validation strategy and thus could not observe the transferability to new unseen functions in our models. In this AAS classification scenario, we can test this to some degree by leaving out entire FIDs as well as creating custom folds of similar FIDs in respect to objective value range as defined in Fig. 2. This means that the two models, based on non-transformed and normalized objective values, are trained and evaluated three times individually, i.e., each time with a different k-fold cross-validation strategy. The IID cross-validation case comprises five folds where each fold consists of all BBOB functions in all dimensions for a particular instance. Cross-validation based on FID refers to folds which only contain a single BBOB function across all instances and dimensions, i.e., 24 folds. The last cross-validation variant is based on six differently sized folds, where each folds consists only of FIDs grouped as in Fig. 2. In each of these three cross-validation strategies, less and less information about the range of objective values can be exploited. In contrast, generalizability becomes increasingly important for competitive results.

Table 2 shows a performance comparison of these six models. The results are structured first into the three different cross-validation strategies based on IIDs, FIDs, and custom groupings and consecutively into the non-transformed and normalized case. Furthermore, the results are aggregated with the arithmetic mean for each combination of problem dimension and BBOB function group as well as an overall aggregation. The displayed values are based on the expected running time (ERT) [3] where each value is divided by the minimal value of each unique combination of problem dimension, function group and cross-validation strategy. To give an example, for dimension 2, BBOB function group 1, and IID cross-validation, the ERT values of the non-transformed and normalized case where divided by the lower ERT value of the two. Here, that is the ERT value of the non-transformed case. Thereby, we can directly measure how many more function evaluations the slower of the two requires in comparison. In this particular example, the normalized case requires 1.19 times more algorithm function evaluations compared to the non-transformed case.

For the IID cross-validation case, we can clearly see that the model based on non-transformed objective values and therefore non-invariant ELA features is superior. However, that does not mean that it is generally better. This cross-validation strategy is particular susceptible to the exploitation of absolute objective values because the model is trained on exactly four instances of any given FID and evaluated on the fifth. We can highlight this by investigating in more detail where the normalized model struggles. This seems to be the case in particular on BBOB function groups 2 and 3. Function group 2 consists of FID 6–9 where 6, 8, and 9 have rather unique ranges of objective values. This is also applicable to the majority of function group 3, ranging from FID 10–14. This underlines that the non-transformed objective values are the source for more discriminative ELA features which are indicative for similar instances of an already seen function. This is not a desirable property of ELA features in general because it limits their capabilities on unseen functions.

The FID cross-validation case is less influenced by absolute objective values. In Table 2, we see that both models reach an equilibrium where none is predominantly better than the other with some exceptions. When we compare the overall performance, we can see that the normalized model is negligibly worse but without real significance.

The last cross-validation case emphasizes the generalization capabilities to new unseen function and value ranges artificially by design. We essentially exclude entire objective value ranges from the training phase to measure the effect this has on the respective model's performance. While the difference is not as large as in the IID cross-validation case, it is apparent that the normalized model performs better than the non-transformed one. Thereby, we argue that the former is more capable to generalize to unseen functions and especially unseen objective value ranges. This highlights the potential of our work especially for real-world problems where objective values can have widely different and arbitrary value ranges.

8   Conclusions and Outlook

This paper highlights the shortcoming of ELA features which are not shift nor scale invariant and simultaneously offers a method w.l.o.g. in single-objective black-box optimization as a remedy. We experimentally illustrated the strong positive effects of our suggested normalization procedure of objective values prior to feature computation, especially in terms of an unbiased feature-based discrimination between BBOB problems as well as more robust setting in AAS without a decrease in competitive performance.

Our normalization methodology naturally generalizes beyond the exemplary focused BBOB benchmark set. It even might become more important in case of AAS models. These are typically built upon a single benchmark set which includes largely varying objective value ranges across all problems. This circumstance is only amplified when multiple benchmark sets are involved.

Hence, future work will indeed focus on a comprehensive AAS study integrating various benchmark problem and ELA feature sets, while systematically analyzing the effect of the suggested objective value normalization. In this endeavour, we will also include alternative transformation techniques such as standardization and robust scaling. Another interesting area for further research is the forcibly induced bounds of the objective space and its theoretical effects on each ELA feature in isolation. Upper and lower bounds of an ELA feature, which is computed on objective values in the interval $[0, 1]$, could be determined and then used to subsequently conduct normalization of ELA features which will improve respective interpretability.

Prospectively, other types of invariance such as against objective space rotation and decision space dimensionality should be investigated and combined with the currently derived valuable insights of this study and related work in combinatorial optimization [5].

References

1. Bossek, J., Doerr, C., Kerschke, P.: Initial design strategies and their effects on sequential model-based optimization: an exploratory case study based on BBOB. In: Proceedings of the 2020 Genetic and Evolutionary Computation Conference, pp. 778–786. GECCO '20, Association for Computing Machinery, New York, NY, USA (2020). https://doi.org/10.1145/3377930.3390155
2. Hansen, N., Auger, A., Ros, R., Mersmann, O., Tušar, T., Brockhoff, D.: COCO: a platform for comparing continuous optimizers in a black-box setting. Optim. Methods Softw. **36**, 114–144 (2021). https://doi.org/10.1080/10556788.2020.1808977

3. Hansen, N., Auger, A., Finck, S., Ros, R.: Real-parameter black-box optimization benchmarking 2010: experimental setup. Research Report RR-7215, INRIA (2010). https://hal.inria.fr/inria-00462481
4. Hansen, N., Finck, S., Ros, R., Auger, A.: Real-parameter black-box optimization benchmarking 2009: noiseless functions definitions. Tech. Rep. RR-6829, INRIA (2009). https://hal.inria.fr/inria-00362633/document
5. Heins, J., Bossek, J., Pohl, J., Seiler, M., Trautmann, H., Kerschke, P.: A study on the effects of normalized TSP features for automated algorithm selection. Theor. Comput. Sci. **940**, 123–145 (2023). https://doi.org/10.1016/j.tcs.2022.10.019
6. Kerschke, P., Hoos, H.H., Neumann, F., Trautmann, H.: Automated algorithm selection: survey and perspectives. Evol. Comput. **27**(1), 3–45 (2019). https://doi.org/10.1162/evco_a_00242
7. Kerschke, P., Preuss, M., Wessing, S., Trautmann, H.: Detecting funnel structures by means of exploratory landscape analysis. In: Proceedings of the 2015 Annual Conference on Genetic and Evolutionary Computation, pp. 265–272. GECCO '15, Association for Computing Machinery, New York, NY, USA (2015). https://doi.org/10.1145/2739480.2754642
8. Kerschke, P., Trautmann, H.: Automated algorithm selection on continuous black-box problems by combining exploratory landscape analysis and machine learning. Evol. Comput. **27**(1), 99–127 (2019). https://doi.org/10.1162/evco_a_00236
9. Kerschke, P., Trautmann, H.: Comprehensive feature-based landscape analysis of continuous and constrained optimization problems using the r-package flacco. In: Bauer, N., Ickstadt, K., Lübke, K., Szepannek, G., Trautmann, H., Vichi, M. (eds.) Applications in Statistical Computing. SCDAKO, pp. 93–123. Springer, Cham (2019). https://doi.org/10.1007/978-3-030-25147-5_7
10. Lunacek, M., Whitley, D.: The dispersion metric and the CMA evolution strategy. In: Proceedings of the 8th Annual Conference on Genetic and Evolutionary Computation. p. 477–484. GECCO '06, Association for Computing Machinery, New York, NY, USA (2006). https://doi.org/10.1145/1143997.1144085
11. Mersmann, O., Bischl, B., Trautmann, H., Preuss, M., Weihs, C., Rudolph, G.: Exploratory landscape analysis. In: Proceedings of the 13th Annual Conference on Genetic and Evolutionary Computation. p. 829–836. GECCO '11, Association for Computing Machinery, New York, NY, USA (2011). https://doi.org/10.1145/2001576.2001690
12. Muñoz Acosta, M.A., Kirley, M., Halgamuge, S.K.: Exploratory landscape analysis of continuous space optimization problems using information content. IEEE Trans. Evol. Comput. (TEVC) **19**(1), 74–87 (2015). https://doi.org/10.1109/TEVC.2014.2302006
13. Muñoz, M.A., Sun, Y., Kirley, M., Halgamuge, S.K.: Algorithm selection for black-box continuous optimization problems: a survey on methods and challenges. Inf. Sci. **317**, 224–245 (2015). https://doi.org/10.1016/j.ins.2015.05.010
14. Pedregosa, F., et al.: Scikit-learn: machine learning in Python. J. Mach. Learn. Res. **12**, 2825–2830 (2011)
15. Prager, R.P.: pflacco: The R-Package flacco in Native Python Code (2022). https://github.com/Reiyan/pflacco, Python Package v1.1.0
16. Prager, R.P., Seiler, M.V., Trautmann, H., Kerschke, P.: Automated algorithm selection in single-objective continuous optimization: a comparative study of deep learning and landscape analysis methods. In: Rudolph, G., Kononova, A.V., Aguirre, H., Kerschke, P., Ochoa, G., Tušar, T. (eds.) Parallel Problem Solving from Nature - PPSN XVII, pp. 3–17. Springer International Publishing, Cham (2022). https://doi.org/10.1007/978-3-031-14714-2_1

17. Renau, Q., Doerr, C., Dreo, J., Doerr, B.: Exploratory landscape analysis is strongly sensitive to the sampling strategy. In: Bäck, T., et al. (eds.) PPSN 2020. LNCS, vol. 12270, pp. 139–153. Springer, Cham (2020). https://doi.org/10.1007/978-3-030-58115-2_10

18. Renau, Q., Dreo, J., Doerr, C., Doerr, B.: Expressiveness and robustness of landscape features. In: Proceedings of the Genetic and Evolutionary Computation Conference Companion. p. 2048–2051. GECCO '19, Association for Computing Machinery, New York, NY, USA (2019). https://doi.org/10.1145/3319619.3326913

19. Renau, Q., Dreo, J., Doerr, C., Doerr, B.: Towards explainable exploratory landscape analysis: extreme feature selection for classifying BBOB functions. In: Castillo, P.A., Jiménez Laredo, J.L. (eds.) EvoApplications 2021. LNCS, vol. 12694, pp. 17–33. Springer, Cham (2021). https://doi.org/10.1007/978-3-030-72699-7_2

20. Škvorc, U., Eftimov, T., Korošec, P.: Understanding the problem space in single-objective numerical optimization using exploratory landscape analysis. Appl. Soft Comput. **90**, 106138 (2020). https://doi.org/10.1016/j.asoc.2020.106138

A Robust Statistical Framework
for the Analysis of the Performances
of Stochastic Optimization Algorithms
Using the Principles of Severity

Sowmya Chandrasekaran$^{(\boxtimes)}$ [ID] and Thomas Bartz-Beielstein [ID]

Institute for Data Science, Engineering, and Analytics, TH Köln, Steinmüllerallee 1,
51643 Gummersbach, Germany
{sowmya.chandrasekaran,thomas.bartz-beielstein}@th-koeln.de

Abstract. Meta-heuristic stochastic optimization algorithms are pre-dominantly used to solve complex real-world problems. Numerous new nature-inspired meta-heuristics are being proposed to address various open challenges. Since many heuristics are stochastic, they could yield different solutions to the same problem for different runs. Hence, there is a need for stringent in-depth statistical analysis of the performances of stochastic optimization algorithms. The proposed severity framework enables researchers and practitioners to define application-specific and meaningful performance evaluation metric that evaluates the magnitude of the performance improvement achieved, which is not only of statistical significance but also of practical relevance.

Keywords: Benchmarking · Statistical comparison · Single objective problems · Evolutionary computation

1   Introduction

There is a continuously growing need for a computationally less expensive as well as efficient solution for numerous problems across various engineering and scientific fields. Hence, there is a requirement to optimize various performance measures such as cost, energy, performance, processing time, etc., with various constraints in terms of available budget, time, etc. This growing need to solve complex real-world problems continuously promotes the development of new optimization techniques.

As heuristic and meta-heuristic optimization algorithms are stochastic in nature, multiple runs of each algorithm are performed, and statistics such as mean, median, variance, best and worst values of the achieved optimum are reported. Whenever a new algorithm is proposed, a common practice is evaluating its performance by comparing it with other existing benchmark algorithms using the many well-known benchmarking suites as in [12,18,24–26,38] which comprises of well-defined benchmark functions.

© The Author(s), under exclusive license to Springer Nature Switzerland AG 2023
J. Correia et al. (Eds.): EvoApplications 2023, LNCS 13989, pp. 426–441, 2023.
https://doi.org/10.1007/978-3-031-30229-9_28

The scope of this paper is an in-depth statistical analysis of the performance of a newly proposed algorithm against an existing benchmark algorithm with hypothesis testing [31] and to identify *the extent* to which we trust the resulting inference using the concept of severity [29]. Inspired by [29], we propose a robust bootstrapping-based hypothesis testing framework, *algCompare*, for comparing the performances of the optimization algorithms. This framework does not rely upon distributional assumptions and can statistically analyse if an algorithm outperforms under a robust and reliable testing procedure.

The paper is organized as follows: Section 2 explains the basics of statistical analysis and the existing statistical tools to evaluate the performances of the optimization algorithms. Section 3 explains the concept of severity. Section 4 summarizes the proposed algorithm comparison framework. Here, we show how severity is integrated into this framework and how it bridges the gap between statistical results and their scientific relevance. In Sect. 5, the Covariance Matrix Adaption Evolutionary Strategy (CMA-ES) and Real Space Particle Swarm Optimization (RSPSO) are arbitrarily chosen as optimization algorithms and are evaluated on the CEC benchmark suites from the years 2013, 2014, 2015, and 2017. Section 6 concludes with summary and outlook. The results presented in this paper are based on an experimental analysis of stochastic optimization heuristics. They can also be applied to any other kind of algorithm comparisons as well, with suitable adaptations.

Reproducibility: The code and the results are made available anonymously at https://anonymous.4open.science/r/algCompare-Paper-B3B9/.

2   Background

Most commonly, descriptive and visualization tools may not be sufficient enough in analyzing the differences in the performances of algorithms, especially when the differences are of smaller magnitude. In such scenarios, a statistical analysis is recommended.

2.1   Statistical Analysis

The experimental results of stochastic optimization algorithms can be statistically analyzed using hypothesis testing [23]. The necessity for the statistical analysis of the performances of the stochastic algorithms is discussed in [1,7,34]. The null hypothesis can be formulated as "There is no statistically significant difference between the compared algorithms", and the alternative hypothesis can be formulated as " There exists a statistically significant difference between the compared algorithms."

Hypothesis testing can be classified into parametric and non-parametric tests. Parametric tests assume a specific type of probability distribution of the data and make inferences about the parameters of the distribution. If applicable, parametric tests are more powerful and require only a smaller sample size than non-parametric tests. Non-parametric tests do not make any *explicit* assumptions about the data or its underlying distributions and are recommended when the assumptions for the safe use of parametric tests are not met. The outcome of both testing procedures is the p-value which turns out to be the measure for deciding whether to retain or reject the null hypothesis. Hypothesis testing will be outlined in Sect. 3.

Drawbacks of Using the p-value. The p-value is showcased as the key measure that determines the fate of every new scientific publication. Despite its significance, popularity, and widespread usage, severe criticism has emerged in recent years, questioning the reliability and interpretation of the p-value [4]. The incapability of the p-value as stronger evidence for any scientific claim was clearly discussed in [5,20,22]. If used without the required caution, the p-value suffers from the *problem of large N*. It cannot be stated whether the p-value obtained using a large number of runs of the experiment offer strong evidence against the p-value obtained using a small number of runs. In conjunction with this criticism, the American Statistical Association (ASA) has questioned the credibility for the p-value by releasing a statement on statistical significance and p-values [37]. This article suggests no single statistic can be claimed as evidence for a scientific conclusion and other appropriate and feasible approaches have to be considered.

2.2   Existing Statistical Tools to Compare Algorithm Performances

In [34], a generalized bootstrapping-based multiple hypothesis testing framework was proposed. Non-parametric statistical tests for the evaluation of the evolutionary algorithms were discussed in [9,11,16]. A deep statistic-based comparison tool was proposed in [15]. In [30], an automatic toolkit to compare the performances of optimization algorithms had been proposed. In [7], a multi-criteria-based statistical comparison method had been discussed. In [14], a practical deep statistical comparison technique has been proposed, which considers practical significance when the performances of the stochastic optimization algorithms was compared. In most of the above-discussed articles, be it parametric or non-parametric tests, like [14,15,30,34], merely the p-value was considered as a key performance measure. Possible alternatives to this p-value testing had been proposed in [6,32], where statistical comparison is performed using Bayesian inferences. It is known that in the Bayesian approach the identification of prior probabilities is a key issue [17]. Hence, in [6], the idea of using informed prior from the previous analysis was suggested. The Bayesian approach can lead to very good results if the prior is correct. However, wrong prior can lead to wrong analysis, and sometimes, can even lead to biased conclusions. Also, in [36], a chess rating System for evolutionary algorithms was proposed, which evaluates the practical significance in terms of confidence intervals.

Key Issues with the Existing Tools. It is desired that any performance improvement claimed is statistically significant and practically relevant. Statistical significance was addressed in all the discussed algorithm comparison schemes. However, in [36], the statistical significance was calculated by the overlap of the confidence intervals, which was claimed to be statistically inconsistent in [14]. Though the practical relevance was addressed in [14,36], the resulting schemes can be conservative. In [14], the practical significance was accounted directly in the hypothesis formulation, which makes the test very conservative as this formulation can increase the Type II error (β), which in turn decreases the power of the test. I.e., even if the newly proposed algorithm outperforms the benchmark algorithm, the ability to detect the performance differences reduces (sometimes significantly) compared to the standard hypothesis testing.

In the proposed framework, both statistical significance and practical relevance are addressed without explicitly introducing additional conservatism by a two-step process. In the first step, the standard hypothesis test is performed to infer a decision based on the p-value. The quality of the decision is stringently evaluated by the concept of severity, the details of which are discussed in the following section.

3   Severity

Let us consider the performances of a benchmark algorithm A, say $\mathbf{a} = (a_1, \ldots, a_n)$, with the new algorithm B, say $\mathbf{b} = (b_1, \ldots, b_n)$. Here, $a_1, \ldots, a_n$ represents the achieved optimum values of the objective function for N runs by the benchmark algorithm. And $b_1, \ldots, b_n$ represents the achieved optimum values of the objective function for N runs by the new algorithm.

Hypothesis Formulation: To explain the concept of hypothesis testing, let us assume Normal, Independent, and Identically Distributed (NIID) data[1]. The hypothesis test is formulated as an upper tail test of the difference between the two means as we are considering an optimization minimization problem. For simplicity, let us consider the pairwise tests with a significance level α. The difference vector $\mathbf{x}$ can be defined as $\mathbf{x} = (x_1, \ldots, x_n)$, where $x_i = a_i - b_i$, $\forall i = \{1, \ldots, n\}$. Let $\bar{\mathbf{x}}$ denote the mean of the vector $\mathbf{x}$. Under the null hypothesis, any observed difference is attributed only to errors in random sampling. Hence, the null hypothesis (H_0) and alternate hypothesis (H_1) are formulated as:
H_0 : B does not outperform A $\implies \bar{\mathbf{x}} \leq 0$.
H_1 : B outperforms A $\implies \bar{\mathbf{x}} > 0$.
The alternate hypothesis implies if the new algorithm, B, seems to be outperforming the benchmark algorithm, A, then $\bar{\mathbf{b}} < \bar{\mathbf{a}}$, (i.e., $\bar{\mathbf{x}} > 0$).

Experimental setup: The hypothesis testing is performed as described in (1).

$$H_0 : \bar{\mathbf{x}} \leq 0; \text{vs.} \ H_1 : \bar{\mathbf{x}} > 0, \tag{1}$$

$$decision = \begin{cases} \text{not-Reject } H_0, & \text{if } d(\mathbf{X}) \leq u_{1-\alpha}, \\ \text{Reject } H_0, & \text{otherwise,} \end{cases}$$

---

[1] Note that in the proposed framework, we do not assume normality of the data. Here it is assumed to simplify the explanation of the concept and without the loss of generality, the concept can be adapted to the cases where the distribution is not known.

where $u_{1-\alpha}$ is the upper tail cut-off point of the normal distribution, which cuts the upper-tail probability of α, and the test statistic $d(\mathbf{X})$ can be represented as

$$d(\mathbf{X}) = \frac{\bar{X} - \mu_0}{\sigma_x}, \tag{2}$$

where μ_0 is the hypothesized mean under H_0, which is zero in our case and standard error $\sigma_x = \frac{\sigma}{\sqrt{N}}$. When a test statistic is observed beyond the cut-off point, we reject the H_0 at a significance level α. The associated power $(1 - \beta)$, significance level (α) and p-value of the test is shown in Table 1.

Pre-data Analysis: Generally, the values for α, β, and hence power $(1-\beta)$ are pre-specified before the experiment is performed. It is recommended to identify the required number of runs, N, based on the desired values of α, β, the difference sought (Δ), and σ. For a one-sided test of size α, the (approximate) formula to identify the required N is

$$N \approx 2 \times (u_{1-\alpha} + u_{1-\beta})^2 \sigma^2 / \Delta^2. \tag{3}$$

Experiment: The benchmark algorithm A and the new algorithm B are tested on a benchmark function for N runs based on (3). And achieved optimum values of the objective function by Algorithm A, $\mathbf{a} = (a_1, \ldots, a_n)$ and achieved optimum values of the objective function by Algorithm B, $\mathbf{b} = (b_1, \ldots, b_n)$ is obtained. The hypothesis testing as described in (1) is performed. Based on the p-value, the H_0 is rejected or retained.

Drawbacks of the Experiment: This hypothesis testing procedure is often criticized as a black and white thinking, as the statistical significance is decided in the form of a yes or no fashion based only on the p-value [4]. The effects of the drawbacks can be minimized by performing a post-data analysis as described below.

Post-data Analysis: Severity, a form of attained power [29], considers the observed experimental result, i.e., $d(\mathbf{x})$ and performs a post-data evaluation after the hypothesis test is performed, and after the decision to either retain or to reject the null hypothesis is obtained. In (1), the test assesses if $d(\mathbf{x})$ falls under the null hypothesis or not. The key difference in severity evaluation is that the test is performed for different values under the alternative hypothesis. This is measured using severity, a probability analogous to the p-value under the alternative hypothesis rather than one under the null [33]. The different values under the alternative hypothesis for which the compatibility is assessed will henceforth be called discrepancy, δ.

The differences between power $(1 - \beta)$, significance level (α), p-value, and severity are shown in Table 1. The key difference between power and severity is that severity depends on the experimental result, i.e., $d(\mathbf{x})$ instead of $u_{1-\alpha}$. Severity is defined separately for the non-rejection (S_{nr}) and the rejection (S_r) of the null hypothesis. Figure 1 depicts the region of severity for the non-rejection when the decision was to retain the null hypothesis and the region of rejection when the decision was to reject the null hypothesis. The S_{nr} values increase

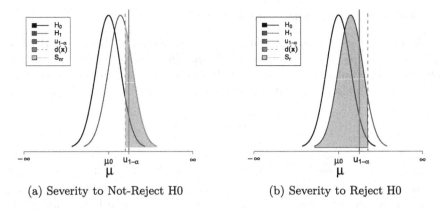

(a) Severity to Not-Reject H0   (b) Severity to Reject H0

Fig. 1. Illustration of the S_{nr} and S_r under the alternate hypothesis: In (a), the green line represents the actual test statistic $d(\mathbf{x})$. As this falls inside the $u_{1-\alpha}$, the decision is to retain the null. The S_{nr} is the area under the H_1 that is beyond $d(\mathbf{x})$ (area shaded in green). In (b), the green line represents the actual test statistic $d(\mathbf{x})$, as this falls beyond the $u_{1-\alpha}$, the decision is to reject the null. The S_r is the area under the H_1 that is within the $d(\mathbf{x})$ (area shaded in green). (Color figure online)

Table 1. Power $(1 - \beta)$, significance level (α), p-value, and severity. P_{H_0} denotes the probability under the assumption that H_0 is true, whereas P_{H_1} denotes the probability under the assumption that H_1 is true.

$1 - \beta$	α	p-value	Severity
$P_{H_1}(d(X) > u_{1-\alpha})$	$P_{H_0}(d(X) > u_{1-\alpha})$	$P_{H_0}(d(X) > d(x))$	S_{nr}: $P_{H_1}(d(X) > d(x))$
			S_r: $P_{H_1}(d(X) \leq d(x))$

monotonically from 0 to 1. The S_r values decrease monotonically from 1 to 0. The closer the value is to 1, the more reliable the decision made with the hypothesis test. A severity value higher than 0.8 is generally considered a reliable support.

Benefits of Severity: Moving beyond the black and white thinking based on the p-value, the severity sheds light in understanding the magnitude of the effect, i.e., considering $\alpha = 0.05$, p-value of 0.00005 or 0.005 or 0.049 are clearly distinguished based on the corresponding S_r value, when the decision is to reject the null hypothesis. Similarly, p-value of 0.051 or 1 is supported by corresponding S_{nr} values. The practical significance and or uncertainty can be accounted using δ. The *problem of large N* is directly addressed using severity by changing the intensity of the severity test. For a larger N, the severity test is less severe. And for a smaller N, the severity test is more severe [29]. In [2], severity is applied to analyze the results of hyper-parameter tuning experiments in the field of machine and deep learning.

3.1   Other Existing Alternative Measures

For the parametric case, two variants of effect size, namely Cohens$_d$ and Hedges$_g$, which are scale-free approaches to quantify the size of the performance difference between the two algorithms, are used. It is the standardized mean difference between the two algorithms [10]:

$$Cohen_d = \frac{\bar{x}_A - \bar{x}_B}{S_p} \qquad (4)$$

where $S_p = \sqrt{\frac{(n_B-1)s_B^2+(n_A-1)s_A^2}{n_A+n_B-2}}$, $\bar{x}_A$ and $\bar{x}_B$ is the sample mean of the algorithms A and B, respectively. The S_p is the pooled standard deviation, n_A, n_B is the sample size of each algorithm, and s_A, s_B is the standard deviation of each method. Cohen suggested effect size as small (0.2), medium (0.5), and large (0.8), along with a strong caution to the applicability in different fields. In [21], Hedges suggested a correction measure as Cohen's d slightly overestimates the standard deviation

$$Hedges_g = 1 - \frac{3}{4(n_A + n_B) - 9} \times g. \qquad (5)$$

For the non-parametric cases, Cliff's delta [27] is commonly used as effect size and is obtained as

$$Cliff_\delta = \frac{\text{no of times}(a_i > b_i) - \text{no of times}(a_i < b_i)}{n_A * n_B}, \qquad (6)$$

where n_A, n_B is the sample size of each algorithm, and a_i, b_i are the results of each run of A and B, respectively. Its value ranges from -1 to 1. A value of +1 or -1 indicates the absence of overlap between A and B, whereas a value of 0 indicates that distributions of A and B overlap completely.

The limitations of Cohens$_d$, Hedges$_g$ is that if both the pooled standard deviation and the mean differences are too high the effect gets canceled out. Also, when the pooled standard deviation is close to zero, the effect size is too high. The Cliff$_\delta$ can be too high even when the observed performance improvement is of negligible magnitude. Detailed analysis of why these measures may not be sufficient to identify the practical significance will be discussed with examples in Sect. 5.

4   Proposed Framework: AlgCompare

With the goal of providing an algorithm comparison framework that does not rely upon distributional assumptions, algCompare is formulated as a bootstrapping-based statistical test to statistically analyse the performances in a robust and reliable manner. Bootstrapping is considered a standard method in statistics and is applied when the data is non-normal, or the distribution is unknown, and a representative sample is available [13, 28]. Since the experimental results obtained from stochastic optimization algorithms are not always normally distributed [16], the generalized hypothesis testing framework discussed in Sect. 3, is adapted for bootstrapping-based t-test. In order to obtain better estimate, we perform sampling with replacement. This procedure also eliminates the dependence of the outcome on the ordering of the optimization runs.

Algorithm 1: Proposed Comparison Framework with Bootstrapping

---

Data: $\mathbf{a} = (a_1, \ldots, a_n)$, $\mathbf{b} = (b_1, \ldots, b_n)$, α, δ, n_b

Result: p, decision, S_{nr}, S_r

Formulate Hypothesis $H_0 : \bar{\mathbf{x}} \leq 0$ vs. $H_1 : \bar{\mathbf{x}} > 0$, $x_i = a_i - b_i$, $i = \{1, 2, .., n\}$

Evaluate observed sample mean difference $t_{obs} = \bar{\mathbf{a}} - \bar{\mathbf{b}}$

Combine $\mathbf{I} = \widehat{\mathbf{ab}}$

repeat

    Draw a bootstrap sample of $2n$ observations with replacement from $\mathbf{I}$

    Let the mean of the first n observation be $\bar{\mathbf{a}}^*$

    Let the last n observations be $\bar{\mathbf{b}}^*$

    Evaluate $t^{*\text{bs}} = \bar{\mathbf{a}}^* - \bar{\mathbf{b}}^*$

    Evaluate $t_s^{*\text{bs}} = \bar{\mathbf{a}}^* - \bar{\mathbf{b}}^* - \delta$

until n_b *times*;

Calculate $p \approx \frac{\#(t^{*\text{bs}} \geq t_{obs})}{n_b}$

if $p \leq \alpha$ **then**

    decision: Reject H_0

    With $t_s^{*\text{bs}}$ obtain $S_r \approx \frac{\#(t_s^{*\text{bs}} \leq t_{obs})}{n_b}$

else

    decision: not-Reject H_0

    With $t_s^{*\text{bs}}$ obtain $S_{nr} \approx \frac{\#(t_s^{*\text{bs}} > t_{obs})}{n_b}$

end

The proposed algCompare framework is described in Algorithm 1. Firstly, the observed sample mean difference t_{obs} is to be calculated. For the bootstrapping procedure, we merge the results of both algorithms into one sample $\mathbf{I}$ of size $2n$. Then we draw a bootstrap sample of $2n$ observations with replacement from $\mathbf{I}$. Then, we evaluate the bootstrap test statistic $t^{*\text{bs}}$. We repeat this procedure based on the bootstrapping re-sample size, n_b, and the desired p-value is then estimated to be the number of times the bootstrap test statistic $t^{*\text{bs}}$ was found to be greater than the t_{obs} in n_b samples. When the p-value is found to be significant, i.e., less than the specified α, then we reject H_0, else we do not reject H_0. In case the decision is to reject H_0, then, S_r, a numeric vector of severity measure for rejecting the null hypothesis for a range of user-defined discrepancies, δ, are obtained. Else, S_{nr}, severity measure for not-rejecting the null hypothesis for a range of δ, is obtained.

Advantages of the Proposed Framework:

The benefits of analysing the performance improvement achieved using algCompare Framework are three-folded.

1. The p-value may be misguiding as the magnitude of the differences observed is not explicitly visible. Severity enables a more transparent analysis by carefully considering the factors like N, δ that might influence the outcome.
2. The claimed performance improvement achieved is more precisely reported. If B outperforms A, the extent to which it outperforms in terms of δ and the corresponding severity support for each discrepancy is obtained. For example, B outperforms A and this decision is supported with 80% severity until a specific δ.
3. The practitioners can analyse the broad spectrum of how strong the performance improvement of B over A is, and which δ is of practical relevance.

5   Casestudy

To showcase the application of the algCompare in practice, this section presents a case study. In [8], SELECTOR is proposed as a reliable and unbiased tool to evaluate the performances of the algorithms. We used the performance data from [8], which is an excellent source of results from several benchmarking suites that are combined under a single roof and available as open-source. It includes the benchmarking suites from the Special Session on Real Parameter Optimization organized at IEEE Congress on Evolutionary Computation (CEC) in the years 2013 [25], 2014 [24], 2015 [26], and 2017 [38], along with the Black-Box Benchmarking (BBOB) suite [18,19]. Both suites provide several single objective problems for benchmarking and have been used in the CEC and BBOB competitions for several years. More precisely, CEC 2013 provides 28 benchmark functions, CEC 2014 provides 30, CEC 2015 with 15, CEC 2017 with 29 benchmark functions of a single instance. And the BBOB suite is with 24 functions from 5 instances. In all suites, dimension 10 is considered. As optimization algorithms, the CMA-ES, Differential Evolution [35] and RSPSO are provided. The experiment consisted of 30 independent runs of each algorithm on each instance of the benchmarking function. Further details about the experimental setup are available in [8].

For our experiments, a subset from [8] is taken. The CMA-ES and RSPSO are arbitrarily chosen as two optimization algorithms and is evaluated on CEC 2013, CEC 2014, CEC 2015, CEC 2017. Nevertheless, we have repeated the experiments for all pairwise combinations of algorithms on all CEC and BBOB performance data, and the results are available in our GitHub repository. Considering page restrictions, we investigate only the performances of CMA-ES vs RSPSO. As other performance metrics for comparison, the Cohens$_d$ and Hedges$_g$ and Cliffs δ are also evaluated using the R package *effectsize* version 0.8.2 [3].

5.1   CMA-ES Outperforms RSPSO

We first formulate the hypothesis to check on which functions the CMA-ES algorithm performs better than the RSPSO algorithm as " H_0: *CMA-ES does not outperform RSPSO*" versus " H_1: *CMA-ES outperforms RSPSO*"

Table 2. Overview of summary statistics for CMA-ES and RSPSO for some specific cases where Cohens$_d$ and Hedges$_g$ are misleading. The winning algorithm is highlighted in bold.

Function	Algorithm	Mean	Median	SD	Cohens$_d$	Hedges$_g$	Cliffs δ
CEC 2013 f2	**CMA-ES**	-1300.00	-1300.00	2.77×10^{-9}			
	RSPSO	1.98×10^5	5.91×10^4	3.55×10^5	0.79	0.78	1
CEC 2013 f3	**CMA-ES**	-1200	-1200	3.21×10^{-9}			
	RSPSO	2.32×10^7	2.5×10^6	5.38×10^7	0.6	0.6	1
CEC2014 f23	**CMA-ES**	2500	2500	1.88×10^{-13}			
	RSPSO	2629.45	2629.45	2.20×10^{-9}	8.3×10^{10}	8.2×10^{10}	1
CEC2014 f26	CMA-ES	2800	2800	1.8×10^{-13}			
	RSPSO	2700.2	2700.17	0.104	1.3×10^3	1.3×10^3	1
CEC2014 f14	CMA-ES	1400.41	1400.44	8.7×10^{-2}			
	RSPSO	1400.29	1400.29	1.65×10^{-1}	0.93	0.94	0.54

The algCompare framework was evaluated for a chosen significance level α of 0.05, $N = 30$ runs, $n_b = 10000$. $\delta \in 10^{-9}, 10^{-6}, 10^{-3}, 10^{-1}, 10^0, 10^1, 10^2, 10^3, 10^6$. The results are shown in Fig. 2. For CEC 2013 benchmark suite, CMA-ES outperformed RSPSO in 8 out of 28 benchmark functions. However, for function 1, the severity support is only 77% at a $\delta = 10^{-9}$. This means the performance improvement achieved by CMA-ES over RSPSO is only in the range of 10^{-9}. CMA-ES outperformed RSPSO in CEC 2014 on 13 out of 30, in CEC 2015 on 7 out of 15 functions, and in CEC 2017 on 11 out of 29 functions. In CEC 2014, for functions 7 and 13, the severity support exists only until a δ of 10^{-3}. Also, in CEC 2015, function 11 is supported by severity of 88% for a δ of 10. Finally, in CEC 2017, for function 2, the severity support exists only until a δ of 10^{-6}. And for function 4, the severity support exists only until a δ of 10^{-3}.

5.2 RSPSO Outperforms CMA-ES

Similarly, the algCompare framework was evaluated conversely to check on which functions RSPSO seems to outperform CMA-ES. And the results are shown in Fig. 3. RSPSO outperformed CMA-ES in CEC 2013 on 15 out of 28 functions, in CEC 2014 on 11 out of 30, in CEC 2015 on 6 out of 15 functions, and in CEC 2017 on 11 out of 29 functions, respectively. For function 22 in CEC2013, the severity offers strong support only until a δ of 0.10. In CEC 2014, for functions 14 and 19, the support stands only until a δ of approx. 0.1. For function 7 in CEC 2015, the severity offers 70% support for a δ of 1.0.

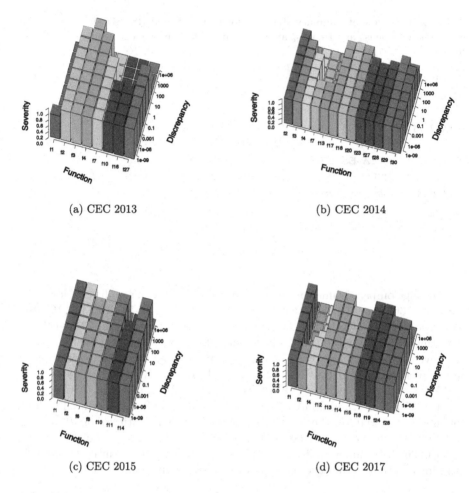

(a) CEC 2013   (b) CEC 2014

(c) CEC 2015   (d) CEC 2017

Fig. 2. CASE 1: Benchmarking functions for which CMA-ES outperforms RSPSO and its corresponding severity support for various discrepancies (δ).

5.3   Evaluating Other Measures

In general, Cohens$_d$ and Hedges$_g$ are recommended only as a part of the parametric test. However, we intended to measure these metrics to evaluate their performances for the data set whose underlying distribution is unknown. Specific

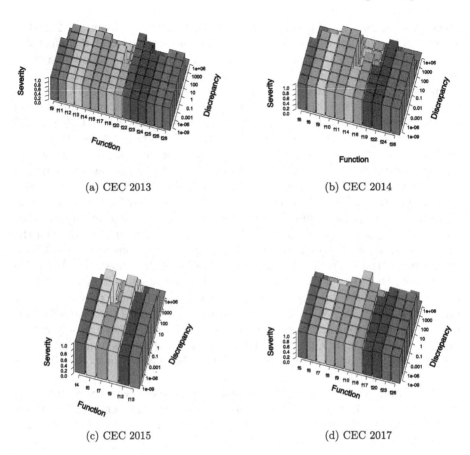

(a) CEC 2013

(b) CEC 2014

(c) CEC 2015

(d) CEC 2017

Fig. 3. CASE 2: Benchmarking functions for which RSPSO outperforms CMA-ES and its corresponding severity support for various discrepancies (δ).

Table 3. Overview of summary statistics for CMA-ES and RSPSO for some specific cases where Cliffs δ is inappropriate. The winning algorithm is highlighted in bold.

Function	Algorithm	Mean	Median	SD	Cohens$_d$	Hedges$_g$	Cliffs$_\delta$
CEC 2013 f10	**CMA-ES**	−499.9904	−499.99260	0.01006582	1.6	1.58	1
	RSPSO	−499.4823	−499.6196	0.4459858			
CEC 2017 f2	**CMA-ES**	200	200	2.08×10^{-9}			
	RSPSO	200	200	9.88×10^{-6}	2.92	2.88	1
CEC 2017 f4	**CMA-ES**	400	400	4.15×10^{-9}			
	RSPSO	403.4718	400.0225	18.85289	0.26	0.257	1

instances are reported in Table 2 to showcase where employing Cohens$_d$ and Hedges$_g$ might be problematic. Our results indicate that in some cases, where both the pooled standard deviation and the mean differences are too high the effect gets canceled out. This is obvious in CEC 2013, for functions 2 and 3, where RSPSO did not converge at all and Cohens$_d$ suggests that the effect size is large (0.79), and medium (0.6). The severity provides strong support for these functions, as illustrated in Fig. 2. In contrast, for function 14 in CEC 2014, where the performances of both the algorithms are in a similar range, the Cohens$_d$ is larger (0.93), and supports RSPSO. The severity support stands for RSPSO only for a δ of 10^{-3}, as shown in Fig. 3. Another possibility where Cohens$_d$ and Hedges$_g$ are inappropriate is shown in the cases of functions 23 and 26 in CEC 2014. Here, the standard deviations of both the algorithms are close to zero. As a result, the values of Cohens$_d$ and Hedges$_g$ are too high. The severity supports CMA-ES until a δ of 100 for function 23. The RSPSO is supported by severity for function 26 until a δ of 10. For the CEC 2014 function 14, the results from Cohens$_d$, Hedges$_g$, and Cliffs$_\delta$ are contradicting. In all these examples, however the Cliffs$_\delta$ is reasonably acceptable considering the ground truth from the summary statistics. However, in Table 3, certain observations where the Cliffs$_\delta$ overestimates the performance improvement achieved are showcased. It can be clearly seen that for function 10 in CEC 2013, and function 2 in CEC 2017 , the magnitude of the differences are very small. But the Cliffs$_\delta$ provides utmost support for CMA-ES, with a maximum attainable value of 1. The severity support for function 10 is until a discrepancy of 1 and for function 2 is only until a discrepancy of 10^{-6}. In CEC 2017, for function 4, the results from Cliffs$_\delta$, and Cohens$_d$, Hedges$_g$ are contradicting. For this, the severity supports CMA-ES only until a δ of 0.001.

Comparing the results, it can be seen that by using severity as the measure to evaluate algorithms, not only do we get statistical significance without conservatism but also stringent evaluation of practical relevance of the obtained results.

6   Summary and Outlook

The algCompare is proposed as a framework for the reliable statistical interpretation of the results from empirically tested algorithms, where a degree of support for the decision is obtained using hypothesis testing for the chosen magnitudes of practical significance. For didactical purposes, the hypothesis testing is formulated considering the mean performance differences among algorithms. Nevertheless, the proposed algCompare framework can be used to compare the median and related performance measures as well. Through the case study, in addition to highlighting the benefits of the algCompare framework, we have shown that employing Cohens$_d$, Hedges$_g$ measures for non-normal data can lead to misleading inferences. Also, it shows that non-parametric Cliffs$_\delta$ may also give an incomplete picture, i.e., it can show that one outperforms the other even in the case where the observed performance improvement is of negligible magnitude.

The concept of severity can be extended for the comparison of multiple algorithms. Multiple pairwise comparisons for comparing the performances of multiple algorithms can be performed by adjusting for the Family-Wise Error Rate

(FWER) using many available correction measures. The framework can be easily integrated with many of the existing algorithm ranking schemes, which are based on parametric or non-parametric tests. This can be done by choosing a relevant and meaningful value for δ and obtaining the severity values along with the p-value. A detailed analysis of which will be published in the near future.

References

1. Bartz-Beielstein, T., et al.: Benchmarking in optimization: best practice and open issues. arXiv preprint arXiv:2007.03488 (2020)
2. Bartz-Beielstein, T., Mersmann, O., Chandrasekaran, S.: Ranking and result aggregation. In: Bartz, E., Bartz-Beielstein, T., Zaefferer, M., Mersmann, O. (eds.) Hyperparameter Tuning for Machine and Deep Learning with R: A Practical Guide, chap. 5, pp. 121–161. Springer Nature (2023). https://doi.org/10.1007/978-981-19-5170-1_5
3. Ben-Shachar, M.S., Lüdecke, D., Makowski, D.: Effectsize: estimation of effect size indices and standardized parameters. J. Open Source Softw. **5**(56), 2815 (2020)
4. Benavoli, A., Corani, G., Demšar, J., Zaffalon, M.: Time for a change: a tutorial for comparing multiple classifiers through Bayesian analysis. J. Mach. Learn. Res. **18**(1), 2653–2688 (2017)
5. Berger, J.O., Sellke, T.: Testing a point null hypothesis: The irreconcilability of p values and evidence. J. Am. Stat. Assoc. **82**(397), 112–122 (1987)
6. Calvo, B., Shir, O.M., Ceberio, J., Doerr, C., Wang, H., Bäck, T., Lozano, J.A.: Bayesian performance analysis for black-box optimization benchmarking. In: Proceedings of the Genetic and Evolutionary Computation Conference Companion, pp. 1789–1797 (2019)
7. Carrano, E.G., Wanner, E.F., Takahashi, R.H.: A multicriteria statistical based comparison methodology for evaluating evolutionary algorithms. IEEE Trans. Evol. Comput. **15**(6), 848–870 (2011)
8. Cenikj, G., Lang, R.D., Engelbrecht, A.P., Doerr, C., Korošec, P., Eftimov, T.: Selector: selecting a representative benchmark suite for reproducible statistical comparison. arXiv preprint arXiv:2204.11527 (2022)
9. Christensen, S., Wineberg, M.: Using appropriate statistics-statistics for artificial intelligence. In: Tutorial Program of the Genetic and Evolutionary Computation Conference, Seattle, WA, pp. 544–564 (2004)
10. Cohen, J.: Statistical power analysis for the behavioral sciences (revised ed.) (1977)
11. Derrac, J., García, S., Molina, D., Herrera, F.: A practical tutorial on the use of nonparametric statistical tests as a methodology for comparing evolutionary and swarm intelligence algorithms. Swarm Evol. Comput. **1**(1), 3–18 (2011)
12. Doerr, C., Ye, F., Horesh, N., Wang, H., Shir, O.M., Bäck, T.: Benchmarking discrete optimization heuristics with IOHprofiler. Appl. Soft Comput. **88**, 106027 (2020)
13. Efron, B., Tibshirani, R.J.: An Introduction to the Bootstrap. CRC Press, Boca Raton (1994)
14. Eftimov, T., Korošec, P.: Identifying practical significance through statistical comparison of meta-heuristic stochastic optimization algorithms. Appl. Soft Comput. **85**, 105862 (2019)
15. Eftimov, T., Korošec, P.: A novel statistical approach for comparing meta-heuristic stochastic optimization algorithms according to the distribution of solutions in the search space. Inf. Sci. **489**, 255–273 (2019)

16. García, S., Molina, D., Lozano, M., Herrera, F.: A study on the use of non-parametric tests for analyzing the evolutionary algorithms' behaviour: a case study on the cec'2005 special session on real parameter optimization. J. Heuristics **15**(6), 617–644 (2009)
17. Gelman, A.: Objections to Bayesian statistics. Bayesian. Analysis **3**(3), 445–449 (2008)
18. Hansen, N., Auger, A., Ros, R., Mersmann, O., Tušar, T., Brockhoff, D.: COCO: a platform for comparing continuous optimizers in a black-box setting. Optim. Methods Softw. **36**(1), 114–144 (2021)
19. Hansen, N., Finck, S., Ros, R., Auger, A.: Real-parameter black-box optimization benchmarking 2009: Noiseless functions definitions. Ph.D. thesis, INRIA (2009)
20. Head, M.L., Holman, L., Lanfear, R., Kahn, A.T., Jennions, M.D.: The extent and consequences of p-hacking in science. PLOS Bio. **13**(3), 1–15 (2015)
21. Hedges, L.V., Olkin, I.: Statistical Methods for Meta-Analysis. Academic Press, New York (1985)
22. Lecoutre, B., Lecoutre, M.P., Poitevineau, J.: Uses, abuses and misuses of significance tests in the scientific community: won't the Bayesian choice be unavoidable? Int. Stat. Rev. **69**(3), 399–417 (2001)
23. Lehmann, E.L., Romano, J.P.: Testing Statistical Hypotheses. Springer, New York (2006). https://doi.org/10.1007/0-387-27605-X
24. Liang, J.J., Qu, B.Y., Suganthan, P.N.: Problem definitions and evaluation criteria for the CEC 2014 special session and competition on single objective real-parameter numerical optimization. In: Computational Intelligence Laboratory, Zhengzhou University, Zhengzhou China and Technical Report, Nanyang Technological University, Singapore, vol. 635, p. 490 (2013)
25. Liang, J.J., Qu, B., Suganthan, P.N., Hernández-Díaz, A.G.: Problem definitions and evaluation criteria for the CEC 2013 special session on real-parameter optimization. In: Computational Intelligence Laboratory, Zhengzhou University, Zhengzhou, China and Nanyang Technological University, Singapore, Technical Report, vol. 201212, iss. 34, pp. 281–295 (2013)
26. Liang, J., Qu, B., Suganthan, P., Chen, Q.: Problem definitions and evaluation criteria for the cec 2015 competition on learning-based real-parameter single objective optimization. In: Technical Report201411A, Computational Intelligence Laboratory, Zhengzhou University, Zhengzhou China and Technical Report, Nanyang Technological University, Singapore, vol. 29, pp. 625–640 (2014)
27. Macbeth, G., Razumiejczyk, E., Ledesma, R.D.: Cliff's delta calculator: a non-parametric effect size program for two groups of observations. Universitas Psychologica **10**(2), 545–555 (2011)
28. Mammen, E., Nandi, S.: Bootstrap and resampling 111.2. Handbook of Computational Statistics: Concepts and Methods, p. 467 (2004)
29. Mayo, D.G., Spanos, A.: Severe testing as a basic concept in a neyman-pearson philosophy of induction. British J. Philos. Sci. **57**(2), 323–357 (2006)
30. Molina, D., LaTorre, A.: Toolkit for the automatic comparison of optimizers: comparing large-scale global optimizers made easy. In: 2018 IEEE Congress on Evolutionary Computation (CEC), pp. 1–8. IEEE (2018)
31. Neyman, J., Pearson, E.S.: On the use and interpretation of certain test criteria for purposes of statistical inference: Part i. Biometrika, pp. 175–240 (1928)
32. Rojas-Delgado, J., Ceberio, J., Calvo, B., Lozano, J.A.: Bayesian performance analysis for algorithm ranking comparison. IEEE Trans. Evol. Comput. **26**(6), 1281–1292 (2022)

33. Senn, S.S.: Statistical issues in drug development, vol. 69. John Wiley & Sons (2008)
34. Shilane, D., Martikainen, J., Dudoit, S., Ovaska, S.J.: A general framework for statistical performance comparison of evolutionary computation algorithms. Inf. Sci. **178**(14), 2870–2879 (2008)
35. Storn, R., Price, K.: Differential evolution-a simple and efficient heuristic for global optimization over continuous spaces. J. Global Optim. **11**(4), 341–359 (1997)
36. Veček, N., Mernik, M., Črepinšek, M.: A chess rating system for evolutionary algorithms: a new method for the comparison and ranking of evolutionary algorithms. Inf. Sci. **277**, 656–679 (2014)
37. Wasserstein, R.L., Lazar, N.A.: The ASA's statement on p-values: context, process, and purpose. Am. Stat. **70**(2), 129–133 (2016)
38. Wu, G., Mallipeddi, R., Suganthan, P.N.: Problem definitions and evaluation criteria for the CEC 2017 competition on constrained real-parameter optimization. National University of Defense Technology, Changsha, Hunan, PR China and Kyungpook National University, Daegu, South Korea and Nanyang Technological University, Singapore, Technical Report (2017)

Towards Constructing a Suite of Multi-objective Optimization Problems with Diverse Landscapes

Andrejaana Andova[1,2]([✉]), Tobias Benecke[3], Harald Ludwig[4], and Tea Tušar[1,2]

[1] Jožef Stefan Institute, Ljubljana, Slovenia
{andreajaana.andova,tea.tusar}@ijs.si
[2] Jožef Stefan International Postgraduate School, Ljubljana, Slovenia
[3] Otto-von-Guericke-University Magdeburg, Magdeburg, Germany
tobias.benecke@ovgu.de
[4] Johannes Kepler University Linz, Linz, Austria
harald.ludwig@jku.at

Abstract. Given that real-world multi-objective optimization problems are generally constructed by combining individual functions to be optimized, it seems sensible that benchmark functions would also follow this procedure. Since the pool of functions to choose from is large and the number of function combinations increases exponentially with the number of objectives, we need a smart way to choose a reasonably sized and diverse collection of function combinations to use in benchmarking experiments. We propose a four-step approach that analyzes the landscape characteristics of all function combinations and selects only the most diverse ones to form a suite of problems. In this initial study, we test this idea on the pool of bbob functions and the case of two objectives. We provide a proof of concept for the proposed approach and its initial results. We also discuss its limitations to be addressed in future work.

Keywords: benchmarking · test problems · multi-objective optimization · exploratory landscape analysis

1 Introduction

One of the purposes of benchmarking is to gain knowledge about algorithm performance on various test problems, which can be applied when solving real-world optimization problems. The latter are often computationally expensive, making it intractable (if not impossible) to perform extensive experiments to find the best algorithm for the given problem. The more realistic scenario in such cases is to find some information about the properties of the real-world problem at hand and solve it using the algorithm that performs best on quick-to-evaluate test problems with similar characteristics.

A. Andova, T. Benecke and H. Ludwig—Contributed equally to this work.

A good choice of benchmark problems is crucial to make this scenario useful in practice. According to [1], a suite of benchmark problems should be *diverse*, containing problems with different characteristics, and *representative* by including difficulties that are found in real-world problems. Its problems should be *scalable and tunable* and, in order to facilitate performance evaluation, have *known optima* or at least some *known best performance*. Ideally, the suite would be *continuously updated* to prevent the over-fitting of algorithms.

In continuous single-objective optimization, the well-known bbob test suite [6] satisfies all these requirements. However, there is no equivalent of such a suite in multi-objective optimization (the bbob-biobj suite [3] is limited to bi-objective problems). The decades old, still most often used benchmark suites in multi-objective optimization, namely ZDT [16], DTLZ [5] and WFG [10], are built following the *bottom-up approach*, meaning they are designed around the desired properties of the Pareto-front, which sacrifices real-world relevance of resulting problems to a simple suite construction procedure and controllable problem properties [4]. Note that because of the prevalence of these suites in multi-objective optimization, algorithms have been overfitting to their problems [11], further motivating the need for a change.

We believe that the alternative approach to problem construction, which uses single-objective test function combinations to create multi-objective problems, more closely resembles the real-world conditions and can be used to create a diverse benchmark suite without the biases brought on by the bottom-up approach. However, simply creating all function combinations from a pool of functions is infeasible due to the high number of resulting problems. With k functions to choose from, we can construct $\binom{k+m-1}{m}$ unique multi-objective function combinations (without including permutations of the same functions), where m is the number of objectives. For example, given $k = 10$ functions, we get 55 function combinations with two objectives, 2002 function combinations with five objectives and 92 378 function combinations with ten objectives. This is not sustainable and a more sophisticated strategy is needed to select good function combinations to form a reasonably sized suite of multi-objective benchmark problems.

In this paper, we propose to use problem landscape characteristics to guide to the selection of a manageable number of function combinations. The idea is to compute problem (dis)similarity using exploratory landscape analysis (ELA). This meets the first requirement for a suite of benchmark problems—its diversity. To make sure that function combinations are also representative of real-world difficulties, we use a pool of functions with this quality, the bbob functions (which are also scalable, tunable, and have known optima). We try this idea in the case of two objectives. Without a formal way to validate such a construction, we look at the resulting problems as well as some intermediate steps in the construction procedure from multiple perspectives to understand the implications of our choices and find ideas for future improvements.

In the following pages, Sect. 2 provides some background on multi-objective optimization problems, the bbob and bbob-biobj(-ext) benchmark suites and the exploratory landscape analysis features used to characterize the problems. Section 3 details the employed four stepped methodology: (i) generating the base set of problems, (ii) computing their ELA features, (iii) selecting a diverse subset of problems, and (iv) generating similar instances. The results of this procedure are presented in Sect. 4, where we first validate our idea on 2-D problems, for which the problem landscape can be visualized, then show the results for all considered dimensions and finally discuss the limitations of our approach and ideas to improve it. Section 5 concludes the paper with final remarks.

2   Background

In this section, we first clarify the terminology around multi-objective optimization problems, followed by a brief introduction to the single-objective bbob functions used to construct the bi-objective problems in this paper. Finally, the ELA features used to characterize the fitness landscapes are shortly presented.

2.1   Multi-objective Optimization Problems

We are concerned with multi-objective optimization problems of the form:

$$\text{minimize}\quad F^{\theta}(x) = (f_1^{\theta_1}(x), \ldots, f_m^{\theta_m}(x)), \tag{1}$$

where $x = (x_1, \ldots, x_n) \in S$ is a search vector from the n-dimensional search space $S \subseteq \mathbb{R}^n$ and $f_i^{\theta_i} : S \to \mathbb{R}$, $i = 1, \ldots, m$, are parameterized objective functions, where $m > 1$ and $\theta = (\theta_1, \ldots, \theta_m) \in \Theta$ parameterizes the function instance.

We use the term *function combination* to denote the non-instantiated m-tuple of functions $F(x) = (f_1(x), \ldots, f_m(x))$, while a *problem* is a particular instance of the function combination, $F^{\theta}(x) = (f_1^{\theta_1}(x), \ldots, f_m^{\theta_m}(x))$. Different instances of a function might be shifted in the decision and/or objective space, can be rotated, etc. and are used to test algorithm (in)variability to these changes.

A problem solution $x \in S$ *dominates* another solution $y \in S$ when $f_i^{\theta_i}(x) \le f_i^{\theta_i}(y)$ for all $i = 1, \ldots, m$ and $f_j^{\theta_j}(x) < f_j^{\theta_j}(y)$ for at least one $j = 1, \ldots, m$. A solution $x^* \in S$ is *Pareto optimal* if there are no solutions $x \in S$ that dominate x^*. All non-dominated solutions represent the *Pareto set* P of the problem, while its image in the objective space is called the *Pareto front*. Additionally, the *ideal point* z^{ideal} in the objective space $\mathbb{R}^m$ is defined as the point whose coordinates equal the optimal values of $f_i^{\theta}(x)$ for each $i = 1, \ldots, m$ independently. That is, $z^{\text{ideal}} = (\inf_{x \in S} f_1^{\theta_1}(x), \ldots, \inf_{x \in S} f_m^{\theta_m}(x))$. Conversely, the *nadir point* z^{nadir} in the objective space $\mathbb{R}^m$ consists in each objective of the worst value obtained by a Pareto optimal solution. That is, $z^{\text{nadir}} = (\sup_{x \in P} f_1^{\theta_1}(x), \ldots, \sup_{x \in P} f_m^{\theta_m}(x))$.

2.2   The bbob Functions

The bbob function suite contains 24 well-known and understood functions in six dimensions (2, 3, 5, 10, 20, and 40) and 15 instances that change through the years [6]. The functions have known optima and incorporate various difficulties found in real-world problems. They were carefully selected to support a variety of research questions and are categorized into five groups based on their properties:

- separable functions (functions f_1 to f_5),
- functions with low or moderate conditioning (f_6 to f_9),
- highly conditioned and unimodal functions (f_{10} to f_{14}),
- multimodal functions with global structure (f_{15} to f_{19}), and
- multimodal functions with weak global structure (f_{20} to f_{24}).

The bbob function suite is implemented in the Comparing Continuous Optimizers (COCO) platform [9], which supports automated benchmarking using these problems. Instances are used to measure robustness of stochastic and deterministic optimizers alike. So in one benchmarking experiment, an algorithm is run once on a total of $24 \cdot 6 \cdot 15 = 2160$ problems.

Two suites of bi-objective problems formed as combinations of the bbob functions were proposed in [3]. The first one, bbob-biobj, contains all function combinations of ten manually chosen bbob functions (two from each group), resulting in 55 function combinations. The second one, bbob-biobj-ext, extends this selection by adding additional function combinations where both functions are different and come from the same group, resulting in 92 function combinations in total. This extension was proposed to increase the diversity of the problems and therefore uses the pool of all but one of the 24 bbob functions (it leaves out the Weierstrass function f_{16} because it has multiple global optima, making the computation of the nadir point intractable). So, in a bi-objective benchmarking experiment, an algorithm needs to optimize $55 \cdot 6 \cdot 15 = 4950$ problems of the bbob-biobj suite or $92 \cdot 6 \cdot 15 = 8280$ problems of the bbob-biobj-ext suite. Both numbers are high and would increase drastically if the same procedure was used to form suites of problems with more objectives.

When bbob functions are combined to form multi-objective problems, two new issues appear [3]. The first is that we no longer know all optimal solutions. The single-objective optima of the functions reveal only the extreme points of the Pareto front, not the entire Pareto front (nor set). This makes performance assessment more challenging. The second issue (addressed in Sect. 3.4) is that the instances of function combinations differ among themselves more than in the case of single-objective bbob problems, especially when one or both of the functions are multimodal. This defies the purpose of instances as 'minor' modifications that should not significantly affect the performance of robust algorithms [8].

2.3 ELA Features

Exploratory landscape analysis is a method that extracts information from the problem landscape [15]. The information is gathered from the objective values of a sample of problem solutions and represented by so-called *features* (numerical values) that characterize the landscape. Hundreds of ELA features can be computed for continuous single-objective problems, see for example the flacco package [12], while the only collection of ELA features for continuous multi-objective problems is (to the best of our knowledge) the one presented in [13]. Its implementation is available from https://gitlab.com/aliefooghe/landscape-features-mo-icops/.

Although the features from [13] were originally proposed for multi-objective interpolated continuous optimization problems, they can also be applied to 'standard' multi-objective optimization problems and are consequently used in this paper. The features are categorized into four types: *global, multimodality, evolvability*, and *ruggedness* features.

The *global landscape features* extract information about the global properties of the multi-objective landscape. Among the calculated properties we find the correlation between two objectives, the average and maximum distance among sampled solutions in the search and the objective space, the proportion of non-dominated solutions, the hypervolume value, etc.

The *features that characterize the landscape multimodality* measure the properties of the local optima. Two different types of local optima are considered: single-objective local optima and multi-objective local optima. The neighborhood of a solution (the set of the closest samples to this solution) is used to detect the single-objective local optima—if no neighbor of the target solution has a better objective value than it, then the target solution is a local optimum. Similarly, a multi-objective local optimum dominates all its neighboring solutions. The multimodality features measure the percentage of solutions that are local optima, their distance, etc.

The *features that characterize the landscape evolvability* measure the proportion of neighbors of a solution that can outperform it. First, the ELA method records for each solution the proportion of neighbors that dominate, are dominated by, or are mutually non-dominated with the corresponding solution. Then, it applies some statistical measures to determine how these numbers differ in the entire sample of solutions.

The last type of features are the *ruggedness features*, which measure the ruggedness of the problem landscape by analyzing the correlation of the evolvability features among neighboring solutions.

3   Methodology

We wish to follow the benchmarking procedure used by the COCO platform [8], which assumes that the function combinations contained in a benchmarking suite are instantiated in several dimensions and instances. The dimensions are needed to test the scalability of the algorithms, while the instances are used to assess their repeatability. This means that we need to compare the properties of problems for multiple dimensions at the same time, as well as produce instances that are very similar to each other (which requires effort, as by default, the multi-objective problem instances can be quite different).

Our procedure to construct a diverse suite of bi-objective problems therefore consists of the following four steps, described in more detail in the rest of this section:

1. Generate 15 instances of all function combinations in all considered dimensions (our base set of problems),
2. Compute the ELA features for all these problems,

3. Select a small suite with diverse problems, and
4. Generate additional similar instances for the selected problems.

3.1 Generating the Base Set of Problems

To generate our base bi-objective problem set, we start with all possible combinations of the bbob functions [6]. However, similarly as in the bbob-biobj-(ext) test suites [3], we remove the Weierstrass function because it has multiple global optima, complicating the computation of the nadir point. Furthermore, we regard both permutations of two functions to be the same, as they produce the same landscapes. This results in 276 unique function combinations.

Instances of the problems are also generated using the same approach as in the bbob-biobj-(ext) test suites [3]. This means that we only consider a bi-objective instance if the single-objective optima have at least the distance of 10^{-4} in the search space, and if the ideal point and the nadir point have at least the distance of 10^{-1} in the non-normalized objective space. These two conditions need to be satisfied for all considered dimensions (2, 3, 5, 10, and 20). We generate 15 different instances for each problem. Considering 276 unique problems with 15 instances in 5 different dimensions results in in 20 700 problems overall.

3.2 Computing the ELA Features

To generate the ELA features described in Sect. 2.3, a sample of solutions is needed. We create it using the Latin Hypercube Sampling method [14], which is one of the most widely used sampling methods in evolutionary computation. To assure stability (avoid that the differences among problem features originate from different samples rather than the different problem landscape properties), we use the same sample for all problems of the same dimension. The sample size is $200 \cdot n$, where n is the search space dimension. We evaluate the samples for each of the five dimensions considered (2, 3, 5, 10, 20) on all problems.

In the next step, we normalize the objective values for each problem using its ideal and nadir points. The ideal point represents the minimal and the nadir point the maximal objective values for normalization. We thus use the min-max normalization on the objective values for each of the solutions in the sample. In this way, the objective values of all problems are comparable. Note however that since the nadir point is used for the normalization maximum, solutions worse than the nadir will have objective values larger than 1.

We then compute 48 features using the code mentioned in Sect. 2.3 with default parameters. This means that the reference point for hypervolume computation equals $(1, 1)$ and only one neighbor is used in neighborhood-based features. For each of the $276 \cdot 15 = 4140$ problem combinations we concatenate the features of all five dimensions to a single feature vector of size 240.

Based on the computed features, we finally filter out some of the unusable features and/or problem instances. The filtering is triggered by one of the following reasons: Firstly, if the feature values are the same for all problems. This happens because some of the features are computed based on the variance of the samples in the search space. Since we use the same sample for all problems of the same dimension, such features are useless in our case and thus removed. Secondly, if the features values cannot be computed because the sample has a single non-dominated solution. Some of the features require multiple non-dominated solutions to compute. Although we are creating problem instances where the Pareto front is not 'too small', this still happens occasionally. As we consider such problems not interesting from the perspective of multi-objective optimization (they are not desirable in the test suite), we remove them from the collection. Lastly, if the feature values cannot be computed because we use only one neighbor (the default setting). Such features are also removed. In this way, we get the feature vector of size 202 and 4030 problem instances.

3.3   Selecting a Diverse Subset of Problems

The general idea of selecting a diverse subset of problems used in this paper is to do so iteratively, always choosing the problem that is most different in terms of ELA features to the ones already selected. Therefore, we divide the set of all problems into the subsets of *selected* and *unselected* problems. Recall that the 48 ELA features for each dimension are combined into a single, large feature vector so that the results of all dimensions are included in this selection process.

Because of the high dimensionality of the feature vector, we compare the problems using cosine similarity instead of some distance metric. Distance metrics would demonstrate stark differences because slight variations in individual dimensions can accumulate over dimensions. When comparing two vectors, cosine similarity only considers their angle rather than their magnitude, making it immune to this issue. To determine the similarity between two bi-objective problem instances, we compute a similarity matrix where the cosine similarity ranges from 0 (completely unrelated) to 1 (identical problems).

The first problem in this iterative process needs to be selected by hand. In our work, we chose the first instance of the double sphere problem as the starting point. The double sphere problem is one of the easiest bi-objective problems to solve and the only one for which we know the entire Pareto set (the line segment between the two single-objective optima) and front (the image of this line segment in the objective space). We believe all suites should contain a well-understood problem like this, so this is a natural starting point for our procedure.

The remaining problems are chosen from the unselected subset as those that are the most different from the already selected. This is achieved as follows. First, we find for each unselected problem the most similar selected one and record their similarity. Then, we compare these similarities and choose the unselected problem with the lowest recorded similarity. In other terms, we try to maximize the minimum similarity between the already selected problems and the ones to be chosen next. This way, we iteratively add the most dissimilar problem to our selected subset.

This approach can generate an arbitrary large subset of benchmark problems. We need to decide when to stop this procedure, i.e., how many function combinations we wish to have in the suite. This is nontrivial and somewhat preference-based, as it requires choosing a trade-off between the suite diversity and representation abilities on the one hand, and the computational cost and evaluation time on the other. If the desired suite size is known, the proposed greedy procedure could also be replaced by a more sophisticated algorithm.

3.4   Generating Similar Instances

Different instances of the same function combination have varying fitness landscapes (and consequently features). This can be problematic when creating a benchmark suite where the instances are supposed to test algorithm repeatability. To solve this problem, we try to find 14 instances that are very similar to the selected problem instance. This way, we can expect similar algorithm performance on all instances of the selected problems.

We generate the problem instances for the N most diverse problems selected (let us call them *target problems*) in a post-processing fashion. For each target problem, we first generate 140 additional instances, the same as in Sect. 3.1. Then, we compute the ELA features as described in Sect. 3.2 on these new instances. This gives us 155 instances for each problem, categorized by the ELA features. Once again, we use cosine similarity to measure the similarity between the target problem and each additional instance. Finally, we select 14 instances that are most similar to the target problem, yielding 15 similar instances.

4   Results and Discussion

In this section, we present the results of applying our methodology on the pool of the bbob functions. We first show the results of a test run on only the 2-D problems as a proof of concept for our approach. Then we present the actual results on multiple dimensions. Finally, we list the limitations of our approach, proposing ideas to improve it in the future.

4.1   Proof of Concept on 2-D Problems

Given that there is no established way to validate our proposed approach, we try to gain a better understanding of its workings by applying it only to 2-D problems, whose landscapes can be visualized. This means that we perform the entire procedure described in Sect. 3, with the exception of the parts that require stacking together the features from problems of different dimensions.

We first select an arbitrary function combination that is also included in the bbob-biobj test suite and therefore has some known properties (see [3] and its supplementary material website at https://numbbo.github.io/bbob-biobj/ for more information). This is the function combination (f_1, f_8), where f_1 is the sphere function and f_8 is the original Rosenbrock function, instantiated with 15 different instances. We use it to visually verify that the way we compute problem features and measure their similarity works as intended.

Figure 1 presents the problem similarity heatmap for the 15 instances of (f_1, f_8) where the rows and columns are sorted in such a way that the similar

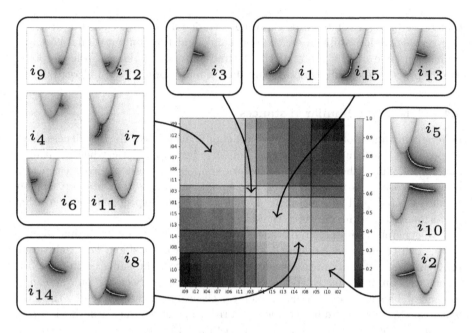

Fig. 1. Problem similarity heatmap for the 15 instances of the 2-D function combination (f_1, f_8) (function F_4 in the `bbob-biobj` suite) and their corresponding landscape visualizations using dominance ranking ratio [3,7] show that the size of the Pareto set (yellow curves) is proportional to the similarity among instance. The instances are numbered as in the `bbob-biobj` suite. (Color figure online)

instances (yellow hues) are placed together. Furthermore, the problem landscape of the different instances is also visualized using dominance rank ratio plots [3,7]. In these plots, the search space is approximated by the grid points and their color denotes the proportion of other grid points that dominate the current one (the more such points, the lighter the color). Additionally, yellow is used to denote the Pareto set approximation (non-dominated grid points).

First, we notice that the problem instances could be grouped into five clusters based on the values of the similarity matrix. Next, we see that the size of the Pareto set is proportional to the similarity among instances. On the one hand we have six mutually similar instances i_9, i_{12}, i_4, i_7, i_6, and i_{11}, with short Pareto sets (top left), and on the other hand, three mutually similar instances i_5, i_{10}, and i_2 with long Pareto sets (bottom right). The instances in the first group are very different from the instances in the last one according to the similarity matrix as well as their landscape visualizations.

The characterization of the `bbob-biob-(ext)` problems in [3], where the first five instances of all 92 problems were visually inspected for five properties (number of Pareto set subsets, number of Pareto front subsets, convex Pareto front, Pareto set outside of $[-5, 5]^2$ and number of basins of attraction),

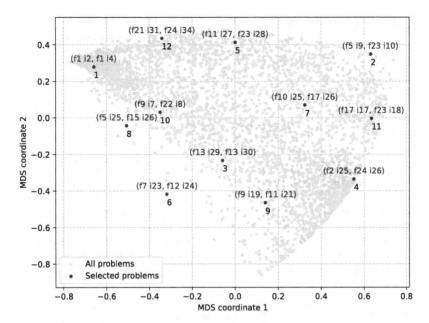

Fig. 2. Multidimensional scaling of the 48-D space of features to a plane for all 2-D problems. The 12 most diverse problems (see Fig. 3) are emphasized and labeled with their consecutive number in black and function instance combination in purple. (Color figure online)

has found no difference between the first five instances of the 2-D function combination (f_1, f_8). In this sense, our approach exhibits higher discriminability powers as it does differentiate between these instances, in fact, they reside in four of the five denoted clusters. We therefore see that the applied ELA features and the cosine similarity metric properly characterize the instances of this function combination, showing promise of this methodology.

If we use the proposed approach to construct a suite of 12 most diverse 2-D function combinations, we get the results shown in Fig. 2. This plot presents a Multidimensional Scaling (MDS) projection of all problems from the 48-D space of ELA features to a plane. MDS chooses a projection so that the distances in the projected space respect the distances well in the original space (we used the default MDS implementation from the `sklearn` Python library to produce this plot). The darker dots denote our selected function instance combinations. We can see that they are relatively uniformly distributed over the entire space, although any such projection needs to be interpreted with caution.

One interesting thing to note is that some `bbob` functions are included in multiple combinations, see for example f_{23}, which was employed three separate times. This is somewhat counter-intuitive, as we could expect the functions to repeat themselves only after (almost) all different available functions have been selected. This might be caused by large differences in some function instances.

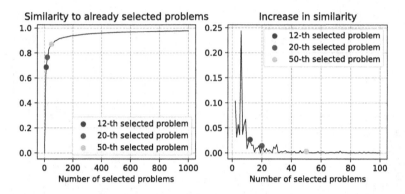

Fig. 3. Similarity to already selected problems when adding new problems to the selection. The plot on the left shows the similarity itself (which has a large increase at first and then flattens around 100 problems), while the plot on the right shows the increase in similarity compared to the previous value. In both plots, the 12th, 20th and 50th problems are emphasized.

4.2   Results on Multiple Dimensions

Deciding on the number of problems to include in a benchmark requires compromising. On the one hand, we want to select many problems which cover a wide range of different characteristics. On the other hand, the benchmark should be of a manageable size to be fit for use in large benchmarking studies. As we can vary the number of selected problems in our approach, it is beneficial to evaluate the impact or usefulness of adding each problem by calculating its similarity to the already selected ones.

Figure 3 shows this in two plots. On the left-hand side plot, we can observe the similarity of a newly selected problem to the already selected ones for the first 1000 problems. The right-hand side plot shows the increase in similarity for each newly selected problem for the first 100 problems. On the left-hand side plot, a logarithmic growth can be observed, increasing rapidly for the first 100 problems selected, then gradually flattening out. On the plot on the right, we can see a rugged decline in the similarity increase for all problems, with a notable peak at the sixth selected problem. Not surprisingly, when only a small number of problems are included in the suite, it is very beneficial to add more. Selecting less than 12 problems, therefore, seems undesirable. We see that the similarity to the already selected problems increases more slowly after the first 12 problems have been selected. However, the data does not provide a clear cut-off point after which selecting more problems becomes less effective.

The number of problems selected cannot be determined clearly as it depends on many factors. In this paper, 12 problems were selected to strike a balance between diversity, representation, and and the ability to execute and show results of benchmarking experiments in the scope of scientific work.

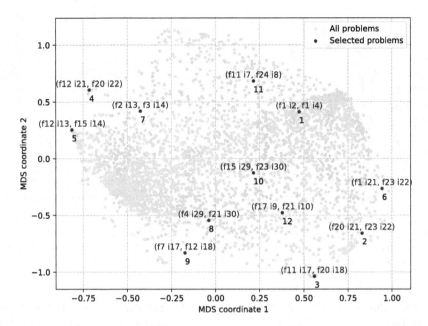

Fig. 4. Multidimensional scaling of the 202-D space of features to a plane for problems with multiple dimensions. The 12 most diverse problems are emphasized and labeled with their consecutive number in black and function instance combination in purple. (Color figure online)

Similar to the results on the 2-D problems in Fig. 2, Fig. 4 shows the multidimensional scaling of the ELA features to a plane for the problems in all used dimensions. We can again notice that the 12 most diverse problems resulting from our approach are rather uniformly distributed over the space, although given that the feature space is 202-D in this case, care must be taken when making assumptions based on such visualizations.

From the point of view of included **bbob** functions, we can see that even more (compared to the previous section) have been selected multiple times. That is, each of f_{11}, f_{12}, f_{20} and f_{23} has been used three times in the top 12 most diverse problems. We are currently unable to explain why some repeated functions are preferred over others that have not yet been included.

Finally, Fig. 5 visualizes with violin plots the distribution of the cosine similarity values between the target problem (one of the 12 problem instances selected to be included in the suite) and all 155 instances of the same problem (in blue) as well as between the target problem and the closest 15 instances of the same problem (in orange). This is done for all 12 target problems. Note that the 155 problem instances are the result of the procedure described in Sect. 3.4.

We can observe that the similarity of the closest instances differs from one function combination to the next. On some combinations, e.g., (f_{12}, f_{20}) and (f_{12}, f_{15}), they are very close, while for others (notably the double sphere function (f_1, f_1), where one of the closest instances is rather far away from the target

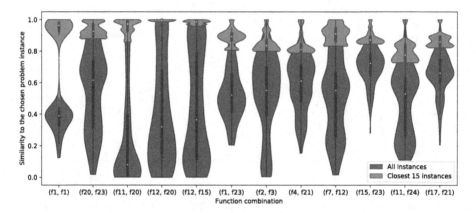

Fig. 5. Distribution of cosine similarity values between each target problem and all 155 corresponding problem instances (in blue), and the target problem and its closest 15 instances (in orange). (Color figure online)

problem) this is not the case. This means that further instances would need to be generated on such problems to achieve better similarity of the top 15 instances.

Additional insights into the fitness landscapes can be gained by analyzing the distributions in the blue violin plots. For example, in the function combination (f_1, f_1) we see two groups forming, one close to the target problem in the similarity range of $[0.8, 1.0]$ and one more distant with a median similarity at around 0.4. This indicates that different instances of this function combination might result in two different problem groups in terms of the ELA features. Interestingly, (f_1, f_{23}), which likewise includes the sphere function f_1, also seems to produce two clusters of problems. On the other hand, some function combinations seem to produce very diverse instances that cover the entire range of similarity to the target problem, like (f_{12}, f_{20}) and (f_{12}, f_{15}). Other combinations, like (f_{15}, f_{23}) and (f_{17}, f_{21}) show only one large group with some outlying instances. A more elaborate evaluation is needed to understand why this happens.

This analysis has shown that a fixed number of function instances is, in general, not enough to achieve close instances. An iterated approach where instances are generated until a closeness threshold has been met might work better (although it could also take a long time).

4.3 Limitations

Our work presents a first attempt at creating a new benchmarking suite by using ELA features to ensure its diversity. While we were able to show that the idea works quite well (especially on the 2-D problems), we acknowledge some limitations of this initial study that need to be addressed in the future.

Our procedure *assumes that all problems are instantiated and characterized* before the selection takes place. Currently, calculating the ELA features is the bottleneck of the proposed methodology. If many more function combinations would have to be explored (for example, in the case of many-objective problems), this could become intractable and an alternative approach that does not require characterizing all function combinations would need to be implemented (perhaps by filtering unpromising function combinations based on the properties of single functions or combinations of two functions).

No quantifiable goal and/or evaluation of the proposed approach exists. Several concepts that we use throughout the study, such as problem diversity, instance closeness, and suite size, are hard to quantify in terms of thresholds that denote a satisfactory value. This means that it is also hard to judge whether our procedure was able to meet its goals. One particular issue that we need to explore is whether the methodology optimizes for outliers. This could result in a suite that is not representative of real-world conditions and therefore go against our goal. One way to evaluate the usefulness of a benchmark suite is to show that it differentiates between algorithms. We will address this in future work.

Sensitivity to Parameters. While we have not tested this extensively yet, our experiments suggest that the resulting problem selection strongly depends on various parameters of our approach, such as the selection of the applied ELA features, the choice of dimensions to be included in the construction, the number and variety of instances, the initial selected problem, etc. This sensitivity would first need to be studied more comprehensively and then decreased where possible (although, of course, the reliance on some parameters cannot be eliminated).

Questionable Scalability in Search Space Dimension. With increasing search space dimension, the sample size used for computing the ELA features increases only linearly, which seems not to be enough for properly categorizing high-dimensional problems. We need to look into this issue more deeply.

Questionable Scalability in Objective Space Dimension. So far, the approach has only been tested on bi-objective problems. In order for the resulting benchmark to be truly representative of the real world, where many problems have more than two objectives [2], the approach should be tested also using three, five and more objectives. This is closely tied with the first limitation.

Using Default ELA Features. While using default ELA features was the 'safe choice' for this set of initial experiments, we have realized that it might have been flawed. For example, a lot of the used features might not contain useful information for predicting algorithm performance, and thus, applying feature selection methods would remove the unnecessary noise in the feature space. Furthermore, we currently used the default setting for calculating the ELA features. However, now the hypervolume reference point coincides with the nadir point, meaning that all non-dominated solutions that are worse than the nadir do not contribute to the hypervolume. We should therefore be looking at alternative hypervolume measures, such as the indicator I_{HV+} from [8], which resolves this issue. Moreover, using only one neighbor might be questionable, we need to explore whether having two or three neighbors works better.

5   Conclusions

Designing multi-objective benchmark problems as combinations of individual single-objective functions closely follows the construction of real-world problems and should therefore be preferred to the bottom-up approach. However, one cannot simply employ all possible function combinations, as they are too many to be usable in practice and a smart way to choose a reasonably sized and diverse collection of function combinations is needed.

In this work, we proposed to use problem landscape characteristics (computed as ELA features) to create a benchmark suite of diverse and representative multi-objective optimization problems of the chosen size. The main idea is to construct the suite by adding problems whose cosine similarity in the ELA feature space to the already selected problems is minimal. The approach was tried on bi-objective combinations of the bbob functions, which are scalable, tunable, and contain difficulties found in real-world problems.

We first used a simplified procedure formulation to prove on 2-D problems that our concept is promising, and then showed results of the actual procedure that uses all considered dimensions. Finally, we listed the limitations of this initial approach that need to be addressed in future work.

Acknowledgements. Special thanks to the Species Society and the organizers of the Species Summer School 2022 for connecting us and making this research possible.

We acknowledge financial support from the Slovenian Research Agency (research project "Constrained multi-objective Optimization Based on Problem Landscape Analysis", young researcher program and research core funding no. P2-0209). This work is also part of the Research Initiative "SmartProSys: Intelligent Process Systems for the Sustainable Production of Chemicals" funded by the Ministry for Science, Energy, Climate Protection and the Environment of the State of Saxony-Anhalt.

References

1. Bartz-Beielstein, T., et al.: Benchmarking in optimization: best practice and open issues. CoRR abs/2007.03488 (2020). https://arxiv.org/abs/2007.03488
2. van der Blom, K., et al.: Towards realistic optimization benchmarks: a question-naire on the properties of real-world problems. In: GECCO 2020: Genetic and Evolutionary Computation Conference, Companion Volume, pp. 293–294. ACM (2020). https://doi.org/10.1145/3377929.3389974
3. Brockhoff, D., Auger, A., Hansen, N., Tušar, T.: Using well-understood single-objective functions in multiobjective black-box optimization test suites. Evol. Comput. **30**(2), 165–193 (2022). https://doi.org/10.1162/evco_a_00298
4. Brockhoff, D., Tušar, T.: GECCO 2022 tutorial on benchmarking multiobjective optimizers 2.0. In: GECCO 2022: Genetic and Evolutionary Computation Conference, Companion Volume, pp. 1269–1309. ACM (2022). https://doi.org/10.1145/3520304.3533635
5. Deb, K., Thiele, L., Laumanns, M., Zitzler, E.: Scalable multi-objective optimization test problems. In: Proceedings of the 2002 Congress on Evolutionary Computation, CEC 2002, pp. 825–830. IEEE (2002). https://doi.org/10.1109/CEC.2002.1007032

6. Finck, S., Hansen, N., Ros, R., Auger, A.: Real-parameter black-box optimization benchmarking 2010: presentation of the noiseless functions. Tech. Rep. 2009/20, Research Center PPE (2009). http://coco.gforge.inria.fr/downloads/download16.00/bbobdocfunctions.pdf
7. Fonseca, C.M.: Multiobjective genetic algorithms with application to control engineering problems, Ph. D. thesis, University of Sheffield (1995)
8. Hansen, N., Auger, A., Brockhoff, D., Tušar, T.: Anytime performance assessment in blackbox optimization benchmarking. IEEE Trans. Evol. Comput. **26**(6), 1293–1305 (2022). https://doi.org/10.1109/TEVC.2022.3210897
9. Hansen, N., Auger, A., Ros, R., Mersmann, O., Tušar, T., Brockhoff, D.: COCO: a platform for comparing continuous optimizers in a black-box setting. Optim. Meth. Softw. **36**(1), 114–144 (2021). https://doi.org/10.1080/10556788.2020.1808977
10. Huband, S., Barone, L., While, L., Hingston, P.: A scalable multi-objective test problem toolkit. In: Coello Coello, C.A., Hernández Aguirre, A., Zitzler, E. (eds.) EMO 2005. LNCS, vol. 3410, pp. 280–295. Springer, Heidelberg (2005). https://doi.org/10.1007/978-3-540-31880-4_20
11. Ishibuchi, H., Setoguchi, Y., Masuda, H., Nojima, Y.: Performance of decomposition-based many-objective algorithms strongly depends on Pareto front shapes. IEEE Trans. Evol. Comput. **21**(2), 169–190 (2017). https://doi.org/10.1109/TEVC.2016.2587749
12. Kerschke, P., Trautmann, H.: Comprehensive feature-based landscape analysis of continuous and constrained optimization problems using the R-Package Flacco. In: Bauer, N., Ickstadt, K., Lübke, K., Szepannek, G., Trautmann, H., Vichi, M. (eds.) Applications in Statistical Computing. SCDAKO, pp. 93–123. Springer, Cham (2019). https://doi.org/10.1007/978-3-030-25147-5_7
13. Liefooghe, A., Vérel, S., Lacroix, B., Zavoianu, A., McCall, J.A.W.: Landscape features and automated algorithm selection for multi-objective interpolated continuous optimisation problems. In: Genetic and Evolutionary Computation Conference, GECCO 2021, pp. 421–429. ACM (2021). https://doi.org/10.1145/3449639.3459353
14. McKay, M.D., Beckman, R.J., Conover, W.J.: A comparison of three methods for selecting values of input variables in the analysis of output from a computer code. Technometrics **42**(1), 55–61 (2000)
15. Mersmann, O., Bischl, B., Trautmann, H., Preuss, M., Weihs, C., Rudolph, G.: Exploratory landscape analysis. In: 13th Annual Genetic and Evolutionary Computation Conference, GECCO 2011, pp. 829–836. ACM (2011). https://doi.org/10.1145/2001576.2001690
16. Zitzler, E., Deb, K., Thiele, L.: Comparison of multiobjective evolutionary algorithms: Empirical results. Evol. Comput. **8**(2), 173–195 (2000). https://doi.org/10.1162/106365600568202

Computational Intelligence
for Sustainability

Using Genetic Programming to Learn Behavioral Models of Lithium Batteries

G. Di Capua[1] ⓘ, C. Bourelly[1] ⓘ, C. De Stefano[1] ⓘ, F. Fontanella[1(✉)] ⓘ,
F. Milano[1] ⓘ, M. Molinara[1] ⓘ, N. Oliva[2] ⓘ, and F. Porpora[1,3] ⓘ

[1] DIEI, University of Cassino and Southern Lazio, Cassino, FR, Italy
{giulia.dicapua,carmine.bourelly,destefano,fontanella,
filippo.milano,m.molinara,francesco.porpora}@unicas.it
[2] DIEM, University of Salerno, Fisciano, SA, Italy
noliva@unisa.it
[3] E-Lectra s.r.l., Cassino, FR, Italy

Abstract. Li-ion batteries play a key role in the sustainable development scenario, since they can allow a better management of renewable energy resources. The performances of Li-ion batteries are influenced by several factors. For this reason, accurate and reliable models of these batteries are needed, not only in the design phase, but also in real operating conditions. In this paper, we present a novel approach based on Genetic Programming (GP) for the voltage prediction of a Lithium Titanate Oxide battery. The proposed approach uses a multi-objective optimization strategy. The evolved models take in input the State-of-Charge (SoC) and provide as output the Charge/discharge rate (C-rate), which is used to evaluate the impact of the charge or discharge speed on the voltage. The experimental results showed that our approach is able to generate optimal candidate analytical models, where the choice of the preferred one is made by evaluating suitable metrics and imposing a sound trade-off between simplicity and accuracy. These results also proved that our GP-based behavioral modeling is more reliable and flexible than those based on a standard machine learning approach, like a neural network.

Keywords: Lithium Batteries · Behavioral Modeling · Genetic Programming · Multi-Objective Optimization

1 Introduction

Over the last decades, Li-ion batteries have become a key energy storage technology in many fields, including e-mobility applications [3]. Lithium based batteries are characterized by a high investment cost in comparison to other energy storage technologies (e.g., lead acid or nickel metal hydride batteries), due to their high energy density and overall performances. However, their performance is highly influenced by temperature, load current, and State-of-Charge (SoC) [27]. Therefore, accurate and reliable models of these batteries are needed, not only in their design phase, but also in their operating conditions. For instance, being able to accurately predict the behavior of the battery

J. Correia et al. (Eds.): EvoApplications 2023, LNCS 13989, pp. 461–474, 2023.
https://doi.org/10.1007/978-3-031-30229-9_30

voltage versus SoC, C-rate and temperature can allow the implementation of efficient battery management systems.

Lithium battery models, like for other power components, fall in two main categories: physical and/or electrochemical models [16, 18], and Equivalent Circuit Models (ECMs) [4, 13]. ECMs are usually preferred and are widely used for their simplicity, since they retain the main electrical characteristics of a battery using electrical components. Electrochemical Impedance Spectroscopy (EIS) is often used to identify the ECM elements and relevant parameters [1]. Typically, the resulting model parameters are valid for a specific battery State-of-Health (SoH). However, during the battery lifetime, the SoH degrades due to a variety of complex aging mechanisms, deriving from electrical, thermal, and mechanical stress. Accordingly, the ECMs need to be recalibrated during battery operation. Since EIS measurements can be time consuming, the recalibration is sometimes not a real affordable solution, especially if the ECM includes many elements. More in general, the ECM is just an instance of the class of the so-called Behavioral Models (BMs), which are analytical functions expressing the global dependence of a system performance on its macroscopic characteristics and operational conditions. The analytical functions of a BM may have a not predetermined mathematical structure, this last being an output of the BM generation method. An ECM is a BM based on a predetermined and fixed mathematical structure associated to the selected equivalent circuit.

Machine Learning (ML) based approaches exhibit enhanced capabilities in identifying correlations that can be used to generate BM of batteries [21]. ML can ensure good trade-off performances in terms of accuracy and applicability. As an inherent disadvantage and limitation, ML-based approaches may require a large amount of training and validation data, and more importantly, they hide the analytical correlation among parameters and performances of batteries. Some limitations of ML can be overcome by Genetic Programming (GP) algorithms [17], which have been recently proposed in literature for BMs generation in power magnetic components [26] and wireless power systems [8] modeling problems. GP has also been used for modeling battery systems. Following this trend, in the last years the effectiveness of GP in finding accurate and reliable models of complex systems has been also used for estimating the SoH of Li-ion batteries [28, 29], or the probability of polluting events caused by Li-ion batteries like cobalt leaching [9]. However, to the best of our knowledge, GP has been never used before to evolve a behavioral model for Lithium Batteries in operating conditions.

Starting from these premises, in this paper we present a novel GP-based technique for modeling Lithium Titanate Oxide (LTO) battery terminal voltage, starting from simulated data. The proposed approach allowed us to discover simplified yet accurate battery performance functions, overcoming the drawbacks of ECM as no a-priori modeling assumptions are required. In order to demonstrate the advantages of the proposed approach, we compared its results with those achieved by using a Neural Network (NN). The comparison results confirmed the effectiveness of the proposed approach.

The paper is organized as follows. Section 2 details an existing ECM for Li-ion batteries, whereas Sect. 3 introduces the GP-based proposed approach and the architecture of the NN used for the comparison. Experimental and comparison results are discussed in Sect. 4, whereas Sect. 5 is devoted to some conclusion remarks.

2   ECMs of a Lithium Titanate Oxide Battery

The adoption of ECM-based strategies to characterize the behavior of Lithium batteries is discussed in several studies [20, 24]. The reliability of the ECMs and the need of conducting appropriate experimental tests to examine the behavior of batteries under various operating situations are both emphasized in [24]. A third or second order ECM is the most popular solution in literature. In the following, the second order ECM of a 13 Ah LTO battery cell is discussed in detail, and implemented in COMSOL Multiphysics® for the generation of a new dataset.

2.1   Adopted ECM for Lithium Titanate Oxide Battery

Figure 1 shows the model adopted in [20]. An Open Circuit Voltage (OCV), an internal resistance (R_S), and two RC branches (R_1-C_1 and R_2-C_2) realize the ECM. When the current is zero ($I_B = 0$), the OCV is equal to the battery terminal voltage. When a charging or discharging current is provided ($I_B \neq 0$), the battery terminal voltage instantly rises or falls, depending on the R_S. The branches R_1-C_1 and R_2-C_2 take into account the dynamics of the battery. Specifically, the R_1-C_1 branch reflects the processes of charge transfer, whereas the R_2-C_2 branch includes the phenomena of diffusion.

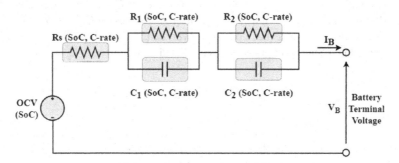

Fig. 1. Second order ECM for lithium adopted in [20].

The amount of charge of a battery in relation to the nominal capacity can be quantified by the SoC, while the C-rate represents the speed used to charge or discharge a battery. The functional relationships of the model parameters are depicted in Fig. 1: the OCV is only SoC-dependent, while all the other parameters are dependent on both SoC and C-rate. When the values of the model parameters are known, the battery voltage V_B is easily obtained from the current I_B applied to the battery. The authors of [20] carried out many experimental tests on a 13 Ah LTO battery to obtain the relevant ECM parameters. Four distinct C-rates (0.25C, 0.5C, 1C, 2C, and 4C) were tested via pulsed conditions, with the SoC value tracked over a range of 5% to 95% with a 5% step.

Figure 2 provides a summary of the resulting ECM parameters. The model was experimentally verified in [20]: given the current, the simulated battery voltage fully matches the measured one. It is crucial to remember that the model validity is only guaranteed within the adopted identification ranges, i.e., C-rate [0.25C, 4C] and SoC [5%, 95%].

2.2 ECM-Based Simulator

The use of a wider dataset than the one provided in [20] can be of help to design a new battery behavioral model. Accordingly, the model described in Section II-A has been implemented in COMSOL Multiphysics®. We used the "Battery Equivalent Circuit" physics [2] to develop the model shown in Fig. 1. Look-up tables have been imported with the values of the circuit parameters, as functions of the SoC and C-rate.

Additional values of SoC and C-rate have been adopted, all given within the initial ranges assumed in [20]. By calculating the battery terminal voltage value with respect to the SoC for each C-rate value, discharge tests have been simulated. It is important to highlight that, though simulations were used to create a new dataset for ML-based algorithms, experimental data in [20] were utilized to determine the starting ECM model. Hence, the new dataset is still based on experimentally-observed battery behavior.

2.3 Dataset

The 13 Ah LTO battery described above has been adopted for the creation of the dataset. The variation of the battery terminal voltage is non-linear with respect to SoC and C-rate. Consequently, the model must provide a sufficiently high resolution over the range of the operating conditions. A total number of $m = 10$ values of C-rate have been considered for COMSOL Multiphysics® runs, yielding different couples (SoC, V_B). A total number of $n = 37$ SoC-V_B values has been obtained over the [5%, 95%] SoC range, with 2.5% step. All simulated SoC and C-rate values are listed in Table 1.

Table 1. LTO battery operating conditions.

Parameters	Values
C-rate	{0.25C, 0.50C, 0.75C, 1.0C, 1.5C, 2.0C, 2.5C, 3.0C, 3.5C, 4.0C}
SoC (%)	{95.0, 92.5, 90.0, 87.5, ...,12.5, 10.0, 7.5, 5.0}

The combinations of all these values result in a dataset of overall $n \times m = 370$ test conditions. From this dataset, we randomly selected 40 samples to create a test set T common to all GP runs. The remaining 330 samples were randomly divided for each run in 270 samples for the training set $\mathbf{T_r}$ and 60 samples for the validation set $\mathbf{V}$.

3 Behavioral Modeling of LTO Battery

3.1 GP-Based Modeling

As mentioned in the Introduction, the goal of our GP algorithm is to express the battery terminal voltage V_B as a function of SoC and C-rate, according to (1):

$$V_{B,gp} = f[\text{SoC}, \mathbf{u}(C - \text{rate})] \tag{1}$$

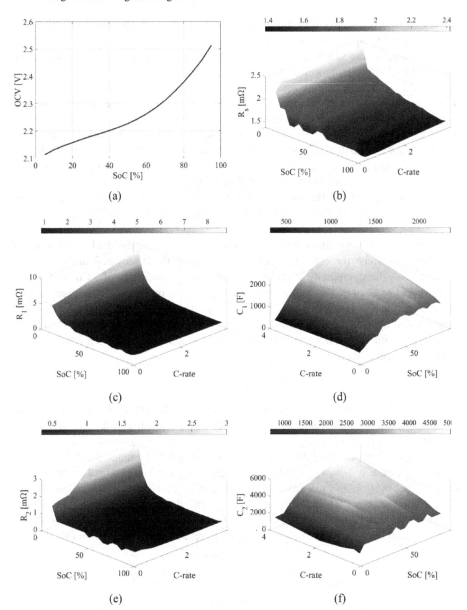

Fig. 2. ECM parameter identification results for the 13 Ah LTO battery used in [20].

where f is a mathematical function and $\mathbf{u}$ is a vector of coefficients, each one expressed as a function of C-rate, and whose number can change for different functions f. According to the canonic expression (1), the SoC has a major effect on V_B, so that it appears as an explicit variable in f. Conversely, the C-rate has a minor effect on V_B, so that it simply influences the coefficients $\mathbf{u}$ of the function f. In other words, SoC is considered as a *primary variable* and C-rate as a *secondary variable*.

According to the effectiveness of genetic operations extensively discussed in the literature [19], our GP algorithm was configured as follows:

- the best 0.5% of the population is subject to elitism and is copied directly into the next generation;
- due to the efficiency and ease of implementation, we use the k-Selection Tournament [25], with $k = 4$, so that 4 individuals are randomly selected from the entire population, and they compete against each other: the individual with the highest fitness wins and is chosen as one of the two next generation parents;
- the remaining part of the population not subject to elitism is generated through crossover and mutation operations, with a percentage of 70% and 30%, respectively.

To evaluate the quality of the model represented by an individual we adopted as objective functions: the error between the outputs of the discovered model and the training data, the complexity of the model and the monotonicity of its $\mathbf{u}$ coefficients with respect to C-rate. This results in a multi-objective optimization problem, whose fitness function is a vector of three elements. Since the GP algorithm works on a single fitness value to perform selection, additional processing is required to transform the fitness vector into a scalar value [5]. The simplest scalar fitness FIT is a weighted sum of three elements [30]:

$$FIT = \alpha_1 F_{rms} + \alpha_2 F_{cmp} + \alpha_3 F_{mnt} \tag{2}$$

where F_{rms}, F_{cmp} and F_{mnt} are the objective functions for the RMS error, the model complexity and the monotonicity of the coefficients, respectively, and α_1, α_2, α_3 are their relative weights (being $\alpha_1 + \alpha_2 + \alpha_3 = 1$). The objective functions F_{rms}, F_{cmp} and F_{mnt} are described in more detail below, and normalized in the range $[0,1]$.

Each element u_i ($i = 1,.., k$) of the coefficients vector $\mathbf{u}$ is obtained by processing the m values u_{ij} ($j = 1,.., m$), calculated by means of the Levenberg-Marquardt NLLS optimization method [22], applied to the m-size C-rate data vector. Given the coefficients vector $\mathbf{u}$, the normalized accuracy objective function F_{rms} of each model is estimated over the training dataset, by calculating the RMS relative error as given in (3):

$$F_{rms} = \sqrt{\frac{1}{m}\frac{1}{n}\sum_{j=1}^{m}\sum_{i=1}^{n}\left(\frac{V_{B,bhv,ij}-V_{B,ij}}{V_{B,ij}}\right)^2} \tag{3}$$

There are many ways to classify the complexity of a model. The most common approach is to simply consider the depth and number of nodes that make up the tree. In this paper, the elements of the external nodes and internal nodes have been assigned different complexity factors. The global complexity of each constructed GP model has been estimated accordingly.

Table 2 summarizes the complexity factors used in our GP algorithm, depending on the type of node, external or internal. Then, the complexity c_f of a model is calculated as:

- each external node (input or coefficient) implies a 0.8 additive contribution to the overall complexity;
- each internal node implies an additive contribution to complexity, depending on the type of algebraic operator or basic function it implements;
- each internal node also implies a further additive contribution equal to the product of the complexity of the function and its argument, with different complexities for the function if the argument is an external or an internal node.

Table 2. External and internal node complexities.

Function	Arity	Description	External nodes complexity	Inner node complexity
sum	2	$h + g$	0.6	1
subtraction	2	$h - g$	0.6	1
multiplication	2	$h \cdot g$	0.75	1.1
division	2	h/g	0.85	1.2
reciprocal	1	$1/h$	0.85	1.2
square root	1	$\sqrt{h}$	1	1.5
square	1	h^2	1	1.5
natural log	1	$\ln(h)$	1	1.5
natural exp	1	e^h	1	1.5
sine	1	$\sin(h)$	1	1.5
cosine	1	$\cos(h)$	1	1.5
tangent	1	$\tan(h)$	1	1.5
inverse sine	1	$\sin^{-1}(h)$	1	1.5
inverse tangent	1	$tan^{-1}(h)$	1	1.5

This choice limits bloat phenomenon [7] and models resulting in involved functions of functions or with many operations on simple functions. The function F_{cmp} refers exclusively to the complexity of the model function f and is associated to its dependence on the primary variable SoC. The normalized complexity objective function F_{cmp} is given by (4):

$$F_{cmp} = \frac{c_f - c_{f,min}}{c_{f,max} - c_{f,min}} \quad (4)$$

The objective function F_{mnt} expresses a qualitative characteristic of the **u** coefficients, namely their monotonicity with respect to C-rate values, and is calculated as in (5)(6), according to the method described in [6]:

$$F_{mnt} = \frac{1}{k} \sum_{i=1}^{k} 2\min\left\{X^{(-)}, X^{(+)}\right\} \quad (5)$$

where:

$$X^{(+)} = \sum_{j=1}^{m-1} \left(C_{rate_{j+1}} - C_{rate_j}\right) \frac{\dot{u}_{i,j+1}^{(+)} + \dot{u}_{i,j}^{(+)}}{2} \tag{6.a}$$

$$X^{(-)} = \sum_{j=1}^{m-1} \left(C_{rate_{j+1}} - C_{rate_j}\right) \frac{\dot{u}_{i,j+1}^{(-)} + \dot{u}_{i,j}^{(-)}}{2} \tag{6.b}$$

$$\dot{u}_{i,j}^{(+)} = \max\left(\frac{u_{i,j} - u_{i,j-1}}{C_{rate_j} - C_{rate_{j-1}}}, 0\right) \tag{6.c}$$

$$\dot{u}_{i,j}^{(-)} = \max\left(\frac{u_{i,j-1} - u_{i,j}}{C_{rate_j} - C_{rate_{j-1}}}, 0\right) \tag{6.d}$$

The GP algorithm was executed over 100 runs, with a population of 80 individuals, evolving over 30 generations for each run. The fitness weights α_1, α_2 and α_3 in (2) were chosen with prevalence to the RMS error objective functions F_{rms}, by setting $\alpha_1 = 0.8$, $\alpha_2 = 0.1$ and $\alpha_3 = 0.1$. This choice was dictated by the preliminary analysis of data, evidencing an inherent smooth behavior of the battery voltage with respect to SoC and C-rate, which fairly yields simple functions f and monotonic u coefficients. At the end of each run, the best GP-based model was selected based on the minimum FIT value.

Figure 3 shows the minimum, maximum, and mean fitness values as function of the generation number, averaged over the 100 runs carried out. From the figure we can observe that there is a slow and steady decrease of the mean fitness. After an initial constant value, this trend is followed also by the minimum (i.e. the best) fitness. This behavior confirms the exploration ability of our GP algorithm, excluding a premature convergence of the best individual toward a local optimum. Even more importantly, from Fig. 3 we can see that there is a wide distance between the mean and the best fitness, until the last generations of evolution. This wide distance confirms that, on average, the individuals in the population are quite different from the best one, therefore we can state that there were no genetic drift phenomena. Overall, we can conclude that our GP-based modeling approach has a good exploration ability in the solution space made of the tree-based models used to express the battery terminal voltage of a LTO battery.

The models from #1 to #4, showing a maximum training percent error lower than 1%, were considered as candidates of interest. Their expressions are listed in Table 3. The corresponding mean (μ_{err}), standard deviation (σ_{err}), and maximum (err_{max}) values of the percent error over the training dataset T_r and the test dataset T are listed in Table 4. All models are expressed as basic analytic functions, with a maximum error of 2.7% on the training dataset and of 4.0% on the validation dataset. Among all these models, some can be preferred, since they are linear combinations of elementary SoC functions, which allow us to distinguish the contribution of each function given its own C-rate coefficients. Accordingly, the model #2 is herein selected. The trend of coefficients u_i (i = 1, 2, 3, 4) for model #2 can be expressed as in (7):

$$u_1 = a_{11} + a_{12}C_{rate} + a_{13}C_{rate}^2 + a_{14}C_{rate}^3 \tag{7.a}$$

$$u_2 = a_{21} + a_{22}C_{rate} + a_{23}C_{rate}^2 + a_{24}C_{rate}^3 \tag{7.b}$$

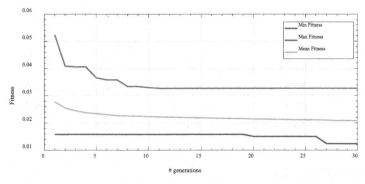

Fig. 3. Evolution of the minimum (best), maximum (worst), and mean fitness values.

$$u_3 = a_{31} \exp(a_{32}C_{rate}) + a_{33} \exp(a_{34}C_{rate}) \tag{7.c}$$

$$u_4 = a_{41} \exp(a_{42}C_{rate}) + a_{43} \exp(a_{44}C_{rate}) \tag{7.d}$$

where $a_{i,1}$, $a_{i,2}$, $a_{i,3}$ and $a_{i,4}$, $i = 1, \ldots, 4$, are listed in Table 5. The final metrics for the testing dataset with coefficients (7) adopted in model #2 are: $\mu_{err} = 0.41$, $\sigma_{err} = 0.47$, and $err_{max} = 2.38$.

Table 3. Expressions of selected models obtained from GP algorithm.

Model	Expression
#1	$u_1 + u_2\left\{1/\,SoC\, -u_3\exp\left[\sin^{-1}(SoC)\right]\right\}$
#2	$u_1 + u_2\{u_4[u_3\tan(SoC)\,-SoC] + \ln(SoC)\}$
#3	$u_1 +u_2 SoC\left[u_3 SoC +\ln(SoC) +u_4\right]$
#4	$u_1 +u_2\left(u_3+\sqrt{SoC}\right)\exp(SoC)$

Table 4. Metrics values for selected GP models.

Model	Training dataset			Testing dataset		
	μ_{err}	σ_{err}	err_{max}	μ_{err}	σ_{err}	err_{max}
#1	0.3	0.3	2.2	0.3	0.3	1.4
#2	0.3	0.3	2.2	0.4	0.6	3.1
#3	0.4	0.4	2.7	0.5	0.8	4.0
#4	0.6	0.5	2.1	0.7	0.6	3.1

Table 5. Coefficient values for the behavioral model #2.

Coefficients	$a_{i,1}$	$a_{i,2}$	$a_{i,3}$	$a_{i,4}$
u_1	2.182E0	−3.971E−2	1.079E−1	−1.546E−2
u_2	2.833E−2	7.373E−3	5.204E−2	−7.200E−3
u_3	1.200E0	−2.958E−1	1.926E−2	5.883E−1
u_4	1.562E + 1	−1.681E0	6.555E0	−8.118E−2

3.2 NN-Based Modeling

To test the effectiveness of the proposed GP-based approach, we compared its results with those achieved by a Multi-Layer Perceptron (MLP) [12], which is a fully connected class of feedforward NN, widely-used both for classification and regression problems. An MLP can be used to approximate any general function, as is well established in many papers [14, 15], where it has demonstrated that a single hidden layer network can approximate any mathematical function with arbitrary precision [23]. Here, we used an MLP to express the battery terminal voltage V_B as a function of SoC and C-rate, according to (8):

$$V_{B,nn} = f[\text{SoC}, \mathbf{u}(C - \text{rate})] \tag{8}$$

where f is an unknown mathematical function.

We used the Weka MLP implementation [11], with 2 input, 1 output and 16 neurons in the hidden layer. After a set of preliminary experiments, we set the values of the hyper-parameters as shown in Table 6. To the point, the global relation between input and output can be summarized as in (9):

$$V_{B,nn} = \sum_{i=1}^{16} w_{Oi}\sigma\left(b_i + w_{i,\text{SoC}}\,\text{SoC} + w_{i,\text{Crate}}\,C_{\text{rate}}\right) + b_O \tag{9}$$

where σ is the sigmoid activation function.

Table 6. Hyper-parameters values of the NN.

Hyper-parameter	Value
Learning rate	0.3
Momentum	0.2
Batch size	10
Number of epochs	5000
Activation function	sigmoid

4 Results and Discussion

Figure 4 shows the comparison between the predicted battery terminal voltage VB values on the 40 test samples for the GP-based model ($V_{B,gp}$, filled red circles) and for the NN-based model ($V_{B,nn}$, filled blue circles). All samples are sorted according to increasing values of C-rate (from 0.25C to 4.0C). Table 7 summarizes the maximum (err_{max}), the mean (μ_{err}) and the standard deviation (σ_{err}) values of the two models percent error distribution, over the test set defined in Sect. 2.

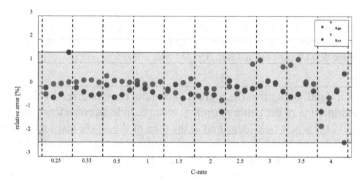

Fig. 4. 13 Ah LTO battery: relative percent errors of GP-based model (red) versus NN-based model (blue) for C-rate from 0.25C to 4.0C. (Color figure online)

Table 7. Comparison on the models percent error distribution.

Metrics	$V_{B,gp}$	$V_{B,nn}$
$|err_{max}|$ [%]	2.38	1.46
μ_{err} [%]	0.41	0.31
σ_{err} [%]	0.47	0.30

From Fig. 4, we can observe that the NN-based model provides, on average, a good estimation of the battery terminal voltage, with a maximum percent error lower than 1.46%. The GP-based model also provides on average a quite reliable battery terminal voltage estimation, with maximum percent error slightly higher (2.38%). In particular, the behavioral model #2 provides good prediction, with quite good accuracy, especially for low C-rate conditions. In fact, by limiting the test dataset from 0.25C to 3.0C, the GP-based model provides $err_{max} = 1.13$, $\mu_{err} = 0.29$ and $\sigma_{err} = 0.25$, while the NN-based model realizes $err_{max} = 1.46$, $\mu_{err} = 0.28$ and $\sigma_{err} = 0.29$.

Finally, Fig. 5 shows the battery terminal voltage obtained by COMSOL simulations (black dotted lines), for different C-rate values, and compares the GP results based on model #2 with the coefficient trend u_i (red dots) and the NN results (blue dots), both achieved on the same 40 test samples mentioned above.

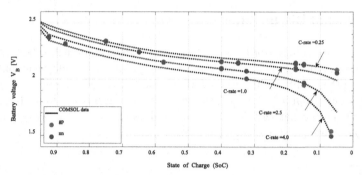

Fig. 5. Battery terminal voltage obtained by COMSOL simulations (black dotted lines), for different C-rate values, GP results based on model #2 with coefficient approximation (red dots), and NN results (blue dots) for the test dataset. (Color figure online)

It is worth highlighting that the highest maximum error realized by the GP is still a quite good estimation of the battery terminal voltage. Indeed, considering the true value $V_B = 1.502$ V, the worst case identified in the data of the whole test set corresponds to $V_{B,gp} = 1.538$ V and $V_{B,nn} = 1.494$ V, thus resulting in maximum absolute errors of 36 mV and 8 mV for the GP-based and NN-based models, respectively.

The huge advantage ensured by our GP-based approach is its ability to express the battery terminal voltage V_B as a simple and accurate function of the SoC and C-rate input variables. This advantage is twofold. From one hand, the GP can find functions (models) which are a linear combinations of simpler terms (as in case of model #2 in Table 3), and can possibly be inverted to estimate the SoC or the C-rate values of the battery, once the battery output voltage VB is known. On the other hand, these models can be easily implemented in commercial simulators and battery management systems. The proposed model is flexible and potentially easy to apply to different families of Li-ion batteries, only requiring a standard curve fitting technique to identify the formula coefficients, based on experimental/simulated data of other Li-ion batteries.

5 Conclusions

In this paper, we have presented a Genetic Programming (GP) approach to evolve behavioral models of Li-ion battery voltage. The proposed method is based on the separation of the influence of the State-of-Charge (SoC) and Charge/discharge rate (C-rate) on the battery output voltage. The discovered model has allowed us to express such a relation through simple analytical functions and can be of help in better predicting the impact of the battery status and operating conditions on the efficacy of energy storage management systems. The results have confirmed that the model identified by the proposed GP approach have achieved satisfactory levels of accuracy, equivalent to those achieved by a Neural Network (NN), which is instead a black-box system. It is worth highlighting that a trained NN, differently from the analytical expression provided by GP, cannot be easily inverted or coded in commercial simulators.

In the future, we plan to extend our study by using our GP algorithm to develop new behavioral models with temperature as an additional input variable. Furthermore, we will test the robustness of our GP-based approach on different types of batteries (including different battery sizes and chemistries).

Acknowledgments. This work was supported by MOST, the Italian National Center for Sustainable Mobility, funded by the Italian Ministry of University and Research (MUR) in 2022–2025, by the "Innovazione per il Controllo avanzato e la gestione di Grid Energetiche" project within the Puglia FESR 2014 – 2020 (CUP B81B22000020007), and by Power4Future S.p.A. under the frame of the grant CDS000944, funded by the Italian Ministry of Enterprises and Made in Italy (MISE).

References

1. Andre, D., Meiler, M., Steiner, K., Wimmer, C., Soczka-Guth, T., Sauer, D.U.: Characterization of high-power lithium-ion batteries by electrochemical impedance spectroscopy. I. Experimental investigation. In: J. Power Sour.**196**(12), 5334–5341 (2011). https://doi.org/10.1016/j.jpowsour.2010.12.102

2. Cai, L., White, R. E.: Mathematical modeling of a lithium ion battery with thermal effects in COMSOL Inc. Multiphysics (MP) software. J. Power Sour. **196**(14), 5985–5989 (2011). https://doi.org/10.1016/j.jpowsour.2011.03.017

3. Catenacci, M., Verdolini, E., Bosetti, V., Fiorese, G.: Going electric: expert survey on the future of battery technologies for electric vehicles. Energy Policy **61**, 403–413 (2013). https://doi.org/10.1016/j.enpol.2013.06.078

4. Chen, M., Rincon-Mora, G.A.: Accurate electrical battery model capable of predicting runtime and I-V performance. IEEE Trans. Energy Conver. **21**(2), 504–511 (2006). https://doi.org/10.1109/TEC.2006.874229

5. Coello Coello, C.A., Lamont, G.B., Van Veldhuizen, D.A.: Evolutionary Algorithms for Solving Multi-Objective Problems. 2nd Edition, Springer, New York (2007). https://doi.org/10.1007/978-0-387-36797-2

6. Davydov, Y., Zitikis, R.: Quantifying non-monotonicity of functions and the lack of positivity in signed measures. Modern Stochast. Theory Appl. **4**(3), 219–231 (2017). https://doi.org/10.15559/17-VMSTA84

7. De Jong, E., Watson, R. A., Pollack, J. B.: Reducing bloat and promoting diversity using multi-objective methods. In: 3rd Annual Conference on Genetic and Evolutionary Computation, pp. 11–18 (2001)

8. Di Capua, G., et al.: Mutual inductance behavioral modeling for wireless power transfer system coils. IEEE Trans. Ind. Electron. **68**(3), pp. 2196–2206 (2021). https://doi.org/10.1109/TIE.2019.2962432

9. Ebrahimzade, H., Khayati, G.R., Schaffie, M.: A novel predictive model for estimation of cobalt leaching from waste Li-ion batteries: Application of genetic programming for design. J. Eenviron. Chem. Eng. **6**(4), 3999–4007 (2018). https://doi.org/10.1016/j.jece.2018.05.045

10. Echevarria, Y., Blanco, C., Sánchez, L.: Learning human-understandable models for the health assessment of li-ion batteries via multi-objective genetic programming. Eng. Appl. Artif. Intell. **86**, 1–10 (2019). https://doi.org/10.1016/j.engappai.2019.08.013

11. Frank, E., Hall, M.A., Witten, I.H.: The WEKA Workbench. Online Appendix for Data Mining: Practical Machine Learning Tools and Techniques, 4th edn.. Morgan Kaufmann (2016)

12. Hastie, T., Tibshirani, R., Friedman, J.: The Elements of Statistical Learning: Data Mining, Inference, and Prediction, 2nd edn. Springer, New York (2009). DOI: https://doi.org/10.1007/978-0-387-84858-7

13. Hentunen, A., Lehmuspelto, T., Suomela, J.: Time-domain parameter extraction method for thévenin-equivalent circuit battery models. IEEE Trans. Energy Conversion **29**(3), 558–566 (2014). https://doi.org/10.1109/TEC.2014.2318205

14. Hornik, K., Stinchcombe, M., White, H.: Multilayer feedforward networks are universal approximators. Neural Networks **2**(5), 359–366 (2003). https://doi.org/10.1016/0893-608 0(89)90020-8

15. Hornik, K.: Approximation capabilities of multilayer feedforward networks. Neural Networks **4**(2), 251–257 (2003). https://doi.org/10.1016/0893-6080(91)90009-T

16. Klein, R., Chaturvedi, N.A., Christensen, J., Ahmed, J., Findeisen, R., Kojic, A.: Electro-chemical model based observer design for a lithium-ion battery. IEEE Trans. Control Syst. Technol. **21**(2), 289–301 (2013). https://doi.org/10.1109/TCST.2011.2178604

17. Koza, J.K.: Genetic programming as a means for programming computers by natural selection. Statist. Comput. **4**, 87–112 (1994). https://doi.org/10.1007/BF00175355

18. Li, Y., et al.: Model order reduction techniques for physics-based lithium-ion battery man-agement: a survey. IEEE Ind. Electron. Mag. **16**(3), 36–51 (2022). https://doi.org/10.1109/MIE.2021.3100318

19. Luke, S., Spector, L.: A comparison of crossover and mutation in genetic programming. Genetic Programming **97**, 240–248 (1997)

20. Madani, S.S., Schaltz, E., Knudsen Kær, S.: An electrical equivalent circuit model of a lithium titanate oxide battery. Batteries (MDPI) **5**(1) (2019). https://doi.org/10.3390/batteries5010031

21. Ng, M.F., Zhao, J., Yan, Q., et al.: Predicting the state of charge and health of batteries using data-driven machine learning. Nat. Mach. Intell. **2**, 161–170 (2020). https://doi.org/10.1038/s42256-020-0156-7

22. Press, W.H., Teukolsky, S.A., Vetterling, W.T., Flannery, B.P.: Numerical Recipes in C: The Art of Scientific Computing, 2nd edn. Cambridge University Press, Cambridge (1992)

23. Rumelhart, D., Hinton, G., Williams, R.: Learning representations by backpropagating errors. Nature **323**, 533–536 (1986). https://doi.org/10.1038/323533a0

24. Schröer, P., van Faassen, H., Nemeth, T., Kuipers, M., Sauer, D.U.: Challenges in modeling high power lithium titanate oxide cells in battery management systems. J. Energy Storage **28** (2020). https://doi.org/10.1016/j.est.2019.101189

25. Shukla, A., Pandey, H.M., Mehrotra, D.: Comparative review of selection techniques in genetic algorithm. In: International Conference on Futuristic Trends on Computational Analysis and Knowledge Management, pp. 515–519 (2015). https://doi.org/10.1109/ABLAZE.2015.715 4916

26. Stoyka, K., Di Capua, G., Femia, N.: A novel AC power loss model for ferrite power inductors. IEEE Trans. Power Electron. **34**(3), 2680–2692 (2019). https://doi.org/10.1109/TPEL.2018. 2848109

27. Stroe, D.I., Swierczynski, M., Stroe, A.I., Knudsen Kær, S.: Generalized Characterization methodology for performance modelling of lithium-ion batteries. Batteries (MDPI) **37**(2) (2016). https://doi.org/10.3390/batteries2040037

28. Yao, H., Jia, X., Wang, B., Guo, B.: A new method for estimating lithium-ion battery capacity using genetic programming combined model. In: 2019 Prognostics and System Health Man-agement Conference, pp. 1–6 (2019). https://doi.org/10.1109/PHM-Qingdao46334.2019.894 2970

29. Yao, H., Jia, X., Zhao, Q., Cheng, Z.-J., Guo, B.: Novel lithium-ion battery state-of-health estimation method using a genetic programming model. IEEE Access **8**, 95333–95344 (2020). https://doi.org/10.1109/ACCESS.2020.2995899

30. Zadeh, L.: Optimality and non-scalar-valued performance criteria. IEEE Trans. Autom. Control **8**(1), 59–60 (1963). https://doi.org/10.1109/TAC.1963.1105511

An Intelligent Optimised Estimation of the Hydraulic Jump Roller Length

Antonio Agresta[1], Chiara Biscarini[1], Fabio Caraffini[2],
and Valentino Santucci[1(✉)]

[1] University for Foreigners of Perugia, Perugia, Italy
{antonio.agresta,chiara.biscarini,valentino.santucci}@unistrapg.it
[2] Department of Computer Science, Swansea University, Computational Foundry,
Swansea SA1 8EN, UK
fabio.caraffini@swansea.ac.uk

Abstract. In this paper, we address a problem in the field of hydraulics which is also relevant in terms of sustainability. Hydraulic jump is a physical phenomenon that occurs both for natural and man-made reasons. Its importance relies on the exploitation of the intrinsic energy dissipation characteristics and on the other hand the danger that might produce on bridges and river structures as a consequence of the interaction with the large vortex structures that are generated. In the present work, we try to address the problem of estimating the hydraulic jump roller length, whose evaluation is inherently affected by empirical errors related to its dissipative nature. The problem is approached using a regression model and exploiting a dataset of observations. Regression is performed by minimising the loss function using ten different black-box optimisers. In particular, we selected some of the most used metaheuristics, such as Evolution Strategies, Particle Swarm Optimisation, Differential Evolution and others. Furthermore, an experimental analysis has been conducted to validate the proposed approach and compare the effectiveness of the metaheuristics.

Keywords: Metaheuristics · Hydraulic jump roller length · Regression · Continuous optimisation · Sustainability

1 Introduction

River flow dynamics [1,2,10] falls under the category of very complex flow evolution, which includes the widest diversity of flow. Such processes are the main damage to river crossing structures.

In the 16th century, the great *Leonardo Da Vinci* observed a visible increase in height in a flowing liquid and, while documenting this phenomenon, associated its occurrence with an abrupt velocity change of its flow from high speed to lower speed. Only later, in the 1800 s, Professor *Giorgio Bidone* (University of Turin) gave a mathematical formulation to the generic description of his illustrious predecessor.

© The Author(s), under exclusive license to Springer Nature Switzerland AG 2023
J. Correia et al. (Eds.): EvoApplications 2023, LNCS 13989, pp. 475–490, 2023.
https://doi.org/10.1007/978-3-031-30229-9_31

The phenomenon *Hydraulic Jump* is a physics problem that has been studied for centuries due to the societal, optical, and technological implications that its solution can bring in terms of sustainability and resilience. Italy is a perfect case study to show this potential, as this country has a territory rich in large rivers and smaller watercourses, creeks, and brooks. However, structures such as bridges and embankments that often date back to decades had poor maintenance over the years. Recently, a consortium of universities named "FABRE" has been created to map and assess the health status, as well as to monitor the procedural schemes of hundreds of bridges to implement emergency recovery interventions.

Among the many dangers that affect bridge piers, the hydraulic jump is one of the most common, as it is intrinsically linked to the nature of the flow and its regime. In this article, we propose an approach based on a learning algorithm optimised with a heuristics technique to evaluate the hydraulic jump characteristics and tackle this problem.

Hydraulic jump is a phenomenon generated by a change in the flow regime from supercritical to subcritical, accompanied by high energy dissipation leading to an increase in the depth of the flow, as shown in Fig. 1a. The increase in dissipation is due to the development of a complex multi-fold turbulent flow structure that causes significant energy losses. Often, hydraulic jump characteristics, analogously to what happens in compressible flow dynamics with the occurrence of shock waves, are affected by the presence of external elements such as obstacles downstream, typically bridge piles, and above all, the characterisation of the bed roughness (see Figs. 1a and 1b).

As in the premisis, hydraulic jumps have been broadly investigated both due to their frequent occurrence in nature and their potential risk for man-made fluvial structures and, on the opposite front, for their possible use as energy dampers for hydraulic structures themselves [15].

From the seminal work between the end and the beginning of the 19th century in terms of both analytical [4,27] and experimental [31] aspects, the analytical form of the characteristics of the hydraulic jump originates from the equation of momentum balance. The early works cited considered the friction forces negligible compared to the others. In subsequent investigations [24] a more generalised study was conducted that included consideration of a more realistic velocity distribution and, therefore, partially took into account the resistance of the boundaries. Rajaratnam [23], Hughes and Flack [19] and Hager and Bremen [14,16] went a little further by implementing the bed shear stress directly from the expression of the momentum equation. Since then, the main issue still is represented by the definition of this "friction term" related in a general sense to the bottom shear stress, but as a matter of fact strictly depending on the different feature of the bottom surface and the flow regime. Quite recently researchers have made a specific effort trying to assess the implementation of the effects of the bed roughness both empirically [5,6,11] and by using teaching-learning-based optimisation techniques [13,20] always distinguishing between the two main characteristics of the hydraulic jump, namely the "Sequent Depth ratio", which is the ratio between the two cross sections with successive depths h_1 and

h_2 as in Fig. 1b and the "Roller Length", defined as the horizontal distance between the toe section with the flow depth h_1 and the roller end corresponding to the cross section h_2. A more specific and mathematically detailed description of those hydraulic jump flow characteristics will be given in the next section.

The scope of the present work is to focus on the Roller Length evaluation, estimating it via a novel form of the roughness contribution, modelled from the "shape" of the roughness function derived from the turbulence charts of the Nikuradze-type diagram [3]. Practically, we design a regression function with unknown parameters, which will be fitted to a dataset of observations by minimising a loss function through a black-box optimiser. In particular, we considered ten popular metaheuristics for black-box continuous optimisation and experimentally compared them through a repeated cross-validation approach.

2   Hydraulical Aspects

The description given in Sect. 1 shows how, due to the inherently dissipative nature of the hydraulic jump phenomenon, the analytical definition of some of its peculiar features is very difficult to obtain and empirical relations are needed based on the principles of the dimensional analysis. In what follows, a brief description of the characteristics of this phenomenon is presented from the hydraulic point of view.

2.1   The Hydraulic Jump over a Rough Surface

The hydraulic jump phenomenon generates a characteristic large vortex, or more correctly a *roller* whose length is directly proportional to the intensity of the phenomenon itself. The length of the roller can be defined as the horizontal distance between the toe section with height h_1 and the section where the roller ends with height h_2 as shown in Fig. 1b.

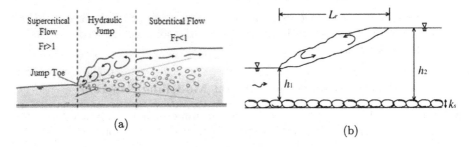

Fig. 1. (a) Hydraulic jump physical schematics and (b) hydraulic jump roller length main characteristics.

Its length can be fairly easily evaluated in experimental tests visualising the flow and its free surface and measuring it, but it is still hard to determine it *a priori* by modelling a functional relation, especially considering the effects of the roughness of the bed. More rigorously, we can identify a dependence of the roller length L_r of the following type:

$$L_r = F(k_s, g, h_1, h_2, U_1, \mu), \tag{1}$$

where: k_s represents the roughness height at the bottom, expressed in *cm*; g is the gravitational acceleration; h_1 and h_2 are, respectively, the depths of the upstream and downstream flow, considering the transition from the supercritical (upstream) to subcritical (downstream) regime; U_1 is the mean flow velocity of the upstream cross section; while μ is the kinematic viscosity of the fluid.

Taking advantage of the bases of dimensional analysis, we can reduce the number of independent variables obtaining the following:

$$\frac{L_r}{h_1} = f_0 \left(\frac{k_s}{h_1}, \frac{U_1 h_1}{\mu}, \frac{h_2}{h_1}, \frac{U_1}{\sqrt{gh_1}} \right). \tag{2}$$

The a-dimensional group $\frac{U_1}{\sqrt{gh_1}}$ is called the "Froude number" (Fr) and represents the relationship between the flow inertial forces and the external gravitational field. It serves as an indicator of the transition between the different flow regimes, namely subcritical, critical, and supercritical, respectively, for $Fr < 1$, $Fr = 1$, and $Fr > 1$. Another ratio between forces acting within the flow, and precisely against inertial forces and viscous ones, is represented by the "Reynolds number" (Re), indicated by the fraction $\frac{U_1 h_1}{\mu}$ which defines the transition from laminar to turbulent regime. In this case, since the Reynolds number is relative to the upstream cross section by the velocity U_1, it is taken for granted that it is in the presence of a fully developed turbulent flow, leading to the possibility of ignoring the viscous effects and hence the dependence on the Reynolds number itself. Therefore, the final form of the functional relation will be:

$$\frac{L_r}{h_1} = f \left(\frac{k_s}{h_1}, \frac{h_2}{h_1}, Fr_1 \right), \tag{3}$$

where the ratio $\frac{h_2}{h_1}$ is defined as "Sequent Depth Ratio" (SDR) and we directly denote the Froude number with the subscript "1", pointing out that it refers to the upstream flow.

2.2   Definition of a Roughness Height Modelling Function

Many have tried for almost a century now to propose valid analytical forms for the relation (3), starting from the very well-known equation for SDR:

$$\frac{h_2}{h_1} = \frac{1}{2} \left(-1 + \sqrt{1 + 8Fr_1^2} \right), \tag{4}$$

obtained in 1828 by the French hydraulic engineer Jean-Baptiste Bélanger from the momentum balance equation. It appears clear right away how this equation does not imply the presence of a rough bottom, hence explicitly, of a shear stress.

Its generalisation including a direct dependence from the velocity distribution was given by Govinda et al. in 1966 [24]:

$$\frac{h_2}{h_1} = \frac{1}{2}\left(-1 + \sqrt{1 + \alpha Fr_1^2}\right),\tag{5}$$

where α also takes into account the condition of zero velocity at the bottom.

To integrate the action of bed shear stress, some authors (Carollo and Ferro, in 2004 [5,6]) have included empirical parameters and, starting from the momentum balance equation, also proposed the following equation

$$\frac{h_2}{h_1} = \frac{1}{2}\left(-1 + \sqrt{1 + 8(1 - \beta)Fr_1^2}\right),\tag{6}$$

where β is empirically defined as $\beta = 0.42\frac{k_s}{h_1}$, Although both α and β are related to the friction of the bed, the first through the velocity profile and the latter through the presence of the height of the roughness of the bed k_s.

Regarding the roller length L_r, Smetana in 1937 was the first to propose a direct correlation with the subsequent depth.

$$\frac{L_r}{h_1} = 6\left(\frac{h_2}{h_1} - 1\right).\tag{7}$$

Hager et Al. [14], via an experimental campaign, suggested the following relation for highly supercritical flows:

$$\frac{L_r}{h_1} = 8(Fr_1 - 1.5).\tag{8}$$

In both the above studies, the proportionality with SDR and Fr is suggested, but in neither case is there any dependence on the roughness of the bed. Hughes and Flack in 1984 [19], Ead and Rajaratnam in 2002 [11] and Carollo and Ferro in 2004 [5], verified the possibility to include such a relation via an empirical coefficient. In particular, taking into consideration both the sequent depth and the Froude number, Carollo and Ferro proposed the following.

$$\frac{L_r}{h_1} = \frac{a}{\left(\frac{h_1}{h_2}\right)^{1.272}}\tag{9}$$

$$\frac{L_r}{h_1} = b(Fr_1 - 1).\tag{10}$$

where the constants a and b, as expected, directly depend on the roughness of the bed. In 2007 the same authors, as a result of an experimental campaign, were appointed. found specific values for empirical coefficients, explicitly linking the roller length L_r to Fr_1 and k_s and simultaneously L_r to SDR, h_2/h_1.

Karbasi followed similar relationships, but with a different approach, in 2016 [20]. He proposed a teaching-learning-based optimisation algorithm to test different regression functions in order to find optimised values for empirical parameters to relate, in this case, the roller length L_r with Fr_1 and h_2/h_1, leaving out the explicit dependence on the roughness of the bed k_s.

In this work, we propose a dimension-based approach where the relation (3) is fully represented using both the results reported above and an evaluation of the roughness of the bed derived from considerations relative to the transition process between a smooth regime and a fully rough one. In the former condition, completely hydraulically smooth, viscosity acts as a damper and cancels out any perturbation caused by the roughness, while in the latter condition, it is the pressure drag on the rough surface that totally leads the process and produces friction. In the middle of these two regimes, a transitional phase occurs where the two processes are present. In order to grossly capture the complex physics described, a modelling function is suggested by Andersson [3] who basically reports the Nikuradse turbulence diagrams in the form of a wall function.

The original idea was to implement such a function, adjusting some of its terms, to the general form of the relation (3). The proposed roughness function equation is as follows:

$$\phi = \begin{cases} 0 & \text{if } k_s^+ \leq K_{SM} \\ \frac{1}{k_s} \ln[C_s k_s^+] \sin\left(\frac{\pi}{2} \frac{\ln k_s^+ - \ln K_{SM}}{\ln K_R - \ln K_{SM}}\right) & \text{if } K_{SM} \leq k_s^+ \leq K_R \\ \frac{1}{k_s} \ln[C_s k_s] & \text{if } k_s^+ > K_R, \end{cases} \tag{11}$$

where the new constants introduced here are K_{SM} and K_R (respectively, "K smooth" and "K rough") that represent the lower and upper bound of the transitionally rough regime. According to literature they have been set as $K_{SM} = 2.25$ and $K_R = 90$. The variable calculated k_s^+ is the so called "Roughness Reynolds number" defined as $k_s^+ = (k_s U_1)/\nu$ with the viscosity of water $\nu = 0.00131$ kg/ms. The constant C_s, called the "Roughness constant" is a numerical re-tuning coefficient and was set at $C_s = 0.5$. To retrieve the mean flow velocity of the cross section U_1, we used the experimental data set extracted from the 2007 work of Carollo and Ferro [7]. In particular, from the values of the Froude number we can compute U_1 as follows:

$$U_1 = Fr_1 \cdot \sqrt{gh_1} \tag{12}$$

to be inserted in Eq. (11).

The idea behind the choice of the type of function (11) was made to best resemble the shape of the functional trend of the roughness function in the three ranges defined by the Froude number, emulating the behaviour of the wall friction function.

By plugging Eq. (11) into Eq. (3), also taking into consideration Equations (9) and (10), we can derive our estimate for the roller length as follows.

$$\frac{L_r}{h_1} = a_1 \cdot \phi + a_2 \cdot \frac{h_2}{h_1} + a_3 \cdot (Fr_1 - 1.5), \tag{13}$$

where the vector $a \in \mathbb{R}^3$ represents three parameters imposed by the design and whose values can be learnt from the data.

3   The Learning Scheme

The roller length estimation function introduced in Equation (13) requires one to identify the values for the parameter vector $a \in \mathbb{R}^3$. With this aim, we can exploit the dataset of observed values provided in [7] in order to learn the a parameters. Therefore, we treat Equation (13) as a regression function and we apply the definition of the parameters a to a regression problem.

We denote by $x \in \mathbb{R}^4$ the vector of the following four values: $x_1 = k_s$, $x_2 = h_1$, $x_3 = h_2$, and $x_4 = Fr_1$. By also noting that ϕ, as defined in Equation (11), is a function of k_s, h_1 and Fr_1, we will use the notation $\phi(x_1, x_2, x_4)$. We can now define the regression function for the roller length as

$$g_a(x) = a_1 \cdot \phi(x_1, x_2, x_4) + a_2 \cdot \frac{x_3}{x_2} + a_3 \cdot (x_4 - 1.5). \tag{14}$$

Furthermore, we denote by $y \in \mathbb{R}$ the roller length values contained in the data set, that is, $y = L_r/h_1$. Therefore, the data set D considered in this work [1] is a set of pairs (x, y) as follows.

$$D = \left\{ (x, y) : x \in \mathbb{R}^3 \text{ and } y \in \mathbb{R} \right\}. \tag{15}$$

It is now clear that the parameters a can be learnt by minimising the following loss function.

$$f(a) = \sum_{(x,y) \in T \subseteq D} \left(g_a(x) - y \right)^2, \tag{16}$$

where the training samples in T are a suitable subset of the entire data set D, which contains a total of 367 observed data samples.

4   Black-Box Optimisers

The loss function introduced in Eq. (16) can be minimised by any black-box optimiser. In this section, we describe the ten metaheuristics considered in this work.

4.1   Random and Quasi-random Search

Trivial random search procedures are considered in this work as baseline methods. In particular, we denote by RS the random search procedure which generates a given number of solutions uniformly at random and, after evaluating them all, selects the best one. Similarly, we also consider the Scrambled Hammersley Search (SHS) method [8] which generates a quasi-random sample of low-discrepancy vectors that homogeneously cover the search space. Although being trivial, RS and SHS are interesting baseline methods for two reasons: all the solutions can be evaluated in parallel (at least in principle), and they allow us to derive indications about the smoothness of the search landscape at hand by comparing their effectiveness with that of other smarter algorithms.

---

[1] The data set is available with the supplementary data at https://doi.org/10.5281/zenodo.7595510.

482   A. Agresta et al.

4.2 Simple Evolution Strategies

Evolution strategies [33] are a family of evolutionary algorithms that evolve
one or more incumbent solutions by means of one or more genetic operators.
In this work we consider two simple evolutionary strategies which adopt the so-
called (1+1) search scheme, i.e., they maintain a single incumbent solution which
is iteratively mutated and the generated mutant becomes the new incumbent
solution if and only if it is fitter than it.

Given the incumbent solution $x \in \mathbb{R}^d$, a mutant $y \in \mathbb{R}^d$ is generated as $y_i \leftarrow x_i + \varepsilon_i$, for any dimension $i = 1, \dots, d$. In this work we term with the acronyms
ES and CES the evolution strategies which perform the perturbation according
to, respectively, a normal distribution or a Cauchy distribution (which is a fat-
tailed variant of the normal distribution). More formally, we have $\varepsilon_i \sim N(0, \sigma_i)$
in ES, and $\varepsilon_i \sim C(0, \gamma_i)$ in CES. In our experiments we adopt the default
configurations of the ES and CES implementations provided in the Nevergrad
library [25], which sets $\sigma_i = \gamma_i = 1$ for any dimension $i = 1, \dots, d$.

4.3 Covariance Matrix Adaptation Evolution Strategies

One of the most sophisticated forms of evolution strategies is the evolution strategy of covariance matrix adaptation (CMA) [17].

Unlike the simple evolution strategies previously described, the CMA main-
tains three entities: a mean vector $m \in \mathbb{R}^d$, a covariance matrix $C \in \mathbb{R}^{d \times d}$, and
a step-size vector $\sigma \in \mathbb{R}^d$. In each iteration, N solutions are sampled from the
(possibly) multivariate normal distribution $N(m, \sigma C)$. The solutions are then
weighted on the basis of their fitness, and then used to update m, C and σ. For
further details, we point the interested reader to [17].

In this work, we consider two standard implementations of this algorithm,
namely CMA and DCMA: The former uses a multivariate covariance matrix,
while the latter maintains a simpler diagonal covariance matrix. Furthermore,
the CMA and DCMA implementations provided in the Nevergrad library [25]
are adopted with their default configurations.

4.4 Differential Evolution

Differential Evolution (DE) [28,29] is a population-based evolutionary meta-
heuristic which was originally proposed in [32].

The DE population is made up of N $d-$dimensional vectors $x_1, \dots, x_N$ that
can be uniformly initialised at random or by using a quasi-random generator
such as the Scrambled Hammersley procedure that produces a low discrepancy
sample of vectors [8]. In this work we adopt both initialization variants, which
are referred to as, respectively, DE and QrDE.

The key operator of the algorithm is the differential mutation that, for each
population individual $x_i \in \mathbb{R}^d$, produces a mutant vector $v_i \in \mathbb{R}^d$ as a linear

combination of some other population individuals. One of the most popular DE mutation strategy is named "current-to-best" and is defined as

$$v_i \leftarrow x_i + F_1(x_{best} - x_i) + F_2(x_{r_1} - x_{r_2}), \tag{17}$$

where: x_{best} is the best population individual so far, F_1 and F_2 are the two scale factor hyperparameters of DE, and x_{r_1}, x_{r_2} are two randomly selected population individuals which are different between them and with respect to x_i.

After the mutation, the vector v_i is recombined with x_i. The most common recombination operator is *binomial crossover* that generates a trial vector $y_i \in \mathbb{R}^d$ by selecting each vector component from either v_i, with probability CR, or x_i, with probability $1 - CR$. The crossover probability $CR \in [0, 1]$ is a hyperparameter of the algorithm. Finally, if v_i is fitter than x_i, it replaces x_i in the next iteration population.

In our experimentation we adopted the default configuration of the DE implementation provided in the Nevergrad library [25], i.e., $N = 30$, $F_1 = F_2 = 0.8$, $CR = 0.5$.

4.5 Particle Swarm Optimization

Particle Swarm Optimization (PSO) [30] is one of the most famous metaheuristic based on swarm intelligence principle, firstly proposed in [21].

PSO maintains a population of N particles. Each particle has a position $x_i \in \mathbb{R}^d$ in the solution space and a velocity $v_i \in \mathbb{R}^d$. The particles are statically connected among them, usually by means of a global topology.

As in DE, the population is evolved during a given number of iterations. At each iteration, the j-th component of x_i is updated according to

$$v_{i,j} \leftarrow \omega v_{i,j} + c_1 r_{1,j}(p_{i,j} - x_{i,j}) + c_2 r_{2,j}(g_j - x_{i,j}), \tag{18}$$

$$x_{i,j} \leftarrow x_{i,j} + v_{i,j}, \tag{19}$$

where: $p_i, g \in \mathbb{R}^d$ are, respectively, the best position ever visited by particle i and the global best position ever visited by the whole swarm; $\omega, c_1, c_2 \in \mathbb{R}^+$ are the so-called *intertial*, *cognitive* and *social* hyperparameters of PSO; while $r_{1,j}, r_{2,j} \in \mathbb{R}$ are randomly generated numbers in $[0, 1]$.

In our experimentation we adopted the default configuration of the PSO implementation provided in the Nevergrad library [25], i.e., $N = 40$, $\omega = 0.72$, $c_1 = c_2 = 1.19$.

4.6 Nelder-Mead

Nelder-Mead (NM) [12] is a mathematical optimization methodology which iteratively updates a simplex of $d+1$ vertices in the solution space. At each iteration, the simplex is updated by trying to replace the worst vertex with a new better solution obtained by means of four operations termed as *reflection*, *expansion*, *contraction*, and *shrink*. We refer the interested reader to [12] for their definitions.

In our experimentation we adopt the standard implementation of NM as provided in the Nevergrad library [25].

A. Agresta et al.

5 Experiments

5.1 Experimental Setup

In order to analyse the proposed methodology, we conducted an experimental comparison among the ten black-box optimisers described in Sect. 4. Following the learning scheme described in Sect. 3, we have performed a k-fold repeated cross-validation [22], where the number of repetitions is $r = 10$ and the number of folds is $k = 3$. Then, for all $r \cdot k = 30$ cross-validation rounds, each algorithm is executed $s = 10$ times with a budget of $10\,000$ objective evaluations. Therefore, a total of $r \cdot k \cdot s = 300$ executions per algorithm were performed.

In any single execution, an algorithm evaluates the loss function by accessing only the data from the training set for the current cross-validation round. However, the optimised a values returned by the algorithm are used to evaluate the regression function on the data of the current test set, thus allowing one to compute the mean percent relative error (MPRE) as follows.

$$MPRE = 100 \cdot \frac{1}{n} \sum_{i=1}^{n} \frac{|\hat{y}_i - y_i|}{y_i}, \tag{20}$$

where: $\hat{y}_i = g_a(x_i)$ is the predicted value computed based on the learnt values a and the observed input data x_i from the test set, y_i is the observed output value of the test set, while n is the number of samples in the test set. With slight abuse of notation, we also denote by MPRE the average mean percentage relative error of an algorithm in all cross-validation rounds. Moreover, we may also use the same measure related to the training set data, but in these cases we will clearly specify that.

The implementations of the selected algorithms available in the popular Python library called *Nevergrad* [25,26] (version 0.5.0, the latest one at the time of writing) were adopted with their default parameters' settings. For the sake of reproducibility, in Table 1 we provide for each algorithm the name of the corresponding Nevergrad class.

Table 1. Names of the Nevergrad classes corresponding to the algorithms adopted in this work. The acronyms of the algorithms are defined in Sect. 4.

Algorithm's acronym	Nevergrad's class
ES	OnePlusOne
CES	CauchyOnePlusOne
CMA	CMA
DCMA	DiagonalCMA
DE	DE
QrDE	QrDE
PSO	RealSpacePSO
NM	NelderMead
RS	RandomSearch
SHS	ScrHammersleySearch

5.2   Experimental Results

First, to validate our approach, we compare the training and test errors of the a values learnt by any single execution of all algorithms considered. In this regard, in Fig. 2 we provide a scatter plot where: the abscissa and ordinate axes represent the percentage relative error on, respectively, the training and test data, any blue point in the graph is the result of a single algorithm execution, while the regression line through the points is drawn in black.

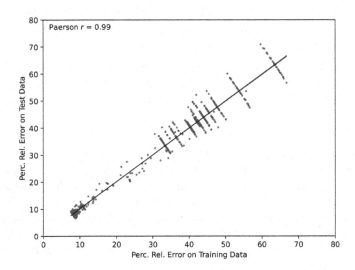

Fig. 2. Correlation between the mean percentage relative training and test errors.

Figure 2 clearly shows that training and test errors are strongly correlated – their Paerson correlation coefficient is 0.99 –, thus validating our approach for learning the regression parameters.

In Table 2 we provide all the statistics about the MPREs – evaluated on the test data – of the best solutions obtained by the ten algorithms in all their executions. The algorithms are ordered by median MPRE and the best results, for each statistic, are provided in bold.

To statistically validate the results, we also performed a statistical analysis on the test MPREs obtained by the algorithms. The omnibus Kruskal-Wallis test [9] rejects the equivalence of effectiveness among the five algorithms with a practically zero p-value (around 10^{-303}). Therefore, a Conover post hoc test has been performed considering the Benjamini-Hochberg adjustment scheme to mitigate the statistical family error rate [18]. The Conover p-values of the comparison between ES and all other algorithms are also provided in Table 2.

Table 2 shows that the eight proper algorithms are clearly better than the two baselines RS (random search) and SHS (a quasi-random sampling method). Moreover, the largest group of algorithms that are not significantly better – under the usual threshold of $\alpha = 0.05$ – than each other is made up of five

Table 2. Statistics about the MPREs evaluated on the test data.

Statistics on the MPREs						
Algorithm	median	pvalue	mean	std	min	max
ES	**8.27**	–	8.18	0.64	6.92	9.18
CES	8.28	0.98	8.18	0.63	6.95	9.16
PSO	8.29	0.90	**8.17**	**0.62**	7.10	**9.10**
NM	8.30	0.14	8.64	2.26	7.12	20.30
QrDE	8.30	0.53	8.24	0.71	**6.73**	11.26
CMA	8.31	8e-03	9.48	5.38	7.12	53.61
DCMA	8.32	0.02	9.44	5.00	7.01	40.21
DE	8.37	2e-03	8.88	2.72	6.89	30.11
SHS	41.84	6e-172	40.01	5.34	30.76	49.58
RS	42.13	6e-184	42.42	13.00	9.88	70.87

algorithms, namely: ES, CES, PSO, NM and QrDE. Among these five algorithms, ES, CES, and PSO have a higher degree of robustness, as indicated by the standard deviation, minimum, and maximum of their MPREs. Their average MPRE is around 8.17%/8.18%. The other statistics also show that these three algorithms are practically equivalent among them. Therefore, our main conclusion is that the two variants of the simple evolution strategy (ES and CES) and the PSO algorithm are to be preferred over the other competitors for the task under examination.

To provide a clearer picture, Fig. 3 shows the box plot graph of the test MPREs obtained by the 300 executions of each algorithm.

Figure 3 mostly confirms the indications previously discussed. In fact, ES, CES and PSO are the only three algorithms which do not have outliers in the graph. This aspect is of particular interest by considering that the ordinate axis is on log scale. Moreover, it is also interesting to observe that the two baseline random search schemes (SHS and RS) are significantly worse than all the other approaches. This suggests that the search landscape of the optimised optimisation problem designed has a structure that is not chaotic.

For the sake of completeness, all pairwise comparisons among the algorithms considered are synthesised in the heat map of Fig. 4. The entries are greenish or reddish when the row-algorithm is, respectively, better or worse than the column-algorithm. The green or red grades are set on the basis of the adjusted Conover p-values, which are also provided in the entries.

Figure 4 is a further confirmation of the conclusions derived, but also provides some additional indications as follows. The two covariance-based evolution strategies CMA and DCMA appear to be less effective than their simpler "cousins" ES and CES. We conjecture that this is probably due to the small search-space dimensionality of the problem under investigation. The plain DE

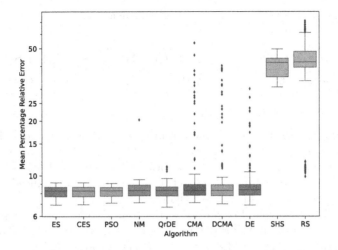

Fig. 3. Boxplot showing the distribution of the MPREs evaluated on the test data.

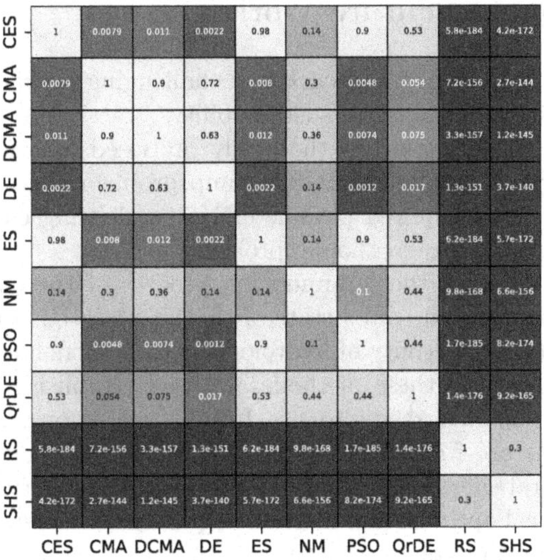

Fig. 4. Conover pvalues of all the pairwise comparisons among the algorithms.

is less effective than QrDE, suggesting that a quasi-random initialisation may improve the resilience to local optima in the DE search scheme.

As a last analysis, for each algorithm – with the exception of the two random search methods RS and SHS – we computed all pairwise Euclidean distances among the 300 a vectors obtained by the different executions carried out in the experimentation. Hence, the maximum Euclidean distances of all algorithms are provided in Table 3.

Table 3. Maximum Euclidean distance observed among the 300 solutions obtained by the different executions.

Algorithm	CES	CMA	DCMA	DE	ES	NM	PSO	QrDE
Max distance	0.67	4.48	4.19	13.39	0.85	6.47	**0.48**	3.81

These data confirm that the most robust methods in terms of MPRE, i.e., ES, CES and PSO, are also the most robust methods when analysed in the search space dimensions. In fact, also noting that the search space range was $[-50, +50]^3$, the maximum distance between the solutions produced by the 300 independent executions is much less than 1.

Finally, for reproducibility purposes, the source code, the full experimental results and the scripts used for the analysis were made publicly available at the following location: https://doi.org/10.5281/zenodo.7595510.

6   Conclusion and Future Work

We proposed a computational approach to a challenging hydraulic problem with important implications in terms of sustainability.

A contribution of this study is the newly proposed function for estimating the roller length of the typical hydraulic jump problem in water courses. This is designed so that the required parameters for evaluating can be learnt from observed data following a regression approach.

Furthermore, we show the minimisation process of the corresponding loss function with the aid of ten tested state-of-the-art black-box optimisation algorithms. To facilitate the activity of hydrology practitioners and researchers who may not be familiar with these methods, we used established algorithms with open implementation and ready to use from a well-known library [25]. The selected algorithms are from different families, such as simple and covariance-based evolution strategies, differential evolution variants, particle swarm optimisation, mathematical optimisation methodologies, and simpler random or quasi-random one-shot optimisers.

With a thorough experimental and validation phase, we show that the proposed approach is robust and effective in terms of the mean percentage of relative error.

Moreover, this investigation allows us to recommend the use of (1+1) evolution strategies (based on both normally and Cauchy distributed mutations) and particle swarm optimisation algorithms over other methods when using the proposed framework. Indeed, amongst the algorithm under investigation, these two classes guaranteed the best balance of both effectiveness and robustness for the problem at hand.

We believe that this is a promising starting point for further research, and we envisage investigating different regression functions, possibly exploiting interpretable machine learning models, in the future. Also, potential ways forward are in the direction of analysing the fitness landscape induced by the regression function and extending the approach to other related problems in the hydraulic field.

Acknowledgement. This work was partially supported by the following research grants from "Università per Stranieri di Perugia": (i) "Finanziamento Dipartimentale alla Ricerca FDR 2022", (ii) "Artificial Intelligence for Education, Social and Human Sciences", (iii) "Progettazione e sviluppo di strumenti digitali per la formazione a distanza".

References

1. Agresta, A., Baioletti, M., Biscarini, C., Caraffini, F., Milani, A., Santucci, V.: Using optimisation meta-heuristics for the roughness estimation problem in river flow analysis. Appl. Sci. **11**(22), 10575 (2021)
2. Agresta, A., Baioletti, M., Biscarini, C., Milani, A., Santucci, V.: Evolutionary Algorithms for Roughness Coefficient Estimation in River Flow Analyses. In: Castillo, P.A., Jiménez Laredo, J.L. (eds.) EvoApplications 2021. LNCS, vol. 12694, pp. 795–811. Springer, Cham (2021). https://doi.org/10.1007/978-3-030-72699-7_50
3. Andersson, J., Oliveira, D.R., Yeginbayeva, I., Leer-Andersen, M., Bensow, R.E.: Review and comparison of methods to model ship hull roughness. Appl. Ocean Res. **99**, 102119 (2020)
4. Bélanger, J.: Essay on the numerical solution of some problems relative to steady flow of water. Paris, France, Carilian-Goeury (1828)
5. Carollo, F., Ferro, V.: Contributo allo studio della lunghezza del risalto libero su fondo liscio e scabro. Rivista di Ingegneria Agraria **35**(4), 13–20 (2004). (in Italian)
6. Carollo, F., Ferro, V.: Determinazione delle altezze coniugate del risalto libero su fondo liscio e scabro. Rivista di Ingegneria Agraria **35**(4), 1–11 (2004)
7. Carollo, F.G., Ferro, V., Pampalone, V.: Hydraulic jumps on rough beds. J. Hydraul. Eng. **133**(9), 989–999 (2007)
8. Cauwet, M.L., et al.: Fully parallel hyperparameter search: reshaped space-filling. In: International Conference on Machine Learning, pp. 1338–1348. PMLR (2020)
9. Derrac, J., García, S., Molina, D., Herrera, F.: A practical tutorial on the use of nonparametric statistical tests as a methodology for comparing evolutionary and swarm intelligence algorithms. Swarm Evol. Comp. **1**(1), 3–18 (2011)
10. Di Francesco, S., Biscarini, C., Manciola, P.: Characterization of a flood event through a sediment analysis: The tescio river case study. Water **8**(7), 308 (2016)
11. Ead, S.A., Rajaratnam, N.: Hydraulic jumps on corrugated beds. J. Hydraul. Eng. **128**(7), 656–663 (2002)
12. Gao, F., Han, L.: Implementing the nelder-mead simplex algorithm with adaptive parameters. Comput. Optim. Appl. **51**(1), 259–277 (2012)
13. Gul, E., Dursun, O.F., Mohammadian, A.: Experimental study and modeling of hydraulic jump for a suddenly expanding stilling basin using different hybrid algorithms. Water Supply **21**(7), 3752–3771 (2021)
14. Hager, W.H., Bremen, R., Kawagoshi, N.: Classical hydraulic jump: length of roller. J. Hydraul. Res. **28**(5), 591–608 (1990)
15. Hager, W.: Energy Dissipators and Hydraulic Jumps, vol. 8. Kluwer Academic Publication, Dordrecht, The Netherlands (1992)

16. Hager, W.H., Bremen, R.: Classical hydraulic jump: sequent depths. J. Hydraul. Res. **27**(5), 565–585 (1989)
17. Hansen, N., Ostermeier, A.: Completely derandomized self-adaptation in evolution strategies. Evol. Comput. **9**(2), 159–195 (2001)
18. Hollander, M., Wolfe, D.A., Chicken, E.: Nonparametric statistical methods, vol. 751. John Wiley & Sons (2013)
19. Hughes, W.C., Flack, J.E.: Hydraulic jump properties over a rough bed. J. Hydraul. Eng. **110**(12), 1755–1771 (1984)
20. Karbasi, M.: Estimation of classical hydraulic jump length using teaching-learning based optimization algorithm. J. Mater. Environ. Sci. **7**, 2947–2954 (2016)
21. Kennedy, J., Eberhart, R.: Particle swarm optimization. In: Proceedings of the IEEE International Conference on Neural Networks, vol. 4, pp. 1942–1948 (1995)
22. Kim, J.H.: Estimating classification error rate: repeated cross-validation, repeated hold-out and bootstrap. Comp. Stat. Data Anal. **53**(11), 3735–3745 (2009)
23. Rajaratnam, N.: The hydraulic jump as a well jet. J. Hydraul. Div. **91**(5), 107–132 (1965). https://doi.org/10.1061/JYCEAJ.0001299
24. Rao, N.S.G.: Ramaprasad: application of momentum equation in the hydraulic jump. La Houille Blanche **52**(4), 451–453 (1966)
25. Rapin, J., Teytaud, O.: Nevergrad - A gradient-free optimization platform. https://GitHub.com/FacebookResearch/Nevergrad (2018)
26. Rapin, J., Bennet, P., Centeno, E., Haziza, D., Moreau, A., Teytaud, O.: Open source evolutionary structured optimization. In: Proceedings of the 2020 Genetic and Evolutionary Computation Conference Companion, pp. 1599–1607 (2020)
27. Safranez, K.: Wechselsprung und die energievernichtung des wassers. Bauingenieur **8**(49), 898–905 (1927)
28. Santucci, V., Baioletti, M., Di Bari, G.: An improved memetic algebraic differential evolution for solving the multidimensional two-way number partitioning problem. Expert Syst. Appl. **178**, 114938 (2021)
29. Santucci, V., Baioletti, M., Milani, A.: An algebraic differential evolution for the linear ordering problem. In: Proceedings of the Companion Publication of the 2015 Annual Conference on Genetic and Evolutionary Computation, pp. 1479–1480 (2015)
30. Santucci, V., Milani, A.: Particle swarm optimization in the EDAs framework. Soft Comput. Ind. Appl. **96**, 87–96 (2011)
31. Smetana, J.: Studi sperimentali sul salto di Bidone libero e annegato. Energ. Elettr. **24**(10), 829–835 (1937)
32. Storn, R., Price, K.: Differential evolution - a simple and efficient heuristic for global optimization over continuous spaces. J. Global Optim. **11**(4), 341–359 (1997)
33. Yao, X., Liu, Y.: Fast evolution strategies. In: Angeline, P.J., Reynolds, R.G., McDonnell, J.R., Eberhart, R. (eds.) EP 1997. LNCS, vol. 1213, pp. 149–161. Springer, Heidelberg (1997). https://doi.org/10.1007/BFb0014808

A Multispectral Image Classification Framework for Estimating the Operational Risk of Lethal Wilt in Oil Palm Crops

Alejandro Peña[1,2], Alejandro Puerta[1], Isis Bonet[1(✉)], Fabio Caraffini[3],
Mario Gongora[4], and Ivan Ochoa[5]

[1] EIA University, Envigado, Colombia
japena@eafit.edu.co, {alejandro.puerta,isis.bonet}@eia.edu.co
[2] Escuela de Administración, Universidad EAFIT, Medellín, Colombia
[3] Department of Computer Science, Swansea University, Swansea, UK
fabio.caraffini@swansea.ac.uk
[4] Institute of Artificial Intelligence, School of Computer Science and Informatics,
De Montfort University, Leicester, UK
mgongora@dmu.ac.uk
[5] UNIPALMA de Los Llanos S.A, Meta, Meta, Colombia
ivan.ochoa@unipalma.com

Abstract. Operational risk is the risk associated with business operations in an organisation. With respect to agricultural crops, in particular, operational risk is a fundamental concept to establish differentiated coverage and to seek protection against different risks. For cultivation, these risks are related to the agricultural business process and to external risk events. An operational risk assessment allows one to identify the limits of environmental and financial sustainability. Specifically, in oil palm cultivation, the characterisation of the associated risk remains a challenge from a technological perspective. To advance in this direction, researchers have used different technologies, including spectral aerial images, unmanned aerial vehicles to construct a vegetation index, intelligent augmented platforms for real-time monitoring, and adaptive fuzzy models to estimate operational risk. In line with these technological developments, in this article we propose a framework for the estimation of the risk assessment associated with the disease of Lethal Wilt (LW) in oil palm plantations. Although our purpose is not to predict lethal wilt, since the framework starts from the result of a prediction model, a model to detect LW in an early stage is used for the demonstration. For the implementation of the prediction model, we use a novel deep learning system based on two neural networks. This refers to a case study conducted at UNIPALMAS. We show that the suitability of our system aims to evaluate operational risks of LW with a confidence level of 99.9% and for a period of 6 months.

Keywords: Oil Palm · Lethal Wilt · Sustainability · Vegetation Index · Unmanned Aerial Vehicles · Convolutional Neural Network · Deep Learning · Augmented Intelligence

© The Author(s), under exclusive license to Springer Nature Switzerland AG 2023
J. Correia et al. (Eds.): EvoApplications 2023, LNCS 13989, pp. 491–506, 2023.
https://doi.org/10.1007/978-3-031-30229-9_32

1 Introduction

The Colombian government considers oil palm a reference crop, as all of its agro-industrial processes generate jobs in rural areas, including those post-conflict [11], where it represents a new lease of life for local communities. In Colombia and throughout South America, diseases such as 'lethal Wilt' (LW) [2,7] cause ecological and economic disasters, making it very difficult to plan activities and run the business in a sustainable way.

Among the most successful characterisation systems for agricultural crops, it is worth mentioning those that use multispectral / hyperspectral images to perform non-destructive phytosanitary diagnoses *on site*. Specifically, in oil palms, some work has been developed [13]. A widely used methodology to exploit the intrinsic information content of these images is the extraction of 'Vegetation Indices' (VIs) [12,13,17,21]. Several works highlight the use of Deep Learning (DL) models [6,10,22]. Some authors have focused on the prediction of oil palm diseases [1,25].

In the literature, researchers are also proposing models specifically to predict the spread of LW [7]. Despite the developments implemented, preventing its transmission is a very challenging task since an infected palm can be contagious already six months before the appearance of symptoms, and is highly contagious four months before they appear, the latter can spread the pathogen readily from diseased to healthy units prior to eradication of visibly infected plants. Due to the high risk of contamination, when it is suspected that a palm has contracted LW, it is ideal to immediately treat it, as well as those that surround it with a radius of two palm units, with chemical pesticides. Furthermore, when the diagnosis is confirmed, the entire hectare must be eradicated [19]. This large elimination of crop areas produces severe losses and the need to replant in the nonproductive stages of a palm tree [11]. In addition, more fertiliser is also needed to accelerate growth and more insecticides are used for soil treatment. The increase in these chemical products carries economic and environmental risks.

For these reasons, the Roundtable Sustainable Palm Oil (RSPO) has provided practitioners with standards and procedures to follow to achieve both the environmental and financial sustainability of such crops [18]. These procedures are essential in South America and worldwide and aim at reducing the need for pesticides and fertilisers while expanding the limits of costs associated with important loss descriptors.

However, estimating the risk of LW-affected plantations is not an easy task, mainly due to the impossibility of having a diagnosis before the occurrence of symptoms by human inspection. Clearly, the solution to this problem is highly dependent on technological advances that make it possible to assess the phytosanitary states of palm units in the very early stages. In this light, we propose a framework for the estimation of operational risk that makes use of predictive models, such as in [6] to function.

Our newly proposed approach to operational risk has the advantage of being applicable to different types of crops and diseases and offers a generic framework to follow to identify the greatest losses.

The remainder of this article has the following structure. In the first section, we will detail the methodology for the analysis and validation of the proposed model. The second section presents the analysis and discussion of the results obtained, according to a series of parameters and metrics that define the general methodology to estimate operational risk with reference to agricultural crops. In the final section, we conclude and present an orientation for future work to improve both environmental and financial sustainability for this type of crop and for agricultural crops in general, when using the concept of computational intelligence.

2 Materials and Methods

This piece of research used real-word data that were carefully collected as part of the project [14] in a Colombian plantation located in the 'Santa Barbara area of Villavicencio' (owned by our collaborative panther Unipalma S.A., with a size of approximately 5000 hectares. [24]). This plantation contains two widely planted oil palm species, namely, the African palm *Elaeis guinensis* and the American palm *Elaeis oleifera - OxG* [19].

The data taken are only representative; the study could not cover all time necessary to collect data successively, but in some ways it is a significant sample to demonstrate the effectiveness of a model of this type. With a small experiment, such as the one described below, it was possible to create a base with diseased palms and healthy palms.

2.1 Acquiring and Preparing the Data Set

A *DJI Phantom 3* series [9] drone equipped with a *Sequoia Parrot* multispectral camera [20] was used to acquire relevant Multispectral Aerial Images (MAIs) from the study area at different heights using four reflectance bands:

- Green (550 nm $\pm$ 40 nm);
- Red (660 nm $\pm$ 40 nm);
- Red Edge (735 nm $\pm$ 40 nm);
- Near Infrared (790 nm $\pm$ 40 nm);

and a resolution of $13\,cm/px$.

Empirically, we determined the optimal flight height of reference to be at $50\,m$, which allows for a coverage of $4200\,m^2$ with a resolution of $1.2\,Mpx$ per band (i.e. $1280\,px \times 960\,px$). Note that weather conditions (e.g., wind, rain, etc.) might have a negative impact on maintaining this height, as well as on the UAV's autonomy, thus preventing each performed flight from being ideal. However, we maintained acceptable heights throughout the acquisition process with sporadic fluctuations never exceeding $120\,m$.

A total of 32 flights were carried out from different locations in the study area over a period of six months. This made sure that the images were taken under different climate conditions. In detail, eight flights were carried out during '*month*

1', 8 more after three months, i.e. 'month 3', and the remaining eight flights were carried out after three more months in 'month 6'. Each flight returned 80 MAIs, thus totalling $8 \times 80 \times 3 = 1920$ images throughout the acquisition period. Therefore, every three months, we stored a sample of approximately 38400 oil palm units, for a total of 115200 units that were considered in this study.

To create the training data set, the collected MAIs were manually segmented so that the cropped images can have a maximum of 50 pixel overlap. This allowed for a varied set of images containing palm units, as well as images containing no palm units at all or no palm units at all. These images are resized to a format of $300\,px \times 300\,px$. This process was performed for all available spectral bands, i.e., red (R), green (G), red edge (RE), near-infrared (NI), and for three vegetation indices. For each MAI and VI, 250 cropped images (CI), where approximately 100 CI correspond to morphologically complete oil palm units (MCOP). The oil palms that form the MCOP set were grouped in healthy palms (ELM), palms with advanced LW (SLM), and palms that are suspicious of LW (ULM).

2.2 Defining Losses and Operational Risk

The Basel II agreement defines Operational Risk (OR) as '... the possibility of incurring in losses due to deficiencies, failures or inadequacies of in human resources, processes, technologies, infrastructure or by the occurrence of external events...' [16].

The aggregate loss distribution (ALD) is one of the most widely used methods for measuring operational risk by financial institutions. It models the aggregate loss of all events that occurred over a time horizon, typically one year. This approach is based on the internal data collected by the business line/risk event. Each risk event can be quantified as the product of the number of loss events (frequency distribution) and the total loss amount (severity distribution) over a period of time (usually one day) [3].

In the Basel III agreement, the ALD distribution is called the Loss Component (LC) [4]. This structure is defined by three risk parameters: Expected Losses (EL), Stress Losses (SL), and Unexpected Losses (UL). These parameters, seen from the point of view of risk of losses due to LW in oil palm, can be interpreted as: EL_{LW} are those losses occurring when oil palm units are correctly characterised by the model as surely being ill before the appearance of LW symptoms - these are treatable and do not require eradication; UL_{LW} relate to those oil palm units for which the model returns a 0.5 ± 0.1 probability of having LW - these uncertain (nonfrequent) cases cannot be classified and could turn into unexpected costs if an expert is requested to check them or if LW symptoms appear in a second moment; SL_{LW} are those losses from oil palm units that were erroneously characterised by the model as healthy eradication will be the only option as soon as advanced LW symptoms are visible.

In agricultural crops affected by the phytosanitary state or agroclimatic events, risk parameters (parametric insurances) are defined through the ALD as follows:

- the '*Operational value at risk*' (OpVar) parameter is a reference value for the financial coverage of OR at which 99.9% of LC is covered;
- The '*Hedge of Expected Losses*' (HEL) parameter establishes the hedge for ELs as shown below:

$$HEL = \frac{EL}{LC} \qquad (1)$$

- the '*Hedge of Unexpected Losses*' (HUL) parameter establishes the hedge for UL as shown below:

$$HUL = \frac{UL}{LC} \qquad (2)$$

In practise, the definitions of the 'hedges express a normalised version of the OpVar value in terms of EL and UL.

2.3 The Employed Vegetation Indices

A Vegetation Index (VI) is a spectral transformation involving at least two frequency bands from a multispectral image. In precision agriculture, these are widely used to highlight relevant vegetation properties and perform spatial and temporal comparisons of terrestrial photosynthetic activity and canopy structural variations [23]. We make use of the following three widely used VIs:

- the Normalised Difference VI (NDVI)

$$NDVI = \frac{\Phi_{NI} - \Phi_{R}}{\Phi_{NI} + \Phi_{R}}; \qquad (3)$$

- the Green Normalised Difference VI (GNDVI)

$$GNDVI = \frac{\Phi_{NI} - \Phi_{G}}{\Phi_{NI} + \Phi_{G}}; \qquad (4)$$

- the Near-Infrared VI (NIVI)

$$NIVI = \frac{\Phi_{NI}}{\Phi_{R}}. \qquad (5)$$

where Φ_{NI}, Φ_{RED}, and Φ_{GREEN} refer to the reflectance of the near-infrared (NI) band, the red band, and the green band, respectively. Note that NDVI, GNVDI $\in [-1, 1]$ and NIVI $\in [0, +\infty]$ - for more details, see [5, 23].

2.4 A Hybrid Neural Deep Learning System

The proposed model for predicting LW consists of the sequence of two neural systems, as shown in Fig. 1. These are designed to automatically isolate palm units from MAIs, which we collected as explained in Sect. 2.1, and detect the presence of LW respectively. In DL, the systems used are termed a Stacked

Neural Network (SNN) and a Convolutions Neural Network (CNN), configured as indicated in Sect. 2.4.

The output of this model allows for the estimation of the OR as explained in Sect. 2.2. Note that our framework for the OR associated with LW is generally valid and does not impose a limitation on the choice of the classifier as long as it returns a probability of classifying the phytosanitary state of a palm. However, to validate our approach, early prediction of LW plays a key role. As this is a challenging task, with the system in Fig. 1, we show a promising strategy to address this problem.

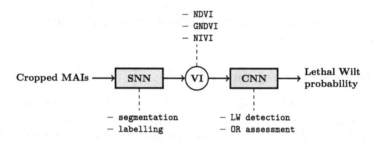

Fig. 1. Scheme of the proposed neural system and roles plated by its two main SNN and CNN components. Note that VIs are extracted in an intermediary step.

Our system first uses unsupervised training to generate a chain of autoencoders capable of extracting features from input cropped MAIs. Then, during the so-called 'stacked' step, the previously trained autoencoders form the hidden layers of the final neural architecture, which is equipped with a last 'softmax' layer to classify complete and incomplete palm units. A RELU activation function is then employed to finalise the network, which is trained through supervised back-propagation with the images prepared as indicated in Sect. 2.

The palm units labelled as 'complete' in the previous step are not being directly fed to the second convolutional block, but transformed into their VI representation. This step is key to extracting relevant vegetation properties, thus obtaining more appropriate input for CNN and more accurate classifications.

For the identification and labelling of MCOPs at a confidence level of 99.9%, Substructure 1 was configured with a total of 1000 learning cycles, for a total of 1000 iterations per cycle. For each learning cycle, the substructure performs a random sampling process with a total of 1000 CI and a total of 20,000 MCOPs that are available in each of the aforementioned databases (7 databases). The structure of the layers was configured for 3 stacked layers (L1, L2, L3) and with the following number of neurons per layer: $L1 : 1000$, $L2 : 500$, $L3 : 250$, and for an FCL that consists of 100 neurons, where a compression rate per layer close to 90% was ensured [15]. The analysis and validation focused on VIs, due to the information that this structure provides on the vigour of the plant in the study zone.

Substructure 2 was configured using the same learning strategy used for the configuration of Substructure 1. For Substructure 2, the input data arise as a result of a random sampling process with reference to a total of 1000 oil palm units, based on the set of MCOPs that were available for month 0, and according to the percent composition that defines the loss structure for Scenario 0, as shown in Table 1. Convolutional patterns were also selected using a similar random sampling process at a confidence level of 99.9%, taking as a reference the set of MCOPs that was available for a period of 6 months. This convolutional process allows for creating a forecast structure to characterise LW in oil palm crops six months before reaching the advanced LW stage. In general terms, the data for each scenario were grouped into 70% for learning (700 MCOPs) and 30% for validation (300 MCOPs) for each learning cycle. The remaining data was used for the general evaluation of the model ($7'000\ MCOPs$).

The obtained probability is normally used for binary classification. However, in the context of our research, it is used to implement the proposed OR estimation approach of Sect. 2.2. This means that we are able to exploit the model to quantify EL_{LW}, UL_{LW} and SL_{LW}, thus being able to build an ALD for the six-month period under investigation.

2.5   Evaluation Metrics

We use established performance metrics to analyse and validate our model.

Categorical Cross Entropy (CCE). CCE is a loss function used to measure how two distributions are different from each other, these being the predicted values $\hat{y}$ and the available values y^D from the D data set in our case. Mathematically, this is defined as

$$CCE = -\frac{1}{|D|} \sum_{k=1}^{|D|} \left[y_k^D \log\left(\hat{y}_k\right) + \left(1 - y_k^D\right) \log\left(1 - \hat{y}_k\right) \right]. \tag{6}$$

Accuracy Weighted Index (ACC$_W$). The ACC$_W$ index is the sum of all fractions of correct predictions for each class, averaged by the total number of samples in the data set. This is calculated as shown below:

$$ACC_W = \frac{1}{|D|} \sum_{i=1}^{C} w_i T_i \tag{7}$$

where

- C is the number of classes;
- T_i is the number correct predictions for the ith class;
- W_i is the frequency of the ith class.

Mean Square Error (MSE) and Root Mean Squared Error (RMSE).
The MSE is a widely used metric mathematically expressed as

$$\text{MSE} = \frac{1}{|D|} \sum_{k=1}^{|D|} \left(y_k^D - \hat{y}_k \right)^2. \tag{8}$$

RMSE is the square root of MSE

3   Experimental Setup and Validation

3.1   Assessment Scenarios

The proposed hybrid model is evaluated twice over the six-month period required to monitor the evolution of LW, as previously explained in Sect. 2.

The first evaluation process is performed on the entire six-month period and takes into consideration two reference risk scenarios based on

- The costs attributed to the natural evolution of LW;
- the costs attributed to the limits of environmental and financial sustainability as indicated in the 'RSPO standards';

respectively. The relevant losses for these two scenarios are quantified per 1,000 units of oil palm (that is, approximately ~ 6.9 hectares) and are shown in Table 1.

Table 1. Losses for the reference scenarios of the RSPO standards and natural evolution based on 1000 oil palm units.

	EL	UL	SL
RSPO Standard	857	142	1
Natural Evolution	607	105	288

For the second evaluation process, we considered four assessment scenarios monitoring the evolution of the phytosanitary states for palm units at different temporal checkpoints. The first scenario assesses the progress of the LW during *month 1* (M1), the second scenario during *month 3* (M3), and the third scenario during *month 6* (M6). The fourth scenario instead considers the available palm units during the entire semester, i.e. months 1–6 (M1-6).

Statistical results for the evaluation of the proposed model are calculated with a confidence level of 99.9% and considering 1000 oil palm units that were randomly chosen from the available ones in the study area at the specific time corresponding to each evaluation scenario.

3.2 Quantifying the Losses

To use and evaluate our model, loss values must be quantified. This is also needed to calculate the ALD and estimate the OR of LW.

An approximation of the costs involved in the management and treatment of oil palm crops affected by LW is provided in Table 2, where the values refer to the USD currency, sometimes given in the form of an expected range where costs can fall (i.e. [min, max]).

Note that costs have been classified in terms of EL, UL and SL values. In detail, all agricultural activities required for the management and treatment of healthy palms with fertilisers have grouped to form the ELs; all fumigation costs, required to treat all palms forming two rings around a palm diagnosed with 'early LW', form the ULs; eradication costs due to the presence of an oil palm unit that is diagnosed with 'advanced LW', which means that one entire hectare of crop will be lost (i.e., 144 palm units), form the SLs.

It should be reported that the SL class includes activities that involve the use of large amounts of pesticides and fertilisers for the treatment of the soil and the eradication of units (Table 2) [11]. Therefore, they contribute the most to pollution and negatively impact the environment.

Table 2. Operational Costs of Oil Palm Crops. All costs are expressed in USD and are estimated over a two-year period. For the sake of clarity, costs are grouped so that (a) displays all ELs, (b) all ULs and (c) all SLs.

(a) Maintenance Costs – EL

Fertilization	274.53
Water for Irrigation	84.65
Management	144.74
Pruning or Ablation	32.19
Other Costs	250.42

(b) Treatment Costs – UL

Sanitary Control	90.34
Weed Control	150.10
Technical Assistance	84.77

(c) Stress Costs – SL

Maintenance Costs (ha.)	[1214.89; 1921.93]
Elimination Costs (ha.)	[688.35; 704.58]
Unproductive Costs (ha.)	[5533.33; 5933.33]
Total Costs (ha.)	[7448.22; 8555.26]
Number of Oil Palms (ha.)	144
Total Costs/Unit	[51.72; 59.41]

4 Experimental Results

The process of transformation of data with the use of vegetation index in combination with this network can extract good characteristics from the representation of the data. An example of the result of this image transformation process is shown in Fig. 2. Traffic light colormaps like the one shown with this index are usually easier to analyze visually. Green indicates healthy and red highlights hotspots or areas of dead or sparse vegetation.

Fig. 2. An example of NDVI representation of a 'completed' plam unit isolated by the SNN block.

Table 3 shows the results that were achieved by the proposed model with regard to the identification & labelling of MCOPs. In this table, the MCOPs, based on the NDVI and GNDVI, show the best results with regard the ACC-w index. This is mainly induced by the lower values that were reached for the CCE and RMSE indices. This confirms the importance of the NDVI and GNDVI indices in the characterisation of agricultural crops in general.

Table 3. Behaviour of the identification substructure vs the identification & labelling of MCOPs

	Training			Validation			Evaluation		
	NDVI	GNDVI	NRVI	NDVI	GNDVI	NRVI	NDVI	GNDVI	NRVI
CCE	0.0665	0.103	8.9563	0.014	0.8506	10.7454	0.0398	0.0643	8.59631
ACC_W	0.9933	0.9733	0.4667	0.7188	0.6667	0.3333	0.9933	0.9933	0.4666
RMSE	0.014	0.0206	0.5333	0.2198	0.2596	0.6667	0.0056	0.0086	0.5333

The results show that the model achieved a significant reduction in LC, close to 90%, starting from a value of $88'765.88USD/ha$ eradicated, down to a value of $8'726.14USD/ha$ eradicated, which places this value slightly above the LC value that defines the RSPO scenario, which reaches $7'309.90USD/ha$ eradicated. The EL, estimated by the proposed model, reached a value of $4'409.92USD/ha$ treated. A value that was slightly below the value established for the EL for Scenario 0, with a value of $5'580.29USD/ha$ treated. This fact shows the good

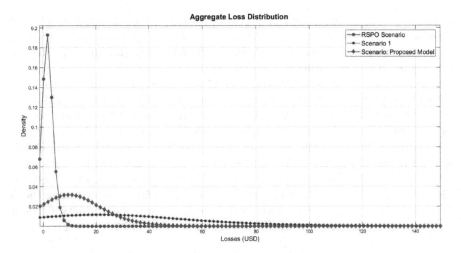

Fig. 3. Aggregate Loss Distribution, ALD - Risk Scenarios

behaviour reached by the proposed model with respect to the sustainability of this type of crops. LC was one of the factors that most affected the behaviour of the proposed model, which can be evidenced through the percentage composition of the coverage. These were below 75% on average ($HEL : 0.50536, HUL : 0.77335$), and below the percentage composition of the coverage established for scenario 0, with an ideal value above 80% on average ($HEL : 0.76438, HUL : 0.92216$).

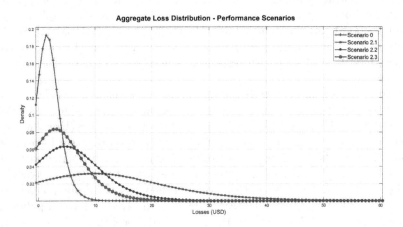

Fig. 4. Aggregate Loss Distribution, ALD - Performance Scenarios

Figure 3 shows the ALDs that represent the loss structure for each of the risk scenarios. The predominant distribution for losses is the logistic distribution, which is in line with the activation function that defines the FCL for the proposed model. This demonstrates the structural and dimensional stability of the proposed model with respect to the loss structure, defined for the RSPO scenario, and for the loss structure defined by the natural evolution of the disease in the study area (Scenario 1).

Figure 4 shows the general behaviour of the proposed model, related to a series of performance scenarios using different convolutional patterns (MCOP) that were obtained over a period of 6 months. In general, these ALDs present the structural characteristics of a long-tail distribution, where the log-logistic distribution stands out. The evolution of the log-likelihood index for this distribution toward larger values demonstrates that this distribution becomes more slender, which is due to a better performance that is characterised by a larger EL value and more extended limits of sustainability. This fact also guarantees both structural stability (the same distribution) and dimensional stability (slender distribution) for the proposed model when characterising different risk scenarios [8].

With regard to the convolutional patterns, we can observe that the proposed model achieved the best performance for the convolutional patterns that were obtained for month 6 (Scenario 2.3). This is due to a better differentiation between healthy oil palms (input data - month 0) and oil palm units that are affected by advanced LW (reference data - month 6). This performance was validated by the coverage, which achieves the highest values for all performance scenarios (HEL : 0.92313, HUL : 069216). However, the low performance was mainly due to the low differentiation between healthy oil palms (month 0) and healthy palms that will develop LW over a period of six months (month 0) (Scenario 2.1). In this scenario, the coverage reached values below 50% on average. It is important to mention that for Scenario 2, the convolutional patterns were selected over a period of 6 months as a result of a random sampling process (mixed temporal pattern), which places the model (ELM:858,ULM:113,SLM:29) slightly below Scenario 2.2. A scenario that shows a medium behaviour (ELM: 860, ULM: 119, SLM: 21) with respect to the SL exhibited by the other risk scenarios.

From a theoretical perspective, Scenario 2.1 constitutes one of the best performing scenarios for the proposed model, since the SLs are assumed as UL (UL = SL), while the UL are added to EL (EL = EL+UL: 11, 754.58 (USD/ha)). This fact considerably extends the sustainability limit. This is due to the identification of the disease in the prior stages, before its appearance, which will clearly allow a differentiated treatment of crop units by the type of risk (parametric risks). For the model, Scenario 2.3 constitutes one of the worst performing scenarios, as oil palms that are part of the UL category can quickly turn into oil palms with advanced LW due to late identification of the disease (SL = SL+UL:9, 837.44 (USD/ha)).

The cumulative distribution functions (CDF - Log-logistic distribution) are grouped by the performance of a scenario. Table 4 shows the parameters that define each CDF, based on a generalised log-logistic structure. The *Beta* parameter (slender parameter) is decreasing for EL and for much more extended CDF

Table 4. Log-logistic structure - Normalised weights

Scenarios	Scenario 2.3			Scenario 2.2			Scenario 2.1		
Parameters	EL	UL	SL	EL	UL	SL	EL	UL	SL
a	−0.83	−1.53	−1.38	−0.90	−1.77	−1.38	−0.82	−1.74	−1.36
α	1.04	0.82	0.99	1.15	1.17	0.99	1.12	0.87	1.02
β	1.95	6.47	14.45	3.32	11.16	14.45	7.12	8.11	9.32

distributions, with wider limits for sustainability, due to better differentiation (Scenario 2.3). However, this parameter is increasing for SL losses, thus generating more slender distributions with less SL, due to better identification of oil palm units with advanced LW. With respect to the parameter a, we can state that this parameter presents the highest values, in general, for the EL category, which exhibits the effect of the weights on the classification of oil palms in the EL category. On the contrary, this value presents low values, in general, for the SL category, which underlines the good classification performance of the proposed model.

The α *factor* (stability factor) remains close to unity, which demonstrates the stability of the behavior of the model when characterizing OR for different loss categories. It is important to note that the UL distributions had the least change. This is due to a better identification of oil palm units with advanced LW, which also favoured the expansion of the sustainability limit (EL limit).

5 Conclusions and Future Work

This paper is a contribution to the assessment of operational risk using an advanced model in order to evaluate the risk taking into account the prediction of losses based on the number of palms that can get a disease. This shows that a model for early prediction of disease in palms can help to better adjust the budget and even lower costs if palms can be treated in time.

The proposed model establishes a general methodology for the configuration of parametric risks in agricultural crops, integrating in a single structure different vegetation indices to measure plant vigour. The model is based on multispectral images and a convolutional mechanism for the characterisation of phytosanitary states at an early stage. Hence, complying with one of the main objectives of parametric insurance, which consists of expanding the limits of environmental and financial sustainability, through a significant reduction in the use of pesticides. This is possible due to the differentiated treatment of crop units that are affected by a pathology in specific areas of the study zone.

The model was compared with other scenarios and can be seen that the OpVar decreases, and in the decomposition of the losses you can see this too. That is exactly what we wanted to demonstrate at the beginning that a model to estimate OpVar having a classifier for diseases such as LW can decrease OpVar, and this is clear in the three components into which it was divided. Regardless of the classifier, if the percentage of correct classification is good, this model will increase by this measure.

Following this investigation, we envisage proposing several more models to be able to select the one achieving better prediction; the extension of this promising system to other plant's diseases that can also cause losses. In addition, we propose the integration of a series of IoB devices (Internet of Beings) to identify the behaviour of the vector (*Haplaxius crudus*) that produces the dispersion of LW in a study zone. This allows for the diagnosis and localisation of the disease in stages before its appearance. Consequently, an augmented intelligence platform could be implemented to monitor oil palm crops that are susceptible to disease for an ostensible improvement of UAV autonomy and to characterise risk in wide areas of land.

Acknowledgements. This piece of research is a part of the project [14] which has received funds from The Royal Academy of Engineering under the Newton Fund Industry Academia Partnership programme (reference: IAPP1\100130).

Furthermore, we want to thank Unipalma S.A. [24] for letting us access their facilities and for all support provided to carry out our project.

References

1. Alaa, H., Waleed, K., Samir, M.S., Tarek, M., Sobeah, H., Abdul, M.: An intelligent approach for detecting palm trees diseases using image processing and machine learning. Int. J. Adv. Comput. Sci. Appl. **11**, 0110757 (2020)
2. Alvarez, E., Mejía, J.F., Contaldo, N., Paltrinieri, S., Duduk, B., Bertaccini, A.: Candidatus phytoplasma asteris strains associated with oil palm lethal wilt in Colombia. Plant Dis. **98**(3), 311–318 (2014)
3. Bank for International Settlements: Operational risk. Supporting document to the new basel capital accord. https://www.bis.org/publ/bcbsca07.pdf (2001)
4. Basel Committee on Banking Supervision: high-level summary of Basel III reforms. Bank for International Settlements (2017). https://www.bis.org/bcbs/publ/d424_hlsummary.pdf
5. Bhandari, A., Kumar, A., Singh, G.: Feature extraction using normalized difference vegetation index (ndvi): a case study of jabalpur city. Procedia Technol. **6**, 612–621 (2012). https://doi.org/10.1016/j.protcy.2012.10.074. 2nd International Conference on Communication, Computing & Security [ICCCS-2012]
6. Bonet, I., Caraffini, F., Pena, A., Puerta, A., Gongora, M.: Oil palm detection via deep transfer learning. In: 2020 IEEE Congress on Evolutionary Computation (CEC), pp. 1–8. IEEE (2020)
7. Fahy, C., Caraffini, F., Gongora, M.: A multi-agent system for modelling the spread of lethal wilt in oil-palm plantations. In: 2020 IEEE Congress on Evolutionary Computation (CEC), pp. 1–7 (2020). https://doi.org/10.1109/CEC48606.2020.9185742
8. Gonzalez-Ruiz, J.D., Peña, A., Duque, E.A., Patiño, A., Chiclana, F., Góngora, M.: Stochastic logistic fuzzy maps for the construction of integrated multi-rates scenarios in the financing of infrastructure projects. Appl. Soft Comput. **85**, 105818 (2019). https://doi.org/10.1016/j.asoc.2019.105818. http://www.sciencedirect.com/science/article/pii/S156849461930599X

9. Industries, D.P.: Phantom 3 Professional. https://www.dji.com/phantom-3-pro/info (2018)
10. Lu, Y., Yi, S., Zeng, N., Liu, Y., Zhang, Y.: Identification of rice diseases using deep convolutional neural networks. Neurocomputing **267**, 378–384 (2017). https://doi.org/10.1016/j.neucom.2017.06.023. http://www.sciencedirect.com/science/article/pii/S0925231217311384
11. Mosquera, M., Valderrama, M., Ruiz, E., Lopez, D., Castro, L.: Costos de producción para el fruto de palma de aceite y el aceite de palma en 2015: estimación en un grupo de productores colombianos. Palmas **38**(2), 11–27 (2017)
12. Paula Ramos, P., Solarte, E., Valdés, C., Sanz, J., Gómez, E.: Características espectrales de la luz reflejada por frutos de café (coffea arabica). Revista de la Sociedad Colombiana de Física (2018)
13. Pena, A., Patino, A., Bonet, I., Gongora, M.: Fuzzy spatial maps to identify oil palm units: spatial fuzzy maps. In: 2018 13th Iberian Conference on Information Systems and Technologies (CISTI), pp. 1–6 (2018). https://doi.org/10.23919/CISTI.2018.8399144
14. Pena, A., Gongora, M., Bonet, I., Caraffini, F.: Intelligent system to improve the sustainability of oil palm crops through the construction of forecasting maps. https://doi.org/10.21253/DMU.11638095 (2020)
15. Peña, A., Bonet, I., Manzur, D., Góngora, M., Caraffini, F.: Validation of convolutional layers in deep learning models to identify patterns in multispectral images. In: 2019 14th Iberian Conference on Information Systems and Technologies (CISTI), pp. 1–6 (2019). https://doi.org/10.23919/CISTI.2019.8760741
16. Peña, A., Bonet, I., Lochmuller, C., Chiclana, F., Góngora, M.: Flexible inverse adaptive fuzzy inference model to identify the evolution of operational value at risk for improving operational risk management. Appl. Soft Comput. **65**, 614–631 (2018). https://doi.org/10.1016/j.asoc.2018.01.024. http://www.sciencedirect.com/science/article/pii/S1568494618300309
17. Peña, P.A., Patiño, P.A., Velásquez, V.J., Góngora, M.: Intelligent system to identify oil palm crop units from multispectral aerial images: Identification of multispectral patterns. In: 2017 12th Iberian Conference on Information Systems and Technologies (CISTI), pp. 1–7 (2017). https://doi.org/10.23919/CISTI.2017.7975991
18. Rspo roundtable for sustainability palm poil: RSPO principles and criteria for the production of sustainable palm oil. PNG Oil Palm Industry Corporation, Department of Agriculture and Livestock (2018)
19. Departamento Agronómico - UNIPALMA S.A.: Guia Técnica - Polinización Asistida en Palma de Aceite (2018). https://www.proamazonia.org/wp-content/uploads/2021/05/Guia-Tecnica-para-polinizaci%C3%B3n-asistida_compressed_web.pdf
20. S.A., P.: Parrot Sequoia - Multispectral Camera. https://www.parrot.com/soluciones-business/profesional/parrot-sequoia (2018)
21. Shamshiri, R.R., Hameed, I.A., Balasundram, S.K., Ahmad, D., Weltzien, C., Yamin, M.: Fundamental research on unmanned aerial vehicles to support precision agriculture in oil palm plantations. In: Zhou, J., Zhang, B. (eds.) Agricultural Robots, chap. 6. IntechOpen, Rijeka (2019). https://doi.org/10.5772/intechopen.80936
22. Sladojevic, S., Arsenovic, M., Anderla, A., Culibrk, D., Stefanovic, D.: Deep neural networks based recognition of plant diseases by leaf image classification computational intelligence and neuroscience. Comput. Intell. Neurosci. **2016**, 11 (2016)

23. Torres-Sánchez, J., López-Granados, F., De Castro, A.I., Peña-Barragán, J.M.: Configuration and specifications of an unmanned aerial vehicle (uav) for early site specific weed management. PLOS ONE **8**(3), 1–15 (2013). https://doi.org/10.1371/journal.pone.0058210
24. Unipalma, S.A.: Unipalma. https://www.unipalma.com/. Accessed 30 Dec 2020
25. Yarak, K., Witayangkurn, A., Kritiyutanont, K., Arunplod, C., Shibasaki, R.: Oil palm tree detection and health classification on high-resolution imagery using deep learning. Agriculture **11**(2), 1–16 (2021)

An AI-Based Support System for Microgrids Energy Management

Alejandro Puerta[1], Santiago Horacio Hoyos[1], Isis Bonet[1(✉)] ⓘ,
and Fabio Caraffini[2] ⓘ

[1] EIA University, Envigado, Colombia
{alejandro.puerta,santiago.hoyos95,isis.bonet}@eia.edu.co
[2] Department of Computer Science, Swansea University, Swansea, UK
fabio.caraffini@swansea.ac.uk

Abstract. Decarbonisation of the economy is the key to reducing greenhouse-effect gas emissions and climate change. One of the ways decarbonisation of economy is electrification of economic sectors. In this case, the implementation of micro-grids in different economic sectors such as households, industry, and commerce is a great mechanism that allows the integration of renewable energies into the electrical power system and to contribute with accelerated energy transition for decarbonisation. However, micro-grids include self-generation through renewable energy and distributed generation, as well as energy efficiency in the consumer. Micro-grids have energetic, economic, and environmental benefits for the user and the power system, but for the security of the energy supply it is necessary to balance the offer and demand of electricity at all times, which in this case must be estimated for the market of the next day. The problem here is how to estimate generation and consume for the next day when the determinant of offer and demand are variable. This paper proposes algorithms of forecasting based on machine learning with high accuracy in a decision support system of management of energy for a micro-grid.

Keywords: micro-grid energy management · artificial intelligence · machine learning · decision support system

1 Introduction

With the arrival of the 4th industrial revolution, technological and regulatory changes, new business models emerge to transform the way companies and people carry out their productive activity. The electricity sector is proof of this, with the adoption of non-conventional sources of renewable energy to ensure reliable supply in an environmentally friendly manner [5].

Due to the rapid adoption of decentralised generation systems in transmission and both distribution and storage systems [1], micro-grids have become an important element of resilience and reliability in this section and, in particular,

J. Correia et al. (Eds.): EvoApplications 2023, LNCS 13989, pp. 507–518, 2023.
https://doi.org/10.1007/978-3-031-30229-9_33

in the field of smart grids. However, as the participation of new market players also continues to increase, this wide use of micro-grids has also raised new challenges for some traditional and regulated utilities.

By definition, a "Micro-Grid is an interconnection system with the capacity to be self-sufficient and operate in isolation if necessary. It includes both generation, storage, and electricity transport, as well as equipment to optimise intelligent energy management by the end user" [12]. A clear example of a micro-grid is a group of houses where each has its own solar system installed to supply their consumption, and they all connect to share their surplus energy. In current times, the Internet of Things (IOT) paradigm is widely used for the implementation of these energy networks, adding more complexity in controlling and balancing between supply and demand. Hence, the need to equip micro-grids with complex systems such as monitoring stations, storage systems, self-generation systems, and integrated demand response capacity in the electrical system also presents technical and economic challenges when it comes to ensuring effective supply to customers.

A clear example of the scenario depicted above is in the city of Medellín (Colombia), where models of certain segments of the city are being generated and applied to micro-grid simulation to understand how the passage of a standard energy structure affects a micro-grid structure [7]. Here, the incorporation of renewable energy into the electricity sector is regulated by technical economic and mechanisms derived from **Law 1715** of 2014 and the most recent **Law 2099** of 2021 on Energy Transition in Colombia, mechanism of renewable integration similar to others in the world [11].

The energy sector is in a process of change, where adaptation has been progressive and increasing on the part of both generators and consumers. Since trends change, consumption patterns adapt to the increase in technology and increasingly seek efficiency in new ways, such as solar energy, so the planning problem is visible. This occurred also with the introduction of micro-grids and distributed generation systems, which change the dynamics of the traditional network, thus providing a space for other technologies to be used.

In the past, energy forecasting has been a very important part of the energy industry and was used to find the correct distribution of resources. The forecasting problem has been approached using mathematical methods, such as long-term [9] and short-term [20] load forecasting, for more than a century. Spatial Load Forecasting has been widely used in transmission and distribution planning as it provides when, where, and how much electricity demand would grow [19]. When the late 2010 s s came, so the adoption of distributed generation by people, turning consumers into consumers, changing the dynamics of the system, therefore affecting the results of the forecasting systems, because of the constantly changing panorama, the forecasting solutions existing to that very moment were not working anymore; thus the need for a more robust and dynamic solution arises [10].

The growth of the economy is associated with a growth in the demand for energy, due to the fact that the different economic activities require energy consumption for their development, in this sense, the impacts of health events such as COVID-19 and climate events such as "El Niño" and "la Niña", generate changes in the load profiles of consumers [2], as well as in the availability and price of the energy supply, putting at risk the supply of energy to end users at reasonable prices.

Self-generation plants are normally based on renewable energies based on non-conventional sources such as photovoltaic solar energy, wind power and small hydroelectric plants. The availability of energy that these generation plants can deliver depends on various climatic factors, such as cloudiness, precipitation, temperature, wind speed, hours of sunshine during the day, and its intensity per unit area, among other factors, which determine variable and intermittent generation, thus causing uncertainty in the amount of energy available to meet demand in the short, medium and long term [13].

Similarly, the energy demand of households and companies is variable [6], either due to the incorporation of new efficient technologies that reduce consumption, due to the entry of new equipment and machinery to support growth, due to the reduction or increase in industrial and commercial production as a result of economic and commercial changes, due to restrictions on mobility due to the effect of a pandemic, among many other factors that explain why the energy demanded by a home or a company varies over time.

Until now, several ways have been used to try to solve the problem of distributed generation forecasting and planning, such as dynamic system control theories, game theories, dynamic forecasting, multi-system control, etc.

With the advancement of computer systems and programming, a great deal of knowledge has been generated in terms of prediction and forecasting systems through mathematical tools and artificial intelligence, where they are being used in all areas of the industry. For example, some authors develop probabilistic forecasting [15]. Neural networks are also very popular to solve this problem [8]. In recent years, deep learning methods have also been used [3, 16].

1.1 Aim and Objectives

The aim of this research is to develop a model to predict generation and consumption patterns. The model must offer an efficient energy management tool in a micro-grid, which requires adequate calculation and estimation of supply and demand, to reach a balance, and to obtain economic and environmental benefits. This balance between supply and demand seeks to minimise energy losses, such as excess supply, where the additional energy generated cannot be used, causing additional costs due to oversizing and opportunity costs if the said excess energy is not sold. However, problems such as blackouts and deficiencies in the energy supply service can arise when there is excess demand, as the supply is not capable of fully supplying the demand at any instant in time. This problem of imbalance between supply and demand usually also appears in micro-grids as a result of an intermittent supply of energy from solar and wind systems, since there is no sun and wind all day, nor is it of the same intensity at the hours of their presence.

We propose a model based on artificial intelligence which is capable of determining the behaviour of the micro-grid and uses those data to make decisions on the energy distribution at each moment.

The structure of the paper is as follows: the next section details the data, machine learning methods, and the metrics used; next, the results obtained are shown and discussed; finally, we draw conclusions and use them to plant future work in this direction.

2   Materials and Methods

2.1   Data

We analysed the literature and identified the predominant types of existing micro-grids. Note that solar energy is classified as a non-conventional source of renewable energy in this context, being a widely used commercial technology of low cost and easy appropriation by users. For this project, the most usable type is a micro-grid connected in alternating current to the national interconnected system, with the possibility of selling surpluses. This is illustrated in Fig. 1.

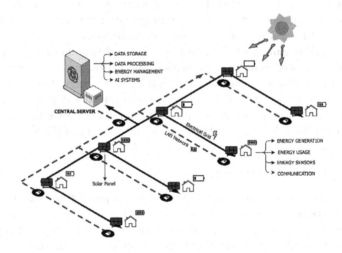

Fig. 1. Micro-grid topology used to test the model

For the consumption curves, we consider the three types of user listed below.

- Residential;
- Commercial;
- Industry.

Most of the information is taken directly from the public information system of the operator of the national interconnected system. The retrieved data allow us to have a minimum analysis horizon of one year, in hourly resolution. Furthermore, technical information provided by the Energy Mining Planning Unit (UPME) is taken from the study [17], where the user determines the load curves as a form of validation mechanism of the load profile obtained for that type of user. Such curves are made for an interval of 24 h, with 24 data per node regardless of the network nature, i.e. residential, commercial, or industrial.

For the operation of the micro-grid, we consider the variable.

- Node consumption;
- Node generation or node self-generation;
- Node storage;
- Prediction of node consumption;
- Prediction of node generation or node self-generation.

For the management of the micro-grid, we propose a generation prediction system. This can be used as input for the management system to make first decisions on to whom and when energy must be delivered.

Using the curve values introduced above, we develop three databases, one for consumption in homes, one for consumption in commercial establishments, and another for industrial entities. For the forecasting model, a data vector is created, with an adjustable window size, to process a future value, either generation or consumption. Data are available hour by hour for each curve in a 24-hour range. For this application, a window of 10 h was selected to predict the consumption of the next hour. The databases are implemented as a set of tables with 400 data each.

2.2 Simulation of Micro-Grid for Validation

Although management of model results is not the objective of this article, the data obtained and simulated from a micro-grid were used to validate future model performance.

For this simulation, we select a university. simulated university was divided into 3 nodes, which are: **Node 1:**

- Solar Generation.
- Consumption of complete block C and laboratories

Node 2:

- Complete Own consumption, with vehicle chargers included.
- Storage

Node 3:

- Consumption of the rest of the university

The data are organised in such a way that they could be analysed by days and hours, in order to perform an average of the data that would deliver a representative consumption curve of a normal consumption in one day in block C of the University. The average is calculated for Tuesdays, Wednesdays and Thursdays, to represent the average consumption of the day when the university is full-time. Note that on holidays or weekends, consumption drops considerably, which would return non-relevant data for our purpose.

The average consumption of Block C is stored hour by hour in KW, with a daily total consumption of 234.89 kWh/d.

The same process was performed on the solar generation data from the photovoltaic system installed at the university.

The data used for the definition of a solar generation curve are the historical ones for the year 2022, where every day was counted to make a curve of an average day, since in this case the system will generate energy in the same way, either on a weekday or weekend.

For the consumption of node 2, there are no historical data, but an average consumption curve was defined with the experts on the subject, based on the study of the Owl and its components, For this node average consumption curve, the assumption was made that consumption started at 5 AM and ended at 6 PM. With the consumption of the Owl, based on the analysis carried out, storage and charging stations for electric vehicles are taken into account. The consumption curve was made in conjunction with the university's expert in electrical matters to ensure that this curve represented as much as possible the real consumption that the electric vehicle cells could be generating.

For the consumption data of node 3, it was defined that it would be carried out as the consumption of the rest of the university, so the average data found is used, adding 85% of the generation of the solar panels, since this is consumed by the same Block C and then the consumption of nodes 1 and 2 was subtracted, so finding the remaining consumption of the rest of the university.

2.3   Machine Learning Algorithms

Support Vector Machine (SVM). This model was proposed by [18]. The key behind SVM is the first phase, where, using the kernel, it transforms the original N-dimensional data into another feature space. After the transformation of the data, the second part is the linear separation of the data. For this, by solving a quadratic equation, they try to find the optimal generalisation by separating the hyperplanes. They are based on a set of data called support vectors, which are the ones that guarantee choosing the best hyperplane. In the case of regression, this method is known as Support Vector Regression.

Artificial Neural Networks (ANN). This model consists of three basic layers: input layer, hidden layer (which can be more than one) and output layer [14]. The layers have interconnected neurons with weights and specific activation functions, which transform the data and pass them to the next layer. Given a network with a fixed set of units and interconnections, this network adjusts the weights using the back-propagation learning process. One of the oldest multilayer networks is the Madaline network, which is no less efficient for that.

eXtreme Gradient Boosting (XGBoost). XGBoost is an ensemble of classifiers based on several tress. It is considered as an optimised gradient tree boosting model, which is based on algorithmic optimisations. What distinguishes this model is the learning algorithm it has to handle sparse data. To optimise the loss function at each level of trees, XGboost uses an objective function, taking into account the results obtained in the previous level. It also adds regularisation to perform the results [4].

2.4   Evaluation Metrics

To measure the results and demonstrate how close the response given by the model was to the desired response, we relied on a measure of accuracy based on the well-known MAPE metric (Eq. 1). Since MAPE gives us an absolute error percentage, we calculate the accuracy as shown in Eq. 2.

$$MAPE = \left| \frac{Yk(t) - Yd(t))}{Yd(t)} \right| \cdot \frac{100}{N} \tag{1}$$

where,

   $Yk(t)$ is the response of the model at an instant of time t.
   $Yd(t)$ is the desired value of at an instant of time t.

$$Accuracy = 100 - MAPE \tag{2}$$

3   Results and Discussion

The results of the best models are shown, which in this case were SVM, Madelaine and XGBoost. Below are the results of these models in the Figs. 2, 3 and 4. In these figures, on the X axis we see the time instants of the series and on the Y axis the value in kilowatts, but normalized between 0 and 1. In red the desired learning value is shown for each instant t and in blue the forecasting of each model at time t.

   The support vector machine was implemented for one of the databases, in this case, home data-set. A hyperparameter adjustment was made, obtaining the best results with a Gaussian kernel, which are shown in Fig. 2.

   The SVM results delivered a success rate of 72.6%, which is not considered sufficient for the application, so the tests were continued in search of another algorithm that would allow greater precision.

   Given the nature of the Madaline, the network is capable of adapting to all kinds of dynamic system behaviour. For this case, there are 10 data inputs, 10 neurons in the internal layer, and one data output. The relevant tests were carried out and the results shown in Fig. 3 were achieved.

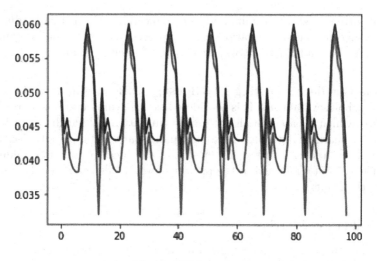

Fig. 2. SVM results

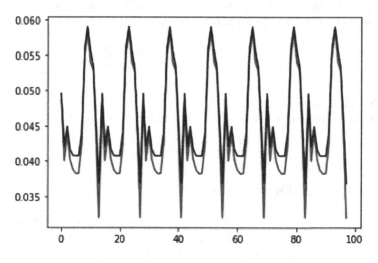

Fig. 3. ANN results

The results obtained are closer to what is sought, with 88.2% accuracy. However, the aim is to optimise the prediction process, so a third test was carried out to seek higher precision.

For the last test, XGBoost was implemented, which is very accurate when performing non-linear regressions of various behaviours. The results obtained are shown in Fig. 4.

The results obtained gave an accuracy of 93%, which is already considered a very good percentage. With the data structure, the tests carried out and the results obtained, it only remains to choose the language and environment to be used.

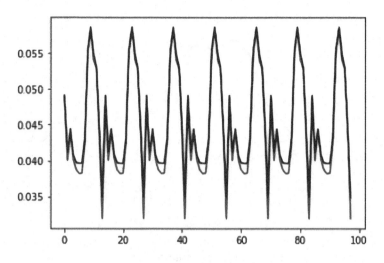

Fig. 4. XGBoost results

Using the data described above in Sect. 2.2 and the XGBoost methods trained as we have shown previously, we obtain the results shown in Figs. 5, 6 and 7. As can be seen, the consumption curve (Blue) is contrasted with the predicted curve (Orange), where it is observed that the prediction begins at 8:00 am.

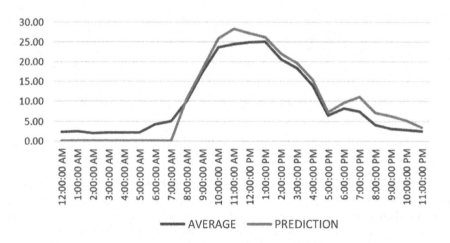

Fig. 5. Forecast result for Node 1

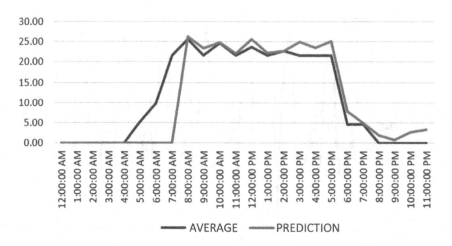

Fig. 6. Forecast result for Node 2

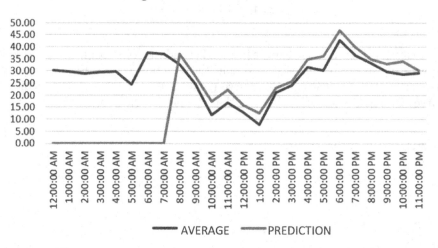

Fig. 7. Forecast result for Node 3

Figures 5, 6 and 7 show in blue the load curve for the average values of energy consumption in each hour of the day (real data) in comparison with the orange line, predicted energy consumption. It can be seen that in the 3 nodes, the prediction system was very successful, since the real curves and those delivered by the neural system mostly coincide with a good percentage accuracy of 88%, which allowed the system to manage the energy in a better way.

For the generation of node 1 (Fig. 5), the neural system also presents very good results, reaching 89% accuracy for the prediction of energy generation by the photovoltaic system.

It can be seen that in the 3 nodes, the prediction system was very successful, since the real curves and those delivered by the neural system mostly coincide with a good percentage accuracy of 88%, which allowed the system to manage the energy in a better way.

For the generation of node 1, the neural system also presents very good results, reaching 89% accuracy for the prediction of energy generation by the photovoltaic system.

It can be concluded with the model applied to the university micro-grid that the prediction model, being trained, is applicable to any type of system. This makes it scalable to any type of system if historical data are available, resulting in an adaptable intelligent management modelling and simulation system.

4 Conclusions and Future Work

The micro-grid model developed makes it possible to capture the characteristics of today's typical connected micro-grids, where the nodes represent users who can also be prosumers if they have solar self-generation systems.

The forecast models defined for this case represent the real behaviour considered for the consumption and generation of solar systems in the micro-grid, with a fidelity level of approximately 88%. These models are useful for planning and programming the operation of the micro-grid.

In the case of the validation on the simulated system of the EIA University micro-grid, it can be concluded that the system works adequately, achieving very promising results in the field of intelligent energy management, since the taking of decisions on how to use energy were correct according to the prediction made in the consumption and generation of each node, with percentages around 89% accuracy.

For the calculation of economic and environmental benefits, it can be concluded that the system works as expected, obtaining adequate results for the amount of energy used from the micro-grid itself according to the simulation itself.

For future work, it is proposed to validate the model in other types of micro-grid, with a greater number of nodes, and to build an energy allocation system according to the estimated prediction.

Acknowledgements. This research is part of the project "Strategy of Transformation of the Colombian Energy Sector in the Horizon 2030" funded by the call 788 of Minciencias: Scientific Ecosystem. Contract number FP44842-210-2018.

References

1. Binder, C.R., Knoeri, C., Hecher, M.: Modeling transition paths towards decentralized regional energy autonomy: the role of legislation, technology adoption, and resource availability. Raumforschung Raumordnung – Spatial Res. Planning **74**(3), 273–284 (2016). https://doi.org/10.1007/s13147-016-0396-5
2. Buechler, E., et al.: Global changes in electricity consumption during COVID-19. iScience **25**(1), 103568 (2022). https://doi.org/10.1016/j.isci.2021.103568

3. Cai, L., Gu, J., Jin, Z.: Two-layer transfer-learning-based architecture for short-term load forecasting. IEEE Trans. Industr. Inf. **16**(3), 1722–1732 (2020). https://doi.org/10.1109/TII.2019.2924326
4. Chen, T., Guestrin, C.: XGBoost: a scalable tree boosting system. In: Proceedings of the 22nd ACM SIGKDD International Conference on Knowledge Discovery And Data Mining, pp. 785–794 (2016)
5. Dogaru, L.: The main goals of the fourth industrial revolution. Renew. Energy Perspect. Procedia Manufact. **46**, 397–401 (2020). https://doi.org/10.1016/j.promfg.2020.03.058
6. EIA University: Sources of energy (2021). https://www.eia.gov/energyexplained/what-is-energy/sources-of-energy.php
7. Ali, F., et al.: Advancing from community to peer-to-peer energy trading in the Medellín-Colombia local energy market trial. IEEE Smart Cities, p. 200 (2022)
8. Hippert, H., Pedreira, C., Souza, R.: Neural networks for short-term load forecasting: a review and evaluation. IEEE Trans. Power Syst. **16**(1), 44–55 (2001). https://doi.org/10.1109/59.910780
9. Hong, T.: Energy forecasting: past, present, and future. foresight. Int. J. Appl. Forecasting **32** 43–48 (2014). https://ideas.repec.org/a/for/ijafaa/y2014i32p43-48.html
10. Hong, T., Pinson, P., Wang, Y., Weron, R., Yang, D., Zareipour, H.: Energy forecasting: a review and outlook. IEEE Open Access J. Power Energy **7**, 376–388 (2020). https://doi.org/10.1109/OAJPE.2020.3029979
11. International energy agency: climate change and energy transition law - policies - iea. https://www.iea.org/policies/13323-climate-change-and-energy-transition-law. Accessed 30 Jan 2023
12. Llano, M.M.: La micro-red inteligente: una ciudad eficiente, en miniatura. Revista universitaria científica, pp. 24–29 (2015). https://www.upb.edu.co/es/documentos/doc-ciudadeficienteminiatura-inv-1464100344537.pdf
13. Ma, J., et al.: Demand and supply-side determinants of electric power consumption and representative roadmaps to 100% renewable systems. J. Clean. Prod. **299**(2006), 126832 (2021). https://doi.org/10.1016/j.jclepro.2021.126832
14. Mitchell, T.M.: Machine Learning. Mcgraw-Hill science. Engineering/Math **1**, 27 (1997)
15. Nowotarski, J., Weron, R.: Recent advances in electricity price forecasting: a review of probabilistic forecasting. Renew. Sustain. Energy Rev. **81**, 1548–1568 (2018). https://doi.org/10.1016/j.rser.2017.05.234. https://www.sciencedirect.com/science/article/pii/S1364032117308808
16. Shi, H., Xu, M., Li, R.: Deep learning for household load forecasting-a novel pooling deep RNN. IEEE Trans. Smart Grid **9**(5), 5271–5280 (2017)
17. UPME: Redes Inteligentes (2019). https://www1.upme.gov.co/DemandayEficiencia/Paginas/Redes-Inteligentes.aspx
18. Vapnik, V.: The Nature Of Statistical Learning Theory. Springer science & Business Media (2013). https://doi.org/10.1007/978-1-4757-3264-1
19. Willis, H., Northcote-Green, J.: Spatial electric load forecasting: a tutorial review. Proc. IEEE **71**(2), 232–253 (1983). https://doi.org/10.1109/PROC.1983.12562
20. Zareipour, H.: Short-term electricity market prices: a review of characteristics and forecasting methods. Handbook of networks in power systems I, pp. 89–121 (2012)

Predicting Normal and Anomalous Urban Traffic with Vectorial Genetic Programming and Transfer Learning

John Rego Hamilton, Anikó Ekárt, and Alina Patelli[(⊠)]

Department of Computer Science, Aston University, Birmingham, UK
{j.connorregohamilton1,a.ekart,a.patelli2}@aston.ac.uk

Abstract. The robust and reliable prediction of urban traffic provides a pathway to reducing pollution, increasing road safety and minimising infrastructure costs. The data driven modelling of vehicle flow through major cities is an inherently complex task, given the intricate topology of real life road networks, the dynamic nature of urban traffic, often disrupted by construction work and large-scale social events, and the various failures of sensing equipment, leading to discontinuous and noisy readings. It thus becomes necessary to look beyond traditional optimisation approaches and consider evolutionary methods, such as Genetic Programming (GP). We investigate the quality of GP traffic models, under both normal and anomalous conditions (such as major sporting events), at two levels: *spatial*, where we enhance standard GP with Transfer Learning (TL) and diversity control in order to learn traffic patterns from areas neighbouring the one where a prediction is needed, and *temporal*. In the latter case, we propose two implementations of GP with TL: one that employs a lag operator to skip over a configurable number of anomalous traffic readings during training and one that leverages Vectorial GP, particularly its linear algebra operators, to smooth out the effect of anomalous data samples on model prediction quality. A thorough experimental investigation conducted on central Birmingham traffic readings collected before and during the 2022 Commonwealth Games demonstrates our models' usefulness in a variety of real-life scenarios.

Keywords: Nature-inspired computing for sustainability · Resilient urban development · AI-driven decision support systems · Intelligent and safe transportation · Urban traffic prediction

1 Introduction

Designing, building and maintaining an accessible, safe and cost-effective urban traffic infrastructure are key milestones on the road to meeting the UN's

This work is supported by the Engineering and Physical Sciences Research Council (Grant Number EP/R512989/1).

© The Author(s), under exclusive license to Springer Nature Switzerland AG 2023
J. Correia et al. (Eds.): EvoApplications 2023, LNCS 13989, pp. 519–535, 2023.
https://doi.org/10.1007/978-3-031-30229-9_34

Sustainable Development goals[1]. The complex decision making involved, particularly at the level of local administration, can be significantly streamlined when robust and reliable predictions of future traffic through key areas of the urban road network are made available by computational means rather than by exclusively relying on human expertise.

To that end, data-driven, Artificial Intelligence methods are widely recognised ways of producing computationally efficient models of dynamic, large-scale transport networks, such as modern day cities [3]. Within that category, evolutionary optimisation techniques, Genetic Programming (GP) in particular, offer an attractive solution to the urban traffic and modelling prediction problem. When combined with Transfer Learning (TL), traditional GP has been shown to predict vehicle flow with competitive accuracy, through urban junctions where traffic readings are either unavailable or unreliable [7,12]. Transferring traffic patterns learnt on areas that are topologically similar to, yet geographically distinct from, the region where a traffic model is needed but difficult or impossible to obtain practically is particularly important when vehicle flow is affected by atypical events (accidents, major sporting events, construction work, etc.). Besides accurately capturing vehicle flow under typical and atypical conditions by *spatial* means, as afforded by TL, traffic anomalies can also be mitigated on a *temporal* level. One way of achieving that is by equipping TL-enhanced prediction algorithms with a lag operator, making it possible to train traffic models on a dynamically adjusted number of past readings, counting backwards from an arbitrarily chosen sample in the training data set. We refer to this approach as *skipping*. Alternatively, the effect of anomalous data on the traffic models' prediction accuracy can be attenuated by applying adequate linear algebra operators that aggregate training samples. We refer to this approach as *smoothing*.

Our goal is to investigate the impact of spatial (TL) and temporal (skipping and smoothing) anomaly handling on the prediction accuracy of GP traffic models. We achieve this by proposing a rigorously validated, robust and competitively accurate approach to the urban traffic modelling and prediction problem. This approach rests on two original contributions:

1. A novel traffic prediction algorithm, $\overline{\text{GENTLER}}$, equipped with TL and smoothing, that produces models transferable to areas different to those where initial training occurs, without compromising prediction accuracy, and is tolerant to anomalous training samples.
2. A rigorous experimental investigation of $\overline{\text{GENTLER}}$'s prediction capability under both normal and anomalous traffic conditions. To enable that, $\overline{\text{GENTLER}}$ is deployed on several areas of the Birmingham city centre and trained on traffic data collected both before and during the 2022 Commonwealth Games. GENTLER [12], a precursor algorithm featuring TL and skipping, is used as a benchmark. The full factorial experiment presented in Sect. 5 demonstrates the competitiveness of the two algorithms in a variety of traffic conditions.

---

[1] Available at https://sdgs.un.org/goals, in particular, goal 11, action 11.2.

A summary of relevant work conducted in the traffic modelling and prediction domain is given in Sect. 2. Section 3 presents a simple yet illuminating example that highlights the value adding potential of incorporating vectorial elements into our existing algorithm. The paper's two original contributions are the topics of Sects. 4 and 5. Section 6 outlines our conclusions.

2 Background

Traffic prediction by means of traditional time series modelling and machine learning has been receiving significant attention [4,11]. Traditional time series modelling, based on the auto-regressive integrated moving average (ARIMA) method and its variants, is well-suited to cases where short-term predictions (a few hours into the future) are sufficient. Conversely, machine learning approaches such as supervised learning underpinned by deep neural networks [1,4,11] and hybrid models [10], have proven successful for short, medium and longer term predictions (up to one week). Yet, they typically require extensive training [1]. This downside can be mitigated by turning to evolutionary algorithms: versions specifically tailored to tackle complex problems related to urban transport have been successfully applied to automate traffic signal management [5,13], design bus route networks with a reduced environmental impact [8], locate electric vehicle charging stations [6], etc.

Another promising approach to traffic prediction is related to transfer learning, a technique that enables transferring models trained on data collected from one area (source) to a neighbouring one (target), where traffic readings may not be available. Li et al. report competitive results when applying transfer learning to highway traffic [9]; their findings are predicated on source and target traffic exhibiting similar patterns, which is less likely when it comes to the complex road networks within major cities. Modelling traffic through those networks requires a more sophisticated version of transfer learning: one example is GENetic programming with Transfer LEarning (GENTLE) [7], an algorithm that produces robust models of urban traffic, transferable from source to target junctions with no need for additional training on the latter. A lag operator ensures that historical traffic flow values can be taken into account in a computationally efficient way, without inflating the terminal set with lagged input terms. GENTLER [12] is an extension of GENTLE whereby, in the case of transfer from multiple sources, the models copied over from one source junction to the next are supplemented with a configurable amount of random trees, in an attempt to achieve a better exploration-exploitation trade-off. When its parameters are optimally configured, GENTLER yields models of significantly better accuracy than GENTLE.

GENTLE and GENTLER use traditional scalar-based symbolic regression. By contrast, Vectorial Genetic Programming (VE-GP), a recent approach specifically designed to model time series [2], utilises vector terminals and vectorial functions, both aggregate and cumulative. VE-GP is applied on a real-life physiological time series prediction problem. When the specifics of the healthcare problem domain are abstracted out, sufficient similarities to traffic prediction remain to justify incorporating certain VE-GP components into GENTLER.

3   Motivating Scenario

The example analysed in this section is a simplified version of a real-world traffic
modelling problem. It uses a basic road layout to explain how anomaly handling
mechanisms operate across space (TL) and time (lag operator enabling skipping
and linear algebra operators facilitating smoothing) in order to produce accurate
traffic predictions that hold under both typical and anomalous conditions.

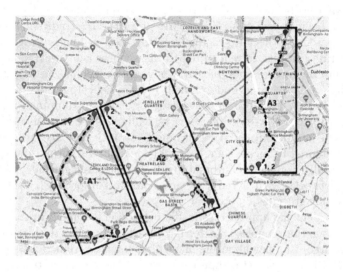

Fig. 1. Three central Birmingham areas monitored before and during the 2022 Commonwealth Games: sensors 1 and 2 measure inflow traffic; the unlabeled sensor captures
outflow traffic.

3.1   Traffic Anomaly Handling over Time: Skipping and Smoothing

Let us consider a topology with two inflow lanes, x_0 and x_1, and one outflow
lane, y. This could be assimilated to any of the three areas shown in Fig. 1, where
sensors 1 and 2 monitor inflow, whilst the remaining sensor records outflow, every
Δt minutes, over a time interval of length L. Assuming $\Delta t = 15$ and $L = 60$,
the traffic readings are stored in matrix R.

$$R = \begin{bmatrix} x_0(t_0) & x_0(t_1) & x_0(t_2) & x_0(t_3) \\ x_1(t_0) & x_1(t_1) & x_1(t_2) & x_1(t_3) \\ y(t_0) & y(t_1) & y(t_2) & y(t_3) \end{bmatrix} \tag{1}$$

The number of rows in matrix R is given by the number of monitored lanes,
whereas the number of columns is equal to the number of traffic readings collected throughout the monitoring interval L. Symbol $x_0(t_0)$ represents the number of vehicles passing through input lane x_0 at the beginning of the monitoring interval, $x_1(t_1)$ stands for the number of vehicles recorded Δt minutes later

through input lane x_1, whilst $y(t_3)$ marks the number of vehicles exiting through outflow lane y at the end of the monitoring interval. The other symbols in matrix R are to be interpreted in a similar fashion.

Model Representation. GENTLER builds models of the traffic flowing through lane y by combining scalar terminal nodes from set T and arithmetic operator nodes from set O, both given in Eq. (2). Operator / stands for protected division, which returns 1 whenever the right hand side operand is 0. Set O also includes a unary *lag* operator that returns the value of its input delayed by one sampling interval, Δt.

$$T = \{x_0, x_1, y\} \quad O = \{+, -, \times, /, lag\} \tag{2}$$

In the case of $\overline{\text{GENTLER}}$, traffic models are represented as combinations of the vector terminals and linear algebra operators given in Eq. (3). The elements of set $\overline{T}$ correspond to the rows in matrix R, in the order given by their indices. The first four elements of set $\overline{O}$ are unary aggregate operators that output, respectively, the mean, sum, maximum and minimum of their input vectors. The following four elements perform the same operations cumulatively, whereas the final four represent the traditional vector addition, subtraction, element-wise product and element-wise protected division. The complete definitions (with examples) of all operators in set $\overline{O}$ are available in [2].

$$\overline{T} = \{\overline{x_0}, \overline{x_1}, \overline{y}\}$$
$$\overline{O} = \{\mathbf{V\_mean, V\_sum, V\_max, V\_min,}$$
$$\mathbf{C\_mean, C\_sum, C\_max, C\_min,}$$
$$\mathbf{VsumW, V\_W, VprW, VdivW}\} \tag{3}$$

Let us consider the tree-like traffic models M and $\overline{M}$ in Fig. 2. The output of the GENTLER tree M represents the predicted vehicle flow, $\hat{y}(t)$, through the output lane.

$$\hat{y}(t) = (x_0(t) + x_1(t-2))/c, t \in [t_0, \dots, t_3], \quad c = const \tag{4}$$

The $\overline{\text{GENTLER}}$ model $\overline{M}$ outputs a vector, the elements of which are predicted outflow values at each time instant in the monitoring interval.

$$[\hat{y}(t_0), \hat{y}(t_1), \hat{y}(t_2), \hat{y}(t_3)] = 1/c \times [x_0(t_0), x_0(t_1), x_0(t_2), x_0(t_3)]$$
$$+1/4c \times (4x_1(t_0) + 3x_1(t_1) + 2x_1(t_2) + x_1(t_3)), \quad c = const \tag{5}$$

Model Evaluation. In order to determine the prediction accuracy (i.e., the fitness) of the traffic models, their outputs are compared against the readings in the third row of matrix R. Assuming that only the first three columns in matrix

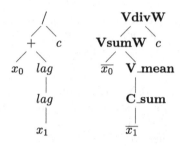

Fig. 2. Model representation in GENTLER (tree M, on the left) and in $\overline{\text{GENTLER}}$ (tree $\overline{M}$, on the right).

R are used for training, the fitness of model M is $\sqrt{\frac{1}{3}\sum_{i=0}^{2}(\hat{y}(t_i) - y(t_i))^2}$, a scalar.
The values $\hat{y}(t), t \in [t_0, \ldots t_2]$ featured in Eq. (6) are those given in Eq. (4).

$$3 \times (fitness(M))^2 = \sum_{i=0}^{2}(\hat{y}(t_i) - y(t_i))^2 =$$

$$
\begin{aligned}
&(x_0(t_0)/c - y(t_0))^2 && : t_0 \\
&+(x_0(t_1)/c - y(t_1))^2 && : t_1 \\
&+((x_0(t_2) + x_1(t_0))/c - y(t_2))^2 && : t_2
\end{aligned}
\tag{6}
$$

By contrast, the fitness of $\overline{M}$ is a vector, as shown in Eq. (7) where values $\hat{y}(t), t \in [t_0, \ldots t_2]$ are given in Eq. (5). However, $\overline{\text{GENTLER}}$ calculates the root mean squared error (RMSE) of candidate models by squaring and averaging the elements of the fitness vector, leading to a scalar value much like in the case of GENTLER.

$$fitness(\overline{M}) = [\hat{y}(t_0) - y(t_0), \hat{y}(t_1) - y(t_1), \hat{y}(t_2) - y(t_2)] \tag{7}$$

The fitness of models M and $\overline{M}$ reveals the different ways in which GENTLER and $\overline{\text{GENTLER}}$ are equipped to handle anomalies in the training data. In cases where traffic is affected by road construction, accidents, popular sporting events, etc., GENTLER manages the disruption by skipping over (potentially) anomalous readings: as shown in Eq. (6), the prediction generated by model M is unaffected by samples $x_1(t_2)$ and $x_1(t_1)$. $\overline{\text{GENTLER}}$ employs a different mechanism: the unary operators in set $\overline{O}$ have a smoothing effect in that data samples are aggregated and cumulated (see second line of Eq. (5)) which reduces the impact of anomalous readings on the prediction accuracy.

3.2 Traffic Anomaly Handling Across Space: Transfer Learning

In a real-world setting, it is often the case that some areas of a city's road network, say A1 and A2 in Fig. 1, are reliably monitored, whilst others are not (the equipment sensing traffic through A3 is faulty, has been incorrectly installed or is missing altogether). Regardless, an accurate model of A3 traffic is needed all the same (e.g., to help decision makers determine the layout of new roads to be built in the area in order to streamline vehicle flow, thus reducing delays, accident rates and pollution). In the absence of native data, A3 traffic models need to be trained on neighbouring junctions A1 and A2 then transferred over to A3 (see [7] for a detailed explanation of the transfer learning process). For the transfer to be successful, the exogenous model (trained on A1 and A2) would need to predict traffic as accurately as an indigenous one (trained on A3, assuming data were available to make that possible). The experimental investigation documented in Sect. 5 demonstrates that this is indeed the case: exogenous models produced by both GENTLER and $\overline{\text{GENTLER}}$, on various combinations of typical and anomalous training data sets predict traffic with an accuracy that is comparable to that of indigenous models.

4 $\overline{\text{GENTLER}}$ Explained

We propose $\overline{\text{GENTLER}}$, vectorial GENetic Programming with Transfer LEarning and Randomisation, that features the following components:

1. Classic Symbolic Regression enhanced with transfer learning and bespoke exploration-exploitation tuning (also found in GENTLER [7]);
2. Vectorial representation (i.e., vector terminals and linear algebra operators enabling smoothing); and
3. Vectorial fitness, customised with a penalty mechanism designed to punish models with scalar outputs (i.e., vectors filled with copies of the same element), thus increasing the selection pressure in favour of trees with (true) vectorial outputs, which are more likely to be accurate traffic predictors.

Algorithm 1 illustrates how GENTLER and $\overline{\text{GENTLER}}$ evolve traffic models. Each row in matrix *models* (line 1) represents the population at a given generation: for GENTLER, the candidate models will be similar to M in Fig. 2 (note the *skipping*-enabling *lag* operator), whereas for $\overline{\text{GENTLER}}$, the trees will look like $\overline{M}$. Line 2 initialises the *estimator*, i.e., the object wrapper for *gp learn*'s *SymbolicTransformer*[2], with all expected evolutionary parameters (see Table 2 caption). On line 3 the *estimator* evolves the population of candidate models, for G_1 generations on training data collected from the first source (src1, which can represent any area in Fig. 1).

The *fit* method in the *gp learn* library implements the classic GP loop; one of the steps involved is fitness calculation. This is presented in Algorithm 2, which runs for every element in $models[G_i]$, where G_i is the current generation. Lines 2 through 5 illustrate the classic RMSE calculation, over all *trn* samples in the training data set (containing readings collected from src1 if $G_i \leq G_1$ or from the second source, src2, if $G_i > G_1$). This is where *smoothing* occurs, as an effect

---

[2] https://gplearn.readthedocs.io/en/stable/reference.html#symbolic-transformer.

Algorithm 1. GENTLER and $\overline{\text{GENTLER}}$ essential logic

1: $models \leftarrow [\,][\,]$
2: $estimator.init()$
3: $models[G_1] \leftarrow estimator.fit(getTrainData(\text{src1}))$

4: ############### **Transfer Learning with Randomisation**
5: **if** $mix == 0$ **then**
6:     $models[G_1 + 1] \leftarrow makeRndModels(N)$
7: **else**
8:     **if** $mix == 1$ **then**
9:         $models[G_1 + 1] \leftarrow models[G_1]$
10:    **else**
11:        $models[G_1 + 1] \leftarrow getBest(models[G_1], hof, cmp, mix \times N)$
                    $\cup\ makeRndModels((1 - mix) \times N)$
12:    **end if**
13: **end if**###

14: $models[G_2] \leftarrow estimator.fit(getTrainData(\text{src2}), models[G_1 + 1])$
15: **return** $getBest(models[G_2])$

of the cumulative and aggregation operator featured in $\hat{y}(t)$. The result, $rmse$, represents the fitness of the model. Fitness calculation in $\overline{\text{GENTLER}}$ implies an additional step (lines 6 through 9): should the output of the current model $\overline{M}$, i.e., the vector containing predicted values $\hat{y}$ for each of the time instants in the training set, contain identical elements (indicating a scalar output), chances are that model will not yield a competitive prediction accuracy, therefore its fitness is increased by a factor of 100 (*pen* on line 8).

Once the population at generation G_1 becomes available, Algorithm 1 continues to the transfer learning stage. This occurs when traffic is to be predicted through a (target or destination) area where historical vehicle flow data are not available (because sensing equipment is absent, faulty or incorrectly installed, or because the target has not yet been built). In these cases, training data have to be collected from adjacent locations (e.g., neighbouring junctions where reliable traffic readings are available), called sources. For illustration, suppose that any one of the areas in Fig. 1 is the destination. Any combination of the remaining two areas (sources) can be used to train the traffic models aimed to predict traffic through the destination; GENTLER assumes two such sources, src1 and src2. In that context, transfer learning consists in training the candidate models on data collected from src1 for G_1 generations (from 1 to G_1) and then on data collected from src2 for $G_2 - G_1$ generations (from $G_1 + 1$ to G_2). Thus, the traffic patterns learnt from src1 data and further refined on src2 data are transferred over to the destination to produce predictions (since all possible destinations shown in Fig. 1 already exist and are monitored, we use those data to validate the transferred model). The selection process of the trees to be transferred between sources (from generation G_1 to $G_1 + 1$) is controlled by three parameters:

Algorithm 2. GENTLER and $\overline{\text{GENTLER}}$ fitness calculation

---

1: $rmse = 0$
2: **for** t **in** $[t_0, \ldots, t_{trn}]$ **do**
3:     $rmse \leftarrow rmse + (\hat{y}(t) - y(t))^2$
4: **end for**
5: $rmse \leftarrow \sqrt{rmse/(trn + 1)}$

6: ##################### penalty ($\overline{\text{GENTLER}}$ only)
7: **if** $\hat{y}(t)$ *identical*, $\forall t$ **in** $[t_0, \ldots, t_{trn}]$ **then**
8:     $rmse \leftarrow rmse \times pen$
9: **end if**######################################

10: **return** $rmse$

- *hof* the number of trees (in order of fitness, starting with the most accurate) to include in the **hall of f**ame and consider for transfer;
- *cmp* the number of least correlated **comp**onents (trees) within the hall of fame to be transferred to src2; and
- *mix* a real number between 0 and 1 representing the proportion of trees transferred from src1 relative to random trees.

Lines 5 through 13 show the three possible ways of putting together the N models in the population at generation $G_1 + 1$, depending on the *mix* value:

- All models are random (lines 5, 6): all traffic patterns learnt on the first source are lost; training starts afresh on src2 (equivalent to pure exploration).
- No models are random (lines 8, 9): all traffic patterns learnt on the first source are transferred over to src2 where training resumes (equivalent to pure exploitation).
- Some models are random (lines 11, 12): $mix \times N$ of the most accurate, least correlated *cmp* trees trained on src1 are transferred over to src2, and the remaining positions are filled with randomly generated models (balance between exploitation and exploration).

The configurable amount of random trees injected at generation $G_1 + 1$ dynamically adjusts the exploration-exploitation ratio: if optimally chosen, the *mix* value will make it possible to leverage the traffic patterns learnt from the first source as well as maintain a healthy amount of genetic diversity.

Once the transfer from the first to the second source is complete, the *fit* function is called again, on line 14, to evolve the models at generation $G_1 + 1$ for $G_2 - G_1$ additional generations, on src2 training data. Method *getBest* (line 15) returns the most accurate model at generation G_2; its output can then be used to predict traffic through the destination (and be validated, if destination traffic readings are available).

5 Experimental Investigation

5.1 Sensor Selection and Traffic Readings Analysis

Transport for West Midland (TfWM) has 178 sensors installed across Birmingham[3]. Out of the ones covering the city centre, we eliminated those with null outputs (most likely due to faults) and those that recorded significantly fewer than 96 samples a day (i.e., captured data less frequently than every 15 min). The remaining sensors monitor vehicle flow through areas A1, A2 and A3 in Fig. 1. A stream of arrows indicates the direction of traffic: inbound vehicles are counted by sensors 1 and 2, whereas outbound ones are captured by the third sensor (unlabeled) in each of the considered areas.

The nine sensors collected data before and during the 2022 Commonwealth Games that took place in Birmingham: readings captured between the 25th of April and the 29th of May are taken to represent traffic under normal conditions, whereas values recorded between the 25th of July and the 28th of August are indicative of traffic under anomalous conditions (i.e., disrupted by restrictions put in place to accommodate for the various sporting events). We conducted full factorial experiments with GENTLER and $\overline{\text{GENTLER}}$, considering all possible combinations of source and destination areas, with relevant models trained and validated (60-40 data split) under normal and anomalous conditions; the results are presented in Table 2.

The TfWM traffic monitoring exercise is young: measurements are of poor quality compared to other major EU municipalities[4]. The two principal challenges are unevenness (standard deviations reported in Table 1 are very high compared to averages, particularly in A2) and scarceness (most acutely in A3, where there are no sensors monitoring traffic through the main road, which is intensely used by motorists going round the Clean Air Zone to avoid fees). Although these data quality related problems are bound to negatively impact the accuracy of the traffic models, we chose not to eliminate outliers during pre-processing, as there is no way to determine whether they are indicative of sensor malfunctions or of spikes in real traffic; their deletion would either deplete the data pool or overly-sanitise it, making it difficult for the ensuing experiments to authentically highlight the strengths of GENTLER and $\overline{\text{GENTLER}}$. Instead, we chose to use the native models (trained and validated on data from the same area) as a baseline, in order to ascertain the relative quality of non-native ones (trained and validated on data collected from different areas), thus showcasing the effectiveness of transfer learning with smoothing and skipping under realistic (in this case, sub-optimal) conditions. We argue that this is a valid experimental approach, as transfer learning is meant to provide models of traffic through areas where native data are not available (roads that are not monitored or yet to be built). For comprehensive evidence that the two algorithms work efficiently on high quality data, the interested reader is referred to [7,12].

---

[3] https://data-tfwm.opendata.arcgis.com.
[4] Such as Darmstadt: https://www.ui.city/en/.

Table 1. Output sensor data used to validate all traffic models: relevant stats.

	Normal			Anomalous		
	A1	A2	A3	A1	A2	A3
Count	1344	1344	1344	1344	1344	1344
Mean	77.73	9.73	160.58	55.60	26.45	150.46
SD	43.00	16.08	83.43	46.28	18.92	80.16
Min	0	0	6	0	0	9
Max	194	100	383	209	173	337

Table 2. GENTLER and $\overline{\text{GENTLER}}$ model accuracy: $hof, cmp, mix = (717, 68, 0.28)$; for all other evolutionary parameters see [12]; RMSE is lowest over 30 runs; % RMSE is the ration of RMSE to the mean of the corresponding validation dataset in Table 1.

	Combination	NN		NA		AN		AA	
		RMSE	% RMSE	RMSE	% RMSE	RMSE	% RMSE	RMSE	% RMSE
GENTLER	A1 -> A1	20.19	25.97	15.59	28.04	20.41	26.26	15.51	27.89
	A2 -> A1	20.58	26.48	16.85	30.31	20.73	26.67	16.92	30.42
	A3 -> A1	415.62	534.67	372.11	669.20	523.20	673.06	240.99	433.40
	A2, A3 -> A1	42.04	54.08	38.35	68.97	21.12	27.17	17.33	31.17
	A3, A2 -> A1	29.78	38.31	24.09	43.33	30.33	39.02	24.57	44.20
	A2 -> A2	5.60	57.54	10.65	40.28	5.56	57.15	10.67	40.35
	A3 -> A2	216.92	2228.22	324.29	1225.66	326.51	3354.00	338.71	1280.16
	A1 -> A2	5.68	58.42	10.75	40.63	5.69	58.46	10.81	40.86
	A3, A1 -> A2	9.72	99.91	16.59	62.70	8.69	89.29	12.67	47.88
	A1, A3 -> A2	6.68	68.68	14.20	53.69	6.68	68.68	14.20	53.69
	A3 -> A3	48.33	30.10	57.46	38.19	50.54	31.47	66.09	43.92
	A2 -> A3	151.90	94.59	140.44	93.34	137.97	85.91	139.49	92.71
	A1 -> A3	150.89	93.96	138.20	91.85	127.33	79.29	126.50	84.07
	A2, A1 -> A3	117.73	73.31	129.76	86.24	124.36	77.44	127.36	84.64
	A1, A2 -> A3	115.82	72.12	127.11	84.48	121.37	75.58	128.44	85.36
$\overline{\text{GENTLER}}$	A1 -> A1	19.28	24.80	15.72	28.27	19.13	24.61	15.34	27.58
	A2 -> A1	21.01	27.03	16.53	29.73	20.69	26.62	16.37	29.45
	A3 -> A1	42.45	54.62	38.68	69.57	47.13	60.63	51.86	93.27
	A2, A3 -> A1	32.02	41.19	24.13	43.39	42.34	54.47	38.62	69.45
	A3, A2 -> A1	30.62	39.39	27.48	49.42	33.28	42.82	24.61	44.26
	A2 -> A2	5.65	58.11	10.40	39.33	5.67	58.31	10.23	38.68
	A3 -> A2	12.77	131.20	20.31	76.78	47.25	485.35	48.88	184.75
	A1 -> A2	5.73	58.87	10.98	41.52	5.65	58.10	10.57	39.95
	A3, A1 -> A2	7.13	73.30	14.34	54.23	12.12	124.59	15.27	57.74
	A1, A3 -> A2	9.85	101.19	13.61	51.43	13.14	135.00	20.26	76.58
	A3 -> A3	48.16	29.99	52.82	35.11	44.80	27.90	53.03	35.24
	A2 -> A3	94.47	58.83	91.18	60.60	94.47	58.83	91.18	60.60
	A1 -> A3	86.28	53.73	97.83	65.01	83.27	51.85	85.76	57.00
	A2, A1 -> A3	110.19	68.62	99.88	66.38	105.45	65.66	96.59	64.19
	A1, A2 -> A3	89.48	55.72	83.12	55.24	100.27	62.44	100.14	66.55

5.2   Results Discussion

We ran GENTLER and $\overline{\text{GENTLER}}$ on all source and destination combinations achievable considering the three areas in Fig. 1. The Combination column in Table 2 lists the source area(s), where training is performed, to the left of the arrow and the destination area, where validation takes place, to the right of the arrow. The experiments we conducted are of two types: homogeneous, where training and validation take place on data collected under matching conditions, and heterogeneous, where training is performed on normal readings and validation on abnormal ones and vice-versa. Experimental results in the former category are presented in the columns headed NN and AA, whereas findings in the latter category are reported in columns NA and AN (the first letter refers to training and the second to validation).

Modelling Traffic Under Normal Conditions. This segment of the experimental analysis is aimed at investigating the impact of transfer learning on model accuracy. It relies on data presented in the two columns of Table 2 under the NN heading.

With an average of 9.7 outgoing vehicles (see Table 1), the flow of traffic through the A2 area under normal conditions is severely limited. Output traffic in the A1 area is significantly higher (77.7 vehicles passing by the Morrisons sensor, on average), with A3 being by far the busiest of the three (160.6 average output flow through the Matalan monitoring point, more than double the volume of traffic out of A1). However vehicle flow through A3 is also the least even, with acute variations in the number of recorded outgoing vehicles, as indicated by a standard deviation of 83.4, twice as high as in the case of A1. These statistics suggest that models trained and validated on data collected from the A1 area are likely to be the most accurate of the three. This is confirmed by the experimental results: out of the three native models produced by GENTLER, the one trained and validated on A1 data has the highest accuracy (relative RMSE of 26%) and transfers over efficiently to A2: the model trained on A1 data and validated on A2 data is of comparable accuracy to the A2 native model (relative RMSE of 58.4% in the case of the former, compared to 57.5% in the case of the latter). The same applies to $\overline{\text{GENTLER}}$ models: 58.9% relative error when training on A1 and transferring to A2, compared to 58.11% achieved by the A2 native model.

The extreme variations in the A3 data make it very difficult to benefit from vehicle flow patterns learnt on the much better behaved A1 traffic: the GENTLER model trained on A1 data and validated on A3 data is more than three times less accurate than the native A3 model, a result that doesn't change when the source junction is A2. When it comes to $\overline{\text{GENTLER}}$, single source transfer learning works better: training models on A1 and, respectively, A2 data results in an accuracy loss (relative to the native A3 model) that is over one order of magnitude smaller than in the case of GENTLER.

The situation improves further when training occurs on A1 and A2 data, subsequently as opposed to separately, before the transfer to A3: the GENTLER models' relative RMSE drops from the 93% – 94% range (which is the case for single source transfer learning), to the 72% – 73% one (achieved for multiple source transfer learning). $\overline{\text{GENTLER}}$ models, single source and multiple source, with A3 as the destination, are all in the 53.7% – 58.8% band, with the exception of the one trained on A2 and then on A1, which is approximately 10% less accurate. As expected, A3 models transfer over very poorly in all single and multiple source combinations. Whilst GENTLER completely fails to apply patterns learnt during A3 training in order to model traffic through A1, $\overline{\text{GENTLER}}$ manages to produce a single source transfer model that is half as accurate as the A1 baseline (the former has a prediction error of 54.6% as opposed to 24.8% in the case of the latter). $\overline{\text{GENTLER}}$ continues to generate single source transfer learning models of superior accuracy to that of GENTLER ones when A3 is the source and A2 the destination. However, both algorithms yield multiple source transfer learning models that significantly outperform single source ones, for all three destinations considered in this study.

This part of the experimental analysis indicates that, under normal conditions (i.e., when vehicle flow through both source and destination junctions is relatively even, as is the case for A1 and A2 but not for A3), transfer learning models and native ones are comparably accurate regardless of overall traffic volume (which is much higher through A1 than A2). Multiple source transfer learning models (that benefit from two bouts of training on different source areas before being validated on the destination one) are consistently superior to single source models. In situations where native models are not available, this experimental conclusion supports our claim that transfer learning models can be be confidently used as competitive substitutes.

Modelling Traffic Under Anomalous Conditions. This segment of the experimental analysis is aimed at comparing transfer learning with skipping (GENTLER) against transfer learning with smoothing ($\overline{\text{GENTLER}}$), in terms of their efficiency at producing traffic models that are competitively accurate, even though they were trained and validated on data affected by anomalies (i.e., vehicle flow disruptions caused by restrictions put in place during the Commonwealth Games). The relevant data are included in the two columns of Table 2 under the AA heading.

Out of the three areas, A2 continues to have the lowest outflow (see columns under the Anomalous heading in Table 1). However, compared to pre-Games readings, traffic exiting A2 through the roundabout at the entrance to the Chinese Quarter is three times as busy. Taken in conjunction with the decreased average flow out of A2 and A3, this suggests that, during the Games, a significant part of Five Ways and Moor St traffic was diverted via the roads around The Mailbox and New St Station. This appears to have had very little effect on the vehicle flow evenness: standard deviation levels under anomalous conditions are comparable to pre-Games ones.

Overall, both skipping and smoothing are efficient at modelling the changes in traffic patterns during the Games:

- A1 predictions stay in the same accuracy band. The A1 native model produced by GENTLER (skipping) has an average RMSE of 27.9% compared to 26% before the Games, whereas the $\overline{\text{GENTLER}}$ model (smoothing) performs at 27.6% compared to 24.8% pre-Games.
- A2 predictions become better: skipping takes the average RMSE down from 57.5% to 40.3%, whereas smoothing achieves a model accuracy increase of almost 20% points. Since A2 is the recipient of traffic diverted from the other two areas during the Games, the fact that both temporal anomaly handling mechanisms we propose are effective at accurately capturing that dynamic is evidence in support of our contributions' value.
- A3 predictions become marginally worse: skipping leads to an increase in the average prediction error from 30.1% pre-Games to 43.9%, whilst smoothing performs better, causing a precision loss of only 5% points.

Transfer learning interacts with skipping in the expected way: GENTLER fails to transfer traffic patterns learnt whilst training on A3 (the area where the output validation data set has the highest standard deviation) to either A1 or A2 destinations. However, when training takes place on A3 and a second source, skipping brings down the average error recorded in the destination area by as much as 52% points compared to pre-Games levels. The transfer learning, smoothing combination leads to single source transfer models comparable to those produced by GENTLER. They are outperformed by $\overline{\text{GENTLER}}$ models trained on multiple sources, most notably when A2 is the destination: there is a drop in prediction error of 15 to 24% points compared to pre-Games levels. This adds to the above mentioned evidence attesting to the efficiency of our temporal anomaly handling mechanisms, in that they are now shown to also be effective in combination with spatial anomaly handling (transfer learning).

Modelling Traffic Under Mixed Conditions. This segment of the experimental analysis is aimed at investigating the efficiency of transfer learning with skipping (GENTLER) and transfer learning with smoothing ($\overline{\text{GENTLER}}$) at predicting anomalous traffic based on patterns learned from normal one and vice-versa. The investigation is based on the data in the four columns at the centre of Table 2 (headed NA and AN).

The performance of the native models produced by GENTLER indicates that skipping is just as effective under heterogeneous conditions (i.e., models are trained on normal traffic and validated on anomalous one or vice-versa) as it is under homogeneous ones (i.e., models are both trained and validated on either normal traffic or abnormal one): the two heterogeneous A1 native models are in the 26.3% – 28% error range, where the homogeneous ones also lie, with a similar conclusion to be drawn for A2 and A3. It is particularly relevant to note that regardless of whether their training occurs under normal or anomalous circumstances, native models predict either type of traffic comparably well. This is most obvious in the case of A2: the AN and NN models are practically equally accurate (approximately 57% average prediction error), whilst the same can be said about the NA and AA models (circa 40% average prediction error). Analysing the performance of $\overline{\text{GENTLER}}$ native models yields similar experimental findings. This implies that, whenever it is necessary to model future traffic anomalies (e.g., predict vehicle flow during the 2026 edition of the Games) for which anomalous training data is yet to become available, skipping and smoothing make it possible for training to be performed on data collected under normal circumstances, without compromising the prediction accuracy.

Discounting combinations that include source A3, where vehicle flow is too uneven to allow for the efficient transfer of learnt patterns across areas, all single and multiple source transfer learning models perform within an accuracy band of approximately 10% points. When combining knowledge learnt by training on A1 and A2 (in either order), skipping produces models capable of predicting normal traffic within a 4% error margin, regardless of whether training took place under normal or anomalous circumstances. When it comes to predicting anomalous traffic, the same source combination, when skipping is applied, brings the margin down to 2.4%. Also note that all four multiple source transfer learning models produced by GENTLER, where A3 is the destination, outperform native models by at least 8% and as much as 21%. Smoothing is less effective than skipping at producing multiple source transfer models that rival the accuracy of native ones (when A3 is the destination, the prediction quality of $\overline{\text{GENTLER}}$ models worsens by as much as twofold). Yet, the observation regarding the relative competitiveness of heterogeneous and homogeneous models continues to be valid in the case of smoothing: $\overline{\text{GENTLER}}$ models trained on A1 and A2 (in either order) are within the 2.1% – 10.3% accuracy band.

6   Conclusions

Traffic modeling and prediction are central to efficient intelligent transportation which, in turn, is a key component of the smart cities initiative and the UN's Sustainable Development strategy. It is thus essential that efficient algorithms be developed to produce traffic predictions with competitive accuracy in a variety of practical settings: the area where traffic is being predicted has not been fitted with sensing equipment, the traffic predictions will inform city planners' decision making concerning the layout of new roads, the traffic readings used for training were collected during sporting events or other kinds of short and medium term disruption, etc.

To cater to such needs, we propose $\overline{\text{GENTLER}}$, a traffic modelling and prediction algorithm that leverages Genetic Programming enhanced with Transfer Learning and randomisation, on the one hand, and presents increased tolerance to training data anomalies, on the other hand. The former quality enables $\overline{\text{GENTLER}}$ to predict vehicle flow through areas where traffic data are not available, by learning from readings collected on neighbouring areas of the road network. The latter feature is afforded by linear algebra functions that mitigate the effect of training outliers via aggregation and cumulation.

$\overline{\text{GENTLER}}$ produces competitive models regardless of whether the training and validation data were collected during typical or anomalous traffic conditions. We support that claim with a comprehensive set of experimental results obtained by running $\overline{\text{GENTLER}}$ on traffic readings obtained before and during the 2022 Birmingham Commonwealth Games. Those results indicate that the prediction accuracy of $\overline{\text{GENTLER}}$ models does not deviate from the GENTLER reference in a statistically significant way. In other words, when native models are not available, or heterogeneous predictions are required, running transfer learning equipped with smoothing and, respectively, skipping, selecting the most accurate of the resulting models and using it to predict traffic through the destination area will yield a level of accuracy that is comparable to baseline.

References

1. Almeida, A., Brás, S., Oliveira, I., Sargento, S.: Vehicular traffic flow prediction using deployed traffic counters in a city. Futur. Gener. Comput. Syst. **128**, 429–442 (2022)
2. Azzali, I., Vanneschi, L., Bakurov, I., Silva, S., Ivaldi, M., Giacobini, M.: Towards the use of vector based GP to predict physiological time series. Appl. Soft Comput. **89**, 106097 (2020)
3. Bliemer, M.C., Ban, X.J., Leclercq, L., Qian, S., Unnikrishnan, A., Yang, X.S.: Special issue on dynamic transportation network modelling, emerging technologies, data analytics and methodology innovations. Transport. Res. Part C: Emerg. Technol. **142**, 103778 (2022)
4. Boukerche, A., Tao, Y., Sun, P.: Artificial intelligence-based vehicular traffic flow prediction methods for supporting intelligent transportation systems. Comput. Netw. **182**, 107484 (2020)
5. Cacco, A., Iacca, G.: Simulation-driven multi-objective evolution for traffic light optimization. In: Castillo, P.A., Jiménez Laredo, J.L., Fernández de Vega, F. (eds.) EvoApplications 2020. LNCS, vol. 12104, pp. 100–116. Springer, Cham (2020). https://doi.org/10.1007/978-3-030-43722-0_7
6. Cintrano, C., Toutouh, J.: Multiobjective electric vehicle charging station locations in a city scale area: malaga study case. In: Jiménez Laredo, J.L., Hidalgo, J.I., Babaagba, K.O. (eds.) EvoApplications 2022. LNCS, vol. 13224, pp. 584–600. Springer, Cham (2022). https://doi.org/10.1007/978-3-031-02462-7_37
7. Ekárt, A., Patelli, A., Lush, V., Ilie-Zudor, E.: GP with transfer learning for urban traffic modelling and prediction. In: 2020 IEEE CEC, pp. 1–8 (2020)
8. Kajihara, S., Sato, H., Takadama, K.: Generating duplex routes for robust bus transport network by improved multi-objective evolutionary algorithm based on decomposition. In: Castillo, P.A., Jiménez Laredo, J.L. (eds.) EvoApplications 2021. LNCS, vol. 12694, pp. 65–80. Springer, Cham (2021). https://doi.org/10.1007/978-3-030-72699-7_5
9. Li, J., Guo, F., Sivakumar, A., Dong, Y., Krishnan, R.: Transferability improvement in short-term traffic prediction using stacked LSTM network. Transport. Res. Part C: Emerg. Technol. **124**, 102977 (2021)
10. de Medrano, R., Aznarte, J.L.: A spatio-temporal attention-based spot-forecasting framework for urban traffic prediction. Appl. Soft Comput. **96**, 106615 (2020)
11. Nagy, A.M., Simon, V.: Survey on traffic prediction in smart cities. Pervasive Mob. Comput. **50**, 148–163 (2018)

12. Patelli, A., Hamilton, J.R., Lush, V., Ekárt, A.: A gentler approach to urban traffic modelling and prediction. In: 2022 IEEE CEC, pp. 1–8 (2022)
13. Wittpohl, M., Plötz, P.-A., Urquhart, N.: Real time optimisation of traffic signals to prioritise public transport. In: Castillo, P.A., Jiménez Laredo, J.L. (eds.) EvoApplications 2021. LNCS, vol. 12694, pp. 162–177. Springer, Cham (2021). https://doi.org/10.1007/978-3-030-72699-7_11

Evolutionary Computation in Edge, Fog, and Cloud Computing

Energy-Aware Dynamic Resource Allocation in Container-Based Clouds via Cooperative Coevolution Genetic Programming

Chen Wang[1]([✉]), Hui Ma[2], Gang Chen[2], Victoria Huang[1], Yongbo Yu[2], and Kameron Christopher[1]

[1] HPC and Data Science Department, National Institute of Water and Atmospheric Research, Wellington, New Zealand
{chen.wang,victoria.huang,kameron.christopher}@niwa.co.nz
[2] School of Engineering and Computer Science, Victoria University of Wellington, Wellington, New Zealand
{hui.ma,aaron.chen,yuyong1}@ecs.vuw.ac.nz

Abstract. As a scalable and lightweight infrastructure technology, containers are quickly gaining popularity in cloud data centers. However, dynamic Resource-Allocation in Container-based clouds (RAC) is challenging due to two interdependent allocation sub-problems, allocating dynamic arriving containers to appropriate Virtual Machines (VMs) and allocating VMs to multiple Physical Machines (PMs). Most of existing research works assume homogeneous PMs and rely on simple and manually designed heuristics such as Best Fit and First Fit, which can only capture limited information, affecting their effectiveness of reducing energy consumption in data centers. In this work, we propose a novel hybrid Cooperative Coevolution Genetic Programming (CCGP) hyper-heuristic approach to automatically generate heuristics that are effective in solving the dynamic RAC problem. Different from existing works, our approach hybridizes Best Fit to automatically designed heuristics to coherently solve the two interdependent sub-problems. Moreover, we introduce a new energy model that accurately captures the energy consumption in a more realistic setting than that in the literature, e.g., real-world workload patterns and heterogeneous PMs. The experiment results show that our approach can significantly reduce energy consumption, in comparison to two state-of-the-art methods.

Keywords: container-based clouds · container allocation · energy efficiency · genetic programming · hyper-heuristic

1 Introduction

Cloud computing has become a pillar of today's software industry by providing on-demand computing resources [4]. A fundamental technology that powers cloud computing is virtualization, e.g., Virtual Machine (VM)-based clouds

© The Author(s), under exclusive license to Springer Nature Switzerland AG 2023
J. Correia et al. (Eds.): EvoApplications 2023, LNCS 13989, pp. 539–555, 2023.
https://doi.org/10.1007/978-3-031-30229-9_35

which enable multiple VMs with different Operating Systems (OS) to share one Physical Machine (PM). However, VM deployment is often considered *heavy-weight* in terms of memory, CPU and start-up time since each VM is essentially a standalone machine with its own OS and a virtual copy of hardware. A recent trend in cloud computing is container-based cloud, which enables multiple applications to be deployed and share resources of the same OS on a VM [4]. Compared with VMs, containers are known for *lightweight* which utilize less resources and can easily scale up.

The widespread use of containers poses new research challenges. One essential issue is the Resource Allocation problem in Container-based clouds (RAC). Different from traditional VM-based cloud which only requires one-level resource allocation (i.e., VM instances are directly allocated to PM instances), RAC involves a two-level decision making process [12]. At the first level, a container needs to be assigned to a VM which involves decisions on VM selection and VM creation when existing VMs may not have enough resources to host the container. The second level allocation is similar to VM-based cloud, which deploys VMs to existing or newly created PMs with varied capacities.

RAC is usually modelled as a vector bin-packing problem [14,15,18,20] which is NP-hard. Thus, existing works usually simplified it by decoupling RAC into two independent single-level optimization without taking consideration of the inter-dependencies between container-VM and VM-PM allocations [10,17,20, 26]. Moreover, RAC has been further simplified by considering only homogeneous PMs [20,22] where the searching space is significantly reduced. As a result, their proposed algorithms may not be suitable for a cloud data center where PMs are equipped with different capacities [25]. Further, existing works [20,22] often assume that application deployment requests arrive to a data center with a simple workload pattern, e.g., one container arrives at the data center every five minutes. However, the number and the time of containers arriving at a data center are changing dynamically [19].

The goal of this paper is to propose an effective approach for energy-aware dynamic resource allocation in container-based clouds with heterogeneous PMs. To achieve this goal, we first propose a problem formulation on the dynamic RAC with the goal of minimizing the overall energy consumption. Different from existing works [10,20,22,25,26], our problem formulation adopts a new energy model and jointly considers heterogeneous VMs and PMs and dynamic container arrivals. Meanwhile, a new Hybrid cooperative co-evolution genetic programming (CCGP) approach is proposed to learn allocation rules hybridized with Best Fit to simultaneously address both container-VM and VM-PM allocations. To evaluate the performance of the hybrid approach in real-world settings, extensive experiments have been conducted on real-world workload patterns. Experiment results show that our proposed hybrid approach can significantly outperform widely-used heuristics and a state-of-the-art algorithm CCGP in reducing the energy consumption.

2   Related Work

Resource allocation in clouds has been widely investigated in the literature [1, 5,8–11,17,20,22,23,26]. However, most studies [1,5,8,9,11,23] only focused on one-level resource allocation where either VM-PM [1,5,8,9,14,23] or container-VM allocation [11] is tackled, neglecting the inter-dependencies between VM-PM and container-VM allocations.

Several recent research works [2,3,10,15,17,20,22,25,26] have studied the problem of two-level resource allocation in container-based clouds as a static problem and proposed meta-heuristic approaches to solve the problem. For example, [20] minimized the energy consumption by maximizing the PM utilization to reduce the number of active PMs. A Two-stage Multi-type Particle Swarm Optimization (TMPSO) algorithm was proposed which combines greedy and heuristic optimization. Similarly, [10] minimized the number of active PMs and instantiated VMs as well as maximized resource utilization. They proposed an Ant Colony Optimization based on Best-Fit (ACO-BF) algorithm where enough VMs are pre-created and assigned to PMs using Best-fit and containers are subsequently allocated to VMs using ACO. However, the above mentioned meta-heuristic approaches consider the resource allocation problem as a static problem where all the requests arrive at a data center at the same time. To handle the dynamic resource allocation problem, a CCGP approach is developed in [22] which simultaneously learns allocation rules for both levels.

From the problem perspective, existing works do not fully capture important features of the problem, e.g., the energy consumption is either not directly optimized or is only measured at a certain time point rather than over a time period [10,20,26]. Evidence has been shown in [22] that even with the same temporal energy consumption, the accumulated energy consumption can be significantly different. Moreover, existing works also tend to simplify the problem by considering homogeneous PMs and simple workload patterns [2,20,22], which cannot be applied in a real-world cloud data center with different types of PMs with different capacities [25].

From the solution perspective, RAC is a challenging two-level optimization problem with a large search space. Existing studies [15,18,20] usually simplified the problem as two independent one-level optimization without taking into account their inter-dependencies [10,17,20,26]. Moreover, a majority solutions relied on human-designed heuristics, e.g., greedy [16,20] and best-fit [10,26]. Their performance heavily relies on the workload patterns [21]. Although different searching algorithms, e.g., ACO, GP, and PSO, have been proposed [10,20–22], GP has shown its promise in effectively handling dynamic resource allocation problems [21,22].

3   Problem Formulation

Dynamic RAC is the process of dynamically allocating a set of arriving containers to VMs and subsequently mapping VMs to PMs according to various constraints.

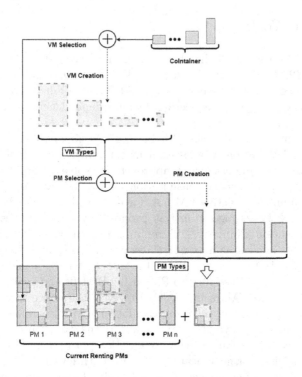

Fig. 1. An overview of dynamic RAC

In the remaining of this paper, we use RAC to refer dynamic RAC. As shown in Fig. 1, RAC follows a decision making process from *VM selection, VM creation, PM selection,* to *PM creation.* VM selection allocates a newly arrived container to a VM chosen from existing VMs. VM creation can be invoked if no existing VMs are preferred for hosting the container. VM creation creates a VM with a selected VM type, which is used for container allocation. When a new VM is created, it is allocated to an exiting PM via PM selection or a new PM via PM creation.

RAC considers a set of $|\mathcal{C}_t|$ containers $\mathcal{C}_t = \{c_1, \ldots, c_{|\mathcal{C}_t|}\}$ arriving at time t. Each container c_i is defined with three attributes, i.e., the CPU $\zeta_{c_i}^{cpu}$, memory $\zeta_{c_i}^{men}$, and OS $\zeta_{c_i}^{os}$ for running container c_i. In general, $\zeta_{c_i}^{os}$ is selected from a given set of $|\Phi|$ OS types $\Phi = \{\psi_1 \ldots, \psi_{|\Phi|}\}$, i.e., $\zeta_{c_i}^{os} \in \Phi$.

When a new container arrives, it needs to be allocated to a VM instance (or VM in short). A VM is defined with an OS $\Omega_j^{vos} \in \Phi$ and a VM type τ_j selected from a set of $|\Gamma|$ VM types provided by cloud providers where $\Gamma = \{\tau_1, \ldots, \tau_{|\Gamma|}\}$. Each VM type τ_j is associated with its CPU capacity $\Omega_{\tau_j}^{vcpu}$, memory capacity $\Omega_{\tau_j}^{vmen}$, CPU overhead $\pi_{\tau_j}^{vcpu}$ and memory overhead $\pi_{\tau_j}^{vmen}$ for creating a VM.

When a new VM is created, it needs to be allocated to a PM. A PM is associated with a PM type $\top_{p_k}$ from a set of $|\Upsilon|$ PM types, $\Upsilon = \{\top_1, \ldots, \top_{|\Upsilon|}\}$. Each PM type $\top_{p_k}$ is defined with a tuple, i.e., $\top_{p_k} = (\Omega_{p_k}^{pcpu}, \Omega_{p_k}^{pmen}, \pi_{p_k}^{pcpu}, \pi_{p_k}^{pmen})$, capturing its CPU and memory capacities, as well as the CPU and memory overhead for creating a new PM.

To calculate the energy consumption for a given PM p at time t, we define a non-linear energy model Eq. (1) based on [6], which can capture energy consumption more accurately than the linear models [22,25].

$$E_{p,t} = E_p^{idle} + (E_p^{full} - E_p^{idle})(2\mu_{p,t}^{cpu} - (\mu_{p,t}^{cpu})^{1.4}) \tag{1}$$

where E_p^{idle} and E_p^{full} represent the energy consumption of the PM p per time unit when it is idle and fully loaded, respectively. Given a number of $|L_t|$ VM instances used at time t, $\mu_{p,t}^{cpu}$ is the CPU utilization level at time t, which can be calculated as

$$\mu_{p,t}^{cpu} = \frac{\sum_{l=1}^{|L_t|}(\sum_{j=1}^{|\Gamma|} \pi_{\tau_j}^{vcpu}\mathfrak{a}_{lj} + \sum_{i=1}^{|C_t|} \zeta_{c_i}^{cpu}\mathfrak{c}_{il})\mathfrak{b}_{lp}}{\sum_{k=1}^{|\Upsilon|}(\pi_{\Upsilon_{p_k}}^{pcpu} + \Omega_{p_k}^{pcpu})\mathfrak{x}_{pk}}$$

where $\mathfrak{a}_{lj}$, $\mathfrak{b}_{lp}$, $\mathfrak{c}_{il}$, and $\mathfrak{x}_{pk}$ are binary variables. For example, $\mathfrak{a}_{lj} = 1$ if the lth running VM instance belongs to type τ_j. $\mathfrak{c}_{il} = 1$ if the ith container is allocated to the lth VM. $\mathfrak{b}_{lp} = 1$ if the lth VM instance is allocated to the pth PM instance. $\mathfrak{x}_{pk} = 1$ if the pth created PM is type of Υ_{p_k}.

RAC aims to find the mapping of containers to VMs (i.e., $\{\mathfrak{a}_{lj}\}$ and $\{\mathfrak{c}_{il}\}$) and VMs to PMs (i.e., $\{\mathfrak{b}_{lp}\}$ and $\{\mathfrak{x}_{pk}\}$) so that the accumulated energy consumption over the allocation period T can be minimized:

$$J = \int_0^T \left(\sum_{p=1}^{|P_t|} E_{p,t}\right) dt \tag{2}$$

where $|P_t|$ a step function that gives the number of PM used at any time t. In line with Eq. (2), RAC is subject to the following constraints:

- A container c_i can only be allocated to one VM instance, i.e., $\sum_{l=1}^{|L_t|} \mathfrak{c}_{il} = 1$.
- A VM can only be allocated to one PM instance, i.e., $\sum_{p=1}^{|P_t|} \mathfrak{b}_{lp} = 1$.
- Each VM and PM instance must have a specific type, i.e., $\sum_{j=1}^{|\Gamma|} \mathfrak{a}_{lj} = 1$ and $\sum_{k=1}^{|\Upsilon|} \mathfrak{x}_{pk} = 1$
- OS compatibility: A container c can only be allocated to a VM instance v if they have the same OS, i.e., $\zeta_{c_i}^{os}\mathfrak{c}_{il} = \Omega_l^{vos}$.
- VM capacity constraint: For a VM instance, the total CPU and memory requirements of all allocated containers must satisfy its capacity, i.e., $\sum_i \zeta_{c_i}^{cpu}\mathfrak{c}_{il} \leq \sum_{j=1}^{|\Gamma|} \Omega_{\tau_j}^{vcpu}\mathfrak{a}_{lj}$ and $\sum_i \zeta_{c_i}^{men}\mathfrak{c}_{il} \leq \sum_{j=1}^{|\Gamma|} \Omega_{\tau_j}^{vmen}\mathfrak{a}_{lj}$.
- PM capacity constraint: For a PM instance, the total CPU and memory requirements of all allocated VMs must not exceed its capacity, i.e., $\sum_{l=1}^{|L_t|}(\sum_{j=1}^{|\Gamma|} \Omega_{\tau_j}^{vcpu}\mathfrak{a}_{lj})\mathfrak{b}_{lp} \leq \sum_{k=1}^{|\Upsilon|} \Omega_{p_k}^{pcpu}\mathfrak{x}_{pk}$ and $\sum_{l=1}^{|L_t|}(\sum_{j=1}^{|\Gamma|} \Omega_{\tau_j}^{vmem}\mathfrak{a}_{lj})\mathfrak{b}_{lp} \leq \Omega_{p_k}^{pmem}\mathfrak{x}_{pk}$.

We summarize the symbols used in this section in Table 1 to facilitate reading.

Table 1. Notations for problem model

Symbol	Meaning
$\mathcal{C}_t$	Set of containers arriving at time t.
c_i	A container of index i, and $c_i \in \mathcal{C}_t$
$\zeta_{c_i}^{cpu}$, $\zeta_{c_i}^{men}$	CPU and memory requirements for container c_i
$\zeta_{c_i}^{os}$	A OS type of container c_i
τ_j	A VM type of index j
Γ	Set of different VM types provided by cloud providers, and $\tau_j \in \Gamma$
Ω_j^{vos}	A OS type of VM τ_j
Φ	Set of different OS types, and $\zeta_{c_i}^{os} \in \Phi$, and $\Omega_j^{vos} \in \Phi$
$\Omega_{\tau_j}^{vcpu}$, $\Omega_{\tau_j}^{vmen}$	CPU capacity and memory capacity of a VM type τ_j.
$\pi_{\tau_j}^{vcpu}$, $\pi_{\tau_j}^{vmen}$	CPU overhead and memory overhead of a VM type τ_j.
T_{p_k}	A PM type of index k
Υ	Set of different PM types provided by cloud providers, and $\mathsf{T}_{p_k} \in \Upsilon$
$\Omega_{p_k}^{pcpu}$, $\Omega_{p_k}^{pmen}$	CPU capacity and memory capacity of a PM type T_{p_k}
$\pi_{p_k}^{pcpu}$, $\pi_{p_k}^{pmen}$	CPU and memory overhead of a PM type T_{p_k}
$E_{p,t}$	Energy consumption of a given PM p at time t
E_p^{idle}, E_p^{full}	Energy consumption of a given PM p per time unit when it is idle and fully loaded
$|L_t|$	Number of VM instances used at time t
$\mu_{p,t}^{cpu}$	CPU utilization level of a given PM p at time t
a_{lj}	$a_{lj} = 1$ if the lth running VM instance belongs to type τ_j, else $a_{lj} = 0$
b_{lp}	$b_{lp} = 1$ if the lth VM instance is allocated to the pth PM instance, else $b_{lp} = 0$
c_{il}	$c_{il} = 1$ if the ith container is allocated to the lth VM, else $c_{il} = 0$
$\mathfrak{x}_{pk}$	$\mathfrak{x}_{pk} = 1$ if the pth created PM is type of T_{p_k}, else $\mathfrak{x}_{pk} = 0$
T	Allocation period
J	Accumulated energy consumption over the allocation period T
$|P_t|$	A step function that gives the number of PM used at any time t

4   A Hybrid Approach for Dynamic RAC

As presented in Sect. 3, the allocation decision includes four components, VM selection, VM creation, PM selection, and PM creation. The proposed hybrid-heuristic approach, aims to automatically generate two rules as solutions for these four decision components. The two rules can be used for four decisions because creating new VM/PM is treated as a special case of VM/PM selection. In particular, one rule r^v is trained for VM selection and creation, and the other hybrid rule, i.e., r^p with the Best Fit, is for PM selection and creation, respectively. A rule is used as a priority function to indicate the preference of selecting a given candidate allocation decision.

Our hybrid-heuristic approach follows the CCGP framework from [22] (shown in Fig. 2) by first randomly initializing two sub-populations. Each sub-population contains candidates for one rule in the form of a tree (see details in Sect. 4.1). The rules are evolved cooperatively by the genetic operators, e.g. crossover and mutation. Based on historical data, a set of training instances are processed and used for fitness evaluation. To evaluate a rule, a pair of rule is formed from each population. The performance of each pair of rule is measured by its fitness value which will be discussed in Sect. 4.2. Based on the fitness values, the best performing rules within each sub-population are selected to generate the next generation. This process repeats until a predefined stopping condition, e.g., maximum generations, is met.

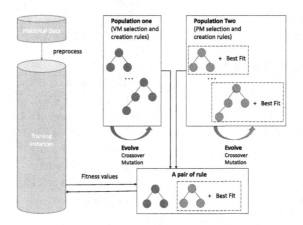

Fig. 2. An overview of the training process of CCGP

4.1 Representation, Terminal and Function Set

We adopt a tree representation for two allocation rules due to its promise in interpretability, expressiveness, and flexibility [22]. To capture features of dynamic arriving container as well as properties of VMs and PMs, a set of terminal nodes and functional nodes are shown in Table 2. As shown in Fig. 3, a simple example of an allocation rule is constructed by a set of terminal nodes, i.e., LeftVMMem, ConCPU and LeftVMCPU, and function/intermediate nodes, i.e., + and /.

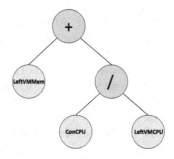

Fig. 3. A simple example of allocation rule for VM selection and creation

Table 2. Terminal and function sets used in our tree-based rules

Terminal Set (VM Selection and Creation Features)	Description
ConMem	Memory requirement of a container
ConCPU	CPU requirement of a container
LeftVMMem	Remaining memory of a VM
LeftVMCPU	Remaining CPU of a VM
VMMemOverhead	Memory overhead of a VM
VMCPUOverhead	CPU overhead of a VM
Terminal Set (PM Selection and Creation Features)	**Description**
VMMem	Memory capacity of a VM
VMCPU	CPU capacity of a VM
LeftPMMem	Remaining Memory of a PM
LeftPMCPU	Remaining CPU of a PM
PMMem	Memory capacity of a PM
PMCPU	CPU capacity of a PM
PMCore	Core requirement of a PM
Function Set (PM/VM Selection and Creation Features)	**Description**
Add, Subtract, Multiply	Arithmetic operations ($+$, $-$, $\times$)
Protected Division	Protected division ($/$) that returns 1 if the denominator is 0.

4.2 Fitness Evaluation

Algorithm 1 shows the process of fitness evaluation on a pair of rules, i.e., VM creation and selection rule r^v, PM creation and selection rule r^p. For every training instance, we start with a data center initialization process, where a group of containers are randomly allocated to suitable VMs, and these VMs are randomly allocated to suitable PMs. Note that the same randomness is used for the data center initialization process in every generation through the use of the same random seed. By doing this, it can ensure the fitness values of every pair of rules in one generation are comparable. Afterwards, we allocated containers one by one based on their arrival time. Note that multiple containers can arrive at the same time t. The arrival time t is also utilized for triggering the process of updating total energy consumption using Eq. (1). For every container, we first filter out running VMs with different OSs from its OS. The remaining running VMs combined with new VMs (one for each VM type) are merged as a candidate VM list, on which VM selection or creation operation is performed based on the calculated priority value using r^v. Whenever a new VM is created, a PM type is selected and PM of the selected type is created using r^p, where the trained heuristic part in r^p is used for the selection of a running PM while Best Fit part in r^p is used for the selection of a new PM instance (one for each PM type) for a PM creation. After allocation all containers in the training instance, we return the total energy consumption J as the fitness value of the pair of rules.

Algorithm 1: Fitness evaluation

Input: Containers (c_t), a pair of rules $(r^v$ and $r^p)$
Output: Accumulated energy consumption (J)

1 **for** *each training instance* **do**
2    $J = 0$;
3   Initialize data center;
4   **for** *each C_t over a period of time* **do**
5   **for** *each c_t in C_t* **do**
6   Update J;
7   Filter out the VMs that require different OS;
8        $vm =$ VMSelectionCreation(c_t, r^v);
9   allocateVM(c,vm);
10   **if** *vm is Newly Created* **then**
11   addVM(vm, the running VM list);
12          $pm =$ PMSelectionCreation(vm, r^p);
13   **if** *pm is Newly Created* **then**
14   addPM(pm, the running PM list);
15   **end**
16   allocatePM(vm, pm);
17   **end**
18   **end**
19   **end**
20 **end**
21 **Return** energy consumption;

---

5   Experiment

To evaluate the proposed approach, we conducted an empirical experiment using real-world datasets in this section. Notably, we compare our approach with a popular baseline used in the recent works [2,9,13,22], i.e., Sub&JustFit/FF rule [15], and a GPHH approach [22] that is recently proposed to solve a similar RAC problem. The problem investigated in this paper is more challenging because it considers real-world workload patterns (instead of using synthetic workload studied in [22]) and heterogeneous PM types. We modify Sub&JustFit/FF rule and GPHH approach [22] to cope with the selection on heterogeneous PM types. In particular, a Best-Fit (BF) rule is added to the Sub&JustFit/FF rule for new PM creation. Meanwhile, GPHH approach [22] is adapted to heterogeneous PM types by considering different PM type options when training the second level allocation rules. In this section, we will use Sub&JustFit/FF&BF and Evo/Evo to represent rules generated by these two modified methods, and use Evo/Hybrid-Evo to represent our hybrid-heuristic approach.

5.1   New Dataset

We designed six scenarios, as summarized in Table 3, for our experimental analysis. In particular, we consider distinct numbers of OS types in the experiment.

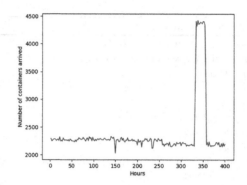

Fig. 4. An example of real-world workload pattern

The proportions of the OS types are determined using the recent market shares of OS reported in [24]. In Table 4, we simulate three OS scenarios where the number of OS increases from 3 to 5. Furthermore, as shown in Tables 5 and 6, we uses popular VM types and PM types settings from Amazon EC2 VM instances and PM reported in [22] as examples in this paper. Note that, we convert CPU Cores of VM to MHz using $VM_{Mhz} = PM_{MHz}/PM_{Core} * VM_{Core}$. Using this formula, we actually create 45 different VM types in Mhz.

To create workload patterns that consider dynamic number and time arrival for containers, we use Bitbrains data [19], i.e., time-serial CPU and memory usage records of applications, to generate 400 instances for each scenario. Each instance is made of two hour records of containers arrived at the data center. This can ensure the number of containers to be large enough for an algorithm to reach a stable status. Note that, we exclude containers that cannot be fulfilled with the largest VM types in Table 5, and the remaining containers are augmented with OS requirements based on the OS distribution in Table 4. We use 100 instances for data center initialization, 200 instances for rule training, and 100 instances for rule testing.

Table 3. Scenarios

Scenarios	Number of OSs	VM types	Container workload patterns
S-1	3	Amazon EC2 VM types	BitBrains Pattern 1
S-2	3	Amazon EC2 VM types	BitBrains Pattern 2
S-3	4	Amazon EC2 VM types	BitBrains Pattern 1
S-4	4	Amazon EC2 VM types	BitBrains Pattern 2
S-5	5	Amazon EC2 VM types	BitBrains Pattern 1
S-6	5	Amazon EC2 VM types	BitBrains Pattern 2

Table 4. OS Distribution

Number of OS	OS distribution (%)
3	50, 30, and 20
4	62.5, 17.5, 15, and 5
5	17.9, 45.4, 23.6, 10.5, and 2.6

Table 5. VM types

VM	CPU (*Core*)	Memory (*GB*)
C5.large	2	4
C5.xlarge	4	8
C5.2xlarge	8	16
C5.4xlarge	16	32
C5.9xlarge	36	72
C5a.8xlarge	32	64
C5a.12xlarge	48	96
C5a.16xlarge	64	128

Note that, different from the dataset used in the very recent work [22], which assumes containers arrive one by one uniformly and a fixed number of contains (i.e., 2500 in [22]) in every instance, our dataset does not make such unrealistic assumptions and consider a varied number of containers can come at any time. See an example of workload pattern in Fig. 4. The new benchmark datasets and the implementation of all approaches in this paper have been made freely available online.[1]

5.2 Parameter Settings

We strictly follow the setting reported in the recent work [22] except the number of generations. The number of generation increases to 200 for the training on challenging new dataset. Other settings remain the same, e.g., number of population is two, each population size is 512, elite size is 3, tournament selection size is 7, crossover rate is 0.8, mutation rate is 0.1, reproduction rate is 0.1, crossover max depth and mutate max depth are 7.

5.3 Performance Comparison

Table 7 shows the comparison of accumulated energy consumption among the three competing rules, i.e., Sub&JustFit/FF&BF, Evo/Evo, and Evo/Hybrid-Evo. Our results show that the Evo/Hybrid-Evo rule learned in all scenarios

---

[1] Online resource allocation benchmarks for container-based clouds and the source code of all discussed approaches are available from https://github.com/ chenwangnida/RAC.

Table 6. PM types

PM (CPU)	CPU (MHz)	Memory (MB)	P_{Idle} (Wh)	P_{Max} (Wh)	Cores
E5-2630	27600	128000	99.6	325	12
E5430	22640	16000	166	265	8
E5507	20264	8000	67	218	8
E5-2620	24000	3200	70	300	12
E5645	28800	16000	63.1	200	12
E5-2650	32000	24000	52.9	215	12
E5-2651	21600	32000	57.5	178	16
E5-2670	41600	24000	54.1	243	12
E5540	10000	72000	151	312	16
E5560	22400	128000	133	288	4
E5-2665	24000	256000	117	314	8
XS5650	31992	64000	80.1	258	12

achieve the lowest accumulated energy consumption. Note that the performance of Evo/Evo does not match the reported results in [22], where Evo/Evo outperforms Sub&JustFit/FF&BF rule. Evo/Evo does not work well under real-world workload pattern settings. This might due to that the big searching space using our dataset.This finding also indicates that Evo/Evo may not be suitable for a cloud data center where different PMs types are considered.

Table 7. The best and mean accumulated energy consumption of three competing rules tested over 100 h (Note: the lower the value the better)

Scenarios	Sub&JustFit/FF&BF	Evo/Evo (best value)	Evo/Hybrid-Evo (best value)	Evo/Evo (mean value)	Evo/Hybrid-Evo (mean value)
S-1	95959	95868	**95496**	125994 ± 22734	**95550 ± 43**
S-2	85296	85128	**84850**	114775 ± 27378	**85049 ± 138**
S-3	95503	95399	**95061**	121537 ± 21237	**95165 ± 137**
S-4	85615	85486	**85181**	116209 ± 24509	**85366 ± 98**
S-5	95442	95289	**94961**	120951 ± 19164	**95018 ± 65**
S-6	85316	85185	**84944**	116336 ± 23293	**85069 ± 60**

To further analyze the differences in energy consumption during the allocation process, we use the deterministic sub&JustFit/FF&BF rule and the two best evolved rules found from Evo/Evo rule, and Evo/Hybrid-Evo in scenario S-1 in Table 3 as an example.

Figure 5 (a) shows that all the rules have comparable energy consumption in the first 10 h. This is because new PMs are barely created, and containers are mainly allocated to the existing running PMs. As time goes by, new PMs must

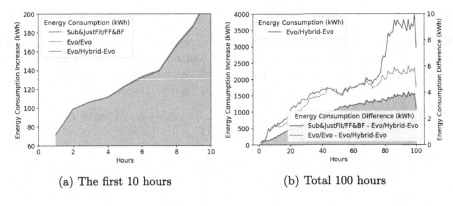

(a) The first 10 hours   (b) Total 100 hours

Fig. 5. Energy consumption during the allocation process for Scenario S-1

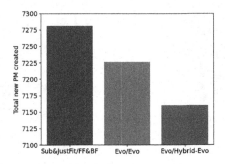

Fig. 6. Total new PMs created

be created to satisfy the resource needs of newly arriving containers. In particular, Evo/Hybrid-Evo rule is expected to create new PMs that result in more opportunities in allocating more containers than the sub&JustFit/FF&BF rule, Evo/Evo rules in the future. Figure 5 (b) shows the average increase in energy consumption over the first 100 h using Evo/Hybrid-Evo rule and the average increase differences between Evo/Hybrid-Evo and other two rules. We can see that the average increase in energy consumption using Evo/Hybrid-Evo rule is much slower than the sub&JustFit/FF&BF and Evo/Evo rules. Meanwhile, sub&JustFit/FF&BF rule is the worst performer in reducing energy consumption.

Figure 6 shows the total number of new PMs created in 100 h using the three rules. The Evo/Hybrid-Evo rule creates the significantly less number of PMs than the other two rules after 100 h. This observation indicates that Evo/Hybrid-Evo rule creates less new PMs in the long run, which could lead to better energy consumption.

5.4   Further Analysis

In this section, we use a simple example to explain the best pair of rule learned from Evo/Hybrid-Evo in 30 runs. The best pair of rule includes a VM selection and creation rule r^v and a PM selection rule, i.e., a trained heuristic part in r^p. These two rules are presented in Eq. (3) and Eq. (4), respectively.

$$score_{r^v} = (((((\text{VMMemOverhead} - \text{CoMem}) - \text{LeftVMMem}) * \text{VMCPUOverhead}) * \text{VMCPUOverhead}) * \text{CoMem}) * \text{CoMem}$$

(3)

$$score_{r^p} = \text{PMCore} - ((\text{VMMem}/((\text{VMMem}/\text{PMMem})/\text{LeftPMCPU})) - ((\text{VMMem}/\text{PMMem})/\text{LeftPMCPU}))$$

(4)

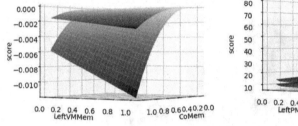 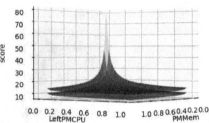

(a) Upper plane τ_1 and lower plane τ_2

(b) Upper plane PM E5540 and Lower plane PM E5-2665

Fig. 7. The 3D contour plots of the best pair of rules learned from Evo/Hybrid-Evo in 30 runs

Let's consider an example of a cloud provider, who provides two types of 8-core VMs (e.g., τ_1 and τ_2) and two types of PMs (e.g., 8-core PM E5-2665 and 16-core PM E5540). For VM type τ_1, its CPU capacity $\Omega_{\tau_1}^{vcpu}$ is equal to 18400 and its memory capacity $\Omega_{\tau_1}^{vmen}$ is equal to 21000. For VM type τ_1, its CPU capacity $\Omega_{\tau_2}^{vcpu}$ is equal to 5000, and memory capacity $\Omega_{\tau_2}^{vmen}$ is equal to 16000. Meanwhile, the overhead on CPU and Memory are 10% of CPU capacity and 200 MB, respectively.

Taking the above numbers into Eq. (3), we can plot $score_{r^v}$, LeftVMMem, and ConMem in a 3D contour figure, shown in Fig. 7 (a) over two planes. In Fig. 7 (a), the upper plane and lower plane are corresponds to VM types τ_1 and τ_2, respectively. This figure indicates that for allocating any incoming container to VMs, VM instance selection or creation is determined by the priority score $score_{r^v}$. Such score is determined not only by VM types, but also by $ConMem$ and $LeftVMMem$. Meanwhile, this score is more sensitive to $ConMem$ than $LeftVMMem$. Moreover, τ_1 is often preferable, compared to τ_2.

In Fig. 7 (b), we assume a VM of type τ_1 is created, and it waits to be allocated to two running PMs instances (e.g., 8-core PM E5-2665 and 16-core PM E5540). Based on this assumption and the PM selection rule in r^p, i.e., Eq. (4), we can plot another 3D contour figure for explaining decisions on PM selection, which is determined by $LeftPMCPU$, $PMMem$, and two planes in Fig. 7 (b). The upper plane and lower plane are corresponds to PM types E5540 and E5-2665, respectively. Similar to Fig. 7 (a), this plot indicates that how the PM selection is affected by $LeftPMCPU$, $PMMem$, and PM types.

6   Conclusions

In this paper we proposed a novel hybrid CCGP hybrid-heuristic approach, named Evo/Hybrid-Evo, to solve a new challenging dynamic RAC problem. We formulated a formal energy consumption model for this problem that that considers heterogeneous PM types. We also evaluated its performance against other promising approaches in a more realistic experimental settings. Our experiment results show that the Evo/Hybrid-Evo achieves significantly lower accumulated energy consumption than the human-designed rule (sub&JustFit/FF&BF rule,) and a state-of-the-art approach (Evo/Evo).

References

1. Abohamama, A.S., Hamouda, E.: A hybrid energy-aware virtual machine placement algorithm for cloud environments. Expert Syst. Appl. **150**, 113306 (2020)
2. Akindele, T., Tan, B., Mei, Y., Ma, H.: Hybrid grouping genetic algorithm for large-scale two-level resource allocation of containers in the cloud. In: Long, G., Yu, X., Wang, S. (eds.) AI 2022. LNCS (LNAI), vol. 13151, pp. 519–530. Springer, Cham (2022). https://doi.org/10.1007/978-3-030-97546-3_42
3. Al-Moalmi, A., Luo, J., Salah, A., Li, K., Yin, L.: A whale optimization system for energy-efficient container placement in data centers. Expert Syst. Appl. **164**, 113719 (2021)
4. Bhardwaj, A., Krishna, C.R.: Virtualization in cloud computing: moving from hypervisor to containerization-a survey. Arab. J. Sci. Eng. **46**(9), 8585–8601 (2021)
5. Bhattacherjee, S., Das, R., Khatua, S., Roy, S.: Energy-efficient migration techniques for cloud environment: a step toward green computing. J. Supercomputing **76**(7), 5192–5220 (2020)
6. Dayarathna, M., Wen, Y., Fan, R.: Data center energy consumption modeling: a survey. IEEE Commun. Surv. Tutorials **18**, 732–794 (2015)
9. Ding, W., Luo, F., Han, L., Gu, C., Lu, H., Fuentes, J.: Adaptive virtual machine consolidation framework based on performance-to-power ratio in cloud data centers. Future Gener. Comput. Syst. **111**, 254–270 (2020)
8. Gharehpasha, S., Masdari, M., Jafarian, A.: Virtual machine placement in cloud data centers using a hybrid multi-verse optimization algorithm. Artif. Intell. Rev. **54**(3), 2221–2257 (2021)

9. Guo, M., Guan, Q., Chen, W., Ji, F., Peng, Z.: Delay-optimal scheduling of VMs in a Queueing cloud computing system with heterogeneous workloads. IEEE Trans. Serv. Comput. 15(1), pp. 110–123 (2022)

10. Hussein, M.K., Mousa, M.H., Alqarni, M.A.: A placement architecture for a container as a service (CAAS) in a cloud environment. J. Cloud Comput. 8(1), 1–15 (2019). https://doi.org/10.1186/s13677-019-0131-1

11. Kaewkasi, C., Chuenmuneewong, K.: Improvement of container scheduling for docker using ant colony optimization. In: 2017 9th International Conference on Knowledge and Smart Technology (KST), pp. 254–259. IEEE (2017)

12. Kanso, A., Youssef, A.: Serverless: beyond the cloud. In: Proceedings of the 2nd International Workshop on Serverless Computing, pp. 6–10 (2017)

13. Li, F., Tan, W.J., Cai, W.: A wholistic optimization of containerized workflow scheduling and deployment in the cloud-edge environment. Simul. Model. Pract. Theory 118, 102521 (2022)

14. Long, S., Wen, W., Li, Z., Li, K., Yu, R., Zhu, J.: A global cost-aware container scheduling strategy in cloud data centers. IEEE Trans. Parallel Distrib. Syst. 33(11), 2752–2766 (2021)

15. Mann, Z.Á.: Interplay of virtual machine selection and virtual machine placement. In: Aiello, M., Johnsen, E.B., Dustdar, S., Georgievski, I. (eds.) ESOCC 2016. LNCS, vol. 9846, pp. 137–151. Springer, Cham (2016). https://doi.org/10.1007/978-3-319-44482-6_9

16. Nardelli, M., Hochreiner, C., Schulte, S.: Elastic provisioning of virtual machines for container deployment. In: Proceedings of the 8th ACM/SPEC on International Conference on Performance Engineering Companion, pp. 5–10 (2017)

17. Piraghaj, S.F., Dastjerdi, A.V., Calheiros, R.N., Buyya, R.: Efficient virtual machine sizing for hosting containers as a service (SERVICES 2015). In: 2015 IEEE World Congress on Services, pp. 31–38. IEEE (2015)

18. Piraghaj, S.F., Dastjerdi, A.V., Calheiros, R.N., Buyya, R.: A framework and algorithm for energy efficient container consolidation in cloud data centers. In: 2015 IEEE International Conference on Data Science and Data Intensive Systems, pp. 368–375. IEEE (2015)

19. Shen, S., Van Beek, V., Iosup, A.: Statistical characterization of business-critical workloads hosted in cloud datacenters. In: 2015 15th IEEE/ACM International Symposium on Cluster, Cloud and Grid Computing, pp. 465–474. IEEE (2015)

20. Shi, T., Ma, H., Chen, G.: Energy-aware container consolidation based on PSO in cloud data centers. In: IEEE CE, pp. 1–8 (2018)

21. Tan, B., Ma, H., Mei, Y.: A genetic programming hyper-heuristic approach for online resource allocation in container-based clouds. In: Mitrovic, T., Xue, B., Li, X. (eds.) AI 2018. LNCS (LNAI), vol. 11320, pp. 146–152. Springer, Cham (2018). https://doi.org/10.1007/978-3-030-03991-2_15

22. Tan, B., Ma, H., Mei, Y., Zhang, M.: A cooperative coevolution genetic programming hyper-heuristics approach for on-line resource allocation in container-based clouds. IEEE Trans. Cloud Comput. 10, 1500–1514 (2022)

23. Tarahomi, M., Izadi, M., Ghobaei-Arani, M.: An efficient power-aware VM allocation mechanism in cloud data centers: a micro genetic-based approach. Clust. Comput. 24(2), 919–934 (2021)

24. Taylor, P.: Global market share held by operating systems for desktop PCs, from Jan 2013 to Dec 2021. Tech. rep. (2022). https://www.statista.com/statistics/218089/global-market-share-of-windows-7

25. Zhang, C., Wang, Y., Wu, H., Guo, H.: An energy-aware host resource management framework for two-tier virtualized cloud data centers. IEEE Access **9**, 3526–3544 (2020)
26. Zhang, R., Zhong, A., Dong, B., Tian, F., Li, R.: Container-VM-PM Architecture: a novel architecture for docker container placement. In: Luo, M., Zhang, L.-J. (eds.) CLOUD 2018. LNCS, vol. 10967, pp. 128–140. Springer, Cham (2018). https://doi.org/10.1007/978-3-319-94295-7_9

A Memetic Genetic Algorithm for Optimal IoT Workflow Scheduling

Amer Saeed[1]([✉])(iD), Gang Chen[1](iD), Hui Ma[1](iD), and Qiang Fu[2](iD)

[1] Victoria University at Wellington, Kelburn, Wellington 6012, New Zealand
{amer.saeed,aaron.chen,hui.ma}@vuw.ac.nz
[2] RMIT, Melbourne, VIC 3000, Australia
qiang.fu@rmit.edu.au
https://www.wgtn.ac.nz/

Abstract. Internet of Things (IoT) devices have become a crucial part
of daily life. Because IoT devices often have small processing capability
and low power supply, two popular technologies, i.e. cloud servers and
fog edges, are increasingly integrated with IoT for workflow execution,
giving rise to the resource allocation and workflow scheduling problem
in hybrid IoT environments, i.e. the IoT workflow scheduling (IoTWS)
problem. To tackle this NP-hard IoTWS problem, a new Genetic Algorithm (GA) called IoTGA has been successfully developed in this paper.
In comparison to state-of-the-art GA approaches from literature, IoTGA
allows fast workflow execution and can explicitly reduce the time and
energy consumption thanks to its use of a newly designed local search
method. Experiments on benchmark IoTWS problems clearly indicate
that IoTGA can significantly outperform several competing GA methods and are more useful in practice.

Keywords: Internet of Things · Workflow Scheduling · Memetic
Genetic Algorithm

1 Introduction

Internet of Things (IoT) starts to play a crucial role in human life [22]. There
has been a rapid rise in demand for IoT connectivity. Experts estimate that the
number of IoT devices will reach 29 billions in 2030 [11]. To help IoT devices meet
users' demand for high-quality ubiquitous service through reducing delay and
energy consumption, fog/edge and cloud servers are increasingly integrated in the
IoT infrastructure [12]. However, the collective use of a myriad of miscellaneous
virtualized computing resources in an IoT environment gives birth to a variety of
technical challenges, including IoT application placement, workflow scheduling,
task offloading, and load balancing [9]. A workflow is a sequence of tasks required
to accomplish a job. Each task in the workflow is preceded and followed by other
tasks, apart from the initial and final tasks [9].

© The Author(s), under exclusive license to Springer Nature Switzerland AG 2023
J. Correia et al. (Eds.): EvoApplications 2023, LNCS 13989, pp. 556–572, 2023.
https://doi.org/10.1007/978-3-031-30229-9_36

In this paper, we focus specifically on the resource allocation and workflow scheduling problem in hybrid IoT environments. To ease discussion, such problems will be referred to as the *IoT workflow scheduling* (IoTWS) problems. Since workflows present a major type of workload for various smart applications in IoT systems, addressing the IoTWS problems enables us to indirectly tackle several related problems, including the task offloading and the load balancing problems [13]. The IoTWS problem therefore has significant research and practical values and deserves more attention in the research community.

In a nutshell, each IoT device in the IoTWS problem can generate a workflow that involves multiple inter-dependent tasks. Each task can be assigned either to an IoT device, a fog server or a remote cloud server for execution. After its execution, a task also needs to transfer its output data to the subsequent dependent tasks/devices. The goal of the IoTWS problem is to find a scheduling solution that assigns every task of all workflows to one device/server for execution so as to minimize the overall makespan required to execute all workflows as well as the total energy consumption. Several existing research works have studied such IoTWS problems (or similar problems) recently [1,10,18]. As a result, multiple algorithms have been successfully developed for IoTWS. These algorithms can be largely divided into four categories: (1) heuristic approaches [4], (2) mathematical optimization approaches [1], (3) local search approaches [25], and (4) evolutionary computation approaches (EC) [10]. Each category of algorithms has its own limitations and drawbacks. Heuristic algorithms can often produce sub-optimal scheduling solutions. Manual design of the heuristic rules can also be labor-intensive and may fail to generalize well to different IoT environments. Mathematical optimization approaches are often accompanied by high computational cost, restricting their applicability only to small scale IoTWS problems. Local search techniques can easily scale to large problem sizes. However, the search process can be easily trapped by poor local optima.

Different from all these algorithms discussed above, evolutionary algorithms, particularly the *genetic algorithm* (GA), adopt a population-based strategy to search for global optimal scheduling solutions [20] and can achieve a desirable balance between algorithm efficiency and effectiveness. A prominent example is the APMA algorithm proposed recently in [10]. APMA utilizes a progressive strategy to divide all tasks to be scheduled into multiple layers. GA is used to schedule tasks of each layer one at a time. A local search operator is also proposed to enhance the performance of the algorithm.

While APMA achieved cutting-edge performance on several IoTWS problems, the layered allocation of tasks requires all tasks in one layer to be completed before executing any tasks in the subsequent layers [10]. This may introduce unnecessary delays to the execution of dependent tasks in a workflow. To address this issue, we develop a new GA method for IoTWS called IoTGA. IoTGA allows immediate execution of any task as long as all tasks it depends upon have been completed. We further develop a new local search operator that can explicitly reduce the communication time and cost required for task execution.

Driven by the goal of developing IoTGA, this paper achieves the following main contributions:

- A new formulation of the IoTWS problem has been presented in this paper. The new formulation allows fast task execution subject to both workflow and allocation dependencies (see Sect. 3).
- A new IoTGA method has been successfully developed in this paper. In comparison to existing algorithms such as APMA [10], IoTGA allows fast task execution and can explicitly reduce the time and energy consumption due to inter-task data communication by using a newly designed local search method.
- Extensive experiments have been conducted in this paper. Our experiment results clearly indicate that IoTGA can significantly outperform APMA and standard GA from literature.

2   Related Works

Workflow execution in IoT environments has many real-world applications. For example, [6] showed that advanced intelligent surveillance can be supported by a large IoT workflow. It is important to carefully schedule the execution of such workflows in order to meet execution time, energy consumption and related performance objectives [14].

Many recently published research works [7] considered the workflow scheduling problem by using only IoT devices. In view of the limited processing capability and power supply of IoT devices, [10,16] further explored the possibility of executing some/all workflow tasks on fog and cloud servers, giving rise to a *hybrid IoT paradigm* that is the central focus of this paper.

Workflow scheduling in hybrid IoT aims to exploit a plethora of virtualized computing resources offered both locally and remotely to achieve the best possible performance in execution time and energy consumption [2]. This is widely known as NP-hard problems in literature [2].

Multiple approaches have been proposed to solve IoTWS problems efficiently and effectively. These approaches as they were explained can be largely divided into four categories: (1) heuristic approaches [4], (2) local search approaches [25], (3) mathematical optimization approaches [1,17], and (4) EC approaches [10,21]. Among these approaches, EC techniques, particularly the APMA algorithm, have achieved the best trade-off between performance and algorithm complexity [10]. While showing promise, APMA only allows layer-by-layer execution of all workflow tasks [10]. Specifically, any task in one layer can not be scheduled for execution until all tasks from the previous layer has been completed, inevitably delaying the execution of some tasks. In addition, any two tasks in the same layer cannot be assigned to the same device for execution in APMA, potentially restricting the number of possible scheduling solutions that the algorithm can produce. In an attempt to achieve better workflow scheduling performance, we propose a new IoTGA algorithm to tackle these issues in this paper.

3 Problem Formulation

We study practical IoTWS problems where multiple workflows generated by various IoT devices can be executed either locally on these IoT devices or partially/completely assigned to cloud servers and/or fog servers. Figure 1 presents an outline of the IoTWS problem. In line with Fig. 1, a workflow WF is defined as a tuple, i.e. $WF = (D(WF), G(WF), T(WF))$, with $D(WF)$ representing the IoT device that generated WF. The final results obtained from executing WF must be sent back to $D(WF)$. $T(WF)$ indicates the time that WF is generated by $D(WF)$. $G(WF)$ is a *Directed Acyclic Graph* (DAG) where each node of $G(WF)$ denotes a separate task $t \in G(WF)$ to be executed and each edge $(t, t') \in G(WF)$ captures the dependency between two tasks $t \in G(WF)$ and $t' \in G(WF)$ in WF (i.e. task t must be executed prior to task t'). Meanwhile, $\forall(t, t') \in G(WF)$, $\rho(t, t')$ measures the amount of data to be transmitted from task t as the input to task t'. $C(t)$ indicates the number of CPU instructions required for executing any task t.

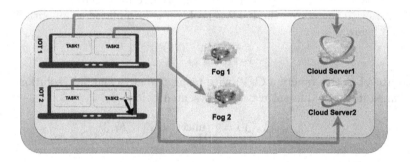

Fig. 1. High-Level Overview of the IoT Workflow Scheduling Problem.

The inputs of the IoTWS problem are defined below:

- $\{WF_1, \ldots, WF_K\}$: the set of workflows to be executed. $\forall i \in \{1, \ldots, K\}$, $D(WF_i) = s^{0,i}$, where $s^{0,i}$ is the i-th IoT device in the IoTWS problem.
- $\{s^{\zeta,I}\}$: the set of computation devices. Each device has its type indicated by ζ and its ID indicated by I. ζ has three optional values: 0: IoT, 1: Fog server, and 2: Cloud Server. We use $C(s^{\zeta,I})$ to denote the *computing capacity* of $s^{\zeta,I}$, i.e., the number of CPU instructions $s^{\zeta,I}$ can process per second.

For any two tasks t and t', t is said to depend on t' if $\exists WF$ s.t. $(t, t') \in G(WF)$. Simultaneously, task t may also depend on t' if they are allocated to the same device $s^{\zeta,I}$ for execution, since any device can only execute its allocated tasks sequentially. The former case is called the *workflow dependency*. The later case is called the *allocation dependency*.

In line with the problem input described above, the solution/output from the IoTWS problem is denoted as $\mathcal{Y} = \{Y_1, \ldots, Y_K\}$, which is a set of *allocation configurations* with respect to all workflows $\{WF_1, \ldots, WF_K\}$. Specifically, $\forall i \in \{1, \ldots, K\}$, Y_i is defined as

$$Y_i = \{(t, s^{\varsigma,I}) | \forall t \in G(WF_i)\}$$

with $(t, s^{\varsigma,I})$ denoting that task t of workflow WF_i is assigned for execution to device $s^{\varsigma,I}$ according to Y_i and hence $\mathcal{Y}$. For the sake of simplicity, we use $\mathcal{Y}(t)$ to indicate this device.

The IoTWS problem defined in this section uses two assumptions commonly considered in literature [10]. *Assumption 1*: each workflow WF has a dummy start task, denoted as $S(WF)$, and a dummy end task, denoted as $E(WF)$. Using the two tasks, we can easily determine the start time of the workflow $ST(WF) = ST(S(WF))$ (i.e. the start time of workflow WF equals to the start time of its start task) and the completion time of the workflow $CT(WF, \mathcal{Y}) = CT(E(WF), \mathcal{Y}))$ (i.e. the completion time of workflow WF equals to the completion time of its end task). Note that the completion time depends on $\mathcal{Y}$ and is expected to be minimized. *Assumption 2*: $\forall i \in \{1, \ldots, K\}, ST(WF_i) = 0$. In other words, all workflows are generated in advance for execution.

In line with the above, the IoTWS problem is formulated as an optimization problem below:

$$\underset{\mathcal{Y}=\{Y_1, \ldots, Y_K\}}{\arg\min} \quad \mathcal{O}(\mathcal{Y})$$

where the objective function $\mathcal{O}(\mathcal{Y})$ with respect to any candidate solution $\mathcal{Y}$ can be defined in three different ways, as presented below:

$$\mathcal{O}_{makespan}(\mathcal{Y}) = \max_{i \in \{1, \ldots, K\}} CT(WF_i, \mathcal{Y}) \tag{1}$$

$$\mathcal{O}_{energy}(\mathcal{Y}) = \sum_{i=1}^{K} Energy(WF_i, \mathcal{Y}) \tag{2}$$

$$\mathcal{O}_{combined}(\mathcal{Y}) = w\tilde{\mathcal{O}}_{makespan}(\mathcal{Y}) + (1-w)\tilde{\mathcal{O}}_{energy}(\mathcal{Y}) \tag{3}$$

where $0 \leq w \leq 1$ controls the relative importance of the normalized makespan $\tilde{\mathcal{O}}_{makespan}$ and the normalized energy consumption $\tilde{\mathcal{O}}_{energy}$. We use the common min-max normalization technique. The maximum makespan and maximum energy consumption used to calculate $\tilde{\mathcal{O}}_{makespan}$ and $\tilde{\mathcal{O}}_{energy}$ are determined by executing all tasks of all workflows locally on IoT devices.

3.1   Makespan Model

This subsection presents the computation of the makespan $\mathcal{O}_{makespan}(\mathcal{Y})$ with respect to any IoTWS solution $\mathcal{Y}$. Following (1), $\mathcal{O}_{makespan}(\mathcal{Y})$ is obtained after calculating $CT(WF, \mathcal{Y})$ of each workflow WF. Meanwhile,

$$CT(WF, \mathcal{Y}) = \max_{t \in G(WF)} CT(t, \mathcal{Y}) \tag{4}$$

Therefore, the focus is on computing the completion time $CT(t, \mathcal{Y})$ of each task $t \in G(WF)$. For any task $t \in G(WF)$, define

$$Pre^1(t) = \{t'|(t', t) \in G(WF)\}$$

to be the set of predecessor tasks that must be executed prior to task t due to the workflow dependency. Further define

$$Pre^2(t) = \{t'|\mathcal{Y}(t') = \mathcal{Y}(t) \text{ and } Ready(t') < Ready(t)\}$$

to be the set of predecessor tasks that must be executed prior to task t due to the allocation dependency. Here the ready time $Ready(t)$ of any task t refers to the time when task t becomes ready for execution [15]. Then $CT(t, \mathcal{Y})$ is calculated as

$$CT(t, \mathcal{Y}) = \frac{C(t)}{C(\mathcal{Y}(t))} +$$
$$\max \left\{ \begin{array}{l} \max\limits_{t' \in Pre^1(t)} \left(CT(t') + Delay(\mathcal{Y}(t), \mathcal{Y}(t')) + \frac{\rho(t', t)}{Band(\mathcal{Y}(t'), \mathcal{Y}(t))} \right), \\ \max\limits_{t' \in Pre^2(t)} CT(t') \end{array} \right\} \quad (5)$$

At the RHS of Eq.(5), the first term computes the time required for executing task t on device $\mathcal{Y}(t)$. The second term determines the earliest ready time of task t. According to Eq.(5), the ready time can never be earlier than the completion time of any task $t' \in Pre^1(t)$ plus the corresponding data propagation and transmission time. Meanwhile, $\forall t' \in Pre^2(t), Ready(t) \geq CT(t')$.

In Eq.(5), $Delay(\mathcal{Y}(t), \mathcal{Y}(t'))$ stands for the propagation delay from device $\mathcal{Y}(t)$ to device $\mathcal{Y}(t')$. It vanishes if $\mathcal{Y}(t) = \mathcal{Y}(t')$. Meanwhile, $Band(\mathcal{Y}(t'), \mathcal{Y}(t))$ gives the communication bandwidth for transmitting data from $\mathcal{Y}(t')$ to $\mathcal{Y}(t)$.

Following Eq.(5), it is straightforward to compute the completion time of every task of any workflow WF, starting from $S(WF)$ till $E(WF)$. The makespan $\mathcal{O}_{makespan}(\mathcal{Y})$ can be determined subsequently for any IoTWS solution $\mathcal{Y}$.

3.2   Energy Consumption Model

This subsection elaborates the computation of the energy consumption $\mathcal{O}_{energy}(\mathcal{Y})$ with respect to any IoTWS solution $\mathcal{Y}$. Following Eq.(2), $\mathcal{O}_{energy}(\mathcal{Y})$ is obtained after calculating $Energy(WF, \mathcal{Y})$ of each workflow WF. Meanwhile,

$$Energy(WF, \mathcal{Y}) = \sum_{t \in G(WF)} Energy(t, \mathcal{Y}) \quad (6)$$

For any task $t \in G(WF)$, its energy consumption is calculated as

$$
Energy(t, \mathcal{Y}) = \begin{cases} \dfrac{C(t)P_{cpu}}{C(\mathcal{Y}(t))} + \dfrac{\rho(t', t)P_{transfer}\theta(\mathcal{Y}(t'), \mathcal{Y}(t))}{Band(\mathcal{Y}(t'), \mathcal{Y}(t))}, & \text{if } \mathcal{Y}(t) = s^{0,i} \\[2ex] \left(CT(t, \mathcal{Y}) - \dfrac{C(t)}{C(\mathcal{Y}(t))}\right)P_{idle} + \\ Delay(\mathcal{Y}(t'), \mathcal{Y}(t))P_{idle} + \\ \dfrac{\rho(t', t)P_{transfer}\theta(\mathcal{Y}(t'), \mathcal{Y}(t))}{Band(\mathcal{Y}(t'), \mathcal{Y}(t))}, & \text{if } \mathcal{Y}(t) = s^{1,I}/s^{2,I} \end{cases}
$$

$$(7)$$

Similar to [10], the IoTWS problem focuses on the energy consumption of the IoT devices. In other words, only the energy consumed by IoT devices and energy consumed for data communication are counted for $\mathcal{O}_{energy}$. There are two cases to calculate $Energy(t, \mathcal{Y})$: case 1 when task t is executed by IoT devices; and case 2 when task t is executed on either fog or cloud servers. Each of the two cases corresponds to the two conditions in Eq.(7) respectively.

For case 1 at the RHS of Eq.(7), P_{cpu} denotes the unit CPU power consumption of an IoT device in the active mode (i.e. when the device is executing any task). $P_{transfer}$ gives the unit data transfer power consumption. $\theta(\mathcal{Y}(t'), \mathcal{Y}(t)) = 1$ only when $\mathcal{Y}(t)$ and/or $\mathcal{Y}(t')$ are IoT devices. It equals 0 otherwise. For case 2, P_{idle} denotes the unit power consumption of an IoT device in the idle mode (i.e. when the device is not executing any task).

Following (6) and (7), it is straightforward to compute the energy consumption of each workflow WF as well as the total energy consumption $\mathcal{O}_{energy}$.

4   Algorithm Design

This section proposes a new IoTGA method to solve the IoTWS problem. We first present an overview of the IoTGA algorithm and then elaborate its building blocks. The design of IoTGA comprises of several major components: solution representation including genotypic solution representation and population initialization (see Subsect. 4.1); fitness evaluation (see Subsect. 4.2); evolution operators including the selection, crossover, mutation operators (see Subsect. 4.3); and the newly proposed local search process (see Subsect. 4.4).

IoTGA starts from a randomly initialized population of genotypic solutions. Each solution is evaluated for fitness. Guide by the evaluated fitness, the selection, crossover and mutation operators are utilized to generate offspring solutions [8]. The best offspring solution is then selected and further improved through a local search process. The solution obtained by local search then replaces the worst offspring solution in the new population for the next generation of IoTGA. The algorithm terminates after it runs for a maximum number of generations.

4.1   Solution Representation and Population Initialization

Each candidate solution evolved by IoTGA is represented by a chromosome as shown in Fig. 2. Subject to the position of each gene in the chromosome, every gene corresponds to a separate task of any given workflow. The value of every gene indicates the device that is assigned to process the respective task.

For example, assuming that there are two workflows WF_1 and WF_2 and each of the two workflows has two separate tasks. Hence a total of four tasks are to be assigned to devices for execution: $t_1 \in G(WF_1)$, $t_2 \in G(WF_1)$, $t_3 \in G(WF_2)$, and $t_4 \in G(WF_2)$. Accordingly, a chromosome comprises four genes, as shown in Fig. 2. The first gene for task t_1 is assigned to device $s^{2,1}$, which is a cloud server. Here the first number 2 in the first gene indicates the device type ζ as explained in Sect. 3. The second number 1 gives the index of that device.

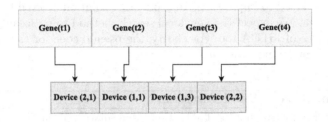

Fig. 2. An example chromosome evolved by IoTGA.

In line with the genotypic representation, the initial population of IoTGA is constructed by randomly creating a total of N chromosomes, where N is the fixed population size. For each chromosome, every gene is given a randomly selected device ID as its value, subject to the condition that any task $t \in G(WF)$ can be assigned to either $D(WF)$ or any cloud/fog servers for execution [10].

4.2 Fitness Evaluation

IoTGA considers three possible fitness functions, as defined respectively in (1), (2), and (3). Regardless of which fitness function is adopted, the process for evolving candidate solutions in IoTGA remains the same. In the experiments, we studied the performance of IoTGA by using (3) as its fitness function, which combines both (1) and (2). It is important to note that (4) and (6) can be determined easily based on the calculated makespan and energy consumption of each workflow. This is achieved by using (5) and (7) respectively. Therefore, IoTGA can assign fitness values to each candidate solution efficiently.

4.3 Evolution Operators

Following many existing works [19,23], we use the popular tournament selection method to select parent solutions in each generation. The tournament size is set to 3 in the experiments.

Some selected solutions are recombined through crossover to generate offspring solutions. Particularly, IoTGA adopts the typical two-point crossover operator to recombine any given pair of selected parent solutions. Similarly, IoTGA uses the common mutation operator to randomly alter one or multiple genes of a parent solution in order to generate its mutated offspring. For each

564   A. Saeed et al.

altered gene, its new value is set to a randomly selected device ID. The frequency of using the crossover and mutation operators is controlled by the *crossover rate* and *mutation rate* respectively. Specific settings of the two parameters can be found in Subsect. 5.2.

4.4 Local Search

In order to significantly enhance the final solution quality, a new local search process is developed to improve the best offspring solution evolved by IoTGA in each generation. We expect the new local search method to noticeably speed up the convergence of IoTGA too. For clarity, the pseudo-code for the local search process has been summarized in Algorithm 1.

Algorithm 1: Local Search Algorithm

Set the current solution $\mathcal{Y}^c$ to the best offspring solution $\mathcal{Y}^o$ in the current generation

Repeat for n=3 steps:

$\hat{\mathcal{Y}}^c = \mathcal{Y}^c$

if $\mathcal{Y}^c$ *is assigned to a new solution* **then**

    Select task $\hat{t}$ with the highest $DT(t)$ defined in (8)

    Identify task $\tilde{t}$ according to (9)

    Update $\hat{\mathcal{Y}}^c$ such that $\hat{\mathcal{Y}}^c(\hat{t}) \leftarrow \hat{\mathcal{Y}}^c(\tilde{t})$

else

    Select task $\hat{t}$ with the next highest $DT(t)$ defined in (8)

    Identify task $\tilde{t}$ according to (9)

    Update $\hat{\mathcal{Y}}^c$ such that $\hat{\mathcal{Y}}^c(\hat{t}) \leftarrow \hat{\mathcal{Y}}^c(\tilde{t})$

end

if $\mathcal{O}(\hat{\mathcal{Y}}^c) \leq \mathcal{O}(\mathcal{Y}^c)$ **then**

    $\mathcal{Y}^c \leftarrow \hat{\mathcal{Y}}^c$

end

return $\mathcal{Y}^c$

According to Algorithm 1, the local search process starts from the offspring solution with the best fitness $\mathcal{Y}^o$. Algorithm 1 sets up first a *current solution* $\mathcal{Y}^c = \mathcal{Y}^o$. Then in each iteration of local search, Algorithm 1 creates a neighboring solution $\hat{\mathcal{Y}}^c$ of $\mathcal{Y}^c$.

In order to for $\hat{\mathcal{Y}}^c$ to have better fitness than $\mathcal{Y}^c$, Algorithm 1 aims to reduce the time and energy consumption caused by data communication. This is because data communication among dependent tasks in workflows have huge impact on the overall makespan and energy consumption of the evolved solutions. Driven by this goal, Algorithm 1 calculates the data communication time required by each task according to $\mathcal{Y}^c$. Specifically, for any task $t \in G(WF)$, its data communication time is determined as

$$DT(t) = \max \left\{ \frac{\rho(t', t)}{Band(\mathcal{Y}^c(t'), \mathcal{Y}^c(t))} \middle| (t', t) \in G(WF) \right\} \qquad (8)$$

Clearly the higher the $DT(t)$ in Eq.(8), more time and energy consumption will be incurred due to data communication requested by task t. Guided by $DT(t)$ in Eq. (8), Algorithm 1 identifies one task $\hat{t}$ with the highest $DT(\hat{t})$ based on $\mathcal{Y}^c$. Subsequently, Algorithm 1 further identifies another task $\tilde{t}$ such that

$$\tilde{t} = \arg\max_{t' \in Pre^1(\hat{t})} \frac{\rho(t', \hat{t})}{Band(\mathcal{Y}^c(t'), \mathcal{Y}^c(\hat{t}))} \tag{9}$$

Clearly, it takes the longest time to transmit data from task $\tilde{t}$ to task $\hat{t}$ among all predecessor tasks belonging to $Pre^1(\hat{t})$. In order to reduce the communication time between task $\tilde{t}$ and task $\hat{t}$, in the resulting neighboring solution $\hat{\mathcal{Y}}^c$, task $\tilde{t}$ and task $\hat{t}$ will be assigned to the same device, thereby reducing the communication time and energy consumption involved. The assignment of other tasks remain identical between $\hat{\mathcal{Y}}^c$ and $\mathcal{Y}^c$. If $\hat{\mathcal{Y}}^c$ has better fitness than $\mathcal{Y}^c$, $\mathcal{Y}^c$ will be updated to $\mathcal{Y}^c \leftarrow \hat{\mathcal{Y}}^c$. Otherwise, $\mathcal{Y}^c$ will remain unchanged. In the latter case, for the next local search iteration, Algorithm 1 selects task $\hat{t}$ with the next highest $DT(\hat{t})$, which will be assigned to a different device for execution to create the neighboring solution. This local search process will be repeated for n iterations to ensure its efficiency and effectiveness. In our experiment, $n = 3$.

Table 1. Benchmark IoTWS problem instances.

Problem	No. IoT devices	WF Configuration	BW$_{fog,IoT}$	BW$_{cloud,IoT}$
P1	6	$\{WF1, WF2, WF3, WF4, WF5, WF6\}$	2000 KB/s	500 KB/s
P2	6	$\{WF1, WF2, WF3, WF4, WF5, WF6\}$	4000 KB/s	1000 KB/s
P3	4	$\{WF1, WF2, WF3, WF4\}$	2000 KB/s	500 KB/s
P4	8	$\{WF1, WF2, WF3, WF4, WF5, WF6, WF5, WF1\}$	2000 KB/s	500 KB/s

5   Experiment Results

We conduct experiments on real-world workflows and compare the results with two benchmark algorithms, i.e. standard GA (or sGA) and APMA [3]. All experimented algorithms are implemented in Python 3.9 and executed on Linux servers.

Table 2. Workflow patterns used in the experiments.

WF ID	WF Pattern	WF Size	Type
WF1	Real-world workflow	7 Tasks	FR
WF2	Real-world workflow	16 Tasks	QR
WF3	Scientific workflows	15 tasks	Montage
WF4	Scientific workflows	20 tasks	CyberShake
WF5	Scientific workflows	20 tasks	Epigenomics
WF6	Scientific workflows	30 tasks	Sipht

Table 3. Common settings across all IoTWS problem instances.

Parameter	Description
Number of Fog/Edge servers	6
Number of Cloud servers	3
Delay between IoT and Fog	0.5 ms
Delay between IoT and Cloud	30 ms
Computing capacity of IoT devices	500 MIPS
Computing capacity of fog servers	3000 MIPS
Computing capacity of cloud servers	4000 MIPS
Idle Power Consumption of IoT device	0.3 W
Active Power Consumption of IoT device	0.9 W
Transmission Power of IoT devices	1.3 W

5.1   Benchmark Problems

Four benchmark IoTWS problem instances are experimented in this study. They are summarized in Table 1. According to this table, problem instances $P1$ and $P2$ considered different communication bandwidth setting [3]. Specifically, $BW_{fog,IoT}$ in Table 1 indicates the communication bandwidth between fog servers and IoT devices and $BW_{cloud,IoT}$ denotes the communication bandwidth between cloud servers and IoT devices. Clearly, problem instance $P2$ has twice the communication bandwidth as the bandwidth available for $P1$. The remaining two problem instances $P3$ and $P4$ allow us to further study the impact of different workflow patterns and IoT system configurations (i.e. the number of IoT devices, fog servers and cloud servers) on the makespan and energy consumption achievable by IoTGA. In particular, our experiments cover up to 8 IoT devices, 6 fog/edge servers, and 3 cloud servers as summarized in Tables 1 and 3.

In addition to the above, following [10], several additional settings remain identical across all benchmark IoTWS problem instances. These include the communication delay, computing capacity of IoT devices, fog servers and cloud servers, as well as the power consumption attributes of IoT devices. According to [10], these settings are typically used in literature and practical applications.

Following [10], both real-world IoT workflows and scientific workflows are utilized in the five benchmark IoTWS problem instances. Specifically, two real-world IoT workflows are considered. The first one (i.e. WF1 in Table 2) supports the IoT Face Recognition (FR) application introduced in [24]. The second one (i.e. WF2 in Table 2) supports the IoT QR code processing application introduced in [26]. Besides IoT workflows, we also experimented four widely used scientific workflows, including Montage, CyberShake, Epigenomics and Sipht (i.e. WF3, WF4, WF5, and WF6 in Table 2) [5].

5.2   Parameter Settings

For a fair comparison, we adopted the recommended (or widely used) parameter settings for sGA and APMA in [10]. Specifically, for sGA the population size, the crossover rate, mutation rate, the maximum number of generations are 50, 0.8, 0.2. and 300 respectively. For APMA, we use the same parameter settings. Following [10], parent selection in APMA is implemented through roulette wheel selection instead of tournament selection in sGA.

Similarly, we set the parameters required by IoTGA as follows. The population size is 50. The crossover rate and mutation rate are 0.8 and 0.2 respectively. The maximum number of generations is 300. The tournament size for tournament selection is 3. For all algorithms, we use (3) as the fitness function with w=0.5 since both makespan and energy consumption are considered equally important. In total $300 \times 50 = 15000$ fitness evaluations were conducted during each algorithm run.

5.3   Results

This subsection reports the final performance achieved by the three competing algorithms in terms of makespan, energy consumption, and their weighted combination on all IoTWS problem instances. We then further analyze the makespan and energy consumption of each workflow in some problem instances. The convergence speed of IoTGA and sGA will also be compared and discussed. Since APMA uses separate GA processes to optimize task execution in each layer [10], it cannot be directly compared with IoTGA in terms of convergence speed.

Table 4. Total makespan, total energy consumption and their weighted combination achieved by all competing algorithms on 4 IoTWS problem instances. The best results verified statistically by using Wilcoxon ranked test with a significance level of 0.05 are bolded.

Problem	Algorithm	Total Makespan (s)	Total Energy Consumption (J)	$\mathcal{O}_{combined}$ ($w = 0.5$)
$P1$	APMA	23.49 ± 4.38	168.62 ± 72.64	192.11 ± 77.02
	sGA	22.37 ± 3.81	42.68 ± 23.92	65.05 ± 27.73
	IoTGA	$\mathbf{16.69 \pm 6.37}$	$\mathbf{10.78 \pm 4.61}$	$\mathbf{27.47 \pm 10.98}$
$P2$	APMA	12.04 ± 1.67	73.56 ± 26.96	85.6 ± 28.63
	sGA	11.28 ± 1.92	21.80 ± 10.86	33.08 ± 12.78
	IoTGA	$\mathbf{8.42 \pm 3.16}$	$\mathbf{6.03 \pm 2.19}$	$\mathbf{14.45 \pm 5.35}$
$P3$	APMA	22.18 ± 6.07	51.85 ± 21.04	74.03 ± 27.11
	sGA	20.17 ± 4.27	10.66 ± 46.22	30.83 ± 50.49
	IoTGA	$\mathbf{15.04 \pm 7.88}$	$\mathbf{8.06 \pm 3.37}$	$\mathbf{23.1 \pm 11.25}$
$P4$	APMA	24.32 ± 4.38	270.83 ± 95.73	295.15 ± 100.11
	sGA	24.18 ± 4.66	107.54 ± 42.56	131.72 ± 47.22
	IoTGA	$\mathbf{17.54 \pm 5.84}$	$\mathbf{13.43 \pm 4.21}$	$\mathbf{0.97 \pm 10.05}$

Table 4 shows the average total makespan and energy consumption along with their weighted combination obtained by three competing algorithms on four IoTWS problem instances. We run every algorithm for 40 indepedent times on each instance. The obtained results are verified statistically by using the Wilcoxon ranked test at a significance level of 0.05. The best results with statistical significance according to the test are bolded. As can be easily verified in Table 4, IoTGA achieved significant improvement on all problem instances, in comparison to both sGA and APMA. For example, on problem instance $P1$, IoTGA reduced the total makespan by 25% and 30%, compared to sGA and APMA respectively. IoTGA also managed to successfully reduce the total energy consumption by almost 4 and 15 times, compared to sGA and APMA. Similar observations have been witnessed on all other problem instances.

Our experiment results show that IoTGA is particularly effective at reducing the total energy consumption. The reason for this achievement is two-fold. As we only consider energy consumed by IoT devices in (7), IoTGA managed to allocate many energy-consuming workflow tasks to fog and cloud servers to significantly save energy, whereas sGA and APMA failed to find such scheduling solutions. Meanwhile, executing tasks on fog/cloud servers may incur extra communication overhead. This is mitigated effectively by IoTGA thanks to its use of the newly designed local search process in Algorithm 1. As a result, IoTGA can noticeably reduce the total makespan too.

We also notice that, although the set of workflows considered in $P2$ is identical to that of $P1$, IoTGA can reduce both the total makespan and energy consumption by roughly a half on $P2$, in comparison to $P1$. This result agrees with our expectation, since $P2$ enjoys twice the communication bandwidth as $P1$, revealing the dominating role of the communication infrastructure on the overall performance of an IoT system.

Table 5. Makespan in seconds achieved by all competing algorithms on each workflow of problem $P1$.

Algorithm	WF1	WF2	WF3	WF4	WF5	WF6
APMA	3.98 ± 5.44	11.82 ± 8.01	6.00 ± 7.66	23.02 ± 5.28	8.77 ± 6.95	3.51 ± 3.68
sGA	5.31 ± 8.51	19.73 ± 7.95	7.44 ± 9.02	21.47 ± 4.10	18.57 ± 6.41	7.44 ± 8.90
IoTGA	$\mathbf{0.43 \pm 0.48}$	$\mathbf{5.76 \pm 8.09}$	$\mathbf{0.58 \pm 0.56}$	$\mathbf{15.41 \pm 7.29}$	$\mathbf{1.95 \pm 3.30}$	$\mathbf{1.60 \pm 2.99}$

Table 6. Energy Consumption in Joules achieved by all competing algorithms on each workflow of problem $P1$.

Algorithm	WF1	WF2	WF3	WF4	WF5	WF6
APMA	6.70 ± 5.62	42.02 ± 17.75	5.73 ± 4.65	27.94 ± 6.95	82.08 ± 57.85	4.14 ± 2.83
sGA	0.22 ± 0.07	15.97 ± 12.20	$\mathbf{0.35 \pm 0.08}$	11.47 ± 5.02	14.00 ± 19.62	0.67 ± 0.45
IoTGA	$\mathbf{0.17 \pm 0.10}$	$\mathbf{0.60 \pm 0.41}$	0.36 ± 0.07	$\mathbf{8.37 \pm 4.45}$	$\mathbf{0.69 \pm 0.19}$	$\mathbf{0.59 \pm 0.22}$

Table 7. Makespan in seconds achieved by all competing algorithms on each workflow of problem $P2$.

Algorithm	WF1	WF2	WF3	WF4	WF5	WF6
APMA	2.16 ± 2.60	4.60 ± 3.44	2.84 ± 3.21	12.04 ± 1.67	4.74 ± 3.75	2.21 ± 2.60
sGA	2.45 ± 4.10	10.20 ± 3.83	3.34 ± 4.41	10.94 ± 2.11	9.35 ± 3.19	3.74 ± 4.37
IoTGA	$\mathbf{0.30 \pm 0.25}$	$\mathbf{3.12 \pm 4.00}$	$\mathbf{0.43 \pm 0.27}$	$\mathbf{7.98 \pm 3.47}$	$\mathbf{1.24 \pm 2.12}$	$\mathbf{0.85 \pm 1.49}$

Table 8. Energy Consumption in Joules achieved by all competing algorithms on each workflow of problem $P2$.

Algorithm	WF1	WF2	WF3	WF4	WF5	WF6
APMA	2.34 ± 2.1	19.11 ± 7.0	2.11 ± 1.8	14.03 ± 2.9	33.47 ± 23.3	2.51 ± 1.6
SGA	0.17 ± 0.07	8.28 ± 6.87	$\mathbf{0.34 \pm 0.08}$	6.14 ± 2.41	6.26 ± 8.77	0.62 ± 0.23
IoTGA	$\mathbf{0.10 \pm 0.10}$	$\mathbf{0.58 \pm 0.30}$	0.35 ± 0.06	$\mathbf{3.78 \pm 2.03}$	$\mathbf{0.68 \pm 0.20}$	$\mathbf{0.55 \pm 0.18}$

Tables 5, 6, 7 and 8 provide detailed makespan and energy consumption achieved by all competing algorithms with respect to each workflow of problem instances $P1$ and $P2$. As evidenced in these tables, IoTGA outperformed sGA and APMA in terms of the energy consumption of WF1, WF2, WF4, WF5 and WF6, since data communication in these workflows can incur huge delay and energy consumption. IoTGA is more effective at improving communication efficiency, in comparison to sGA and APMA. In fact, many folds of energy savings have been achieved on some workflows (e.g. WF5 in Tables 6 and 8). Consequently, although IoTGA did not reduce energy consumption of workflow WF3, the total energy consumption was significantly reduced in Table 4. The same as energy consumption, IoTGA consistently outperformed sGA and APMA regarding the makespan of each workflow for both problem instances $P1$ and $P2$. Similar observations have also been witnessed on other problem instances. In view of the results in Tables 5, 6, 7 and 8, we found that IoTGA not only can improve the overall performance of an IoT system, each workflow executed in the system can also benefit from using the algorithm (IoTGA does not sacrifice the performance of any individual workflow).

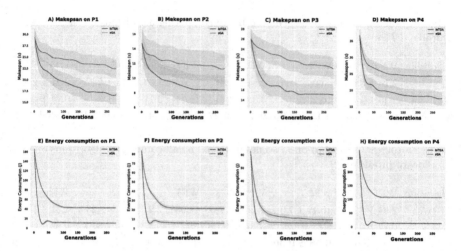

Fig. 3. Convergence of makespan and energy consumption on 4 problem instances.

Figure 3 compares the convergence speed of IoTGA and sGA in terms of both makespan and energy consumption on all problem instances. As evidenced in this figure, for all problems, IoTGA can converge significantly faster than sGA. Since the key difference between IoTGA and sGA is the local search process, Fig. 3 confirms that Algorithm 1 proposed in this paper can significantly increase the chance for IoTGA to find high-quality scheduling solutions and hence its convergence speed.

6   Conclusions

In this paper, a new GA approach called IoTGA has been successfully developed to tackle the IoT workflow scheduling (IoTWS) problem in hybrid IoT environments. IoTGA was designed to jointly minimize the total makespan of executing all requested workflows and the corresponding total energy consumption. In view of the strong influence of data communication on the performance of IoT systems, we proposed a new local search process that can effectively improve the best evolved solution in each generation of IoTGA by explicitly reducing the communication overhead involved in workflow execution. Experiments have been conducted on four benchmark IoTWS problem instances. Our experiment results clearly showed that IoTGA can significantly outperform sGA and APMA by a large margin in term of both the total makespan and energy consumption. IoTGA was also shown to converge much faster than sGA.

References

1. Abd Elaziz, M., Abualigah, L., Ibrahim, R.A., Attiya, I.: Iot workflow scheduling using intelligent arithmetic optimization algorithm in fog computing. In: Computational Intelligence and Neuroscience 2021 (2021)
2. Abualigah, L., Diabat, A., Elaziz, M.A.: Intelligent workflow scheduling for big data applications in IoT cloud computing environments. Cluster Comput. **24**(4), 2957–2976 (2021)
3. Aburukba, R.O., AliKarrar, M., Landolsi, T., El-Fakih, K.: Scheduling internet of things requests to minimize latency in hybrid fog-cloud computing. Future Gen. Comput. Syst. **111**, 539–551 (2020)
4. Aburukba, R.O., Landolsi, T., Omer, D.: A heuristic scheduling approach for fog-cloud computing environment with stationary IoT devices. J. Network Comput. Appl. **180**, 102994 (2021)
5. Ahmad, Z., et al.: Scientific workflows management and scheduling in cloud computing: taxonomy, prospects, and challenges. IEEE Access **9**, 53491–53508 (2021)
6. Alsurdeh, R., Calheiros, R.N., Matawie, K.M., Javadi, B.: Hybrid workflow provisioning and scheduling on edge cloud computing using a gradient descent search approach. In: 2020 19th International Symposium on Parallel and Distributed Computing (ISPDC), pp. 68–75. IEEE (2020)
7. Chen, X., Cai, Y., Shi, Q., Zhao, M., Champagne, B., Hanzo, L.: Efficient resource allocation for relay-assisted computation offloading in mobile-edge computing. IEEE Internet Things J. **7**(3), 2452–2468 (2019)
8. Eiben, A., Smith, J.: Introduction to Evolutionary Computing (Natural Computing Series). Springer, Heidelberg (2008)
9. Ghobaei-Arani, M., Souri, A., Rahmanian, A.A.: Resource management approaches in fog computing: a comprehensive review. J. Grid Comput. **18**(1), 1–42 (2020)
10. Goudarzi, M., Wu, H., Palaniswami, M., Buyya, R.: An application placement technique for concurrent IoT applications in edge and fog computing environments. IEEE Trans. Mob. Comput. **20**(4), 1298–1311 (2020)
11. Knuth, D.: Number of Internet of Things (IoT) connected devices worldwide from 2019 to 2021, with forecasts from 2022 to 2030 kernel description. https://www.statista.com/statistics/1183457/iot-connected-devices-worldwide/. Accessed 30 Sept 2010
12. Laroui, M., Nour, B., Moungla, H., Cherif, M.A., Afifi, H., Guizani, M.: Edge and fog computing for IoT: a survey on current research activities & future directions. Comput. Commun. **180**, 210–231 (2021)
13. Li, S., Zhai, D., Du, P., Han, T.: Energy-efficient task offloading, load balancing, and resource allocation in mobile edge computing enabled IoT networks. Sci. China Inf. Sci. **62**(2), 1–3 (2019)
14. Li, Z., Ge, J., Hu, H., Song, W., Hu, H., Luo, B.: Cost and energy aware scheduling algorithm for scientific workflows with deadline constraint in clouds. IEEE Trans. Serv. Comput. **11**(4), 713–726 (2015)
15. Liu, Y., et al.: Dependency-aware task scheduling in vehicular edge computing. IEEE Internet Things J. **7**(6), 4961–4971 (2020)
16. Miao, Y., Wu, G., Li, M., Ghoneim, A., Al-Rakhami, M., Hossain, M.S.: Intelligent task prediction and computation offloading based on mobile-edge cloud computing. Fut. Gener. Comput. Syst. **102**, 925–931 (2020)

17. Mohammadi, S., Pedram, H., PourKarimi, L.: Integer linear programming-based cost optimization for scheduling scientific workflows in multi-cloud environments. J. Supercomput. **74**(9), 4717–4745 (2018). https://doi.org/10.1007/s11227-018-2465-8
18. Mokni, M., Yassa, S., Hajlaoui, J.E., Chelouah, R., Omri, M.N.: Cooperative agents-based approach for workflow scheduling on fog-cloud computing. J. Ambient Intell. Hum. Comput. 1–20 (2021)
19. Pan, L., Liu, X., Jia, Z., Xu, J., Li, X.: A multi-objective clustering evolutionary algorithm for multi-workflow computation offloading in mobile edge computing. IEEE Trans. Cloud Comput. (2021)
20. Sriraghavendra, M., Chawla, P., Wu, H., Gill, S.S., Buyya, R.: DoSP: a deadline-aware dynamic service placement algorithm for workflow-oriented IoT applications in fog-cloud computing environments. In: Tiwari, R., Mittal, M., Goyal, L.M. (eds.) Energy Conservation Solutions for Fog-Edge Computing Paradigms. LNDECT, vol. 74, pp. 21–47. Springer, Singapore (2022). https://doi.org/10.1007/978-981-16-3448-2_2
21. Sulaiman, M., Halim, Z., Lebbah, M., Waqas, M., Tu, S.: An evolutionary computing-based efficient hybrid task scheduling approach for heterogeneous computing environment. J. Grid Comput. **19**(1), 1–31 (2021)
22. Tahsien, S.M., Karimipour, H., Spachos, P.: Machine learning based solutions for security of internet of things (IoT): a survey. J. Network Comput. Appl. **161**, 102630 (2020)
23. Tan, B., Ma, H., Mei, Y.: A group genetic algorithm for resource allocation in container-based clouds. In: Paquete, L., Zarges, C. (eds.) EvoCOP 2020. LNCS, vol. 12102, pp. 180–196. Springer, Cham (2020). https://doi.org/10.1007/978-3-030-43680-3_12
24. Wu, H., Knottenbelt, W.J., Wolter, K.: An efficient application partitioning algorithm in mobile environments. IEEE Trans. Parallel Distrib. Syst. **30**(7), 1464–1480 (2019)
25. Xing, L., Zhang, M., Li, H., Gong, M., Yang, J., Wang, K.: Local search driven periodic scheduling for workflows with random task runtime in clouds. Comput. Ind. Eng. **168**, 108033 (2022)
26. Yang, L., Cao, J., Yuan, Y., Li, T., Han, A., Chan, A.: A framework for partitioning and execution of data stream applications in mobile cloud computing. ACM SIGMETRICS Perform. Eval. Rev. **40**(4), 23–32 (2013)

Multi-objective Location-Aware Service Brokering in Multi-cloud - A GPHH Approach with Transfer Learning

Yuheng Chen[1], Tao Shi[2(✉)], Hui Ma[1], and Gang Chen[1]

[1] Victoria University of Wellington, Wellington, New Zealand
{chenyuhe,hui.ma,aaron.chen}@ecs.vuw.ac.nz
[2] Qingdao Agricultural University, Qingdao, China
shitao@qau.edu.cn

Abstract. With the increasing number of cloud services in multi-cloud, it has been a challenging task to choose suitable cloud services in consideration of multiple conflicting objectives. Multi-objective location-aware service brokering in multi-cloud aims to find a set of trade-off solutions that minimize both the cost and latency. To achieve this goal, existing approaches either manually design brokering heuristics or automatically generate heuristics via Genetic Programming Hyper-Heuristics (GPHH) for each problem domain from scratch. However, manually designing heuristics takes a long time and requires domain knowledge. Also, knowledge learnt from one problem domain can be helpful for solving another problem domain. To effectively and efficiently generate heuristics for any new problem domain, we propose three novel GPHH-based approaches with transfer learning to automatically generate a group of Pareto-optimal heuristics. Experimental evaluations on real-world datasets demonstrate that our proposed GPHH with transfer learning approaches can outperform existing approaches.

Keywords: Multi-objective optimization · Multi-cloud · Service brokering · Genetic programming · GPHH · Transfer learning

1 Introduction

Service brokering in multi-cloud plays an important role in helping application providers find services in multi-cloud to deploy their applications [23]. It has become a challenging task in finding the suitable services, as there are numerous data centers for application providers to select. As brokers, they have the problem to recommend a list of candidate services from different cloud providers to meet practical requirements. On the one hand, cloud services in different regions can have significantly varied prices. Take m6g.large from Amazon Web Service (AWS) as an example, the price of the service in North Virginia is $0.077, while the price of the same service provided in Sao Paulo is $0.1224. On the other hand,

© The Author(s), under exclusive license to Springer Nature Switzerland AG 2023
J. Correia et al. (Eds.): EvoApplications 2023, LNCS 13989, pp. 573–587, 2023.
https://doi.org/10.1007/978-3-031-30229-9_37

the network latency between consumers and data centers may affect the performance of cloud applications. In order to provide cloud services with low latency, cloud providers set up their data centers across different locations. Therefore, we need to consider the balance between the performance in terms of network latency and the cost of cloud services. In this paper, we study the problem of multi-objective location-aware service brokering (MOLSB) in multi-cloud that simultaneously consider both cost and performance of services.

There are many existing works on multi-cloud service brokering problems. Most of them assume that all the requests are assigned at the same time, which is static problem [8,18,19,25]. A genetic algorithm (GA) approach was proposed in [19], merging the two objectives, i.e., the cost and network latency, by a weighted sum function. Other works assume that the requests are assigned dynamically once they arrived. In [4], a multi-objective genetic programming hyper-heuristic (GPHH) approach, named GPHH-MOLSB, was proposed to generate automatically designed non-dominating rules for assigning requests. The rules designed by GPHH-MOLSB are effective in solving problem instances of the same problem domain. However, new rules need to be evolved by GPHH-MOLSB from scratch for new problem domains, ignoring the fact that the knowledge learned from other problem domains can be used. Therefore, it is desirable to design effective transfer learning methods to effectively transfer the knowledge learned from previous domains while designing new rules. Some transfer learning methods have been proposed for single-objective combinatorial optimisation problems, e.g., [1] for arc routing problems, [9] for workflow scheduling problems. However, to the best of our knowledge, there is no transfer learning method proposed for GPHH in solving multi-objective combinatorial optimisation problems, in particular for the MOLSB probem.

The aim of this paper is to develop new GPHH approaches with transfer learning techniques that can automatically design heuristics for the MOLSB problem. To achieve this, we will

1. Propose three multi-objective GPHH approaches with different transfer learning techniques. Each of the approach can generate a Pareto Front of heuristics;
2. Evaluate the proposed approaches and compare their performance with an existing GPHH based approach, using datasets collected from the real world;
3. Analyse the results of the experiments in terms of Hypervolume (HV) and Inverted Generational Distance (IGD).

The remaining sections of this paper are organised as follows. In Sect. 2, we review the related existing work in this area. The problem is described and modeled in Sect. 3. In Sect. 4, we present the GPHH approaches with transfer learning techniques to solve the dynamic MOLSB problem. Section 5 discusses the design of the experiments and analyses the experimental results. Section 6 concludes the paper.

2   Related Works

Various heuristic and meta-heuristic approaches have been proposed to solve the static MOLSB problem. In [12], a greedy algorithm was proposed to solve the problem. Since the greedy algorithm only allows requests arriving at the same time, it will get stuck in local optima. To improve the algorithm proposed in [12], a genetic algorithm (GA) approach was proposed to optimise the static MOLSB problem in [19]. These approaches assume all the application deployment requests are known and arrive at the same time.

Hyper-heuristic methods are used to automatically generate heuristic rules to solve dynamic combinatorial optimisation problems. Genetic programming (GP), as an evolutionary computation technique, aims to evolve and generate rules to solve a specific task. In each GP generation, genetic operators, including selection, crossover and mutation, are applied to evolve GP rules [16]. Through many generations of evolution, a final rule will be selected by the algorithm. Hyper-heuristic algorithms aim to explore in heuristic space, rather than the solution space [3]. For dynamic optimisation problem, GPHH has been applied in many areas [29,30] for single-objective problems. To solve dynamic multi-objective multi-cloud brokering problem, [4] proposed a GPHH approach, named GPHH-MOLSB, with a newly designed terminal set, to automatically generate a set of trade-off heuristics for users to choose according to their QoS preferences. The rules generated by GPHH-MOLSB significantly outperformed the heuristics that were human-designed.

Transfer learning is a technique that can solve novel tasks by using the knowledge learned from previous tasks [6]. With the help of transfer learning techniques, evolutionary algorithms may obtain a better initial performance, reduce computational time and achieve better results in the target domain, compared to the algorithms without transfer learning [9,27]. In [6], the authors listed three main transfer approaches in genetic programming in solving symbolic regression problems, including transferring the full tree from the final generation of the source domain, transferring selected sub-trees with high fitness from the final generation of the source domain and transferring the best individuals at each generation of evolution performed in the source domain. In [9], a new transfer learning approach of GPHH was proposed for dynamic multi-workflow scheduling problem. In order to reduce the computational cost on target domain, a randomly initialized population is trained on a simpler source domain for a predefined number of generations. Then the population is evolved further in the tougher target domain for another predefined number of generations. The transfer learning approach saves the overall computation time by reducing the number of generations to be performed in the target domain. In [1], the authors proposed a novel genetic programming approach with knowledge transferring to solve the uncertain capacitated arc routing problem. The author compared the proposed method with several existing transfer learning approach, including DDGP [2], FullTree [6], GATL [15], SubTree [6] and TLGPC [13].

Most of the existing transfer learning approaches for GPHH as summarized above were proposed for solving single objective problems. In multi-objective problems, existing approaches are not applicable or effective. For example, the TLGPC approach selects the subtrees of the individuals which are better than the mean value of the fitness in the final generation. This approach cannot be applied to multi-objective problems since the final generation provides a Pareto Front of individuals, and each of them is non-dominated. To satisfy the requirement for application deployment with different QoS preferences, we need to generate a set of trade-off solutions for users to choose. Therefore, we need to design GPHH approaches with transfer learning that can be applied to solve multi-objective problems.

3   Problem Definition

The MOLSB problem aims to allocate dynamically arriving applications to suitable cloud resources, i.e., VM instances in data centers, so as to minimize both the application deployment cost and the network latency. In this section, we present a formal model of the MOLSB problem. Constrains and assumptions of this problem are also formulated. The key notations to be used for problem definition are listed in Table 1.

For dynamically arriving application deployment requests S, a broker selects VMs at different locations from multiple cloud providers. Let $\mathcal{R}$ denotes the set of different region of data centres. In each region $r \in \mathcal{R}$, different types of VM instances are provided by cloud providers, denoted by V. Each VM type $v \in V$ has different price $C_{v,r}$ in different region $r \in R_v$. We use G_v and M_v to denote the capacities of CPU and memory that are provided by a VM type v. The set of all regions that provide VM type v is denoted as R_v

During a time period T (e.g., one day), a broker receives a sequence of requests. We use N to denote the total number of received requests. Each request i from user location u_i has two types of resource requirements, i.e., CPU g_i and memory m_i, and the time period t_i it will use the VM instance for. Once a new request i arrives at time T_i, the broker will assign it to an instance of VM type v in region r. Here, $L_{i,r}$ denotes the network latency between user location u_i and data center region r.

Following [12], we have the following assumptions about VM instances in this paper.

- The price of a VM instance can be affected by different service providers, different VM types, and different regions of data centers.
- The configuration of a VM instance cannot be modified if the VM instance has a request assigned to.
- Any single VM instance of type $v \in V$ can only have one request assigned to it at a time.
- Each request i must be assigned to exactly one VM instance of any type $v \in V$.

Table 1. Mathematical notations

Notation	Definition
V	Set of VM types
$\mathcal{R}$	Set of regions that multi-cloud data centers span
G_v	CPU capacity of VM type v
M_v	memory capacity of VM type v
R_v	Set of available regions of VM type v
$C_{v,r}$	The unit price of VM type v in region r
T	Time span of an application deployment
N	Total number of requests during time span T
T_i	arrival time of Request i
u_i	user location of Request i
g_i	CPU requirement of Request i
m_i	memory requirement of Request i
t_i	VM usage time for request i
$L_{i,r}$	Network latency between request i and region r
$x_{i,v,r}$	Binary variable indicating whether request i is assigned to VM type v in region r
TC	Total cost of the selected VMs
ANL	Average network latency of the selected VMs

The following constraints should be satisfied if a request i is assigned to a VM instance v.

$$g_i \leqslant G_v,$$
$$m_i \leqslant M_v. \tag{1}$$

Equation (1) implies that the VM instance must satisfy the resource requirement of the assigned request. A VM type is *capacity-feasible* to a request i, if the capacity of CPU is greater than or equal to g_i and the capacity of memory is greater than or equal to m_i.

The total cost of VM instances (TC) and the average network latency between users and data centres (ANL) can be calculated as follows:

$$TC = \sum_{i=1}^{N} \sum_{v \in V} \sum_{r \in R_v} C_{v,r} t_i x_{i,v,r}$$

$$ANL = \frac{1}{N} \sum_{i=1}^{N} \sum_{v \in V} \sum_{r \in R_v} L_{i,r} x_{i,v,r}, \tag{2}$$

where $x_{i,v,r} \in \{0, 1\}$ determines whether request i is assigned to an instance of VM type $v \in V$ in region $r \in R_v$.

Therefore, for application deployment requests MOLSB aims to find best resources available from multiple cloud with two objectives, minimizing TC and minimizing ANL, as defined in Eq. (3):

$$\min \quad TC,$$
$$\min \quad ANL. \tag{3}$$

4   GPHH with Transfer Learning for MOLSB

To effectively evolve brokering rules for MOLSB, we propose a GPHH algorithm with three different transfer learning approaches.

An overview of the GPHH with transfer learning approaches is shown in Fig. 1. The transfer learning approaches initialise a population of brokering rules from the previously trained rules in a source domain. In each generation, these rules are evolved by genetic operators, and evaluated by a set of training instances generated from the target domain. After a predefined number of iterations, a set of brokering rules are generated to solve the dynamic MOLSB problem on the target domain.

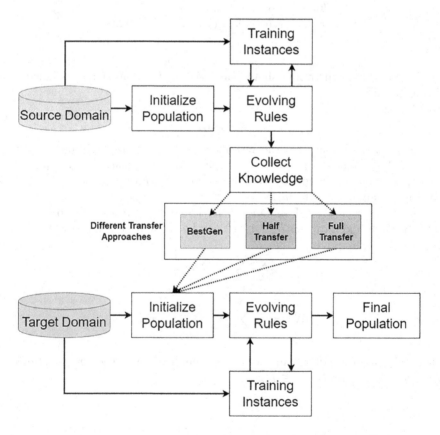

Fig. 1. The training progress of GPHH transfer learning approaches

4.1 Representation and Terminal Set

We use trees to represent the mathematical expressions of brokering rules, where the leaves are terminals and intermediate nodes are functions. To evolve brokering rules for MOLSB, we need a set of problem related features for terminal nodes as well as a set of arithmetic functions for internal nodes. We use the same set of features for terminals as in [4], since they are important features of the problem. The terminal set and function set are summarized in Table 2. Note that we use the same terminal and function set for both source and target domains.

Table 2. Terminal Set And Function Set

Terminal	Symbol	Definition
CPU	g_i	Request i's CPU requirement
Memory	m_i	Request i's memory requirement
Time	t_i	VM usage time for request i
Latency	$L_{i,r}$	The latency of request i from the data center in region r to the user
BestPrice	$min(C_{v,r})$	The minimum price of VM type v in region r that satisfies the constraints in Eq. (1).
Function set	$+, -, \times$, protected division, maximum, minimum, cosine and sine	

An example of multi-cloud brokering rule represented as a GP tree using the terminal and function set is shown in Fig. 2.

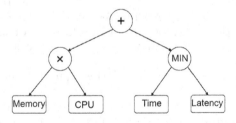

Fig. 2. An example of the tree-based representation

4.2 Transfer Approaches

To investigate using transfer learning in GPHH for solving the Multi-Objective Location-aware Service Brokering (MOLSB) problem, we propose three different transfer approaches in this paper, including *BestGen*, *Half-Transfer* and *Full-Transfer*. In Fig. 3, the training process of GPHH with three different approaches on the source domain is presented. The training process on the target domain of the three transfer approaches is further summarized in Algorithm 1.

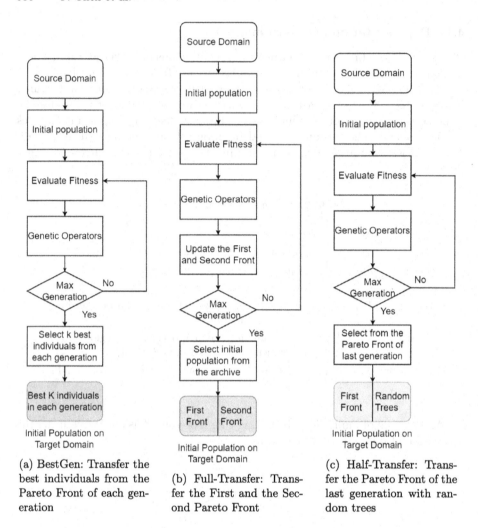

(a) BestGen: Transfer the best individuals from the Pareto Front of each generation

(b) Full-Transfer: Transfer the First and the Second Pareto Front

(c) Half-Transfer: Transfer the Pareto Front of the last generation with random trees

Fig. 3. The training process of three different approaches on source domain

BestGen. BestGen in Algorithm 1 is a transfer learning approach that transfers some best individuals from the source domain to the target domain. It is different from the approach in [6] which was proposed for the single-objective optimisation problem. The transfer learning approach in [6] collects from 10% to 20% of the best individuals in each generation on the source domain, which cannot be used to solve our multi-objective optimisation problem. In this paper we propose to select a pre-defined number of individuals from the Pareto Front of each generation. Since the number of individuals in the Pareto Front can be very large, the crowding distance [5] is applied to ensure our selected individuals keep most of the diversity of the Pareto Front. The selected individuals are then used to generate the initial population of the the target domain. The training process starts from a population with knowledge collected from the source domain.

Algorithm 1. Training process of GPHH with transfer learning

Input: a list of training instance on source domain S_1, a list of training instance on target domain S_2,

Output: a set of non-dominating heuristics on target domain

1: Randomly initialize the population P_1
2: **while** max generation not reached **do**
3:     Evolve offspring through selection, crossover and mutation
4:     Evaluate the fitness of all evolved offspring on source domain
5: **end while**
6: Identify the set of individuals I to transfer from source domain to target domain

$$I = \begin{cases} \text{Best of } k \text{ individuals in each generation,} & \textbf{BestGen} \\ 50\% \text{ from First Front} + 50\% \text{ from Second Front,} & \textbf{Full-Transfer} \\ 50\% \text{ from First Front rules,} & \textbf{Half-Transfer} \end{cases}$$

7: Initialize the population P_2 using a combination of I and randomly generated rules

8: **while** max generation not reached **do**
9:     Evolve offspring through selection, crossover and mutation
10:     Evaluate the fitness of all evolved offspring on target domain
11: **end while**
12: Return the Pareto Front of the last evolved population

Full-Transfer. Full-Transfer approach in Algorithm 1 randomly selects half of the first Pareto Front and half of the second Pareto Front over one single run on source domain. The second Pareto Front refers to the evolved rules that are only dominated by the rules in the first Pareto Front and are not dominated by other rules in the second Pareto Front. In the training process on the source domain, we keep the individuals in the first Pareto Front and the second Pareto Front. The archive is updated by each generation. After the training process on the source domain, the initial population is generated from the archive that consists of the First and Second Pareto Front of the training process.

Half-Transfer. Half-Transfer in Algorithm 1 is to transfer half of the final population from the source domain to the target domain. To solve our multi-objective problem we design an approach to selectively choose the solutions on the front from the final generation evolved in the source domain. The Half-Transfer approach selects the Pareto Front over one single run on source domain. The approach then randomly selects half of evolved brokering rules from the Pareto Front obtained on the source domain to create the initial population for evolution on the target domain. The other half of the initial population is randomly generated to maintain the diversity.

Note that Full-Transfer approach uses the second Pareto Front to keep the diversity. Compared to the Half-Transfer approach, it transfers more knowledge from the source domain, since the second Pareto Front is significantly better than the randomly generated rules.

4.3   Crossover and Mutation

Three genetic operators, crossover, mutation, and selection, are applied in our transfer learning methods. In the crossover process, each parent individual is cut-off by a randomly selected point. Then the offspring can be generated by swapping the two parts starting from the cut-off points from their parents. To maintain the diversity, the mutation operation randomly generates a new branch at the mutation point of the parent individuals.

4.4   Fitness Evaluation

We use fitness values to evaluate each individual. Each pair of fitness values is computed based on Eq. (3). Each individual can be represented as a mathematical function. Given the arrived requests, we can apply the function of each individual to sort the priority of candidate regions list and select the VM in the region that has the highest priority in the region list (see [4] for more details). After all requests are assigned, we calculate the sum of the cost TC and the average latency ANL as the fitness values for this individual.

5   Experiments

In this section, the experimental design is introduced. We perform the experiments to simulate the cloud environment in the real world. We also present our evaluation results on the generated rules using Hypervolume (HV) [28] and Inverted Global Distance (IGD) [14].

In our experiments, hypervolume (HV) describes the area that covers all area dominated by our proposed rules and the reference point (1,1) [10]. A higher value of HV represents a better diversity of our proposed solutions [20].

IGD describes the average distance from our proposed rules to the true Pareto Front. The true Pareto Front in this problem can be approximated by obtaining the Pareto Front from all non-dominated solutions over 30 independent runs. A lower value of IGD indicates a better diversity and convergence of our proposed solutions [17].

5.1   Simulation and Datasets

We use the simulator developed in [24] for multi-cloud service brokering to evaluate our proposed transfer learning approaches.

The simulator has the following features:

- For each VM type, the number of VM instances available is infinite.
- Requests arrive sequentially in a fixed timespan (one day).
- Once a request arrived, the request will be assigned immediately.
- All requests are equally important.
- Multiple VM types and multiple datacenters are available in the simulator.

The dataset that we used for training and testing are real world data, as used in [4]. 15 different VM types are included in our experiments, 5 from each service provider, Alibaba Elastic Compute Service (ECS), Amazon Web Services (AWS) and Microsoft Azure. As in [21,22], we adopt 82 user locations in the Sprint IP Network[1] to simulate the global user community. To determine the network latency between users and assigned VM instances, we collect real-world observations of network latency from Sprint IP backbone network databases[2].

The source domain of the MOLSB problem with 8 different data centers, including Dublin, Singapore, Sydney, North Virginia, Mumbai, Tokyo, North California and Sao Paulo. Following [7], User requests arrive at the data centers with 1% arrival rate of the Microsoft Azure dataset. The target problem domain is the MOLSB problem with 15 different data centers, with 7 more regions including Frankfurt, Hong Kong, London, Paris, Seoul, Stockholm and Montreal.

To evaluate our proposed GPHH with transfer learning, we created 5 test cases of the target domain. Each test case contains one test set, which is unseen in training process. The performance of each test cases and the average performance of 5 test sets are calculated.

5.2 Baseline Algorithm

As mentioned in Sect. 1, there is no transfer learning method proposed for the MOLSB problem. To evaluate our proposed GPHH with three transfer learning approaches, namely BestGen, Half-Transfer, and Full-Transfer, we compare the transfer learning approaches with GPHH-MOLSB [4]. GPHH-MOLSB is trained on target domain directly without any knowledge transferred from the source domain.

5.3 Parameter Settings

In our experiments, we follow the GPHH parameters settings in the existing work [4] and [26]. In both the source and target domain, the population size is 1024 and the generation size is 100. The crossover rate and the mutation rate are 90% and 10%. The maximum depth is 7. In order to approximate the true Pareto Front, we train 30 independent runs with different random seeds. All transfer learning approaches are implemented using DEAP [11].

5.4 Results

Tables 3 and 4 show the average HV and IGD results of GPHH with different transfer techniques and the baseline method without knowledge transferred, i.e., GPHH-MOLSB.

---

[1] https://www.sprint.net.
[2] https://www.sprint.net/tools/ip-network-performance.

Table 3. Average HV result of 5 test cases

Test Case	GPHH-MOLSB [4]	BestGen	Half-Transfer	Full-Transfer
1	0.9800 ± 0.0005	0.9809 ± 0.0003	0.9810 ± 0.0003	**0.9811 ± 0.0002**
2	0.9798 ± 0.0006	0.9802 ± 0.0004	0.9803 ± 0.0003	**0.9805 ± 0.0003**
3	0.9796 ± 0.0005	0.9801 ± 0.0004	0.9803 ± 0.0002	**0.9804 ± 0.0002**
4	0.9779 ± 0.0008	0.9787 ± 0.0005	0.9788 ± 0.0003	**0.9790 ± 0.0002**
5	0.9786 ± 0.0006	0.9794 ± 0.0003	0.9795 ± 0.0003	**0.9796 ± 0.0002**
Average	0.9791 ± 0.0006	0.9798 ± 0.0004	0.9800 ± 0.0003	**0.9801 ± 0.0002**

As can be seen from Table 3, all three transfer techniques have better HV results than GPHH-MOLSB. For example, in test case 1, GPHH-MOLSB has a HV result of 0.9800. The HV results of BestGen, Half Transfer and Full Transfer approaches are 0.9809, 0.9810 and 0.9811. Similarly, in Table 4, the average IGD results of BestGen, Half Transfer and Full Transfer approaches are 0.00069, 0.00069 and 0.00065, while the average IGD of GPHH-MOLSB is 0.00112. In Table 3, the Full Transfer approach achieves the highest HV in all 5 test cases, while the BestGen approach has the lowest HV result.

Table 4. Average IGD result of 5 test cases

Test Case	GPHH-MOLSB [4]	BestGen	Half-Transfer	Full-Transfer
1	10.2223 ± 2.4358e−4	6.2131 ± 0.5317e−4	6.2117 ± 0.4607e−4	**5.8971±0.5317e−4**
2	11.5345 ± 3.0194e−4	6.7907 ± 0.4728e−4	6.8737 ± 0.4598e−4	**6.3855±0.5289e−4**
3	11.2997 ± 2.8345e−4	6.8132 ± 0.6295e−4	6.8763 ± 0.5525e−4	**6.4603±0.5971e−4**
4	11.1337 ± 3.2152e−4	7.0447 ± 0.5916e−4	7.1584 ± 0.5642e−4	**6.6599±0.4936e−4**
5	12.0743 ± 3.3086e−4	7.5021 ± 0.4838e−4	7.4144 ± 0.4997e−4	**7.2684±0.5037e−4**
Average	11.2475 ± 2.9627e−4	6.8727 ± 0.5419e−4	6.9069 ± 0.5074e−4	**6.5342±0.5310e−4**

In Table 4, the Full-Transfer approach has the lowest IGD in all 5 test cases, BestGen has the highest IGD in most of the test cases except test case 4. All three transfer learning approaches have achieved better performance than GPHH-MOLSB. This demonstrates that the three transfer learning methods are able to keep the diversity of the population, which is important for evolutionary processes.

The results in Tables 3 and 4 show that the heuristics generated by three transfer learning approaches outperform the GPHH-MOLSB without transfer knowledge in terms of the IGD and HV results.

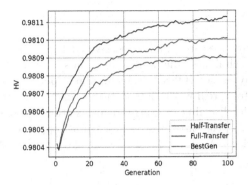

Fig. 4. Convergence curve of HV for test case 1 on 15 data centers

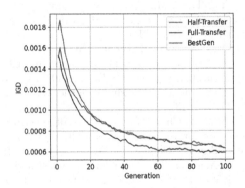

Fig. 5. Convergence curve of IGD for test case 1 on 15 data centers

To further investigate the effectiveness of the three transfer learning approaches during the training process of the source domain, we analyze the convergence curves of the three approaches. Figures 4 and 5 present the convergence curves regarding the average HV and IGD obtained by three transfer learning approaches on the test case 1 with 15 data centers. As seen in the two figures, all three approaches start with a high HV and a low IGD results. The HV results of all three transfer approaches increase rapidly, and the IGD results decrease significantly from the first generation to the 40-th generation. From the 80th generation to the last generation, the HV and IGD results change slightly with fluctuation in all three transfer approaches.

As seen in Figs. 4 and 5, among the three transfer learning methods, GPHH with Full-Transfer method performs the best through all the generations. The results demonstrate that knowledge obtained from the source domain can help GPHH to generate high quality rules in the target domain.

6   Conclusion

In this paper, we propose three multi-objective GPHH approaches with different transfer learning techniques, including Full-Transfer, Half-Transfer and Best-Gen, to solve the dynamic MOLSB problem. Our experimental evaluation using datasets collected from real world demonstrates that all the three approaches of GPHH with transfer learning outperform an existing approach without using transfer learning. All the three transfer learning approaches generate better heuristics with higher HV and lower IGD than the existing GPHH method.

References

1. Ardeh, M.A., Mei, Y., Zhang, M.: Genetic programming with knowledge transfer and guided search for uncertain capacitated arc routing problem. IEEE Trans. Evol. Comput. **26**(4), 765–779 (2021)
2. Ardeh, M.A., Mei, Y., Zhangz, M.: Diversity-driven knowledge transfer for GPHH to solve uncertain capacitated arc routing problem. In: 2020 IEEE Symposium Series on Computational Intelligence (SSCI), pp. 2407–2414. IEEE (2020)
3. Burke, E.K., et al.: Hyper-heuristics: a survey of the state of the art. J. Oper. Res. Soc. **64**(12), 1695–1724 (2013)
4. Chen, Y., Shi, T., Ma, H., Chen, G.: Automatically design heuristics for multi-objective location-aware service brokering in multi-cloud. In: 2022 IEEE International Conference on Services Computing (SCC), pp. 206–214. IEEE (2022)
5. Deb, K., Pratap, A., Agarwal, S., Meyarivan, T.: A fast and elitist multiobjective genetic algorithm: NSGA-II. IEEE Trans. Evol. Comput. **6**(2), 182–197 (2002)
6. Dinh, T.T.H., Chu, T.H., Nguyen, Q.U.: Transfer learning in genetic programming. In: 2015 IEEE Congress on Evolutionary Computation (CEC), pp. 1145–1151. IEEE (2015)
7. Du, B., Wu, C., Huang, Z.: Learning resource allocation and pricing for cloud profit maximization. In: Proceedings of the AAAI Conference on Artificial Intelligence, vol. 33, pp. 7570–7577 (2019)
8. Durillo, J.J., Fard, H.M., Prodan, R.: Moheft: A multi-objective list-based method for workflow scheduling. In: 4th IEEE International Conference on Cloud Computing Technology and Science Proceedings, pp. 185–192. IEEE (2012)
9. Escott, K.-R., Ma, H., Chen, G.: Transfer learning assisted GPHH for dynamic multi-workflow scheduling in cloud computing. In: Long, G., Yu, X., Wang, S. (eds.) AI 2022. LNCS (LNAI), vol. 13151, pp. 440–451. Springer, Cham (2022). https://doi.org/10.1007/978-3-030-97546-3_36
10. Fonseca, C.M., Paquete, L., López-Ibánez, M.: An improved dimension-sweep algorithm for the hypervolume indicator. In: 2006 IEEE international conference on evolutionary computation, pp. 1157–1163. IEEE (2006)
11. Fortin, F.A., De Rainville, F.M., Gardner, M.A.G., Parizeau, M., Gagné, C.: Deap: Evolutionary algorithms made easy. J. Mach. Learn. Res. **13**(1), 2171–2175 (2012)
12. Heilig, L., Buyya, R., Voß, S.: Location-aware brokering for consumers in multi-cloud computing environments. J. Netw. Comput. Appl. **95**, 79–93 (2017)
13. Iqbal, M., Xue, B., Al-Sahaf, H., Zhang, M.: Cross-domain reuse of extracted knowledge in genetic programming for image classification. IEEE Trans. Evol. Comput. **21**(4), 569–587 (2017)

14. Ishibuchi, H., Masuda, H., Tanigaki, Y., Nojima, Y.: Modified distance calculation in generational distance and inverted generational distance. In: Gaspar-Cunha, A., Henggeler Antunes, C., Coello, C.C. (eds.) EMO 2015. LNCS, vol. 9019, pp. 110–125. Springer, Cham (2015). https://doi.org/10.1007/978-3-319-15892-1_8
15. Koçer, B., Arslan, A.: Genetic transfer learning. Expert Syst. Appl. **37**(10), 6997–7002 (2010)
16. Koza, J.R., Poli, R.: Genetic programming. In: Burke, E.K., Kendall, G. (eds.) Search Methodologies, pp. 127–164. Springer, Boston (2005). https://doi.org/10.1007/0-387-28356-0_5
17. Ma, H., da Silva, A.S., Kuang, W.: NSGA-II with local search for multi-objective application deployment in multi-cloud. In: 2019 IEEE Congress on Evolutionary Computation (CEC), pp. 2800–2807. IEEE (2019)
18. Mansouri, Y., Toosi, A.N., Buyya, R.: Brokering algorithms for optimizing the availability and cost of cloud storage services. In: 2013 IEEE 5th International Conference on Cloud Computing Technology and Science, vol. 1, pp. 581–589 (2013)
19. Shi, T., Ma, H., Chen, G.: A genetic-based approach to location-aware cloud service brokering in multi-cloud environment. In: 2019 IEEE International Conference on Services Computing (SCC), pp. 146–153. IEEE (2019)
20. Shi, T., Ma, H., Chen, G.: Seeding-based multi-objective evolutionary algorithms for multi-cloud composite applications deployment. In: 2020 IEEE International Conference on Services Computing (SCC), pp. 240–247. IEEE (2020)
21. Shi, T., Ma, H., Chen, G., Hartmann, S.: Location-aware and budget-constrained application replication and deployment in multi-cloud environment. In: 2020 IEEE International Conference on Web Services (ICWS), pp. 110–117. IEEE (2020)
22. Shi, T., Ma, H., Chen, G., Hartmann, S.: Location-aware and budget-constrained service deployment for composite applications in multi-cloud environment. IEEE Trans. Parallel Distrib. Syst. **31**(8), 1954–1969 (2020)
23. Shi, T., Ma, H., Chen, G., Hartmann, S.: Cost-effective web application replication and deployment in multi-cloud environment. IEEE Trans. Parallel Distrib. Syst. **33**(8), 1982–1995 (2021)
24. Shi, T., Ma, H., Chen, G., Hartmann, S.: Location-aware and budget-constrained service brokering in multi-cloud via deep reinforcement learning. In: Hacid, H., Kao, O., Mecella, M., Moha, N., Paik, H. (eds.) ICSOC 2021. LNCS, vol. 13121, pp. 756–764. Springer, Cham (2021). https://doi.org/10.1007/978-3-030-91431-8_52
25. Simarro, J.L.L., Moreno-Vozmediano, R., Montero, R.S., Llorente, I.M.: Dynamic placement of virtual machines for cost optimization in multi-cloud environments. In: International Conference on High Performance Computing Simulation, pp. 1–7 (2011)
26. Tan, B., Ma, H., Mei, Y.: A hybrid genetic programming hyper-heuristic approach for online two-level resource allocation in container-based clouds. In: 2019 IEEE Congress on Evolutionary Computation (CEC), pp. 2681–2688. IEEE (2019)
27. Weiss, K., Khoshgoftaar, T.M., Wang, D.D.: A survey of transfer learning. J. Big Data **3**(1), 1–40 (2016). https://doi.org/10.1186/s40537-016-0043-6
28. While, L., Hingston, P., Barone, L., Huband, S.: A faster algorithm for calculating hypervolume. IEEE Trans. Evol. Comput. **10**(1), 29–38 (2006)
29. Zhang, F., Mei, Y., Nguyen, S., Zhang, M.: Evolving scheduling heuristics via genetic programming with feature selection in dynamic flexible job-shop scheduling. IEEE Trans. Cybern. **51**(4), 1797–1811 (2020)
30. Zhang, F., Mei, Y., Nguyen, S., Zhang, M., Tan, K.C.: Surrogate-assisted evolutionary multitask genetic programming for dynamic flexible job shop scheduling. IEEE Trans. Evol. Comput. **25**(4), 651–665 (2021)

Evolutionary Machine Learning

Evolving Neural Networks for Robotic Arm Control

Anthony Horgan and Karl Mason[(✉)][iD]

School of Computer Science, University of Galway, Galway, Ireland
{a.horgan5,karl.mason}@universityofgalway.ie

Abstract. Developing effective and adaptive robotic arm controllers is crucial for many industries, e.g. manufacturing. Traditional pre-programmed controllers cannot adapt to changing environments. This study investigates how neuroevolution can be used to develop robotic arm controllers and addresses key gaps in the existing literature, such as incorporating expert demonstrations and analyzing the robustness of evolved controllers. In addition to addressing these questions, this work compares different controller architectures and training algorithms. The proposed evolutionary neural network motion controller can accurately complete the random target reaching task, moving to within 1.7 cm from the target on average. An evolutionary supervisor neural network approach is also proposed to solve the pick-and-place task. The proposed method achieves a high successful completion rate, 927 out of 1000 trials.

Keywords: Neural Networks · Robot Arm · Evolutionary Algorithms · Neuroevolution

1 Introduction

Robotic arms are used to perform an extensive range of tasks in a number of real-world environments. Traditionally, they have been manually programmed to perform predefined tasks. Often, tasks which are too dangerous, require extreme accuracy or are too tedious for humans, can be performed quickly and safely by robotic arms. Some of these tasks include: picking and placing objects, welding, machining, cutting and assembling. The traditional approach of manually programming robotic arms is suitable for narrowly-defined tasks in static environments and is less suitable for dynamic environments.

Machine learning is a promising paradigm to enable robots to operate in dynamic environments. Recent studies have successfully applied machine learning techniques to robotic arm control tasks, e.g. reinforcement learning (RL) [1,7]. Neuroevolution is a machine learning approach that has been demonstrated to perform competitively with RL [23,27]. It can address some of the limitations present with RL, e.g. difficulty in dealing with sparse rewards and long time horizons [23]. There are several examples of using neuroevolution to control a robotic arm in the literature [4,11,21,28].

© The Author(s), under exclusive license to Springer Nature Switzerland AG 2023
J. Correia et al. (Eds.): EvoApplications 2023, LNCS 13989, pp. 591–607, 2023.
https://doi.org/10.1007/978-3-031-30229-9_38

There are multiple research topics that remain relatively unexplored in the literature relating to neuroevolution and robotic arm control, e.g. studies relating to expert demonstrations and controller robustness. This work utilizes expert demonstrations to improve the evolutionary process when evolving neural networks for robotic arm control. There are no examples of such a study in the literature.

This paper makes the following contributions:

1. To compare the performance of multiple well studied meta-heuristics for robotic arm control.
2. To evaluate the impact of single versus multiple robotic arm joint angle updates on learning.
3. To investigate the impact of expert demonstrations on the evolved robotic arm controller.
4. To assess the robustness of the evolved robotic arm controller to joint damage.
5. To propose a robotic arm supervisor neuroevolution architecture to perform a pick and place task.

The rest of this paper is structured as follows. Section 2 discusses work related to this paper. Section 3 outlines the experimental setup for the simulations conducted. Section 4 presents the results of the experiments conducted and provides a discussion of these. Section 5 draws conclusions from this research.

2   Related Work

2.1   Neuroevolution

Neuroevolution uses evolutionary algorithms (EA) to evolve neural networks (NN). Many neuroevolution approaches evolve the weights of neural networks with a fixed network topology. Each gene in the genotype corresponds to the value for a specific weight parameter in the neural network. Each set of network parameters θ_i are evaluated on some task, e.g. robotic arm control. The fitness of each set of parameters is determined using a fitness function $F(\theta_i)$. The next generation of the population is then created based on the fittest individuals in the current generation.

Neuroevolution has been shown to be a competitive alternative for RL [23, 27]. Salimans et al. implemented a type of evolution strategy called a natural evolution strategy to evolve the weights of deep NNs to play Atari 2600 games and to solve continuous robotic control tasks at a competitive level compared to state-of-the-art Deep Reinforcment Learning (DRL) methods [23]. Such et al. use a gradient-free evolutionary algorithm to evolve very large networks to play Atari games, outperforming DRL methods such as DQN and A3C on a number of games [27]. Mason and Grijalva show that Neuroevolution is a competitive alternative to RL methods for control of heating, ventilation and air-conditioning systems [20].

This paper will compare three algorithms to train a neural network for robotic arm control. These include two evolutionary algorithms, Differential Evolution (DE) and Covariance Matrix Adaptation Evolution Strategy (CMA-ES), and one swarm based meta-heuristic, Particle Swarm Optimization (PSO). Including PSO as a comparison will provide further context to the performance of the two evolutionary methods. Each of these algorithms are outlined briefly in the next sections.

Differential Evolution (DE) is a commonly used and well established evolutionary algorithm [26]. In DE, individuals are perturbed according to the scaled difference between other randomly selected individuals. Pseudocode outlining the DE Algorithm is illustrated in Algorithm 1. DE has been used before in neuroevolution in a variety of contexts [3, 14, 19, 24].

Algorithm 1. Differential Evolution

1: Randomly initialize each individual $x_i \in \mathbb{R}^n$
2: **while** termination criterion not met **do**
3:     **for** each individual x in population **do**
4:         Randomly select three individuals a, b, c from population which are distinct from each other and distinct from x
5:         Randomly select $R \in \{1, 2, ..., n\}$
6:         $y \leftarrow x$
7:         **for** each dimension $d = 1, 2, ..., n$ **do**
8:           Generate random number $r_d \sim U(0, 1)$
9:           **if** $r_d < CR$ or $d = R$ **then**
10:             $y_d \leftarrow a_d + F(b_d - c_d)$
11:         **if** $cost(y) \leq cost(x)$ **then**
12:           $x \leftarrow y$

Covariance Matrix Adaptation Evolution Strategy (CMA-ES) is one of the most prominent and powerful evolutionary algorithms [10]. In CMA-ES, candidate solutions are sampled from a multivariate normal distribution which represents a statistical model of the population. The best performing solutions are used to update the distribution parameters. Algorithm 2 demonstrates how the CMA-ES algorithm operates.

CMA-ES has been shown to be an effective method of optimizing the weights of neural networks for RL tasks [13, 18, 19]. Igel and Christian use CMA-ES to evolve networks to solve several variants of the pole-balancing problem [13]. Mason et al. use CMA-ES to evolve networks to predict power demand [18].

Hansen proposed a method of injecting external solutions into the CMA-ES evolution process [9]. The CMA-ES algorithm is altered to make it robust to the inclusion of external solutions in the update process. Injecting external candidate solutions in this way can lead to improved convergence speed.

Algorithm 2. Covariance Matrix Adaptation Evolution Strategy

1: Initialize $C = I, p_c = 0, p_\sigma = 0, \sigma = \sigma_0, m = m_0$
2: **while** termination criterion not met **do**
3:    Sample solutions: $x_i = m + \sigma y_i$, where $y_i \sim \mathcal{N}_i(0, C)$ for $i = 1, ..., \lambda$
4:    Update mean: $m \leftarrow m + c_m \sigma y_w$, where $y_w = \Sigma_{i=1}^{\mu} w_{rk(i)} y_i$
5:    Update σ path: $p_\sigma \leftarrow (1 - c_\sigma)p_\sigma + \sqrt{1 - (1 - c_\sigma)^2}\sqrt{\mu_w}C^{-\frac{1}{2}}y_w$
6:    Update C path: $p_c \leftarrow (1 - c_c)p_c + 1_{[0,2n]}\{\|p_\sigma\|^2\}\sqrt{1 - (1 - c_c)^2}\sqrt{\mu_w}y_w$
7:    Update σ: $\sigma \leftarrow \sigma \times \exp\left(\frac{c_\sigma}{d_\sigma}\left(\frac{\|p_\sigma\|}{\mathbb{E}\|\mathcal{N}(0,\,I)\|} - 1\right)\right)$
8:    Update C: $C \leftarrow C + c_\mu \Sigma_{i=1}^{\lambda} w_{rk(i)}(y_i y_i^T - C) + c_1(p_c p_c^T - C)$

Particle Swarm Optimization (PSO) is a swarm intelligence meta-heuristic which represents candidate solutions as particles traversing the solution space [17]. The particle's velocity is influenced by the best position seen by that particle as well as the best position seen by any particle. Algorithm 3 describes how PSO operates.

There have been many approaches which use PSO and its variants to train the weights [6,8] and to train the weights and architectures [29] of neural networks. Gudise et al. show that PSO can converge faster than backpropagation for optimizing network weights [8]. Espinal et al. show that DE has better computational performance than PSO but that PSO exhibits better classification performance for evolving shallow networks [5].

Algorithm 3. Particle Swarm Optimization

1: **for** each particle $i = 1, 2, ..., popsize$ **do**
2:    Randomly initialize particle position x_i
3:    Randomly initialize particle velocity v_i
4:    Initialize particle's personal best position to current position $p_i \leftarrow x_i$
5:    **if** $cost(x_i) < cost(g)$ **then**
6:        Update global best position $g \leftarrow x_i$
7: **while** termination criterion not met **do**
8:    **for** each particle $i = 1, 2, ..., popsize$ **do**
9:        **for** each dimension $d = 1, 2, ..., N$ **do**
10:           Generate random numbers $r_p, r_g \sim U(0, 1)$
11:           Update particle velocity $v_{i,d} \leftarrow wv_{i,d} + C_1 r_p(p_{i,d} - x_{i,d}) + C_2 r_g(g_d - x_{i,d})$
12:       Update particle position $x_i \leftarrow x_i + v_i$
13:       **if** $cost(x_i) < cost(p_i)$ **then**
14:           Update particle's personal best position $p_i \leftarrow x_i$
15:           **if** $cost(p_i) < cost(g)$ **then**
16:               Update global best position $g \leftarrow p_i$

2.2   Robotic Arm Controllers

A traditional controller for a robotic arm involves solving the inverse kinematic (IK) equations of the arm [25, pp. 85–98] or using path planning algorithms [25, pp 149–170]. There are drawbacks to both approaches. Inverse kinematics controllers do not take collisions with obstacles into account. Path planning algorithms are computationally expensive. These approaches are unsuitable for changing environments.

There have been several applications of neuroevolution to robotic arm control tasks [4,11,21,28]. Wen et al. evolve a neural network controller for a planar musculoskeletal arm [28]. Neuroevolution of Augmenting Topologies (NEAT) is used to evolve two separate controllers to perform random target reaching (one which controls the delta/change in joint angles and another which controls the activation of human-like muscles which move the arm joints). Controllers are evaluated according to a pre-computed training set of joint configurations.

Huang et al. use limited human supervision to evolve a NN controller which can grasp objects of various shapes with a human-like robotic hand [11]. NEAT is used to evolve a controller which takes depth-sensor data as input, and outputs the joint configurations necessary to grasp the object.

Moriarty uses neuroevolution to evolve a controller that can reach a target whilst avoiding obstacles [21]. A primary network is evolved to move the end effector to the target while avoiding the obstacle and a secondary network is evolved to make more precise movements to reach the target position. Combining these networks produces a robust control policy which integrates both obstacle avoidance and target reaching.

The previous study most similar to the work presented in this paper is the work of D'Silva and Miikkulainen [4]. The authors evolve neural network controllers to reach random targets in three different environments: **I**: environments with no obstacles, **II**: environments with stationary obstacles, **III**: environments with moving obstacles. The NEAT algorithm is used to evolve controllers. The network output determines how much to move each joint during each time step of the episode. By using neuroevolution, the authors evolved a controller which is robust to a dynamic environment with moving obstacles. This cannot be achieved using inverse kinematics or path finding controllers.

Learning from Demonstration (LfD) and Imitation Learning refer to the paradigm of machine learning techniques that learn to mimic/imitate an expert [12]. LfD is often used in learning robotic controllers [2]. Karpov shows that expert demonstrations can be used alongside neuroevolution to solve robotic control tasks in a video game environment [16]. The research outlined in this paper will combine LfD with neuroevolution for robotic arm control.

3   Experimental Setup

3.1   Simulated Robotic Arm

A robot arm with eight degrees of freedom is simulated using the CoppeliaSim simulator [22]. This is illustrated in Fig. 1. The parameters of the arm joints are outlined in Table 1. The arm is approximately 80 cm from base to tip.

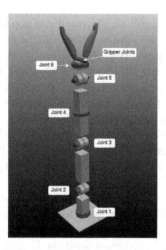

Fig. 1. Simulated Robot in CoppeliaSim

Table 1. Simulated Robotic Arm Joints

Label	Joint Name	Min Angle (deg)	Max Angle (deg)
J1	Rotating Base	−90	90
J2	Shoulder Joint	0	120
J3	Elbow Joint	0	120
J4	Forearm Rotational	−90	90
J5	Wrist Vertical	−120	120
J6	Wrist Rotational	−90	90
J7	Left Gripper	0	90
J8	Right Gripper	0	90

PyRep is a tailored framework that can run on top of CopelliaSim [15]. It is used for all experiments presented in this paper. The evolved networks move the arm using joint position control. The networks output the desired position for each joint and do not need to take the physics of the environment into account.

3.2   Experiments

A total of 5 experiments are conducted in this paper. The experiments are run for 100,000 fitness function evaluations each. These are outlined below.

1. Comparing the performance of CMA-ES, DE and PSO when training a network to move the end effector to a target location.
2. Comparing the performance of the evolved network using the single-step approach vs the multi-step approach. Comparing the performance when the network output is used to set the final arm joint angles in a single movement/time-step with when the output is used to incrementally update the joint angle over multiple time-steps.
3. Incorporating expert demonstrations when evolving a NN for robotic arm control.
4. Assessing the impact of disabled joints on evolved NN controller performance.
5. Evaluating the performance of a CMA-ES evolved supervisor network to perform a pick-and-place task.

3.3   Evolving Robotic Arm Controller

Figure 2 illustrates how the evolutionary neural network algorithm interacts with the simulated robotic environment.

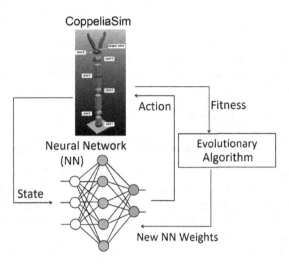

Fig. 2. Evolutionary Neural Network Interaction with Robotic Arm Environment

The objective of the robotic arm in Experiments 1–4 is to move the end effector to the target location. The fitness of the network in this movement task is calculated as the euclidean distance between the end effector and the target at the end of the episode, as presented in Eq. 1.

$$Fitness = \sqrt{(target_x - tip_x)^2 + (target_y - tip_y)^2 + (target_z - tip_z)^2} \quad (1)$$

The training dataset consists of fifty target positions which are placed within a semi-circle region on the floor in front of the arm. The z co-ordinate of the target position is always 0. The target positions are evenly dispersed over the target region.

Two different NN outputs for robotic arm control were considered:

1. Single-Step: NN ouputs the desired final joint angle (α) in a single action.
2. Multi-Step: NN outputs an incremental change to the joint angles ($\Delta\alpha$) over 20 time-steps.

Single-Step. The single-step controller outputs the joint angles/positions necessary to place the end effector at the target. The network input in these simulations are the x, y co-ordinates of the target position.

Each of the output nodes determines the angle (α_i) for a corresponding joint i. This approach provides the NN with the lowest amount of state information and aims to set the joint angles in a single action. The purpose of running this experiment is to establish how the NN can perform with minimal state information.

Multi-Step. This approach evolves the NN to adjust the joint angles over multiple time-steps. The NNs output the change in joint angle $(\Delta\alpha_i)$ for each joint i at each time-step. The NNs are provided with the following normalized state information:

- Six joint angles that represent current joint state;
- x, y co-ordinates of target position;
- x, y, z co-ordinates of target position relative to the end effector.

The joint deltas are normalized to be in the range $[-5, +5]$ degrees. The network has an additional output node which controls when to stop the episode. The episode begins by placing the target and setting the joints to their starting positions. The episode ends if the value outputted by the stop node is greater than 0.5 or after twenty time-steps are completed.

Expert Demonstrations. The goal of this experiment is to evolve a network to learn from a set of expert examples which demonstrate how to move the arm in order to accurately complete the single-step movement task.

The dataset consists of 50 expert demonstrations. This number was selected as it is a reasonable number of demonstrations that would be available from a pre-programmed robotic controller. Each entry consists of the x, y co-ordinates of the target positions along with the joint angles which place the end effector at these targets. Inverse kinematics are used to find an appropriate joint configuration for each target position. The fitness here is equal to the mean squared error (MSE) between joint angles outputted by the network (α) and the expert joint angles (α') as outlined in Eq. 2.

$$Fitness = MSE(\alpha, \alpha') \tag{2}$$

The purpose of this simulation is to determine how expert demonstrations can impact performance. This converts the problem into a supervised learning task.

An additional simulation is conducted that utilizes expert demonstrations. This simulation involves utilizing the evolved network, trained from expert demonstrations in a supervised learning manner, as a seed network when evolving the robotic arm controller in CoppeliaSim. This seeding takes the form of setting the initial CMA-ES mean value as the weights of the imitation seed network. In addition to this, the seed network parameters are injected into the population until parameters with a lower distance to target are found.

Disabling Joints. This experiment involves disabling each of the robotic arm joints to test the robustness of the evolved controllers. As each joint is disabled, the NN controller is evaluated based on its ability to reach a target location using Eq. 1.

Evolving Supervisor Network for Pick and Place Task. The final experiment is to develop a controller to complete the pick and place task. The task can be broken down into four steps or sub-tasks:

1. move the end effector to a randomly placed object
2. grasp the object
3. move the object to a randomly chosen target location
4. place the object.

Moving the end effector to the object (sub-tasks 1) and target location (sub-tasks 3) are the challenging sub-tasks for pick-and-place. A NN can be evolved to achieve sub-tasks 1 and 2 using the same approach taken in previous experiments. Sub-tasks 2 and 4 are relatively straightforward tasks which can be performed using a pre-programmed routine instead of utilizing neuroevolution to open and close a gripper.

Supervisor Network. The evolved motion control networks are used to perform the pick-and-place task by way of a supervisor network. The supervisor network is evolved to predict which of the four sub-tasks is to be performed at each step of the pick-and-place task. Each sub-task is performed by a subroutine. Sub-tasks 1 and 3 (moving to object, moving to target) are performed by a motion control network. This is illustrated in Fig. 3.

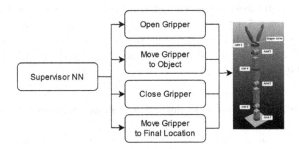

Fig. 3. Supervisor Neural Network for Pick and Place

At the start of an episode, the arm joints are set to their starting positions and the object and target are placed in front of the arm. Throughout the episode, the supervisor network must select which sub-task to perform at each of the 4 environment interactions. The supervisor network must call each of the four sub-task controllers in the correct order to successfully complete the pick and place task. The supervisor network is given the state information s= [x and y co-ordinates of object, x and y co-ordinates of target, all eight joint angles].

The supervisor network is evolved with a supervised learning approach using a dataset of 20,000 (state information, sub-task label) pairs gathered by performing each of the four sub-tasks in the correct order. The fitness of a supervisor network is calculated as the mean squared error between the network output of sub-task selection and the correct sub-task selection.

For this task, the sub-tasks are always performed in the same order. For other robotic control tasks this might not be the case. Therefore, it is important to evolve a NN which can select the sub-task based on the state of the environment.

3.4   Hyperparameters

A comprehensive parameter sweep resulting in the following parameters for the network architecture, CMA-ES, DE and PSO.

- **Neural Network.** All evolved networks consist of a single hidden layer of 20 nodes.
- **Covariance Matrix Adaptation Evolution Strategy.** Hyperparameters: $\sigma_0 = 0.1$, initial $m = 0$, $\lambda = 19$ for single-step and 21 for multi-step
- **Differential Evolution.** Hyperparameters: population size = 200, differential weight = 0.2, crossover probability = 0.9.
- **Particle Swarm Optimization.** Hyperparameters: population size = 100, momentum = 0.4, local influence = 1.5, global influence = 2.

4   Results

4.1   Comparing CMA-ES, DE and PSO for Single-Step Task

Figure 4 (Left) visualizes the spread of single-step network solutions found by each algorithm over 5 runs. Table 2 shows the average distance to target of solutions evolved by each algorithm on training and test sets. Figure 4 (Right) shows the distance to target of the best solution found at each generation for each of DE, PSO and CMA-ES.

Table 2. Performance on single-step task averaged over five runs

Algorithm	Mean Train Target Distance	Std. Train Target Distance	Mean Test Target Distance	Std. Test Target Distance
CMA-ES	0.0195	0.0037	0.0272	0.0039
DE	0.0822	0.0109	0.0881	0.0094
PSO	0.1042	0.0173	0.1194	0.0141

Figures 4 (Left) and (Right) clearly demonstrate that CMA-ES performs significantly better than DE and PSO. Two-tailed Wilcoxon Rank-Sum tests with

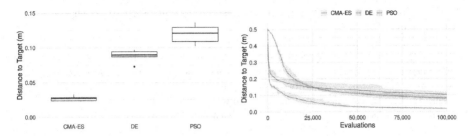

Fig. 4. (Left) Boxplot of distance to target for DE, PSO, CMA on unseen target locations. (Right) Convergence of each algorithm.

a significance level of 0.01 conclude that CMA-ES achieves significantly lower cost (distance to target) than both PSO and DE. CMA-ES is therefore selected to train all networks in all subsequent experiments.

4.2 Comparing Single-Step with Mult-Step Control

A multi-step network, both with and without the position of the target relative to the end effector as NN input, is compared to the single-step network. Figures 5 (Left) and (Right) illustrate the distance to target and convergence achieved by single and multi-step control approaches.

Table 3 demonstrates the positive impact that relative position information has on the multi-step control approach. This controller significantly outperforms all other controllers when compared using a two-tailed Wilcoxon Rank-Sum test ($\alpha = 0.01$).

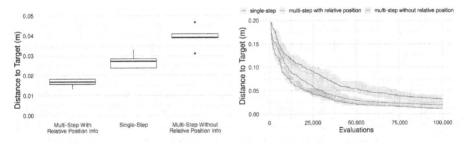

Fig. 5. (Left) Boxplot of distance to target for single-step and multi-step control. (Right) Convergence of single-step and multi-step controllers.

This result demonstrates that progressively updating the arm joint angles over 20 time-steps is more effective than simply setting the final joint angles in one time-step. There is a trade off in terms of computational cost however. An evaluation of a single-step network takes ≈ 150 ms. An evaluation of a multi-step network takes ≈ 500 ms.

Table 3. Performance of CMA-ES on single-step and multi-step approaches averaged over five runs

Algorithm	Mean Train Target Distance	Std. Train Target Distance	Mean Test Target Distance	Std. Test Target Distance
Single-step	0.0195	0.0037	0.0272	0.0039
Multi-step (relative position)	0.0122	0.0016	0.0165	0.0021
Multi-step (no relative position)	0.0322	0.0067	0.0394	0.0056

4.3 Learning from Expert Demonstrations

Figure 6 compares the spread of controller fitness trained utilizing the single-step learning approach with the controller trained utilizing supervised learning from expert demonstrations. Both approaches are capable of evolving a network controller to move to the same fifty target positions. The accuracy of the controller trained in simulation without expert demonstrations results in a lower distance to target when compared to using expert demonstrations through supervised learning.

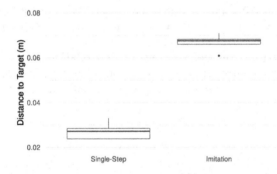

Fig. 6. Learning from environment interactions vs learning from expert demonstrations

The next simulation involved utilizing the neural network controller, pre-trained using expert demonstrations, as a seed for the controller evolved in simulation. Figure 7 shows the convergence of the distance to target using CMA-ES on the single-step movement task trained with and without this expert demonstration pre-trained seed network. When the seed is utilized, low-cost solutions are found earlier and the algorithm converges with less variation. The final distance to target achieved both with and without a seed network is not significantly different when tested using the Wilcoxon test ($\alpha = 0.01$). This implies that if expert demonstrations are available, they can be utilized to speed up training however they are not mandatory for good performance.

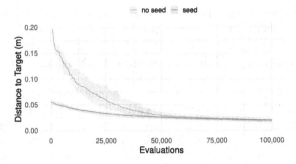

Fig. 7. Convergence with and without expert demonstration seed network

4.4   Movement with Disabled Joints

When robotic arms are deployed in real-world environments they can suffer damage which can impair their ability to perform as intended. The results presented in this section investigate how disabling a joint of the robotic arm impacts the performance of the motion control networks.

The motion control networks were originally evolved with full control over all six main joints. The networks were then evaluated on the test set to see how their motion control is impacted when a single joint was disabled (i.e. the network will try to change the joints positions as usual but one joint cannot be moved).

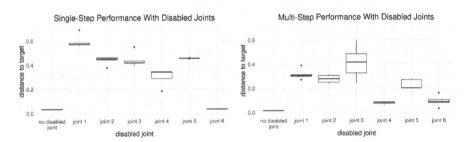

Fig. 8. Boxplot of distance to target achieved by single-step network (Left) and multi-step network (Right) with different joints disabled.

Figure 8 (Left) shows the performance spread of five single-step networks and Fig. 8 (Right) shows the performance spread of five multi-step networks. The single-step network and the multi-step network are affected by disabled joints in different ways. The single-step network does not significantly suffer when joint 6 is disabled. This is because joint 6 only rotates the end effector and while this changes the orientation of the end effector, it does not change its position relative to the target location. This is not true for the multi-step network whose performance suffers when joint 6 is disabled. The suspected reason for this is that the multi-step network receives the joint positions at each time-step.

Disabling joint 6 would therefore cause the network to receive state information which it is unfamiliar with. This then results in sub-optimal joint movement.

On the other hand, as can be seen in Fig. 8, when joints 1, 2, 4 or 5 are disabled the performance of the single-step network degrades much more than the performance of the multi-step network.

In the case of both the single-step and the multi-step networks, disabling a single joint usually has a considerable negative impact on the networks ability to move to the target. In part, this is because disabling most joints (namely joints 1, 2, 3 and 5) reduces the reachable workspace of the arm. However, even when losing control of a joint of lesser importance (e.g. joint 4), the motion control networks struggle to reach the target.

4.5   Pick and Place Task Using Supervisor Network

In order to tackle the pick and place task, the supervisor network was evolved for 400,000 evaluations. When evaluated on 4000 unseen test cases, the evolved network selects the correct sub-task controller with an accuracy of 99.5%.

The evolved supervisor network was evaluated by performing 1000 episodes of the pick-and-place task using a multi-step network. Out of the 1000 episodes, 927 were completed successfully, 33 failed due to the supervisor network selecting the wrong sub-task controller and 51 failed because the object could not be properly grasped due to the end effector being place too far away. The combination of the evolved supervisor network and evolved multi-step network results in a system which can accurately perform the pick-and-place task.

5   Conclusion and Future Work

This work has demonstrated that neuroevolution is an effective approach to motion control of a robotic arm compared to a DRL approach based on performance on the random target reaching and pick-and-place tasks. The primary findings of this research are:

1. CMA-ES is more effective than DE and PSO for training a neural network robotic arm controller. CMA-ES achieves a mean distance to target of 0.027 m. This is significantly lower than DE (0.088 m) and PSO (0.119 m).
2. Evolving a multi-step neural network controller provides increased performance when compared to a single-step controller in terms of distance between end effector and target location. The multi-step network achieves a mean distance to target of 0.017 m, compared to the single-step network (0.027 m). For context, the approach used by Gu. et al. consistently achieves a distance to target of below 0.05 m on a similar task. There is an increased computational cost ($\approx 3\times$) required to train the multi-step network.

3. Training a neural network controller in a supervised learning manner from expert demonstrations does not lead to increased performance when compared to the controller evolved through interacting with the environment. The simulator trained controller performs significantly better. Utilizing a pre-trained neural network as a seed for the network evolved in simulation speeds up the training process, however the final controller is not significantly better than without using a pre-trained seed.
4. Disabling robotic arm joints has a negative impact on the performance of the evolved controller, as expected. The multi-step controller is more adversely impacted by disabling the wrist rotation joint, when compared to the single-step network which is negligibly impacted by disabling this joint. Conversely, the negative impact of disabling most other joints is less severe when using the multi-step controller.
5. Evolving a supervisor neural network is an appropriate technique for solving the pick-and-place task. The evolved supervisor network selects the correct sub-task controller with an accuracy of 99.5% in unseen test cases. When evaluated on 1000 episodes of pick-and-place, the supervisor network successfully completed 927 out of 1000 trials.

There are many avenues for future work that have arisen from this research, e.g. investigating different supervisor network architectures for the pick-and-place task and evolving controllers that can adapt when joints are disabled.

References

1. Amarjyoti, S.: Deep reinforcement learning for robotic manipulation-the state of the art. arXiv preprint arXiv:1701.08878 (2017)
2. Argall, B.D., Chernova, S., Veloso, M., Browning, B.: A survey of robot learning from demonstration. Robot. Auton. Syst. **57**(5), 469–483 (2009)
3. Baioletti, M., Di Bari, G., Poggioni, V., Tracolli, M.: Differential evolution for learning large neural networks (2018)
4. D'Silva, T., Miikkulainen, R.: Learning dynamic obstacle avoidance for a robot arm using neuroevolution. Neural Process. Lett. **30**(1), 59–69 (2009)
5. Espinal, A., et al.: Comparison of PSO and de for training neural networks. In: 2011 10th Mexican International Conference on Artificial Intelligence, pp. 83–87. IEEE (2011)
6. Grimaldi, E.A., Grimaccia, F., Mussetta, M., Zich, R.: PSO as an effective learning algorithm for neural network applications. In: Proceedings. ICCEA 2004. 2004 3rd International Conference on Computational Electromagnetics and Its Applications, 2004, pp. 557–560. IEEE (2004)
7. Gu, S., Holly, E., Lillicrap, T., Levine, S.: Deep reinforcement learning for robotic manipulation with asynchronous off-policy updates. In: 2017 IEEE International Conference on Robotics and Automation (ICRA), pp. 3389–3396. IEEE (2017)
8. Gudise, V.G., Venayagamoorthy, G.K.: Comparison of particle swarm optimization and backpropagation as training algorithms for neural networks. In: Proceedings of the 2003 IEEE Swarm Intelligence Symposium. SIS 2003 (Cat. No. 03EX706), pp. 110–117. IEEE (2003)

9. Hansen, N.: Injecting external solutions into cma-es. arXiv preprint arXiv:1110.4181 (2011)
10. Hansen, N., Ostermeier, A.: Completely derandomized self-adaptation in evolution strategies. Evol. Comput. **9**(2), 159–195 (2001)
11. Huang, P.C., Lehman, J., Mok, A.K., Miikkulainen, R., Sentis, L.: Grasping novel objects with a dexterous robotic hand through neuroevolution. In: 2014 IEEE Symposium on Computational Intelligence in Control and Automation (CICA), pp. 1–8. IEEE (2014)
12. Hussein, A., Gaber, M.M., Elyan, E., Jayne, C.: Imitation learning: a survey of learning methods. ACM Comput. Surv. (CSUR) **50**(2), 1–35 (2017)
13. Igel, C.: Neuroevolution for reinforcement learning using evolution strategies. In: The 2003 Congress on Evolutionary Computation, 2003. CEC 2003, vol. 4, pp. 2588–2595. IEEE (2003)
14. Ilonen, J., Kamarainen, J.K., Lampinen, J.: Differential evolution training algorithm for feed-forward neural networks. Neural Process. Lett. **17**(1), 93–105 (2003)
15. James, S., Freese, M., Davison, A.J.: Pyrep: Bringing v-rep to deep robot learning. arXiv preprint arXiv:1906.11176 (2019)
16. Karpov, I.V., Valsalam, V.K., Miikkulainen, R.: Human-assisted neuroevolution through shaping, advice and examples. In: Proceedings of the 13th Annual Conference on Genetic and Evolutionary Computation, pp. 371–378 (2011)
17. Kennedy, J., Eberhart, R.: Particle swarm optimization. In: Proceedings of ICNN 1995-International Conference on Neural Networks, vol. 4, pp. 1942–1948. IEEE (1995)
18. Mason, K., Duggan, J., Howley, E.: Forecasting energy demand, wind generation and carbon dioxide emissions in Ireland using evolutionary neural networks. Energy **155**, 705–720 (2018)
19. Mason, K., Duggan, M., Barrett, E., Duggan, J., Howley, E.: Predicting host CPU utilization in the cloud using evolutionary neural networks. Futur. Gener. Comput. Syst. **86**, 162–173 (2018)
20. Mason, K., Grijalva, S.: Building HVAC control via neural networks and natural evolution strategies. In: 2021 IEEE Congress on Evolutionary Computation (CEC), pp. 2483–2490. IEEE (2021)
21. Moriarty, D.E., Miikkulainen, R.: Evolving obstacle avoidance behavior in a robot arm. In: From animals to animats 4: Proceedings of the Fourth International Conference on Simulation of Adaptive Behavior, vol. 4, p. 468. MIT Press (1996)
22. Rohmer, E., Singh, S.P., Freese, M.: V-rep: a versatile and scalable robot simulation framework. In: 2013 IEEE/RSJ International Conference on Intelligent Robots and Systems, pp. 1321–1326. IEEE (2013)
23. Salimans, T., Ho, J., Chen, X., Sidor, S., Sutskever, I.: Evolution strategies as a scalable alternative to reinforcement learning. arXiv preprint arXiv:1703.03864 (2017)
24. Slowik, A., Bialko, M.: Training of artificial neural networks using differential evolution algorithm. In: 2008 Conference on Human System Interactions, pp. 60–65. IEEE (2008)
25. Spong, M.W., Hutchinson, S., Vidyasagar, M., et al.: Robot Modeling and Control, vol. 3. Wiley, New York (2006)
26. Storn, R., Price, K.: Differential evolution-a simple and efficient heuristic for global optimization over continuous spaces. J. Global Optim. **11**(4), 341–359 (1997)
27. Such, F.P., Madhavan, V., Conti, E., Lehman, J., Stanley, K.O., Clune, J.: Deep neuroevolution: Genetic algorithms are a competitive alternative for training deep neural networks for reinforcement learning. arXiv preprint arXiv:1712.06567 (2017)

28. Wen, R., Guo, Z., Zhao, T., Ma, X., Wang, Q., Wu, Z.: Neuroevolution of augment-ing topologies based musculor-skeletal arm neurocontroller. In: 2017 IEEE Inter-national Instrumentation and Measurement Technology Conference (I2MTC), pp. 1–6. IEEE (2017)

29. Zhang, C., Shao, H., Li, Y.: Particle swarm optimisation for evolving artificial neural network. In: SMC 2000 Conference Proceedings. 2000 IEEE International Conference on Systems, Man and Cybernetics'.Cybernetics Evolving to Systems, Humans, Organizations, and Their Complex Interactions'(cat. no. 0. vol. 4, pp. 2487–2490. IEEE (2000)

Centroid-Based Differential Evolution with Composite Trial Vector Generation Strategies for Neural Network Training

Sahar Rahmani[1], Seyed Jalaleddin Mousavirad[2(✉)], Mohammed El-Abd[3],
Gerald Schaefer[4], and Diego Oliva[5]

[1] Department of Computer Engineering, Amirkabir University of Technology,
Tehran, Iran
[2] Universidade da Beira Interior, Covilhã, Portugal
jalalmoosavirad@gmail.com
[3] College of Engineering and Applied Sciences, American University of Kuwait,
Salmiya, Kuwait
melabd@auk.edu.kw
[4] Department of Computer Science, Loughborough University, Loughborough, UK
gerald.schaefer@ieee.org
[5] Depto. de Innovación Basada en la Información y el Conocimiento,
Universidad de Guadalajara, CUCEI, Guadalajara, Mexico
diego.oliva@cucei.udg.mx

Abstract. The learning process of feedforward neural networks, which determines suitable connection weights and biases, is a challenging machine learning problems and significantly impact how well neural networks work. Back-propagation, a gradient descent-based method, is one of the most popular learning algorithms, but tends to get stuck in local optima. Differential evolution (DE), a popular population-based meta-heuristic algorithm, is an interesting alternative for tackling challenging optimisation problems. In this paper, we present Cen-CoDE, a centroid-based differential evolution algorithm with composite trial vector generation strategies and control parameters to train neural networks. Our algorithm encodes weights and biases into a candidate solution, employs a centroid-based strategy in three different ways to generate different trial vectors, while the objective function is based on classification error. In our experiments, we show Cen-CoDE to outperform other contemporary techniques.

Keywords: feedforward neural network · learning · classification · differential evolution

1 Introduction

In the domains of artificial intelligence, machine learning, and deep learning, neural networks enable computer programs to discover patterns and solve various

J. Correia et al. (Eds.): EvoApplications 2023, LNCS 13989, pp. 608–622, 2023.
https://doi.org/10.1007/978-3-031-30229-9_39

problems. Artificial neural networks (ANNs) have been extensively used to tackle complex problems in medicine [16], finance [38], food quality evaluation [32], and business [7], among others. The human brain serves as the inspiration for ANNs, which have demonstrated good performance in both supervised and unsupervised learning.

Feedforward neural networks (FFNNs) are the most widely employed type of ANNs. Finding the best weights and biases is the goal of learning in FFNNs in order to reduce the differences between expected and actual outputs. Among the most popular learning algorithms are those based on gradient descent, including back-propagation (BP). These however also suffer from certain issues including sensitivity to starting weights and a tendency to get stuck in local optima [19,29].

Population-based metaheuristic algorithms (PBMHs), such as differential evolution (DE) [37], particle swarm optimisation (PSO) [36], genetic algorithm (GA) [40], and artificial bee colony algorithm (ABC) [12] have been shown to be competitive alternatives to gradient descent-based algorithms and can increase convergence speed and reduce the likelihood of encountering local optima.

DE is a well-established PBMH that has been demonstrated to perform effectively and efficiently on a wide range of issues from various domains [10,25]. To guide its population toward the global optimum, it employs mutation, crossover, and selection operators in every generation [39]. Several DE algorithms have been presented for FFNN learning. A DE algorithm with a modified optimal mutation operator is suggested in [14]. Studies on three clinical classification datasets confirm the effectiveness of the proposed approach. To identify the best connections in FFNNs, [4] introduces the adaptive Cauchy DE (ACDE) and self-adaptive DE (SaDE) and show the proposed algorithm to be better than standard DE. A novel training technique based on quasi-opposition-based learning is proposed in [28], demonstrating better performance in classification applications, while [30] proposes an RDE-OP algorithm, which combines a region-based DE algorithm with quasi-opposition-based learning. The present population is divided using a clustering approach that RDE-OP uses to create crossover operators at each cluster centre. Other DE-based training algorithms include [20,24,26].

One of the state-of-the-art DE algorithms, introduced in [39], is CoDE, which enhances DE by mixing several trial vector generation algorithms and with various control parameter settings. In particular, CoDE generates trial vectors by a random combination of three well-studied trial vector generation schemes and three control parameter settings.

The technique of centroid-based sampling can be employed to enhance PBMH algorithms [23,27]. [33] demonstrates that the likelihood of individuals being nearer an unknown solution is higher toward the centre of the search space compared to randomly-located candidate solutions. A center-based DE is proposed in [9] as a method to solve complex problems. They use the base vector in the DE's mutation operator, which is the centre of the points. By replacing the base vector with a centre point, a similar method is used to enhance the SHADE (success-history based parameter adaptation) algorithm [8]. [15] uses centroid-based sampling for the initialisation of a cooperative co-evolutionary algorithm and develops three schemes: centre-based normal distribution sampling (CNS), central golden region (CGR), and hybrid random-centre normal distribution sampling (HRCN). Centre-based Latin hypercube sampling is used in [23] to address deception in optimisation problems.

In this paper, we present a novel DE-based training algorithm for FNNS. The two main contributions of the paper are: (1) a novel strategy based on the CoDE algorithm is proposed to find optimal weights in a neural network, and (2) a novel centre-based strategy for CoDE, Cen-CoDE, is introduced to improve the results. Since CoDE employs three distinct generation strategies, Cen-CoDE incorporates its centroid-based scheme in three different ways into the CoDE algorithm. Cen-CoDe uses a vector-based representation, while classification error is used as objective function. Our extensive experiments indicate that Cen-CoDE supports excellent NN-based pattern classification, outperforming various other techniques.

The remainder of the paper is organised as follows. Some background is provided in Sect. 2, while Sect. 3 describes our proposed algorithm in detail. Section 4 presents experimental findings, and Sect. 5 concludes the paper.

2   Background

2.1   Feedforward Neural Networks

A feedforward neural network, as a non-parametric machine learning algorithm, is defined by a set of simple nodes called neurons. FFNNs are supervised artificial neural networks where information moves in only one direction -forward-from the input nodes, through the hidden nodes, and to the output nodes. Each link in the network has a weight that represents the strength of the connection between two nodes. In this paper, we use the sigmoid function

$$\delta(net) = \frac{1}{1 + e^{-net}},\tag{1}$$

as activation function to obtain a neuron's output, where net refers to the input signal. The performance of FFNNs is heavily influenced by the weights. As a result, one of the essential tasks in FFNNs is determining appropriate weights during the example-based training phase.

2.2   Composite Differential Evolution

CoDE (differential evolution with composite trial vector generation strategies and control parameters) [39] is a DE algorithm that has shown considerable performance in solving complex optimisation problems. CoDE generates new trial vectors by combining various trial vector generation algorithms at random with various control parameter values. To create the pool of candidates, CoDE uses three trial vector generation strategies:

– **rand/1/bin** :

$$v_i = x_{r1} + F(x_{r2} - x_{r3})\tag{2}$$

and

$$u_{i,j} = \begin{cases} v_{i,j}, & rand(0,1) \leq CR \; or \; j == j_{rand} \\ x_{i,j}, & \text{otherwise} \end{cases}, \tag{3}$$

where x_{r1}, x_{r2}, and x_{r3} are three different randomly selected individuals, and F is a scaling factor.

- **rand/2/bin**

$$v_i = X_{r1} + F(X_{r2} - X_{r3}) + F(X_{r4} - X_{r5}) \tag{4}$$

and

$$u_{i,j} = \begin{cases} v_{i,j}, & rand(0,1) \leq CR \; or \; j == j_{rand} \\ x_{i,j}, & \text{otherwise} \end{cases}, \tag{5}$$

where x_{r1}, x_{r2}, x_{r3}, x_{r4}, and x_{r5} are five different randomly selected candidate solutions from the current population.

- **current-to-rand/1**

$$\vec{u}_i = X_i + rand(X_{r1} - X_i) + F(X_{r2} - X_{r3}) \tag{6}$$

which does not use binomial crossover.

Three control parameter settings are chosen in a parameter candidate pool as $[F= 1.0, C_r= 0.1]$, $[F= 1.0, C_r= 0.9]$, and $[F= 0.8, C_r= 0.2]$, respectively.

Each generation uses a different trial vector generation method from the strategy candidate pool along with a control parameter setting selected at random from the parameter candidate pool. Consequently, for each target vector, three trial vectors are created. The best one enters the next generation if it is better than its target vector.

2.3 Centroid-Based Sampling Strategy

The idea of centre-based sampling is first proposed in [33], which compares the likelihood of randomly generated points and centre points being close to an unknown solution, and finds that the likelihood of points being close to an unknown solution is significantly higher in the centre of the search space.

For a search space in the interval of $[a, b]$, the centre is defined as

$$c_i = \frac{a_i + b_i}{2}, \tag{7}$$

where $i=1,\ldots, D$ and D defines the search space dimensions.

3 Proposed Cen-CoDE Algorithm

Our proposed Cen-CoDE algorithm combines a centroid-based approach with differential evolution, composite trial vector generating strategies and control parameters. It is a combination of a centroid-based strategy and the CoDE algorithm. In the following, we first describe the main components of Cen-CoDE and then explain how these fit together to form the algorithm.

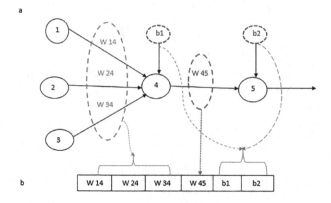

Fig. 1. (a) An FFNN with 3-1-1 architecture. (b) the corresponding representation.

3.1 Representation

We encode the connection weights and biases using a real-valued representation. The length of each candidate solution is the sum of all connections and biases. Figure 1 shows a typical example to illustrate this representation.

3.2 Objective Function

In this paper, we perform pattern classification, and therefore use an objective function based on classification error, defined as

$$E = 100 \sum_{p=1}^{P} \xi(\sum_{n=1}^{P} \overrightarrow{p}/P), \tag{8}$$

with

$$\xi(\overrightarrow{p}) = \begin{cases} 1 & \text{if } \overrightarrow{o_p} \neq \overrightarrow{d_p} \\ 0 & \text{otherwise} \end{cases}, \tag{9}$$

where for each P-dimensional input vector in the training data, the corresponding desired output is $\overrightarrow{d_p}$, while $\overrightarrow{o_p}$ is output predicted by the network. Finding optimal connection weights and biases in an FFNN is the main objective of learning to reduce classification error.

3.3 Centroid-Based CoDE

In our proposed algorithm, in each iteration, the average of the N best candidate solutions is calculated as

$$\overrightarrow{x_{center}} = \frac{\overrightarrow{x_{b1}} + \cdots + \overrightarrow{x_{bi}} + \cdots + \overrightarrow{x_{bN}}}{N}, \tag{10}$$

where $\overrightarrow{x_{bi}}$ is the i-th best candidate solution. The centre of a 1-D problem is shown in Fig. 2 using an example with 5 candidate solutions (circular points). For $N = 3$, the three best candidate solutions are $B = \{\overrightarrow{x_2}, \overrightarrow{x_3}, \overrightarrow{x_4}\}$, their corresponding locations $P = \{3, 5, 8\}$, and the centre-based candidate solution $\overrightarrow{x_{center}} = 5.33$. Since CoDE uses three generation strategies, Cen-CoDE employs the centroid-based strategy in three ways as follows:

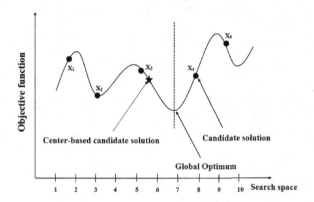

Fig. 2. Example of center-based candidate solution for a 1-D problem.

– **centre/rand/1/bin** :

$$v_i = x_{mean} + F(x_{r1} - x_{r2}) \tag{11}$$

and

$$u_{i,j} = \begin{cases} v_{i,j}, & rand(0,1) \leq CR \, or \, j == j_{rand} \\ x_{i,j}, & \text{otherwise} \end{cases} \tag{12}$$

– **centre/rand/2/bin**

$$v_i = X_{mean} + F(X_{r1} - X_{r2} + F(X_{r3} - X_{r4}), \tag{13}$$

and

$$u_{i,j} = \begin{cases} v_{i,j}, & rand(0,1) \leq CR \, or \, j == j_{rand} \\ x_{i,j}, & \text{otherwise.} \end{cases} \tag{14}$$

– **centre/Current-to-rand/1**

$$\vec{U}_i = X_i + rand(X_{mean} - X_i) + F(X_{r1} - X_{r2}) \tag{15}$$

where X_{mean} is the average of the N best candidate solutions.

Cen-CoDE generates new trial vectors by combining the centroid-based generation strategies with various control parameters.

3.4 Algorithm

Our proposed Cen-CoDE algorithm, is presented, in the form of pseudo-code, in Algorithm 1.

Input : N_P: population size.

Max_{FEX}: maximum number of function evaluations.

Strategy candidate pool:"rand/1/bin","rand/2/bin", and "current-to-rand/1".

Parameter candidate pool: $[F = 1.0, C_r = 0.1]$, $[F = 1.0, C_r = 0.9]$, and $[F = 0.8, C_r = 0.2]$.

Output: Individual with the best objective function value in the population.

$G = 0$;

Generate initial population $P_0 = \{x_{1,0}, \ldots, x_{NP,0}\}$ by uniformly and randomly sampling from feasible solution space;

Evaluate objective function values $F(x_{1,0}), \ldots, F(x_{NP,0})$ using Eq. (8);

$FES = N_P$;

while $FES < Max_{FES}$ **do**

$\quad P_{G+1} = \emptyset$;

$\quad$ **for** $i \leftarrow 1$ **to** NP **do**

$\qquad$ Select N best candidate solutions as $\overrightarrow{x_{b1}}, \overrightarrow{x_{b2}}, \ldots, \overrightarrow{x_{bN}}$;

$\qquad x_{mean} = \frac{\overrightarrow{x_{b1}} + \overrightarrow{x_{b2}} + \cdots + \overrightarrow{x_{bN}}}{N}$;

$\qquad$ Use the three trial vector generation strategies, each with a control parameter setting randomly selected from the parameter candidate pool, to generate three trial vectors u_{i1}, u_{i2}, and u_{i3} for the target vector x_i using Eqs. (11) to (15);

$\qquad$ Evaluate objective function values of u_{i1}, u_{i2}, and u_{i3} using Eq. (8);

$\qquad$ Choose u_i as best trial vector from u_{i1}, u_{i2}, and u_{i3};

$\qquad P_{G+1} = P_{G+1} \cup \text{Select}(x_i, u_i)$;

$\qquad FES = FES + 3$;

$\quad$ **end**

$\quad G = G + 1$;

end

Algorithm 1: Cen-CoDE algorithm.

4 Experimental Results

For our experiments, we use several complex benchmark classification datasets from different domains, and compare our Cen-CoDE algorithm to 14 other algorithms. Table 1 lists the main characteristics of the used datasets. We perform k-fold cross-validation, with $k = 10$, where each dataset is randomly divided into k folds, with one fold acting as the test set and the remaining $k - 1$ folds as training data, and repeating the process so that each fold is used as test set. Each connection weight and bias has lower and upper bound of -10 and 10, respectively. Since here we do not focus on the structure of neural networks, we use the recommendation from [21,22,24] to define the number of neurons in the (single) hidden layer as $2 \times N + 1$, where N is the total number of features in the dataset.

Table 1. Main characteristics of the datasets.

	# instances	# features	# classes
Iris	150	4	3
Breast Cancer	699	9	2
Liver	345	7	2
Pima	768	2	2
Seed	210	7	3
Vertebral	310	6	3

We compare our proposed Cen-CoDE algorithm with both conventional and PBMH algorithms. Conventional algorithms include gradient descent with momentum backpropagation (GDM) [31], gradient descent with adaptive learning rate backpropagation (GDA) [3], gradient descent with momentum and adaptive learning rate backpropagation (GDMA) [35], resilient backpropagation (RP) [34], scaled conjugate gradient backpropagation (SCG) [18], and Bayesian regularisation backpropagation(BR) [5]. The PBMHs used for comparison include particle swarm optimisation (PSO) [6], artificial bee colony (ABC) [11], grey wolf optimiser [17], dragonfly algorithm (DA) [13], sine cosine algorithm (SCA) [2], whale optimisation algorithm (WOA) [1], DE, and cuckoo search [41]. Also, a comparison between CoDE and Cen-CoDE is performed to assess the proposed centre-based strategy. All PBMH algorithms have a population size and a function evaluation count of 50 and 25.000, respectively. For the other algorithms, we employ default parameters values from the cited publications.

The Iris dataset which classifies samples into 3 types (setosa, versicolor, and virginica), is evaluated on networks of structure 4-9-3. The obtained results, given in Table 2, show that Cen-CoDE outperformes the other methods, yielding the best classification accuracy of 98.00 with a low standard deviation of 3.22. CoDE is ranked second, followed by SCG, BR, and PSO.

The well-known Breast Cancer dataset contains 699 samples and 9 features, leading to networks of structure 9-19-1 with 210 decision variables, From the results in Table 3, we can see that CenCoDE again performs best, followed by CoDE and GWO (tied).

The Liver Disorders Dataset is a clinical dataset with 345 samples divided into 2 labels and 7 features. The resulting 7-15-1 networks thus comprise 105 decision variables. According to Table 4, CenCoDE is top ranked with an accuracy rate of 75.1, while the second- and third-ranked algorithms are CoDE and PSO, respectively. Again, we can see a significant improvement over CoDE due to the proposed centre-based approach.

Table 2. Classification results for Iris dataset.

	mean	std.dev.	rank
GDM	92.00	8.20	10.5
GDA	94.67	5.26	7
GDMA	82.67	16.98	15
RP	95.33	5.49	6
SCG	96.00	8.43	4
BR	96.00	7.17	4
PSO	96.00	5.62	4
ABC	84.67	9.45	14
GWO	93.33	4.44	8
DA	92.67	5.84	9
SCA	90.67	7.83	12
WOA	87.33	8.58	13
DE	92.00	5.26	10.5
CS	55.00	7.73	16
CoDE	97.10	7.73	2
Cen-CoDE	98.00	3.22	1

Table 3. Classification results for Breast Cancer dataset.

	mean	std.dev.	rank
GDM	92.99	7.6	14
GDA	95.9	2.04	13
GDMA	90.47	5.94	16
RP	96.05	2.47	12
SCG	96.63	2.09	11
BR	91.81	3.82	15
PSO	97.95	1.72	4.5
ABC	97.95	1.03	4.5
GWO	98.10	1.39	2.5
DA	97.51	1.83	6
SCA	97.08	1.82	8
WOA	97.07	1.96	9
DE	97.36	2.06	7
CS	97.00	1.58	10
CoDE	98.10	1.25	2.5
Cen-CoDE	98.38	1.61	1

Table 4. Classification results for Liver dataset.

	mean	std.dev.	rank
GDM	59.47	15.05	15
GDA	58.24	6.53	16
GDMA	60.66	13.99	14
RP	65.82	5.21	10
SCG	66.97	9.6	9
BR	70.71	6.93	6
PSO	73.36	6.28	3
ABC	70.75	6.47	5
GWO	73.01	9.74	4
DA	70.42	7.01	7
SCA	65.5	5.96	11
WOA	62.87	6.4	12
DE	67.81	8.21	8
CS	61.4	7.89	13
CoDE	74.4	6.92	2
Cen-CoDE	75.1	6.66	1

One of the most difficult datasets is the Pima Indian Diabetes dataset for which networks of structure 9-19-2 are trained, thus involving 171 decision variables. The results are shown in Table 5. As we can see from there, SCG gives the best results here, followed by ABC and CenCoDE, which again outperforms CoDE.

The Seed dataset contains 210 samples of seven geometrical features of wheat kernels, including compactness, perimeter, and area. From the classification results in Table 6, BR shows the best performance with an accuracy of 90.95, while Cen-CoDE, CoDE, GDA, and SCG are all tied in the next place.

The Vertebral dataset categorises orthopaedic patients into three groups based on six bio-mechanical parameters. We solve for 105 decision variables based on 13 neurons in the hidden layer and report the results in Table 7. Here, PSO yields the best performance, followed by Cen-CoDE, CoDE and DE which are all tied.

Table 8 compares the algorithms based on all their rankings in a single table. As can be seen, Cen-CoDE archieves the best average rank, confirming very competitive performance of our proposed algorithm. By a wide margin, CoDE ranks second, indicating the notable capability of the proposed centre-based strategy. Also comparing the results of DE and CoDE, with mean ranks of 9.42 and 5.67, respectively, we can conclude that the composite trial vector strategy can improve the performance of neural network training. GDM, GDMA and CS perform worst among the algorithms.

Table 5. Classification results for Pima dataset.

	mean	std.dev.	rank
GDM	67.98	13.03	15
GDA	75.91	4.52	9
GDMA	73.2	6.74	13
RP	76.83	4.5	8
SCG	78.52	3.05	1
BR	75.89	5.37	10
PSO	77.6	3.24	6
ABC	78.26	4.45	2
GWO	67.45	2.79	16
DA	77.85	5.4	5
SCA	74.47	4.2	12
WOA	76.95	3.65	7
DE	70.94	4.97	14
CS	75.78	3.08	11
CoDE	77.90	4.24	4
Cen-CoDE	77.99	4.12	3

Table 6. Classification results for Pima dataset.

	mean	std.dev.	rank
GDM	47.62	29.95	16
GDA	82.38	6.75	3.5
GDMA	80.00	16.47	7
RP	80.48	9.38	6
SCG	82.38	9.54	3.5
BR	90.95	7.6	1
PSO	78.1	11.92	8.5
ABC	72.38	8.03	10
GWO	78.1	10.09	8.5
DA	70.48	7.03	12.5
SCA	71.43	8.98	11
WOA	70.48	8.92	12.5
DE	70.00	11.01	14
CS	62.85	4.37	15
CoDE	82.38	8.33	3.5
Cen-CoDE	82.38	8.1	3.5

Table 7. Classification results for Pima dataset.

	mean	std.dev.	rank
GDM	76.13	9.15	14
GDA	80.65	5.27	11
GDMA	72.9	17.62	15
RP	78.39	5.05	13
SCG	82.26	8.5	7.5
BR	84.52	6.94	5
PSO	86.45	8.02	1
ABC	82.9	5.7	6
GWO	81.94	7.93	9.5
DA	81.94	5.31	9.5
SCA	82.26	10.67	7.5
WOA	79.03	10.09	12
DE	85.16	5.31	3
CS	65.81	6.12	16
CoDE	85.16	6.51	3
Cen-CoDE	85.16	6.48	3

Table 8. Rankings of all algorithms over all datasets.

	Iris	Cancer	Liver	Pima	Seed	Vertebral	mean rank	overall rank
GDM	10.5	14	15	15	16	14	14.08	16
GDA	7	13	16	9	3.5	11	9.97	11
GDMA	16	15	14	13	7	15	13.33	14
RP	6	12	10	8	6	13	9.17	9
SCG	4	11	9	1	3.5	7.5	6.00	4
BR	4	15	6	10	1	5	7.83	6
PSO	4	4.5	3	6	8.5	1	4.50	2
ABC	14	4.5	5	2	10	6	6.92	5
GWO	8	2.5	4	16	8.5	9.5	8.08	7
DA	9	6	7	5	12.5	9.5	8.17	8
SCA	12	8	11	12	11	6.5	10.08	12
WOA	13	9	12	7	12.5	12	10.92	13
DE	10.5	7	8	14	14	3	9.42	10
CS	16	10	13	11	15	16	13.50	15
CoDE	2	2.5	2	4	3.5	3	5.67	3
Cen-CoDE	1	1	1	3	3.5	3	2.08	1

5   Conclusions

Finding optimal values for connection weights and biases is a challenging task in neural network training. The most popular learning algorithms use gradient-based methods and thus suffer from drawbacks of getting stuck in a local minimum. In this paper, we have proposed a differential evolution-based algorithm for network training. In particular, we combine with the CoDE algorithm with a centroid-based strategy to generate new candidate solutions in three ways. The effectiveness of our proposed Cen-CoDE algorithm is confirmed by experimental results on a variety of difficult classification problems, on which it is shown to outperform a variety of other techniques. In future work, we intend to expand this work to other network types and to simultaneously solve for network structure and weights.

References

1. Aljarah, I., Faris, H., Mirjalili, S.: Optimizing connection weights in neural networks using the whale optimization algorithm. Soft. Comput. **22**(1), 1–15 (2018)
2. Bairathi, D., Gopalani, D.: Salp Swarm Algorithm (SSA) for Training Feed-Forward Neural Networks. In: Bansal, J.C., Das, K.N., Nagar, A., Deep, K., Ojha, A.K. (eds.) Soft Computing for Problem Solving. AISC, vol. 816, pp. 521–534. Springer, Singapore (2019). https://doi.org/10.1007/978-981-13-1592-3_41
3. Beale, H.D., Demuth, H.B., Hagan, M.: Neural Network Design. PWS, Boston (1996)
4. Choi, T.J., Ahn, C.W.: Adaptive Cauchy differential evolution with strategy adaptation and its application to training large-scale artificial neural networks. In: He, C., Mo, H., Pan, L., Zhao, Y. (eds.) BIC-TA 2017. CCIS, vol. 791, pp. 502–510. Springer, Singapore (2017). https://doi.org/10.1007/978-981-10-7179-9_39
5. Foresee, F.D., Hagan, M.T.: Gauss-Newton approximation to Bayesian learning. In: International Conference on Neural Networks, vol. 3, pp. 1930–1935 (1997)
6. Gudise, V.G., Venayagamoorthy, G.K.: Comparison of particle swarm optimization and backpropagation as training algorithms for neural networks. In: IEEE Swarm Intelligence Symposium, pp. 110–117 (2003)
7. Heo, W., Lee, J.M., Park, N., Grable, J.E.: Using artificial neural network techniques to improve the description and prediction of household financial ratios. J. Behav. Exp. Financ. **25**, 100273 (2020)
8. Hiba, H., El-Abd, M., Rahnamayan, S.: Improving SHADE with center-based mutation for large-scale optimization. In: IEEE Congress on Evolutionary Computation, pp. 1533–1540 (2019)
9. Hiba, H., Mahdavi, S., Rahnamayan, S.: Differential evolution with center-based mutation for large-scale optimization. In: IEEE Symposium Series on Computational Intelligence, pp. 1–8 (2017)
10. Jalaleddin, M.S., Shahryar, R., Gerald, S.: Many-level image thresholding using a center-based differential evolution algorithm. In: Congress on Evolutionary Computation (CEC). IEEE (2020)
11. Karaboga, D., Akay, B., Ozturk, C.: Artificial bee colony (ABC) optimization algorithm for training feed-forward neural networks. In: International Conference on Modeling Decisions for Artificial Intelligence, pp. 318–329 (2007)

12. Karaboga, D., Basturk, B.: A powerful and efficient algorithm for numerical function optimization: artificial bee colony (ABC) algorithm. J. Global Optim. **39**(3), 459–471 (2007)
13. Khishe, M., Safari, A.: Classification of sonar targets using an MLP neural network trained by dragonfly algorithm. Wireless Pers. Commun. **108**(4), 2241–2260 (2019)
14. Leema, N., Nehemiah, H.K., Kannan, A.: Neural network classifier optimization using differential evolution with global information and back propagation algorithm for clinical datasets. Appl. Soft Comput. **49**, 834–844 (2016)
15. Mahdavi, S., Rahnamayan, S., Deb, K.: Center-based initialization of cooperative co-evolutionary algorithm for large-scale optimization. In: IEEE Congress on Evolutionary Computation, pp. 3557–3565 (2016)
16. Minnema, J., van Eijnatten, M., Kouw, W., Diblen, F., Mendrik, A., Wolff, J.: Ct image segmentation of bone for medical additive manufacturing using a convolutional neural network. Comput. Biol. Med. **103**, 130–139 (2018)
17. Mirjalili, S.: How effective is the grey wolf optimizer in training multi-layer perceptrons. Appl. Intell. **43**(1), 150–161 (2015)
18. Møller, M.F.: A scaled conjugate gradient algorithm for fast supervised learning. Aarhus University, Computer Science Department (1990)
19. Moravvej, S.V., Mousavirad, S.J., Moghadam, M.H., Saadatmand, M.: An LSTM-based plagiarism detection via attention mechanism and a population-based approach for pre-training parameters with imbalanced classes. In: Mantoro, T., Lee, M., Ayu, M.A., Wong, K.W., Hidayanto, A.N. (eds.) ICONIP 2021. LNCS, vol. 13110, pp. 690–701. Springer, Cham (2021). https://doi.org/10.1007/978-3-030-92238-2_57
20. Moravvej, S.V., Mousavirad, S.J., Oliva, D., Schaefer, G., Sobhaninia, Z.: An improved de algorithm to optimise the learning process of a Bert-based plagiarism detection model. In: 2022 IEEE Congress on Evolutionary Computation (CEC), pp. 1–7. IEEE (2022)
21. Mousavirad, S.J., Bidgoli, A.A., Ebrahimpour-Komleh, H., Schaefer, G., Korovin, I.: An effective hybrid approach for optimising the learning process of multi-layer neural networks. In: International Symposium on Neural Networks, pp. 309–317 (2019)
22. Mousavirad, S.J., Bidgoli, A.A., Komleh, H.E., Schaefer, G.: A memetic imperialist competitive algorithm with chaotic maps for multi-layer neural network training. Int. J. Bio-Inspired Comput. **14**(4), 227–236 (2019)
23. Mousavirad, S.J., Bidgoli, A.A., Rahnamayan, S.: Tackling deceptive optimization problems using opposition-based de with center-based Latin hypercube initialization. In: 14th International Conference on Computer Science & Education, pp. 394–400 (2019)
24. Mousavirad, S.J., Gandomi, A.H., Homayoun, H.: A clustering-based differential evolution boosted by a regularisation-based objective function and a local refinement for neural network training. In: IEEE Congress on Evolutionary Computation, pp. 1–8 (2022)
25. Mousavirad, S.J., Oliva, D., Chakrabortty, R.K., Zabihzadeh, D., Hinojosa, S.: Population-based self-adaptive generalised masi entropy for image segmentation: A novel representation. Knowl.-Based Syst. **245**, 108610 (2022)
26. Mousavirad, S.J., Oliva, D., Hinojosa, S., Schaefer, G.: Differential evolution-based neural network training incorporating a centroid-based strategy and dynamic opposition-based learning. In: IEEE Congress on Evolutionary Computation, pp. 1233–1240 (2021)

27. Mousavirad, S.J., Rahnamayan, S.: CenPSO: a novel center-based particle swarm optimization algorithm for large-scale optimization. In: IEEE International Conference on Systems, Man, and Cybernetics, pp. 2066–2071 (2020)
28. Mousavirad, S.J., Rahnamayan, S.: Evolving feedforward neural networks using a quasi-opposition-based differential evolution for data classification. In: IEEE Symposium Series on Computational Intelligence, pp. 2320–2326 (2020)
29. Mousavirad, S.J., Schaefer, G., Jalali, S.M.J., Korovin, I.: A benchmark of recent population-based metaheuristic algorithms for multi-layer neural network training. In: Genetic and Evolutionary Computation Conference Companion, pp. 1402–1408 (2020)
30. Mousavirad, S.J., Schaefer, G., Korovin, I., Oliva, D.: RDE-OP: A region-based differential evolution algorithm incorporation opposition-based learning for optimising the learning process of multi-layer neural networks. In: International Conference on the Applications of Evolutionary Computation, pp. 407–420 (2021)
31. Phansalkar, V., Sastry, P.: Analysis of the back-propagation algorithm with momentum. IEEE Trans. Neural Networks 5(3), 505–506 (1994)
32. Rad, S.J.M., Tab, F.A., Mollazade, K.: Classification of rice varieties using optimal color and texture features and BP neural networks. In: 7th Iranian Conference on Machine Vision and Image Processing, pp. 1–5 (2011)
33. Rahnamayan, S., Wang, G.G.: Center-based sampling for population-based algorithms. In: IEEE Congress on Evolutionary Computation, pp. 933–938 (2009)
34. Riedmiller, M., Braun, H.: A direct adaptive method for faster backpropagation learning: the RPROP algorithm. In: IEEE International Conference on Neural Networks, pp. 586–591 (1993)
35. Scales, L.: Introduction to Non-linear Optimization. Macmillan International Higher Education (1985)
36. Shi, Y., Eberhart, R.: A modified particle swarm optimizer. In: IEEE International Conference on Evolutionary Computation, pp. 69–73 (1998)
37. Storn, R., Price, K.: Differential evolution-a simple and efficient heuristic for global optimization over continuous spaces. J. Global Optim. 11(4), 341–359 (1997)
38. Tsai, C.F., Wu, J.W.: Using neural network ensembles for bankruptcy prediction and credit scoring. Expert Syst. Appl. 34(4), 2639–2649 (2008)
39. Wang, Y., Cai, Z., Zhang, Q.: Differential evolution with composite trial vector generation strategies and control parameters. IEEE Trans. Evol. Comput. 15(1), 55–66 (2011)
40. Whitley, D.: A genetic algorithm tutorial. Stat. Comput. 4(2), 65–85 (1994)
41. Yi, J.H., Xu, W.H., Chen, Y.T.: Novel back propagation optimization by cuckoo search algorithm. Scientific World J. 2014 (2014)

AMTEA-Based Multi-task Optimisation for Multi-objective Feature Selection in Classification

Jiabin Lin[✉], Qi Chen, Bing Xue, and Mengjie Zhang

School of Engineering and Computer Science, Victoria University of Wellington,
PO Box 600, Wellington 6140, New Zealand
{jiabin,qi.chen,bing.xue,mengjie.zhang}@ecs.vuw.ac.nz

Abstract. Feature selection is important nowadays due to many real-world datasets usually having a large number of features. Evolutionary multi-objective optimisation algorithms have been successfully used for feature selection which usually has two conflicting objectives, i.e., maximising the classification accuracy and minimising the number of selected features. However, most of the existing evolutionary multi-objective feature selection algorithms tend to address feature selection tasks independently, even when these feature selection tasks are related. Multi-task optimisation, which aims to improve the performance of multiple tasks by sharing common knowledge among them, has been used in many areas. However, there is not much work on utilising multi-task optimisation for feature selection. In this work, we develop a new multi-task multi-objective feature selection algorithm. This algorithm aims to address multiple related feature selection tasks simultaneously and facilitate knowledge capturing and transferring among the related tasks. Furthermore, a method is developed for transferring knowledge between related feature selection tasks having different features. This method can avoid transferring information between the unique features of tasks by transforming the probability models of them. We compare the proposed algorithm with the single-task multi-objective feature selection algorithm on seven sets of related feature selection tasks. Experimental results show that the proposed algorithm achieves better classification performance than the single-task algorithm with the help of knowledge transferring among related feature selection tasks. Further analysis shows that the features selected by our proposed algorithm can be more relevant to the classification tasks.

Keywords: Feature selection · Multi-task optimisation ·
Multi-objective optimisation · Evolutionary computation

1 Introduction

Many real-world datasets contain a large number of features, which can be a challenge for classification. Feature selection which removes irrelevant and redundant features, can improve classification performance and speed up the learning

J. Correia et al. (Eds.): EvoApplications 2023, LNCS 13989, pp. 623–639, 2023.
https://doi.org/10.1007/978-3-031-30229-9_40

process. Feature selection usually has two objectives, i.e. minimising the number of selected features and minimising the classification error rate. Most existing multi-objective feature selection algorithms tend to address feature selection tasks independently. However, many real-world feature selection tasks are related and have common knowledge. For example, the IMDB movie review dataset and the MR movie review dataset have many common features about movie reviews, which indicates that feature selection on these two datasets are related and have common knowledge. Therefore, sharing common knowledge across the related feature selection tasks can potentially improve their performance.

Multi-task optimisation, which aims to address multiple related learning tasks simultaneously and transfer common knowledge across them has been used in many areas, such as the traveling salesman problem [14] and job-shop scheduling problems [21,22] and symbolic regression problems [20] in recent years. However, the research on multi-task optimisation for feature selection has not been much. In [4,5], multi-task optimisation was used for addressing high-dimensional feature selection tasks. In their work, an assistant feature selection task with a lower dimensionality is generated based on the main feature selection task. During the evolutionary process, knowledge is transferred from the assistant task to the main task. By doing this, the performance of the main task can be improved. One of the limitations of these two algorithms is that they only focus on the performance of the main task. However, multi-task optimisation aims to improve the performance of all the tasks. Furthermore, using existing multi-task optimisation algorithms for feature selection is still challenging. Many existing methods are based on transferring a number of solutions across multiple related tasks [8,11]. The selection of transferred solutions is usually based on learning the relationships between tasks, which is usually achieved by analyzing the relatedness between the solutions of the multiple tasks, which can be challenging for feature selection.

An adaptive model-based transfer-enabled evolutionary algorithm (AMTEA) [6], which is an evolutionary sequential transfer learning, was proposed to address a target task with the help of knowledge learned from the experience of addressing the other tasks from the source domains. In AMTEA, analyzing the relationship between tasks is achieved by analyzing the relationships between the probabilistic distributions of solutions of them, which can be effective and efficient for feature selection. Multi-task optimisation can be treated as conducting multiple evolutionary sequential transfer learning simultaneously. Therefore, in this work, we aim to develop a new multi-task otpimisation algorithm based on AMTEA for feature selection. Different from AMTEA, our new multi-task optimisation algorithm focuses on addressing multiple related feature selection tasks simultaneously rather than addressing only one main task at a time. In AMTEA, an offline knowledge pool containing probabilistic models from the source domain is built for a target task. However, for our multi-task optimisation algorithm, we propose a method for building an online knowledge pool containing probabilistic models for multiple tasks. Although AMTEA can be useful for feature selection, it is challenging for AMTEA to deal with case feature selection tasks that have different features. In AMTEA, the search spaces of the source tasks and target tasks are transformed into a unified search space with the range of [0, 1]. In this way, information may be transferred between unique features of different tasks, which may not be effective for feature selection since the meanings of different features are different. To address this challenge, we propose a strategy for transforming the probabilistic models when the related feature selection tasks have different features.

The main objectives of this paper can be summarised as follows.

- Develop a new multi-task optimisation algorithm for feature selection, which addresses multiple feature selection tasks simultaneously rather than just focusing on the main task at a time.
- Develop a method for building an online knowledge pool that provides knowledge for multiple tasks simultaneously.
- Develop a method for transforming the probabilistic models for the case multiple tasks have different features.
- Examine the effectiveness of the developed algorithm by comparing it with a single-task feature selection algorithm and testing on a set of benchmark datasets having different features.

2   Related Work

2.1   Evolutionary Multi-objective Feature Selection

In recent years, evolutionary multi-objective optimisation has achieved great success in feature selection. Recently, a new multi-objective wrapper method was developed for feature selection in image classification [10]. In their work, new representations and breeding operators are developed to enhance the performance of multi-objective feature selection methods. Wang et al. [17] develop an enhanced multi-objective feature selection algorithm based on the sampling strategy to reduce the computational cost and improve the performance. In their work, K-means clustering based on differential selection and a ladder-like sample utilization strategy is proposed to reduce the size of training samples, improving feature selection's efficiency. In summary, many of these algorithms have shown their effectiveness in solving feature selection tasks. However, they tend to address each feature selection task separately, while many real-world feature selection tasks connect to each other and share common knowledge for problem-solving. It is worth exploring methods to utilize this knowledge to enhance the related feature selection tasks.

2.2   Multi-task Optimisation for Feature Selection

Multi-task optimisation, which aims to address multiple related optimisation tasks simultaneously and facilitate knowledge transfer across them, has been successfully used in many areas in recent years. However, research on multi-task optimisation for feature selection has still been limited. In [4], a multi-task optimisation-based algorithm is proposed to address high-dimensional feature selection problems. In their work, a knee point selection scheme is developed to distinguish promising features from all features. An assisted task is generated by selecting a subset of features from the target task. The target task and the assisted task are addressed simultaneously, and a new crossover operator is proposed to transfer information. It has been shown that the performance on the target task can be significantly improved with the help of information from the assisted task. One of the limitations of their work is that their method only focuses on the performance of the main feature selection task rather than the performance of all the tasks. Although these recently proposed algorithms have shown the potential of multi-task optimisation for feature selection, it is still challenging to use existing algorithms for feature selection. The main reason is that many existing multi-task optimisation algorithms are based on transferring a number of solutions across multiple related tasks [8, 11]. In these existing algorithms, transferring useful solutions across tasks helps achieve positive transfer. The selection of transferred solutions is usually based on learning the relationships between tasks, which is achieved by learning the relatedness between the current populations of these tasks. However, feature selection tasks usually have many features, making it hard to analyze the relationship between tasks in feature selection. Furthermore, the complex interactions between features can also make it challenging to the relationships between features selection tasks.

2.3 Adaptive Model-Based Transfer-Enabled Evolutionary Algorithm–AMTEA

The adaptive model-based transfer-enabled evolutionary algorithm (AMTEA), an evolutionary sequential transfer learning algorithm, was proposed in [6]. AMTEA aims to improve the performance on a target task with the help of knowledge from source tasks that have been addressed. Assume the total number of target and source tasks is K, including one target task and $K-1$ source tasks. In AMTEA, a probabilistic model is used to describe the solution distribution of one task. Thus, the knowledge for solving the task is stored in the probabilistic model. Compared with storing all the solutions, storing probabilistic models can save a huge amount of computational memory. $K-1$ probabilistic models for the $K-1$ addressed (source) tasks are built at the beginning of the algorithm. Then, at each generation, a probabilistic model is built to describe the solution distribution of the target task. Furthermore, a mixture probabilistic model is built based on the combination of the K models, including the target and $K-1$ source models. The mixture model is built based on the relatedness between the source and the target tasks. If the source task is highly related to the target task, its probabilistic model will greatly impact the mixture model. Compared with learning the relationships between the solutions of different tasks, learning the relationships between the probabilistic models of different tasks can be more effective and efficient. After the mixture model is built, new solutions will be generated based on it and transferred to the target task. By doing this, knowledge for problem-solving is transferred from the source tasks to the target task.

AMTEA has a good potential to be developed as a multi-task optimisation algorithm for feature selection. However, it is still challenging. Firstly, AMTEA is based on the assumption that all the source tasks have been addressed. A knowledge pool containing probability models is built based on the solutions of the source tasks. However, in multi-task optimisation, all the tasks are addressed simultaneously, which makes it challenging to build a knowledge pool for them. To address this issue, a method for building an online knowledge pool for knowledge transferring among multiple tasks simultaneously is developed. Furthermore, it is challenging for AMTEA to transfer knowledge between feature selections having different features. Using the transferring strategy in AMTEA, information may be transferred between the unique features from different tasks, which does not work for feature selection since different features have different meanings. To address this issue, a method for transferring knowledge across related feature selection tasks having different features is proposed. The proposed algorithm is expected to achieve better performance than state-of-the-art feature selection algorithms without significantly increasing computational time.

3   Proposed Method

In this work, we develop a new multi-task multi-objective feature selection algorithm based on AMTEA named FSMTO, which includes two main components: first, a method to build an online knowledge pool storing the probabilistic models which describe the distributions of solutions in multiple tasks during the evolutionary process; second, a method for transforming probabilistic models for the case tasks have different features. This algorithm aims to address multiple related feature selection tasks having common features simultaneously. The online knowledge pool is used for storing the captured knowledge of addressing the multiple related tasks during the evolutionary process. In this paper, knowledge is transferred across the common features of the multiple related feature selection tasks. Therefore, the probabilistic models are transformed when the multiple tasks have different features. This framework can work with many evolutionary multi-objective feature selection algorithms such as NSGA-II [7], IBEA [24], and MOEA/D [23]. In the following sections, the framework of the proposed algorithm is described first. Then, the details of the two components will be introduced.

3.1   The Framework of FSMTO

Figure 1 shows the workflow of the proposed algorithm. As shown in Fig. 1, K feature selection tasks represented by $T_k, k = 1, 2, ..., K$ are addressed simultaneously. An online knowledge pool containing K probabilistic models represented by $M_k^t, k = 1, 2, ..., K$ is built for the multiple tasks at the tth generation. At the $t - 1^{th}(t > 0)$ generation, K probabilistic models are built based on the populations of the K feature selection tasks. Then, knowledge is transferred from the knowledge pool to each of the K feature selection tasks at the tth generation by using the knowledge transferring method in AMTEA.

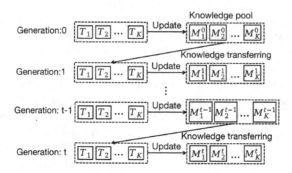

Fig. 1. The workflow of the proposed algorithm FSMTO.

Algorithm 1 shows the pseudocode of the proposed multi-task multi-objective feature selection algorithm FSMTO. As can be seen, K populations are randomly initialized for the K tasks at the beginning of FSMTO. Each population has a size of N. During the evolutionary process, new solutions for each task are generated either based on knowledge transfer across tasks or genetic operators such as cross-over and mutation, which is controlled by the transfer interval Δ between the two closest knowledge transfers. When Δ is large, most of the new solutions for each task are generated by genetic operators. Otherwise, most of the new solutions are generated based on knowledge transfer. After the new solutions are generated, a set of solutions with better objective values will be selected for the next generation under our proposed framework. Finally, the proposed algorithm will select K sets of solutions (feature subsets) for the K feature selection tasks.

As mentioned previously, the key idea of AMTEA is to build a mixture model based on the source probabilistic models for the target task. Since FSMTO is a multi-task optimisation algorithm, knowledge transferring based on AMTEA is conducted K times simultaneously for K feature selection tasks. Here, we introduce how to transfer knowledge from the knowledge pool to one of the K feature selection tasks. The rest $K-1$ tasks follow the same way. Following AMTEA, a mixture model is built to approximate the distribution of optimal solutions for the target task. Let FS represent a set of optimal solutions of the feature selection task, $P^*(\mathbf{x})$ represents the probabilistic distribution of FS. The $K-1$ source probabilistic models are used to approximate the true latent probabilistic model $P^*(\mathbf{x})$ of the target task. At the tth generation, the approximated true latent probabilistic model for the target task is represented by $Q_t(\mathbf{x})$. Let ϕ_K represent the probabilistic model built based on the population of the target task at the $t^t h$ generation. They satisfy the following relationship.

$$Q_t(\mathbf{x}) \approx \Sigma_{l=1}^{K} \alpha_l^t \phi_l(\mathbf{x}) \tag{1}$$

where $\phi_l(\mathbf{x}), l = 1, 2, ..., K-1$ are the $K-1$ source probabilistic models. α_l^t represents the coefficient of the lth model. At the tth generation, the current population of the target task can be represented by U^t. To approximate $Q_t(\mathbf{x})$, the α_l^ts that maximise the probability of observing the data U_t should be found.

Since $Q_t(\mathbf{x})$ is used to approximate $P^*(\mathbf{x})$, the following relationship will be satisfied.

$$P^*(\mathbf{x}) \approx lim_{t \to +\infty} Q_t(\mathbf{x}) \tag{2}$$

$\phi_l(\mathbf{x}), l = 1, 2, ..., K - 1$ represent the $K - 1$ probabilistic models of the source tasks, while $\phi_K(\mathbf{x})$ is a model describing the target dataset U_t. Obtaining α_l^t of the mixture model is equal to maximising Eq. (3).

$$\log L = \Sigma_{i=1}^{N} \log \Sigma_{l=1}^{K} \alpha_l^t \phi_l(\mathbf{x}) \tag{3}$$

where, N is the number of instances of U_t. The log function helps to transform the products in the objective into sums, which is easier to deal with. $\mathbf{x} \in U_t$. Transfer coefficients α_l^t are related to the similarity between the solution distributions of the source task and the target task. The value of α_l^t is in the range of [0,1]. If α_l^t is close to 1, the lth source task is similar to the target task.

After the kth mixture model is built, new solutions are sampled based on the kth mixture model and transferred to the kth feature selection task, which achieves knowledge transferring.

3.2   Build an Online Knowledge Pool

In FSMTO, multiple feature selection tasks are addressed parallelly at the same time. At each generation of the evolutionary process, knowledge of solving each task can be captured from the population and stored in a probabilistic model and then shared across tasks. An online knowledge pool that contains these probabilistic models is built. The probability model used in this work is based on the assumption that features are independent. K probabilistic models are built for the K feature selection tasks at each generation.

The probabilistic model, which describes the current population of a feature selection task, is shown as follows.

$$P(o_1 = k_{o_1}, o_2 = k_{o_2}, ..., o_D = k_{o_D}) = \prod_{i=1}^{D} p_i^{k_{o_i}} (1 - p_i)^{1 - k_{o_i}} \tag{4}$$

where, o_i represents whether the ith feature is selected. $k_{o_i} \in \{0, 1\}$. p_i is the probability of selecting the ith feature, which can be calculated based on the solutions of the feature selection task. For example, if the population of a feature selection task consists of $c_1 = [1, 1, 0, 0], c_2 = [1, 0, 0, 0], c_3 = [0, 1, 0, 1]$. Then, the probabilistic model is:

$$P(o_1 = k_{o_1}, o_2 = k_{o_2}, o_3 = k_{o_3}, o_4 = k_{o_4}) =$$
$$((\frac{2}{3})^{k_{o_1}} (\frac{1}{3})^{1-k_{o_1}})(\frac{2}{3}^{k_{o_2}} \frac{1}{3}^{1-k_{o_2}})(0^{k_{o_3}} 1^{1-k_{o_3}})((\frac{1}{3})^{k_{o_4}} (\frac{2}{3})^{1-k_{o_4}}) \tag{5}$$

At the current generation, K newly generated probabilistic models replace the K old models and update the knowledge pool.

Algorithm 1. Pseudocode of FSMTO

Input:
  K related feature selection tasks and a transfer interval Δ;
Output:
  Solution(s) to all the K feature selection tasks;
  Set $t = 1$;
  Generate K initial populations $P_1(t), ..., P_K(t)$ for the K tasks, the size of each
  population is N;
  Evaluate the individuals based on their objective functions
  while t is not larger than the maximum generation **do**
    Update the probabilistic models in the knowledge pool based on the K current
    populations.
    if mod (t+1, Δ) == 0 **then**
      for k=1:K **do**
        Build a mixture model $q_k(\mathbf{x}|t)$ for the kth task according to Eq (3);
        Randomly sample the offspring population $P_k^c(t)$ from the mixture model
        $q_k(\mathbf{x}|t)$;
    else
      for k=1:K **do**
        Randomly select individuals from $P_k(t)$ to form a parent population $P_k^s(t)$;
        Use genetic operators including crossover and mutation to generate an off-
        spring population $P_k^c(t)$ based on the parent population $P_k^s(t)$;
    for k=1:K **do**
      Evaluate individuals in $P_k^c(t)$ based on their objective functions;
      Select next generation $P_k(t+1)$ from $P_k(t) \cup P_k^c(t)$ based on multi-objective
      optimisation algorithms;
      Set $t = t + 1$;

3.3   Transform the Probabilistic Models

When the search spaces of related tasks are different, the probabilistic model
of one task can not be transferred to another task without any transformation.
Since different features have different meanings, knowledge needs to be transferred across common features of the related feature selection tasks. In this work,
each task will be treated as both the target task and the source task; thus knowledge transferring is conducted K times among these tasks.

   We introduce how to transform the probabilistic model of a source task.
The rest tasks follow the same way. Let $V_{target} = \{o_1^t, ..., o_a^t, u_1^t, ..., u_b^t\}$ rep-
resent the feature vector of a target feature selection task, while $V_{source} = \{o_1^s, ..., o_a^s, z_1^s, ..., z_c^s\}$ represents the feature vector of a source feature selection
task where $o_i^t = o_i^s, i = 1, ..., a$. Here $o_1^t, ..., o_a^t$ and $o_1^s, ..., o_a^s$ represent the com-
mon features of the target and the source tasks. $u_1^t, ..., u_b^t$ and $z_1^s, ..., z_c^s$ represent the unique features from the target task and the source task, respectively. In this case, the target probabilistic model can be split into two parts
as shown in Eq. (6), where the first part is $P(o_1^t = k_{o_1^t}, ..., o_a^t = k_{o_a^t}) = \prod_{i=1}^a p_{o_i^t}^{k_{o_i^t}} (1 - p_{o_i^t})^{1-k_{o_i^t}}$ and the second part is $P(u_1^t = k_{u_1^t}, ..., u_c^t = k_{u_c^t}) = \prod_{i=1}^c p_{u_i^t}^{k_{u_i^t}} (1 - p_{u_i^t})^{1-k_{u_i^t}}$.

$$P(o_1^t = k_{o_1^t}, ..., o_a^t = k_{o_a^t}, u_1^t = k_{u_1^t}, ..., u_c^t = k_{u_c^t}) =$$
$$\prod_{i=1}^{a} p_{o_i^t}^{k_{o_i^t}} (1 - p_{o_i^t})^{1-k_{o_i^t}} \cdot \prod_{i=1}^{c} p_{u_i^t}^{k_{u_i^t}} (1 - p_{u_i^t})^{1-k_{u_i^t}} \tag{6}$$

Similarly, the source probabilistic model is represented by

$$P(o_1^s = k_{o_1^s}, ..., o_a^s = k_{o_a^s}, z_1^s = k_{z_1^s}, ..., z_c^s = k_{z_c^s}) =$$
$$\prod_{i=1}^{a} p_{o_i^s}^{k_{o_i^s}} (1 - p_{o_i^s})^{1-k_{o_i^s}} \cdot \prod_{i=1}^{c} p_{z_i^s}^{k_{z_i^s}} (1 - p_{z_i^s})^{1-k_{z_i^s}} \tag{7}$$

As can be seen, the differents between the target probabilistic model and the source probabilistic model are

$$P(u_1^t = k_{u_1^t}, ..., u_c^t = k_{u_c^t}) = \prod_{i=1}^{c} p_{u_i^t}^{k_{u_i^t}} (1 - p_{u_i^t})^{1-k_{u_i^t}} \tag{8}$$

and

$$P(z_1^s = k_{z_1^s}, ..., z_c^s = k_{z_c^s}) = \prod_{i=1}^{c} p_{z_i^s}^{k_{z_i^s}} (1 - p_{z_i^s})^{1-k_{z_i^s}} \tag{9}$$

Since features $z_1^s, ..., z_c^s$ are not in the target probabilistic model, Eq. (9) of the source probabilistic model should be replaced when the source probabilistic model is transferred to the target task. Due to the fact that the source task does not contain the information about the unique features in the target task $u_1^t, ..., u_b^t$, The probabilities of selecting features $u_1^t, ..., u_b^t$ in the source tasks are all 0. Equation (9) of the source task will be replaced with

$$P(u_1^t = k_{u_1^t}, ..., u_b^t = k_{u_b^t}) = \prod_{i=1}^{b} 0^{k_{u_i^t}} 1^{1-k_{u_i^t}} \tag{10}$$

The transformed source probabilistic model which can be used for the target task is represented as follows.

$$P(o_1^s = k_{o_1^s}, ..., o_a^s = k_{o_a^s}, u_1^t = k_{u_1^t}, ..., u_b^t = k_{u_b^t}) =$$
$$\prod_{i=1}^{a} p_{o_i^s}^{k_{o_i^s}} (1 - p_{o_i^s})^{1-k_{o_i^s}} \cdot \prod_{i=1}^{b} 0^{k_{u_i^t}} 1^{1-k_{u_i^t}} \tag{11}$$

4   Experiment Design

In this paper, improving the strength pareto evolutionary algorithm (SPEA2), which is the commonly used multi-objective framework [24], is used as a baseline. We compare SPEA2 with SPEA2 cooperating with the proposed algorithm, named SPEA2-FSMTO. Firstly, we compare the hypervolume values obtained by the two algorithms on the training sets of the benchmark datasets. Secondly, we compare the hypervolume values obtained by the two algorithms on the test sets of the benchmark datasets. Furthermore, we compare the computational time cost of the two algorithms on the benchmark datasets. Finally, further analysis of the selected features of the two algorithms is conducted.

Table 1. Datasets

Dataset	Instance	Features	Class	Dataset	Instance	Features	Class
Wine-1	1599	12	2	Dermatology-1	358	34	2
Wine-2	4898			Dermatology-2			
Mushroom-1	3516	22	2	Dermatology-3			
Mushroom-2	4608			Dermatology-4			
Magic-1	2212	10	2	Dermatology-5			
Magic-2	16808			Dermatology-6			
Letter-1	1151	16	2	Waveform-1	1331	40	3
Letter-2	1543			Waveform-2	3669		
Letter-3	1509			News-1	3749	200	4
Letter-4	1575			News-2	3743		
Letter-5	1536			News-3	3702		
Letter-6	1573			News-4	3669		

4.1 Benchmark Datasets

Multiple related datasets, which have common features but different distributions, and similar tasks, have been used in this paper. The feature selection tasks on these datasets are related and have common knowledge, which can be used for evaluating the performance of our proposed algorithm. Their properties, such as the number of features, instances, and classes of these datasets, are summarised in Table 1.

- **Wine**: The Red Wine dataset and the White Wine dataset, which have the same features but different data distributions, are the two representative datasets in the field of wine quality classification [2].
- **Mushroom**: Mushroom is used to classify whether a mushroom is edible or poisonous. Two related datasets generated from the original Mushroom dataset are widely used in transfer learning [15].
- **Magic:** The Magic dataset contains image data of hadronic showers. This dataset is used for the classification of hadronic showers caused by primary gammas or upper atmospheres. The original dataset can be divided into two related sub-datasets with the same feature space, and different data distributions [16].
- **Letter:** Letter Recognition dataset contains image data of letters from A to Z [9]. This dataset also can be divided into several binary classification datasets, which have the same search space but different data distributions. Based on this, six sub-datasets 'I vs T', 'E vs F', 'C vs G', 'M vs N', 'Q vs O', and 'X vs Y' are generated, respectively. 'I vs T' denotes as a sub-dataset with the task of classification between letter 'I' and letter 'T', which contains the instances with a label of 'I' or 'T'. This also applies for the other five sub-datasets.

- **Dermatology:** Dermatology contains instances of six dermatological diseases. This dataset has 33 clinical and histopathological attributes. Following [1], Dermatology can be divided into six sub-datasets which are used for binary classification tasks. These sub-datasets have the same search space but different data distributions, which can be used for transfer learning.
- **Waveform:** The Waveform dataset, which is usually used for transfer learning, contains the data of three classes of waves. This original dataset can be transformed into two related datasets where Waveform-1 contains the instances which have the first feature larger than 0.15 and the second feature larger than 0, and Waveform-2 contains the remaining instances.
- **20 Newsgroups:** The 20 Newsgroups dataset contains nearly 20,000 newsgroup documents, evenly across 20 different news categories [3]. In this research, we create four related sub-datasets based on the original dataset, which are named News-1, News-2, News-3, and News-4. The properties of these four generated sub-datasets, including the document categories, are summarised in Table 2. As can be seen, the search spaces of the feature selection tasks are similar but different, which is sensible to evaluate the effectiveness of our proposed method for transforming the probabilistic models across tasks having different features.

Table 2. Document Categories in News Datasets

Name	Documents Categories	Name	Documents Categories
News-1	alt.atheism,comp.graphics, rec.autos,sci.space	News-3	alt.atheism,soc.religion.christian, talk.politics.guns,rec.motorcycles
News-2	alt.atheism,comp.graphics, misc.forsale,rec.motorcycles	News-4	alt.atheism,rec.sport.hockey, talk.politics.guns,sci.crypt

4.2 Parameter Settings

In the experiments, each dataset is randomly split with 70% forms the training set, while the rest 30% will be put into the test set. The classification error rate, the second objective value of multi-objective feature selection, of candidate feature subsets is calculated based on the Decision Tree with 10-fold cross-validation on the training set [13].

Following [12], in this work, the population size is set to the number of features in each task. For the purpose of having enough computational resources for the feature selection algorithms to find a set of good solutions, the maximal number of generations is set to five times the population size but will not be larger than 200. Each algorithm will run 30 independent times on each dataset. The one-point crossover operator and mutation are used in this research. The probability of mutation is set to $1/D$, where D is the number of features [12]. The transfer interval of AMTEA is set to 2. Table 3 shows the summary of the parameter settings of this paper.

4.3 Result Analyses and Discussions

In this section, the comparisons between the performance of SPEA2-FSMTO and SPEA2 are used to verify the effectiveness of the proposed multi-task optimisation algorithm. Hypervolume (HV) [18], which is one of the most popular performance indicators in evolutionary multi-objective optimisation is used for comparisons. A larger HV value indicates better multi-objective optimisation performance. Since the two objectives of feature selection are both in the range of [0,1], the reference point of HV is set as (1.1, 1.1) in this paper [19]. The Wilcoxon test with a significance level of 0.05 is used for the performance comparison in this paper. The sign of '+' means that SPEA2-FSMTO is significantly better than SPEA2, while the sign of '−' means that SPEA2-FSMTO is significantly worse than SPEA2. The sign of '≈' means that SPEA2-FSMTO ties SPEA2. Finally, the features selected by the two compared algorithms on one typical dataset, Magic, are analyzed.

Table 3. The parameter setting

Parameters	Settings
Population size	N
Maximum generation	Min(200, 5*N)
Dimension of solutions	D
Probability of mutation	1/D
Transfer interval	2
Reference point for hypervolume	(1.1,1.1)

Table 4. Hypervolume on the training sets

	SPEA2	SPEA2-FSMTO			SPEA2	SPEA2-FSMTO	
	Mean (Std)	Mean (Std)	W-test		Mean (Std)	Mean (Std)	W-test
Wine-1	0.956(0.033)	**0.976(0.004)**	+	Dermatology-1	1.171(0.003)	**1.173(0.001)**	+
Wine-2	0.909(0.024)	**0.919(0.003)**	+	Dermatology-2	1.15(0.011)	1.153(0.003)	≈
Mushroom-1	1.153(0.009)	**1.157(0.0)**	+	Dermatology-3	1.178(0.0)	**1.178(0.0)**	+
Mushroom-2	1.159(0.002)	**1.16(0.0)**	+	Dermatology-4	1.159(0.003)	1.159(0.002)	+
Magic-1	0.972(0.037)	**0.987(0.023)**	+	Dermatology-5	1.176(0.003)	1.176(0.003)	≈
Magic-2	0.867(0.035)	**0.873(0.032)**	+	Dermatology-6	1.175(0.002)	**1.177(0.001)**	+
Letter-1	1.091(0.036)	**1.122(0.005)**	+	Waveform-1	0.921(0.01)	**0.927(0.009)**	+
Letter-2	1.116(0.024)	**1.13(0.002)**	+	Waveform-2	0.903(0.02)	**0.914(0.01)**	+
Letter-3	1.053(0.028)	**1.073(0.005)**	+	News-1	0.789(0.016)	**0.909(0.005)**	+
Letter-4	1.078(0.039)	**1.104(0.006)**	+	News-2	0.814(0.016)	**0.94(0.004)**	+
Letter-5	1.037(0.06)	**1.082(0.004)**	+	News-3	0.779(0.015)	**0.882(0.006)**	+
Letter-6	1.113(0.023)	**1.127(0.001)**	+	News-4	0.82(0.02)	**0.94(0.005)**	+

Comparisons on Hypervolume Values on the Training Sets and the Test Sets: Tables 4 and 5 show the averages and standard deviations of the HV values calculated based on the final fronts obtained by the two algorithms, SPEA2-FSMTO and SPEA2, over the 30 independent runs on the training sets and test sets. Based on the statistical test results, SPEA2-FSMTO has a significantly better training performance than SPEA2 on 22 of the 24 benchmark datasets. Furthermore, the proposed algorithm has a better testing performance than SPEA2 on 18 of the 24 tasks. Overall, with the help of knowledge transferring among related feature selection tasks, the proposed algorithm can obtain a better performance than SPEA2 on the benchmark datasets.

Table 5. Hypervolume on the test sets

	SPEA2	SPEA2-FSMTO			SPEA2	SPEA2-FSMTO	
	Mean (Std)	Mean (Std)	W-test		Mean (Std)	Mean (Std)	W-test
Wine-1	0.951(0.036)	**0.972(0.008)**	+	Dermatology-1	1.176(0.003)	1.176(0.003)	≈
Wine-2	0.909(0.025)	0.919(0.004)	≈	Dermatology-2	1.114(0.017)	1.119(0.007)	≈
Mushroom-1	1.153(0.009)	**1.157(0.0)**	+	Dermatology-3	**1.16(0.006)**	1.158(0.0)	−
Mushroom-2	1.159(0.003)	**1.16(0.0)**	+	Dermatology-4	**1.152(0.007)**	1.145(0.009)	−
Magic-1	0.956(0.036)	**0.973(0.023)**	+	Dermatology-5	1.178(0.0)	1.178(0.0)	≈
Magic-2	0.868(0.035)	**0.874(0.032)**	+	Dermatology-6	1.173(0.006)	**1.175(0.004)**	+
Letter-1	1.091(0.036)	**1.122(0.005)**	+	Waveform-1	0.896(0.013)	0.902(0.019)	≈
Letter-2	1.116(0.024)	**1.13(0.002)**	+	Waveform-2	0.89(0.022)	**0.901(0.014)**	+
Letter-3	1.053(0.028)	**1.073(0.005)**	+	News-1	0.74(0.015)	**0.869(0.007)**	+
Letter-4	1.078(0.039)	**1.104(0.006)**	+	News-2	0.785(0.015)	**0.925(0.006)**	+
Letter-5	1.037(0.06)	**1.082(0.004)**	+	News-3	0.695(0.017)	**0.796(0.01)**	+
Letter-6	1.111(0.023)	**1.125(0.003)**	+	News-4	0.758(0.019)	**0.875(0.007)**	+

Table 6. Computational time (in Seconds) of the algorithms on the training sets

	SPEA2	SPEA2-FSMTO	
	Mean(Std)	Mean(Std)	W-test
Wine	51.296(3.068)	**47.048(2.474)**	+
Mushroom	126.609(2.897)	**113.095(3.624)**	+
Magic	139.418(11.916)	**117.064(5.654)**	+
Letter	207.59(24.883)	**180.798(2.852)**	+
Dermatology	794.868(22.465)	**690.16(5.253)**	+
Waveform	805.292(41.735)	**671.073(25.318)**	+
News	5679.773(663.745)	5461.877(1344.209)	≈

Table 7. Features selected by SPEA2 and SPEA2-FSMTO on Magic-1 dataset.

SPEA2 on Magic-1	
Classification accuracy	Selected features
0.899844994	fLength, fSize, fConc, fAsym, fM3Long, fAlpha
0.896606619	fLength, fM3Long, fAlpha
0.860410557	fM3Long, fAlpha
SPEA2-FSMTO on Magic-1	
Classification accuracy	Selected features
0.914046921	fLength, fWidth, fSize, fM3Long, fAlpha
0.896606619	fLength, fM3Long, fAlpha
0.873338919	fLength, fAlpha
0.757246025	fLength, fWidth, fAlpha

Comparisons on Computational Time: To investigate the efficiency of the proposed feature selection framework, we also compare the computational cost of the two feature selection algorithms on the training sets. Table 6 shows the average and standard deviation of the computational time (in seconds) of the two algorithms over 30 runs. The results show that SPEA2-FSMTO spends less computational time than SPEA2 for feature selection. There are several possible reasons: first, building the mixture model in AMTEA does not consume many computational resources [6]; second, nearly half of the new solutions are generated by probabilistic models in SPEA2-FSMTO, which may be faster than generating all the new solutions by genetic operators in SPEA2; Finally, SPEA2-FSMTO selects a smaller number of features than SPEA2 (see the next subsection), which helps to save computational cost in fitness evaluations. Therefore, FSMTO can improve the performance of SPEA2 with better efficiency for most datasets.

Table 8. Features selected by SPEA2 and SPEA2-FSMTO on Magic-2 dataset.

SPEA2 on Magic-2	
Classification accuracy	Selected features
0.796940798	fWidth, fSize, fConc, fAlpha, fDist
0.790144579	fLength, fWidth, fSize, fAlpha
0.757246025	fLength, fWidth, fAlpha
SPEA2-FSMTO on Magic-2	
Classification accuracy	Selected features
0.797450569	fLength, fWidth, fSize, fM3Long, fAlpha
0.790144579	fLength, fWidth, fSize, fAlpha
0.765918555	fWidth, fSize, fAlpha
0.757246025	fLength, fWidth, fAlpha

Further Analysis on the Selected Features: To further understand the effectiveness of the proposed method, in this subsection, we compare the features selected by the single-task multi-objective feature selection algorithm and the proposed multi-task multi-objective feature selection algorithm. The Magic datasets contain ten features, which is feasible for our analysis. Furthermore, the meaning of each feature of the Magic datasets is known, which is helpful in understanding why the selected features result in good classification performance. Therefore, we use the two Magic datasets as an example for our analysis. Table 7 and Table 8 show the features selected by the two algorithms on the run that has the median hypervolume values. As can be seen, SPEA2-FSMTO obtains better classification performance and selects a smaller number of features than SPEA2. SPEA2 selects features including {*fLength, fSize, fCon, fAsym, fM3Long, fAlpha*} and obtain a classification accuracy of 0.899 on Magic-1, while SPEA2-FSMTO selects features {*fLength, fWidth, fSize, fM3Long, fAlpha*} and obtains a classification accuracy of 0.914 on Magic-1. The main difference between them is that SPEA2-FSMTO selects the feature {*fWidth*}, but SPEA2 does not, which means the feature {*fWidth*} is useful to the classification task of the Magic-1 dataset. Furthermore, both SPEA2 and SPEA2-FSMTO select the feature {*fWidth*} on the Magic-2 dataset. Since SPEA2-FSMTO addressed the two related feature selection tasks, Magic-1 and Magic-2, together, the knowledge of selecting the feature {*fWidth*} can be transferred across them. However, the knowledge of selecting the feature {*fWidth*} can not be transferred from the feature selection task Magic-2 to Magic-1 in SPEA2, which results in lower classification accuracy on Magic-1 than SPEA2-FSMTO. Overall, SPEA2-FSMTO can select good features and achieve improved classification performance with the help of knowledge transfer across related feature selection tasks.

5   Conclusions

In this work, a new multi-task optimisation framework is developed for feature selection based on AMTEA. The proposed algorithm addresses multiple related feature selection tasks simultaneously rather than addressing one main task at a time. A method is developed to build an online knowledge pool that contains knowledge for multiple tasks. Furthermore, a method is developed to transform the probabilistic models for the case tasks with different features. To investigate the effectiveness of the proposed algorithm, related feature selection tasks having different features have been tested. Experimental results show that SPEA2 cooperating with the proposed algorithm selects a smaller number of features and achieves a better classification performance than that of SPEA2 based feature selection without knowledge transfer across tasks. Furthermore, the experimental results verified the effectiveness of the proposed for addressing multiple tasks having different features simultaneously.

References

1. Argyriou, A., Evgeniou, T., Pontil, M.: Convex multi-task feature learning. Mach. Learn. **73**(3), 243–272 (2008)
2. Cao, B., Pan, S.J., Zhang, Y., Yeung, D.Y., Yang, Q.: Adaptive transfer learning. In: Proceedings of the AAAI Conference on Artificial Intelligence, vol. 24, pp. 407–412 (2010)
3. Chandra, A.: Comparison of feature selection for imbalance text datasets. In: 2019 International Conference on Information Management and Technology (ICIMTech), vol. 1, pp. 68–72. IEEE (2019)
4. Chen, K., Xue, B., Zhang, M., Zhou, F.: An evolutionary multitasking-based feature selection method for high-dimensional classification. IEEE Trans. Cybern. **52**, 7172–7186 (2020)
5. Chen, K., Xue, B., Zhang, M., Zhou, F.: Evolutionary multitasking for feature selection in high-dimensional classification via particle swarm optimisation. IEEE Trans. Evol. Comput. **26**, 446–460 (2021)
6. Da, B., Gupta, A., Ong, Y.S.: Curbing negative influences online for seamless transfer evolutionary optimization. IEEE Trans. cybern. **49**(12), 4365–4378 (2018)
7. Deb, K., Pratap, A., Agarwal, S., Meyarivan, T.: A fast and elitist multiobjective genetic algorithm: NSGA-II. IEEE Trans. Evol. Comput. **6**(2), 182–197 (2002)
8. Feng, L., et al.: Explicit evolutionary multitasking for combinatorial optimization: a case study on capacitated vehicle routing problem. IEEE Trans. Cybern. **51**(6), 3143–3156 (2020)
9. Gonçalves, A.R., Das, P., Chatterjee, S., Sivakumar, V., Von Zuben, F.J., Banerjee, A.: Multi-task sparse structure learning. In: Proceedings of the 23rd ACM International Conference on Conference on Information and Knowledge Management, pp. 451–460 (2014)
10. González, J., Ortega, J., Damas, M., Martín-Smith, P., Gan, J.Q.: A new multi-objective wrapper method for feature selection-accuracy and stability analysis for bci. Neurocomputing **333**, 407–418 (2019)
11. Lin, J., Liu, H.L., Xue, B., Zhang, M., Gu, F.: Multiobjective multitasking optimization based on incremental learning. IEEE Trans. Evol. Comput. **24**(5), 824–838 (2019)
12. Nguyen, B.H., Xue, B., Andreae, P., Ishibuchi, H., Zhang, M.: Multiple reference points-based decomposition for multiobjective feature selection in classification: static and dynamic mechanisms. IEEE Trans. Evol. Comput. **24**(1), 170–184 (2019)
13. Nguyen, B.H., Xue, B., Zhang, M.: A constrained competitive swarm optimiser with an SVM-based surrogate model for feature selection. IEEE Trans. Evol. Comput. (2022). https://doi.org/10.1109/TEVC.2022.3197427
14. Osaba, E., Del Ser, J., Martinez, A.D., Lobo, J.L., Nebro, A.J., Yang, X.S.: MO-MFCGA: multiobjective multifactorial cellular genetic algorithm for evolutionary multitasking. In: 2021 IEEE Symposium Series on Computational Intelligence (SSCI), pp. 1–8. IEEE (2021)
15. Segev, N., Harel, M., Mannor, S., Crammer, K., El-Yaniv, R.: Learn on source, refine on target: a model transfer learning framework with random forests. IEEE Trans. Pattern Anal. Mach. Intell. **39**(9), 1811–1824 (2016)
16. Shi, Y., Lan, Z., Liu, W., Bi, W.: Extending semi-supervised learning methods for inductive transfer learning. In: 2009 Ninth IEEE International Conference on Data Mining, pp. 483–492. IEEE (2009)

17. Wang, X.H., Zhang, Y., Sun, X.Y., Wang, Y.L., Du, C.H.: Multi-objective feature selection based on artificial bee colony: an acceleration approach with variable sample size. Appl. Soft Comput. **88**, 106041 (2020)
18. While, L., Hingston, P., Barone, L., Huband, S.: A faster algorithm for calculating hypervolume. IEEE Trans. Evol. Comput. **10**(1), 29–38 (2006)
19. Xu, H., Xue, B., Zhang, M.: A duplication analysis-based evolutionary algorithm for biobjective feature selection. IEEE Trans. Evol. Comput. **25**(2), 205–218 (2020)
20. Xu, Q., Wang, N., Wang, L., Li, W., Sun, Q.: Multi-task optimization and multi-task evolutionary computation in the past five years: a brief review. Mathematics **9**(8), 864 (2021)
21. Zhang, F., Mei, Y., Nguyen, S., Zhang, M.: Multitask multiobjective genetic programming for automated scheduling heuristic learning in dynamic flexible job-shop scheduling. IEEE Transactions on Cybernetics (2022). https://doi.org/10.1109/TCYB.2022.3196887
22. Zhang, F., Mei, Y., Nguyen, S., Zhang, M., Tan, K.C.: Surrogate-assisted evolutionary multitask genetic programming for dynamic flexible job shop scheduling. IEEE Trans. Evol. Comput. (2021). https://doi.org/10.1109/TEVC.2021.3065707
23. Zhang, Q., Li, H.: MOEA/D: a multiobjective evolutionary algorithm based on decomposition. IEEE Trans. Evol. Comput. **11**(6), 712–731 (2007)
24. Zitzler, E., Laumanns, M., Thiele, L.: Spea 2: improving the strength pareto evolutionary algorithm. TIK-report 103 (2001)

Under the Hood of Transfer Learning for Deep Neuroevolution

Stefano Sarti[1]($\boxtimes$)(iD), Nuno Laurenço[2](iD), Jason Adair[1](iD), Penousal Machado[2](iD), and Gabriela Ochoa[1](iD)

[1] University of Stirling, Scotland, UK
{stefano.sarti,jason.adair,gabriela.ochoa}@stir.ac.uk
[2] University of Coimbra, CISUC, DEI, Coimbra, Portugal
{naml,machado}@dei.uc.pt

Abstract. Deep-neuroevolution is the optimisation of deep neural architectures using evolutionary computation. Amongst these techniques, Fast-Deep Evolutionary Network Structured Representation (Fast-DENSER) has achieved considerable success in the development of Convolutional Neural Networks (CNNs) for image classification. In this study, variants of this algorithm are seen through the lens of Neuroevolution Trajectory Networks (NTNs), which use complex network modelling and visualisation to uncover intrinsic characteristics. We examine how evolution uses previously acquired knowledge on some datasets to inform the search for new domains with a specific focus on the architecture of CNNs. Results show that the transfer learning paradigm works as intended as networks mutate, incorporating layers from the best models trained on previous datasets. The use of specifically designed NTNs in this analysis enabled us to perceive the architectural characteristics that evolution favours in the design of CNNs. These findings provide novel insights that may inform the future creation of Deep Neural Networks (DNNs).

Keywords: Dynamic-Structured Grammatical Evolution ·
Neuroevolution · Neuroevolution Trajectory Networks · Computer Vision

1 Introduction

Neuroevolution is the application of Evolutionary Computation to the design of Artificial Neural Networks (ANNs) - replacing the time-consuming, trial-and-error process of hand-designing those networks. Neuroevolution can be subdivided into systems that: evolve and optimise neural structures; optimise the learning parameters; and optimise both the learning and the architectures simultaneously. This paper is based on a successful and efficient algorithm from the latter group, Fast-DENSER [1].

Fast-DENSER evolves both structures and learning parameters of Convolutional Neural Networks (CNNs) through a Dynamic Structured Grammatical Evolution (DSGE) encoding, coupled with a $(1+\lambda)$ Evolution Strategy (ES)

J. Correia et al. (Eds.): EvoApplications 2023, LNCS 13989, pp. 640–655, 2023.
https://doi.org/10.1007/978-3-031-30229-9_41

approach. It has been successfully deployed for Transfer Learning (using incremental development) where models evolved in a classification dataset are used to seed the evolution of models for a subsequent dataset [2]. The rationale is that knowledge from previously generated models could be reused to inform the search and optimisation of models for other problem domains - avoiding random initialisation.

This work focuses on the modelling and visualisation of the evolutionary processes of Fast-DENSER using Neuroevolution Trajectory Networks (NTNs) [13–15]. Specific to neuroevolution, this analysis has been derived from Search Trajectory Networks (STNs), a technique that uses complex networks to depict the search process of metaheuristics. Here, NTNs are used to reveal intrinsic characteristics of several variants of Fast-DENSER, examining the search space and comparing variants to the original version of the algorithm. We analyse two algorithms: a variant in which the Evolutionary Strategy (ES) includes both crossover and mutation (with elitism), and a local search variant.

In the incremental development setup, we broaden the current analysis by inspecting the knowledge transferred from one classification dataset to the next. We design a version of NTNs to characterise the differences between solutions and determine which key components (layers and learning) are deemed valuable by evolutionary dynamics and therefore inherited as previous knowledge. This offers a novel contribution through the global visualisation of network characteristics inherited from multiple independent runs. These new insights can serve to inform decisions for both neuroevolution algorithms and expert Deep Neural Networks hand designers. The key contributions of this paper are as follows.

- Examination of Fast-DENSER variants for transfer learning developed to include crossover and local search
- An in-depth analysis of intrinsic characteristics of said variants when evolving CNNs for benchmark problems using recently developed NTNs techniques
- Augmentation of NTNs tools to present the knowledge transferred during incremental development of CNNs, to globally assess the key components selected by evolution

The rest of this paper is organised as follows. Section 2 outlines the background knowledge. Our methodology is outlined in Sect. 3, presenting the rationale behind the proposed analysis. Section 4 provides an overview of the statistical results for each evaluated variant. Section 5 reveals the creation and analysis of the NTN models. Finally, Sect. 6 summarises our main findings and suggestions for future work.

2 Background Knowledge

2.1 Grammar-based Neuroevolution

DENSER - DSGE Powered Neuroevolution. DENSER is a neuroevolution technique that delivers human-comparable results [3] by using two evolutionary algorithms (EAs) simultaneously: a Genetics Algorithm (GA) and Dynamic

Structured Grammatical Evolution (DSGE) [7]. In DENSER the GA level is responsible for encoding the macrostructure of the networks, which requires the allowed network structures to be defined. For instance, in the development of CNNs the allowed GA structure will be: [(features, 1, 10), (classification, 1, 2), (softmax, 1, 1), (learning, 1, 1)], with each tuple being the valid starting symbols, and the minimum and maximum number of times these can be used. Therefore, based on the specified criteria, the GA can evolve structures with up to ten convolutional/pooling layers, followed by up to two fully-connected layers, and the classification layer (softmax); the learning tuple instead is specified for codifying parameters to train the network.

While the GA is responsible for evolving the network structure based on the allowed configuration, the DSGE evolves the parameters inherent to the evolved layers. These parameters and allowed values are specified by the user in the *grammar* (for example, see [1]).

This multi-EAs combination proved to be effective for neuroevolution, based on a grammar encoding [3]. The GA encapsulates the genetic information for the layers, making it easier to apply variation operators. Using the DSGE allows this technique to be generalisable to other problem domains, as changing the grammar configurations can explicitly redirect the search towards other network types.

Fast-DENSER. Although effective in generating CNNs for image classification [3], DENSER is slow in its evolution and training process due to large populations. In response, Fast-DENSER [1] was introduced. This is 20 times faster without loss in performance. It is achieved by a $(1 + \lambda)$ Evolutionary Strategy (ES), where lambda is set to generate four offspring from a single parent (the elite) in each generation. This drastic reduction in population size corresponds to lower training and evaluation time, less computational resources, resulting in a fraction of its running time compared to DENSER.

Fast-DENSER offers further improvements over DENSER in that it: improves the flexibility of DENSER by allowing the layers of the evolved networks to connect to any of the previous layers - known as *skip connections*; maps layers to inputs (similar to a direct acyclic graph); limits the number of previous layers that can be used as inputs (a maximum number of levels back constraint). By enforcing the rule in which layers always have to be connected to previous layers the algorithm avoids invalid solutions that could be caused by disjoint graphs.

The initialisation of Fast-DENSER is achieved in a gradual and constructive manner. In the initial population, a lower upper bound on the maximum number of layers is enforced (on the outer-level encapsulation); thus ensuring that the network is initially simple, deepening the networks through generations and preventing excessively deep architectures saturating the training progress early on.

In Fast-DENSER, several types of mutation operators are included - with the notable exception of *crossover*. There are four operators included in the outer encapsulation layer: **add evolutionary unit, remove evolutionary unit, add connection, remove connection**. The inner-level mutation operators are as described in [3], which alter real values/integers related to the grammatical expansion possibilities.

2.2   Incremental Development/Transfer Learning

The concept of incremental development, specific to Fast-DENSER was introduced in [2]. The initial objective was to evaluate whether substantial benefits could be achieved from transferring knowledge between DNNs trained on related classification problems. In turn, this would correspond to a reduction in the number of generations necessary to reach comparative fitness levels as those initialised with a random population. Additionally, the authors aimed to assess if the knowledge gained from previous datasets worked at every evolutionary stage (not just at initialisation) - demonstrating that incrementally developed DNNs are more resilient and capable of generalising better that their non-incrementally evolved equivalents on similar classification datasets.

In [2], the neuroevolution algorithm was configured with a specific grammar for the construction of Convolutional Neural Networks; these architectures have been successful in image classification/object detection [16]. At the end of each run, the best individual was selected from the final population and used to spawn (based on the ES criteria) the new population deployed to solve the subsequent benchmark problem. Four benchmarks for image classification/object detection were used, in the following order: MNIST, SVHN, CIFAR10 and Fashion-MNIST. This approach resulted in faster convergence to the optimal fitness, thanks to the incorporation of previous knowledge.

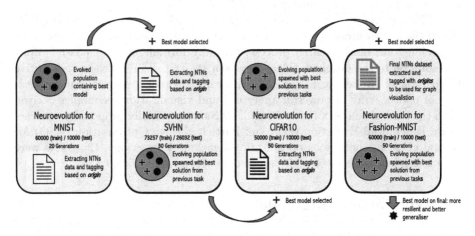

Fig. 1. Fast-DENSER with Incremental Development. During the incremental development of CNNs for benchmark classification problems, information essential to the NTNs creation is extracted and labelled based on its *origin*. Train and test split used for each dataset were based on [2].

As shown in Fig. 1, after a population is evolved based on the simpler benchmark MNIST, Fast-DENSER begins a new search, evolving for the new domain SVHN, with the incorporation of the best found model from MNIST. The incorporation of this knowledge begins at population inception but is also achieved during the mutation stage. At inception, the following can occur to generate individuals: (i) include all feature extraction layers but randomly generate the classification layers; (ii) random generation of feature extraction layers but port classification layers from previous best models; (iii) transfer knowledge of learning evolutionary unit but generate all other components from scratch; (iv) randomly initialise every evolutionary units at random, without incorporating any previous knowledge.

Mutation can leverage previous knowledge too. The operators are unchanged from those in [2]: addition, removal, and/or duplication of any evolutionary unit. In incremental development, the operator can select to adopt information of the best models from previous datasets or forgo this and use the knowledge in the current population.

2.3   Neuroevolution Trajectory Networks

Neuroevolution Trajectory Networks [13] are Search Trajectory Networks specific to the study of neuroevolution dynamics. As their derivation, many of the concepts that constitute STNs are applicable and transferable to NTNs. Nonetheless, there are exceptions to these that specifically characterise unique neuroevolution criteria. Thus far, these have been successfully applied in the field of Neuroevolution to study the intrinsic dynamics related to crossover, diversity, novelty and topological complexity of ANNs [13–15].

STNs are graph-based, data-driven models which represent defined trajectories of the search process of a given algorithm. In these graphs, nodes determine a specific state of the search process, typically formulated as a possible optimal solution to the problem. Edges represent the transition between consecutive search states. Modelling a system as a graph (network) favours the application of a vast array of powerful analytical and visualisation techniques derived from the field of complex networks [10]. STNs were initially proposed to characterise Differential Evolution (DE) and Particle Swarm Optimisation (PSO) for several classical continuous optimisation benchmark functions [12]. STNs were later extended to be successfully applied to other realms [5,6,9,11].

3   Experimental Methodology

In our study, two Fast-DENSER search algorithms were compared: a variant where mutation and *single-point crossover* were implemented at the ES level and a simple *local search* where the initial solution is the best of five initial solutions.

The rationale behind these variants is twofold. The first variant tests the hypothesis that the use of crossover does not produce any significant improvements to the algorithm, similar to observations in previous neuroevolution analyses [14,15]. The second variant is inspired by the same motivations that led to the development of Fast-DENSER [1] in which the authors' objective was to optimise DENSER to search for good performing solutions, with limited resources and time frames. This leads to our development of a local search variant of

Fast-DENSER to further reduce testing and training times while preventing degradation of performance. This is further corroborated by the evidence that local search can be a simple yet powerful algorithm [8].

To examine these variants we compare the performance of incrementally developed solutions derived from the same four datasets as in [2] (MNIST, SVHN, CIFAR10, and Fashion-MNIST); averaging over 10 independent runs. The analysis takes into account the depth of the networks, the training times and both evolutionary and testing accuracy to identify how knowledge is transferred between problems. To accomplish this we deploy a specific configuration of Neuroevolution Trajectory Networks (NTNs) that focus on highlighting how the architecture, in terms of layers and learning evolutionary units, morphed and developed from the starting dataset (MNIST). This is accomplished through tagging the evolutionary units with their *origin*. A form of identification to describe from which dataset the specific evolutionary unit originated. These tags are later parsed and used to construct and decorate the networks. Details are given in Fig. 3 of Sect. 5.

The experiment focuses on the last dataset solved in terms of performance analysis and NTNs visualisation. That is because the last dataset should incorporate knowledge gained from the previous three datasets. In the performance analysis section, we examine how the variants have performed from the perspective of Fashion-MNIST; with particular attention to the size of the evolved DNNs using the trainable parameters metric as a proxy. Furthermore, we assess the time taken for each variant to train to explore whether reducing Fast-DENSER's five individuals population to a single-point local search can be advantageous in speeding up the algorithm without impairing performance.

The parameters used to configure Fast-DENSER are the same as those in [2] with the exception of the variant-specific ones introduced by this analysis. In the *crossover* variant the ES has a 50/50 chance of choosing between mutation and single-point crossover. The *local search* uses a single individual and is configured as a $(1, \lambda)$, where $\lambda = 1$. Finally, to simplify the visualisations the number of maximum generations for Fashion-MNIST was reduced from 100 to 50. All variants studied include elitism and retraining. Specific to Fast-DENSER, as described in [1], retraining allows the elite to have the same training time as the parent selected for mutation or crossover.

4   Performance Analysis

Figure 2 illustrates the evolutionary performance of the tested variants on the last classification problem: Fashion-MNIST. These solutions were incrementally developed and knowledge was transferred through evolution from the previous classification datasets. In this enquiry, we focus on the fitness value, i.e. the *accuracy* of the classifications, the number of trainable parameters that the CNNs possess as a proxy of the *depth* of the neural network, and the amount of *training time* required. These are relevant metrics to support our hypotheses.

As can be derived from the plots in Fig. 2 and results in Table 1, the performance in terms of accuracy of the local search variant is comparable to Fast-DENSER where $\lambda = 4$. The trainable parameters are higher on average, meaning deeper networks but much more stationary throughout evolution, despite the higher standard deviation. Noteworthy, as predicted, local search achieved the lowest average training time amongst the variant without significant loss in performance. Furthermore, it appeared that a drastic loss in the average accuracy of the original algorithm occurred around the 8th generation. Conversely, local search appears to produce a much more stable performance across generations.

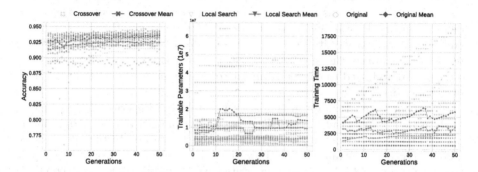

Fig. 2. Convergence plots of the evolutionary performance of all variants based on the last dataset (fashion-MNIST). The analysis focuses on classification accuracy, depth of the CNNs in terms of trainable parameters, and training time spent.

This demonstrates that, from the evolutionary performance perspective, local search is comparable in performance with respect to accuracy and superior in training time (see Table 1). Despite this, local search's accuracy in *testing* conditions did not perform as well with an average of 92.73% compared to the original with an average of 93.27%. The crossover variant, as seen in the evolutionary set-up, performed the worst with an average testing accuracy of 91.87%. Since the distributions appear Gaussian, we proceed to perform a T-test to evaluate the accuracy of local search vs the original Fast-DENSER. It is found that both evolutionary ($p = 0.478$) and testing accuracy ($p = 0.0736$) are comparable and not significantly different.

Table 1. Average evolutionary and testing performance for the 10 runs on the Fashion-MNIST benchmark classification problem.

	Train	Test	Time(sec.)	Trainable Parameters
Local Search$^{(std)}$	$0.9316^{(3.5e-3)}$	$0.9273^{(7.2e-3)}$	$2236^{(479)}$	$1.996318e7^{(8.048529e6)}$
Original$^{(std)}$	$0.9329^{(2.6e-3)}$	$0.9327^{(5.2e-3)}$	$5852^{(1166)}$	$1.406319e7^{(1.703391e6)}$
Crossover$^{(std)}$	$0.9236^{(3.6e-3)}$	$0.9187^{(1.46e-2)}$	$3234^{(412)}$	$1.191039e7^{(1.779603e6)}$

5   NTNs Analysis

Neuroevolution Trajectory Networks (NTNs) are Search Trajectory Networks (STNs) applied to neuroevolution dynamics. In this work, we tailor STNs to neural architecture search by focusing on the learned units that are transferred from one dataset to the next. The model definitions reported below are those of STNs, which are largely applicable to NTNs, can be found in [11]; additionally we highlight the specific differences that constitute NTNs for this research.

These network models require the following definitions.

Representative Solution. A solution (i.e. a CNN structure) that represents the status of the search process (by generation) in the space of potential architectures (layers) allowed by the defined grammar of the DSGE apparatus. The solution with the best fitness in the population at a given iteration is chosen as the representative solution. Fitness is calculated by classification accuracy.

Location. A non-empty subset of solutions that results from retrieving types of DNN layer found while running the algorithm. Each solution in the search space is mapped to one location. Several similar solutions are generally mapped to the same location, as the locations represent a partition of the architecture's search space.

Trajectory. Given a sequence of representative solutions in the order in which they are encountered during neuroevolution. A neuroevolution trajectory is a search sequence, mapping solutions to their corresponding location.

Node. A location in a neuroevolution trajectory. The set of nodes is denoted by N. Nodes have attributes which are used to decorate the network, maximising the conveyance of information about transfer learning.

Edges. Edges are directed and connect two consecutive locations in the trajectory. Edges are weighted with the number of times a transition between two given nodes occurred during the NTN modelling process. The set of edges is denoted by E.

Neuroevolution Trajectory Network. A directed graph $NTN = (N, E)$, with node set N, and edge set E as defined above.

5.1   Sampling and Model Construction

The network models are constructed from a data log derived by running the studied variants on the last benchmark classification dataset (Fashion-MNIST); a network is representative of 10 runs of each variant. The process works as follows: we incrementally develop solutions for all datasets starting from MNIST and ending with Fashion-MNIST. Although these logs are created on all the datasets we perform an analysis in retrospect, only focusing on generating NTNs networks based on the last dataset; to examine what knowledge was transferred in terms of CNNs architectures. The focus on the final dataset allows us to determine from which of the previous datasets (if any) the learning originated and was transferred.

To achieve NTNs logs for network generation a sequence of steps is recorded from evolution; that is two adjacent representative solutions in the search process. Solutions (evolved CNNs structures) are represented by a unique signature reflecting the allowed architecture of the networks discovered; from features extraction to classification.

The link between these consecutive steps is what forms edges in the networks. Each representative solution is made to store essential attributes such as the solution's classification accuracy (fitness), the origin of the layers and learning units, as well as the depth (size) of the CNNs in terms of trainable parameters. All of these attributes will be used according to different decoration styles; which are described below. After this log file is created, a post-processing stage maps solutions to locations based on their origin, and models the network object as per the aforementioned definitions.

In Fig. 3 we illustrate how this complex multidimensional space visualisation for transfer learning is achieved. As previously stated due to the numerous components to be represented, the decision was to compartmentalise the visualisation of networks in groups based on the allowed layers of the grammar configuration for CNNs (features and classification). This removes the other components, leaving only the DNNs' architectural components of interest (see *Evo. units filter* in Fig. 3). We create 3 filter groups, each containing 2 layers types: group 1 for *fully connected* and *dropout* layers, group 2 describing *pooling average* and *pooling max* layers and group 3 for *batch normalisation* and *convolution* layers. These groups have their associated colours as decorators representing them. Three learning optimisers are shown (Adam, Rmsprop, Gradient Descent) and carry the same colour throughout the groups.

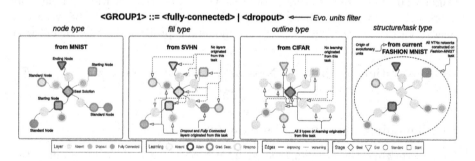

Fig. 3. Illustration of how the NTNs visualisation of this complex multidimensional space is achieved. To highlight the knowledge transferred from different datasets it was decided to compartmentalise the visualisation based on layer types (Evo. units filter) and distinguish by the problem in which the layers and learning units were adopted. Layer and learning that are *absent* from a problem are represented both with a neutral colour.

The compartmentalised (by layer group) visualisations are related to the network constructed using the representative solutions (architectures comprised by layer types) for the last dataset Fashion-MNIST, hence why the structure and layout of the networks are the same throughout. To further enhance the interpretability, the seed for *force-directed* layout [4] is fixed to reproduce the same visualisation, allowing for ease of comparison of knowledge transferred between datasets. The NTNs' visualisations are subdivided by classification problems. If no evolutionary units were derived from any dataset, a pale colour was used to indicate their absence (see Fig 3).

In Fig. 3 each dataset visualisation describes the contributions of that dataset to the final network. The nodes' *shape* represent the different stages that are possible in a trajectory. A representative solution that starts a trajectory (start of a run) is depicted as a square. Standard nodes are circular. The end of a trajectory (end of a run) is a triangle and the best solution identified (best architecture in terms of accuracy) is represented by a diamond (exemplified by the text inside "from MNIST").

The *size* of the shapes is dictated by the normalised metric of *trainable parameters* for the CNN, which is a proxy of the depth of the DNN.

The *fill* of the nodes is dependent on whether the specific layers of the filter group originated from that dataset (exemplified by the text inside "from SVHN").

The *outline* colour is consistent throughout groups as the available optimisers for all variants are: Adam, Rmsprop and Gradient Descent (exemplified by the text inside "from CIFAR").

5.2 NTNs Analysis: Revealing Transfer Learning

In the illustrations (Figs. 4, 5 and 6) the NTNs network constructed from the last classification problem (Fashion-MNIST) depicts the dynamic of the variants of this neuroevolution algorithm over 10 runs of 50 generations each. Each variant produces a different network, comprising different representative solutions. The network structure within each group for the same variant does not change between datasets to allow for a clear comparison of the learning that was transferred.

General Observations. The predominant observation that we can obtain by looking at these network models is that, although the allowed sequence of layers specified by the grammar is relatively constrained, all variants of Fast-DENSER seem to traverse the architectural search space in diverse ways, without generating many points of convergence to specific solutions. Some solutions are re-visited multiple times by the same trajectories indicating a reiteration of accepted CNNs' architectures.

NTNs (as with STNs) are good indicators of the diversity that a given algorithm produces. With this in mind, we can observe that the original Fast-DENSER, although it only leverages the mutation operator in the ES, is capable of producing diverse solutions, both in layers and learning (84 nodes, 45 improving edges and 34 worsening ones). Conversely, crossover does not have the same characteristic of diversity, noticeable from the reduced amount of nodes in the NTNs (39 nodes, 22 improving edges and 12 worsening ones). Finally, local search has the least amount of diversity (22 nodes, 17 improving edges and 0 worsening ones) with one of the trajectories showing a point of convergence that leads to multiple ending solutions. It is notable that this variant is more resilient as it does not produce any worsening transitions between representative solutions.

In line with observations from Fig. 2, we can infer from the NTNs that the size of the neuroevolved CNNs (node size) is much greater throughout the local search NTN than in the other variants. In particular, we can also identify which of the layers, acquired through which of the datasets, are responsible for inflating the size; as it is known: primarily *fully-connected* and *convolution* layers.

With respect to the learning units ported, we observe that the vast majority have been derived from CIFAR, closely followed by Fashion-MNIST. The most popular choice of optimiser transferred overall was *gradient-descent*, ported from almost every datasets and present in nearly every variant.

Finally, the length of trajectories varies greatly between variants. The original (being the most diverse) has longer trajectories on average, indicating greater architectural exploration, which may explain the longer training times. Crossover trajectories are less explorative than the original but longer than local search which has the shortest paths, which is explainable by the intrinsic characteristics of this variant.

Group-Specific Observations

Fast-DENSER Group 1: Fully-Connected and Dropout Layers. In Fig. 4 we can infer the following.

- Fast-DENSER evolutionary process has correctly deemed the drop-out layer to be a fundamental architectural component in CNNs. This is visible both in local search (Fig. 4c) and the original variants (Fig. 4a) as the vast majority of nodes possess the units. On the other hand crossover does not present a high percentage of these, which may explain the lower performance.
- Fully connected layers are also predominant in all variants. Similarly, crossover does not have as many, which are mainly acquired during CIFAR and Fashion-MNIST. The original variant has acquired these units almost evenly on all datasets for most of the trajectories.
- Interestingly a trajectory in Fig. 4a on the lower left corner appears to have gained and retained *dropout* layers from almost all the datasets except from SVHN where all the solutions have acquired *fully-connected* units.

Fast-DENSER Group 2: Pooling Average and Pooling Max Layers. In Fig. 5 we can observe the following.

- All variants show that there is a general lack of these evolutionary units. This is noteworthy but not unforeseen as this is a type of layer that provides

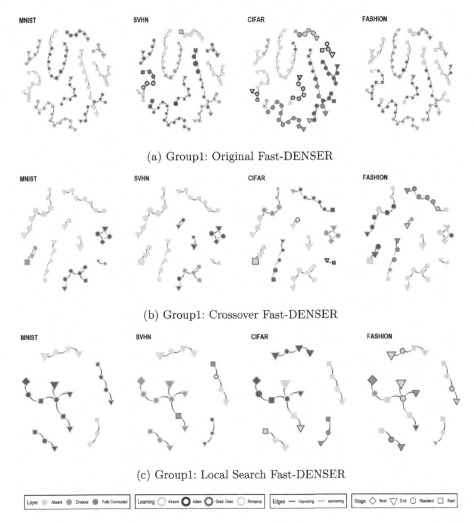

(a) Group1: Original Fast-DENSER

(b) Group1: Crossover Fast-DENSER

(c) Group1: Local Search Fast-DENSER

Fig. 4. In this visualisation we present the NTNs of *group 1* for all the 3 analysed variants.

a summarisation of the feature map from convolutional layers to reduce the number of parameters to learn from. A useful component that this algorithm and its variants do not view as indispensable.

- In particular in local search, Fig. 5c these layers are almost nonexistent. Noteworthy, only one of the trajectories possesses these in the lower right area.
- Generally, CIFAR and the current dataset (Fashion-MNIST) are the location of origin of these. With the exception of the original variant (Fig. 5a) which transported these units since the very initial dataset (MNIST).

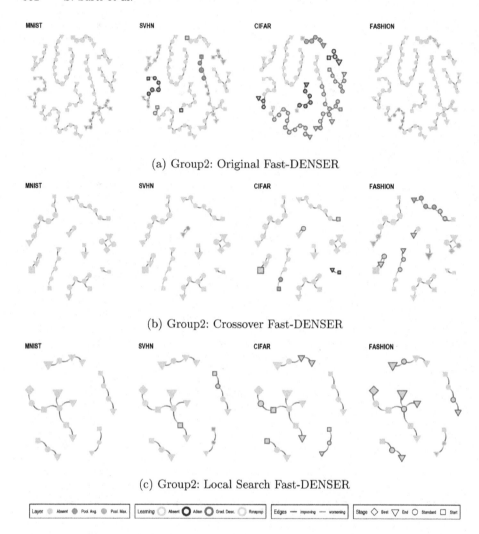

(a) Group2: Original Fast-DENSER

(b) Group2: Crossover Fast-DENSER

(c) Group2: Local Search Fast-DENSER

Fig. 5. In this visualisation we present the NTNs of *group 2* for all the 3 analysed variants.

Fast-DENSER Group 3: Batch Normalisation and Convolution Layers. In Fig. 6 we can observe the following.

– The NTNs models confirm to different extents that these evolutionary units are essential components of the CNNs architectures. In Fig. 6a it is visible that such units have originated from all of the preceding datasets, highlighting their importance. The original variant appears to be offering a vast variety of these.
– The convolutional units are predominant in all of the variants as these are the essential building blocks for the creation of our networks of interest. In

the original version, the majority of these appear to be originating in CIFAR and the current dataset. Meanwhile, local search has these originating further into the past, in the first two datasets (see Fig 6c).

– In local search the four trajectories that converge to a representative solution have batch normalisation layers deriving from MNIST, and the convolution ones originated from SVHN, the following dataset. Meanwhile, only two solutions from a single trajectory have batch-normalisation coming from the current problem, proving that past evolution is sufficient in the construction of networks for the current dataset.

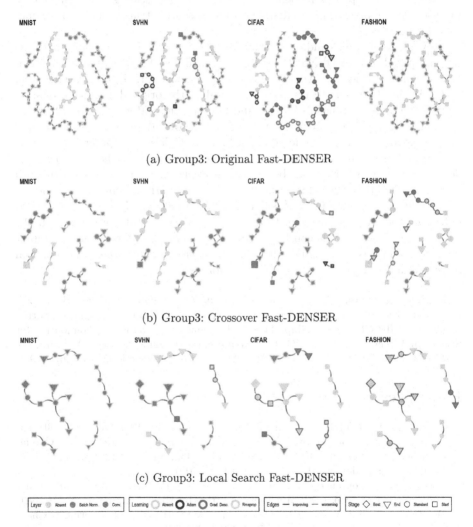

(a) Group3: Original Fast-DENSER

(b) Group3: Crossover Fast-DENSER

(c) Group3: Local Search Fast-DENSER

Fig. 6. In this visualisation we present the NTNs of *group 3* for all the 3 analysed variants.

6   Conclusion

The principal enquiry of this paper was to visually examine the dynamics of transfer learning for deep neuroevolution. In this work, two variants of the Fast-DENSER algorithm were evaluated: adding crossover and using a simple local search. The motivations were: to observe if decreasing the population to a single individual could produce accurate performance with a further reduction in training time; and to test the intuition that an ES equipped with crossover would *not* improve the original algorithm, which only used mutation. These results on incremental development helped to validate both hypotheses; confirming that local search is a powerful paradigm and that crossover struggles to improve neuroevolution.

Key NTNs findings indicated that the variants differed in solution diversity compared to Fast-DENSER. Namely, in local search, convergence and fewer nodes signalled a lesser diversity. Trajectory lengths varied greatly, with local search having the shortest trajectories. Notably, the original algorithm demonstrated the highest transfer of layers and learning that often originated from all previous datasets. The absence of transfer and the overall presence of the pooling layers, which evolution considered superfluous in CNNs, were points of interest.

We propose that this form of analysis could be used to help shape DNN design and that this tool can be crucial in explaining intrinsic traits in both hand-crafted and neuroevolution models. In future work, we aim to examine the neuroevolution of other network structures (i.e. Recurrent Neural Networks) in different problem domains (i.e. time series prediction, real-world images). We envision that the analysis can be extended to maximise information. An intent for future analysis is to signal when combinations of layer types, derived from incremental development, can improve performance.

Acknowledgements. This research is the result of the collaboration with the Departamento de Engenharia Informática of the University of Coimbra, Portugal; supported by the 2022 SPECIES scholarship. This work is funded by the FCT - Foundation for Science and Technology, I.P./MCTES through national funds (PIDDAC), within the scope of CISUC R&D Unit - UIDB/00326/2020 or project code UIDP/00326/2020.

References

1. Assunção, F., Lourenço, N., Machado, P., Ribeiro, B.: Fast DENSER: efficient deep neuroevolution. In: Sekanina, L., Hu, T., Lourenço, N., Richter, H., García-Sánchez, P. (eds.) EuroGP 2019. LNCS, vol. 11451, pp. 197–212. Springer, Cham (2019). https://doi.org/10.1007/978-3-030-16670-0_13
2. Assunção, F., Lourenço, N., Ribeiro, B., Machado, P.: Incremental evolution and development of deep artificial neural networks. In: Hu, T., Lourenço, N., Medvet, E., Divina, F. (eds.) EuroGP 2020. LNCS, vol. 12101, pp. 35–51. Springer, Cham (2020). https://doi.org/10.1007/978-3-030-44094-7_3
3. Assunção, F., Lourenço, N., Machado, P., Ribeiro, B.: DENSER: deep evolutionary network structured representation. Genet. Program Evolvable Mach. **20**(1), 5–35 (2018). https://doi.org/10.1007/s10710-018-9339-y

4. Fruchterman, T.M.J., Reingold, E.M.: Graph drawing by force-directed placement. Softw. Pract. Exper. **21**(11), 1129–1164 (1991)
5. Lavinas, Y., Aranha, C., Ochoa, G.: Search trajectories networks of multiobjective evolutionary algorithms. In: Jiménez Laredo, J.L., Hidalgo, J.I., Babaagba, K.O. (eds.) EvoApplications 2022. LNCS, vol. 13224, pp. 223–238. Springer, Cham (2022). https://doi.org/10.1007/978-3-031-02462-7_15
6. Lavinas, Y., Ladeira, M., Ochoa, G., Aranha, C.: Component-wise analysis of automatically designed multiobjective algorithms on constrained problems. In: Proceedings of the Genetic and Evolutionary Computation Conference. p. 538–546. GECCO 2022, Association for Computing Machinery, New York, NY, USA (2022). https://doi.org/10.1145/3512290.3528719
7. Lourenço, N., Assunção, F., Pereira, F.B., Costa, E., Machado, P.: Structured grammatical evolution: a dynamic approach. In: Handbook of Grammatical Evolution, pp. 137–161. Springer International Publishing, Cham (2018). https://doi.org/10.1007/978-3-319-78717-6_6
8. Lourenço, H.R., Martin, O.C., Stützle, T.: Iterated local search: framework and applications. In: Handbook of Metaheuristics, pp. 129–168. Springer International Publishing, Cham (2019). https://doi.org/10.1007/978-3-319-91086-4_5
9. Narvaez-Teran, V., Ochoa, G., Rodriguez-Tello, E.: Search trajectory networks applied to the cyclic bandwidth sum problem. IEEE Access **9**, 1 (2021). https://doi.org/10.1109/access.2021.3126015
10. Newman, M.E.J.: Networks: an introduction. Oxford University Press, Oxford; New York (2010)
11. Ochoa, G., Malan, K.M., Blum, C.: Search trajectory networks: A tool for analysing and visualising the behaviour of metaheuristics. Appl. Soft Comput. **109**, 107492 (2021). https://doi.org/10.1016/j.asoc.2021.107492
12. Ochoa, G., Malan, K.M., Blum, C.: Search trajectory networks of population-based algorithms in continuous spaces. In: Castillo, P.A., Jiménez Laredo, J.L., Fernández de Vega, F. (eds.) EvoApplications 2020. LNCS, vol. 12104, pp. 70–85. Springer, Cham (2020). https://doi.org/10.1007/978-3-030-43722-0_5
13. Sarti, S., Adair, J., Ochoa, G.: Neuroevolution trajectory networks of the behaviour space. In: Jiménez Laredo, J.L., Hidalgo, J.I., Babaagba, K.O. (eds.) EvoApplications 2022. LNCS, vol. 13224, pp. 685–703. Springer, Cham (2022). https://doi.org/10.1007/978-3-031-02462-7_43
14. Sarti, S., Adair, J., Ochoa, G.: Recombination and novelty in neuroevolution: a visual analysis. SN Comput. Sci. **3**(3), 185 (2022). https://doi.org/10.1007/s42979-022-01064-6
15. Sarti, S., Ochoa, G.: A NEAT visualisation of neuroevolution trajectories. In: Castillo, P.A., Jiménez Laredo, J.L. (eds.) EvoApplications 2021. LNCS, vol. 12694, pp. 714–728. Springer, Cham (2021). https://doi.org/10.1007/978-3-030-72699-7_45
16. Valueva, M., Nagornov, N., Lyakhov, P., Valuev, G., Chervyakov, N.: Application of the residue number system to reduce hardware costs of the convolutional neural network implementation. Math. Comput. Simul. **177**, 232–243 (2020). https://doi.org/10.1016/j.matcom.2020.04.031

Feature Selection on Epistatic Problems Using Genetic Algorithms with Nested Classifiers

Pedro Carvalho[1]([✉]) [iD], Bruno Ribeiro[2] [iD], Nuno M. Rodrigues[3] [iD],
João E. Batista[3] [iD], Leonardo Vanneschi[4] [iD], and Sara Silva[3] [iD]

[1] University of Coimbra, Centre for Informatics and Systems of the University
of Coimbra, Department of Informatics Engineering, Coimbra, Portugal
`pfcarvalho@dei.uc.pt`
[2] GECAD - Research Group on Intelligent Engineering and Computing for Advanced
Innovation and Development, LASI - Intelligent Systems Associate LAboratory,
Polytechnic of Porto, Rua Dr. António Bernardino de Almeida, 431,
4249-015 Porto, Portugal
`brgri@isep.ipp.pt`
[3] LASIGE, Department of Informatics, Faculty of Sciences, University of Lisbon,
Lisbon, Portugal
`{nmrodrigues,jebatista,sara}@fc.ul.pt`
[4] NOVA Information Management School (NOVA IMS), Universidade Nova de
Lisboa, Campus de Campolide, 1070-312 Lisboa, Portugal
`lvanneschi@novaims.unl.pt`

Abstract. Feature selection is becoming an essential part of machine
learning pipelines, including the ones generated by recent AutoML tools.
In case of datasets with epistatic interactions between the features, like
many datasets from the bioinformatics domain, feature selection may
even become crucial. A recent method called SLUG has outperformed the
state-of-the-art algorithms for feature selection on a large set of epistatic
noisy datasets. SLUG uses genetic programming (GP) as a classifier
(learner), nested inside a genetic algorithm (GA) that performs feature
selection (wrapper). In this work, we pair GA with different learners, in
an attempt to match the results of SLUG with less computational effort.
We also propose a new feedback mechanism between the learner and the
wrapper to improve the convergence towards the key features. Although
we do not match the results of SLUG, we demonstrate the positive effect
of the feedback mechanism, motivating additional research in this area
to further improve SLUG and other existing feature selection methods.

Keywords: Feature Selection · Epistasis · Genetic Algorithms ·
Genetic Programming · Decision Trees · Machine Learning ·
Genome-Wide Association Studies

Supplementary Information The online version contains supplementary material
available at https://doi.org/10.1007/978-3-031-30229-9_42.

J. Correia et al. (Eds.): EvoApplications 2023, LNCS 13989, pp. 656–671, 2023.
https://doi.org/10.1007/978-3-031-30229-9_42

1   Introduction

Feature selection is becoming progressively more important as Machine Learning (ML) addresses increasingly difficult problems involving large and noisy datasets. Particularly in the bioinformatics domain, datasets tend to be intractable in terms of dimensionality, and ML pipelines normally perform a very strong feature selection prior to applying any learning algorithm. One such example is the Tree-based Pipeline Optimization Tool (TPOT) [13,14], an automated ML (AutoML) system based on Genetic Programming (GP) [17], where feature selection is given paramount importance. More recent variants, called TPOT-MDR [22], and TPOT-FSS [10], have specialized in genetics research while introducing in their pipelines yet other elements dedicated to feature selection. MDR stands for multi-factor dimensionality reduction, while FSS stands for feature set selector. TPOT and its variants have been evaluated on simulated Genome-Wide Association Studies (GWAS) datasets generated by GAMETES [24], as well as in real-world datasets. GAMETES is an open-source software package designed to generate GWAS datasets with pure epistatic interactions between the features. With the increasing amount of feature selection, TPOT and its variants have shown a growing ability to detect and model the predictive features of these very difficult datasets.

The GAMETES datasets have recently been used to evaluate other classification methods. In [9], a multiclass classifier called M4GP was compared with several off-the-shelf ML methods on six GAMETES datasets, proving to be the only method able to find and model the epistatic relationships, except for the hardest dataset where all the methods failed. M4GP is based on GP and improves upon its predecessor M3GP [11] by using a stack-based representation [16] for the individuals and lexicase selection [23] for choosing parents. In [2], M4GP was further evaluated and compared with TPOT-MDR as well as other methods, including extreme gradient boosting and a deep neural network. Additionally, expert knowledge filter (EKF) algorithms were coupled with both M4GP and TPOT-MDR, and all the variants were compared with each other. The EKF algorithms are pre-processing filters sensitive to interactions, based on the Relief family of feature selection algorithms [8]. M4GP+EKF used ReliefF [8] to select the top 10 features before M4GP was run and achieved comparable results to the much heavier TPOT variants on 16 GAMETES problems. In [20], one GAMETES problem was used to test a new ensemble GP system (eGP), with interesting results that showed the importance of pre-selecting features before running the classifier. Finally, a new method called SLUG [19] has recently been compared to standard GP, M3GP, M4GP, and M4GP+EKF, surpassing all of them on ten GAMETES problems. SLUG uses a Genetic Algorithm (GA) as a wrapper to GP, where each GA individual selects a subset of features and takes as fitness the accuracy of the best individual of a GP run that uses those features. However, despite the good results achieved, SLUG is faithful to its name and runs extremely slowly. Coupling GA with classifiers that are computationally cheaper than GP was also attempted [19], but without much success. Among decision trees (DT) [18], random forests (RF) [1] and extreme gradient boosting (XGBoost) [4], only the pairing of GA with DT suggested that it might be able to find the key features needed to solve the problem.

In this work, we explore a new set of classifiers to pair with the GA feature selection wrapper. We begin by evaluating different classifiers on datasets that contain only the key features, thus establishing baselines of what each method can do with the right features. Then we pair GA with different classifiers, focusing mainly on cheap and fast DT-based methods. Finally, we present a novel feedback mechanism between the classifier and its GA wrapper, enabling enriched communication between the two components. The proposed mechanism extracts the features used in the best classifier model and creates a new GA individual with the same features. We perform extensive comparisons and analyze whether the new feedback mechanism speeds the GA evolution towards the minimal set of key features.

In Sect. 2 we explain our methodology; then we describe the GAMETES datasets, we define the baseline results and we describe all the classifiers used and the new feedback mechanism. In Sect. 3 we detail our experimental setup, including the parameters used in the experiments. The results are presented in Sect. 4, while Sect. 5 concludes and outlines directions for future work.

2   Methodology

Most of the previously mentioned methods were evaluated on both synthetic and real-world datasets. However, it was their ability to solve the synthetic GAMETES problems that set them apart from each other and have been driving the research towards better methods. We evaluate our methods exclusively on these difficult datasets, which are described in Sect. 2.1.

It is known that the major difficulty of the GAMETES datasets is finding the key features to model. However, what performance is expected from different ML methods once those features are found is not widely reported in the literature. In order to establish baselines, we evaluate several classifiers on datasets where only the key features remained, in Sect. 2.2.

We propose pairing GA with different classifiers besides the ones already evaluated in the literature. Since we use some GAMETES datasets that had not been used before, we describe both published and newly proposed pairings in Sect. 2.3.

Finally, we introduce a novel feedback mechanism between the classifier and GA, which allows the GA to know which features were actually used by the best model returned by the classifier. We describe this method in Sect. 2.4.

2.1   Datasets

The GAMETES [24] datasets are produced by a tool that embeds epistatic gene-gene interactions into noisy n-locus genetic datasets. For each dataset, an n-way epistatic interaction is present that is predictive of the classification, but this is masked by the presence of confounding features and noisy classifications. These datasets have different degrees of difficulty according to the number of epistatic pairs, the total number of features, and the signal-to-noise ratio, where lower ratios mean harder problems.

In this work we use 16 different GAMETES datasets, using [2, 3] epistatic pairs, [10, 100, 1000] total features, and [0.05, 0.1, 0.2, 0.4] signal-to-noise ratio values. Due to computational and time constraints, we could not perform experiments on all the possible combinations of the number of features and signal-to-noise ratio, or go beyond 1000 features. All the datasets contain 2000 observations, perfectly balanced between the two classes.

2.2 Defining Baselines

Different classifiers were given only the key features of each used GAMETES dataset, allowing us to establish performance baselines for different ML methods, which represent the expected performance once the key features are found. Besides the methods already mentioned (GP, M4GP, DT), we also evaluated the Mahalanobis Distance Classifier (MDC), Support Vector Machines (SVM) [5], and MultiLayer Perceptron (MLP) [21].

Table 1. Median test accuracy of MDC, GP, M4GP, SVM, MLP, and DT classifiers (sorted by average value) over 90 runs (30 for each of the 10, 100 and 1000 feature GAMETES datasets), using only the key features. GP and M4GP use 30 generations and population size of 100, like the GP part of the SLUG algorithm. In later sections of this work, we will consider that an algorithm obtains good results if its median test accuracy surpasses the 90% DT baseline (last row).

Classifier	2w_0.05	2w_0.1	2w_0.2	2w_0.4	3w_0.1	3w_0.2	Average
MDC	62.83	67.50	71.33	77.67	50.33	51.25	63.49
GP	63.33	68.83	72.25	65.00	64.00	62.16	65.93
M4GP	61.67	68.75	71.67	79.00	58.25	59.16	66.42
SVM	62.75	68.83	71.92	79.17	63.08	65.41	68.53
MLP	62.67	68.83	71.75	79.25	64.75	67.58	69.13
DT	63.25	68.83	72.33	79.25	64.75	68.50	69.49
Baseline	**56.93**	**61.95**	**65.10**	**71.33**	**58.28**	**61.65**	

Table 1 displays the results, in terms of accuracy on unseen data (test accuracy), obtained on different sets of GAMETES problems, where each value is the median obtained by each method on the three problems (with 10, 100, 1000 features) with the same number of epistatic pairs and signal-to-noise ratio. In the table, datasets are represented using the notation Nw_M, where N is the number of epistatic pairs and M is the signal-to-noise ratio. We recall that, although the original problems had different numbers of features, the amount of key features is the same for all 2-way (2w) or 3-way (3w) datasets. The rows are sorted by the average result of each method.

We do not observe any major differences between the methods, except for MDC, which falls behind on the 3w datasets, and GP, which underperforms on the 2w_0.4 datasets. Despite the success of the SLUG method, these results suggest that GP may not even be the best classifier to pair with GA since it adds computational costs without achieving the best results. They also suggest that the DT classifiers obtain the best results overall, despite requiring fewer computational resources. The last row of the table contains values calculated as 90% of the values obtained by the best method, DT. In later sections of this work, we will consider that an algorithm obtains good results if its median test accuracy surpasses this baseline.

From now on, we focus only on the GP-based methods already used on GAMETES problems in the literature (GP, M4GP) and on DT-based methods, therefore disregarding MDC, SVM and MLP.

2.3   Classifiers

Different classifiers were used as standalone or paired with GA and evaluated on the selected GAMETES problems. Some had already been used in the literature (mostly GP-based methods), while others are being proposed here for the first time (mostly those with names beginning with "GA"). Most of the new methods are based on DT classifiers, for their lightweight computational demands and good baseline results shown above. Any considerations regarding the previous performance of published methods refer exclusively to results obtained on GAMETES datasets.

GP. This is standard GP, unpaired with any wrapper, using the same settings as the GP part of the SLUG method. Already used for comparison in two works [19, 20], it was successful only on the easiest problems.

M3GP. The M3GP method, predecessor of M4GP, was also used for comparison in two works [19,20]. It performed well on the easiest problems, better than M4GP.

M4GP. The M4GP method has already been studied in different works [2,9,19]. It performed worse than M3GP in the SLUG work, even on the easiest problems.

M4GP-E. This is the M4GP+EKF method that uses the ReliefF feature selection before running M4GP. Already used in different works [2,19], it was one of the strongest methods, only surpassed by SLUG.

SLUG. This is the GA+GP pairing that was named SLUG [19]. It is a nested evolutionary method that performs a full GP run for each GA individual. The GA is the wrapper, where each individual is a binary string of length equal to the number of features in the original dataset, indicating whether each feature is selected or not. When evaluating a GA individual, the selected features are given to GP, the learner. A complete run of GP is performed on the training set with the restricted set of features selected by the GA, and the best fitness achieved in the GP run becomes the fitness of the GA individual. GP does not have to use all the features selected by the GA since it also performs its own feature selection. This means that the GA only has to pre-select a sufficiently small subset that includes the key features, and GP will do the rest. This has been pointed out as one of the main reasons for the success of SLUG, which outperformed all the other methods listed here so far.

GAM4GP. This is a new pairing, GA+M4GP. The SLUG work showed that the superiority of SLUG over M4GP-E was due to GA being a better wrapper than the EKF feature selection (and not because M4GP was better than GP). Therefore, we pair GA with M4GP, following the same process as described for SLUG, with the expectation that it may perform better than any of the previous combinations.

GADT. This is the pairing GA+DT. It was already tested as a faster alternative to GA+GP in the SLUG work [19], but only on one problem (of 1000 features), without success. However, since DT is the method with the best baseline, we give GA+DT another try and thoroughly evaluate it (as well as its new variants below) in the complete set of selected problems.

GAGADT. This is a new variant of GADT where, instead of nesting a DT in each GA individual, a complete GADT method is nested in each GA individual (GA+GA+DT). This approach adds a new level of computational complexity but also provides a new layer of feature selection, where the second GA further selects features from the ones provided by the first GA. Since DT performs well once the key features are found, the effort of this method is mainly on strengthening the feature selection.

GAMDT. This is another variant of GADT where, instead of nesting a single DT in each GA individual, multiple decisions trees (MDT) are created and nested in each GA individual (GA+MDT). Each tree of this population is built using only a random subset of the n features pre-selected by the GA, where each subset contains m features with $1 \leq m \leq n$. This way, each tree has its own subset of allowed features. This approach is similar to the one used in the eGP work, with the same goal of promoting the creation of trees with a large variety of features, but the similarities with evolutionary or ensemble methods end here. In MDT, all the trees are independently evaluated on the training set, but only the best one is kept as the classifier without any population evolution or ensemble voting.

2.4   Adding Feedback

A surprising behavior was observed during preliminary testing, affecting all the wrapper+learner methods described above. During the GA evolution, once the key features are found, they may be lost again.

It may be tempting to think that, in order to avoid this, the GA elitism should guarantee that the best individual is never lost from the population. However, the problem is not a lack of elitism, but rather a lack of communication between the GA and the learner. In the specific case of SLUG, the GA informs GP of what features GP can use, and GP informs the GA of what fitness (accuracy) was obtained by the best model. However, GP never reports to the GA which of

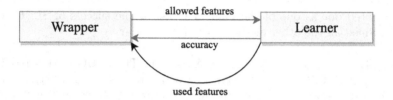

Fig. 1. Using the feedback mechanism proposed, the wrapper communicates the allowed features and the best learner communicates the used features in addition to the accuracy.

the allowed features it actually used in the model. Therefore, even if elitism was used to always keep the best GA individual in the population, this individual does not know which of its features are important or not. Each time a new GA individual is evaluated, a new GP run is performed, with new random initial models and a final best model that may or may not be using the right features.

Hence, a new feedback mechanism is used to address this limitation, where the learner informs the wrapper which features it effectively used. Figure 1 represents a wrapper and a learner with the three types of communication between them: the wrapper communicates the allowed features; the learner communicates the accuracy; the learner communicates the used features. This way, the wrapper has more information about the problem space, which may lead to faster evolution. We apply this mechanism to the GAMDT method described above, obtaining yet another classifier:

GAMDT+. This method expands on GAMDT by incorporating the additional feedback mechanism. Every time a GA individual is evaluated, a new population of trees is created, and the best tree is chosen, just like in GAMDT. The difference is that this tree reports back to the GA not only its accuracy but also the features it uses. The GA always keeps track of the best set of features, i.e., the ones used by the best tree among all the trees created from the beginning of the run. Each new GA generation then includes one individual whose binary string only allows this best set of features.

We chose to apply the feedback mechanism to a DT-based method because they all run faster than either SLUG or GAM4DT. Another reason is that the deterministic nature of the DT learners causes higher selective pressure on the features. In fact, the same subset of allowed features always produces a model with the same features. Among the DT-based methods, we chose GAMDT because it performs better than GADT and runs faster than GAGADT.

3   Experimental Setup

To validate our approach, we compare all the proposed classifiers with SLUG and the other listed GP-based methods, using the parameters of the original work [19] where applicable.

The GA runs for 50 generations with a population of 100 individuals. Parents are selected using roulette selection. GA uses a standard crossover operator that switches a same-sized segment between the two parents with a 70% probability. For mutation, individuals have a 1% probability of suffering a bit flip mutation, where each bit is flipped with a $1/m$ probability, where m is the length of the binary string (the number of features in the data).

When using a GP-based classifier (SLUG, GAM4GP), we use population size 100 for 30 generations. For SLUG, the selection method is tournament selection with tournaments of 5 individuals. Elitism is also used, ensuring the best individual in the population is always copied to the next generation. The variation operators are subtree crossover and mutation, both applied with a 50% probability. The official SLUG repository [12] is used in these experiments. In GAM4GP, with lexicase selection and no elitism, a detailed explanation of the variation operators used in M4GP can be found in the cited works. For GAM4GP, we interface the existing implementation of M4GP [3] with the GA feature selection module of the official SLUG codebase. For the DT-based methods, we use the scikit-learn [15] implementation with a maximum depth of 6 and Gini impurity as the split criterion. The MDT approaches create 100 different trees for each GA individual, each with a random number of features as previously described.

As standalone methods, GP, M3GP, M4GP and M4GP-E all run for 100 generations with 500 individuals in the population. For GP and M3GP, we use the implementation from StdGP [7] and the official M3GP codebase [6]. Similar to M4GP, we refer the reader to the cited works on each of these approaches for details on their respective operators.

We repeat all experiments 30 times, randomly splitting the dataset into 70% training and 30% test for each repetition. Each run uses a different random seed, to ensure the repeatability of the experiments.

4   Results

We evaluate all the methods on all the datasets, measuring the overall accuracy. We present the results as boxplots (training and test) of the 30 runs and tables with the (test) medians. We remark that these are not the results on the last generation, but rather the results of the best model of the run, because of the loss of key features reported in Sect. 2.4. To assess the statistical significance of the results, we perform one-way non-parametric ANOVA analysis by means of Kruskal-Wallis, using 0.01 as the significance threshold. The Supplementary Material contains the p-values obtained in all the pairwise comparisons.

Given the large amount of datasets, we split the results in three parts, obtained on the GAMETES datasets with 10, 100 and 1000 features, respectively. Then we compare the GAMDT and GAMDT+ methods directly, paying special attention to their convergence speed and dispersion of test accuracy across runs.

4.1   10 Feature Datasets

These results are displayed in Table 2 and Fig. 2. In the vast majority of the cases, the methods obtain a median test accuracy above the baseline (see Table 1), which means they probably find the key features. Nevertheless, there are statistically significant differences. SLUG and GAM4GP are always among the best, while the DT-based methods only achieve this in four of the six datasets, and not the same for all methods. One interesting observation is that, although the 2w_0.4 dataset has a higher signal-to-noise ratio than the other 2w datasets, standalone GP performs worse precisely on this dataset. This was already observed in Table 1 and also agrees with the results reported in the SLUG work [19].

Table 2. Median test accuracy on the 10 feature datasets over 30 runs. The best results for each problem are identified in green, more than one if there is no statistically significant difference between them. Results better than the baseline are underlined.

10 features	2w_0.05	2w_0.1	2w_0.2	2w_0.4	3w_0.1	3w_0.2
GP	62.83	68.17	71.00	66.33	62.42	62.75
M3GP	62.17	67.67	69.17	79.58	61.08	64.50
M4GP	61.67	67.50	69.17	79.17	64.25	65.58
M4GP-E	61.25	66.50	69.92	78.42	64.25	64.25
SLUG	62.92	68.17	70.83	79.67	63.75	65.75
GAM4GP	63.00	68.25	71.00	80.08	64.42	66.58
GADT	62.08	66.92	68.67	79.50	63.75	66.33
GAGADT	61.92	66.92	68.83	79.50	63.42	66.33
GAMDT	62.00	67.33	69.00	78.75	63.58	66.17
GAMDT+	62.25	66.67	69.17	79.50	63.67	66.17
Baseline	**56.93**	**61.95**	**65.10**	**71.33**	**58.28**	**61.65**

4.2   100 Feature Datasets

These results are displayed in Table 3 and Fig. 3. Given the much higher difficulty in finding two/three key features in these datasets, it is not surprising that most classifiers fail to find them on the 2w_0.05 dataset as well as on both 3w datasets. On the remaining 2w datasets, only GP, M3GP and GADT struggle to find the key features. Once again, the SLUG method is the winner, followed closely by GAM4GP and M4GP-E.

4.3   1000 Feature Datasets

These results are displayed in Table 4 and Fig. 4. The first thing to notice is that none of the methods is able to reach the baseline values on the 3w datasets, in

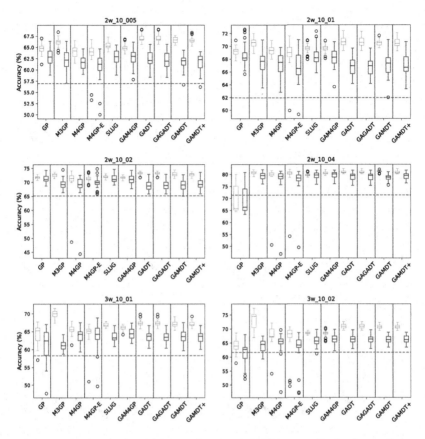

Fig. 2. Performance on the GAMETES 10 feature datasets. Each plot contains, for each method, the results on the training (left, in blue) and test (right, in red) sets. The dashed horizontal lines represent the baselines reported in Table 1. (Color figure online)

Table 3. Median test accuracy on the 100 feature datasets over 30 runs. The best results for each problem are identified in green, more than one if there is no statistically significant difference between them. Results better than the baseline are underlined.

100 features	2w_0.05	2w_0.1	2w_0.2	2w_0.4	3w_0.1	3w_0.2
GP	52.08	53.50	50.92	51.00	49.50	50.33
M3GP	51.33	63.67	68.00	53.67	49.58	51.17
M4GP	56.08	66.08	68.08	75.92	49.67	50.67
M4GP-E	60.67	67.25	70.92	78.08	49.08	50.50
SLUG	61.67	68.08	72.25	77.67	56.58	59.42
GAM4GP	60.00	67.25	70.92	77.33	50.58	50.92
GADT	50.17	59.58	62.83	65.67	49.33	50.50
GAGADT	52.33	63.33	67.33	75.58	50.33	51.58
GAMDT	50.67	63.83	68.08	76.50	49.92	52.17
GAMDT+	52.00	64.83	68.25	76.08	49.67	50.83
Baseline	**56.93**	**61.95**	**65.10**	**71.33**	**58.28**	**61.65**

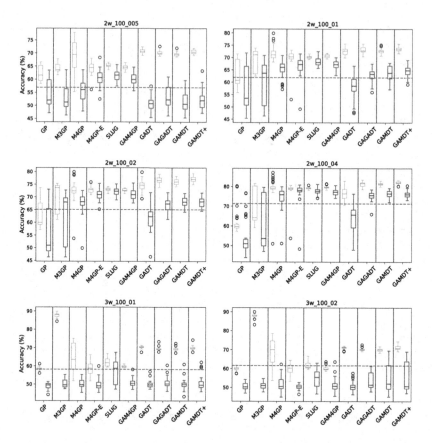

Fig. 3. Performance on the GAMETES 100 feature datasets. Each plot contains, for each method, the results on the training (left, in blue) and test (right, in red) sets. The dashed horizontal lines represent the baselines reported in Table 1. (Color figure online)

any run. On the 2w datasets, both SLUG and M4GP-E reach the baseline, and notably also GAMDT+ is able to reach it on the 2w_0.4 dataset, with GAMDT extremely close on this same problem. GAM4GP shows a high dispersion of values, where many of them also reach the baseline on the 2w datasets, although the median remains below the line. SLUG and M4GP-E are the best methods on 2w_0.2 and 2w_0.4, respectively, and all methods are basically equally bad on the 3w datasets. GAM4GP is, once again, the closest contender.

4.4   Effect of the Feedback Mechanism: GAMDT vs GAMDT+

The GAMETES datasets are difficult problems that require a strong feature selection. Either the methods are able to find the key features, or they will not be able to learn the training data without overfitting. The new feedback mechanism implemented in GAMDT, giving rise to GAMDT+, helps to find

Table 4. Median test accuracy on the 1000 feature datasets over 30 runs. The best results for each problem are identified in green, more than one if there is no statistically significant difference between them. Results better than the baseline are underlined.

1000 features	2w_0.2	2w_0.4	3w_0.1	3w_0.2
GP	50.25	49.50	50.33	49.67
M3GP	49.00	50.73	49.67	51.08
M4GP	50.00	51.08	50.17	50.83
M4GP-E	<u>69.17</u>	<u>77.50</u>	49.83	49.00
SLUG	<u>72.00</u>	<u>75.31</u>	49.75	50.42
GAM4GP	53.92	63.00	50.00	50.25
GADT	50.83	52.92	50.08	50.25
GAGADT	51.17	61.50	49.58	49.75
GAMDT	51.00	70.17	49.58	50.00
GAMDT+	49.58	<u>75.42</u>	49.92	50.08
Baseline	**65.10**	**71.33**	**58.28**	**61.65**

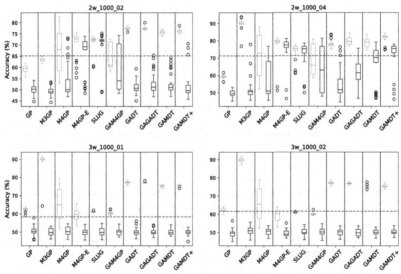

Fig. 4. Performance on the GAMETES 1000 feature datasets. Each plot contains, for each method, the results on the training (left, in blue) and test (right, in red) sets. The dashed horizontal lines represent the baselines reported in Table 1. (Color figure online)

the key features in some cases. One of these is the 2w_1000_04 dataset, where the difference between GAMDT and GAMDT+ is significant with a p-value of 0.0004 (all the p-values can be found in the Suplementary Material).

Figure 5 shows the evolution of accuracy on the 2w_1000_04 dataset. The lines represent the medians of the 30 runs, while the boxplot-like representations

668 P. Carvalho et al.

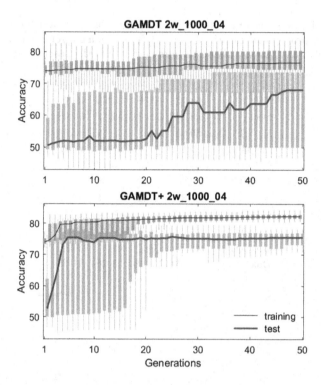

Fig. 5. Evolution of accuracy on the 2w_1000_04 dataset with GAMDT and GAMDT+. The lines represent medians of the 30 runs, training (black) and test (red); the lighter boxplots behind them represent the respective distribution of accuracy values. (Color figure online)

behind the lines represent the distributions of the accuracy values in the different runs. In this figure, we can see that GAMDT+ (the bottom plot) finds the key features much earlier in the run, with the median test accuracy peaking around generation 5 and having a low dispersion of results before generation 20. In comparison, GAMDT (the top plot) clearly struggles to find the key features, maintaining a large dispersion of test accuracy and a highly oscillating median until the end of the run. Also in training, GAMDT+ reveals better and less disperse results.

In other datasets (not shown), the effect is also visible in terms of reducing the dispersion of values, and consequently also the oscillations of the median, even when the key features are not found.

5 Conclusions

We have studied the application of several classification methods that include feature selection to the GAMETES datasets. Our goal was to match the results of the recent state-of-the-art SLUG method, but with a lower computational cost.

SLUG uses genetic programming (GP) as a classifier (learner), nested inside a genetic algorithm (GA) that performs feature selection (wrapper).

We have proposed pairing GA with different types of classifiers, most of them based on decision trees (DT) for their low computational cost. We have also introduced a new feedback mechanism that improves the communication between the wrapper and the learner, with the goal of speeding the feature selection towards the few key features required to solve the GAMETES problems. With the additional feedback, the best learner reports back to the wrapper which features it effectively used. Tested on one DT-based method, the feedback mechanism proved to improve convergence once the key features are found, but did not prove to be very helpful in finding them in the first place.

Overall, the results obtained with GP-based learners were better than the ones obtained with DT-based learners, and none of the methods was able to surpass SLUG. Despite this, the positive effects of the feedback mechanism motivate further research in this area. As future work, we will apply the feedback mechanism also to the GP-based methods, and continue researching this type of feedback for improving other existing feature selection methods.

Acknowledgments. This work was supported by FCT, Portugal, through funding of LASIGE Research Unit (UIDB/00408/2020, UIDP/00408/2020) and CISUC (UID/CEC/00326/2020); projects AICE (DSAIPA/DS/0113/2019), from FCT, and RETINA (NORTE-01-0145-FEDER-000062), supported by Norte Portugal Regional Operational Programme (NORTE 2020), under the PORTUGAL 2020 Partnership Agreement, through the European Regional Development Fund (ERDF). The authors acknowledge the work facilities and equipment provided by GECAD research center (UIDB/00760/2020) to the project team. The authors were also supported by their respective PhD grants, Pedro Carvalho (UI/BD/151053/2021), Nuno Rodrigues (2021/05322/BD), João Batista (SFRH/BD/143972/2019).

References

1. Breiman, L.: Random forests. Mach. Learn. **45**(1), 5–32 (2001). https://doi.org/10.1023/A:1010933404324
2. Cava, W.L., Silva, S., Danai, K., Spector, L., Vanneschi, L., Moore, J.H.: Multidimensional genetic programming for multiclass classification. Swarm Evol. Comput. **44**, 260–272 (2019). https://doi.org/10.1016/j.swevo.2018.03.015
3. cavalab: cavalab/ellyn: python-wrapped version of ellen, a linear genetic programming system for symbolic regression and classification. https://github.com/cavalab/ellyn
4. Chen, T., Guestrin, C.: XGBoost: a scalable tree boosting system. In: Proceedings of the 22nd ACM SIGKDD international conference on knowledge discovery and data mining, pp. 785–794. KDD '16, ACM, New York, NY, USA (2016). https://doi.org/10.1145/2939672.2939785
5. Cortes, C., Vapnik, V.: Support-vector networks. Mach. Learn. **20**(3), 273–297 (1995)
6. Jespb: Jespb/python-m3gp: an easy-to-use scikit-learn inspired implementation of the multidimensional multiclass genetic programming with multidimensional populations (m3gp) algorithm. https://github.com/jespb/Python-M3GP

7. Jespb: Jespb/python-stdgp: an easy-to-use scikit-learn inspired implementation of the standard genetic programming (stdgp) algorithm. https://github.com/jespb/Python-StdGP

8. Kononenko, I.: Estimating attributes: analysis and extensions of RELIEF. In: Bergadano, F., De Raedt, L. (eds.) ECML 1994. LNCS, vol. 784, pp. 171–182. Springer, Heidelberg (1994). https://doi.org/10.1007/3-540-57868-4_57

9. La Cava, W., Silva, S., Vanneschi, L., Spector, L., Moore, J.: Genetic programming representations for multi-dimensional feature learning in biomedical classification. In: Squillero, G., Sim, K. (eds.) EvoApplications 2017. LNCS, vol. 10199, pp. 158–173. Springer, Cham (2017). https://doi.org/10.1007/978-3-319-55849-3_11

10. Le, T.T., Fu, W., Moore, J.H.: Scaling tree-based automated machine learning to biomedical big data with a feature set selector. Bioinformatics 36(1), 250–256 (2020)

11. Muñoz, L., Silva, S., Trujillo, L.: M3GP – Multiclass Classification with GP. In: Machado, P., et al. (eds.) EuroGP 2015. LNCS, vol. 9025, pp. 78–91. Springer, Cham (2015). https://doi.org/10.1007/978-3-319-16501-1_7

12. NMVRodrigues: Nmvrodrigues/slug: an easy-to-use scikit-learn inspired implementation of the feature selection using genetic algorithms and genetic programming (slug) algorithm. https://github.com/NMVRodrigues/SLUG

13. Olson, R.S., Bartley, N., Urbanowicz, R.J., Moore, J.H.: Evaluation of a tree-based pipeline optimization tool for automating data science. In: Proceedings of the Genetic and Evolutionary Computation Conference 2016, pp. 485–492. GECCO '16, ACM, New York, NY, USA (2016). https://doi.org/10.1145/2908812.2908918

14. Olson, R.S., Urbanowicz, R.J., Andrews, P.C., Lavender, N.A., Kidd, L.C., Moore, J.H.: Automating biomedical data science through tree-based pipeline optimization. In: Squillero, G., Burelli, P. (eds.) EvoApplications 2016. LNCS, vol. 9597, pp. 123–137. Springer, Cham (2016). https://doi.org/10.1007/978-3-319-31204-0_9

15. Pedregosa, F., et al.: Scikit-learn: machine learning in Python. J. Mach. Learn. Res. 12, 2825–2830 (2011)

16. Perkis, T.: Stack-based genetic programming. In: Proceedings of the First IEEE Conference on Evolutionary Computation. IEEE World Congress on Computational Intelligence, pp. 148–153, vol. 1 (1994). https://doi.org/10.1109/ICEC.1994.350025

17. Poli, R., B. Langdon, W., Mcphee, N.: A field guide to genetic programming. Lulu Enterprises, UK Ltd (01 2008)

18. Quinlan, J.R.: Induction of decision trees. Mach. Learn. 1(1), 81–106 (1986). https://doi.org/10.1023/A:1022643204877

19. Rodrigues, N.M., Batista, J.E., Cava, W.L., Vanneschi, L., Silva, S.: SLUG: Feature selection using genetic algorithms and genetic programming. In: Lecture Notes in Computer Science, pp. 68–84. Springer International Publishing (2022). https://doi.org/10.1007/978-3-031-02056-8_5

20. Rodrigues, N.M., Batista, J.E., Silva, S.: Ensemble genetic programming. In: Hu, T., Lourenço, N., Medvet, E., Divina, F. (eds.) EuroGP 2020. LNCS, vol. 12101, pp. 151–166. Springer, Cham (2020). https://doi.org/10.1007/978-3-030-44094-7_10

21. Rumelhart, D.E., Hinton, G.E., Williams, R.J.: Learning representations by back-propagating errors. Nature 323(6088), 533–536 (1986)

22. Sohn, A., Olson, R.S., Moore, J.H.: Toward the automated analysis of complex diseases in genome-wide association studies using genetic programming. In: Proceedings of the Genetic and Evolutionary Computation Conference, pp. 489–496. GECCO '17, Association for Computing Machinery, New York, NY, USA (2017). https://doi.org/10.1145/3071178.3071212

23. Spector, L.: Assessment of problem modality by differential performance of lexicase selection in genetic programming: a preliminary report. In: Proceedings of the 14th Annual Conference Companion on Genetic and Evolutionary Computation, pp. 401–408. GECCO '12, Association for Computing Machinery, New York, NY, USA (2012). https://doi.org/10.1145/2330784.2330846

24. Urbanowicz, R., Kiralis, J., Sinnott-Armstrong, N., et al.: GAMETES: a fast, direct algorithm for generating pure, strict, epistatic models with random architectures. BioData Mining 5(16) (2012). https://doi.org/10.1186/1756-0381-5-16

Grammar-Guided Evolution of the U-Net

Mahsa Mahdinejad[1]([✉]) , Aidan Murphy[2] , Michael Tetteh[1] ,
Allan de Lima[1] , Patrick Healy[1] , and Conor Ryan[1]

[1] University of Limerick, Limerick, Ireland
{Mahsa.Mahdinejad,Michael.Tetteh,Allan.Delima,Patrick.Healy,
Conor.Ryan}@ul.ie
[2] University College Dublin, Dublin, Ireland
Aidan.Murphy@ucd.ie

Abstract. Deep learning is an effective and efficient method for image segmentation. Several neural network designs have been investigated, a notable example being the U-Net which has outperformed other segmentation models in different challenges. The Spatial Attention U-Net is a variation of the U-Net, which utilizes *DropBlock* and an attention block in addition to the typical U-Net convolutional blocks, which boosted the accuracy of the U-Net and reduced over-fitting. Optimising neural networks is costly, time-consuming and often requires expert guidance to determine the best mix of hyper-parameters for a particular problem. We demonstrate that grammatical evolution (GE) can be used to create U-Net and Spatial Attention U-Net architectures and optimise its choice of hyper-parameters. Our results show improved performance over state-of-the-art models on the Retinal Blood Vessel problem, increasing both *AUC*, from 0.978 to 0.979 and *Accuracy*, from 0.964 0.966, from the base models. Crucially, GE can achieve these improvements while finding a model which is 10 times smaller than the base models. A smaller model would enable its use in smart devices, such as smart phones, or in edge computing.

Keywords: Deep Learning · Evolutionary Algorithm · Image
Segmentation · Grammatical Evolution

1 Introduction

Specialists and clinicians are increasingly using image processing and machine learning technology to help them investigate and identify anomalies. Diabetes is a serious and widespread disease that, if untreated, can cause blindness [5] or death [19]. Damage and a swelling of blood vessels in the eye are the main contributors to diabetic retinopathy (DR), a form of blindness brought on by diabetes. One of the main causes of blindness is specifically the formation of harsh exudates (a type of fluid) around the fovea (a crucial portion of the neurosensory retina). Early diagnosis and laser photocoagulation, however, is able to slow the growth of DR in the retina. Early detection of DR requires a manual inspection

J. Correia et al. (Eds.): EvoApplications 2023, LNCS 13989, pp. 672–686, 2023.
https://doi.org/10.1007/978-3-031-30229-9_43

of retinal images, as it does not become evident until after a diabetes diagnosis. The Retinal Blood Vessel Segmentation task is the process of extracting blood vessels from retinal images to diagnose DR [27].

The identification of blood vessels in images of eyes is a difficult task, however cutting-edge Neural Networks (NNs) approaches are employed to resolve this issue. A convolutional NN (CNN) [13], an improved variation of the traditional NN, has emerged as the most successful method for tackling image segmentation problems. Due to the availability of powerful supercomputers for training larger, more complex CNNs, a wider range of image processing tasks are being effectively accomplished by CNNs. These achievements have come at the expense of complexity, despite the fact that many typical CNN architectures are so expensive to train. As a result, developing a CNN architecture, training it, testing it and selecting the appropriate hyper-parameters for a task are becoming increasingly challenging and costly. We use Grammatical Evolution (GE) to find the best CNN architecture and parameters automatically [22].

In contrast to our earlier work [12,20] and [16], we use GE to choose both the hyper-parameters and evolve the architecture of a CNN rather than a using a Genetic Algorithm (GA) [11] which optimises the hyper-parameters.

Our results show that GE prefers a Spatial Attention U-Net (SA-UNet) over a U-Net and that the SA-UNet evolved achieves higher *Accuracy* than manual designs. The SA-UNet representation improves the U-Net by adding an attention block and also using *DropBlock* instead of *DropOut*. Further, a SA-UNet reduces over-fitting and concentrates on important aspects of the image while suppressing unnecessary ones. In other words, the SA-UNet cleans up the images by removing noise and unimportant background elements.

To explore if we can improve the performance of our models one step more, we decide to change our fitness function in the second sets of experiments.

We first create a grammar and conduct experimentation on it. Results indicated that the performance of the SA-UNet exceeded that of the U-Net, so further experimentation was conducted with special attention fixed in each model. Lastly, with real world applications in mind, we investigate the use of different fitness functions, first *Val-accuracy* followed by *F2-Score*, on the evolution of models.

In Sect. 2, we provide background information on our methodology. In Sect. 3, the experimental setup is explained and discussed in depth, and Sect. 4 provides the results. Section 5 presents our conclusions as well as our ideas for further work.

2   Background

2.1   Image Segmentation

A significant field of research in computer vision and machine learning is image segmentation [17]. Medical image processing, autonomous driving, and urban navigation all demand precise, dependable, and efficient segmentation algorithms.

In many clinical processes, medical image segmentation, such as diagnosis, planning, and executing treatments is a crucial step. These visual analyses are often carried out by trained therapists. Therefore, creating reliable and strong image segmentation algorithms is a key requirement for medical image analysis. Image segmentation is the process of breaking down a digital image into many subgroups in order to reduce the image's complexity and make future processing or analysis easier.

2.2   Convolutional Neural Networks

A branch of machine learning known as deep learning (DL) focuses on artificial neural networks (ANNs), particularly networks with many complex layers. CNNs have made significant strides in a variety of pattern identification fields, from voice recognition to image processing. CNN's greatest benefit is their ability to decrease the number of parameters in ANNs by using filters and kernels in convolution blocks [2]. This success has encouraged academics and developers to think about ever-larger models in order to complete difficult tasks that were previously unsolvable with conventional ANNs. However, solving a problem a CNN requires that the images do not have spatially dependent features. Another key aspect of CNNs is their ability to identify and use abstract features when inputs pass through many deep layers.

2.3   Autoencoders

An autoencoder is an unsupervised learning strategy for NNs that trains the network to ignore signal "noise" in order to develop efficient data representations. This is referred to as encoding. Autoencoders have been used on many different segmentation tasks [7]. They can be divided into two main parts, an encoder and a decoder 1. The encoder is the portion which encodes input data from the train-validate-test set into a substantially smaller representation than the input data (often by several orders of magnitude). Convolutional blocks are used in the encoder, which are followed by pooling blocks. The middle section contains the compressed knowledge representations. The decoder, is the part which assists the network in "decompressing" knowledge representations and reconstructing data from its encoded form. Finally, the output is contrasted with a ground truth. Data denoising [9] and dimensionality reduction [26] for data visualization are two practical applications of autoencoders.

2.4   The U-Net and SA-UNet

The most recent strategies for addressing the blood vessel segmentation challenge are all NN-based methods. The original U-Net was developed by [21]. According to the authors, the U-Net was a completely CNN variant that could provide good prediction using a minimal training dataset. A U-net design consists of a U-shaped (Fig. 2) convolutional autoencoder that has extra connections between

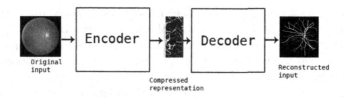

Fig. 1. An auto-encoder acting on an image.

the encoder and decoder components. The contracting path and the expanding path, which are located on the left and right sides of a U-Net, respectively, separate a U-Net into two pieces. The expanding path uses up-sampling, whereas the contracting path uses down-sampling to extract features from input images.

We have previously optimized a U-Net design and produced a smaller model than the standard U-Net [12,20]. This allowed the model to be trained more quickly and affordably, and as a result, the models were more suited to running on compact or embedded systems. However, this more compact form came at the expense of accuracy. This trade-off is no longer justifiable when the accuracy gap between the models widens as newer, state-of-the-art models have improved performance on benchmarks to ever-higher levels. To improve the performance of the network, several U-Net modifications have been suggested. The design of a U-Net has undergone some changes, including the Dual Encoding U-Net (DEU-Net) [25], which employs two encoders and attention blocks. The U-Net++ [29] is another example, which seeks to close the feature map gap between the encoder and decoder, as well as other modifications.

In our experimentation, we also use a spatial attention U-Net (SA-UNet) [10], an improved version of a U-Net. An SA-UNet has two key differences compared to a U-Net, as shown in Fig. 2. First, it uses spatial attention at the bottom of a U-Net [28], in the middle section, and second, it uses *DropBlock* rather than *DropOut* [8] (see below).

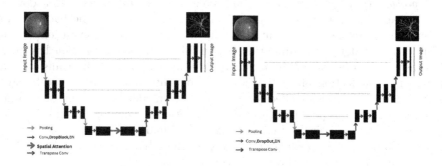

Fig. 2. Comparison SA-UNet (left) and U-Net (right) architecture.

Small training sets are frequently encountered in supervised machine learning applications. This might be due to a lack of available data, the cost of gathering, or a variety of other factors. Most CNN models require massive quantities of data to learn well to avoid over-fitting [6] and exhibit poor performance when trained on small datasets.

A regularization approach called *DropOut* [4] ignores or "drops out" some neurons stochastically during the training process. The *dropout rate* is a hyper-parameter dictating the chance of training a particular node in a layer. A value of 0.0 denotes no drop out, while 1.0 drops out the entire output layer. A suitable dropout value in a hidden layer is usually between 0.4 and 0.6.

DropBlock is another CNN regularization approach that, like *DropOut*, ensures that no units are dependant on each other throughout the training process by changing the input's units to 0. It differs from *DropOut* by removing continuous areas from a layer's feature map rather than individual random units. This difference is demonstrated in Fig. 3.

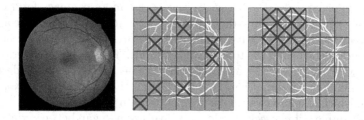

Fig. 3. An example contrasting *DropOut* and *DropBlock's* methods for disregarding units in an input image.

DropBlock has been shown to efficiently minimize network over-fitting, with small training datasets. *DropBlock* needs the user to provide the values of two parameters: *block-size*, which defines the amount of pixels in each block, and *keep-probability*, which determines the likelihood of the block being shut down.

The lack of contrast between the blood vessel region and the backdrop in retinal fundus images is a significant difficulty in retinal segmentation. Much effort has been done to address this lack of differentiation. *Attention* has been demonstrated to be one of the most powerful approaches [24] by translating a query and a collection of crucial data into output to help the network learn better. In order to provide effective outcomes, the model has to learn structural information, which is possible with spatial attention. This information is learned by emphasizing key units and minimizing background noise. The channel axis is used to apply the max-pooling and average-pooling blocks first, which are then concatenated to create an effective feature detector.

2.5 Evolutionary Algorithms

Population-based optimization methods called evolutionary algorithms (EAs) are motivated by natural selection.

A population of solutions is first created in an EA experiment along with a fitness function. This population is then evolved across a number of generations, with the strongest individuals surviving and carrying their traits to next generations, while the weaker individuals will disappear. An individual in our experiments is a CNN architecture and its corresponding hyper-parameters.

2.6 Grammatical Evolution

GE is popular variant of Genetic Programming (GP). It is an evolutionary method that uses a grammar to map a string, called a genotype, to an executable program, called a phenotype. GE has been used to a variety of test problems and real-world applications, with satisfactory results [3,15]. GE often uses a context free grammar, written in Backus-Naur-Form (BNF). A BNF grammar, G, is represented by the four-tuple $<N, T, P, S>$, where N is a set of non-terminals, T is a set of terminals, P is a set of production rules that map the components of N to T, and S (a member of N) is the start symbol.

Deciding on a suitable EA is challenging. Our previous works used a GA for optimising and tuning hyper-parameters of our CNN. GE has some advantages compared to the GA [18], as it gives more freedom in selecting hyper-parameters which can lead to more complex or simpler models, as seen in Figs. 9 and 10, when compared to a fixed-length GA representation. Furthermore, GE can modify the architecture of CNNs and tune the hyper-parameters. The use of a grammar simplifies the addition and expansion of hyper-parameters, enabling solving more difficult tasks on complex networks. We explain our proposed grammar in detail in Sect. 3.4.

3 Experimental Setup

Our experiments started first by using *Val-Accuracy* as the fitness function and asking GE to find both the best architecture and corresponding hyper-parameters.

After analysing the results of the first set of experiments, it was decided to run the next set of experiments with the attention layer fixed in the model, as a SA-UNet architecture. In the final sets of experiments, we modified the fitness function to *F2-Score* to see if it can improve the model's performance.

3.1 Dataset

In order to facilitate comparative study on blood vessel segmentation in retinal images for the diagnosis, screening, treatment, and assessment of many disorders including diabetes, the DRIVE (Digital Retinal Images for Vessel Extraction) [23] database was established. Experts identify and employ morphological characteristics of retinal blood vessels such as length, width, texture, and branching patterns. Research into the relationship between vascular deformability and hypertensive retinopathy, vessel diameter measurement in connection to hypertension diagnosis, and computer-assisted laser surgery may all benefit from automatic identification and study of the vasculature. The retinal vascular branch has also been demonstrated to be particular to each individual and may be used for biometric identification. Two sets of 20 images each were created from the 40 total images, a training set (Fig. 4) and a test set (Fig. 5). After applying an augmentation method on the training set we had 260 total images which we randomly picked 26 to create the validation set.

For the training images, there is only one manual segmentation of the vasculature available. The test cases include two manual segmentations. While the other can be used to compare computer-generated segmentation to those made by a human observer, the first serves as the standard. Additionally, a mask image showing the region of interest is provided for each retinal scan. All human observers who manually segmented the vasculature were led and instructed by an experienced ophthalmologist.

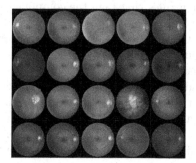

Fig. 4. DRIVE training images.

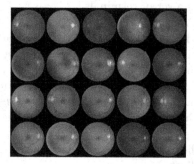

Fig. 5. DRIVE test images.

3.2   Hardware and Software

The experiments were carried out on a single system configured with an Nvidia Quadro RTX 8000 GPU. The networks were trained using TensorFlow [1] and GRAPE (Grammatical Algorithms in Python for Evolution) [14] was the GE framework used.

3.3   Evaluation Metrics

Our models were evaluated using *Recall, Specificity*, area under the curve (*AUC*), positive predictive value (*Precision*), negative predictive value (*NPV*), *F2-Score* and, Matthew's Correlation Coefficient (*MCC*). We demonstrate how different evaluation rates work in the segmentation process in Fig. 6.

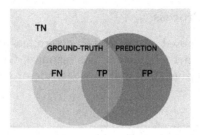

Fig. 6. Example of evaluations in segmentation.

For example, a True-Positive (TP) is when a pixel's value correctly matches the predicted . False-Positive (FP) pixels are those that are found in the prediction but are not part of the ground truth, and so on. *Recall* is the rate of TPs which can also be considered to be the test's sensitivity to noticing small changes. *Specificity* is the rate of TNs. The relevant formulae are:

$$Recall = \frac{TP}{TP + FN}$$

$$Specificity = \frac{TN}{TN + FP}$$

$$Precision = \frac{TP}{TP + FP}$$

$$NPV = \frac{TN}{TN + FN}.$$

MCC 3.3, is a measure for calculating the difference between predicted and actual results. *MCC* is a reliable statistical rate that returns a high score only if the prediction was successful in each of the four sections of the confusion matrix (TP, FN, TN, and FP)

$$MCC = \frac{TP \times TN - FP \times FN}{\sqrt{(TP + FP)(TP + FN)(TN + FN)(TN + FN)}}.$$

F2-Score 3.3 is another evaluation metric which we considered in our experiments. *F2-Score*, is a special case of the *Fβ-Score* metric 3.3, where $\beta = 2$

$$F\beta-Score = \frac{(1 + \beta^2) \times Precision \times Recall}{(\beta^2) \times Precision + Recall}$$

With *F2-Score*, *Recall* becomes more significant and *Precision* becomes less important. Therefore, *F2-Score* is better suited in applications where correctly classifying as many positive samples as possible is more important than maximising the number of correct classifications. In other words, we want the blue circle in Fig. 6 to dominate the yellow circle. In real world settings, it may be preferable to have a very high TP rate, as an incorrect FN may have dire consequences.

3.4   Evolutionary Process

Figure 7 shows our proposed grammar. GE determines the architecture of the model as well as its hyper-parameters. For instance, GE may choose whether to use *DropOut* or *DropBlock*, as well as what *rate* or *block-size* and *keep-prob* to use.

```
<encoder>   ::= <down> | <encoder>\n\t<down>
<decoder>   ::= <up> | <decoder>\n\t<up>
<spatial>   ::= <sp> | <conv>
  <layer>   ::= <conv>\n\t<drop_type>\n\t<Batch>
   <down>   ::= <layer>\n\t<pool>
     <up>   ::= <transpose>\n\t<drop_type>\n\t<Batch>
     <sp>   ::= x = spatial_attention(x)
<transpose> ::= x = Conv2DTranspose(<filters>,(2, 2),<padding>)(x)
   <conv>   ::= x = Conv2D(<filters>,<kernel>,<activation>, padding = <padding>)(x)
   <pool>   ::= x = <pooling>((2, 2), padding='same' )(x)
  <Batch>   ::= x = BatchNormalization()(x)
<DropBlock> ::= x = DropBlock2D(<block_size>, <keep_prob>)(x)
<Dropout>   ::= x = Dropout(<rate>)(x)
<drop_type> ::= <Dropout> | <DropBlock>
<activation> ::= 'relu' | 'sigmoid' | ...
 <pooling>  ::= MaxPooling2D | AveragePooling2D
    <rate>  ::= 0.4 | 0.5 | 0.6
<optimisation> ::= 'Adam' | 'sgd' | 'adamax' | ...
 <filters>  ::= 16 | 32 | 64 | 128 | 256
  <kernel>  ::= (1, 1) | (2, 2) | ...
<block_size> ::= 7 | 9
<keep_prob> ::= 0.8 | 0.9
 <padding>  ::= "valid" | "same"
```

Fig. 7. Proposed GE Grammar.

Table 1. List of the parameters used to run GE.

Parameter	Value
Fitness Function	*Val-Accuracy* and *F2-Score*
Runs	5, 5 and 3
Total Generations	15
Population Size	100
Crossover Rate	0.9
Mutation Rate	0.5
Epochs (Training)	15
Epochs (Best)	500

Table 1 shows our GE parameters. As we employ a non-semantic grammar, we had a large number of invalid individuals with mismatched output dimensions. Decoder layers were supposed to come up depending on how many layers our encoder had descended, however in some individuals, this didn't happen.

Therefore, each experiment had a population of 100, which is higher than in our prior studies, and it was run for 15 generations. One-point crossover and bit-flip mutation were utilized in each experiment, together with random initialization to create the initial population. During the evolutionary run, each individual was trained for 15 epochs with a batch size of 4. The top-performing model was trained for an additional 500 epochs.

We used two different fitness functions. *Validation-accuracy* was the fitness function in our first and second sets of experiments. We repeated first sets 5 times and then 5 more just for optimising hyper-parameters of SA-UNet. Validation-accuracy helps in preventing over-fitting, which is important for our small training dataset. Next, we used *F2-Score* as the fitness function to create a model which is optimised to maximise True-Positive rate. We performed 3 runs for the second fitness function.

4   Results

In Table 2 we show the hyper-parameters of our best models. *Exp1* and *Exp2* are the two best models found during our first set of experiments (5 runs). *Exp1* is a SA-UNet and *Exp2* is a U-Net. The results from our first set of experiments showed that GE chose a SA-UNet more frequently than a U-Net architecture.

Our second set of experiments used a SA-UNet as a base model for the evolutionary process. The hyper-parameters of the best model which GE discovered in this set of experiments is shown in *Exp3*.

The best model from the last set of our experiments, in which we used *F2-Score* as the fitness function, is seen in *Exp4*. Interestingly, this is a U-Net.

As can be seen, all of the models have the combination of *DropBlock* and *DropOut*, a feature rarely seen in human designed networks. We can see some similarity in U-Net and SA-UNet models. All the U-Net models have $Depth = 2$, $Kernel = (5, 5)$, $Padding = same$ and $Pooling = AveragePooling$. In SA-UNet models we have $Depth = 2$, $Kernel = (3, 3)$ and $Kernel = (5, 5)$, $Padding = same$ and $Padding = valid$, while both *AveragePooling* and *MaxPooling* were used in SA-UNet. $Filter\text{-}Size = 16$ is the most frequent one amongst all models. *'adam'* and *'Nadam'* are the most popular *optimisers*. We have a good combination among choosing an *Activation* function.

Table 3 shows the full results of our experimentation on test dataset, with the best results compared with the standard U-Net[1] and standard SA-UNet[2], as found online. *Exp1* obtained a higher AUC, (0.976), than *Exp2*, 0.963.

Exp3 achieved better AUC, 0.978, compared to the standard SA-UNet, 0.977. It's *size*, 3.8M, is also much smaller than the standard SA-UNet, 17.1M.

Indeed, GE evolved more accurate and much smaller models, the smallest of which, *Exp2*, was a comparatively tiny 0.9M when compared to the standard U-Net size, 42.4M, and standard SA-UNet size, 17.1M. This highlights the great benefits GE can provide.

---

[1] https://github.com/zhixuhao/unet.

[2] https://github.com/clguo/SA-UNet.

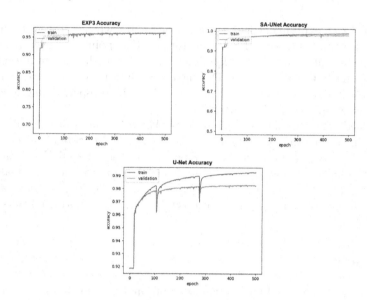

Fig. 8. Different models' training and validation plots for 500 epochs.

Exp4 achieved best *AUC*, 0.979, *Accuracy*, 0.966 and *Specificity*, 0.988 compared to the original models and those found during our evolutionary experiments.

It is also worth noting that our model has less over-fitting than the SA-UNet and U-Net, as demonstrated in Fig. 8.

We also demonstrate the arrangement of the hyper-parameters in two of our models in Fig. 9 and Fig. 10.

The results of our best models compare to U-Net and SA-UNet for two images are shown in Fig. 11. Our model discovered more details in both images.

Table 2. Hyper-parameters of the best models.

Name	Exp1	Exp2	Exp3	Exp4
Model	SA-UNet	U-Net	SA-UNet	U-Net
Depth	3	2	3	2
Filter	16–64–128	32–64–128	16–32–64–256	16–256
Kernel	3–5	5	3–5	5
Drop-type	Block-Out	Block-Out	Block-Out	Block-Out
Padding	same-valid	same	same-valid	same
Optimiser	adam	adam	Nadam	Nadam
Activation	selu-tanh-sigmoid	realu-tanh	relu-softsign-sigmoid	elu-sigmoid
Pooling	Avg-Max	Avg	Avg-Max	Avg

Table 3. Results of test set for both fitness functions. The bold numbers are the best results and underlined numbers are the best results which GE found.

Name	U-Net	SA-UNet	Exp1	Exp2	Exp3	Exp4
Accuracy	0.964	0.963	0.956	0.957	0.964	**0.966**
AUC	0.978	0.977	0.976	0.963	0.978	**0.979**
NPV	0.980	0.983	0.967	**0.993**	0.986	0.975
Specificity	0.980	0.977	0.984	0.961	0.974	**0.988**
Recall	0.791	0.760	**0.839**	0.583	0.734	0.737
Precision	0.799	0.812	0.707	**0.889**	0.836	0.853
F2-Score	**0.793**	0.770	0.768	0.627	0.752	0.757
MCC	**0.776**	0.766	0.747	0.700	0.764	0.774
Size	42.4 M	17.1 M	4 M	**0.9 M**	3.8 M	10.3 M

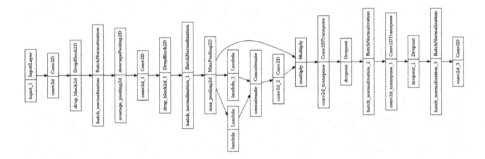

Fig. 9. Exp3 Architecture.

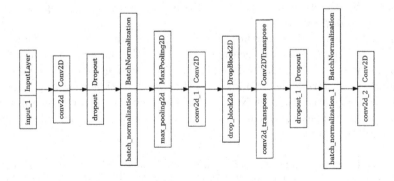

Fig. 10. Exp2 Architecture.

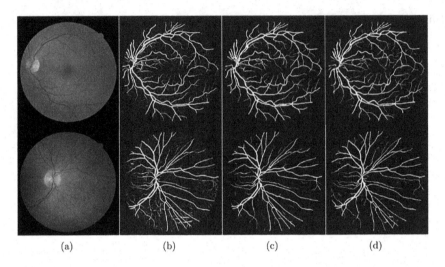

(a)   (b)   (c)   (d)

Fig. 11. Comparison the results of 2 images. b is Exp4, c is U-Net and d is SA-UNet.

5   Conclusion

The architecture and hyper-parameters of a CNN used in image segmentation were optimized by GE. Previously, the U-Net and SA-UNet were employed as base models for a GA to optimize their hyper-parameters separately. In this study, we utilized GE to create and optimise diverse architectures, at the same time. We employed several drop types and attention blocks to decrease noise in order to solve over-fitting and enhance the performance of CNN models.

Our results demonstrated that we outperformed the original models in terms of *AUC* and *Accuracy*. Furthermore, our models are over ten times smaller than U-Net and SA-UNet. A smaller model has the advantage of requiring less training time and computation costs.

We plan to undertake further experimentation with semantic grammars in order to further increase the accuracy of our models; these will remove all morphologically invalid individuals. Further training for longer epochs may also lead to improved performance. Including further modification that have shown to improve the performance of U-Nets is also an avenue for future research. Finally we wish to investigate the use of a multi objective fitness function when evolving U-Net architectures.

Acknowledgements. The Science Foundation Ireland (SFI) Centre for Research Training in Artificial Intelligence (CRT-AI), Grant No. 18/CRT/6223, and the Irish Software Engineering Research Centre (Lero), Grant No. 16/IA/4605, both provided funding for this study. The second author was supported by Science Foundation Ireland grant 20/FFP-P/8818.

References

1. Abadi, M., et al.: Tensorflow: a system for large-scale machine learning. In: 12th USENIX Symposium on Operating Systems Design and Implementation (OSDI 2016), pp. 265–283 (2016)
2. Albawi, S., Mohammed, T.A., Al-Zawi, S.: Understanding of a convolutional neural network. In: 2017 International Conference on Engineering and Technology (ICET), pp. 1–6. IEEE (2017)
3. Assunçao, F., Lourenço, N., Machado, P., Ribeiro, B.: Automatic generation of neural networks with structured grammatical evolution. In: 2017 IEEE Congress on Evolutionary Computation (CEC), pp. 1557–1564. IEEE (2017)
4. Baldi, P., Sadowski, P.J.: Understanding dropout. In: Advances in Neural Information Processing Systems, vol. 26 (2013)
5. Ciulla, T.A., Amador, A.G., Zinman, B.: Diabetic retinopathy and diabetic macular edema: pathophysiology, screening, and novel therapies. Diabetes Care **26**(9), 2653–2664 (2003)
6. Dietterich, T.: Overfitting and undercomputing in machine learning. ACM Comput. Surv. (CSUR) **27**(3), 326–327 (1995)
7. Doersch, C.: Tutorial on variational autoencoders. arXiv preprint arXiv:1606.05908 (2016)
8. Ghiasi, G., Lin, T.Y., Le, Q.V.: Dropblock: a regularization method for convolutional networks. In: Advances in Neural Information Processing Systems, vol. 31 (2018)
9. Gondara, L.: Medical image denoising using convolutional denoising autoencoders. In: 2016 IEEE 16th International Conference on Data Mining Workshops (ICDMW), pp. 241–246. IEEE (2016)
10. Guo, C., Szemenyei, M., Yi, Y., Wang, W., Chen, B., Fan, C.: SA-UNet: spatial attention U-Net for retinal vessel segmentation. In: 2020 25th International Conference on Pattern Recognition (ICPR), pp. 1236–1242. IEEE (2021)
11. Holland, J.H.: Adaptation in Natural and Artificial Systems. University of Michigan Press, Ann Arbor (1975). 2nd edn. 1992
12. Houreh, Y., Mahdinejad, M., Naredo, E., Dias, D.M., Ryan, C.: HNAS: hyper neural architecture search for image segmentation. In: ICAART (2), pp. 246–256 (2021)
13. LeCun, Y., Bengio, Y., et al.: Convolutional networks for images, speech, and time series. Handb. Brain Theory Neural Netw. **3361**(10), 1995 (1995)
14. de Lima, A., Carvalho, S., Dias, D.M., Naredo, E., Sullivan, J.P., Ryan, C.: Grape: grammatical algorithms in python for evolution. Signals **3**(3), 642–663 (2022)
15. Lima, R., Pozo, A., Mendiburu, A., Santana, R.: Automatic design of deep neural networks applied to image segmentation problems. In: Hu, T., Lourenço, N., Medvet, E. (eds.) EuroGP 2021. LNCS, vol. 12691, pp. 98–113. Springer, Cham (2021). https://doi.org/10.1007/978-3-030-72812-0_7
16. Mahdinejad., M., Murphy., A., Healy., P., Ryan., C.: Parameterising the SA-UNet using a genetic algorithm. In: Proceedings of the 14th International Joint Conference on Computational Intelligence - ECTA, pp. 97–104. INSTICC, SciTePress (2022)
17. Minaee, S., Boykov, Y.Y., Porikli, F., Plaza, A.J., Kehtarnavaz, N., Terzopoulos, D.: Image segmentation using deep learning: a survey. IEEE Trans. Pattern Anal. Mach. Intell. (2021)

18. Nyathi, T., Pillay, N.: Comparison of a genetic algorithm to grammatical evolution for automated design of genetic programming classification algorithms. Expert Syst. Appl. **104**, 213–234 (2018)
19. Ogurtsova, K., et al.: IDF diabetes atlas: global estimates for the prevalence of diabetes for 2015 and 2040. Diabetes Res. Clin. Pract. **128**, 40–50 (2017)
20. Popat, V., Mahdinejad, M., Cedeño, O.D., Naredo, E., Ryan, C.: GA-based U-Net architecture optimization applied to retina blood vessel segmentation. In: IJCCI, pp. 192–199 (2020)
21. Ronneberger, O., Fischer, P., Brox, T.: U-net: convolutional networks for biomedical image segmentation. In: Navab, N., Hornegger, J., Wells, W.M., Frangi, A.F. (eds.) MICCAI 2015. LNCS, vol. 9351, pp. 234–241. Springer, Cham (2015). https://doi.org/10.1007/978-3-319-24574-4_28
22. Ryan, C., Collins, J.J., Neill, M.O.: Grammatical evolution: evolving programs for an arbitrary language. In: Banzhaf, W., Poli, R., Schoenauer, M., Fogarty, T.C. (eds.) EuroGP 1998. LNCS, vol. 1391, pp. 83–96. Springer, Heidelberg (1998). https://doi.org/10.1007/BFb0055930
23. Staal, J.: DRIVE: digital retinal images for vessel extraction (2018). https://www.isi.uu.nl/Research/Databases/DRIVE
24. Vaswani, A., et al.: Attention is all you need. In: Advances in Neural Information Processing Systems, vol. 30 (2017)
25. Wang, B., Qiu, S., He, H.: Dual encoding U-net for retinal vessel segmentation. In: Shen, D., et al. (eds.) MICCAI 2019. LNCS, vol. 11764, pp. 84–92. Springer, Cham (2019). https://doi.org/10.1007/978-3-030-32239-7_10
26. Wang, W., Huang, Y., Wang, Y., Wang, L.: Generalized autoencoder: a neural network framework for dimensionality reduction. In: Proceedings of the IEEE Conference on Computer Vision and Pattern Recognition Workshops, pp. 490–497 (2014)
27. Winder, R.J., Morrow, P.J., McRitchie, I.N., Bailie, J., Hart, P.M.: Algorithms for digital image processing in diabetic retinopathy. Comput. Med. Imaging Graph. **33**(8), 608–622 (2009)
28. Woo, S., Park, J., Lee, J.-Y., Kweon, I.S.: CBAM: convolutional block attention module. In: Ferrari, V., Hebert, M., Sminchisescu, C., Weiss, Y. (eds.) ECCV 2018. LNCS, vol. 11211, pp. 3–19. Springer, Cham (2018). https://doi.org/10.1007/978-3-030-01234-2_1
29. Zhou, Z., Rahman Siddiquee, M.M., Tajbakhsh, N., Liang, J.: UNet++: a nested U-net architecture for medical image segmentation. In: Stoyanov, D., et al. (eds.) DLMIA/ML-CDS -2018. LNCS, vol. 11045, pp. 3–11. Springer, Cham (2018). https://doi.org/10.1007/978-3-030-00889-5_1

Explaining Recommender Systems by Evolutionary Interests Mix Modeling

Piotr Lipinski[✉]

Computational Intelligence Research Group, Institute of Computer Science,
University of Wroclaw, Wrocław, Poland
`lipinski@cs.uni.wroc.pl`

Abstract. This paper focuses on explaining the results of Recommender Systems, that aim at suggesting, for a given user, the most accurate products, among a given set of available products, and modeling how different types of user activities, such as based on user interests in different categories of products, affect the results of the recommender system. It proposes an evolutionary approach to interests mix modeling that defines the relation between the characteristic of the user ratings and the composition of the list of the recommended products. Computational experiments, performed on some selected benchmarks derived from the MovieLens dataset, confirmed the accuracy and efficiency of the proposed approach.

Keywords: Recommender Systems · Explainable Artificial Intelligence · Evolutionary Algorithms · Multi-objective Optimization

1 Introduction

Machine Learning approaches gain more and more popularity due to their accuracy and efficiency in solving various real-world problems. However, most of them are usually based on very large and complex models extremely difficult to interpret by human experts responsible for decision making processes. As this constitutes a serious bottleneck in applying Machine Learning models, more and more recent research focuses on Explainable Artificial Intelligence that aims at learning interpretable, but sometimes weaker, models.

In this paper, we focus on Recommendation Systems [10], that aim at suggesting, for a given user, the most accurate products, among a given set of available products, and, instead of constraining models to make them interpretable, we try to explain in some extend the behavior of regular models. We consider how different types of user activities, such as based on user interests in different categories of products, affect the results of the recommender system, such as the categories of the recommended products. We introduce an interest mix modeling approach, similar to marketing mix modeling [5] that aims at explaining the revenue by the outlays on particular marketing channels, that tries to model the characteristic of the composition of the list of the recommended products

© The Author(s), under exclusive license to Springer Nature Switzerland AG 2023
J. Correia et al. (Eds.): EvoApplications 2023, LNCS 13989, pp. 687–699, 2023.
https://doi.org/10.1007/978-3-031-30229-9_44

by user interests in particular categories of products expressed by the ratings of products.

This paper is structured in the following manner: Sect. 2 presents an overview of popular Recommender Systems. Section 3 defines the problem. Section 4 presents an evolutionary approach to interests mix modeling. Section 5 describes the computational experiments. Section 6 concludes the paper.

2   Recommender Systems

Recommender systems aim at suggesting, for a given user, the most accurate products, among a given set of available products. Although various types of recommendation systems exist [8,11,12], related to various types of real-world applications [10], in this paper we focus on the collaborative filtering approach [10], especially with the Singular Value Decomposition (SVD) algorithm [10], but our studies may be easily applied also to other types of recommender systems.

Let $U_1, U_2, \ldots, U_n$ denote the users under consideration. Let $P_1, P_2, \ldots, P_m$ denote the products under consideration. Let $R \in \mathbb{R}^{n \times m}$ denote the rating matrix of elements r_{ij}, for $i = 1, 2, \ldots, n$ and $j = 1, 2, \ldots, m$, representing the rating given by the user U_i to the product P_j (or null if the user U_i did not rate the product P_j). In the collaborative filtering approach, for each user U_i, the recommender system computes a sequence of recommendations based on the rating matrix R (and, in some approaches, on some meta-data concerning products and users, such as categorization of the products).

One of the popular recommendation algorithms in the collaborative filtering approach is the SVD algorithm based on the rating matrix factorization. First, it tries to approximate the rating matrix R by

$$\tilde{R} = U \cdot V^T, \tag{1}$$

where $U \in \mathbb{R}^{n \times k}$ is a matrix of vectors representing the users (so-called user embeddings), $V \in \mathbb{R}^{m \times k}$ is a matrix of vectors representing the products (so-called product embeddings), so

$$\tilde{r}_{ij} = u_i \cdot v_j, \tag{2}$$

where u_i is the i-th column of U (the embedding of i-th user) and v_j is the j-th column of V (the embedding of j-th product). It constructs $\tilde{R}$ by minimizing the error

$$\sum |r_{ij} - \tilde{r}_{ij}|, \tag{3}$$

where the sum is over all i and j for each r_{ij} is not null (i.e. for each the user U_i rated the product P_j). Second, it uses the user and products embeddings to evaluate an approximated rating of each product for each user. Finally, for each user, it returns the products with the highest approximated ratings.

Although the rest of the studies uses such particular approach, this paper focuses mainly on explaining the behaviour and the results of the recommender system, regardless of the recommendation algorithm itself, so the proposed approach may be easily applied to any type of the recommender system.

3   Understanding the Recommendations

From a certain point of view, a recommender system may be perceived as a type of a black box, because it is unclear how different types of user activities (e.g. related to user interests in different categories of products) affect the results of the recommender system (the lists of the recommended products). Some practical questions may arise: Does high rating of a certain category of products causes more products of this category in recommendations? How important are high ratings of some products of a category for filling the recommendations with other products of the category? For a certain list of the recommended products, what types of user interests contributed the most to suggesting such a composition of the list? What is the impact of a high rating of one category on recommending the other?

In this paper, we propose an approach to user interests mix modeling that may be helpful in explaining the black box behavior of the recommender system and answering the above questions.

In order to illustrate the basic concepts, we focus on one of the most popular recommendation system benchmarks, the MovieLens dataset [7], which contains ratings of movies by spectators.

3.1   Simple Regression Model

For the first illustration, we consider two genres, Action and Drama, and a selected user. Figure 1 presents the number of movies of each genre rated by the user in successive time instants of the period under consideration. Figure 2 presents the number of Action movies among the 100 most recommended movies to the user in the successive time instants. One may see that the plot is highly noised, as the recommender system is non-deterministic, but it is also visible that the relation between the number of movies rated and the number of movies recommended is not direct. Moreover, the number of recommended movies may be also influenced by other genres (e.g. the user rates a movie of the Action genre, but the movie may be also a comedy, so it may also increase the number of comedies in the list of the recommended movies getting one place for a possible action movie).

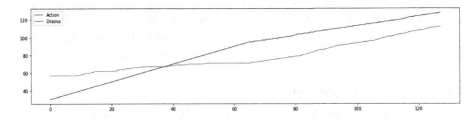

Fig. 1. The number of movies of each genre rated by the user in successive time instants

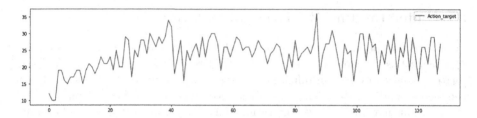

Fig. 2. The number of Action movies among the 100 most recommended movies to the user by the Recommender System in successive time instants

In order to explain how user interests in different categories of products affect the composition of the list of the recommended products, a simple regression model, with the numbers of Action and Drama movies rated by the user as the input data and the number of Action movies among the 100 most recommended movies to the user as the output data, may be used. Figure 3 presents the prediction given by the linear regression model for the dataset presented in Fig. 1 and 2 (the Mean Average Percentage Error (MAPE) of the prediction is 0.1836). It is worthy noting that the input time series were denoised with a short-term moving average before applying the regression model.

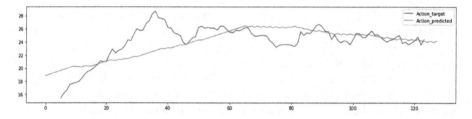

Fig. 3. The predicted number of Action movies among the 100 most recommended movies and the observed values. Prediction based on the linear regression model with input data defined by Action and Drama ratings. The input time series were denoised with a short-term moving average before applying the regression model

Such a regression model may be enriched with additional input data concerning more genres than Action and Drama only. Figure 4 presents the similar prediction given by the linear regression model for the dataset with additional Adventure, Romance and Crime genres (the MAPE error of the prediction is 0.1714).

3.2   Saturation Regression Model

A further extension of the linear regression model from the previous section may consider non-linear saturation of the user interests, where we assume that the user interest is not directly proportional to the number of movies rated, but defined by a saturation curve, such as the popular Hill curve [6].

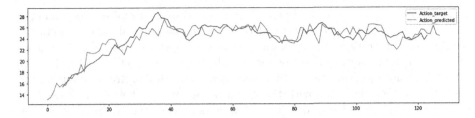

Fig. 4. The predicted number of Action movies among the 100 most recommended movies and the observed values. Prediction based on the linear regression model with input data defined by Action, Drama, Adventure, Romance and Crime ratings. The input time series were denoised with a short-term moving average before applying the regression model

Figure 5 presents the saturation curve of the user interests defined by the Hill curve

$$\phi(x; \theta) = \frac{x^n}{k + x^n}, \tag{4}$$

where $\theta = [k, n]$ is the vector of two parameters of the Hill curve.

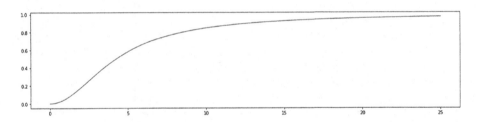

Fig. 5. An example of a saturation curve of the user interests defined by the Hill curve. The horizontal axis corresponds to the number of movies of a given genre rated, the vertical axis corresponds to the level of the saturation of the user interests by the given genre

3.3   Interpreting the Saturation Regression Model

Despite the simple formulation, our research aims at delivering valuable, interpretable results. Therefore, the goal is not only to provide an accurate regression model (there may be many regression models with comparable accuracy), but also to select such a model that reliably explain the black box of the recommender systems. Therefore, one of the important factors are the regression coefficients, which may be perceived unrealistic if they deviate from some real-world expectations.

Figures 6 compares some of possible models. The horizontal axis presents the accuracy error of the model (the MAPE error), the vertical axis presents the deviation of the ratio of the regression coefficients from the ratio of rated movies of corresponding genres (described in details in the next section). One may see that there are many models with similar accuracy, ranging from about 0.16 to about 0.24, but the model with the lowest accuracy error has a high coefficient ratio deviation. Looking at details, such a model has unrealistic regression coefficients, impossible to interpret, and suggesting that the prediction results are mainly influenced by the intercept of the regression with only a weak participation of the input variables. Although such a model gains a low accuracy error, it cannot explain the behavior of the recommender system. Recalling that the recommendations are somehow noised, the lowest accuracy error of the model, might not mean that the model reflect the behaviour perfectly, it may just overfit to the noise. Therefore, there is a reason to study other models, even if they have a little higher accuracy error, such as the models in the center or even on the right side of the plot, that has worse accuracy, but better regression coefficients and they seem to better reflect the behaviour of the recommendation system.

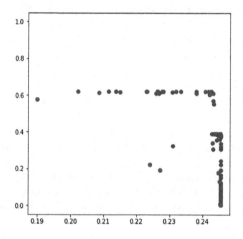

Fig. 6. The comparison of 100 possible models. The horizontal axis corresponds to the accuracy error of the model, the vertical axis corresponds to the deviation of the ratio of the regression coefficients from the ratio of rated movies of corresponding genres

3.4   Problem Definition

Let $C_1, C_2, \ldots, C_k$ denote the categories of the products (e.g. movie genres in the MovieLens dataset). Let $x_1, x_2, \ldots, x_k$ denote the number of products liked by the user in each category (e.g. the number of movies, in each genre, highly rated by the user). Let $\phi_1, \phi_2, \ldots, \phi_k$ be the saturation function for each category transforming the number of movies liked into a number between 0 and 1 defining the user interests in the genre (such as the Hill saturation curve described in the previous sections). The aim of the research is to find the parameters of the following model

$$\alpha_1\phi_1(x_1; \theta_1) + \alpha_2\phi_2(x_2; \theta_2) + \ldots + \alpha_k\phi_k(x_k; \theta_k) + \alpha_0 = y, \tag{5}$$

where y is the number of products of a given category among the 100 most recommended products for the user.

As the training dataset usually consists of data recorded for the selected user in successive time instants, we introduce a time-related notation: for $t = 1, 2, \ldots, T$, let $x_i^{(t)}$ denote the number of products liked by the user in the i-th category at time t and let $y^{(t)}$ denote the number of products of a given category among the 100 most recommended products for the user at time t.

4 Evolutionary Interests Mix Modeling Algorithm (EIMMA)

Our approach to solving the problem defined reformulates finding the parameters of the model as an multi-objective optimization problem and uses an evolutionary algorithm based on an evolutionary strategy [1,3] with a multi-criteria selection coming from NSGA-II [4].

4.1 Search Space and Objective Functions

Candidate solutions Θ to the optimization problem are encoded in chromosomes in the form of continuous vectors whose successive coordinates correspond to successive parameters $\theta_1, \theta_2, \ldots, \theta_k$ of the preprocessing functions $\phi_1, \phi_2, \ldots, \phi_k$ (based on the saturation curves) for the successive categories $C_1, C_2, \ldots, C_k$ of the products included in the model.

Each such chromosome Θ defines the parameters of the preprocessing functions and the remaining parameters of the entire interest mix model, i.e. $\alpha_0, \alpha_1, \ldots, \alpha_k$, are defined by the regression coefficients obtained in the learning process for linear regression models on a given training dataset (we use the regular Ordinary Least Squares learning algorithm for the linear regression model and its implementation available in the scikit-learn package [2,9]).

Having the entire interest mix model parametrized, we may evaluate its performance on the given training dataset and calculate its accuracy defined by the MAPE error. It constitutes the first criterion for evaluating the model and leads to the primary objective function used in the proposed algorithm, i.e.

$$F_1(\Theta) = \frac{1}{T}\sum_{t=1}^{T}\left|\frac{y^{(t)} - \hat{y}^{(t)}}{y^{(t)}}\right|^2, \tag{6}$$

where $t = 1, 2, \ldots, T$ corresponds to the successive time instants from the training dataset, $y^{(t)}$ denotes the number of recommended products from the given category under study at the time t (the observed value) and $\hat{y}^{(t)}$ denotes the predicted value for the time t. Although the notation does not include directly the chromosome Θ, $\hat{y}^{(t)}$ comes from the interest mix model with parameters defined by the chromosome Θ.

In order to avoid overfitting and promote more realistic models (based on some practitioner's heuristics), beside the primary objective function, an auxiliary one is used in the proposed algorithm, and the optimization problem is considered as multi-objective optimization problem. The auxiliary objective function aims at keeping the regression coefficients proportional to the number of products rated in each category, i.e.

$$F_2(\Theta) = \sum_{i=1}^{k} \left| \frac{\alpha_k}{\alpha^*} - \frac{x_k}{x^*} \right|^2, \tag{7}$$

where $\alpha^* = \alpha_1 + \alpha_2 + \ldots + \alpha_k$ and $x^* = x_1 + x_2 + \ldots + x_k$, where

$$x_i = \frac{1}{T} \sum_{t=1}^{T} x_i^{(t)}.$$

The idea behind it is to make the model more realistic and assume that if an user evaluates many products of a category, the category must be of a high importance for the user interests and a good recommender system should reflect it in its results. Certainly, this is only a heuristic assumption, but it is just an additional criterion for the optimization and may be more or less ignored when selecting the final model from the Pareto front. It constitutes the second criterion that measures the practical relevance of the model.

4.2   Crossover and Mutation

As in regular Evolutionary Strategies [1], each individual (Θ, σ) consists of a chromosome Θ representing a candidate solution to the optimization problem and an auxiliary chromosome σ representing the mutation parameter.

In EIMMA, the crossover operator recombines two individuals in such a way that parameters of the corresponding preprocessing functions are matched with each other and recombined one by one so that the corresponding Hill curves were averaged. The offspring individual encodes a Hill curve being somewhere between the Hill curves of the first and the second parent individuals, which is performed by sampling a number of random points from the first and the second Hill curve, averaging them, and interpolating a new Hill curve.

After recombination, for each offspring individual, the mutation operator adds a random noise drawn from the Gaussian distribution $\mathcal{N}(0, \sigma)$.

The auxiliary chromosomes σ are recombined by a one-point crossover and a mutation adding a random noise drawn from the Gaussian distribution $\mathcal{N}(0, \tau)$, where τ is a constant parameter of the algorithm.

4.3 Selection

In EIMMA, ranking selection is used with non-dominated sorting operator as in regular NSGA-II [4]. First, the population is split into successive Pareto fronts. Next, for each individual in the population, the crowding distance is evaluated. Finally, the selection is made on the basis of the Pareto front number and the crowding distance.

4.4 EIMMA

Algorithm 1 presents the overview of EIMMA. It start with generating an initial population, composed of N random individuals. Afterwards, the population evolves under the influence of the crossover and mutation operators.

Algorithm 1. EIMMA

$P_1 = \text{Initial-Population}(N)$
for $i = 1 \to K$ **do**
    $\text{Evaluate}(P_i)$
    $P_i' = \emptyset$
    for $j = 1 \to M$ **do**
        $\text{Parents} = \text{Parent-Selection}(P_i)$
        $\text{Offprings} = \text{Crossover}(\text{Parents})$
        $\text{Offprings} = \text{Mutation}(\text{Offsprings})$
        $P_i' = P_i' \cup \{\text{Offprings}\}$
    end for
    $P_{i+1} = \text{Selection}(P_i \cup P_i')$
end for

---

5 Experiments

5.1 Dataset

In the experiments, we focus on the MovieLens dataset [7]. In order to evaluate the approach proposed, we select a number of users existing in the dataset and simulate their behavior in time by adding a sequence of random ratings in the successive time instants and recording the recommendations evaluated by the recommendation system.

5.2 Case Study

Figure 1 presents a case study of one user. In the original dataset (time 0), he rated 31 action movies and 58 drama movies. In each of the successive 64 time instances, he evaluated one more, randomly chosen, action movie. Then, in each of the successive 64 time instances, he evaluated one more, randomly chosen, drama movie. As some action movies was also dramas, as well as some dramas was also action movies, the plots on Fig. 1 are not flat.

In each of such 128 time instances, the recommender system was run (trained on the current data) and the recommendations were recorded, Fig. 2 presents the number of action movies among the 100 most recommended movies in each time instant. It is worth noticing that the recommender system is not deterministic, so the changes on Fig. 2 are noised and before the further modeling the target time series were denoised with a short-term moving average and the input time series were denoised with Exponential Weighted Moving Average (EWMA) with the halflife parameter included in the preprocessing function parameters θ for each genre separately (Fig. 7).

Figures 8 and 9 presents a Pareto front of candidate solutions for the multi-objective optimization problem under study. Each candidate solution represents an interest mix model, where the parameters of the preprocessing functions were encoded in the chromosome and the remaining parameters are defined by the regression coefficients obtained in the learning process for linear regression models on the given training dataset.

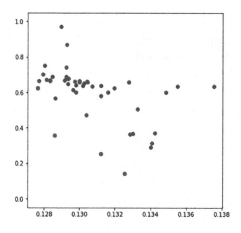

Fig. 7. The initial Pareto Front (in red) of 50 possible models. The horizontal axis corresponds to the accuracy error of the model, the vertical axis corresponds to the deviation of the ratio of the regression coefficients from the ratio of rated movies of corresponding genres (Color figure online)

Using such a Pareto front, a human analyst may pick an appropriate reliable model, taking into consideration the accuracy (the first objective) and the auxiliary heuristic reality measure (the second objective), and use the selected model to explain and interpret the behavior of the recommender system. Equation (5) enables him to estimate what is the impact of a high rating of one category on recommending the other, etc.

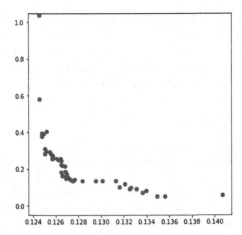

Fig. 8. The final Pareto Front (in red) of 50 possible models. The horizontal axis corresponds to the accuracy error of the model, the vertical axis corresponds to the deviation of the ratio of the regression coefficients from the ratio of rated movies of corresponding genres (Color figure online)

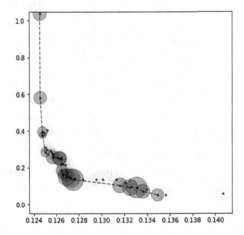

Fig. 9. The final Pareto Front of 50 possible models. The horizontal axis corresponds to the accuracy error of the model, the vertical axis corresponds to the deviation of the ratio of the regression coefficients from the ratio of rated movies of corresponding genres. The size of the circle corresponds to the σ mutation parameters encoded in the candidate solution, which approximately illustrates the possible region of expected mutations

Introducing the second objective and discovering the entire Pareto front, instead of one single solution optimal according to the first objective only, assures that the analyst has the possibility of rejecting some unrealistic models, perhaps being some effects of overfitting to the noised training data, such as the models with unreliable regression coefficients (e.g. significantly different from the proportions of the numbers of products highly rated in individual categories).

For instance, the most accurate model on Fig. 9 has the negative regression coefficient related to the same category as we consider in the output (in other words - the more highly rated products in the category under study, the less recommended products of that category - which makes the model unreliable despite the high accuracy, assuming that the recommender system is fit).

5.3 Summary of Results

Although the final selection of the final model depends on the particular choice of the human experts responsible for decision making processes, the majority of the discovered models outperform the baseline model. Table 1 presents the summary of statistics of these models (the time series cross validation approach with the train and the following test datasets) for the proposed approach (EIMMA) described in Sect. 4 and the simple regression model (SRM) described in Sect. 3.

Table 1. Summary of results

Experiment	F_1 of SRM	F_1 of EIMMA	F_2 of EIMMA
0	0.161966	0.134900	0.050123
1	0.162260	0.124420	1.759589
2	0.160671	0.124537	1.037544
3	0.169212	0.124572	0.580250
4	0.165599	0.124774	0.395716
5	0.176182	0.125013	0.283046
6	0.175665	0.124797	0.376827
7	0.168785	0.126477	0.183480
8	0.160123	0.126314	0.248945
9	0.169062	0.131617	0.102384

6 Conclusions

This paper proposes the interest mix modeling approach and Evolutionary Interests Mix Modeling Algorithm (EIMMA) for explaining how different types of user activities (e.g. related to user interests in different categories of products) affect the results of the recommender system (the lists of the recommended products). Computational experiments, performed on some selected benchmarks derived from the MovieLens dataset, confirmed the accuracy and efficiency of the proposed approach.

However, further research may concern studies on mutual relations between the categories of the products and extending interests mix modeling with additional components related to correlations between categories. Moreover, our current research assumes that the interest mix model is stationary and does not changes with time, which may not be true in larger time horizons, so introducing time-related coefficients may also improve the approach.

Acknowledgment. This work was supported by the Polish National Science Centre (NCN) under grant OPUS-18 no. 2019/35/B/ST6/04379. Calculations have been carried out using resources provided by Wroclaw Centre for Networking and Supercomputing, Grant No. 405.

References

1. Beyer, H.G., Schwefel, H.P.: Evolution strategies - a comprehensive introduction. Nat. Comput. **1**(1), 3–52 (2002). https://doi.org/10.1023/A:1015059928466
2. Buitinck, L., et al.: API design for machine learning software: experiences from the scikit-learn project. In: ECML PKDD Workshop: Languages for Data Mining and Machine Learning, pp. 108–122 (2013)
3. Bäck, T.: Introduction to evolution strategies. In: Proceedings of the Companion Publication of the 2014 Annual Conference on Genetic and Evolutionary Computation. GECCO Comp 2014, pp. 251–280. Association for Computing Machinery (2014). https://doi.org/10.1145/2598394.2605337
4. Deb, K., Pratap, A., Agarwal, S., Meyarivan, T.: A fast and elitist multiobjective genetic algorithm: NSGA-II. IEEE Trans. Evol. Comput. **6**(2), 182–197 (2002). https://doi.org/10.1109/4235.996017, Conference Name: IEEE Transactions on Evolutionary Computation
5. Gatignon, H.: Chapter 15 Marketing-mix models. In: Handbooks in Operations Research and Management Science, Marketing, vol. 5, pp. 697–732. Elsevier (1993). https://doi.org/10.1016/S0927-0507(05)80038-6
6. Gesztelyi, R., Zsuga, J., Kemeny-Beke, A., Varga, B., Juhasz, B., Tosaki, A.: The Hill equation and the origin of quantitative pharmacology. Arch. Hist. Exact Sci. **66**(4), 427–438 (2012). https://doi.org/10.1007/s00407-012-0098-5
7. Harper, F.M., Konstan, J.A.: The MovieLens datasets: history and context. ACM Trans. Interact. Intell. Syst. **5**(4), 19:1–19:19 (2015). DOI: https://doi.org/10.1145/2827872
8. He, X., Liao, L., Zhang, H., Nie, L., Hu, X., Chua, T.S.: Neural collaborative filtering. In: Proceedings of the 26th International Conference on World Wide Web, pp. 173–182 (2017). https://doi.org/10.1145/3038912.3052569
9. Pedregosa, F., et al.: Scikit-learn: machine learning in Python. J. Mach. Learn. Res. **12**, 2825–2830 (2011)
10. Ricci, F., Rokach, L., Shapira, B., Kantor, P.B. (eds.): Springer, Boston (2011). https://doi.org/10.1007/978-0-387-85820-3
11. Tian, C., Xie, Y., Li, Y., Yang, N., Zhao, W.X.: Learning to denoise unreliable interactions for graph collaborative filtering. In: Proceedings of the 45th International ACM SIGIR Conference on Research and Development in Information Retrieval, pp. 122–132 (2022). https://doi.org/10.1145/3477495.3531889
12. Zhang, S., Yao, L., Sun, A., Tay, Y.: Deep learning based recommender system: a survey and new perspectives. ACM Comput. Surv. **52**(1), 5:1–5:38 (2019). https://doi.org/10.1145/3285029

Machine Learning and AI in Digital Healthcare and Personalized Medicine

Multi-objective Evolutionary Discretization of Gene Expression Profiles: Application to COVID-19 Severity Prediction

David Rojas-Velazquez[1,2]($\boxtimes$) (iD), Alberto Tonda[3,4] (iD), Itzel Rodriguez-Guerra[5],
Aletta D. Kraneveld[1] (iD), and Alejandro Lopez-Rincon[1,2] (iD)

[1] Division of Pharmacology, University of Utrecht, Universiteitsweg 99,
3584 CG Utrecht, The Netherlands
{e.d.rojasvelazquez,A.D.Kraneveld,a.lopezrincon}@uu.nl

[2] Department of Data Science, Julius Center for Health Sciences and Primary Care,
University Medical Center Utrecht, Utrecht, The Netherlands

[3] UMR 518 MIA-PS, INRAE, Université Paris-Saclay, 91120 Palaiseau, France
alberto.tonda@inrae.fr

[4] Institut des Systèmes Complexes de Paris Île-de-France (ISC-PIF) - UAR 3611
CNRS, Paris, France

[5] Centro de Investigaciones en Ciencias Microbiológicas, Instituto de Ciencias,
Benemérita Universidad Autónoma de Puebla IC 11 Puebla, Puebla, Mexico
Guadalupe.rodriguezgue@alumno.buap.mx

Abstract. Machine learning models can use information from gene expressions in patients to efficiently predict the severity of symptoms for several diseases. Medical experts, however, still need to understand the reasoning behind the predictions before trusting them. In their day-to-day practice, physicians prefer using *gene expression profiles*, consisting of a discretized subset of all data from gene expressions: in these profiles, genes are typically reported as either over-expressed or under-expressed, using discretization thresholds computed on data from a healthy control group. A discretized profile allows medical experts to quickly categorize patients at a glance. Building on previous works related to the automatic discretization of patient profiles, we present a novel approach that frames the problem as a multi-objective optimization task: on the one hand, after discretization, the medical expert would prefer to have as few different profiles as possible, to be able to classify patients in an intuitive way; on the other hand, the loss of information has to be minimized. Loss of information can be estimated using the performance of a classifier trained on the discretized gene expression levels. We apply one common state-of-the-art evolutionary multi-objective algorithm, NSGA-II, to the discretization of a dataset of COVID-19 patients that developed either mild or severe symptoms. The results show not only that the solutions found by the approach dominate traditional discretization based on statistical analysis and are more generally valid than those obtained through single-objective optimization, but that the candidate Pareto-optimal solutions preserve the sense-making that practitioners find necessary to trust the results.

J. Correia et al. (Eds.): EvoApplications 2023, LNCS 13989, pp. 703–717, 2023.
https://doi.org/10.1007/978-3-031-30229-9_45

Keywords: Gene Expressions · Patient Profiles · Multi-Objective
Evolutionary Algorithms · COVID-19

1   Introduction

The information available up to November 16, 2022 indicates that the SARS-
CoV-2 pandemic continues, with 640 million cases and over 6 million deaths [37].
Due to the magnitude of the viral outbreak, one of the great problems that
humanity faces is the lack of medical equipment to care for infected patients.
Several studies have tried to predict the difference between the severity of the
cases using machine learning to analyze datasets with Chest X-Ray images, but
each dataset needs to be validated by experts, annotated with the corresponding
lesions of lung diseases, and features extracted based on recommendations of
medical personnel [1] resulting in the information taking time to be available
for analysis. One alternative to avoid the problem with Chest X-Ray images is
to use omics data, for example using DNA methylation [24,31], mRNA gene
expression [16,32] and/or microRNA [44] data, to quickly anticipate if a patient
will be in need of intensive care, and efficiently distribute the available medical
resources. Several sources [7,39] consider the correct management of beds and
the resources available in hospitals to be a crucial factor in reducing mortality
rates from COVID-19 in patients with severe infections.

Although there have been varying degrees of success with the use of multi-
omic data for diagnostic and prognostic purposes in general, one of the challenges
lies in translating the results into meaningful diagnostic tests or biomarkers
for clinical practice [11]. Nowadays multiple mRNA gene expression datasets
are available in public repositories: typically gene expression data will include
thousands of genes (features) related to just a few samples, leading to several
challenges for finding meaningful biomarkers. As humans cannot process the
information contained in thousands of genes due to the complexity of the data, it
is necessary to use computational tools such as machine learning (ML) techniques
to obtain reliable predictions [26,27,36].

mRNA gene expression datasets are matrices generated by sequencing data,
where typically each column corresponds to a specific variant of a gene, and
each row to a sample from a tissue (e.g. peripheral blood mononuclear cells,
peripheral blood leukocytes, whole blood, etc.). The values in the matrix are the
expression levels of the given variant of a gene for a patient sample [3].

Nevertheless, to make sense of the data, medical practitioners often resort to
the creation of *gene expression profiles*, discretizations of gene expressions, where
each continuous value is assigned to a discrete category, for example under- or
over-expressed gene. In this work, patient profile is defined as a set of gene expression values that uniquely characterize the patient, often discretized to make it
more readable by domain experts. Two patients have the same patient profile if
their discretized values are the same for the expression of each considered gene.
Categories are usually evaluated using thresholds based on the mean values of
gene expressions from healthy controls as a baseline. While this procedure can
help the sensemaking of the experts, such a discretization leads to loss of information and could potentially impair classification performance. Furthermore,

relying on control groups for discretizing gene expressions can lead to the wrong conclusions, as gene expression variability can be high, and control groups are usually comparatively small.

In this paper, we frame the problem of discretizing gene expression profiles from patients as a multi-objective task: on the one hand, the aim is to deliver a discretization that can be easily interpreted by a domain expert, ideally minimizing the different types of patient profiles; on the other hand, the loss of information resulting from the discretization should be minimal. While counting the number of different patient profiles resulting from the discretization is trivial, the loss of information can be assessed through the performance of a classifier in cross-validation on the discretized patient profiles. While approaches for the automated discretization of patient profiles have already been proposed in literature [30], the problem was previously conceived as single-objective optimization, with an arbitrary choice of weights to find compromises between conflicting objectives.

To test the proposed approach, real data from 138 participants, including information from 60,671 genes, were used and compared with a classical discretization approach based on mean values of gene expression from the group of healthy controls, and the single best solution found by a previously presented single-objective automated approach. Experimental results show that the proposed methodology is effective, identifying 12 genes highly correlated with the response to treatment and being able to discretize their gene expression levels into gene expression profiles. This helps to increase the performance of classifiers, and at the same time provides a human-interpretable explanation of the development of mild or dire symptoms from a COVID-19 infection. An expert analysis performed by domain experts provides a final validation of our approach.

2 Background

This section provides the minimal information needed to introduce the scope of our work.

2.1 Feature Selection

In machine learning (ML), feature selection (FS) is defined as the process of choosing the features of a dataset in order to obtain a minimal, informative subset. Features may not be part of this subset for two main reasons: they might be unrelated to the underlying nature of the problem, just adding noise; or they might be heavily correlated with other features, adding no relevant information for the task. Applications range from face recognition [43] to medicine [49], and approaches can be divided into two categories [19]: filters that score features according to a criterion (often a statistical test); and recursive procedures (forward or backwards) that attempt to reduce the features to a small set of non-redundant ones [10, 25].

In the scope of this work, we focus on recursive FS algorithms, in particular Recursive Ensemble Feature Selection (REFS). The method is a variation of Recursive Feature Elimination (RFE) [20], scoring the features in a 10-fold cross-validation scheme, using 8 different classifiers: Gradient Boosting [18], Passive Aggressive Classifier [13], Logistic Regression [12], Support Vector Machines Classifier [35], Random Forest [5], Stochastic Gradient Descent on Linear Models [48], Ridge Classifier [21] and Bagging [4]. The 10-fold cross-validation scheme was implemented following the nested cross-validation approach described in [42], which proves to be an effective approach to avoid the overfitting of machine learning models when working with data sets with a small number of samples [42]. The lowest-scoring features are removed from the analysis and the process is repeated until the overall classification accuracy drops below a given threshold. The use of an ensemble of classifiers reduces the effects of the inherent bias in each ML algorithm, thus delivering a more objective feature ranking. This technique has been applied successfully for problems involving both mRNA [27] and miRNA [26], featuring number of features ranging from 1,046 to 54,675.

2.2 Gene Expression Profiles

For diagnostic purposes it is not uncommon to generate heatmaps via computational techniques such as clustering or univariate analysis to find genetic expression profiles [14,28,33], using healthy controls as a baseline to obtain discretization thresholds. A visual representation of the transformation from a gene expression dataset to gene expression profiles is shown in Fig. 1.

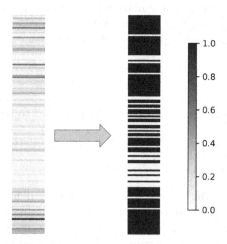

Fig. 1. Example of discretization of gene expression data to gene expression profiles for one gene, using a threshold value to distinguish between over-expressed (dark green) and under-expressed (light green) genes. (Color figure online)

Nevertheless, the generated genetic profiles or found biomarkers are difficult to translate into clinical practice [11]. One of the reasons is that not every time it is possible to include healthy controls as a baseline in the studies, and therefore reference values from other studies should be used which could be affected by the batch effect [36]. Furthermore, control groups are usually small in size, while the variability of gene expression can be considerable. Automated discretization could be a viable alternative, but approaches in literature [30] only frame the problem as single-objective, while in reality there are two conflicting objectives: maximizing classification performance and minimizing the number of different patient profiles to be analyzed by a human expert.

3   Proposed Approach

In this section, a novel approach for the multi-objective discretization of gene expression data to obtain gene expression profiles, interpretable by domain experts is presented. After performing a step of feature selection to identify the most relevant genes, these identified genes are discretized using thresholds optimized by a multi-objective evolutionary algorithm (MOEA), The conflicting objectives to be optimized are the classification performance and the number of different profiles (rows with different values) in the discretized dataset.

3.1   Feature Selection

For the feature selection step, the REFS algorithm was executed. REFS is an algorithm that uses the feedback of an ensemble of classifiers to rank each feature depending on its use capacity for the process of classification. The lowest-scoring features are removed, and the classification/ranking is repeated, until the average classification accuracy is below to a threshold defined by the user. The objective of this process is to select the most meaningful genes to correctly predict and model COVID-19 patients' severity (mild/severe).

3.2   Multi-objective Evolutionary Discretization

After running the REFS algorithm, a small set of relevant features is obtained. In this case, the values associated to the relevant features are complex and difficult to read for medical decision making. So, instead of showing each feature as a continuous value, we use a MOEA to categorize the values into underexpression and overexpression, optimizing the thresholds for each selected feature (gene). Given the reduced set of features $F = \{f_0, f_1, f_2, ..., f_n\}$, given by the REFS algorithm, EAs was used to transform input variables into *over* and *under* expressed values, labeled as 0 and 1, respectively: that is, EA will generate a vector of thresholds $I = \{t_0, t_1, t_2, ..., t_n\}$ to discretize each variable. Two criteria were used to optimize the discretization: number of different profiles, to be minimized to support the sensemaking of domain experts; and classification performance (given the labels of the dataset, corresponding to mild or severe symptoms), to be maximized, as a proxy of information loss. Consequently, the fitness functions for an individual I are given by:

$$f_1(I) = \frac{1.0}{1.0 + F1_{cv}(X_i, y)}$$

$$f_2(I) = n_p$$

(1)

where X_i is the dataset discretized according to the thresholds of individual I, y is the vector with the labels (mild/severe symptoms) for each sample, $F1_{cv}(X, y)$ is the F1 score in a cross-validation, n_p is the number of different profiles in the dataset after discretization. Previous works on automated discretization [30] employed accuracy as part of the evaluation, but F1 (a number between 0 and 1, representing the harmonic mean between precision and recall) is a preferable metric in case of unbalanced datasets, where samples from one class are more prevalent, as is often the case for medical data. The fitness function should be minimized, since the ideal candidate solution presents a high F1 and a low number of different profiles, to facilitate the understanding of domain experts.

4   Experimental Evaluation

The proposed approach is implemented in Python 3, relying on the `inspyred`[1] package for NSGA-II and the `scikit-learn` [34] package for classification. All the code and data needed to reproduce the experiments is freely available in the GitHub repository: https://github.com/to-be-disclosed/after-review.

4.1   Data

The dataset GSE169687 was selected from the gene expression omnibus (GEO) repository[2]. This dataset contains 138 samples of mRNA from peripheral blood of recovered COVID-19 patients at different time points, and 14 healthy controls. Only the 138 samples from patients with either mild/moderate (n=109) or severe/critical (n=29) symptoms were considered in the experiments, while the information from the 14 healthy controls was later used to compute an expert candidate solution to compare against. There are 60,671 ensemble genes (features) for each sample, and the dataset was divided into 2 classes, where label 0 was assigned to patients with mild/moderate symptoms and label 1 to patients with severe/critical symptoms.

4.2   Feature Selection

REFS algorithm was run 10 times, and a reduction from 60,671 to 12 features (highlighted as the optimal trade-off in Fig. 2) was ultimately obtained. This translates to the expression levels shown in Fig. 3.

---

[1] https://pythonhosted.org/inspyred/.
[2] https://www.ncbi.nlm.nih.gov/geo/.

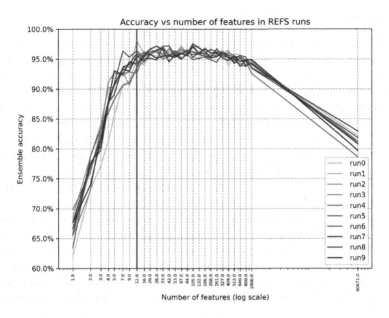

Fig. 2. 10 runs of the REFS algorithm. The solution considered as the best compromise between accuracy and number of features is marked with a red line (n = 12). (Color figure online)

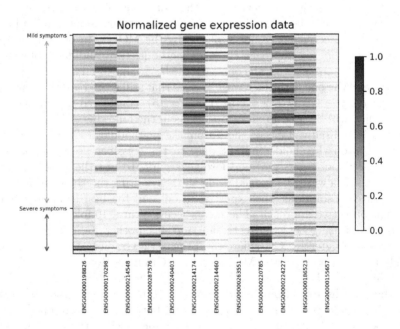

Fig. 3. Heatmap of normalized gene expression data, showing the values of each patient for the 12 most important genes (labeled as ENSG*) selected by the REFS algorithm.

4.3   Profile Generator

The MOEA selected for the profile generator is the NSGA-II [15], considered among the state of the art for multi-objective optimization with a relatively low number of objectives. After a few trial runs, the algorithm is set with the following parameters: $\mu = 200$, $\lambda = 350$, $p_c = 0.8$, $p_m = 0.2$, stop condition after 100 generations, and Logistic Regression [12] as the classifier chosen to compute classification performance $F1_{cv}$ for the fitness function described in Eq. 1. The choice of Logistic Regression is motivated by its effectiveness and training speed, making it one of the most suitable algorithms for our scenario. The classifier is run in a 10-fold cross-validation at each evaluation, in order to obtain a more reliable estimate of F1. The whole evolutionary optimization process is repeated 30 times, to assess the variance in the final results.

To provide a comparison, profiles based on a classical technique of the domain were compared, using as a reference the gene expression levels of the healthy controls to discretize the gene expressions of the patients: in other words, if a patient has the gene expression for a given gene higher than the mean of the 14 healthy controls in the dataset, that gene expression will be categorized as over-expressed (label 1); otherwise, it will be categorized as under-expressed (label 0).

The results of the experiments are shown in Fig. 4: it is clear that the point corresponding to the classical discretization technique (orange triangle) is dominated by the candidate solutions found by the proposed approach. On the other hand, it is interesting to observe that the point corresponding to the single-objective automated discretization [30] (red cross) is Pareto-optimal, and covers a part of the objective space not explored by the MOEA. In order to properly compare the proposed approach to other methods, two candidate solutions on the Pareto front are selected: the one with the best value of $F1_{cv}$ ($F1_{cv} = 0.9457$, $n_p = 40$); and the one with the fewest profiles that still scored above an arbitrary threshold $F1_{cv} \geq 0.9$ ($F1_{cv} = 0.9257$, $n_p = 15$), considered acceptable by an expert. They will be used in all comparisons in the following.

A possible disadvantage of the automatic methodologies is the generation of discretization thresholds that fit only to the classifier used for the fitness function (Logistic Regression) and lose generality. To test the generality of the method, the data was transformed using all the resulting thresholds from the two MOEA-selected solutions, the expert solution, and the best solution found by the single-objective approach. The mean accuracy in a 10-fold cross-validation was computed using Logistic Regression and seven other state-of-the-art classifiers: Passive Aggressive Classifier [13], Stochastic Gradient Descent on Linear Models [48], Support Vector Machines [35], Gradient Boosting [18], Random Forest [5], Ridge Classifier [21] and Bagging [4]. The results presented in Table 1 show that the discretizations found through automated approaches all perform better than the expert solution. Furthermore, both solutions obtained through the proposed MOEA approach on average perform better than the one found through single-objective optimization, hinting that the latter might be overfitted to the performance of Logistic Regression.

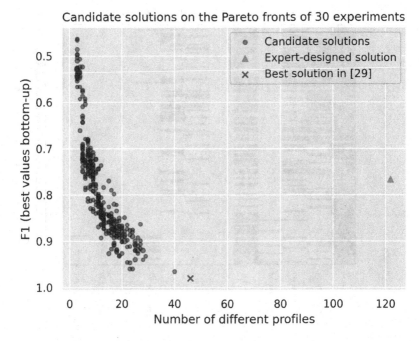

Fig. 4. Experimental results, with number of profiles on the x-axis, best values towards the left, and F1 on the y-axis, best values towards the bottom. Points in blue represent candidate solutions found during the runs. The orange triangle corresponds to the expert discretization based on gene expressions of healthy controls ($F1_{cv} = 0.77$, $n_p = 122$). The red cross corresponds to the best value found by the single-objective method presented in [30] ($F1_{cv} = 0.98$, $n_p = 46$). (Color figure online)

Table 1. Discretization strategies compared, using the F1 from different state-of-the-art classifiers, in a 10-fold cross-validation. **MOEA Highest F1** is the candidate solution with the highest F1 on the Pareto front. **MOEA Fewest Profiles** is the Pareto-optimal solution with the fewest profiles. **Best single-objective** is the performance of the best individual produced by the single-objective optimizer in [30]. **Expert solution** indicates the discretization performed using healthy controls as reference for the thresholds in the gene expression levels.

Classifier	MOEA Highest F1 (40 profiles)	MOEA Fewest profiles (15 profiles)	Best single-objective (46 profiles)	Expert solution (122 profiles)
BaggingClassifier	0.8990 ± 0.1344	0.9114 ± 0.0908	0.8883 ± 0.1414	0.6633 ± 0.2406
GradientBoostingClassifier	0.9314 ± 0.0859	0.9257 ± 0.0923	0.8362 ± 0.1543	0.6190 ± 0.2376
LogisticRegression	0.9457 ± 0.0842	0.9257 ± 0.0923	0.9800 ± 0.0600	0.7667 ± 0.1633
PassiveAggressiveClassifier	0.8324 ± 0.1976	0.7267 ± 0.3072	0.8455 ± 0.1213	0.6943 ± 0.2615
RandomForestClassifier	0.9314 ± 0.0859	0.9457 ± 0.0842	0.8924 ± 0.1165	0.6824 ± 0.2475
RidgeClassifier	0.8467 ± 0.3027	0.8657 ± 0.2967	0.9000 ± 0.1000	0.7667 ± 0.1633
SVC	0.9514 ± 0.0756	0.9257 ± 0.0923	0.9457 ± 0.0842	0.8029 ± 0.0977
Mean F1	0.9252	0.9059	0.8892	0.6990

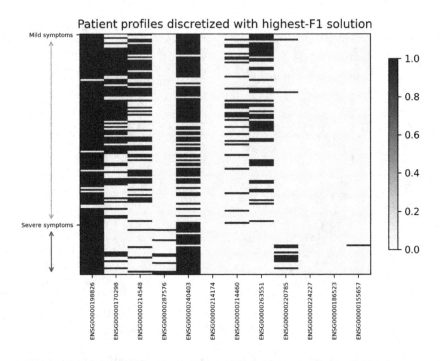

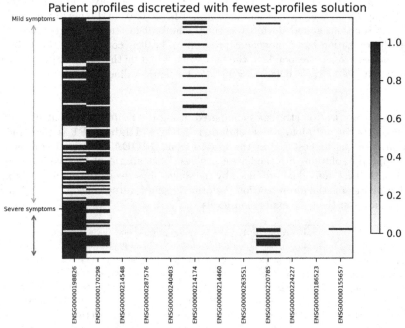

Fig. 5. Heatmaps produced by applying the set of thresholds identified by the two selected Pareto-optimal solutions: highest F1 (top) and fewest profiles (bottom).

Figure 5 shows the heatmaps obtained by discretizing the original dataset using the two solutions selected on the Pareto front: note how some of the discretized features are homogeneous, with all or nearly all values assigned to the same class. Analyzing the values of the thresholds reported in Table 2, it is interesting to notice how the MOEA, in order to reduce the number of profiles, set some of the thresholds to 1.0, *de facto* assigning almost all values of that feature (values between 0.0 and 1.0) to the same class. From a certain point of view, it's as if the algorithm were performing a second feature selection. An inspection of all the Pareto-optimal solutions found during the 30 experiments found that the most common thresholds set to 1.0 are for ENSG00000198826 (gene ARHGAP11A), ENSG00000214174 (gene AMZ2P1), and ENSG00000186523 (gene FAM86B1).

Table 2. Generated thresholds for each of the genes.

Ensemble ID	Gene ID	Thresholds (highest F1)	Thresholds (fewest profiles)	Thresholds (single objective)
ENSG00000198826	ARHGAP11A	0.0342	1.0000	0.2869
ENSG00000170298	LGALS9B	0.1654	0.5804	0.2817
ENSG00000214548	MEG3	0.1592	1.0000	0.6731
ENSG00000287576	-	0.5799	0.5858	0.4580
ENSG00000240403	KIR3DL2	0.0846	0.2320	0.1622
ENSG00000214174	AMZ2P1	1.0000	1.0000	0.2013
ENSG00000214460	TPT1P6	0.4441	1.0000	0.1101
ENSG00000263551	-	0.2366	0.1977	0.2379
ENSG00000220785	MTMR9LP	0.6698	0.9265	0.6528
ENSG00000224227	OR2L1P	1.0000	0.0267	0.9902
ENSG00000186523	FAM86B1	1.0000	1.0000	0.9191
ENSG00000155657	TTN	0.6623	0.7013	0.9880

5   Discussion

The results described in Table 1 show that the proposed approach outperforms the expert-driven discretization methodology from both the classification performance, and the number of profiles produced after discretization. Additionally, the high accuracy results obtained in a 10-fold cross-validation for several classifier provide evidence that using only Logistic Regression as element of the fitness function does not produce overfit in the discretization thresholds to a single classifier, contrary to the single-objective discretization.

Two of the 12 genes selected by the proposed approach are novel transcripts: ENSG00000263551 and ENSG00000287576. These two genes are listed as lncRNA (long, non-coding RNA) in the gene cards database [38], and there is no information available related to the subject as well as with ENSG00000214460 (TPT1P6 gene). These findings could be an important lead for new research on the subject, as they have never been associated with any particular biological function in literature, to our knowledge.

AMZ2P1, KIR3DL2and LGALS9B genes are directly related to the severity of symptoms in COVID. AMZ2P1 was found to be over-expressed in healthy retesting-positive COVID-19 patients [17]. KIR3DL is a killer cell immunoglobulin-like receptor gene and it's over-expression in presence of HLA-B is associated with moderate COVID-19 [2]. LGALS9 was identified as a COVID-19 severity protein biomarker, further evaluation of this gene provided evidence that it is also implicated in the apoptosis-associated cytokine Fas cell surface death receptor as a causal mediator of severe COVID-19 [22]. Although, not related to COVID-19 severity, we found two genes that are directly related to COVID-19 symptoms: TTN and OR2L1P. TTN encodes a large abundant protein of striated muscle. This gene is related to the molecular mechanisms behind muscle loss in COVID and this is associated with altered regulation of several cytokines [6], this could explain the muscle fatigue present in COVID-19 patients. OR2LP1 encodes an olfactory receptor, which interacts with odorant molecules in the nose, connected to the inflammatory reaction in the nasal cavity due to COVID, that leads to a temporary anosmia, where the odors are not able to reach the olfactory receptor neurons [40]; loss of the sense of smell is a common COVID-19 symptom [23].

Genes MEG3, FAM86B1, MTMR9LP and ARHGAP11A do not appear as biomarkers directly related to COVID-19. However, the over-expression of MEG3 may be favorable to virus infection, as evidenced in influenza A, via ADAR over-expression [8]. Since ADAR has been found as a controlling element in cellular response to viral infections, its regulation by MEG3 and interactions with lncR-NAs in SARS-CoV-2 infected cells may influence progression of the disease [41]. MTMR9LP is a lncRNA that is under-expressed in cryptococcal meningitis patients in comparison of healthy controls, related with cytokine expression and immune response triggered by cryptococcal infection [47].

ARHGAP11A is a member of the Rho GTPase-activating protein (RhoGAP) subfamily: this gene is under-expressed in tumors and is usually associated with malignant progression, this may be due the ability of ARHGAP11 to physically bind to p53 and promote its function, to induce cell-cycle arrest and apoptosis [45]. Interestingly, it has been found that a highly expressed ARHGAP11 is a sign for bad prognosis and poor survival rate in lung adenocarcinoma [9]. Also, this gene is under-expressed in pulmonary arterial hypertension, and this could be related to its role as a regulator of cell cycle-dependent motility [29]. FAM86B1 is a gene proposed as a "dark gene" directly linked with the survival of patients in complex diseases, specifically in bladder urothelial carcinoma (BLCA) [46]. Nevertheless, ARHGAP11A, AMZ2P1 and FAM86B1 probably provide less information, as inferred from the thresholds set by the MOEA, that are often set to 1.0 for these genes among the Pareto-optimal solutions.

6   Conclusions and Future Works

In this paper, a novel multi-objective evolutionary approach to the discretization of gene expression data was presented to obtain interpretable gene expression profiles, this approach can also lead to good classification accuracy when used with ML classifiers. The results on a real-world dataset related to COVID-19 (with patients exhibiting either mild or severe symptoms) seem promising, showing that the proposed technique performs better than a more classical approach based on a comparison with healthy controls, and produces results with better generality than a previous single-objective approach. In addition, we generated a set of rules given 12 specific genes to be used as a guide to decide whether a patient will present severe symptoms. An expert analysis reveals that the genes identified by our approach are known in literature, and the experts are satisfied by the discretization options provided, with a preference for solutions producing fewer profiles. While the initial results are promising, tests are needed in other real-world databases related to COVID-19 patients. Additionally, in order to claim generality, the proposed approach still needs to be evaluated on datasets related to different diseases.

References

1. Alghamdi, H.S., Amoudi, G., Elhag, S., Saeedi, K., Nasser, J.: Deep learning approaches for detecting Covid-19 from chest x-ray images: a survey. IEEE Access **9**, 20235–20254 (2021)
2. Bernal, E., et al.: Activating killer-cell immunoglobulin-like receptors are associated with the severity of Covid-19. J. Infect. Diseases (2021)
3. Brazma, A., Vilo, J.: Gene expression data analysis. FEBS Lett. **480**(1), 17–24 (2000)
4. Breiman, L.: Pasting small votes for classification in large databases and on-line. Mach. Learn. **36**(1–2), 85–103 (1999)
5. Breiman, L.: Random forests. Mach. Learn. **45**(1), 5–32 (2001)
6. Cantu, N., et al.: Synergistic effects of multiple factors involved in Covid-19-dependent muscle loss. Aging Disease, 9 (2021)
7. Cavallo, J.J., Donoho, D.A., Forman, H.P.: Hospital capacity and operations in the coronavirus disease 2019 (COVID-19) pandemic—planning for the nth patient. JAMA Health Forum **1**(3), e200345 (2020). https://doi.org/10.1001/jamahealthforum.2020.0345
8. de Chassey, B., et al.: The interactomes of influenza virus ns1 and ns2 proteins identify new host factors and provide insights for adar1 playing a supportive role in virus replication. PLoS Pathog. **9**(7), e1003440 (2013)
9. Chen, S., Duan, H., Xie, Y., Li, X., Zhao, Y.: Expression and prognostic analysis of rho gtpase-activating protein 11a in lung adenocarcinoma. Ann. Transl. Med. **9**(10) (2021)
10. Chien, Y., Fu, K.S.: On the generalized karhunen-loève expansion (corresp.). IEEE Trans. Inf. Theory **13**(3), 518–520 (1967)
11. Chin, L., Gray, J.W.: Translating insights from the cancer genome into clinical practice. Nature **452**(7187), 553–563 (2008)
12. Cox, D.R.: The regression analysis of binary sequences. J. Roy. Stat. Soc. Ser. B (Methodol.), 215–242 (1958)
13. Crammer, K., Dekel, O., Keshet, J., Shalev-Shwartz, S., Singer, Y.: Online passive-aggressive algorithms. J. Mach. Learn. Res. **7**(Mar), 551–585 (2006)

14. Cruz-Rodriguez, N., Quijano, S.M., Enciso, L.J., Combita, A.L., Zabaleta, J.: Gene expression signature predicts induction treatment response and clinical outcome in adult Colombian patients with acute lymphoblastic leukemia (2016)

15. Deb, K., Pratap, A., Agarwal, S., Meyarivan, T.: A fast and elitist multiobjective genetic algorithm: NSGA-II. IEEE Trans. Evol. Comput. **6**(2), 182–197 (2002)

16. Delorey, T.M., et al.: Covid-19 tissue atlases reveal SARS-COV-2 pathology and cellular targets. Nature, 1–8 (2021)

17. Fang, K.Y., et al.: Screening the hub genes and analyzing the mechanisms in discharged Covid-19 patients retesting positive through bioinformatics analysis. J. Clin. Lab. Anal. **36**(7), e24495 (2022)

18. Friedman, J.H.: Greedy function approximation: a gradient boosting machine. Ann. stat., 1189–1232 (2001)

19. Guyon, I., Elisseeff, A.: An introduction to variable and feature selection. J. Mach. Learn. Res. **3**, 1157–1182 (2003)

20. Guyon, I., Weston, J., Barnhill, S., Vapnik, V.: Gene selection for cancer classification using support vector machines. Mach. Learn. **46**(1–3), 389–422 (2002)

21. Hoerl, A.E., Kennard, R.W.: Ridge regression: biased estimation for nonorthogonal problems. Technometrics **12**(1), 55–67 (1970). https://doi.org/10.1080/00401706.1970.10488634

22. Klaric, L., et al.: Mendelian randomisation identifies alternative splicing of the FAS death receptor as a mediator of severe Covid-19. medRxiv (2021)

23. Klopfenstein, T., et al.: Features of anosmia in Covid-19. Medecine et maladies infectieuses **50**(5), 436–439 (2020)

24. Konigsberg, I.R., et al.: Host methylation predicts SARS-COV-2 infection and clinical outcome. Commun. Med. **1**(1), 1–10 (2021)

25. Lewis, P.: The characteristic selection problem in recognition systems. IRE Trans. Inf. Theory **8**(2), 171–178 (1962)

26. Lopez-Rincon, A., Martinez-Archundia, M., Martinez-Ruiz, G.U., Schoenhuth, A., Tonda, A.: Automatic discovery of 100-mirna signature for cancer classification using ensemble feature selection. BMC Bioinform. **20**(1), 480 (2019)

27. Lopez-Rincon, A., et al.: Machine learning-based ensemble recursive feature selection of circulating mirnas for cancer tumor classification. Cancers **12**(7), 1785 (2020)

28. Lu, Y., et al.: Dynamic edge-based biomarker non-invasively predicts hepatocellular carcinoma with hepatitis b virus infection for individual patients based on blood testing. J. Mol. Cell Biol. **11**(8), 665–677 (2019)

29. Ma, Y., Chen, S.S., Feng, Y.Y., Wang, H.L.: Identification of novel biomarkers involved in pulmonary arterial hypertension based on multiple-microarray analysis. Biosci. Rep. **40**(9) (2020)

30. Mouhrim, N., Tonda, A., Rodríguez-Guerra, I., Kraneveld, A.D., Rincon, A.L.: An evolutionary approach to the discretization of gene expression profiles to predict the severity of COVID-19. In: Proceedings of the Genetic and Evolutionary Computation Conference Companion. ACM, July 2022. https://doi.org/10.1145/3520304.3529001

31. de Moura, M.C., et al.: Epigenome-wide association study of Covid-19 severity with respiratory failure. EBioMedicine **66**, 103339 (2021)

32. Ng, D.L., et al.: A diagnostic host response biosignature for Covid-19 from RNA profiling of nasal swabs and blood. Sci. Adv. **7**(6), eabe5984 (2021)

33. Paiva, B., et al.: Phenotypic and genomic analysis of multiple myeloma minimal residual disease tumor cells: a new model to understand chemoresistance. Blood J. Am. Soc. Hematol. **127**(15), 1896–1906 (2016)

34. Pedregosa, F.,et al.: Scikit-learn: machine learning in Python. J. Mach. Learn. Res. **12**, 2825–2830 (2011)

35. Platt, J.: Others: probabilistic outputs for support vector machines and comparisons to regularized likelihood methods. Adv. Large Margin Classifiers **10**(3), 61–74 (1999)

36. Rincon, A.L., Kraneveld, A.D., Tonda, A.: Batch correction of genomic data in chronic fatigue syndrome using CMA-ES. In: Proceedings of the 2020 Genetic and Evolutionary Computation Conference Companion, pp. 277–278. ACM, Cancun, July 2020. https://doi.org/10.1145/3377929.3389947

37. Roser, M.: Covid-19 data explorer (2022). https://ourworldindata.org/explorers/coronavirus-data-explorer

38. Safran, M., et al.: Genecards version 3: the human gene integrator. Database 2010 (2010)

39. Sussman, N.: Time for bed(s): hospital capacity and mortality from Covid-19. COVIDEconomics, pp. 116–129 (2020)

40. Torabi, A., et al.: Proinflammatory cytokines in the olfactory mucosa result in Covid-19 induced anosmia. ACS Chem. Neurosci. **11**(13), 1909–1913 (2020)

41. Turjya, R.R., Khan, M.A.A.K., Mir Md. Khademul Islam, A.B.: Perversely expressed long noncoding RNAs can alter host response and viral proliferation in SARS-COV-2 infection. Future Virol. **15**(9), 577–593 (2020)

42. Vabalas, A., Gowen, E., Poliakoff, E., Casson, A.J.: Machine learning algorithm validation with a limited sample size. PLoS ONE **14**(11), e0224365 (2019)

43. Vignolo, L.D., Milone, D.H., Scharcanski, J.: Feature selection for face recognition based on multi-objective evolutionary wrappers. Expert Syst. Appl. **40**(13), 5077–5084 (2013)

44. Wilson, J.C., et al.: Integrated mirna/cytokine/chemokine profiling reveals severity-associated step changes and principal correlates of fatality in Covid-19. Iscience, 103672 (2021)

45. Xu, J., et al.: RhoGAPs attenuate cell proliferation by direct interaction with p53 tetramerization domain. Cell Rep. **3**(5), 1526–1538 (2013)

46. Yan, J., Li, P., Gao, R., Li, Y., Chen, L.: Identifying critical states of complex diseases by single-sample jensen-shannon divergence. Front. Oncol. **11**, 1824 (2021)

47. Zhang, L., et al.: Long noncoding RNA expression profile from cryptococcal meningitis patients identifies dpy19l1p1 as a new disease marker. CNS Neurosci. Therapeutics **25**(6), 772–782 (2019)

48. Zhang, T.: Solving large scale linear prediction problems using stochastic gradient descent algorithms. In: Twenty-First International Conference on Machine Learning - ICML 2004. ACM Press (2004). https://doi.org/10.1145/1015330.1015332

49. Zhou, Z., Li, S., Qin, G., Folkert, M., Jiang, S., Wang, J.: Multi-objective-based radiomic feature selection for lesion malignancy classification. IEEE J. Biomed. Health Inform. **24**(1), 194–204 (2020). https://doi.org/10.1109/jbhi.2019.2902298

Interactive Stage-Wise Optimisation of Personalised Medicine Supply Chains

Andreea Avramescu[(✉)][iD], Manuel López-Ibáñez[iD],
and Richard Allmendinger[iD]

Alliance Manchester Business School, University of Manchester,
Manchester M15 6PB, UK
{andreea.avramescu,manuel.lopez-ibanez,
richard.allmendinger}@manchester.ac.uk

Abstract. Personalised medicine (PM) is a new area of healthcare that has shown promising results in offering treatments for rare and advanced stage diseases. Nonetheless, their supply chain differs from the traditional healthcare model by adding a high level of complexity through product individualisation. The patient is now also the donor, and becomes part of a complex manufacturing and delivery system. PM biopharmaceuticals are cryopreserved (frozen) and re-engineered at specialised facilities, before being returned to the same patient. The corresponding facility location problem (FLP) consists of both manufacturing and cryopreservation sites, having a set of constraints that are not met in other healthcare supply chains. As a result, the FLP in the context of PM has only been partially analysed and additional research is still necessary to reach optimal network configurations. In this paper, we extend the solution methods previously proposed for PM FLP by using a stage-wise approach in which the (constrained binary multi-objective) problem is divided into smaller, logical parts. We approach the problem from the perspective of the decision maker (DM) and use the R-NSGA-II algorithm to find a set of desirable solutions. By optimising only part of the decision space at a time, we reduce the complexity of the problem and allow the DM to compare the objective values obtained between different supply chain configurations. Our results suggest that allowing the DM to interact with the optimisation process can lead to good and desirable solutions in a shorter computational time, and more flexible network configurations.

Keywords: personalised biopharmaceuticals · supply chain · R-NSGA-II · interactive multi-objective optimisation

1 Introduction

A sub-optimal supply chain can lead to considerable costs for businesses. However, for some fields, the consequences go beyond the economic perspective and influence the customers directly. A prominent example that has received increased attention is the field of personalised medicine (PM). PM focuses particularly on the pharmaceuticals and biopharmaceuticals developed as a result

J. Correia et al. (Eds.): EvoApplications 2023, LNCS 13989, pp. 718–733, 2023.
https://doi.org/10.1007/978-3-031-30229-9_46

of this new healthcare paradigm [5]. These treatments are among the products with the highest re-engineering process, starting with raw cells from the patient and, after complex manufacturing, they are returned to the same individual [23]. It is then for the first time when the donor and the recipient are the same person and become part of the healthcare supply chain.

The market value of personalised medicine biopharmaceuticals (PMBs) is expected to surpass $2 billion [12], with about 40 products approved for commercialisation (i.e. after proceeding through three clinical stages and obtaining the regulatory permissions) [11] and over 2000 in any of the three clinical trials in Europe [21]. Regardless of both economic and social promises of PMBs [10], the companies are struggling to offer the products globally. Causes of failure include the inability to ensure a timely and cost-efficient production such that it meets the quality standards reached in clinical trials. Market authorisation for medical products is a lengthy procedure that, once obtained, comes with a series of regulatory practices and quality assurance measurements that need to be met for each patient. Not meeting the constraints of the market authorisation can lead to partial or complete market withdrawal [1,16].

The current configuration for the personalised medicine biopharmaceuticals (PMBs) delivery is following strategical and organisational decisions that have been used for traditional pharmaceutical. Nevertheless, PMBs use novel technology and are among the newest products on the healthcare market. Research around its supply chain from an operations research perspective is still incipient. A sub-optimal supply chain will increase further the gap in access to revolutionary healthcare and not advance the usage of personalised medicine beyond the current one, namely when everything else has failed.

The scope of this paper is to contribute to the development of more holistic supply chain models and solution methods that are specifically designed for the optimisation of PMBs. We start from the mathematical formulations of a centralised (i.e. the demand and supply nodes are always different) introduced in [2] and an integrated supply chain (demand and supply points can coincide) presented in [3]. Unlike the previous papers whose focus was on the network design, we focus on the impact of the problem dimension and the ability to overcome it by limiting the search space with the support of a decision maker (DM). In both cases, the problem is formulated as a multi-objective facility location problem (FLP) consisting of one mandatory facility, which must process each product, and a helper facility that can reduce the impact of the constraints.

The FLP is a known class of NP-hard problems. In the context of PM, exact solution methods have only been successfully applied for simple formulations, with a small decision space. If we are to consider the real-world demand for PMBs with global coverage, the formulation leads to hundreds of demand nodes and thousands of candidate locations for multiple facility types. Moreover, the mathematical models proposed so far in the literature for PMBs were formulated as mixed-integer non-linear programming, and thus, metaheuristics were the preferred solution method.

In this paper, we analyse the impact of the DM, usually a biopharmaceutical company, in the optimisation process. We divide the problem into six stages and reduce the number of decision variables that are optimised at once. This approach offers the DM the flexibility to create different supply chain configurations and stop the optimisation process when desirable solutions had been obtained. We use a reference-point based algorithm, R-NSGA-II [9], to obtain a set of non-dominated solutions for each of the six stages.

The rest of the paper is organised as follows. A brief summary of the PMB decisions at the strategic level of the supply chain are introduced in Sect. 2. Some of the related academic research covering PMB supply chain optimisation is presented in Sect. 3. Section 4 describes the exact problem formulation applied in this paper and the algorithm used to solve it, while the case study used is described in Sect. 5. The results are presented in Sect. 6 before the paper ends with conclusions and directions for future research in Sect. 7.

2   Supply Chain and Problem Formulation

The PMBs supply chain follows a 1-to-1 model where each product is created individually, having the patient as starting and end points. The demand locations are usually hospitals, where the first step of the supply chain, leukapheresis is conducted. Through leukapheresis, the cells required for the creation of the PMB are collected while the rest of the blood is returned to the patient. This initial step represents one of the reasons while the success rate of the supply chain is of high importance. The cells go through a manufacturing process where they are re-engineered to fight a specific disease. As with any production, the manufacturing can also fail at different steps for various reasons. If a PMB manufacturing fails, the supply chain of that patient needs to be restarted. However, depending on the individual's health condition, collecting a second set of cells might not always be possible. The FLP problems corresponding to the PMBs supply chain can be divided into two broad categories, referred here as *centralised* and *integrated* configurations.

Centralised Supply Chain. The manufacturing process is taking place at highly specialised manufacturing facilities (MFs) to which the cells are delivered from the hospital. After the manufacturing is finished, they are returned to the same patient. Usually, assuming the current workflow of the public health-care, this will coincide with the same hospital from which the cells were initially collected. The cells are time and temperature sensitive, with PMB being considered some of the most sensitive medical products. As a way of mitigating the short shelf-life and risking the quality of the cells, a cryopreservation technique is applied. The frozen cells then have an extended shelf-life that is generally not imposing a time restriction anymore unless the product needs to be stored. The shelf-life of frozen cells is long enough for the delivery between any two points to be met.

The process of cryopreservation is conducted at the MFs once the re-engineering process is finished, and at independent cryopreservation facilities (CFs) after leaving the hospital. A more detailed description and the corresponding mathematical model for the above were defined in [2].

Integrated Supply Chain. Opening new MFs and CFs is expensive and has to consider a number of constraints, like obtaining regulatory approvals for each location. Moreover, each PMB is targeting diseases with a low global demand. A high initial investment might turn out to be economically inefficient for the biopharmaceutical companies. Additionally, a large default delivery time is expected, since placing a high number of such facilities around the globe is not feasible. As an alternative, different levels of decentralisation had been proposed [13,14]. We analyse the impact of smaller manufacturing and cryopreservation units (MUs and CUs) as possible integration at demand points (i.e. hospitals). In this scenario, either the cryopreservation or the manufacturing can take place directly at the hospital, providing a potential solution for locations with high demand or extremely isolated. An integrated CU avoids the shelf-life constraint. An integrated MU allows the completion of the 1-to-1 supply chain within the demand point. A more detailed description of the integrated model and its formulation were defined in [3] and a simplified mathematical model is described below.

In line with previous research, the problem is formulated as a multi-objective facility location-allocation problem where we aim to:

1. *maximise the number of hospitals that can accept patients for PMB* from a pre-defined list, because they have a level of integration or they are within the shelf-life radius of a facility;
2. *minimise the waiting average delivery time per patient*, calculated as the total delivery and production time (excluding leukapheresis) and dividing by a known total demand;
3. *minimise the total cost of the supply chain*, calculated as the cost of building the MFs and CFs, the integration cost at hospitals of MUs and CUs, and the production cost which is product dependent;

Objective 1 is calculated as the number of hospitals that have covered demand. These correspond to those hospitals that have an MU or are allocated to an MF. Partial demand coverage is not allowed and it is assumed that the entire demand from a hospital is covered. Objective 2 is the average delivery time for all hospitals. The hospitals that have an MU will have no delivery time, while those with CU will only have the direct delivery to an allocated MF. Objective 3 is the sum of the construction cost for MFs and CFs and the cost of integration in the case of MUs and CUs.

FR_j (Eq. 5) is the failure rate associated to each location. Equations (6) and (7) are classic FLP constraints and restrict one facility type to be opened at each location. This applies for both demand nodes and candidate locations. Similarly, if a hospital has an MU or CU, it cannot be allocated to an MF or

CF (8). All hospitals that are allocated to a CF must be allocated to an MF (9). Hospitals allocation can only be done at open MFs or CFs (10). Equation (11) is the shelf-life constraint. If the demand from a hospital is delivered to an MF via a CF, then the shelf-life constraint between CF and MF is deactivated.

Indices and Parameters

$i \in I$ Demand nodes (hospitals).

$j \in J$ Candidate locations for placing an MF/CF.

$k \in K$ Production modes of MU/MF such as manual, semi-automatic, and automatic $(k > 1)$; the value $k = 1$ denotes cryopreservation (either CU or CF).

r_k Failure rate associated to each production mode $(k > 1)$.

$d_{ij}, d_{ij'}, d_{ij'}$ Distances between facilities.

c_{ik}, c_{jk} Cost for building an MU $(k > 1)$ or CU $(k = 1)$, and cost for building an MF $(k > 1)$ or CF $(k = 1)$ respectively.

θ Shelf life.

T Very large number.

Decision Variables

h_{ik} 1 if an MU or CU is built at hospital $i \in I$, 0 otherwise; where $k \in K$ represents either CU $(k = 1)$ or MU with various production modes $(k > 1)$.

x_{jk} 1 if a MF or CF is placed at location $j \in J$, 0 otherwise; where $k \in K$ has the same meaning as above but now refers to independent facilities.

y_{ij} 1 if hospital i is allocated to either an MF or CF at location j, 0 otherwise.

The following helper variables are used to check if hospital i has an MU (h_i^{M}), it is allocated to an MF (z_i^{M}) or to a CF (z_i^{C}):

$$h_i^{\mathrm{M}} = \sum_{k=2}^{K} h_{ik} \qquad x_j^{\mathrm{M}} = \sum_{k=2}^{K} x_{jk} \qquad z_i^{\mathrm{M}} = \sum_{j \in J} y_{ij} x_j^{\mathrm{M}} \qquad z_i^{\mathrm{C}} = \sum_{j \in J} y_{ij} x_{j1}$$

Objectives

$$\text{Minimize} \quad 1 - \frac{\sum_{i \in I} h_i^{\mathrm{M}} + z_i^{\mathrm{M}}}{|I|} \tag{1}$$

$$\text{Minimize} \quad \frac{TotalTime}{\sum_{i \in I} h_i^{\mathrm{M}} + z_i^{\mathrm{M}}} \tag{2}$$

$$\text{Minimize} \quad \sum_{i \in I} \sum_{k=1}^{K} c_{ik} h_{ik} + \sum_{j \in J} \sum_{k=1}^{K} c_{jk} x_{jk} \tag{3}$$

where:

$$TotalTime = \sum_{i \in I} (1 - h_i^{\mathrm{M}}) \sum_{j \in J} y_{ij} x_j^{\mathrm{M}} \Big[d_{ji} + (1 + FR_j)$$

$$\cdot \Big((1 - z_i^{\mathrm{C}}) d_{ij} + z_i^{\mathrm{C}} \sum_{j' \in J} y_{ij'} x_{j'1} (d_{ij'} + d_{j'j}) \Big) \Big] \tag{4}$$

$$FR_j = \sum_{k=2}^{K} r_k x_{jk} \tag{5}$$

$$\sum_{k=1}^{K} h_{ik} \leq 1 \quad \forall i \in I \tag{6}$$

$$\sum_{k=1}^{K} x_{jk} \leq 1 \quad \forall j \in J \tag{7}$$

$$h_i^{\mathrm{M}} + z_i^{\mathrm{M}} \leq 1, \quad h_{i1} + z_i^{\mathrm{C}} \leq 1 \quad \forall i \in I \tag{8}$$

$$z_i^{\mathrm{C}} \leq z_i^{\mathrm{M}} \leq 1 \quad \forall i \in I \tag{9}$$

$$y_{ij} \leq \sum_{k=1}^{K} x_{jk} \quad \forall i \in I, \forall j \in J \tag{10}$$

$$\sum_{j \in J} y_{ij} x_j^{\mathrm{M}} d_{ij} \leq z_i^{\mathrm{C}}(24\,\mathrm{h}) + (1 - z_i^{\mathrm{C}})\theta + h_{i1} \cdot T \quad \forall i \in I \tag{11}$$

3   Related Work

The FLP in PMBs supply chain has been only recently approached from an optimisation perspective. Nonetheless, the papers that focused on either the mathematical modelling of the problem or the extension of solution methods have highlighted the impact of the supply chain on the availability of PMBs [20, 22].

The majority of the papers on PMBs FLP focus on the optimisation of the centralised model. The first paper to propose a mathematical formulation is [25]. They use a small case-study of a CAR-T therapy, with only 11 demand points and 6 candidate locations for the MFs. Their formulation is a classical uncapacitated FLP, solved using exact methods. In most of the papers there has been little emphasise on the development of solution methods and more on the extension of the problem formulation, such as in the extension presented in [19] on centralised model, and [4] who introduce intermediate storage units as a decentralisation method.

Using an exact solver, such as CPLEX, was feasible as the formulations in previous papers had a low number of decision variables, a small decision space, and linear formulations. Nonetheless, none of these papers considered the location of the CFs as part of the optimisation problems, and did not look at integration levels for either MFs or CFs. Karakostas et al. [17] presented a decentralisation of the PMBs supply chain that, taking inspiration from supply networks for blood transfusion bags, proposes the use of local treatment facilities alongside specialised hospitals, as well as mobile medical units for the treatment delivery. They use a General Variable Neighborhood Search (GVNS) to tackle a realistic-sized CAR-T case study. A different integrated model was proposed by Avramescu et al. [3] that considers both the optimisation of the CFs and the hospitals facility integration. To the best of our knowledge, the impact of the DM in the optimisation process has not yet been studied for this problem.

3.1 Problem Stages

The integrated configuration (IC) is an extension of the centralised one (CC). Even though there may be a solution where the demand could be completely covered by MUs, for it to be feasible, at least an MF needs to exist for purposes of quality check and quality assurance. In other words, the IC will always comprise the model of the CC. It is then possible to allow the DM to stop the optimisation process at different stages, with a different number of facility types, by dividing the problem into its logical sub-components, as follows:

I **Stage C1:** At this stage in the optimisation process we only optimise the location of the MFs, reducing the problem to the classic uncapacitated FLP [7]. This is the only mandatory stage as the production of a PMB cannot happen unless the cells are re-engineered at one of the MFs. A default production mode chosen by the DM is set.

II **Stage C2:** After the optimisation of the MFs, the DM can decide whether to accept any of the non-dominated solutions obtained in the previous step or add CFs. The CFs will increase the overall delivery time for a patient since the cells will need to be first frozen and then delivered to an MF. However, the CFs can reduce the cost of the supply chain by reducing the number of MFs while maintaining or increasing the number of hospitals that can accept PMBs.

III **Stage C3:** The final stage of the CC optimises the production modes of the MFs. Each mode of production has an associated construction and production cost, and failure rate. If the production of a PMB fails, the entire supply chain for that patient needs to be restarted from leukapheresis, thus increasing the total time (Eq. 4).

The graphical representation of the three stages is shown in Fig. 1.

Once the CC is created, the DM has the choice of adding different levels of integration to the supply chain. In this paper, we have created the integration stages by mirroring the centralised ones to simplify the process for the DM. At the same time, this distribution creates the possibility of interchangeably switching between CC or IC at the beginning of the optimisation process. However, alternative levels of integration or order of stages could also be applied. The mirrored integrated stages are as follows:

IV **Stage I1:** Add MUs at hospitals. If a hospital has an MU, the cells can be processed on-site and no delivery is necessary. The obvious advantage in this case would be a considerable decrease in delivery time for the patient and the elimination of the cryopreservation process. Even though the cost of a single MU is cheaper than an MF, this cost would be unprofitable for hospitals with low demand.

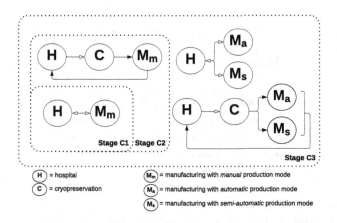

Fig. 1. Facilities and network direction for the different stages in the centralised scenario where integration model is not considered. Each of the higher level stages will contain all combinations at the lower level stages. Filled (resp. unfilled) arrows correspond to frozen (resp. fresh) delivery of cells.

V **Stage I2:** Add CUs at hospitals. In this case the cells can be transported to any MF without having to pass by a CF first. To simplify the Chain of Identity and Chain of Custody, we restrict the cells that come from a hospital with CUs to only be processed at MFs and not hospitals with MUs.

VI **Stage I3:** this final stage only optimises the production modes of the hospitals with MUs.

Figure 2 shows the route of the integrated hospitals. We consider that integration is always built on top of centralisation, hence stages I1, I2, and I3 would also contain stages C1, C2, and C3.

4   Solution Methods

In order to set an evaluation criterion for the solutions obtained at different stages, we use a reference point for each of the three objectives. Since the only requirement for a supply chain of PMB to be feasible is to have an MF, the DM can decide at which point in the optimisation process to stop the evaluations and save a set with non-dominated solutions. A solution is considered to be non-dominated if there is no other solution that can obtain a better value for any of the objectives without worsening another objective function.

We use R-NSGA-II as solution method. R-NSGA-II [9] is a reference point based variant of the well-known multi-objective evolutionary algorithm NSGA-II [8]. The classic evolutionary multi-objective optimisation (EMO) methods try to approximate the entire Pareto front by distributing evenly the non-dominated solutions across the objectives. In comparison, a preference-based EMO allows the DM to set ideal values for the objective functions and the non-dominated

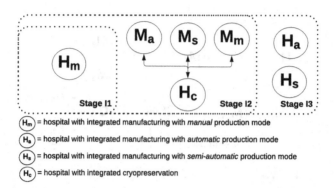

Fig. 2. Facilities and network direction for the different stages in the integrated scenario. Any integration step is always optimised on top of the solutions obtained after the last stage of the centralised scenario. Each of the higher level stages will contain all combinations at the lower level stages. Filled (resp. unfilled) arrows correspond to frozen (resp. fresh) delivery of cells. No connection means that there is no delivery taking place and the facility acts as both demand and supply nodes.

solutions returned by the algorithm will then be concentrated around that particular region. R-NSGA-II is among the most popular EMO algorithms and differs from the classic NSGA-II in the crowding distance, which is calculated using the Euclidean distance from the reference points.

R-NSGA-II is used as the solution method for all six stages of the problem. At the first stage, for both CC and IC, the initial population is created randomly by placing MFs with p probability in the candidate locations. At the second stage, the initialisation is done by adding CFs at empty locations or by changing some of the existing MFs. This approach was preferred as compared to just randomly placing CFs to avoid a drop in the solution quality at the first generation of the following stage. Adding facilities at random to an already optimised solution will, in most cases, lead to worse objective function values. As a consequence, any progress is stagnating until the values of the objectives reach at least the one obtained at the end of the previous stage. As the final stage is only optimising the production modes of existing MFs or MUs, the initialisation will also switch the mode rather than adding new facilities. The entire logic of the solution method is shown in Fig. 3.

5   Case Study

To test the stage-wise approach proposed, we solved the problem using a case study of a PMB with market authorisation. Kymriah is one of the popular CAR-T therapies used for the treatment of blood cancer. The PMB obtained initially authorisation for the treatment of paediatric and young adult patients suffering from relapsed or refractory (r/r) B-cell precursor acute lymphoblastic leukaemia (ALL) in 2017, and in 2022 it obtained a treatment designation expansion for

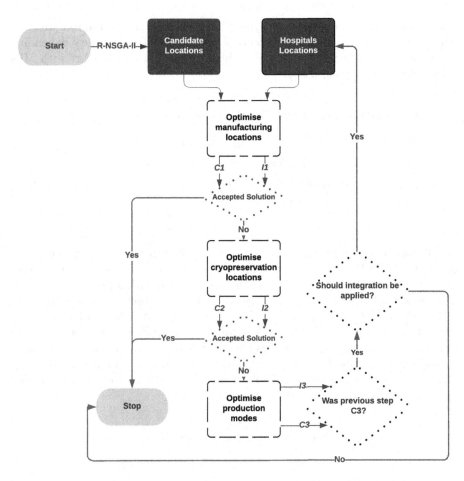

Fig. 3. Flowchart of solution method highlighting the problem stages and the steps at which the DM can stop or continue the optimisation process.

adults with r/r follicular lymphoma (FL), after at least two lines of alternative treatment [6]. It is worth mentioning that the market authorisations usually take place gradually in different jurisdictions around the world, however, for simplicity, we assume the current demand for Kymriah and do not consider time-windows in the problem.

The set of demand locations is comprised of all hospitals that can accept paediatric or adult patients, or both. The demand allocated to each hospital was divided equally based on the type of patients it can accept, using estimated data obtained from the Institute for Health Metrics and Evaluation (IHME) [15]. The set of candidate locations corresponds to the 1000 biggest cities in the world.

The production modes were proposed in [18] of manual, semi-automatic, and automatic. The corresponding percentage of failed PMB per facility depending on the production mode is shown in Table 1. Stages C1 and I3 start with MFs with manual production mode.

Table 1. Lower and upper limits for the failure rate of each production mode.

	Lower Limit	Upper Limit
Manual	5%	15%
Semi-automatic	3%	10%
Automatic	1%	5%

To obtain new solutions, we have used well-known mutation and crossover operators for FLP. The mutation removes and adds facilities with a probability of 0.3 each, or moves a facility to a different location with a probability of 0.4. The mutation is applied for all facility types. We also apply a uniform crossover [24] to the vectors of hospitals and locations separately. The results presented were obtained by running R-NSGA-II for 6000 generations ($50 \cdot 6000$ solution evaluations), with a population size of 50, over 20 independent runs. The stage to be optimised changes after every 1000 generations ($50 \cdot 1000$ solution evaluations). The reference points set by the DM are shown in Table 2. The reference points are used in two separate experiments.

Table 2. Reference point values for the three objectives.

	Time	Cost	Coverage
Reference point #1	0	£5 × 10^7	100%
Reference point #2	2	£2 × 10^7	50%

6   Results

In a multi-objective problem, the result is a set of non-dominated solutions, rather than a single optimal value. The DM would see a set of solutions that are considered to be equally good. Figures 4 and 5 show the non-dominated solutions obtained throughout the optimisation process of the six stages. The colours represent the stage in which each solution was obtained. The axes of the plots were intentionally left unstandardised for better visualisation of the solutions distribution.

As expected, none of the stages fully dominates the solutions of another one. Even though the solutions are getting closer to the reference points throughout

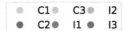

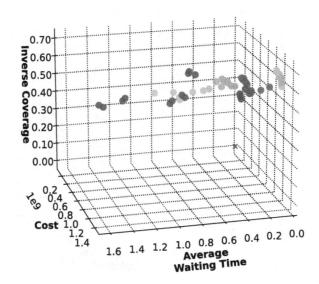

Fig. 4. Non-dominated solutions over all 6 stages for reference point #1 (red ×, avg. time of 0 h, total cost of £5 × 10⁷, and 100% of the hospitals to be covered). (Color figure online)

the optimisation process, there is no clear distinction between the layers of the different stages. Stage C1 is limited in placing only MFs, which are the most expensive facilities and have a small radius of hospitals they can cover due to the shelf-life. As a result, this stage is almost entirely overtaken by the rest of the solutions in relation to the distance to the reference points. However, the behaviour of the following stages indicates that the switch stagnates and even reverses some of the progress previously made. As the new facilities are initially added randomly at the beginning of each stage, it is likely that some of the resulted solutions are further away from the reference point. A possible solution could be the implementation of a heuristic progressing the problem based on each objective and allowing the new facilities to be placed only if they improve any of the solutions found in the previous generation.

In regard to the values obtained for the three objectives, the progress throughout the entire run is not considerable. The exact delivery times, cost, and demand coverage for the different stages are shown in Tables 3 and 4. While different parameters, such as a larger population or a higher number of generations, could obtain better values, the progress will be marginal. This is because there is a high number of configurations that can lead to exactly the same or very similar objectives. For example, the difference in the number of hospitals covered in

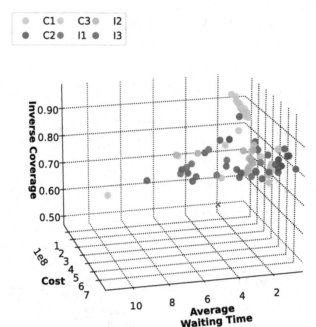

Fig. 5. Non-dominated solutions over all 6 stages for reference point #2 (red ×, avg. time of 2 h, total cost of £2 × 10^7, and 50% of the hospitals to be covered). (Color figure online)

the solutions is 5% and 10% respectively for the two chosen reference points. Even though this translates into a high number of patients, it is still far from the reference point. Two alternatives would lead to faster convergence. Firstly, more aggressive mutations and crossover, could be applied to force the solutions further away in the decision space. Nonetheless, creating problem specific operators would always require an in-depth understanding of the algorithm and the problem formulation, which in many cases is not feasible for the DM.

A better alternative to the above would be to alternate the order of the stages based on the knowledge of the DM in relation to the location of the hospitals and the candidate locations, as well as the construction costs, and the product specific shelf-life. As noticeable in the results shown here, starting with manufacturing, a balance between cost and coverage needs to be obtained to respect the DMs preference of the reference point. By adding helper facilities, such as CFs or any of the integration facilities, the cost can increase more gradually, which is not possible when adding MFs.

The stage-wise approach to the PBM FLP has great potential when interpreting it from the perspective of the DM. In particular, the DM may decide to not run all stages according to trade-off between potential improvement in objectives and computational cost of additional stages. Moreover, dividing the

Table 3. Minimum, maximum, and average values for the three objectives for each of the six stages for reference point with an average time $0\,$h, a total cost of $£5 \times 10^7$, and 100% of the hospitals to be covered.

Stage	Minimum			Maximum			Average		
	Delivery Time	Total Cost	Uncovered Demand	Delivery Time	Total Cost	Uncovered Demand	Delivery Time	Total Cost	Uncovered Demand
Stage C1	0.09	£1.299.589.746	63%	0.14	£1.537.998.717	68%	0.10	£1.366.436.596	67%
Stage C2	0.31	£1.181.083.004	66%	2.22	£1.390.225.244	71%	0.46	£1.341.230.335	67%
Stage C3	0.36	£1.223.012.953	62%	0.85	£1.407.694.906	67%	0.50	£1.373.040.643	63%
Stage I1	0.12	£1.231.458.286	62%	0.21	£1.416.062.239	67%	0.17	£1.360.299.763	63%
Stage I2	0.17	£1.217.884.781	60%	0.70	£1.468.876.481	67%	0.21	£1.369.973.041	63%
Stage I3	0.19	£1.336.629.378	61%	0.79	£1.455.158.975	64%	0.47	£1.400.055.662	62%

Table 4. Minimum, maximum, and average for the three objectives for each of the six stages for reference point with an average time of $2\,$h, a total cost of $£2 \times 10^7$, and 50% of the hospitals to be covered.

Stage	Minimum			Maximum			Average		
	Delivery Time	Total Cost	Uncovered Demand	Delivery Time	Total Cost	Uncovered Demand	Delivery Time	Total Cost	Uncovered Demand
Stage C1	1.45	£301.108.858	89%	1.81	£454.522.604	92%	1.73	£366.708.333	91%
Stage C2	3.69	£511.937.523	83%	8.07	£699.453.456	87%	4.64	£667.094.623	83%
Stage C3	3.07	£538.528.353	79%	7.04	£752.448.241	85%	4.04	£680.533.178	81%
Stage I1	3.32	£565.582.583	80%	4.80	£735.704.812	84%	4.69	£691.437.126	81%
Stage I2	3.70	£488.838.005	77%	165.46	£835.378.150	87%	17.59	£673.242.906	81%
Stage I3	1.58	£663.663.922	79%	11.04	£761.911.651	82%	5.09	£712.235.625	80%

problem into stages that lead to different supply chain configurations, allows the DM to gain an understanding of the benefits or disadvantages of having different facility types. Each facility usually comes with a set of required regulatory approvals, thus it is useful that each stage finds solutions restricted to particular types of facilities.

7   Conclusion and Future Research Directions

The overarching goal of this paper was to introduce the DM in the optimisation process of the supply chain of personalised medicine. We did so by dividing the problem into logical stages that can be optimised independently and allowing the DM to set as reference points expected values for the objective functions. The problem was solved using a popular preference-based EMO algorithm, R-NSGA-II. Our results suggest that the supply chain of PMBs could benefit from interactive optimisation methods. If the problem is further extended, for example to include other decentralisation options, the involvement of the DM might avoid the need for implementation of problem specific heuristics.

For simplicity, we did not allow the DM to change the order of the stages. However, this is something that will be investigated in further research. Since the only constraint in relation to the facilities available in a configuration is to have an MF or an MU, the DM might decide to skip C2 and the optimisation of CFs, and just optimise the productions modes in I2. Furthermore, it is also possible that a different order of the stages, even though partially leading to infeasible solution (e.g., starting with C2), can lead overall to a faster convergence.

The level of complexity for the problem can also differ by adding additional stages. We have restricted the supply chain to only centralised and integrated configurations, however more decentralised networks have been proposed in the literature. Although in the current available practices the decentralisation levels are not applied, with the expected increase in demand, it is argued that smaller facilities that can act as either demand or supply points should be used for the PMB supply chain.

Finally, more experiments are necessary to analyse the behaviour of the problem. The impact of each facility type on the objectives, for each of the six stages, as well as the order in which they are added to the optimisation is still unclear. Moreover, we have run each stage for an equal number of generations, regardless of their decision space. Variable population size and generations per stage could benefit the algorithm and allow it to add additional facilities when the progress between generations is not significant. It would also be useful to study the impact of the DM changing the reference point between the stages.

References

1. Abou-El-Enein, M., Elsanhoury, A., Reinke, P.: Overcoming challenges facing advanced therapies in the EU market. Cell Stem Cell **19**(3), 293–297 (2016)
2. Avramescu, A., Allmendinger, R., López-Ibáñez, M.: A multi-objective multi-type facility location problem for the delivery of personalised medicine. In: Castillo, P.A., Jiménez Laredo, J.L. (eds.) EvoApplications 2021. LNCS, vol. 12694, pp. 388–403. Springer, Cham (2021). https://doi.org/10.1007/978-3-030-72699-7_25
3. Avramescu, A., Allmendinger, R., López-Ibáñez, M., Adriana, L.: Composite facility location problems: a case study of personalised medicine. In: 19th IEEE International Conference in Computational Intelligence in Bioinformatics and Computational Biology (2022)
4. Bernardi, A., Sarkis, M., Triantafyllou, N., Lakelin, M., Shah, N., Papathanasiou, M.M.: Assessment of intermediate storage and distribution nodes in personalised medicine. Comput. Chem. Eng. **157**, 107582 (2022)
5. Branke, J., Farid, S.S., Shah, N.: Industry 4.0: a vision for personalized medicine supply chains? Cell Gene Therapy Insights **2**(2), 263–270 (2016)
6. CBER: KYMRIAH (tisagenlecleucel) (2022). https://www.fda.gov/vaccines-blood-biologics/cellular-gene-therapy-products/kymriah-tisagenlecleucel. Accessed 15 Nov 2022
7. Cornuéjols, G., Nemhauser, G., Wolsey, L.: The uncapacitated facility location problem. Technical report, Cornell University Operations Research and Industrial Engineering (1983)
8. Deb, K., Pratap, A., Agarwal, S., Meyarivan, T.: A fast and elitist multi-objective genetic algorithm: NSGA-II. IEEE Trans. Evol. Comput. **6**(2), 182–197 (2002)
9. Deb, K., Sundar, J.: Reference point based multi-objective optimization using evolutionary algorithms. In: Cattolico, M., et al. (eds.) GECCO, pp. 635–642. ACM Press, New York (2006)
10. Di Sanzo, M., et al.: Clinical applications of personalized medicine: a new paradigm and challenge. Curr. Pharm. Biotechnol. **18**(3), 194–203 (2017)

11. Eder, C., Wild, C.: Technology forecast: advanced therapies in late clinical research, EMA approval or clinical application via hospital exemption. J. Mark. Access Health Policy **7**(1), 1600939 (2019)

12. Grand View Research: Advanced Therapy Medicinal Products Market Size, Share & Trends Analysis Report by Therapy Type (CAR-T, Gene, Cell, Stem Cell Therapy), by Region (North America, Europe, APAC, ROW), and Segment Forecasts, 2021–2028 (2021). https://www.grandviewresearch.com/industry-analysis/advanced-therapy-medicinal-products-market. Accessed 07 Apr 2021

13. Harrison, R.P., Qasim, A., Medcalf, N.: Centralised versus decentralised manufacturing and the delivery of healthcare products: a United Kingdom exemplar. Cytotherapy **20**, 873–890 (2018)

14. Harrison, R.P., Ruck, S., Rafiq, Q., Medcalf, N.: Decentralised manufacturing of cell and gene therapy products: learning from other healthcare sectors. Biotechnol. Adv. **36**, 345–357 (2018)

15. Institute for Health Metrics and Evaluation (IHME): IHME data (2022). http://ghdx.healthdata.org/ihme_data. Accessed 15 Nov 2022

16. Jarosławski, S., Toumi, M.: Sipuleucel-T (Provenge®)-autopsy of an innovative paradigm change in cancer treatment: why a single-product biotech company failed to capitalize on its breakthrough invention. BioDrugs **29**(5), 301–307 (2015)

17. Karakostas, P., Panoskaltsis, N., Mantalaris, A., Georgiadis, M.C.: Optimization of CAR T-cell therapies supply chains. Comput. Chem. Eng. **139**, 106913 (2020)

18. Lopes, A.G., Noel, R., Sinclair, A.: Cost analysis of vein-to-vein CAR T-cell therapy: automated manufacturing and supply chain. Cell Gene Therapy Insights **6**(3), 487–510 (2020)

19. Moschou, D., Papathanasiou, M.M., Lakelin, M., Shah, N.: Investment planning in personalised medicine. In: Computer Aided Chemical Engineering 30th European Symposium on Computer Aided Process Engineering, pp. 49–54 (2020)

20. Papathanasiou, M.M., Stamatis, C., Lakelin, M., Farid, S., Titchener-Hooker, N., Shah, N.: Autologous CAR T-cell therapies supply chain: challenges and opportunities? Cancer Gene Ther. **27**(10–11), 799–809 (2020)

21. Alliance for Regenerative Medicine: Clinical Trials in Europe: Recent Trends in ATMP Development. Techncial report, Alliance for Regenerative Medicine, October 2019

22. Sarkis, M., Bernardi, A., Shah, N., Papathanasiou, M.M.: Decision support tools for next-generation vaccines and advanced therapy medicinal products: present and future. Curr. Opin. Chem. Eng. **32**, 100689 (2021)

23. Sarkis, M., Bernardi, A., Shah, N., Papathanasiou, M.M.: Emerging challenges and opportunities in pharmaceutical manufacturing and distribution. Processes **9**(3), 457 (2021)

24. Syswerda, G.: Uniform crossover in genetic algorithms. In: Schaffer, J.D. (ed.) Proceedings of the Third International Conference on Genetic Algorithms, pp. 2–9. Morgan Kaufmann Publishers, San Mateo (1989)

25. Wang, X., Kong, Q., Papathanasiou, M.M., Shah, N.: Precision healthcare supply chain design through multi-objective stochastic programming. In: 13th International Symposium on Process Systems Engineering (PSE 2018) Computer Aided Chemical Engineering, pp. 2137–2142 (2018)

Resilient Bio-inspired Algorithms

Artificial Neural Algorithms

Further Investigations on the Characteristics of Neural Network Based Opinion Selection Mechanisms for Robotics Swarms

Ahmed Almansoori[1,2]([✉])(iD), Muhanad Alkilabi[1,2](iD), and Elio Tuci[1](iD)

[1] University of Namur, Namur, Belgium
{ahmed.almansoori,elio.tuci}@unamur.be
[2] University of Kerbala, Karbala, Iraq
{Ahmed.kamil,muhanad.hayder}@uokerbala.edu.iq
https://www.unamur.be/info/

Abstract. Collective decision-making is a process that allows a group of autonomous agents to make a decision in a way that once the decision is made it cannot be attributed to any agent in the group. In the swarm robotics literature, collective decision-making mechanisms have generally been designed using behaviour-based control structures. That is, the individual decision-making mechanisms are integrated into modular control systems, in which each module concerns a specific behavioural response required by the robots to respond to physical and social stimuli. Recently, an alternative solution has been proposed which is based on the use of dynamical neural networks as individual decision-making mechanisms. This alternative solution proved effective in a perceptual discrimination task under various operating conditions and for swarms that differ in size. In this paper, we further investigate the characteristics of this neural model for opinion selection using three different tests. The first test examines the ability of the neural model to underpin consensus among the swarm members in an environment where all available options have the same quality and cost (i.e., a symmetrical environment). The second test evaluates the neural model with respect to a type of environmental variability related to the spatial distribution of the options. The third test examines the extent to which the neural model is tolerant to the failure of individual components. The results of our simulations show that the neural model allows the swarm to reach consensus in a symmetrical environment, and that it makes the swarm relatively resilient to major sensor failure. We also show that the swarm performance drops in accuracy in those cases in which the perceptual cues are patchily distributed.

Keywords: Swarm robotics · Collective decision making · Symmetry breaking · Collective perception

J. Correia et al. (Eds.): EvoApplications 2023, LNCS 13989, pp. 737–750, 2023.
https://doi.org/10.1007/978-3-031-30229-9_47

1   Introduction

Swarm robotics is a particular type of multi-robot system in which each robot
has its own controller, perception is local, and communication is based on spatial
proximity [10]. Among the most studied mechanisms in swarm robotics, there
are those that allow the swarm to make a decision collectively. Collective decision
making refers to a situation in which robots collectively make a choice among
two or more alternative options in a way that, when the decision is made, it is no
longer attributable to any single individual robot [28]. According to [8], collective
decision-making is considered one of the key behaviours in swarm robotics and
can be classified into: task allocation and consensus achievement. Task allocation
refers to a process by which the group's performance is increased by splitting the
swarm members into multiple subgroups, each of which is dedicated to solving a
particular task. Consensus achievement refers to a process which allows all the
swarm members to share the same opinion with respect to alternative options.
When the number of available options is finite, the consensus problem, referred
to as the "best-of-n" problem [28], requires the swarm to reach a consensus
on the best among the available option, when options differ in qualities, or on
any option, when options have equal quality or the same utility for the swarm
members [19]. In other words, when confronted with the "best-of-n" type of
problems, the swarm must avoid separating into two or more distinct subgroups
in which robots of a group share an opinion different from the robots of another
group.

In the swarm robotics literature, the mechanisms underpinning collective
decision-making have been generally designed using behaviour-based type of
control structures. That is, the individual decision-making mechanisms are integrated into modular control systems, in which each module takes care of developing a specific behavioural response required by the robots to respond to physical
and social stimuli. Control structures developed with these principles have been
demonstrated to be effective in supporting the collective decision-making process
in a variety of scenarios [23,26,27]. However, the capability of these swarms to
adapt to different sources of variability tend to be limited to those circumstances
that have been clearly predicted by the designer, leaving the robots potentially
unprepared to overcome unexpected and unpredictable events that may occur
in any complex natural settings. As stated in [17], further research is needed to
design opinion selection mechanisms that fit the needs of swarm robotics systems
to allow them to mimic natural swarms in terms of robustness, scalability, and
flexibility.

In [1], the authors proposed an alternative approach to the classic hand-
designed controller, based on the use of artificial neural networks synthesised
using evolutionary computation techniques as individual opinion selection mechanisms [25]. They have tested this design method on a type of best-of-n problem
originally described in [26]. This problem is characterised by two or more options
whose quality concerns the relative proportions with which they are distributed
in the environment. The swarm's task is to reach a consensus on which option
is the most represented in the environment. Given that every single robot can

only explore a small portion of the environment, only the swarm, by relying on collective intelligence, can correctly evaluate the options' quality and eventually choose the best one. By investigating the behaviour of a swarm engaged in this collective perceptual discrimination task, they have shown that artificial neural networks synthesised using evolutionary computation techniques can be an effective design method to allow the robots to reach a consensus on the best available options in the environment.

With extensive comparative experimental work, the authors in [2] have shown that the neural network based opinion selection mechanisms (also referred to, in this paper, as the neural model) is more effective than the classic hand-designed Voter model (see [26]) in a set of environmental conditions generated by varying the level of difficulty of the perceptual discrimination task, by varying the maximum distance for robot-robot communication, and also by dynamically varying the option quality when the swarm has already reached a consensus. The study also showed that the performances of a swarm controlled by the neural model are less touched by variations in the swarm size than those of a swarm controlled by the hand-coded Voter model. The neural model is more effective than alternative hand-designed solutions because it allows each robot to integrate physical and social evidence in a more adaptive and effective way than classic hand-designed approaches in which these different sources of information are treated following some designer "imposed" principles (see [2]).

In this paper, we further explore the characteristics of the neural model with respect to its effectiveness in underpinning opinion selection in a swarm of robots engaged in the above mentioned collective perceptual discrimination task. In particular, we describe the results of three different evaluation tests on the best evolved neural model, designed as illustrated in the previous research work (see [1,2], for details). The first test is related to the ability of the swarm to reach a consensus in a perfectly symmetrical environment where options have the same quality [28]. The capability to break environmental symmetries is an important feature of opinion selection mechanisms because it allows the swarm to keep on operating as a coherent unit and to overcome the individual limitations with the group responses even when feedback modulation based decision processes are not triggered by environmental structures [3,4]. This is the reason why the symmetry-breaking process has been studied in a different types of best-of-n scenarios, such as the prey-hunting scenario [29], the double-bridge scenario [16], and the aggregation scenario [9,12–15,24]. The second test evaluates the neural model with respect to a type of environmental variability related to the spatial distribution of the options. In particular, we use the benchmarks proposed in [5] to evaluate the neural model in eight different environments in which the options are more patchily distributed than the environment experienced by the swarm during the control system design phase. These different spatial distributions of the perceptual cues have also been used in [6] to test the robustness of decision making mechanisms for a swarm of robots controlled by a statistically grounded algorithm against spatial correlations in an unknown environment. The third set of tests examines the extent to which the neural model is tolerant to the failure

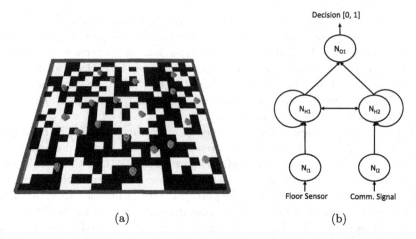

Fig. 1. (a) The simulated arena with the robots engaged on the perceptual discrimination task. (b) The dynamic neural network that underpins the opinion selection in each simulated robot.

of individual components. In particular, we investigate the robustness of the neural model against failure of the floor sensor in a progressively higher number of robots within the swarm. The floor sensor is of fundamental importance in this task since it allows the robots to evaluate the quality of available options individually.

The results of these tests contribute to generating a more informative estimation of the effectiveness of the neural model within the swarm robotics community. Although our tests are limited to the collective perceptual discrimination task, we show that swarms in which robots are controlled by the neural model can converge to consensus in spite of environmental symmetries. Moreover, they can keep on operating effectively even when more than 40% of the robots suffer from a major sensor failure. As in other opinion selection models, we show that a swarm controlled by the neural model undergoes a significant performance drop in most environments in which the perceptual cues are patchily distributed. The significance of these findings will be further elaborated in the Sect. 4.

2   Methods

This study is run in a simulation environment which models the wheeled mobile robot e-puck2, a robotic platform commonly used in swarm robotics experiments [22]. The robot sensory apparatus used in this experiment includes eight infra-red sensors to measure the proximity of obstacles, a floor sensor to perceive the colour of the floor in binary format (i.e., the sensor reads 0 if the robot is on a black tile and 1 if on a white tile), and the range&bearing board for local communication. In particular, each robot emits a binary signal which refers to its opinion about which colour covers the majority of the arena floor (i.e., 1 for white and 0 for black). The maximum robot-robot communication distance is set to 50 cm. This communication system can be reliably implemented on the physical e-puck2 robot with the range&bearing board. The robot movements are computed with a differential drive kinematic model [11]. To compensate for the simulation-reality gap, 10% uniform noise is added to all sensor readings, the motor outputs and the position of the robot.

The simulation environment is characterised by a close arena of 2×2 m with the floor covered with black and white tiles, 10×10 cm each, distributed randomly on the floor (see Fig. 1a). 20 robots are randomly initialised within the arena. During the evaluation, they move randomly, while avoiding obstacles (i.e., the arena walls and other robots) for 400 s, corresponding to the length of an evaluation trial. As in [26], the task of the 20 robots swarm is to reach a consensus (i.e., all robots sharing the same opinion) about which colour (black or white) covers the largest portion of the arena floor. At each simulation update cycle, the robots sample the arena floor underneath their body and listen to the closest neighbour's opinion. After that, they disseminate their current opinion and update their positions. Given the robots' pseudo-random walk and the random distribution of black and white tiles on the arena floor, we estimated, by simulating multiple times the task, that each robot explores, on average, only about 18% of the arena floor during each evaluation period. Thus, a consensus has to be reached by exploiting collective intelligence through local communication for opinion exchanges.

A hand-coded algorithm makes the robots moving within the arena according to an isotropic random walk, with a fixed step length (5 s, at 20 cm/s), and turning angles chosen from a wrapped Cauchy probability distribution characterised by the following probability density function (PDF):

$$f(\theta, \mu, \rho) = \frac{1}{2\pi} \frac{1 - \rho^2}{1 + \rho^2 - 2\rho \cos(\theta - \mu)}, \quad 0 < \rho < 1, \tag{1}$$

where $\mu = 0$ is the average value of the distribution, and ρ determines the distribution skewness (see [20]). For $\rho = 0$, the distribution becomes uniform and provides no correlation between consecutive movements, while for $\rho = 1$, a Dirac distribution is obtained, corresponding to straight-line motion. In this study $\rho = 0.5$. While moving around, the robots continuously perform an obstacle avoidance behaviour. To perform obstacle avoidance, first a robot detects an obstacle with its infra-red sensors, then stops and keeps on changing its headings of a randomly chosen angle uniformly drawn in $[0, \pi]$ until no obstacles are perceived from the front four sensors.

The process underpinning the development of the individual opinion is regulated by a continuous time recurrent neural network (CTRNN) [7], synthesised using evolutionary computation techniques. The neural network has a multi-layer topology, as shown in Fig. 1b: neurons $N_{I,1}$ and $N_{I,2}$ take input from the robot's floor sensor and the eventual communication signal (1 for white-dominant, 0 for black-dominant, and 0.5 whenever there is no other robots at less than 50 cm from the receiver), neuron $N_{O,1}$ is used to set the robot opinion, and neurons $N_{H,1}$ and $N_{H,2}$ form a fully recurrent continuous time hidden layer. The input neurons are simple relay units, while the output neuron is governed by the following equations:

$$o = \sigma(O + \beta^O), \qquad (2)$$

$$O = \sum_{i=1}^{2} W_i^O \, \sigma(H_i + \beta_i^H), \qquad (3)$$

$$\sigma(z) = (1 + e^{-z})^{-1}, \qquad (4)$$

where, using terms derived from an analogy with real neurons, O and H_i are the cell potentials of respectively output neuron and hidden neuron i, β^O and β^H are bias terms, W_i^O is the strength of the synaptic connection from hidden neuron i to output neuron, and $\sigma(H_i + \beta_i)$ are the firing rates. The hidden units are governed by the following equation:

$$\tau_j \dot{H}_j = -H_j + \sum_{i=1}^{2} W_{ij}^H \sigma(H_i + \beta_i^H) + \sum_{i=1}^{2} W_{ij}^I I_i, \qquad (5)$$

where τ_j is the decay constant, W_{ij}^H is the strength of the synaptic connection from hidden neuron i to hidden neuron j, W_{ij}^I is the strength of the connection from input neuron i to hidden neuron j, and I_i is the intensity of the sensory perturbation on neuron i. Cell potentials are set to 0 each time a network is initialised or reset. State equations are integrated using the forward Euler method with an integration step-size of 0.1 seconds. Neuron $N_{O,1}$ is used to set the robot opinion, which corresponds to 1 (i.e., white-dominant) when the neuron firing rate is above the threshold 0.5, and 0 (i.e., black-dominant) otherwise.

The network parameters, that is, the weights of the connections between neurons, the bias terms, and the decay constants, are genetically encoded parameters, set using simple tournament-based selection evolutionary algorithms as illustrated as illustrated in [1,2]. The swarm is homogeneous; that is, the neural network in charge of the individual opinion selection process is cloned in each of the 20 robots of the swarm.

3   Results

In this section, we illustrate the performances of a swarm of simulated robots engaged in three different tests related to the robustness of collective decision making strategies in a collective perceptual discrimination task. In all these tests, the robots' individual decisions are underpinned by neural network-based mechanisms synthesised using evolutionary computation techniques as described in Sect. 2.

In all tests, the swarm is evaluated in condition A (the hard scenario), in which the most represented colour covers 55% of the floor, and in condition B (the simple scenario), in which the most represented colour covers 66% of the arena floor. In test I, the swarm is also evaluated in condition S (the symmetry case), in which each colour covers exactly 50% the arena floor.

Since the operational principles underpinning the individual opinion selection process are not functionally symmetric with respect to the dominant colour, in each test conditions, the swarm undergoes 50 trials in a black-dominant and 50 trials in a white-dominant environment. A trial starts with the 20 robots randomly positioned in the arena and terminates after 400 s. During this time, each robot performs a random walk while avoiding obstacles, and it interacts with the other robots using communication signals, as illustrated in Sect. 2.

To evaluate the group performance, we employ two metrics. The first metric is the *accuracy* of the decision, which corresponds to the proportion of trials (over 50, for each type of environment) in which the swarm reached consensus on the opinion corresponding to the currently most represented colour on the arena floor. In test I, for condition S (the symmetry case), accuracy refers to the proportion of trials (over 50, for each type of environment) in which the swarm reached consensus on any opinion. Consensus refers to the circumstance in which all robots share the same opinion for at least 10 s within a trial. The second metric is the *time*, within a trial, required to the swarm to converge to a consensus state. This metric is calculated on the successful trials only.

3.1   Test I: The Symmetry-Breaking Test

The first test is related to the ability of the swarm to reach a consensus in a perfectly symmetrical environment. As we mentioned above, symmetry-breaking indicates the ability of the swarm to converge on a single shared opinion (instead of multiple opinions) even in those cases in which the alternative options offered by the environment have equal quality [28]. In our collective perceptual discrimination task, the symmetry case corresponds to environments in which the floor is covered by the same proportion of black and white tiles. The results of the symmetry-breaking tests are illustrated in Fig. 2, which depicts the accuracy of the decision (see Fig. 2a) and the distribution of times required for the swarm to reach a consensus (see Fig. 2b) in condition S (i.e., the symmetrical environment). The graphs in Fig. 2 also show the performances of the swarm in condition A and B, as terms of comparison. In Fig. 2a, for condition S, the black and the white bars refer to the proportion of trials (over 50 trials) in which the swarm reaches a consensus on the black and on the white opinion, respectively. In Fig. 2b, for condition S, the black and white boxes refer to the distribution of times to convergence to consensus to the black opinion and to the white opinion, respectively. For condition A and B, the black bars/boxes refer to performances (i.e., accuracy in Fig. 2a, and time to convergence in Fig. 2b) in black-dominant environments, while the white bars/boxes to performances in white-dominant environments. These results show that, in condition S, the swarm always reaches a consensus (i.e., the black plus the white bar in Fig. 2a, condition S, add to accuracy 1). We also note that, due to the stochastic nature of the decision process, the frequency of convergence to opinion black is only

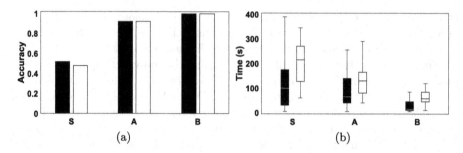

Fig. 2. The symmetry breaking test. In (a), the bars refers to the accuracy of the collective decision, while in (b) the boxes refers to the distribution of times to convergence to consensus. In both graphs, the label S refers to evaluation trials in condition S, that is in symmetrical environments with 50% of black and 50% of white tiles; the A labels refers to evaluation trials in condition A, while the B label refers to trials in condition B. In (a), for condition S, the black and the white bars refer to the proportion of trials (over 50) in which the swarm reach a consensus on the black and on the white opinion, respectively. In (b), for condition S, the black and white boxes refer to the distribution of times to convergence to consensus to the black opinion and to the white opinion, respectively. For condition A and B, the black bars/boxes refer to performances (i.e., accuracy in graph a, and time to convergence in graph b) over 50 trials in black-dominant environments, while the white bars/boxes to performances over 50 trials in white-dominant environments.

marginally higher than the frequency of opinion white. The accuracy in perfectly symmetrical environments is higher than the accuracy in condition A, where the swarm's performance attains about 90% accuracy in both the black-dominant and the white-dominant environment (see Fig. 2a, condition A, black and white bars). As already discussed in [2], for progressively simpler perceptual discrimination tasks, the swarm's performance tends to the 100% accuracy (see Fig. 2a, condition B, black and white bars). For what concerns the distribution of times to convergence, symmetrical environments are those that require longer time to the swarm to reach a consensus (see Fig. 2b, condition S). This finding is in line with previous similar research studies in collective decision making illustrated in [18,19,21]. The results of our symmetry-breaking test also show that the swarm requires longer time to reach consensus in the white-dominant than in the black-dominant environment (see Fig. 2b, white boxes). As explained in [2], this is due to a genetic bias, by which each robot starts an evaluation trial with opinion white. Paradoxically, this bias delays the convergence to consensus to white. This is due to the fact that the swarm has to go through a series of global states in which the initial genetically-induced consensus to white is progressively lost, and subsequently recovered through a genuine collective decision process triggered by the interactions between the robots (see [2], for details). In summary, the symmetry-breaking test tells us that the neural network based decision making mechanisms are extremely effective in underpinning consensus among the members of the swarm also in a perfectly symmetrical environments.

However, the symmetry case requires longer times to the swarm to reach consensus than the non-symmetrical cases. The distribution of times to consensus become progressively shorter in progressively simpler perceptual discrimination tasks (i.e., those in which the proportion of floor taken by the dominant colour is progressively larger).

3.2   Test II: The Tiles Distribution Test

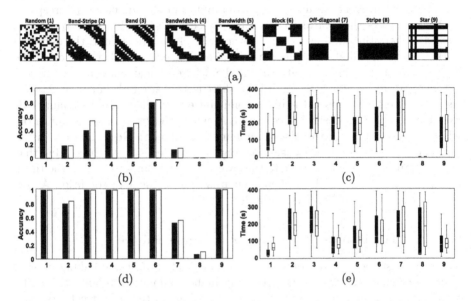

Fig. 3. Tiles distribution test. (a) The nine floor patterns. In the first patterns on the left, black and white tiles are randomly distributed. In all other patterns, the tiles are distributed as illustrated in [5]. Graphs in (b) and (d) show the accuracy of the collective decision for each pattern in condition A and in condition B, respectively. Graphs in (c) and (e) show the distribution of times to convergence to consensus for each pattern in condition A and in condition B, respectively. In (b), (c), (d), and (e) the black bars/boxes refers to the performances in black-dominant environment, while the white bar/boxes refers to the performances in white-dominant environment.

The second test is related to the effectiveness of the swarm to reach consensus in environments in which the perceptual evidence related to the different options is not randomly distributed. Black and white tiles are arranged in order to form specific patterns which may influence the capability of the swarm to reach consensus given the fact that each robot can only explore a limited portion of the arena floor. The patters we used for this test are those originally illustrated in [5] and shown in Fig. 3a.

We remind the reader that pattern 1 (Random) is the one used during the design phase of the neural model. The graphs in Fig. 3b, and 3d, show the accuracy of the collective decision for each pattern in condition A and in condition B, respectively. In condition A (see Fig. 3b), it turns out that the characteristics of each pattern have a substantial effect on the accuracy of the swarm's collection decision. In general, the swarm performance tend to degrade when the perceptual evidence is spatially arranged in distinctive clusters or patches (see patterns 2, 3, 4, 5, 7, 8 in Fig. 3b). The less patchy the distribution of perceptual evidence, the higher the swarm accuracy. This can be accounted for by considering that patchy environments facilitate the emergence of an alignment of opinion among spatially proximal robots randomly wandering within specific clusters. The local alignment on different opinions, supported by the predominant local perceptual evidence hinders the swarm to reach the consensus state. In condition B (see Fig. 3d), where the dominant colour covers 66% of the arena floor, the influence of the patchy distribution of the perceptual evidence has smaller impact on the swarm accuracy except for pattern 7 and 8 where a clear performance drop is observed. Moreover, for both conditions, there is a substantial similarity between the swarm performances in the black-dominant and in the white-dominant environment except for pattern 4 in condition A, where the swarm does better in the white-dominant than in the black-dominant environment. Regarding the distribution of times to reach consensus (see Fig. 3c, and 3e) it is worth noticing that any pattern requires longer time to the swarm to reach consensus than the times recorded for patterns 1. This phenomenon is more evident in condition A (see Fig. 3c) than in condition B (see Fig. 3e).

3.3 Test III: Robots' Floor Sensor Readings Failure

The third test focuses the robustness of the collective decision-making process under conditions in which a progressively higher number of robots within the swarm suffers from failures of the floor sensor, which instead of correctly reading the colour of the floor, it returns randomly generated binary values. The graphs in Fig. 4 show the swarm performances (i.e., accuracy of the group decision, see Fig. 4a, and 4c, and time to convergence to consensus, see Fig. 4b, and 4d) for different numbers of robots (from 0 to 10) suffering from failures of the floor sensor. Figure 4a, and 4b refer to condition A, while Fig. 4c, and 4d refer to condition B. We recall the reader that the swarm size is 20. The results of these tests unequivocally indicate that the collective decision-making progress is relatively robust to this type of disruption. In condition B, the swarm attains 100% accuracy in both types of environment even when 50% of the swarm suffers from failure of the floor sensor (see Fig. 4c). In condition A, only when more than 35% of the swarm suffers from floor sensor failure (see Fig. 4a, for 7 robots), the accuracy starts to progressively drop only for the white-dominant environment. For the black-dominant environment, the swarm accuracy tend to remain above 80% even with 10 robots suffering from the floor sensor failure (see Fig. 4a. Regarding time to convergence to consensus, the results of the test clearly indicate that the higher the number of robots with floor sensor failure the longer the time to reach consensus in white-dominant environments. This trend is observed both in condition A and in condition B (see Fig. 4b, and 4c).

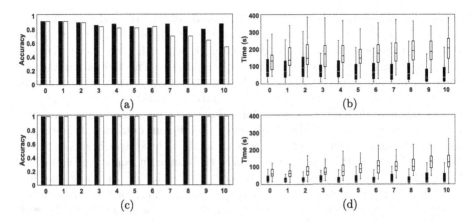

Fig. 4. Floor sensor readings failure test. Graphs in (a) and (c) show the accuracy of the collective decision for each pattern in condition A and in condition B, respectively. Graphs in (b) and (d) show the distribution of times to convergence to consensus for each pattern in condition A and in condition B, respectively. In all graphs, black bars/boxes refers to performances in black-dominant environments, while white bars/boxes to performances in white-dominant environments; the labels on the x-axes indicate the number of robots affected by the floor sensor failure.

4   Conclusion

In this paper, we further investigated the characteristics of the neural model described in [1,2] with respect to its effectiveness in underpinning consensus in a swarm of robots engaged in a collective perceptual discrimination task. In particular, we run three tests. The first test focused on the swarm's ability to establish a consensus in a symmetrical environment where all alternatives are of equal quality. The findings of this test indicate that the decision-making processes based on neural networks are exceptionally successful at supporting consensus among the swarm members, even in situations with perfect symmetry. As expected, it takes longer for the swarm to reach consensus in the symmetrical than in the non-symmetrical environments. The third set of tests examined the extent to which the neural model is tolerant to the failure of individual components. More precisely, we investigated the neural model's robustness against floor sensor's failure in a progressively higher number of robots within the swarm. The results of this test unequivocally indicated that the collective decision-making process is relatively robust to this type of disruption. In condition B, the swarm attains 100% accuracy even when 50% of the swarm suffers from a failure of the floor sensor. In the condition A, where the dominant colour covers 55% of the arena floor, the accuracy remains over 80% even when more than 35% of the swarm suffered from a floor sensor failure.

The second test evaluated the neural model with respect to a type of environmental variability related to the spatial distribution of the options. We employed the benchmarks proposed in [5] to evaluate the neural model in eight different environments in which the options are more patchily distributed than the environment experienced by the swarm during the design phase. This test showed that the swarm's performance tends to degrade when the perceptual evidence is spatially arranged in distinctive clusters or patches. The less patchy the distribution of perceptual evidence, the higher the swarm accuracy. This can be accounted for by considering that patchy patterns facilitate the emergence of an alignment of opinion among spatially proximal robots randomly wandering within specific clusters. The local alignment on different opinions, supported by the predominant local perceptual evidence, hinders the swarm from reaching a consensus state. In the condition B, where the dominant colour covers 66% of the arena floor, the influence of the patchy distribution of the perceptual evidence has a smaller impact on the swarm's accuracy in most studied patterns. There are several elements on which we plan to act in the future, to try to overcome this limitation. One idea is to develop exploration strategies that adapt to the characteristics of the environment. For example, instead of moving with the same pseudo-random walk as in this study, the robots could adaptively mix Lévy flight type of random walk with Brownian motion to sample more distant portion of the environment. Alternatively, we are planning to integrate into the neural model the mechanisms underpinning the robots movements. This would allow us to exploit the adaptivity of the neural model also with respect to the exploration strategies. More effective swarm decision strategies in patchy environments can also come from an increased flexibility of the individual mechanisms in charge of mixing perceptual cues and social evidence.

Acknowledgments. Ahmed Almansoori is funded by a CERUNA grant from the University of Namur (BE).

References

1. Almansoori, A., Alkilabi, M., Colin, J.N., Tuci, E.: On the evolution of mechanisms for collective decision making in a swarm of robots. In: Artificial Life and Evolutionary Computation, pp. 109–120. Springer, Cham (2022)
2. Almansoori, A., Alkilabi, M., Tuci, E.: A comparative study on decision making mechanisms in a simulated swarm of robots. In: 2022 IEEE Congress on Evolutionary Computation (CEC), pp. 1–8. IEEE (2022)
3. Amé, J., Halloy, J., Rivault, C., Detrain, C., Deneubourg, J.: Collegial decision making based on social amplification leads to optimal group formation. PNAS (2006)
4. Amé, J., Rivault, C., Deneubourg, J.: Cockroach aggregation based on strain odour recognition. Animal Behaviour (2004)
5. Bartashevich, P., Mostaghim, S.: Benchmarking collective perception: new task difficulty metrics for collective decision-making. In: Moura Oliveira, P., Novais, P., Reis, L.P. (eds.) EPIA 2019. LNCS (LNAI), vol. 11804, pp. 699–711. Springer, Cham (2019). https://doi.org/10.1007/978-3-030-30241-2_58
6. Bartashevich, P., Mostaghim, S.: Multi-featured collective perception with evidence theory: tackling spatial correlations. Swarm Intell. **15**(1), 83–110 (2021)

7. Beer, R.D.: A dynamical systems perspective on agent-environment interaction. Art. Intell. **72**, 173–215 (1995)

8. Brambilla, M., Ferrante, E., Birattari, M., Dorigo, M.: Swarm robotics: a review from the swarm engineering perspective. Swarm Intell. **7**(1), 1–41 (2013)

9. Campo, A., Garnier, S., Dédriche, O., Zekkri, M., Dorigo, M.: Self-organized discrimination of resources. PLoS ONE **6**(5), e19888 (2010)

10. Dorigo, M., Şahin, E.: Guest editorial. special issue: swarm robotics. Autonomous Robots **17**(2–3), 111–113 (2004)

11. Dudek, G., Jenkin, M.: Computational Principles of Mobile Robotics. Cambridge University Press, Cambridge (2000)

12. Firat, Z., Ferrante, E., Gillet, Y., Tuci, E.: On self-organised aggregation dynamics in swarms of robots with informed robots. Neural Comput. Appl. **32**(17), 13825–13841 (2020). https://doi.org/10.1007/s00521-020-04791-0

13. Francesca, G., Brambilla, M., Brutschy, A., Trianni, V., Birattari, M.: Automode: a novel approach to the automatic design of control software for robot swarms. Swarm Intell. **8**(2), 89–112 (2014)

14. Francesca, G., Brambilla, M., Trianni, V., Dorigo, M., Birattari, M.: Analysing an evolved robotic behaviour using a biological model of collegial decision making. In: Ziemke, T., Balkenius, C., Hallam, J. (eds.) SAB 2012. LNCS (LNAI), vol. 7426, pp. 381–390. Springer, Heidelberg (2012). https://doi.org/10.1007/978-3-642-33093-3_38

15. Garnier, S., Gautrais, J., Asadpour, M., Jost, C., Theraulaz, G.: Self-organized aggregation triggers collective decision making in a group of cockroach-like robots. Adapt. Behav. **17**(2), 109–133 (2009)

16. Garnier, S., Tache, F., Combe, M., Grimal, A., Theraulaz, G.: Alice in pheromone land: an experimental setup for the study of ant-like robots. In: 2007 IEEE Swarm Intelligence Symposium, pp. 37–44. IEEE (2007)

17. Hamann, H.: Swarm Robotics: A Formal Approach. Springer, Cham (2018)

18. Hamann, H., Meyer, B., Schmickl, T., Crailsheim, K.: A model of symmetry breaking in collective decision-making. In: Doncieux, S., Girard, B., Guillot, A., Hallam, J., Meyer, J.-A., Mouret, J.-B. (eds.) SAB 2010. LNCS (LNAI), vol. 6226, pp. 639–648. Springer, Heidelberg (2010). https://doi.org/10.1007/978-3-642-15193-4_60

19. Hamann, H., Schmickl, T., Wörn, H., Crailsheim, K.: Analysis of emergent symmetry breaking in collective decision making. Neural Comput. Appl. **21**(2), 207–218 (2012)

20. Kato, S., Jones, M.: An extended family of circular distributions related to wrapped cauchy distributions via brownian motion. Bernoulli **19**(1), 154–171 (2013)

21. Khaluf, Y.: The emergence of collective response to decisions in a group of physical agents. In: ALIFE 2021: The 2021 Conference on Artificial Life. MIT Press (2021)

22. Mondada, F., et al.: The e-puck, a robot designed for education in engineering. In: Proceedings of the 9th International Conference on Autonomous Robot Systems and Competitions, vol. 1, pp. 59–65 (2009)

23. Scheidler, A., Brutschy, A., Ferrante, E., Dorigo, M.: The k-unanimity rule for self-organized decision-making in swarms of robots. IEEE Trans. Cybern. **46**, 1175–1188 (2016)

24. Sion, A., Reina, A., Birattari, M., Tuci, E.: Controlling robot swarm aggregation through a minority of informed robots. In: Dorigo, M., et al. (eds.) Swarm Intelligence, pp. 91–103. Springer, Cham (2022). https://doi.org/10.1007/978-3-031-20176-9_8

25. Trianni, V., Tuci, E., Ampatzis, C., Dorigo, M.: Evolutionary swarm robotics: a theoretical and methodological itinerary from individual neuro-controllers to collective behaviours. In: Vargas, P.A., Paolo, E.D., Harvey, I., Husbands, P. (eds.) The Horizons of Evolutionary Robotics, pp. 153–178. MIT Press, Cambridge (2014)
26. Valentini, G., Brambilla, D., Hamann, H., Dorigo, M.: Collective perception of environmental features in a robot swarm. In: Dorigo, M., et al. (eds.) ANTS 2016. LNCS, vol. 9882, pp. 65–76. Springer, Cham (2016). https://doi.org/10.1007/978-3-319-44427-7_6
27. Valentini, G., Hamann, H., Dorigo, M.: Efficient decision-making in a self-organizing robot swarm: on the speed versus accuracy trade-off. In: Proceedings of the 2015 International Conference on Autonomous Agents and Multiagent Systems (AAMAS), pp. 1305–1314. International Foundation for Autonomous Agents and Multiagent Systems (2015)
28. Valentini, G., Ferrante, E., Dorigo, M.: The best-of-n problem in robot swarms: formalization, state of the art, and novel perspectives. Front. Robot. AI 4, 9 (2017)
29. Wessnitzer, J., Melhuish, C.: Collective decision-making and behaviour transitions in distributed ad hoc wireless networks of mobile robots: target-hunting. In: Banzhaf, W., Ziegler, J., Christaller, T., Dittrich, P., Kim, J.T. (eds.) ECAL 2003. LNCS (LNAI), vol. 2801, pp. 893–902. Springer, Heidelberg (2003). https://doi.org/10.1007/978-3-540-39432-7_96

Soft Computing Applied to Games

Deep Reinforcement Learning for 5 × 5 Multiplayer Go

Brahim Driss[1(✉)], Jérôme Arjonilla[1], Hui Wang[1], Abdallah Saffidine[2], and Tristan Cazenave[1]

[1] LAMSADE, Université Paris Dauphine - PSL, CNRS, Paris, France
brahim.driss0@dauphine.eu, {hui.wang,Tristan.Cazenave}@dauphine.psl.eu
[2] University of New South Wales, Sydney, Australia

Abstract. In recent years, much progress has been made in computer Go and most of the results have been obtained thanks to search algorithms (Monte Carlo Tree Search) and Deep Reinforcement Learning (DRL). In this paper, we propose to use and analyze the latest algorithms that use search and DRL (AlphaZero and Descent algorithms) to automatically learn to play an extended version of the game of Go with more than two players. We show that using search and DRL we were able to improve the level of play, even though there are more than two players.

Keywords: Multi-agent · Deep reinforcement learning · Go

1 Introduction

Due to a huge game tree complexity, the game of Go has been an important source of work in the perfect information setting. In 2007, search algorithms have been able to increase drastically the performance of computer Go programs [8–11]. In 2016, AlphaGo has been able to beat a strong professional player for the first time [15]. This great success has been achievable thanks to a combination of two key elements: Search (Monte Carlo Tree Search [1]) and Learning (Reinforcement Learning) methods [15–17]. Currently, the level of play of such algorithms is far superior to those of any human player.

Even though, the game of Go has been given great interest, less has been done on variants of the game. In practice, there exist many variants of the game of Go such as Blind Go [5] (where the players cannot see the board), Phantom Go [2] (where the players cannot see the opponent stones) or Capture Go (where the game is finished when the first player to capture a stone wins). In this paper, we study the variant Multiplayer Go. As the name suggests, Multiplayer Go is a variant of the game of Go where there are more than two players. Going from two-player Go to Multiplayer Go makes the game even more complex.

In this paper, we propose to apply and analyze the latest developments in the game of Go to the game of Multiplayer Go. More specifically, we are using search and reinforcement learning method such as AlphaZero [15] and Descent [6,7]

J. Correia et al. (Eds.): EvoApplications 2023, LNCS 13989, pp. 753–764, 2023.
https://doi.org/10.1007/978-3-031-30229-9_48

for the game of Multiplayer Go. In past work, [3] has described multiple UCT algorithms with different multi-agent behaviors (coalitions, paranoid or with alliance) for the game of Multiplayer Go and in [4], they successfully improve past performances using GRAVE, a heuristic method for MCTS algorithms. An adaptation of AlphaZero to multiplayer games was also used in [14] for the multiplayer versions of Tic-Tac-Toe and Connect 4.

The paper is organized as follows: the second section presents the game of Multiplayer Go, section three presents the algorithms we have been using to analyze the game, section four presents our results and the last section summarizes our work and future work.

2 Multiplayer Go

The game of Go is a strategic board game with perfect information, played by two players. Each player aims at capturing more territory than their opponent by placing stones on the board. One is playing black stones and the second is playing white stones. At each turn, one player is acting and placing a stone on a vacant intersection of the board. After being placed, a stone cannot be moved or removed by the player. Nevertheless, a player stone can be removed by its opponent if the latter successfully surrounds the stone on all orthogonally adjacent points. The game ends when no player is able to make a move or until none of them wishes to move.

For the scoring, there exist multiple rules, in our case, we have used the Chinese rule *i.e.,* the winner of the game is defined by the number of stone that a player has on the board, plus the number of empty intersections surrounded by that player's stones and komi (bonus added to the second player as a compensation for playing second).

Multiplayer Go is a variant of the game of Go with more than two players. In our case, we have added a third player which is playing a third color, red. An example is provided in Fig. 1. By adding a third player, one must be wary to not create a queer game [13].

A queer game is when, in some positions, no player can force a win. As an example, in some positions, even if a player is sure to lose, they can still have an impact on the winner by adapting this strategy. As a consequence, coalitions can arise in order to help defeat another opponent.

As a preliminary matter, we have compared the performance of two different type of rewards, one that uses the winning player as an objective and another one that tries to maximize the score, using Chinese rules. Maximizing the score allows us to reduce the problem of queer games. This analysis had already been carried out in [3]. For the remainder of the paper, we thus use the objective for each player to maximize his score using Chinese rules.

Even tough the rules are relatively simple, the game of Go is known as an extremely complex one in comparison to other board game such as Chess. With a larger board (common board have a size of 19×19), the number of legal board positions has been estimated to be 2.1×10^{170}. Moreover, with a large number

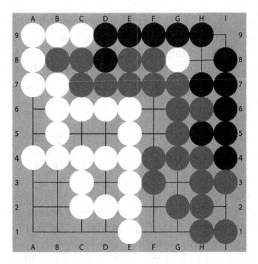

Fig. 1. A game of Multiplayer Go.

of possible actions and longer games on average, the complexity is much greater than that. Worse than that, in our case, the addition of a three-player game adds a significant layer to the complexity of the game. As a consequence, in an effort to study the learning in a multiplayer setup rather than the difficulty of the game, we studied a simplified version of Go on a 5×5 board. Nevertheless, even being smaller, the version of the game is very complex.

3 Deep Reinforcement Learning

In this section, we described the algorithms used for addressing Multiplayer Go. The Subsect. 3.1 presents Monte Carlo Tree Search (MCTS) and its variant UCT, the Subsect. 3.2 presents AlphaZero and the Subsect. 3.3 presents Descent. All the hyperparameters used are explained and defined in Table 1 and in Table 2.

3.1 Monte Carlo Tree Search

In our experimentation, we have been using UCT (Upper Confidence bounds applied to Trees), which is a variant of MCTS, as a baseline. Before explaining UCT, we must explain MCTS. Monte Carlo Tree Search [1] is the state of the art in perfect information games. MCTS is a tree search algorithm which works as follows (i) **selection**—select a path of node based on the exploitation policy (ii) **expansion**—expand the tree by adding a new child node (iii) **playout**—estimate the child node by using an exploration policy (iv) **backpropagation**—backpropagate the result obtained from the playout through the nodes chosen during the selection phase.

UCT is a variant of MCTS where the selection phase is decided by UCB (Upper Confidence Bounds), a bandit algorithm and where the playout use a random policy. The UCB formula is decomposed on two parts, the first part represents the exploitation *i.e.*, it attempts to play the best action observed so far, and the second part represents the exploration *i.e.*, it attempts to play an action less visited.

The formula is defined as follow:

$$UCT(s,a) = Q(s,a) + c\sqrt{\frac{ln[N(s)]}{N(s,a)}} \tag{1}$$

where the best action is the one that maximizes the upper confidence bound $UCT(s,a)$, s denotes the state of the game, a is an action possible from the set $A(s)$ which represents all the actions possibles in the state s. Q represents the value when playing the action a in the state s, $N(s,a)$ is the number of times that the action a has been visited in the state s, $N(s)$ represents the number of times that the state s has been visited and c is a variable that help controlling the exploration.

Furthermore, as we are in a multiplayer context, we must use Multiplayer UCT [18] which is the same algorithm as UCT where the only difference being the score representation. In a multiplayer setup, we get an array instead of a single value, containing the results of the different players.

In our experimentation, we use the following hyper-parameters (i) $n = 180$ the number of rollout *i.e.*, the number of times the playout is repeated in order to obtain a better approximation of the child node (ii) $c = 0.8$.

3.2   AlphaZero

As a famous deep reinforcement learning paradigm, combining online Monte Carlo Tree Search (MCTS) and offline neural network has been widely applied to solve game-related problems, especially known as AlphaGo series programs [15–17]. MCTS is used to enhance the policy and the neural network provides the state estimation.

The neural network based MCTS employs PUCT formula to balance the exploration and exploitation as following:

$$PUCT(s,a) = Q(s,a) + c_{puct}.P_\theta(s,a)\frac{\sqrt{N(s)}}{1 + N(s,a)} \tag{2}$$

where the best action is the one that maximizes the upper confidence bound $PUCT(s,a)$, s denotes the state of the game, a is an action possible from the set $A(s)$ which represents all the actions possibles in the state s. Q represents the value when playing the action a in the state s, $N(s,a)$ is the number of times that the action a has been visited in the state s, $N(s)$ represents the number of times that the state s has been visited, c_{puct} is a variable that help controlling the exploration and $P_\theta(s,a)$ is the estimation of taking action a in the state s according to the policy of the network θ.

The network architecture is similar to the original AlphaZero one, having the board as input and producing two outputs: a probability distribution over moves (policy head) and a vector of score prediction for every player (value head).

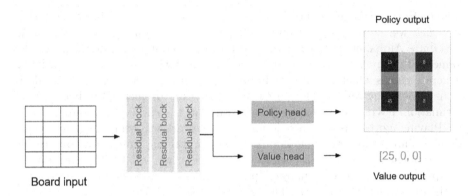

Fig. 2. AlphaZero network architecture

Network Architecture. The general architecture of AlphaZero can be found in Fig. 2. All hyperparameters and differences can be found in Table 2.

In order to encode the board, we are using a 5×5 matrix. The input of the state is represented by 6 channels where each channel is a 5×5 matrix. The first 3 channels represent the position of one player's stones on the board, and the last 3 channels represent the current player who has to play.

After the encode of the game, 8 residual block of width 128 are placed one after the other. The first residual block take the game encoded as an input. Each residual block are using convolution kernel of size 3 with the activation function being ReLU and as we are using a residual block, the input of the layer is also for the next residual block.

The policy head give the probability of playing each action for the current player *i.e.,* for the 25 actions, the policy head return a value between 0 and 1. Thus, the policy head is composed of a 1×1 convolution, outputting a policy distribution in logits.

The value head outputs 3 values as described in multiplayer UCT [18] which estimated the value that each player will obtain *i.e.,* for the 3 players, the value head returns a score between 0 to 25. Thus, the value head is a fully connected with 3 parallel hidden layers, each one connected to an output layer of size 26 (possible integer scores for each player from 0 to 25 points).

Usually, in the game of Go, a player has the possibility of passing. However, when tested, and even with the goal of maximizing the score, the agent starts passing even in positions where it is still possible to continue gaining more points. In order to fix this performance issue, we removed the pass from the network, and only allowed passing when no more moves are possible (or only moves that fill the eyes, where an eye is a single empty space inside a group).

Warm-Start Self-play. For training our network, we have been using self-play *i.e.,* we compete against ourselves, save the data collected and train using this data. AlphaZero self-play starts with randomly initialized networks for the value and policy. Both of them are used, combined with tree search, in order to generate games which are used to improve the networks, leading to better decisions by learning MCTS selected moves (policy improvement) and better value estimation, having access to the games results at the end (value improvement).

Playing moves based on a random policy network will generate games where decisions are almost random, leading to longer training time before observable improvement. In order to accelerate this process, we add UCT agents during self-play, replacing (one or many) AlphaZero agents randomly with a decreasing probability ϵ, where $\epsilon = max((0.5 - \frac{iteration\_number}{n\_updates}), 0.05)$.

As a result, in earlier iterations, every AlphaZero agent (each color) has a 50% probability in every game to be replaced by a UCT agent, this probability decreases every iteration, reaching 5% by iteration 50 where it stops decreasing.

Table 1. Training cycle hyperparameters

Hyperparameter	Description	Value
n	Number of rollouts	180
c	Exploration constant in UCT/PUCT	0.8
n_updates	Number of network updates (iterations)	50
n_games	Number of self-play games per update	1000
n_envs	Number of parallel workers	8
buffer_size	Size of replay buffer	2000
N	Total games played ($n\_updates * n\_games$)	50000

3.3   Descent

The second deep reinforcement learning algorithm is Descent [6,7]. Descent is a recent algorithm which has shown great success in international competitions such as the 2021 Computer Olympiad of the ICGA. Descent is not based on MCTS but on Unbounded MinMax [12].

Table 2. Neural network hyperparameters

Hyperparameter	Description	AlphaZero Value	Descent Value
n_res	Number of residual blocks	8	8
res_filters	Number of output filters in convolutions in residual blocks	128	128
res_kernel_size	Convolution kernel size in residual blocks	3	3
res_activation	Activation in residual blocks	ReLU	ReLU
policy_filters	Number of output filters in policy head	1	None
policy_kernel_size	Convolution kenel size in policy head	1	None
policy_activation	Activation in the last layer of the policy head	Softmax	None
value_activation	Activation function in the last layer of value head	Softmax	Linear
kernel_regularizer	L2 regularization applied to all weights	0.0001	0.0001
policy_loss	Loss function used for the policy head	Categorical crossentropy	None
value_loss	Loss function used for the value head	Categorical crossentropy	Mean squared error
optimizer	Training optimizer	SGD	Adam
lr	Training learning rate	0.0001	0.0001

At the difference of Minimax, Unbounded Minimax explores the game tree in a non-homogeneous way where the exploration is a best-first approach at each iteration. In descent, the exploration is a best-first approach, that is recursively applied until the end of the game. This allows to backpropagate the values of terminal states more efficiently through the nodes chosen. Furthermore, during the exploration phase, the best move is determined by the utilization of a neural network.

Descent architecture is the same as AlphaZero but, in AlphaZero there are two output networks (policy head and value head) and in descent, there is only one output network (value head). Furthermore, in the value head of descent, we are using a linear for the activation function as a regression for player scores.

The neural network has been trained with the value obtained from the minimax values of the trees built during the game. In addition, each state which has been explored during the game (not just for the sequences of states of the played games) is learned. As a result, all the information acquired during the search is used during the learning process. We use the same network and hyperparameters as described in the original article [7] and train for 120 h, the same duration as AlphaZero.

In the original paper, the authors have achieved better performance that AlphaZero on multiple game and more quickly.

4   Experimental Results

The experiences were made on 2 NVIDIA GeForce RTX 2080 TI. Each test have been experimented on 500 games. All neural networks have been trained for 120 h.

Table 3. Average number of point when all players are using UCT. The test has been run on 500 game with 95% confidence interval.

	Black	White	Red
Point	11.5 ± 1.0	7.1 ± 0.9	6.3 ± 0.9

As baseline, we are using the Table 3. In this table, we observe the average number of points when all player are using UCT. As we are not using komi during our experimentation, it makes sense to observe that the black player has an advantage against white and red. However, as we can see, white does not have a significant advantage in comparison to the red player.

4.1 Training of AlphaZero and Descent

In this subsection, we are analyzing the performance of AlphaZero and Descent against UCT for all players. In Fig. 3, we can observe the evolution of the performance according to the training time where we tested the performance every 12 h.

AlphaZero

For AlphaZero, the improvement of the performance are available in a, c and e of the Fig. 3. As we can observer, AlphaZero improves its performance for the three different colors using the same network. At the end of the training, we observe that the average points obtained are close to 16, 10, 11 for the black, white and red players respectively.

Most of the improvement in AlphaZero has been done during the first 60 h of training. Even tough, we observe a stagnation after 60 h, the performance of AlphaZero are superior to UCT. The average score per game increased from 10 to 16, from 7 to 10 and 6 to 11, respectively for the black, white and red players.

In addition, when playing black and red, AlphaZero is able to outperform all other opponents, and when playing white, obtains a score almost equal to black while being at a disadvantage.

Descent

For Descent, the improvement of the performance are available in b, d and f of the Fig. 3. The figure shows that Descent also leads to performance improvement in a multiplayer setup. All the players converge to the same average score, around 11 (12, 12, 11 for black, white, red respectively) and each of them having at least better or equal performance to UCT. Having different network update strategies and network architecture (Descent does not use a policy network to guide move selection), the two methods do not converge to the same performance.

In past work of Descent, the authors had better and faster results than AlphaZero. However, as we can observe, after 120 h, AlphaZero has better scores for the black and red players but Descent is better for the white player. Nevertheless, we can observe that AlphaZero is not improving a lot after the 60 h mark whereas Descent did not start showing stagnation in its curves and this even after 120 h of training.

4.2 Black Against White and Red

AlphaZero and Descent use different planning methods. This can result in different strength and strategies while playing different positions in a game. Remember that each method use a single network for all the different positions. The UCT baseline in Table 3 confirm that Black has an advantage when playing first,

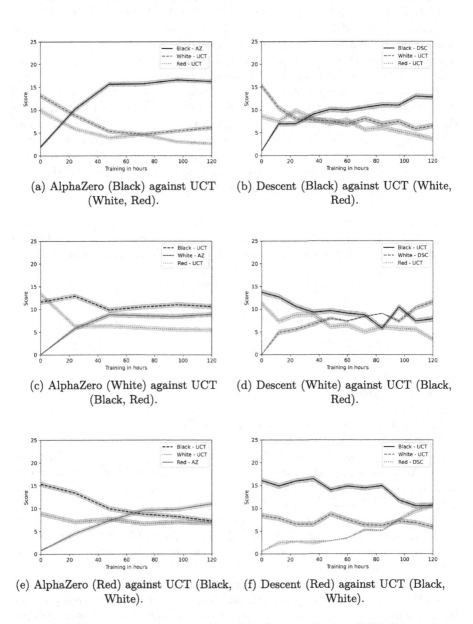

(a) AlphaZero (Black) against UCT (White, Red).

(b) Descent (Black) against UCT (White, Red).

(c) AlphaZero (White) against UCT (Black, Red).

(d) Descent (White) against UCT (Black, Red).

(e) AlphaZero (Red) against UCT (Black, White).

(f) Descent (Red) against UCT (Black, White).

Fig. 3. Left/Right figure represents AlphaZero/Descent against UCT. In y-axis we observe the average points obtained and in x-axis, we observe the training in hours.

which is an expected results since it's the case in Go. Without using komi, we will only focus on how both methods play when having this advantage. Table 4 shows the results of 500 games testing Black strength at attacking each other weaker positions.

We notice that AlphaZero tends to be more aggressive when playing Black against itself, which is what it learn during self-play, achieving an average score of 13.3 points. The same aggression does not work effectively against Descent defenses (White and Red) since it only gets 11 points on average. Looking at Descent scores, playing as Black against AlphaZero defenses achieves an overage score of 12.1 points, which is between both AlphaZero scores as Black. Playing against itself only show a small difference in score going to 11.9.

In both cases, Descent is stronger when playing positions at disadvantage, and does not show a bigger difference playing against AlphaZero as Black, meaning that Descent is more balanced in strengh between all the different positions. AlphaZero on the other hand, will try to be more aggressive against medium defenses (16 points against UCT and 13.3 points against itself) but this also mean that it can be slightly weaker playing White and Red and that the same strategy will not be effective against better defenses (only 11 points against Descent).

Table 4. Average number of point when black is against the others players for different algorithms. The test has been run on 500 games with 95% confidence interval.

		UCT	White and Red AlphaZero	Descent
Black	AlphaZero	16.2 ± 0.3	13.3 ± 0.4	11 ± 0.2
	Descent	12.7 ± 0.5	12.1 ± 0.6	11.9 ± 0.3

5 Conclusion

In this paper, we used and analyzed Deep Reinforcement Learning for one of the variants of the game of Go, the game of Multiplayer Go. We have been using AlphaZero and Descent, which have been showing great success in recent years. We demonstrate that both algorithms are applicable in Multiplayer Go and both of them are able to learn in the context of multiplayer game which is more complex than two players.

Both of the algorithms have been able to beat or equalizes UCT in all players positions (Black, White and Red). In addition, against UCT, the two algorithms obtain very close results in a short training time and neither of the two has been able to beat the other in all cases.

In addition to this, we analyze the impact of the black player using a Deep RL algorithm against the other Deep RL algorithm for the white and red position. In this context, we show that Descent is more balanced in strength between different positions than AlphaZero which result in a better defense, but that

AlphaZero can achieve better performance against medium and weaker defenses (himself or UCT) than Descent.

In future work, we expect to use Deep Reinforcement Learning on other multiplayer games, to increase the number of agents and to use it on larger boards. Furthermore, we have observed that AlphaZero stops improving after 60 h of training which is not the case for Descent. As a consequence, we are interested in making more and longer experiments in order to compare more accurately the two DRL algorithms.

References

1. Browne, C.B., Powley, E., Whitehouse, D., Lucas, S.M., Cowling, P.I., Rohlfshagen, P., Tavener, S., Perez, D., Samothrakis, S., Colton, S.: A survey of Monte Carlo tree search methods. IEEE Trans. Comput. Intell. AI Games 4(1), 1–43 (2012)
2. Cazenave, T.: A phantom-go program. In: van den Herik, H.J., Hsu, S.-C., Hsu, T., Donkers, H.H.L.M.J. (eds.) ACG 2005. LNCS, vol. 4250, pp. 120–125. Springer, Heidelberg (2006). https://doi.org/10.1007/11922155_9
3. Cazenave, T.: Multi-player go. In: van den Herik, H.J., Xu, X., Ma, Z., Winands, M.H.M. (eds.) CG 2008. LNCS, vol. 5131, pp. 50–59. Springer, Heidelberg (2008). https://doi.org/10.1007/978-3-540-87608-3_5
4. Cazenave, T.: Generalized rapid action value estimation. In: Proceedings of the Twenty-Fourth International Joint Conference on Artificial Intelligence, IJCAI 2015, Buenos Aires, Argentina, July 25–31, 2015, pp. 754–760 (2015)
5. Chou, P.C., et al.: Computational and human intelligence in blind go. In: Computational Intelligence and Games. Seoul, North Korea (Aug 2011). http://hal.inria.fr/inria-00625849
6. Cohen-Solal, Q.: Learning to play two-player perfect-information games without knowledge. ArXiv abs/2008.01188 (2020)
7. Cohen-Solal, Q., Cazenave, T.: Minimax strikes back. Reinforcement Learning in Games at AAAI (2021)
8. Coulom, R.: Efficient selectivity and backup operators in Monte-Carlo tree search. In: van den Herik, H.J., Ciancarini, P., Donkers, H.H.L.M.J. (eds.) CG 2006. LNCS, vol. 4630, pp. 72–83. Springer, Heidelberg (2007). https://doi.org/10.1007/978-3-540-75538-8_7
9. Coulom, R.: Computing elo ratings of move patterns in the game of Go. ICGA J. 30(4), 198–208 (2007)
10. Gelly, S., Silver, D.: Monte-Carlo tree search and rapid action value estimation in computer Go. Artif. Intell. 175(11), 1856–1875 (2011)
11. Kocsis, L., Szepesvári, C.: Bandit based Monte-Carlo planning. In: Fürnkranz, J., Scheffer, T., Spiliopoulou, M. (eds.) ECML 2006. LNCS (LNAI), vol. 4212, pp. 282–293. Springer, Heidelberg (2006). https://doi.org/10.1007/11871842_29
12. Korf, R.E., Chickering, D.M.: Best-first minimax search. Artif. Intell. 84(1), 299–337 (1996)
13. Loeb, D.E.: Stable winning coalitions. Games No Chance 29, 451–471 (1996)
14. Petosa, N., Balch, T.R.: Multiplayer alphazero. ArXiv abs/1910.13012 (2019)

15. Silver, D., et al.: Mastering the game of go with deep neural networks and tree search. Nature **529**, 484–489 (2016)
16. Silver, D., et al.: Mastering chess and shogi by self-play with a general reinforcement learning algorithm. CoRR abs/1712.01815 (2017). arxiv.org/abs/1712.01815
17. Silver, D., et al.: A general reinforcement learning algorithm that masters chess, shogi, and go through self-play. Science **362**(6419), 1140–1144 (2018)
18. Sturtevant, N.R.: An analysis of UCT in multi-player games. In: van den Herik, H.J., Xu, X., Ma, Z., Winands, M.H.M. (eds.) CG 2008. LNCS, vol. 5131, pp. 37–49. Springer, Heidelberg (2008). https://doi.org/10.1007/978-3-540-87608-3_4

Genetic Programming and Coevolution to Play the Bomberman™ Video Game

Robert Gold[1] , Henrique Branquinho[2] , Erik Hemberg[1] ,
Una-May O'Reilly[1] , and Pablo García-Sánchez[3](✉)

[1] ALFA, MIT CSAIL, Cambridge, USA
{robertgold,hembergerik,unamay}@csail.mit.edu
[2] Centre for Informatics and Systems of the University of Coimbra,
Department of Informatics Engineering, University of Coimbra, Coimbra, Portugal
branquinho@dei.uc.pt
[3] Department of Computer Architecture and Computer Technology,
University of Granada, Granada, Spain
pablogarcia@ugr.es

Abstract. The field of video games is of great interest to researchers in computational intelligence due to the complex, rich and dynamic nature they provide. We propose using Genetic Programming with coevolution and lexicographic fitness to generate an agent that plays the *Bomberman*^{TM} game. We investigate two sets of Genetic Programming building blocks: one contains conditions relative to movement, and the other does not. We aim to see whether the benefits of these movement-related conditions outweigh the negatives caused by increased search space size. We show that the benefits gained do not outweigh the increase in search space size.

Keywords: *Bomberman*^{TM} · Genetic Programming · Video Games · Lexicographical Fitness · Artificial Intelligence

1   Introduction

Video games are interesting for research in artificial intelligence (AI). They provide complex problems, rich human-computer interaction, and a large amount of data due to their popularity [18]. Competitions for agents playing video games have become a popular venue to compare methods for creating intelligent agents.

There are many applications of AI techniques in video games, such as neural networks and reinforcement learning. In addition to other methods, evolutionary algorithms have proven helpful tools for creating game-playing agents. For example, Genetic Algorithms (GAs) optimise the hard-coded parameters of an agent [15]. However, this limits the agent's behaviour to the knowledge of the human

© The Author(s), under exclusive license to Springer Nature Switzerland AG 2023
J. Correia et al. (Eds.): EvoApplications 2023, LNCS 13989, pp. 765–779, 2023.
https://doi.org/10.1007/978-3-031-30229-9_49

who programmed it. Other methods, such as Genetic Programming (GP), can build upon this. One application of Genetic Programming is to generate source code that determines an agent's behaviour based on all possible actions the agent can perform according to all possible inputs. Thus GP can result in better agents than those optimised with GA [7].

This paper looks at the video game *BomberMan*[TM]. In particular, we look at an open-source version called *BomberLand*, which is used by AI competitions in video games[1]. *BomberLand*[TM]is interesting because, unlike other games, the attacks are asynchronous; an agent can place a bomb and activate it at any point. In addition, an agent can die from its bombs. The game's goal is to defeat the enemy by escaping all bomb explosions and using items that provide advantages while attacking the enemy with bombs.

No other AI agents that play *BomberLand* were available during the development of this work. Thus, we decided to use a competitive coevolutionary [16] approach to assigning fitness to individuals. Coevolution calculates an agent's fitness by having it battle against other individuals (agents) within the population. The goal is to create an agent that performs well against many opponents instead of just one opponent.

A GP coevolutionary approach can be effective when a problem contains uncertainty and dynamic information, and there is no explicit objective function to optimise, only performing well in the game. Thus, we compare two search spaces with and without movement-related conditions. The conditions, called *canMove*, measure whether an agent can safely move in a given cardinal direction (up, down, left, right). Without these conditions, movement actions may either do nothing due to being unable to move in the provided direction or result in damage if they move into something that will damage them. These conditions may result in more effective agent behaviours, but they will also increase the search space size. We aim to determine whether the benefits gained through the *canMove* conditions outweigh the negatives caused by increased search space size.

The experimental setup is as follows: GP generates individuals, then converts them to source code for evaluation by the *BomberLand* environment. The evaluation results are fed back into the GP, and the GP assigns fitness values to each individual. This process continues until a stopping criterion is reached. See Fig. 1 for a visual. The code repository used to obtain the results presented in this paper is available at [11].

The rest of the paper is structured as follows. First, we discuss the state-of-the-art. Then we discuss the game environment used in Sect. 3. A description of agent behaviours follows this, and the algorithm used to optimise them. Then, in Sect. 5, results are discussed. Finally, conclusions and future work are presented.

---

[1] https://www.gocoder.one/bomberland.

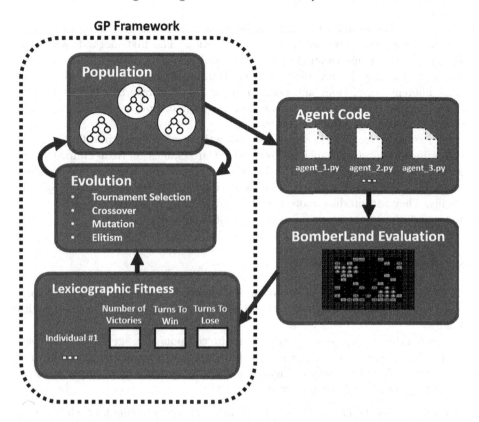

Fig. 1. Overview of the experimental setup. GP generates individuals, then converts them to source code for evaluation by the *BomberLand* environment. The evaluation results are fed back into the GP, and the GP assigns fitness values to each individual. This process continues until a stopping criterion is reached.

2   Related Work

There are applications of GP in different types of video games, such as Real-Time Simulators (RTS) [9], First Person Shooters (FPS) [3], or Game Theory [10]. In particular, GP has been successfully used to create high-performing agents in other games such as *StarCraft*™[9], and *Super Mario Brothers*™[6].

GP can obtain better results than other AI techniques and even human players [7]. Victory-based lexicographical fitness techniques, as opposed to other aggregation-based fitness methods, have shown promising results because they produce more aggressive agents [4].

Coevolution includes interactions between individuals in one or more populations. We discuss two methods of coevolution. The first method, which we use in this paper, uses a single population and performs the scoring process by matching individuals according to a selection mechanism. The second method uses different populations and tries to mimic an arms race between simultaneously evolving populations. Both methods use competition, so this technique can also be called *competitive* coevolution [2].

There are several works which use competitive coevolution in different types of gaming environments, such as Collectable Card Games [8], Real-Time Strategy games [17], and board games [1]. These examples do not need to train against an external enemy. Therefore, there is no risk of overfitting against a specific enemy. They can produce more versatile individuals, as the enemies (the rest of the population) encompass a larger space of behaviours.

3   Game Environment

The goal of *BomberLand* is to defeat the enemy by escaping all bomb explosions and using items which provide advantages while attacking the enemy with bombs.

In *BomberLand*, there are several main components: agents, actions performed by agents, bombs, power-ups, blocks, power-ups, the ring of fire, and game ticks (simulation steps). Figure 2 shows a screenshot of a sample *Bomber-Land* map with these components.

Agents. In *BomberLand*, two agents (or players) compete against each other on a randomly generated map of a fixed size. Agents have a complete vision of the entire map. In other words, agents know what is in each spot on the map.

Each agent controls three units (represented as images or knights, depending on the player). Each unit has health points (HP) and an inventory that contains bombs. An agent wins when all of the opponent's units are dead. If all units reach 0 HP in the same game tick, the game will be considered a tie. A unit begins with 3 HP and is defeated when its HP reaches 0.

Agent Actions. Units can move around the map in any cardinal direction (up, left, down, right), plant bombs, and remotely detonate them. No unit moves if two units try to move to the same spot on the same game tick. In addition, they can also do nothing. We call the actions an agent can perform *up, down, left, right, bomb, detonate,* and *noop*.

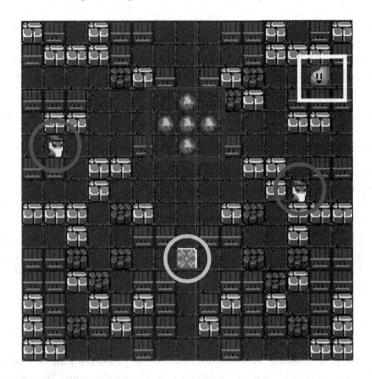

Fig. 2. A sample *BomberLand* map. In the yellow circle is a freeze power-up, in the pink circles are units, in the blue square is a bomb explosion, and in the white square is a blast power-up. Scattered around the map are varying different types of blocks. (Color figure online)

Bombs. A unit can place a bomb and manually activate it at anytime greater than five ticks after being placed. A consequence is that a unit can die due to being within the blast radius of its bomb. If a unit gets caught in a bomb's blast range, it loses one HP. After a unit gets hit by a bomb, it becomes invulnerable to damage for five game ticks.

Units passing through an explosion will lose 1 HP unless they are invulnerable. Bomb blasts last for five ticks. Bombs automatically explode 30 ticks after being placed. Lastly, bombs detonate if another exploding bomb hits them.

Power-Ups. Power-ups are items units can pick up to give them an advantage over their opponents. Power-ups randomly spawn when blocks are destroyed but disappear after 40 ticks. There are three types of power-ups:

– Blast – increases the bomb blast radius of a unit
– Freeze—temporarily freezes a random enemy unit for 15 ticks;
– Ammunition – adds a bomb to a unit's inventory.

The Ring of Fire. After 200 game ticks, the map gradually fills up with flames that destroy all blocks and power-ups. The fire also detonates bombs. The fire starts in the top-left and bottom-right corners of the map and spirals towards the centre of the map. More fire spawns every two ticks until one or no units are alive.

Blocks. Blocks are scattered throughout the game map and act as obstacles. They can contain power-ups that appear upon block destruction. There are three types of blocks:

- Wooden blocks—can be destroyed by a bomb in 1 hit;
- Ore blocks—can be destroyed by a bomb in 3 hits;
- Metal blocks – cannot be destroyed, except by the ring of fire.

Game Ticks (Simulation Steps). *BomberLand* measure time in ticks, i.e., simulation steps. Agents perform one single action every game tick. Invalid actions do nothing. An example of an invalid action is to move left if there is a block to the left of an agent.

4   GP Setup

This section describes the building blocks used to generate individuals and the coevolutionary algorithm we use to perform fitness evaluations. We use a version of the strongly-typed GP framework proposed in [12] and [13] called STGP-Sharp [14]. We use the standard two-point crossover, mutation, and 5-tournament selection operators. Lastly, we used ramped population initialisation. A balance must be found between search space size and search space expressiveness to optimise the GP search algorithm [5]. We compare two versions of our search space which we call the *canMove* and *notCanMove* search spaces. The *canMove* search space contains GP operators to determine if the given agent can move in a specified direction. We call these *canMove* conditions. The *notCanMove* search space does not contain the *canMove* conditions. The latter may result in actions that will do nothing or be damaged (e.g., an agent may try to move up even if there is a block, bomb explosion, or ring of fire directly above it), but the search space will be smaller, which may improve results. See Subsect. 4.1 for more information on the *canMove* conditions.

4.1   Building Blocks

An individual represents an agent's behaviour, modelled by a binary decision tree. We define two types of building blocks: conditionals and agent actions. We define agent actions in Subsect. 3. Internal nodes represent conditionals, and leaf nodes represent agent actions. Figure 3 shows an example of an individual.

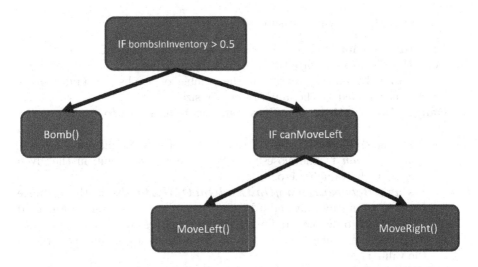

Fig. 3. An example of an individual. Internal nodes represent conditionals within a binary decision tree. Leaf nodes represent actions for an agent to perform. This example is a manually created individual, not an individual generated by our code.

Conditionals are based on measures within the game that an agent perceives. These measures are then compared to a threshold to determine whether to factor in the measure. For example, the $distanceToNearestBombLeft$ measure checks if an agent can move left based on whether there is a block, bomb explosion, or ring of fire within a certain distance to the agent's left. Let $Bombs$ be the set of bombs to the left of a given unit currently on the map, map_w be the width of the current map, $unit_x$ be the x-coordinate of a given unit, and b_x be the x-coordinate of a given bomb. Note that since all $b \in Bombs$ are to the left of a given unit, then all $b_x < unit_x$. We can then say

$$distanceToNearestBombLeft = \begin{cases} \max_{b \in Bombs} 1 - \dfrac{unit_x - b_x}{map_w}, & \text{if } Bombs \neq \emptyset \\ 0, & \text{otherwise} \end{cases}$$

At each game tick, we collect several measures from the game state for each game unit. All measures, except $canMove$ conditions, are represented by floating point numbers between zero and one. All $canMove$ measures are represented by booleans. We define measures representing distances as the absolute distance to a unit, normalised with respect to the map's dimensions, subtracted from one. Thus, the closer something is to a unit, the closer the measure's value will be to one.

Let the syntax $<Direction>$ mean to replace the bracketed string with a given cardinal direction. For example, if $Direction$ is $Left$, then $distanceTo$ $NearestBomb<Direction>$ is equivalent to $distanceToNearestBombLeft$.

We collect the following measures at each game tick:

- **Unit Information:** Information related to unit health and inventory.
  - *HP* represents unit health normalised by max HP;
  - **bombsInInventory** represents the number of bombs in a unit's inventory normalised by the unit's inventory size.
- **Danger Information:** Quantifies distances from a unit to dangerous elements.
  - *distanceToNearestBomb<Direction>* returns the distance in direction *Direction* to the closest bomb. If there is no bomb in the given direction, the value is zero.
  - *distanceBetweenEnemyUnitAndOneOfMyBombs* is the distance between an enemy unit and a bomb planted by a given unit, normalised with respect to the blast radius. If multiple enemies are in the blast radius, it returns the smallest distance. If no enemy units are within blast range, the value is zero.
  - *distanceToClosestEnemyUnit* is the distance between a given unit and its closest enemy unit in any direction. This measure does not consider any other map elements (blocks, fire, bombs).
- **Power-Up Information:** Quantifies distances between a unit and power-ups. We separate into two types of power-ups: ammunition and freeze or blast.
  - *distanceToAmmoPowerUp<Direction>* measures the distance between a unit and the closest ammunition power-up in direction *Direction*. The value is zero if there is no ammunition power-up in the given direction.
  - *distanceToFreezeOrBlastPowerUp<Direction>* measures the distance between a unit and the closest freeze or blast power-up in direction *Direction*. The value is zero if there is no freeze or blast power-up in the given direction.
- **Movement Information:** Booleans which represent whether a unit can move in a given direction. Blasts do nothing to an agent if it is invulnerable. These measures are only used in the *canMove* version of the GP search space, as outlined previously.
  - *canMove<Direction>* represents whether a unit is able to move in direction *Direction* in the current game tick.

4.2   Coevolutionary Setup

Each generation GP creates a population and pits individuals within that population together pairwise to simulate battles. We do this for each pair N times. Due to the stochastic nature of *BomberLand*, this method better characterises individual fitness, as explained in [8]. Note that N is the same as the parameter *number of battles per fitness evaluation* discussed in Sect. 4.4.

Our implementation re-evaluates individuals in each generation since fitness values obtained in one generation are not comparable to those obtained in other generations due to populations across generations being different.

We use lexicographic fitness to compare individuals [8]. For each individual, we generate a fitness vector once per generation. The fitness vector is a composition of the following metrics:

- The number of times that individual i wins across all battles. We call this V_i.
- The sum of turns across all battles in which individual i wins, normalised by the number of battles where individual i wins. We call this TW_i.
- The number of turns across all battles in which individual i loses, normalised by the number of battles where individual i loses. Let us denote this as TL_i.

Note that an agent's fitness vector will not be affected if a battle results in a tie. For an individual i, we define a fitness vector F_i for individual i to be

$$(V_i, TW_i, TL_i).$$

Let F_i^d denote the dth dimension of a fitness vector F_i for individual i.

Formally, given a population P, let

$$Wins : P \times P \to [0, N]$$

be a function that returns how many times one individual wins against another. Then, each individual competes in $N \cdot (|P| - 1)$ battles per generation. In total, there are $N \cdot (|P| - 1)^2$ battles per generation across all individuals.

We can also say that given some individual $i \in P$,

$$V_i = \sum_{j \in P \setminus \{i\}} Wins(i, j).$$

In addition, let

$$Turns : P \times P \to \mathbb{N}$$

be a function that returns the sum of turns it takes for one individual to beat another across all battles between them. Then,

$$TW_i = \frac{1}{V_i} \sum_{j \in P \setminus \{i\}} Turns(i, j).$$

Lastly, let B_i be the number of battles involving individual i, and let L_i be the number of battles in which individual i loses. Thus,

$$B_i = N \cdot (|P| - 1), \text{ and}$$

$$L_i = B_i - V_i.$$

Therefore

$$TL_i = \frac{1}{L_i} \sum_{j \in P \setminus \{i\}} Turns(j, i).$$

Note the use of $Turns(j, i)$ instead of $Turns(i, j)$. $Turns(i, j)$ concerns battles in which individual i wins. Therefore, $Turns(j, i)$ concerns battles in which individual i loses.

4.3   Fitness Vector Comparison

The game's goal is to win; thus, the number of victories an individual achieves is the highest priority. In other words, an individual has higher fitness than another if it wins more times, independently of turns taken to win or lose. If two individuals have the same number of victories, then we compare the number of turns needed to win. Fewer turns-to-win means the individual wins faster. Thus, we minimise turns-to-win. If the turns-to-win metric is the same for both individuals, we compare turns-to-lose. More turns-to-lose means it takes longer for an individual to lose. Thus, we maximise turns-to-lose. In particular, we maximise V_i, and TL_i, while we minimise TW_i.

In addition, we compare individuals' fitness vectors lexicographically, from left to right. Each dimension of the fitness vector is either minimised or maximised independently. For example, we could minimise the first dimension of the fitness vector and maximise the second. Let

$$D = \{n \in \mathbb{N} : 0 \le n < dim(F_i)\}.$$

Then we say

$$M : D \to \{\text{Max, Min}\}$$

represents a mapping of each dimension $d \in D$ of a fitness vector F_i to either the symbol Max or Min, which we use to represent whether we will maximise or minimise that dimension. Therefore, for our experiment, we say

$$M(d) = \begin{cases} Max, & \text{if } d = 0 \\ Min, & \text{if } d = 1 \\ Max, & \text{if } d = 2 \end{cases}$$

We define Algorithm 1, which we use to determine which individual, i or j, performs better according to our fitness vector F described previously. If they have the same fitness, we return $None$.

Algorithm 1. Lexicographic Fitness Comparison

Require: $dim(F_i) = dim(F_j)$
Require: $F_i = (V_i, TW_i, TL_i)$
Require: M := (Max, Min, Max)
  for $0 \leq d < dim(F_i)$ **do**
    if $F_i^d = F_j^d$ **then**
      continue
    else if $M(d)=$ Max **then**
      if $F_i^d > F_j^d$ **then**
        return i
      else
        return j
      end if
    else
      if $F_i^d < F_j^d$ **then**
        return i
      else
        return j
      end if
    end if
  end for
return $None$

4.4 Parameters

We set the population size to 10, which has been used effectively in previous papers [8]. We set the max depth of genomes to 5 and crossover and mutation probabilities to 0.8 and 0.2, respectively. Parents are selected using a five-tournament selection. The stop criterion is 20 generations. Each individual faces every other individual in the population 10 times. Lastly, we run the GP search algorithm 8 times for both the *canMove* and *notCanMove* search spaces. We summarise all parameters in Table 1.

Table 1. Summary of the parameters used during the evolutionary process.

Parameter Name	Value
Population Size	10
Max Depth	5
Crossover Probability	0.8
Mutation Probability	0.2
Number of Generations	20
Tournament Size	5
Elite size	5
Number of Battles Per Fitness Evaluation	10
Number of Runs of the GP Algorithm	8

5   Results

We perform experiments to compare the results of running the GP search algorithm using the *canMove* and *notCanMove* search spaces. We analyse whether reducing the search space results in better-performing agents, even if it results in some move actions either doing nothing or resulting in damage due to a bomb explosion or ring of fire.

Figures 4, 5, and 6 show the distribution of the three dimensions of the fitness vector per generation across all GP runs for both the *canMove* and *notCanMove* search spaces. These graphs are similar to the graphs presented in [8] such that there is high variability in the first generations, stabilising in later generations. Although fitness appears to plateau rapidly, we do not compare fitness across generations due to the reasons discussed in the first paragraph of Sect. 4.2. For example, obtaining ten wins in generation 10 may be more difficult than obtaining ten wins in generation 2 since individuals in generation 10 may perform better than individuals in generation 2.

As previously said, individuals are re-evaluated each generation in an attempt to avoid the presence of sub-optimal individuals that survive by chance but also let higher performing individuals remain in the population as long as they are high performing.

To compare the *canMove* and *notCanMove* versions, we evaluate the best eight individuals of both methods against the best of the other. We do this ten times.

The mean number of wins for the version with *canMove* conditions was 33.87 ± 10.24, and without is 45.85 ± 6.97. The two samples do not follow a normal distribution, as shown by the p-value for the Kolmogorov-Smirnov test being 0.7223. Thus, we run a Kruskall-Wallis test, indicating significant differences (p-value = 0.02). Thus, we show that the *notCanMove* version generally results in better-performing individuals.

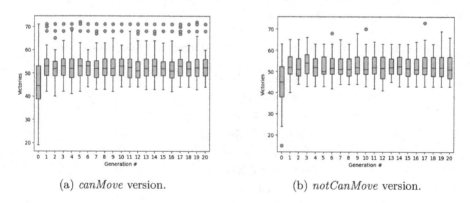

(a) *canMove* version.   (b) *notCanMove* version.

Fig. 4. Box plot of the distribution of the first dimension of the fitness vector (number of victories) for individuals across all runs per generation.

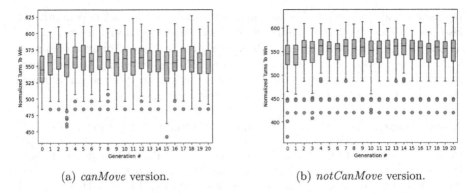

(a) *canMove* version.   (b) *notCanMove* version.

Fig. 5. Box plot of the distribution of the second dimension of the fitness vector (normalised number of turns-to-win in all the battles) for individuals across all runs per generation.

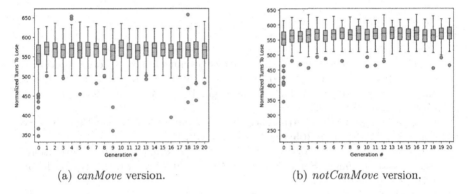

(a) *canMove* version.   (b) *notCanMove* version.

Fig. 6. Box plot of the distribution of the third dimension of the fitness vector (normalised number of turns-to-lose in all battles) for individuals across all runs per generation.

6   Conclusions

This paper describes the use of genetic programming to generate intelligent agents that play the *Bomberman* video game using the open source *Bomber-Land* environment. The algorithm considers the video game's characteristics (all actions and inputs from the environment) and its stochastic nature.

The results show that reducing the search space by removing GP building blocks which check whether an agent can move in a given direction, results in generally better-performing agents. This result is in spite that these GP building blocks prevent an agent from performing a move action when either it is incapable of doing so or will damage the agent.

For future work, we intend to perform experiments in which we mix the best agents from the literature with the GP population and test with varying GP parameters, such as population size, number of generations, and max depth. We also plan to add stochastic components to the generated decision tree, e.g. by adding probabilities of choosing one branch or another instead of fixed distances. This strategy could result in more difficult-to-predict agents and potentially more challenging to defeat. Lastly, it would be interesting to analyse the variability in the behaviour of generated agents using a clustering method.

Acknowledgements. This work is funded by national funds through the FCT - Foundation for Science and Technology, I.P., within the scope of the project CISUC - UID/CEC/00326/2020 and by European Social Fund, through the Regional Operational Program Centro 2020. This work is also supported by the Ministerio español de Economía y Competitividad under project PID2020-115570GB-C22 (DemocratAI::UGR). The second author is funded by Foundation for Science and Technology (FCT), Portugal, under the grant 2022.11314.BD. This work started as a project at the first SPECIES Summer School 2022 (https://species-society.org/summer-school-2022/), for which we would also like to thank the organisers.

References

1. Angeline, P.J., Pollack, J.B.: Competitive environments evolve better solutions for complex tasks. In: Forrest, S. (ed.) Proceedings of the 5th International Conference on Genetic Algorithms, Urbana-Champaign, IL, USA, June 1993, pp. 264–270. Morgan Kaufmann (1993)

2. Dawkins, R., Krebs, J.R.: Arms races between and within species. Proc. Roy. Soc. Lond. Ser. B. Biol. Sci. **205**(1161), 489–511 (1979)

3. Esparcia-Alcázar, A., Moravec, J.: Fitness approximation for bot evolution in genetic programming. Soft. Comput. **17**(8), 1479–1487 (2013). https://doi.org/10.1007/s00500-012-0965-7

4. Fernández-Ares, A., et al.: It's time to stop: a comparison of termination conditions in the evolution of game bots. In: Mora, A.M., Squillero, G. (eds.) EvoApplications 2015. LNCS, vol. 9028, pp. 355–368. Springer, Cham (2015). https://doi.org/10.1007/978-3-319-16549-3_29

5. Forrest, S., Mitchell, M.: Relative building-block fitness and the building block hypothesis. In: Whitley, L.D. (ed.) Proceedings of the Second Workshop on Foundations of Genetic Algorithms. Vail, Colorado, USA, 26–29 July 1992, pp. 109–126. Morgan Kaufmann (1992). https://doi.org/10.1016/b978-0-08-094832-4.50013-1

6. de Freitas, J.M., de Souza, F.R., Bernardino, H.S.: Evolving controllers for Mario AI using grammar-based genetic programming. In: 2018 IEEE Congress on Evolutionary Computation, CEC 2018, Rio de Janeiro, Brazil, 8–13 July 2018, pp. 1–8. IEEE (2018). https://doi.org/10.1109/CEC.2018.8477698

7. García-Sánchez, P., Fernández-Ares, A., Mora, A.M., Castillo, P.A., González, J., Guervós, J.J.M.: Tree depth influence in genetic programming for generation of competitive agents for RTS games. In: Esparcia-Alcázar, A.I., Mora, A.M. (eds.) EvoApplications 2014. LNCS, vol. 8602, pp. 411–421. Springer, Heidelberg (2014). https://doi.org/10.1007/978-3-662-45523-4_34

8. García-Sánchez, P., Tonda, A.P., Leiva, A.J.F., Cotta, C.: Optimizing hearthstone agents using an evolutionary algorithm. Knowl. Based Syst. **188** (2020). https://doi.org/10.1016/j.knosys.2019.105032

9. García-Sánchez, P., Tonda, A.P., Mora, A.M., Squillero, G., Guervós, J.J.M.: Towards automatic starcraft strategy generation using genetic programming. In: 2015 IEEE Conference on Computational Intelligence and Games, CIG 2015, Tainan, Taiwan, 31 August–2 September 2015, pp. 284–291. IEEE (2015). https://doi.org/10.1109/CIG.2015.7317940

10. Gaudesi, M., Piccolo, E., Squillero, G., Tonda, A.P.: TURAN: evolving non-deterministic players for the iterated prisoner's dilemma. In: Proceedings of the IEEE Congress on Evolutionary Computation, CEC 2014, Beijing, China, 6–11 July 2014, pp. 21–27. IEEE (2014). https://doi.org/10.1109/CEC.2014.6900564

11. Gold, R., Branquinho, H., García-Sánchez, P.: BomberLand-Evostar-2023 (2023). https://github.com/18goldr/BomberLand-Evostar-2023

12. Gold, R., Grant, A.H., Hemberg, E., Gunaratne, C., O'Reilly, U.M.: GUI-based, efficient genetic programming for unity3D. In: Proceedings of the Genetic and Evolutionary Computation Conference Companion. GECCO 2022, pp. 2310–2313. Association for Computing Machinery, New York (2022). https://doi.org/10.1145/3520304.3534022

13. Gold, R., Grant, A.H., Hemberg, E., Gunaratne, C., O'Reilly, U.M.: GUI-based, efficient genetic programming and AI planning For Unity3D. In: Trujillo, L., Winkler, S.M., Silva, S., Banzhaf, W. (eds.) Genetic Programming Theory and Practice XIX. Genetic and Evolutionary Computation. Springer, Singapore (2023). https://doi.org/10.1007/978-981-19-8460-0_3

14. Gold, R., Haydn Grant, A., Hemberg, E., Gunaratne, C., O'Reilly, U.M.: STGP-Sharp (2023). https://github.com/ALFA-group/STGP-Sharp

15. Katoch, S., Chauhan, S.S., Kumar, V.: A review on genetic algorithm: past, present, and future. Multim. Tools Appl. **80**(5), 8091–8126 (2021). https://doi.org/10.1007/s11042-020-10139-6

16. Potter, M.A., De Jong, K.A.: A cooperative coevolutionary approach to function optimization. In: Davidor, Y., Schwefel, H.-P., Männer, R. (eds.) PPSN 1994. LNCS, vol. 866, pp. 249–257. Springer, Heidelberg (1994). https://doi.org/10.1007/3-540-58484-6_269

17. Smith, G., Avery, P., Houmanfar, R., Louis, S.J.: Using co-evolved RTS opponents to teach spatial tactics. In: Yannakakis, G.N., Togelius, J. (eds.) Proceedings of the 2010 IEEE Conference on Computational Intelligence and Games, CIG 2010, Copenhagen, Denmark, 18–21 August, 2010, pp. 146–153. IEEE (2010). https://doi.org/10.1109/ITW.2010.5593359

18. Yannakakis, G.N., Togelius, J.: Artificial Intelligence and Games. Springer, Cham (2018). https://doi.org/10.1007/978-3-319-63519-4

Surrogate-Assisted Evolutionary Optimisation

A Surrogate Function in Cellular GA for the Traffic Light Scheduling Problem

Andrea Villagra[1(✉)] and Gabriel Luque[2]

[1] Universidad Nacional de la Patagonia Austral, Caleta Olivia, Argentina
avillagra@uaco.unpa.edu.ar
[2] ITIS Software, Universidad de Málaga, Málaga, Spain
gluque@uma.es

Abstract. The traffic light scheduling problem is undoubtedly one of the most critical problems in a modern traffic management system. Appropriate traffic light planning can improve traffic flows, reduce vehicles' emissions, and provide benefits for the whole city. Metaheuristics, notably the Cellular Genetic Algorithm (cGA), offer an alternative way of solving this optimization problem by providing "good solutions" to adjust the traffic lights to mitigate traffic congestion. However, one of the unresolved issues is these methods use very time-consuming operations. Specifically, the evaluation is a complex process since a simulator should be executed to get the quality of the solutions. In this work, we focus on this topic and propose using an artificial neural network (as a surrogate system) to tackle this problem. Our experiments show very promising results since our proposal can significantly reduce the execution time while maintaining (and even, in some scenarios, improving) the quality of the solutions.

Keywords: Genetic algorithms · Surrogate systems · Traffic light scheduling problem

1 Introduction

With the increase in urbanization, traffic congestion is a serious issue [6,15,34]. This domain is becoming more important since traffic congestion caused by the increasing number of people imposes extra costs, which limits the city's development and causes more emissions and health problems. There is an urgent need to develop an intelligent approach to address traffic congestion at signalized road intersections.

Traffic lights are straightforward and essential elements used in urban environments to organize transit, mainly vehicles [25], to avoid accidents and to improve the traffic flow. Researchers in transport systems and traffic control have

This research is partially funded by the Universidad Nacional de la Patagonia Austral, the Universidad de Málaga, and the project PID 2020-116727RB-I00 (HUmove) funded by MCIN/AEI/10.13039/501100011033; and TAILOR ICT-48 Network (No. 952215) funded by EU Horizon 2020 research and innovation programme.

J. Correia et al. (Eds.): EvoApplications 2023, LNCS 13989, pp. 783–797, 2023.
https://doi.org/10.1007/978-3-031-30229-9_50

suggested that autonomous vehicles are the future of transportation [17,23,31]. Despite this might be true for some developed countries, a fully autonomous vehicle is still far from being achieved for those who are on their way to development. Governments and transport researchers aim to find solutions to traffic congestion problems. Their main objective is to eliminate traffic congestion. Although many efforts have been made and different strategies have been implemented to reduce traffic in big cities, congestion problems still exist, especially in developing countries.

The problem of area traffic control is one of the most challenging problems in traffic flow control and requires the use of advanced techniques to solve even the smallest cases. In contrast to intelligent traffic light control techniques, which require new infrastructure, location of sensors, and modifications to already structured civil works, artificial intelligence techniques in optimizing traffic light cycles are presented as a viable, fast, efficient, and more economical tool.

Current research efforts in traffic signal control methods can be basically categorized into two classes, i.e. fixed time methods [8,12] and traffic responsive methods [35,36]. The former aims to find suitable fixed-time signal plans based on historical traffic demand, while the latter dynamically adjusts the signal state according to the traffic information detected in real-time. Although traffic-responsive methods are technically sound, their performance depends heavily on real-time sensor systems and they are generally difficult to apply to the whole city owing to the high operational cost [22,33]. Besides, the majority of traffic lights in real-world work under fixed signal timing plans and the traffic flows tend to repeat similar patterns like morning and evening peaks. Therefore, developing efficient fixed–time traffic signal optimization methods is of great practical significance.

In this work, we focus on the control of traffic lights, using the cellular genetic algorithm (cGA) [1] that has had outstanding results in the past [32], achieving not only a reduction in the travel time of vehicles but also a reduction in the gases emitted compared to other well-known population metaheuristics such as Particle Swarm Optimization (PSO) and Difference Evolution (DE). One of the most critical problems in this approach is the execution time. This technique relies on utilizing a very time-consuming microscopic simulator to get the quality of the solution. Therefore, this work aims at reducing the computational effort of this algorithm by combining it with an artificial neural network (ANN) used as a surrogate function without lowering the quality of the solutions.

The rest of this paper is organized as follows. Section 2 presents some previous studies in this domain. Later, we describe the problem and our approach (Sect. 3). The experiments and their results are discussed in Sect. 4. Finally, in Sect. 5, we conclude by summarizing our main achievements and by giving some hints about the following analyses which can extend this study.

2   Related Work

There is no doubt that a good traffic light control plan can improve traffic performance. Thus, finding a suitable solution for this problem has been actively investigated for years. Metaheuristics approaches are among the most widely used techniques in this domain. In [19], the authors presented a hybrid solution algorithm for optimization of the traffic light control problem based on Simulated Annealing (SA) and Genetic Algorithm (GA). In [7], they applied a discrete harmony search algorithm to the scheduling traffic light problem, whereas in [9], five metaheuristics were implemented. The authors in [5,11,29] developed GA to optimize this problem setting of the respective networks and objective functions. A Bee Colony algorithm is used in [13] in under-saturated and over-saturated traffic conditions. A recent study [28] provides a comprehensive survey that discusses different ways of using swarm intelligence and evolutionary algorithms to solve different kinds of traffic light problems.

However, most of these approaches present the same issue: they require a very long execution time to get a suitable scheduling plan for the traffic lights. These long runs are mainly due to the necessity of using a simulator and other very time-consuming processes. Several approaches have been proposed in the literature to alleviate these problems, such as using parallel methods [2,27], reducing the number of evaluations by analyzing the properties of the search space [18], among others.

One of the most promising approaches to reducing fitness evaluation times is using surrogate systems [4]. These generally replace the simulator with a machine-learning technique. Artificial Neural Networks (ANNs) are widespread approaches that act as surrogate systems due to their ability to approximate any function [21,26]. To the best of our knowledge, this kind of approach has never been applied to the problem solved in this work.

3   Algorithmic Proposal

This section groups the definition of the problem to be solved, the characteristics of the basic algorithm used to solve the problem, and finally the representation of the solutions.

3.1   Problem Definition

The problem of traffic light control has been investigated in different studies for years [3,10,20,25]. This problem is, typically aimed at optimizing several objectives for a particular region in a specific duration at a time, such as maximizing the number of vehicles arriving at their destination within a specific period, maximizing the average velocity of all the vehicles in the region, minimizing the waiting time of the vehicles at intersections, and minimizing the length of the waiting queue, just to name a few. To mathematically describe the study problem, the representation used by [10] is employed. This problem can be seen as a multi-objective problem. It aims to maximize the number of vehicles that arrive at their destination (NV_D), or at minimizing the number of vehicles that do not arrive at their destination (NV_{ND}), during the simulation time (T_S); minimize the total vehicle travel time (TT_v) from arrival at the simulation to arrive at the destination; minimizing the stop and waiting for time of all vehicles (TT_{EP}); maximize the radius P of green and red colours in each phase state of all intersections, defined as (Eq. 1):

$$P = \sum_{i=0}^{inter} \sum_{j=0}^{fs} d_{i,j} \frac{g_{i,j}}{r_{i,j}}, \tag{1}$$

in which *inter* denotes the number of all intersections, *fs* represents the number of phases in each intersection, and $g_{i,j}$ and $r_{i,j}$ are the number of green and red signal colors, respectively, at intersection i and at phase state j, with duration $d_{i,j}$.

Then, all objectives are combined into a single one as follows:

$$F = \frac{TT_v + TT_{EP} + NV_{ND} \times T_S}{NV_D^2 + P}, \tag{2}$$

The problem consists of the minimization of Eq. 2. This aggregation function off all objectives has been used in different articles with promising results [10, 24, 32].

3.2   Characterizing Cellular Genetic Algorithms

The cGA is a particular class of Evolutionary Algorithm (EA), and a subclass of Genetic Algorithms with a spatially structured population, i.e. individuals in the population can be paired with their neighbours [1]. These small overlapping neighbourhoods help to explore the search space due to slow diffusion of solutions allowing for exploration and exploitation to take place in each neighbourhood through the genetic operators. Algorithm 1 presents the pseudo-code of a cGA.

In a cGA, the population is usually structured on a 2-dimensional toroidal grid (gridcGA). As a family of a GA, the cGA encodes the decision variables of a search problem into finite-length strings of variables of some alphabet of specific cardinality. The strings are candidate solutions and are called "chromosomes". In the same way as in a GA, each of the variables that form the chromosome is called "gene" and "allele" to the different values the genes can take. Once the problem to be solved has been encoded through one or several chromosomes (also called "individuals") and the fitness function has been defined, the solutions to the problem are evolved, i.e. the population of solutions, taking into account the following steps: Initialisation (line 1), selection (lines 5–7), recombination (line 8), mutation (line 9), evaluation (line 10) and replacement (line 11).

Algorithm 1. Pseudo-code of a cGA

```
 1: GeneratePop(gridcGA);
 2: for s ← 1 to MAX_STEPS do
 3:    for x ← 1 to WIDTH do
 4:       for y ← to HEIGHT do
 5:          nList ← ComputeNeigh (gridcGA,individual(x,y));
 6:          parent1 ← individual(x,y);
 7:          parent2 ← Selection(nList);
 8:          offspring ← Recombination (gridcGA, Pc, parent1, parent2);
 9:          offspring ← Mutation(gridcGA, Pm, offspring)
10:          Evaluation(offspring);
11:          Add(gridcGA, individual(x,y), offspring);
12:       end for
13:    end for
14: end for
```

3.3   Representation and Solution Strategy

Within the semaphore cycle, a possible solution to the problem is represented by a vector of positive integers, where each integer represents the duration (in seconds) of each semaphore phase of each intersection. An instance of an urban scenario in SUMO [16] (an open-source, highly portable, microscopic road traffic simulation package designed to handle large road networks) is composed of: intersections, traffic lights, roads, and vehicles moving along their previously specified routes. Traffic lights are located at intersections and control vehicles' flow by following their color signals and cycle duration programs. In this context, all traffic lights located at the same intersection are governed by a common program (*tl-logic*), since they must be synchronized for traffic safety. In addition, to avoid vehicle collisions and accidents for all traffic signals at an intersection, the combination of colour signals during a cycle period always remains valid [18] and must follow the specific traffic rules for intersections. Figure 1 illustrates the elements that form a traffic light cycle program. The *tl-logic* with id = "i" corresponds to an intersection of a SUMO map. The *tl-logic* contains four phases with a duration of "50", "4", "13", and "4" s. The phase surrounded by the ellipse is the first phase and it contains the state $GrGr$ meaning that two traffic lights are green (G) while the other two are red (r) for 50s. The following phase changes the state of the traffic lights $yryr$ (y is yellow) for 4s, and so on. When the last phase is executed, the cycle starts all over again. This is a repetitive process until the end of the simulation. A vector of candidate solutions

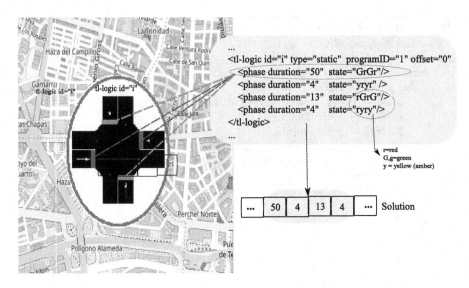

Fig. 1. Málaga intersection and phase duration (cycles) encoded inside a tentative solution [32]

(semaphore program) is sent to SUMO, which calculates its target value after a simulation procedure.

Regarding the representation, as mentioned above, according to SUMO speci-fication for cycle scheduling, the possible solutions are encoded as integer vectors. Each vector element represents a phase duration of the traffic lights involved at a given intersection.

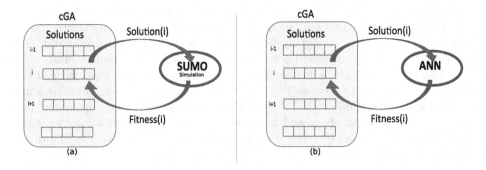

Fig. 2. Optimisation strategy (a) using SUMO (b) using ANN

The optimisation strategy used is the following (see Fig. 2(a)). When the cGA generates a new solution, it is immediately used to update the cycle program. SUMO is then run to simulate an instance scenario with streets, addresses, obstacles, traffic lights, vehicles, routes, etc., with the new state of the cycle program. After the simulation, it returns the global information needed to calculate the quality of the solution (fitness). In addition, it also returns information for the following gases CO_2 (carbon dioxide), CO (carbon monoxide), NO_x (nitrogen oxides), PM_x (particulate matter), and HC (hydrocarbons).

3.4 Surrogate Model

To reduce the computational effort of cGA, an ANN is used as a surrogate fitness function (see Fig. 2(b)), i.e., instead of evaluating the solution by doing the simulation with SUMO, a trained artificial neural network predicts the fitness value. This ANN receives a solution obtained by cGA and returns the corresponding fitness function (without evaluating the solution with SUMO).

ANN is a system derived from our biological nervous system. The purpose of ANN is to make calculations between input and output values. There are two basic operations in ANN. One of them is the training process. The other is the test. If the objective is that the ANN establishes a connection between the input and the output, a well-defined training data set must first pass the method. The training data set should include the outputs expected from the ANN to respond. There must be data on all possible conditions in the training data set. The stronger the training data set, the stronger the output of the ANN will be. ANN consists of layers and the neurons located in these layers, including the entrance, exit, and hidden layers. The ANN model starts the training and testing process with input values. Input values are pre-treated with weight ratios. Then, an ANN output is produced with the collection and activation function. The effectiveness of ANN directly depends on its type, internal structure, and problem characteristics. In the next section, these characteristics are described.

4   Experiments and Results

This section is divided into two subsections. The first subsection presents the dataset, the parameter settings used, and the development environment. The second section describes the experiments and analysis of results.

4.1 Datasets, Parameter Settings, and Environment

This section presents the characteristics of the instances used (datasets), the setting of the cGA parameters, the construction of the ANN model, details of its training, and finally, the design of the experiments.

In order to analyse the performance of cGA with ANN, two problem instances located in two large metropolitan areas (Málaga and Paris) are used. The scenarios considered were created by extracting information from real digital maps.

The Málaga instance has 190 traffic lights and covers an area of $2.55 \, \text{km}^2$; meanwhile, the Paris instance has 378 traffic lights and covers $5.5 \, \text{km}^2$. The traffic flow was set at 1,200 vehicles in both scenarios.

Regarding the simulation procedure, each vehicle started its own trip from a starting point to a destination. The speed limit for the vehicles is $50 \, \text{km/h}$, which is the typical maximum value in urban areas. The simulation time was set to 2,200 s for the Málaga instance, and 3,400 s for the Paris instance, as it consisted of a larger number of traffic lights. These times have been chosen to gather realistic traffic information in a reasonable execution time. The simulation was conducted by executing the traffic simulator SUMO release 0.31.0. for Linux.

In [32], the authors applied cGA to the same set of instances that are used in this paper. It is essential to clarify that the aim of this article is not to analyse the cGA algorithm but to analyse the behavior of the surrogated function. For this reason, we use their setting and is the following: (a) the population is composed of 100 solutions (individuals); (b) for the recombination, one of the parents is the current individual considered in the loop of cGA, and the other one is obtained by using Binary Tournament Selection in its neighborhood; (c) the DPX1 operator (the one with the largest part form the best parent) is always applied (probability of recombination, $Pc = 1.0$); (d) integer mutation with a probability of mutation (Pm) is set to $1/L$ where L = length of the solution. Finally, (e) the replacement criterion is "if not worse". The maximum number of evaluations was set to 30,000 function evaluations overall.

The model and training of the ANN were implemented in Python 3.9.7 using the KerasRegressor functionality. Keras is a neural network library for Python [14] capable of using the TensorFlow and Theano backend. We used a dataset formed by 30,000 elements (for each instance) obtained from results of cGA consisting of the time for each light (190 or 378) and the fitness value. This dataset was divided into 80% elements for training and 20% for testing. We know that the quality of our learning model is directly proportional to the quality of the data. For this reason, to avoid over-fitting or under-fitting, in both data sets (training and testing) we performed stratified cuts of the data to have in both sets representatives of each stratum looking for a balance between bias and variance.

Based on the outcome of several experiments of our dataset, and executing an exhaustive search over hyper-parameters by applying scikit-learn's Grid-SearchCV on the training set. GridSearchCV allows systemic evaluation and selection of model parameters. By specifying a model and the parameters to be tested, you can evaluate the performance of the former in terms of the latter by means of cross-validation. GreenSerchCV was applied to a representative set (20%) of the 30,000-element dataset, using StratifiedShuffleSplit. This is a technique that mixes Stratified Kfold + Shuffle Split in order to obtain a frequency as proportional as possible of all the classes we have in the dataset and to be able to apply to this resulting subset the exhaustive search of hyperparameters: Grid Search with the aim of reducing the computational time required to obtain the best parameterization. It is appropriate to point out that the computation for this type of exhaustive method grows exponentially as the number of data increases.

We constructed the following ANN model for Málaga. This was done with the optimum values formed by an input layer with 190 neurons, by an output layer with one neuron, by three hidden layers with 190, 90 and 45 neurons, respectively, by a number of epochs in 150, and by a batch size in 100. The optimizer was set to "adam", the activation function to "Sigmoid", and mean squared error (MSE) was used as the loss function. The model of ANN for Paris has the same structure and values as Málaga except for the number of neurons for the input and hidden layers. In this case, the input layer was set to 378 neurons, and the hidden layers to 378, 189, and 94 neurons, respectively. The trained models were used in cGA as a surrogate function.

All the algorithms were implemented in Java and each execution was run on a machine with an Intel Xeon Gold 6240R with two processors, 48 cores at 2.40 GHz and 220 GB. The cluster was managed by HTCondor 8.2.7 [30], which allowed us to perform parallel independent executions to reduce the overall experimentation time. We performed 30 independent runs for all algorithms.

Throughout the paper, the best values have been marked in bold. In all the experiments, we performed ANOVA or Kruskal Wallis tests, depending on the normality distributions, to assess their statistical significance (p-value is set to 0.05) using RStudio 1.1.463. Statistically significant differences between the algorithms are shown with the symbol ($\bullet$), while non-significant differences are shown with ($-$).

To analyze the performance of the cGA with a surrogate function, we defined different ways to apply the ANN model. The first one, called cGA-s1, applies the ANN in 100% of the evaluations, the second one, called cGA-s2, applies the ANN in the first 50% of the evaluations, and the third one, called cGA-s3, uses SUMO in the first 50% of evaluations and for the rest the ANN model. We also compared these new approaches against the original cGA, which only uses SUMO.

4.2 Results and Analysis

To analyze the performance of the algorithms using the surrogate function, we paid attention to the quality of the found solutions and the work-clock time to reach these solutions. The reported solution quality is always calculated using the SUMO simulator to achieve a fair comparison among the different approaches (using or not the simulator). As we commented before, in this work we also report the statistical comparison among the algorithms. This result is shown in A|K-W column in the tables where A stands for an ANOVA test while K-W represents Kruskal-Wallis test.

Table 1 displays mean best values and mean time obtained by the original cGA and our versions of cGA with surrogate function (cGA-s1, cGA-s2 and cGA-s3) for the Málaga instance. From the solution quality point of view, we can observe that all the approaches using surrogate models outperform the cGA only applying SUMO. In fact, the best technique for this scenario is the algorithm that does not use SUMO at all: cGA-s1. Our hypothesis about this behaviour is due to the use of approximate values instead of the exact ones offered by the simulator. Our surrogate model obtained very accurate results (the MSE in the test dataset was 0.001). Still, there are minimal differences with the actual value, and these discrepancies can help the algorithm to better explore the search space for this scenario. Of course, we should also consider the training of our surrogate system, but the training time is very small (around 690 s), and it is negligible compared to the cGA times.

Regarding the running time, as expected, the surrogate model reduces the execution time of the technique in a very significant way. The most accurate model, our proposal cGA-s1, reduces the time by 99.5% when compared with that needed for the original cGA using SUMO. According to our statistical analysis, both differences (in quality and time) are significant.

Table 1. Mean values obtained for Málaga instance

	Málaga				
	cGA	cGA-s1	cGA-s2	cGA-s3	A|K-W
Fitness value	0.4308	**0.4065**	0.4180	0.4181	(•)
Time (s)	162,009	**862**	61,707	62,619	(•)

Figure 3 exhibits the mean values (fitness value and time) obtained by the algorithms. These results clearly show the difference between the results obtained by our cGA with the surrogate function with respect to cGA. They reached lower values for the fitness and for the time needed to reach them in Málaga instance.

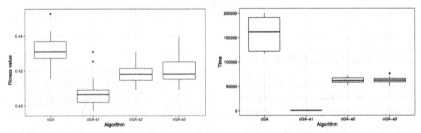

(a) Mean fitness values obtained by the algorithms

(b) Mean time in seconds (s) needed by the algorithms to reach their best fitness values

Fig. 3. Box-plots of the mean best values (fitness and time) obtained by the algorithms for Málaga instance

Table 2 shows the same results as Table 1, but for the Paris instance. For this scenario, and similarly to the previous one, our approaches using a surrogate system significantly reduce the execution time compared to the original cGA. However, in this case, the quality of the solutions found by our techniques is slightly worse. There are, nonetheless, several options to improve the quality of our approaches. Firstly, we are currently using the ANN with the same configuration as in the first scenario (the Málaga one). We can configure the surrogate system specifically for the Paris instance. Secondly, since our approaches are very fast, we can run the technique for a larger number of steps. This can help improve the quality of the solutions, and the resulting method would continue to be faster than the original cGA. We will analyze these possibilities in future studies.

Table 2. Mean values obtained by cGA, cGA-s1, cGA-s2, and cGA-s3 for Paris instance

	Paris				
	cGA	cGA-s1	cGA-s2	cGA-s3	A|K-W
Fitness value	**0.6844**	0.7151	0.7217	0.7166	(•)
Time (s)	287,924	**722**	124,015	111,362	(•)

In Fig. 4(a) we can see that cGA finds the lowest fitness value, but we also see that the time required by our approaches (Fig. 4(b)) is up to 99.75% lower than the one needed for the original cGA.

In our previous analyses, we do not consider the effect of the training time of the model, which is approximately 690 s for Málaga and 880 s for Paris. However, it's important to note that the improvement in time is reported for every execution of the cGA. A cGA requires many executions for tuning the parameters or gathering statistical data since it's a non-deterministic process and several runs are mandatory. Since the training process is only executed once, the larger the number of runs of cGA, the smaller the effect of training time. In our experiments, we ran each algorithm 30 times. An approximation time of the whole experiments using a variant is the training time (if it is used) plus 30 times the execution time. The time for training the models is negligible with respect to the running time of most of the variants (a few minutes against tens of hours). This time is only significant in cGA-s1 variants, but the difference concerning the other versions is vast, even adding this training time to the running time.

Figure 5 shows the evolution of one execution of each algorithm. Similar to previous studies, we compare the actual fitness (calculated with the simulator) to provide a fair comparison. Several conclusions can be drawn from these figures. First, the approach using only the surrogate model cGA-s1 (the green line) provides a rapid convergence, but it can also make small improvements in further iterations. These small improvements can be seen at iteration 220 in the Málaga scenario and around iteration number 280 in Paris. In this latter scenario, a larger number of iterations could improve the results. In contrast, the original model using only the SUMO simulator, cGA (the red line), provides a slower convergence that seems more appropriate for the Paris scenario. While

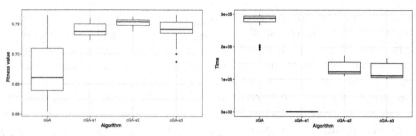

(a) Mean fitness values obtained by the algorithms

(b) Mean time in seconds (s) needed by the algorithms to reach their best fitness values

Fig. 4. Box-plots of the mean best values (fitness and time) obtained by the algorithms for Paris instance

in the Paris scenario it already converged to a local optimal (flat line in the last iterations), more improvements are expected in the Málaga instance with a larger number of iterations. However, since the evaluation using SUMO simulator is a very time-consuming process, larger executions could not possible. The alternation between SUMO and the surrogate model does not seem to provide any advantage. We can observe that fitness does not improve when the way of evaluating the solutions is changed (in generation 150).

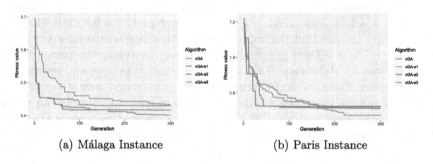

(a) Málaga Instance

(b) Paris Instance

Fig. 5. Evolution of mean fitness value for all algorithms

5   Conclusions

This work has focused on using a surrogate function to decrease the computational time of the cGA to find the optimal value. This metaheuristic has been taken as a basis, which has had satisfactory results in previous works. In the original cGA, the SUMO simulator was used to evaluate the solution. The simulator provides a very realistic evaluation of the proposed traffic light planning, but this is at the expense of an enormous computational cost. We propose to replace the use of the simulator with a surrogate model. For the surrogate function, we have trained an artificial neural network and used it to predict the value of the evaluation function instead of using SUMO. We applied this surrogate model for all the evaluations (cGA-s1) or 50% of the evaluations (cGA-s2 and cGA-s3) in two real-world instances (Málaga and Paris).

With the experiments carried out, we have verified that using a surrogate function reduces the computational effort (in time) to obtain the best fitness value for the instances analyzed. We have reduced the execution time between 60% in the cGA-s2 and cGA-s3 (only partially using the surrogate model) and 99% in the cGA-s1 approach (which does not use the simulator). This indicates that we have entirely fulfilled the primary goal of this work. It is also important to highlight that the work clock time reduction is achieved by finding solutions with similar quality to the original cGA in the Paris instance, and by improving the results in the Málaga scenario.

In future work, there are several issues we intend to address, such as using other surrogate functions or optimization strategies like the Bayesian one and, also, investigating other ways of combining the SUMO evaluations with surrogate ones. Furthermore, with regard to the reduction of initial parameters, it would be interesting to apply a more complex ETL and feature selection processes in order to reduce the search space and to analyze the behaviour of the algorithm. In addition, to use a pre-trained dynamically fitting artificial neural network model that will allow us to test new techniques and more experiments (other routes). Also in terms of dataset generation, study different dataset sizes and different ways of generating solutions (random, Latin Hypercube sampling, among others). Finally, define the objective function as a multi-objective function.

References

1. Alba, E., Dorronsoro, B.: Cellular genetic algorithms, vol. 42. Springer Science & Business Media (2009)
2. Alba, E., Luque, G., Nesmachnow, S.: Parallel metaheuristics: recent advances and new trends. Int. Trans. Oper. Res. **20**(1), 1–48 (2013)
3. Araghi, S., Khosravi, A., Creighton, D.: A review on computational intelligence methods for controlling traffic signal timing. Expert Syst. Appl. **42**(3), 1538–1550 (2015)
4. Bhosekar, A., Ierapetritou, M.: Advances in surrogate based modeling, feasibility analysis, and optimization: a review. Comput. Chem. Eng. **108**, 250–267 (2018)
5. Bie, Y., Cheng, S., Liu, Z.: Optimization of signal-timing parameters for the intersection with hook turns. Transport **32**(2), 233–241 (2017)
6. Drop, N., Garlińska, D.: Evaluation of intelligent transport systems used in urban agglomerations and intercity roads by professional truck drivers. Sustainability **13**(5), 2935 (2021)
7. Gao, K., Zhang, Y., Sadollah, A., Su, R.: Optimizing urban traffic light scheduling problem using harmony search with ensemble of local search. Appl. Soft Comput. **48**, 359–372 (2016)

8. Gao, K., Zhang, Y., Su, R., Yang, F., Suganthan, P.N., Zhou, M.: Solving traffic signal scheduling problems in heterogeneous traffic network by using meta-heuristics. IEEE Trans. Intell. Transp. Syst. **20**(9), 3272–3282 (2018)
9. Gao, Y., Liu, Y., Hu, H., Ge, Y.: Signal optimization for an isolated intersection with illegal permissive left-turning movement. Transportmetrica B: transport dynamics (2018)
10. Garcia-Nieto, J., Olivera, A.C., Alba, E.: Optimal cycle program of traffic lights with particle swarm optimization. IEEE Trans. Evol. Comput. **17**(6), 823–839 (2013)
11. Guo, J., Kong, Y., Li, Z., Huang, W., Cao, J., Wei, Y.: A model and genetic algorithm for area-wide intersection signal optimization under user equilibrium traffic. Math. Comput. Simul. **155**, 92–104 (2019)
12. Hu, W., Wang, H., Qiu, Z., Nie, C., Yan, L.: A quantum particle swarm optimization driven urban traffic light scheduling model. Neural Comput. Appl. **29**(3), 901–911 (2018)
13. Jovanović, A., Teodorović, D.: Pre-timed control for an under-saturated and over-saturated isolated intersection: a bee colony optimization approach. Transp. Plan. Technol. **40**(5), 556–576 (2017)
14. Keras, C.F.: Github repository. http://github.com/fchollet/keras. Accessed 2020-04-01 (2015)
15. Khan, M.U., Saeed, S., Nehdi, M.L., Rehan, R.: Macroscopic traffic-flow modelling based on gap-filling behavior of heterogeneous traffic. Appl. Sci. **11**(9), 4278 (2021)
16. Krajzewicz, D., Bonert, M., Wagner, P.: The open source traffic simulation package sumo. RoboCup 2006 (2006)
17. Kuutti, S., Fallah, S., Katsaros, K., Dianati, M., Mccullough, F., Mouzakitis, A.: A survey of the state-of-the-art localization techniques and their potentials for autonomous vehicle applications. IEEE Internet Things J. **5**(2), 829–846 (2018)
18. Lee, E.H., Eriksson, D., Perrone, V., Seeger, M.: A nonmyopic approach to cost-constrained bayesian optimization. In: Uncertainty in Artificial Intelligence, pp. 568–577. PMLR (2021)
19. Li, Z., Schonfeld, P.: Hybrid simulated annealing and genetic algorithm for optimizing arterial signal timings under oversaturated traffic conditions. J. Adv. Transp. **49**(1), 153–170 (2015)
20. Liu, Z.: A survey of intelligence methods in urban traffic signal control. IJCSNS Int. J. Comput. Sci. Network Secur. **7**(7), 105–112 (2007)
21. Miriyala, S.S., Subramanian, V.R., Mitra, K.: Transform-ann for online optimization of complex industrial processes: casting process as case study. Eur. J. Oper. Res. **264**(1), 294–309 (2018)
22. Mousavi, S.S., Schukat, M., Howley, E.: Traffic light control using deep policy-gradient and value-function-based reinforcement learning. IET Intel. Transport Syst. **11**(7), 417–423 (2017)
23. Olayode, I., Tartibu, L., Okwu, M., Uchechi, U.: Intelligent transportation systems, un-signalized road intersections and traffic congestion in johannesburg: a systematic review. Procedia CIRP **91**, 844–850 (2020)
24. Olivera, A.C., García-Nieto, J.M., Alba, E.: Reducing vehicle emissions and fuel consumption in the city by using particle swarm optimization. Appl. Intell. **42**(3), 389–405 (2015)
25. Papageorgiou, M., Diakaki, C., Dinopoulou, V., Kotsialos, A., Wang, Y.: Review of road traffic control strategies. Proc. IEEE **91**(12), 2043–2067 (2003)
26. Samarasinghe, S.: Neural networks for applied sciences and engineering: from fundamentals to complex pattern recognition. Auerbach publications (2016)

27. Segredo, E., Luque, G., Segura, C., Alba, E.: Optimising real-world traffic cycle programs by using evolutionary computation. IEEE Access **7**, 43915–43932 (2019)
28. Shaikh, P.W., El-Abd, M., Khanafer, M., Gao, K.: A review on swarm intelligence and evolutionary algorithms for solving the traffic signal control problem. IEEE Trans. Intell. Transp. Syst. **23**(1), 48–63 (2020)
29. Tan, M.K., Chuo, H.S.E., Chin, R.K.Y., Yeo, K.B., Teo, K.T.K.: Optimization of traffic network signal timing using decentralized genetic algorithm. In: 2017 IEEE 2nd International Conference on Automatic Control and Intelligent Systems (I2CACIS). pp. 62–67. IEEE (2017)
30. Thain, D., Tannenbaum, T., Livny, M.: Distributed computing in practice: the condor experience. Concurrency and computation: practice and experience **17**(2–4), 323–356 (2005)
31. Van Brummelen, J., O'Brien, M., Gruyer, D., Najjaran, H.: Autonomous vehicle perception: The technology of today and tomorrow. Transportation research part C: emerging technologies **89**, 384–406 (2018)
32. Villagra, A., Alba, E., Luque, G.: A better understanding on traffic light scheduling: New cellular gas and new in-depth analysis of solutions. Journal of Computational Science **41**, 101085 (2020)
33. Wei, H., Zheng, G., Yao, H., Li, Z.: Intellilight: A reinforcement learning approach for intelligent traffic light control. In: Proceedings of the 24th ACM SIGKDD International Conference on Knowledge Discovery & Data Mining. pp. 2496–2505 (2018)
34. Xu, H., Zhuo, Z., Chen, J., Fang, X.: Traffic signal coordination control along oversaturated two-way arterials. PeerJ Computer Science **6**, e319 (2020)
35. Yang, S., Yang, B., Wong, H.S., Kang, Z.: Cooperative traffic signal control using multi-step return and off-policy asynchronous advantage actor-critic graph algorithm. Knowl.-Based Syst. **183**, 104855 (2019)
36. Yao, Z., Shen, L., Liu, R., Jiang, Y., Yang, X.: A dynamic predictive traffic signal control framework in a cross-sectional vehicle infrastructure integration environment. IEEE Trans. Intell. Transp. Syst. **21**(4), 1455–1466 (2019)

Surrogate-Assisted (1 + 1)-CMA-ES with Switching Mechanism of Utility Functions

Yutaro Yamada[1(✉)], Kento Uchida[1], Shota Saito[1,2],
and Shinichi Shirakawa[1]

[1] Graduate School of Environment and Information Sciences,
Yokohama National University, Kanagawa, Japan
{yamada-yutaro-dw,uchida-kento-nc,saito-shota-bt}@ynu.jp
shirakawa-shinichi-bg@ynu.ac.jp
[2] SkillUp AI Co. Ltd., Tokyo, Japan

Abstract. The invariance to any monotonically increasing transformation of the objective function is an essential property of the covariance matrix adaptation evolution strategy (CMA-ES). However, the surrogate-assisted CMA-ES often loses this invariance because the performance of the surrogate model is influenced by such transformation. In this paper, we propose a surrogate-assisted (1 + 1)-CMA-ES with the Gaussian process regression (GPR) possessing the robustness for such transformation. We introduce two utility functions based on the ranking and Lebesgue measure of candidate solutions, which are invariant to such transformation. We train GPR with the utility function values instead of the objective function values to estimate the ranking of the candidate solutions. Moreover, we propose switching Gaussian process CMA-ES (SGP-CMA-ES), which contains a mechanism selecting the suitable target variable of GPR from utility function values in addition to the objective function values to improve the performance on the objective functions that GPR can estimate easily. The experimental results show that SGP-CMA-ES is superior to existing methods on the objective function with several transformations, while maintaining the performance comparable to that of existing methods on the objective function without transformation.

Keywords: Covariance Matrix Adaptation Evolution Strategy ·
Surrogate Model · Gaussian Process Regression · Invariance

1 Introduction

The covariance matrix adaptation evolution strategy (CMA-ES) [8] is one of the state-of-the-art black-box optimization methods for continuous search space. CMA-ES employs a multivariate Gaussian distribution as the law of the candidate solutions and updates the distribution parameters to generate superior candidate solutions. CMA-ES possesses several invariance properties to some transformations of the search space and objective function. These invariance properties ensure that the behaviors of CMA-ES are identical on the objective functions in a certain class, which is important to realize efficient optimization performance on a wide class of objective functions. The invariance properties of CMA-ES are summarized as follows:

1. Invariance to any invertible affine transformation of the search space (including shifting, rotation, and scale transformation).
2. Invariance to any monotonically increasing transformation of the objective function.

The first invariance property is primarily achieved by the update of the covariance matrix, called the covariance matrix adaptation. The second invariance property is owned by the ranking transformation of the evaluation values of the candidate solutions on the objective function. Several variants of CMA-ES that inherit these invariance properties have been proposed, such as $(\mu/\mu_{\mathrm{w}}, \lambda)$-CMA-ES [8] and $(1 + 1)$-CMA-ES [10].

In real-world problems, the evaluation of the objective function is often expensive, and the evaluation of numerous candidate solutions involves high cost. To reduce the number of evaluations on the objective function, the surrogate model has been applied to estimate the evaluation values. Several kinds of models have been incorporated into CMA-ES to estimate the evaluation values precisely, such as support vector machine (SVM) [20], weighted linear regression [6,12], and Gaussian process regression (GPR) [16,19]. Particularly, GP-CMA-ES [19], which incorporates GPR to $(1+1)$-CMA-ES as a surrogate model, shows promising optimization performance on unimodal objective functions.

The surrogate model-assisted CMA-ES is desired to inherit the invariance properties of CMA-ES. Some mechanisms have been introduced into the regression models to inherit the invariance to any invertible affine transformations of the search space, such as transformation to the full quadratic form of input variable [6] and Mahalanobis distance based on the covariance matrix of the sampling distribution [14,19]. However, most of the surrogate-assisted CMA-ESs lose the invariance to any monotonically increasing transformation of the objective function because popular regression models are not invariant to such transformation. To inherit this invariance, Loshchilov et al. [13,14] incorporated the rank-based SVM into $(\mu/\mu_{\mathrm{w}}, \lambda)$-CMA-ES. Unfortunately, the optimization performance of the surrogate-assisted CMA-ESs with this invariance was limited compared to those without such invariance when not considering the monotonically increasing transformation of the objective function.

For CMA-ES with the elitist strategy, Abbasnejad and Arnold [2] applied the warp function to GPR in GP-CAM-ES to construct a robust surrogate model for monotonically increasing transformations of the objective function. Their method, termed ws-CMA-ES, successfully maintained the optimization performance under transformations using power functions with multiple exponent settings. However, performance deterioration may occur with other kinds of transformation because their design of the warp function is also based on the power function, as confirmed in Sect. 5.4.

In this paper, we propose *switching Gaussian process CMA-ES* (SGP-CMA-ES) to make GP-CMA-ES robust against a wide range of monotonically increasing transformations of the objective function. SGP-CMA-ES utilizes the ranking- and Lebesgue measure-based utility functions $u_{\text{rank}}, u_{\text{Leb}}$ that possess the invariance to any monotonically increasing transformation of the objective function. They were applied to Bayesian optimization in [18]. Then, SGP-CMA-ES selects the suitable target variable of GPR from three utility function values, including the ranking-based utility function values $u_{\text{rank}}(\mathbf{x})$, the Lebesgue measure-based utility function values $u_{\text{Leb}}(\mathbf{x})$, and the objective function values $u_{\text{original}}(\mathbf{x}) = f(\mathbf{x})$. SGP-CMA-ES trains GPR to estimate the utility function values as the surrogate model. The numerical experiments show that SGP-CMA-ES achieved at least almost the same optimization performance as GP-CMA-ES and ws-CMA-ES in both cases with and without transformations used in our experiment. Moreover, we confirmed the advantage of SGP-CMA-ES over ws-CMA-ES when applying a transformation not used in the reference of ws-CMA-ES [2].

2   Backgrounds

2.1   CMA-ES with Surrogate Models

Several kinds of regression models, such as SVM [20], weighted linear regression [6,12], and GPR [16,19], have been applied as surrogate models for CMA-ES. In popular update rules of surrogate-assisted $(\mu/\mu_{\text{w}}, \lambda)$-CMA-ES, a part of λ candidate solutions are selected and evaluated on the objective function, and others are evaluated on the surrogate model. Then the surrogate model learns the objective function through the selected candidate solutions. There are several selection methods, including selecting the superior candidate solutions on the surrogate model [6] and selecting the candidate solutions with high uncertainty criteria on GPR [16]. Moreover, the adaptation mechanisms of the number of selected candidate solutions also have been proposed [6,12]. For example, lq-CMA-ES [6] uses the Kendall rank correlation test between the rankings on the objective function and the surrogate model. As another mechanism, lq-CMA-ES also injects the best candidate solutions on the surrogate model, which is given by the quadratic model and realizes an efficient optimization, especially on the convex quadratic functions.

Several variants of surrogate-assisted CMA-ES apply surrogate models that predict the objective function value. Unfortunately, they lose the invariance to any monotonically increasing transformation of the objective function because their models are not invariant to such transformation. To maintain this invariance, a rank-based SVM is applied as the surrogate model of CMA-ES [14,17]. The rank-based SVM does not break the invariance properties of the original CMA-ES because it predicts the ranking of inputs. Particularly, ^{s*}ACM-ES [14] performs another CMA-ES to update the hyperparameters of the rank-based SVM and shows a significant improvement on unimodal objective functions.

There are several variants of $(1 + 1)$-ES with the surrogate model [11,21]. These methods evaluate the candidate solution on the surrogate model first, and then, the candidate solution is evaluated on the objective function if the candidate solution is estimated to be superior to the current elitist solution.

GP-CMA-ES [19] additionally introduces covariance matrix adaptation, which uses both of the rank-one update in the original $(1 + 1)$-CMA-ES [10] and the rank-μ update obtained on the surrogate model. Furthermore, ws-CMA-ES [2] applies the wrap function to GPR in GP-CMA-ES to improve the robustness for monotonically increasing transformations of the objective function. The details of GP-CMA-ES and ws-CMA-ES are explained in Sect. 3.

2.2 Utility Transformations of Objective Function Value

The utility transformation is the transformation of the objective function value of candidate solutions and is introduced to achieve the invariance to any monotonically increasing transformation of the objective function. The ranking-based weights in the update of CMA-ES can be considered as the estimation of the utility function values based on the quantiles of the candidate solutions [15]. On the other hand, Akimoto [3] proposed a utility function based on the Lebesgue measure of the candidate solution, which was used to investigate the convergence rate of the rank-μ update CMA-ES on the monotonic convex-quadratic-composite functions. The details of these utility functions are shown in Sect. 4.1. The ranking- and Lebesgue measure-based utility functions are also applied to Bayesian optimization to achieve the optimization performance invariant to monotonically increasing transformation of the objective function [18].

3   Surrogate-Assisted CMA-ES with Elitist Strategy

3.1   Gaussian Process CMA-ES

Gaussian process CMA-ES (GP-CMA-ES) [19] is a variant of surrogate-assisted $(1+1)$-CMA-ES with GPR. The state of GP-CMA-ES consists of the elitist solution $\mathbf{m} \in \mathbb{R}^n$, the step-size $\sigma \in \mathbb{R}_{>0}$, the covariance matrix $\mathbf{C} \in \mathbb{R}^{n \times n}$, and the evolution path $\mathbf{s} \in \mathbb{R}^n$. GP-CMA-ES also possesses the archive of m candidate solutions $\mathcal{A} = \{(\mathbf{x}_k, f(\mathbf{x}_k))\}_{k=1}^m$ evaluated on the objective function $f : \mathbb{R}^n \to \mathbb{R}$ in prior iterations. The single update of GP-CMA-ES in the minimization of f is as follows, which is depicted in Algorithm 1. First, GP-CMA-ES constructs a GPR f_ϵ using the archive $\mathcal{A}$. Then, λ candidate solutions are generated as $\mathbf{x}_i = \mathbf{m} + \sigma \mathbf{A} \mathbf{z}_i$ using samples from n-dimensional standard Gaussian distribution $\mathbf{z}_i \sim \mathcal{N}(\mathbf{0}, \mathbf{I})$ and the square root of the covariance matrix $\mathbf{A} = \mathbf{C}^{1/2}$. GP-CMA-ES evaluates these solutions on f_ϵ and assigns the weight w_i for i-th best sample $\mathbf{x}_{i:\lambda}$ for $i = 1, \cdots, \lambda$. The weights are given by $w_1 = 1$ when $\lambda = 1$, and are set as the default weight values of $(\mu/\mu_w, \lambda)$-CMA-ES in [5] when $\lambda > 1$. Then, new candidate solution is obtained as $\mathbf{x}_{\text{new}} = \mathbf{m} + \sigma \mathbf{A} \mathbf{z}$ using the weighted sum of $\mathbf{z}_i$ as $\mathbf{z} = \sum_{i=1}^{\lambda} w_i \mathbf{z}_{i:\lambda}$. If the evaluation value of $\mathbf{x}_{\text{new}}$ on f_ϵ is worse than the best objective function value so far, the step-size is decreased by the factor of $e^{-d_1/D}$. Otherwise, $\mathbf{x}_{\text{new}}$ is evaluated on the objective function and the archive is updated. If $\mathbf{x}_{\text{new}}$ is inferior to the elitist solution $\mathbf{m}$ on f, the step-size is decreased by the factor of $e^{-d_2/D}$. Otherwise, GP-CMA-ES updates the elitist solution as $\mathbf{m} = \mathbf{x}_{\text{new}}$, increases the step-size by the factor of $e^{d_3/D}$ and updates the evolution path and the covariance matrix. If $\lambda > 1$, the update rules of the evolution path and the covariance matrix read

$$\mathbf{s} \leftarrow (1-c)\mathbf{s} + \sqrt{c(2-c)\mu_{\text{eff}}}\,\mathbf{A}\mathbf{z} \tag{1}$$

$$\mathbf{C} \leftarrow (1 - c_1 - c_\mu)\mathbf{C} + c_1\mathbf{s}\mathbf{s}^{\text{T}} + c_\mu \mathbf{A}\left(\sum_{j=1}^{\lambda} w_j^\circ \mathbf{z}_{j:\lambda}(\mathbf{z}_{j:\lambda})^{\text{T}}\right)\mathbf{A}^{\text{T}}, \tag{2}$$

respectively. The settings of hyperparameters $c, c_1, c_\mu, \mu_{\text{eff}}$ and weights w_i° are referred to [5]. The weights w_i° are determined by the ranking on f_ϵ. If $\lambda = 1$, then the update rules of $\mathbf{s}$ and $\mathbf{C}$ are the same way as $(1+1)$-CMA-ES introduced by [10] with its default hyperparameter settings. The hyperparameters d_1, d_2 and d_3 are also set depending on λ: they are $d_1 = 0.2, d_2 = 1.0$ and $d_3 = 1.0$ for $\lambda > 1$, while $d_1 = 0.05, d_2 = 0.2$ and $d_3 = 0.6$ for $\lambda = 1$. The parameter D is set to $\sqrt{n+1}$ independently from λ.

The kernel function of GP-CMA-ES is designed based on the Mahalanobis distance using the covariance of the search distribution $\sigma^2\mathbf{C}$ as

$$k(\mathbf{x}, \mathbf{x}') = \exp\left(-\frac{(\mathbf{x}-\mathbf{x}')^{\text{T}}\mathbf{C}^{-1}(\mathbf{x}-\mathbf{x}')}{2\,h^2\sigma^2}\right), \tag{3}$$

where the length scale parameter is set as $h = 8n$. GP-CMA-ES assumes the mean of the prior distribution as $f(\mathbf{m})$, and, denoting $\mathbf{f}_m = (f(\mathbf{x}_1), \cdots, f(\mathbf{x}_m))^{\text{T}}$ and $\mathbf{k}_m(\mathbf{x}) = (k(\mathbf{x}_1, \mathbf{x}), \cdots, k(\mathbf{x}_m, \mathbf{x}))^{\text{T}}$, it computes the evaluation value on the surrogate model as

$$f_\epsilon(\mathbf{x}) = f(\mathbf{m}) + \mathbf{k}_m(\mathbf{x})^{\text{T}}\mathbf{K}_m^{-1}(\mathbf{f}_m - f(\mathbf{m})\mathbf{1}_m), \tag{4}$$

where $\mathbf{1}_m = (1, \cdots, 1)^{\text{T}}$ is a vector consisted of m ones and $\mathbf{K}_m \in \mathbb{R}^{m \times m}$ is a Gram matrix whose (i, j) element is given by $(\mathbf{K}_m)_{i,j} = k(\mathbf{x}_i, \mathbf{x}_j)$. GP-CMA-ES uses the latest $m = (n+2)^2$ data in the archive to construct the surrogate model to reduce the computational cost. Moreover, as initialization phase, GP-CMA-ES does not use the surrogate model in the first $2n$ iterations and updates as the model-free $(1+1)$-CMA-ES with $d_2 = 0.25$ and $d_3 = 1.0$.

3.2   Warped Surrogate CMA-ES

Warped surrogate CMA-ES (ws-CMA-ES) is a variant of GP-CMA-ES, which improves the robustness to monotonically increasing transformations of the objective function. In the construction of GPR, ws-CMA-ES transforms the observed objective function values using the warp function defined as

$$\Omega_{\langle p,q\rangle}(y) = (y-q)^p, \tag{5}$$

where p and q are the shift and power of the warp, respectively. The predicted value of the objective function value applied to the wrap function is obtained as

$$f_\epsilon(\mathbf{x}) = \Omega_{\langle p,q\rangle}(f(\mathbf{m})) + \mathbf{k}_m(\mathbf{x})^{\text{T}}\mathbf{K}_m^{-1}\tilde{\mathbf{f}}_m, \tag{6}$$

Algorithm 1. Single update in GP-CMA-ES

Require: Parameters $\mathbf{m}, \sigma, \mathbf{C}, \mathbf{s}$ of CMA-ES, archive $\mathcal{A} = \{(\mathbf{x}_k, f(\mathbf{x}_k))\}_{k=1}^{m}$
1: Construct a surrogate model f_ϵ using $\mathcal{A}$.
2: **for** $i = 1, \ldots, \lambda$ **do**
3:    Generate $\mathbf{x}_i = \mathbf{m} + \sigma \mathbf{A} \mathbf{z}_i$ using $\mathbf{z}_i \sim \mathcal{N}(\mathbf{0}, \mathbf{I})$ and $\mathbf{A} = \mathbf{C}^{1/2}$
4:    Evaluate $\mathbf{x}_i$ on the surrogate model f_ϵ.
5: **end for**
6: Compute $\mathbf{x}_{\text{new}} = \mathbf{m} + \sigma \mathbf{A} \mathbf{z}$ using $\mathbf{z} = \sum_{i=1}^{\lambda} w_i \mathbf{z}_{i:\lambda}$
7: Evaluate $\mathbf{x}_{\text{new}}$ on the surrogate model f_ϵ.
8: **if** $f_\epsilon(\mathbf{x}_{\text{new}}) > f(\mathbf{m})$ **then**
9:    Decrease the step-size as $\sigma \leftarrow \sigma e^{-d_1/D}$
10: **else**
11:    Evaluate $\mathbf{x}_{\text{new}}$ on the objective function f and add $(\mathbf{x}_{\text{new}}, f(\mathbf{x}_{\text{new}}))$ to $\mathcal{A}$.
12:    **if** $f(\mathbf{x}_{\text{new}}) > f(\mathbf{m})$ **then**
13:       Decrease the step-size as $\sigma \leftarrow \sigma e^{-d_2/D}$
14:    **else**
15:       Update the elitist solution as $\mathbf{m} \leftarrow \mathbf{x}_{\text{new}}$
16:       Increase the step-size as $\sigma \leftarrow \sigma e^{d_3/D}$
17:       Update $\mathbf{s}$ and $\mathbf{C}$ using (1) and (2).
18:    **end if**
19: **end if**

---

where the i-th element of $\tilde{\mathbf{f}}_m$ is given by $(\tilde{\mathbf{f}}_m)_i = \Omega_{\langle p,q \rangle}(f(\mathbf{x}_i)) - \Omega_{\langle p,q \rangle}(f(\mathbf{m}))$. With reasonable settings of p and q, the rankings of transformed evaluation values using the warp function are the same as those of the original evaluation values. The parameters p, q are adaptively updated by a simple coordinate search so that the Kendall rank correlation coefficient τ between the observed objective function values and the predicted values is maintained over 0.9. The details are found in [1] and [2].

4   Switching Gaussian Process CMA-ES

We modify the update rule of GP-CMA-ES and propose switching Gaussian process CMA-ES (SGP-CMA-ES) to improve the robustness to monotonically increasing transformations of the objective function. We incorporate two components to GP-CMA-ES, the utility functions which are invariant to any monotonically increasing transformation of the objective function and a switching mechanism of the target variable of GPR.

4.1   Utility Functions

We introduce three utility functions based on the objective function value, the ranking, and the Lebesgue measure of the candidate solutions. These are used to calculate the candidates of the target variable of GPR. The ranking- and Lebesgue measure-based utility functions were applied to Bayesian optimization in [18].

Untransformed Utility Function: We use the original objective function value as a candidate of utility function value as $u_{\text{original}}(\mathbf{x}_i) = f(\mathbf{x}_i)$. This untransformed utility aims to maintain the performance of SGP-CMA-ES almost the same as that of GP-CMA-ES on easily predictable objective functions by GPR.

Ranking-Based Utility Function: The ranking-based utility is used in some well-known probabilistic model-based black-box methods, including CMA-ES, and is related to the estimation of quantiles on f [15]. Given a probability distribution with its probability density function p, the quantile of $\mathbf{x}$ on f is the probability of sampling a superior sample to $\mathbf{x}$, which reads

$$q(\mathbf{x}) = \int \mathbb{I}\{f(\mathbf{x}) \le f(\mathbf{x}')\}p(\mathbf{x}')d\mathbf{x}', \tag{7}$$

where $\mathbb{I}$ is the indicator function. Because the quantile cannot be obtained analytically on black-box functions, the existing methods estimate it by Monte Carlo approximation using multiple samples $\{\mathbf{x}_j\}$ and their ranking as

$$\hat{q}(\mathbf{x}_i) = \frac{1}{m}\sum_{j=1}^{m} \mathbb{I}\{f(\mathbf{x}_i) \le f(\mathbf{x}_j)\} = \frac{\text{rk}(\mathbf{x}_i)}{m}, \tag{8}$$

where m and rk are the number of samples and the ranking of the given sample, respectively. Then, introducing a non-decreasing function $w : \mathbb{R} \to \mathbb{R}$, the ranking-based utility function value is obtained as $w(\hat{q}(\mathbf{x}))$. Although there are many possible choices of w, we simply set $w(z) = z$ and obtain the utility function as

$$u_{\text{rank}}(\mathbf{x}_i) = \frac{\text{rk}(\mathbf{x}_i)}{m}. \tag{9}$$

Lebesgue Measure-Based Utility Function: We also introduce another utility function based on the Lebesgue measure of the set consisting of the elements superior to a given sample on the objective function [3], which is given by

$$V_f(\mathbf{x}) = (\mu_{\text{Leb}}\{\mathbf{x}' : f(\mathbf{x}) \le f(\mathbf{x}')\})^{2/n} = \left(\int \mathbb{I}\{f(\mathbf{x}) \le f(\mathbf{x}')\}d\mathbf{x}'\right)^{2/n}. \tag{10}$$

We perform Monte Carlo approximation using m samples $\{\mathbf{x}_j\}$ to estimate it on black-box functions. If the likelihood p of the samples is known, the utility function value is obtained as

$$u_{\text{Leb}}(\mathbf{x}_i) = \left(\frac{1}{m}\sum_{j=1}^{m} \frac{\mathbb{I}\{f(\mathbf{x}_j) \le f(\mathbf{x}_i)\}}{p(\mathbf{x}_j)}\right)^{2/n}. \tag{11}$$

We note that the likelihood of samples in the archive cannot be obtained analytically in the context of GP-CMA-ES because the distribution parameters are updated iteratively, and the update of the archive is performed only for the better candidate solution than the current elitist solution. Following [18], we apply a kernel density estimation to obtain the estimated value $\hat{p}(\mathbf{x})$ of the likelihood $p(\mathbf{x})$. The detail of our estimation method is explained in Sect. 4.3.

4.2   Switching Mechanism of Utility Function

When using the ranking- and Lebesgue measure-based utility functions described in Sect. 4.1, the performance may deteriorate on the objective function that GPR can estimate easily instead of obtaining the invariance for any monotonically increasing transformation of the objective function. We propose a switching mechanism using priorities $\theta \in [-1, 1]^M$ of M utility functions, including the untransformed utility function u_{original}, to prevent the performance deterioration by selecting the suitable target variable of GPR. We use three utility functions introduced in Sect. 4.1, i.e., $M = 3$ in the experiment.

The proposed SGP-CMA-ES uses the utility function u with the largest priority to construct the GPR. Given the same kernel function as GP-CMA-ES in (3), the evaluation value on the surrogate model is obtained using $\mathbf{u}_m = (u(\mathbf{x}_1), \cdots, u(\mathbf{x}_m))^{\text{T}}$ as[1]

$$f_\epsilon(\mathbf{x}) = u(\mathbf{m}) + \mathbf{k}_m(\mathbf{x})^{\text{T}} \mathbf{K}_m^{-1} (\mathbf{u}_m - u(\mathbf{m}) \mathbf{1}_m). \tag{12}$$

The priorities are initialized as $\theta = (0, \cdots, 0)^{\text{T}}$ and updated when the candidate solution is evaluated on the objective function. Given the update direction $\Delta\theta \in \mathbb{R}^M$, the update rule of the priorities reads

$$\theta \leftarrow (1 - c_\theta) \theta + c_\theta \Delta\theta, \tag{13}$$

where c_θ is set to $(n + 4)^{-1}$. Let us denote the index of the selected utility function as i_{select}. When the evaluated candidate solution is superior to the elitist solution, the selected utility function is considered to be reliable and set the update direction as $\Delta\theta_{i_{\text{select}}} = 1$. Otherwise, the selected utility function is unreliable and set $\Delta\theta_{i_{\text{select}}} = -1$. The priorities for other utility functions are set as $\Delta\theta_i = 0$ to approach to the neutral state 0. Selecting the surrogate model with the highest priority maintains the surrogate model accurately.

4.3   Implementation of SGP-CMA-ES

Algorithm 2 shows the single update of SGP-CMA-ES. The modifications from GP-CMA-ES are as follows. In Lines 1–3, SGP-CMA-ES selects the utility with the largest priority and trains GPR using the utility function values of solutions in the archive. In Line 10, SGP-CMA-ES compares $f_\epsilon(\mathbf{x}_{\text{new}})$ with $f_\epsilon(\mathbf{m})$ computed by (12) instead of $f(\mathbf{m})$. In Line 14, the priorities are updated using (13).

In the calculation of the Lebesgue measure-based utility value (11), we use the estimated likelihood $\hat{p}(\mathbf{x})$, instead of $p(\mathbf{x})$, obtained by the kernel density estimation with the Gaussian kernel. We compute the estimated likelihood of $\mathbf{x}_i$ using the candidate solutions $\{\mathbf{x}_j\}_{j=1}^m$ in the archive $\mathcal{A}$ as

$$\hat{p}(\mathbf{x}_i) = \frac{1}{mb\sqrt{2\pi\sigma^2|\mathbf{C}|}} \sum_{j=1}^m \exp\left(-\frac{(\mathbf{x}_i - \mathbf{x}_j)^{\text{T}} \mathbf{C}^{-1} (\mathbf{x}_i - \mathbf{x}_j)}{2\sigma^2 b^2}\right), \tag{14}$$

where the bandwidth is set to $b = m^{-1/(n+4)}$.

---

[1] Since the evaluations on u_{rank} and u_{Leb} require the ranking of input, we replace $u(\mathbf{m})$ in (12) with the smallest value in $\{u(\mathbf{x}_1), \cdots, u(\mathbf{x}_m)\}$ if $\mathbf{m}$ is not in the archive and the original objective function is not selected as the target variable of GPR.

Algorithm 2. Single update in SGP-CMA-ES. The steps colored in blue are modified and additional steps to GP-CMA-ES in Algorithm 1.

Require: Parameters of CMA-ES $\mathbf{m}, \sigma, \mathbf{C}, \mathbf{s}$, archive $\mathcal{A} = \{(\mathbf{x}_k, f(\mathbf{x}_k))\}_{k=1}^m$
1: Select the utility function u with the largest priority in θ.
2: Calculate $u(\mathbf{x}_k)$ for $\mathbf{x}_k$ using $\mathcal{A}$ and obtain $\mathcal{A}' = \{(\mathbf{x}_k, u(\mathbf{x}_k))\}_{k=1}^m$.
3: Construct a surrogate model f_e using $\mathcal{A}'$.
4: **for** $i = 1, \ldots, \lambda$ **do**
5:     Generate $\mathbf{x}_i = \mathbf{m} + \sigma \mathbf{A} \mathbf{z}_i$ using $\mathbf{z}_i \sim \mathcal{N}(\mathbf{0}, \mathbf{I})$ and $\mathbf{A} = \mathbf{C}^{1/2}$
6:     Evaluate $\mathbf{x}_i$ on the surrogate model f_e.
7: **end for**
8: $\mathbf{x}_{\text{new}} = \mathbf{m} + \sigma \mathbf{A} \mathbf{z}$, $\mathbf{z} = \sum_{i=1}^{\lambda} w_i \mathbf{z}_{i:\lambda}$
9: Evaluate $\mathbf{x}_{\text{new}}$ and $\mathbf{m}$ on the surrogate model f_e.
10: **if** $f_e(\mathbf{x}_{\text{new}}) > f_e(\mathbf{m})$ **then**
11:     Decrease the step-size as $\sigma \leftarrow \sigma e^{-d_1/D}$.
12: **else**
13:     Evaluate $\mathbf{x}_{\text{new}}$ on the objective function f and add $(\mathbf{x}_{\text{new}}, f(\mathbf{x}_{\text{new}}))$ to $\mathcal{A}$.
14:     Update the priorities θ using (13).
15:     Update the smoothed success rate p_{succ} using (15).
16:     **if** $f(\mathbf{x}_{\text{new}}) > f(\mathbf{m})$ **then**
17:         Decrease the step-size as $\sigma \leftarrow \sigma e^{-d_2/D}$.
18:     **else**
19:         Update the elitist solution as $\mathbf{m} \leftarrow \mathbf{x}_{\text{new}}$.
20:         Increase the step-size as $\sigma \leftarrow \sigma e^{d_3/D}$.
21:         Update $\mathbf{s}$ and $\mathbf{C}$ using (16) and (17).
22:         $\mathbf{C} \leftarrow (n/\text{Tr}(\mathbf{C}))\mathbf{C}$.
23:     **end if**
24: **end if**

Furthermore, we modify the update rules of the evolution path $\mathbf{s}$ and covariance matrix $\mathbf{C}$ following [10] to stabilize the covariance matrix adaptation when the utility function is used. We introduce the smoothed success rate p_{succ} which is updated as

$$p_{\text{succ}} \leftarrow (1 - c_p)p_{\text{succ}} + c_p h_{\text{succ}}, \tag{15}$$

where h_{succ} is set as $h_{\text{succ}} = 1$ when the elitist solution is updated, and $h_{\text{succ}} = 0$ otherwise. This update (15) is performed when the candidate solution is evaluated on the objective function.

Then, denoting $h = \mathbb{I}\{p_{\text{succ}} < p_{\text{thresh}}\}$ and $\delta(h) = (1 - h)c_1 c(2 - c)$, $\mathbf{s}$ and $\mathbf{C}$ are updated by

$$\mathbf{s} \leftarrow (1 - c)\mathbf{s} + h\sqrt{c(2 - c)\mu_{\text{eff}}}\,\mathbf{A}\mathbf{z} \tag{16}$$

$$\mathbf{C} \leftarrow (1 - c_1 - c_\mu + \delta(h))\mathbf{C} + c_1 \mathbf{s}\mathbf{s}^{\mathrm{T}} + c_\mu \mathbf{A}\left(\sum_{j=1}^{\lambda} w_j^\circ \mathbf{z}_{j:\lambda} \mathbf{z}_{j:\lambda}^{\mathrm{T}}\right)\mathbf{A}^{\mathrm{T}}. \tag{17}$$

When p_{succ} is smaller than a threshold p_{thresh}, i.e., $h = 1$, these update rules are the same as (1) and (2). We use the default settings of p_{thresh}, c_p and the initial value of p_{succ} provided in [10]. Additionally, we normalized the trace of the covariance matrix [4] as in Line 22.

5   Experiment

5.1   Experimental Setting

We evaluate the performance of SGP-CMA-ES by comparing the following methods. Note that $(\mu/\mu_w, \lambda)$-CMA-ES and $(1+1)$-CMA-ES have no surrogate model.

- $(\mu/\mu_w, \lambda)$-CMA-ES; we used `pycma 3.1.0` [7] as the implementation.
- $(1 + 1)$-CMA-ES; we implemented according to [10].
- $s*$ACM-ES; we use the implementation using `Matlab` at loshchilov.com.
- lq-CMA-ES; we implemented using `pycma 3.1.0` according to [6].
- GP-CMA-ES; we implemented according to [19].
- ws-CMA-ES; we implemented according to [1,2]
- Ranking; GP-CMA-ES with ranking-based utility function (9).
- Lebesgue; GP-CMA-ES with Lebesgue measure-based utility function (11).

We set the hyperparameters of those methods as their default settings. For GP-CMA-ES, ws-CMA-ES, Lebesgue, Ranking, and SGP-CMA-ES, we set the sample size as $\lambda = 10$. All methods except for lq-CMA-ES are invariant to any invertible affine transformation of the search space. We note that lq-CMA-ES switches the linear model, coordinate-wise quadratic model, and full quadratic model by turn as the number of observed data increases, and it achieves the invariance to any invertible affine transformation of the search space when using the full quadratic model. Moreover, $(\mu/\mu_w, \lambda)$-CMA-ES, $(1 + 1)$-CMA-ES, $s*$ACM-ES, Lebesgue, and Ranking also possess the invariance to any monotonically increasing transformation of the objective function. Although SGP-CMA-ES does not possess such invariance, we expect that its search performance is not severely degraded compared with GP-CMA-ES and ws-CMA-ES when applying monotonically increasing transformation to the objective function.

We evaluated those methods using the following parametrized families of benchmark functions introduced in [19].

- Spherically symmetric functions $f(\mathbf{x}) = \|\mathbf{x}\|^{\alpha}$. When $\alpha = 2$, this recovers the sphere function.
- Convex quadratic functions $f(\mathbf{x}) = \mathbf{x}^{\mathrm{T}}\mathbf{B}\mathbf{x}$ with a diagonal matrix $\mathbf{B}$ whose i-th diagonal element is given by $B_{i,i} = \beta^{(i-1)/(n-1)}$. When $\beta = 10^6$, this recovers the ellipsoid function.
- Quartic functions $f(\mathbf{x}) = \sum_{i=1}^{n} \gamma(x_{i+1} - x_i^2)^2 + (1 - x_i)^2$. When $\gamma = 100$, this recovers the Rosenbrock function.

We regarded a trial as successful when the best evaluation value was less than $(10^{-8})^{\alpha/2}$ on the spherically symmetric functions and less than 10^{-8} on the other functions. The initial value of the mean vector $\mathbf{m}$ is set uniformly at random in $[-4, 4]^n$. The initial values of the step-size and covariance matrix are set as

$\sigma = 2$ and $\mathbf{C} = \mathbf{I}$, respectively. On the quartic functions with $n \geq 4$, there is a local minimum and CMA-ES sometimes unsuccessfully converges around it if undesired initial distribution parameters were given. We discarded such trials and collected 15 successful trials for each experimental setting because the aim of the experiment is the evaluation of the search performance on the unimodal objective functions. We note that our experimental settings follow the reference [2,19].

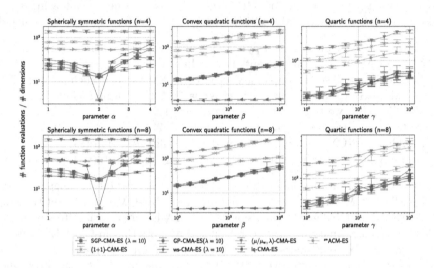

Fig. 1. The number of evaluations on the parametrized benchmark functions. The medians and interquartile ranges of 15 trials are plotted.

5.2   Results on Parametrized Benchmark Functions

First, we evaluated the optimization performance on the parameterized benchmark functions changing the parameters. We set the parameters as $\alpha \in [1, 4]$, $\beta \in [1, 10^6]$, and $\gamma \in [1, 100]$ for the spherically symmetric function, convex quadratic function, and quartic function, respectively. We conducted the experiment with $n = 4$ and $n = 8$. Figure 1 shows the medians and interquartile ranges of the number of evaluations on the objective function over 15 trials. On the convex quadratic function and quartic function, SGP-CMA-ES showed a competitive search performance to GP-CMA-ES and ws-CMA-ES. This implies that our mechanism does not impose performance degradation on the objective function without a monotonically increasing transformation of objective function values. On spherically symmetric function, SGP-CMA-ES outperformed the GP-CMA-ES with a large value of α, while ws-CMA-ES realized more efficient optimization. We suppose that the setting of the warp function of ws-CMA-ES was suitable for the transformation using the power function, and some transformations may degrade its performance, which will be investigated in Sect. 5.4.

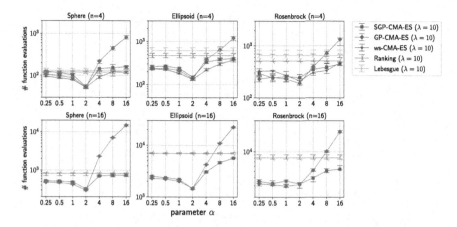

Fig. 2. The number of evaluations on the benchmark functions applied the transformation using power function (18) with $\alpha \in [0.25, 16]$. The medians and interquartile ranges of 15 trials are plotted.

The methods with the invariance to any monotonically increasing transformation of the objective function were worse than the others. On the other hand, lq-CMA-ES showed the most efficient search performance on the spherically symmetric function with $\alpha = 2$ and the convex quadratic functions because the full quadratic model of lq-CMA-ES can estimate the landscape of the objective function perfectly. However, the performance of lq-CMA-ES was significantly degraded on the spherically symmetric function with $\alpha \neq 2$, which indicated the limitation of the search performance of lq-CMA-ES.

5.3   Results with Transformation Using Power Function

Next, we compared the search performance on the sphere, ellipsoid, and Rosenbrock functions with the transformation using the power function as

$$g_{\text{power}}(y) = y^{\alpha/2}, \tag{18}$$

where $\alpha > 0$ is the exponent parameter. In this experiment, we set the evaluation value of a sample $\mathbf{x}$ as $g_{\text{power}}(f(\mathbf{x}))$ instead of $f(\mathbf{x})$. We regard a trial as successful if the best evaluation value on f is less than 10^{-8}. We varied the parameter α in $[2^{-2}, \ldots, 2^4]$ and compared SGP-CMA-ES with GP-CMA-ES, ws-CMA-ES, Ranking, and Lebesgue with $n = 4$ and $n = 16$.

Figure 2 shows the medians and interquartile ranges of the number of evaluations on the objective function over 15 trials. For Ranking and Lebesgue, the results with $\alpha = 2$ are displayed because their performance is not influenced by the setting of α. In the case of $n = 4$, the performance of GP-CMA-ES was degraded with large α on all of the functions. In contrast, SGP-CMA-ES achieved

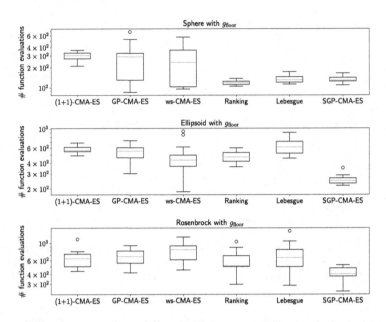

Fig. 3. The number of evaluations on the benchmark functions ($n = 4$) applied the transformation using floor function (19).

the competitive performance to the best method among GP-CMA-ES, Ranking, and Lebesgue for all settings of α. Moreover, SGP-CMA-ES was superior to all of them on the ellipsoid and Rosenbrock functions with α in $[4, 16]$. This implies that the suitable utility function changed in the optimization process, and SGP-CMA-ES successfully selected it. We note that ws-CMA-ES is the best method for all the settings except for Rosenbrock with small α. A possible reason is that the warp function of ws-CMA-ES is specialized for the transformation using the power function.

Next, we focus on the result with $n = 16$. We do not display the results of Lebesgue and ws-CMA-ES because of the following reasons: Lebesgue failed the optimization of the ellipsoid and Rosenbrock more than 3 times on average, and the computational cost for tuning the parameters p and q of ws-CMA-ES is too high due to the increase of the archive size when $n = 16$. Focusing on SGP-CMA-ES, it is superior to GP-CMA-ES and Ranking on the ellipsoid and Rosenbrock functions with large α, while it showed almost the same performance with the best method in the rest cases. We note that SGP-CMA-ES contains the Lebesgue measure-based utility as a candidate for utility. This implies that SGP-CMA-ES successfully selected the suitable utility for the given objective function.

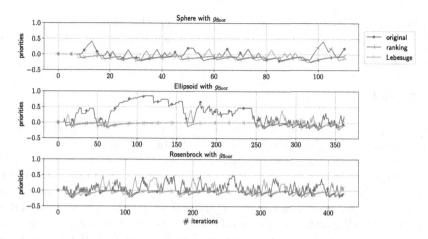

Fig. 4. The transition of priorities of target variables for GPR on the benchmark functions with the floor function transformation (19). We plot the result of a typical trial for each benchmark function.

5.4 Results with Transformation Using Floor Function

Finally, we considered the other kind of monotonically increasing transformation to investigate the limitation of ws-CMA-ES. We selected a discontinuous monotonically increasing transformation as

$$g_{\text{floor}}(y) = y + 10\lfloor \ln y \rfloor. \tag{19}$$

Same as in the previous experiment, we evaluate a sample $\mathbf{x}$ as $g_{\text{floor}}(f(\mathbf{x}))$ instead of $f(\mathbf{x})$. A trial was regarded as successful when the best evaluation value on f reached less than 10^{-8}.

Figure 3 shows the box plots of the number of evaluations on the objective function over 15 trials with $n = 4$. Focusing on GP-CMA-ES and ws-CMA-ES, the results contain large variances on the sphere function, and their medians were almost the same as that of $(1+1)$-CMA-ES on all functions. SGP-CMA-ES achieves the best performance on the ellipsoid and Rosenbrock functions, while it is second best on the sphere function. Especially, SGP-CMA-ES was significantly better than all compared methods on ellipsoid and Rosenbrock functions. This implies that the utilities were appropriately switched during the optimization process. Figure 4 displays the transitions of the priorities of utility functions in typical trials. On the ellipsoid function, the original objective function was mainly used at first, and then the ranking- and Lebesgue measure-based utilities were selected. A possible reason for such behavior is that the second term of (19) can be ignored when the values of $y = f(\mathbf{x})$ are significantly large, and modeling the original objective function by GPR is appropriate until the f values of generated candidate solutions decrease. This behavior is notable on the ellipsoid function due to its scale. We also confirmed that the transitions of priorities differ depending on the objective functions, which implies the utility function was adaptively selected to accelerate the optimization process.

6   Conclusion

We have proposed a novel surrogate-assisted $(1 + 1)$-CMA-ES, which is an improvement of GP-CMA-ES and switches the target variable of GPR among two invariant utilities and the original objective. The proposed SGP-CMA-ES adopts the ranking- and Lebesgue measure-based utility functions as the utility candidates. Thanks to the invariance of those utility functions to the monotonically increasing transformation of the objective function, the optimization performance of SGP-CMA-ES does not significantly degrade by such transformations. In the experiment with the transformation using the power function, we observed that SGP-CMA-ES performed as well as the best methods among GP-CMA-ES, Ranking, and Lebesgue for all settings. Moreover, SGP-CMA-ES was significantly better than compared methods on the ellipsoid and Rosenbrock functions with transformation using the floor function because it adaptively changed the utility function during the optimization process.

Introducing sophisticated utility functions regardless of the invariance, such as the warp function in ws-CMA-ES, is a possible future work to improve the optimization performance. Another direction is to improve the selection mechanism of the suitable utility function. For Bayesian optimization, portfolio strategies [9] are applied to select the suitable acquisition function. Combining such strategies with our method is an important future work.

References

1. Abbasnejad, A., Arnold, D.V.: Adaptive function value warping for surrogate model assisted evolutionary optimization. Master's thesis, Dalhousie University (2021)
2. Abbasnejad, A., Arnold, D.V.: Adaptive function value warping for surrogate model assisted evolutionary optimization. In: Rudolph, G., Kononova, A.V., Aguirre, H., Kerschke, P., Ochoa, G., Tušar, T. (eds.) PPSN 2022. LNCS, vol. 13398, pp. 76–89. Springer, Cham (2022). https://doi.org/10.1007/978-3-031-14714-2_6
3. Akimoto, Y.: Analysis of a natural gradient algorithm on monotonic convex-quadratic-composite functions. In: Proceedings of the 14th Annual Conference on Genetic and Evolutionary Computation Conference (GECCO), pp. 1293–1300. ACM Press (2012). https://doi.org/10.1145/2330163.2330343
4. Akimoto, Y., Sakuma, J., Ono, I., Kobayashi, S.: Functionally specialized CMA-ES: a modification of CMA-ES based on the specialization of the functions of covariance matrix adaptation and step size adaptation. In: Proceedings of the 10th Annual Conference on Genetic and Evolutionary Computation (GECCO), pp. 479–486. ACM Press (2008). https://doi.org/10.1145/1389095.1389188
5. Hansen, N.: The CMA evolution strategy: a tutorial. CoRR abs/1604.00772 (2016)

6. Hansen, N.: A global surrogate assisted CMA-ES. In: Proceedings of the Genetic and Evolutionary Computation Conference (GECCO), pp. 664–672. ACM Press (2019). https://doi.org/10.1145/3321707.3321842

7. Hansen, N., Akimoto, Y., Baudis, P.: CMA-ES/pycma on github (2019)

8. Hansen, N., Ostermeier, A.: Adapting arbitrary normal mutation distributions in evolution strategies: the covariance matrix adaptation. In: Proceedings of IEEE International Conference on Evolutionary Computation, pp. 312–317 (1996). https://doi.org/10.1109/ICEC.1996.542381

9. Hoffman, M., Brochu, E., de Freitas, N.: Portfolio allocation for Bayesian optimization. In: Proceedings of the Twenty-Seventh Conference on Uncertainty in Artificial Intelligence, pp. 327–336. AUAI Press (2011)

10. Igel, C., Suttorp, T., Hansen, N.: A computational efficient covariance matrix update and a (1+1)-CMA for evolution strategies. In: Proceedings of the 8th Annual Conference on Genetic and Evolutionary Computation (GECCO), pp. 453–460. ACM Press (2006). https://doi.org/10.1145/1143997.1144082

11. Kayhani, A., Arnold, D.V.: Design of a surrogate model assisted (1 + 1)-ES. In: Auger, A., Fonseca, C.M., Lourenço, N., Machado, P., Paquete, L., Whitley, D. (eds.) PPSN 2018. LNCS, vol. 11101, pp. 16–28. Springer, Cham (2018). https://doi.org/10.1007/978-3-319-99253-2_2

12. Kern, S., Hansen, N., Koumoutsakos, P.: Local meta-models for optimization using evolution strategies. In: Runarsson, T.P., Beyer, H.-G., Burke, E., Merelo-Guervós, J.J., Whitley, L.D., Yao, X. (eds.) PPSN 2006. LNCS, vol. 4193, pp. 939–948. Springer, Heidelberg (2006). https://doi.org/10.1007/11844297_95

13. Loshchilov, I., Schoenauer, M., Sebag, M.: Comparison-based optimizers need comparison-based surrogates. In: Schaefer, R., Cotta, C., Kołodziej, J., Rudolph, G. (eds.) PPSN 2010. LNCS, vol. 6238, pp. 364–373. Springer, Heidelberg (2010). https://doi.org/10.1007/978-3-642-15844-5_37

14. Loshchilov, I., Schoenauer, M., Sebag, M.: Self-adaptive surrogate-assisted covariance matrix adaptation evolution strategy. In: Proceedings of the 14th Annual Conference on Genetic and Evolutionary Computation (GECCO), pp. 321–328. ACM Press (2012). https://doi.org/10.1145/2330163.2330210

15. Ollivier, Y., Arnold, L., Auger, A., Hansen, N.: Information-geometric optimization algorithms: a unifying picture via invariance principles. J. Mach. Learn. Res. 18(1), 564–628 (2017)

16. Pitra, Z., Bajer, L., Holeňa, M.: Doubly trained evolution control for the surrogate CMA-ES. In: Handl, J., Hart, E., Lewis, P.R., López-Ibáñez, M., Ochoa, G., Paechter, B. (eds.) PPSN 2016. LNCS, vol. 9921, pp. 59–68. Springer, Cham (2016). https://doi.org/10.1007/978-3-319-45823-6_6

17. Runarsson, T.P.: Ordinal regression in evolutionary computation. In: Runarsson, T.P., Beyer, H.-G., Burke, E., Merelo-Guervós, J.J., Whitley, L.D., Yao, X. (eds.) PPSN 2006. LNCS, vol. 4193, pp. 1048–1057. Springer, Heidelberg (2006). https://doi.org/10.1007/11844297_106

18. Shirakawa, S.: Impact of invariant objective for order preserving transformation in Bayesian optimization. In: 2016 IEEE Congress on Evolutionary Computation (CEC), pp. 1432–1437 (2016). https://doi.org/10.1109/CEC.2016.7743958

19. Toal, L., Arnold, D.V.: Simple surrogate model assisted optimization with covariance matrix adaptation. In: Bäck, T., et al. (eds.) PPSN 2020. LNCS, vol. 12269, pp. 184–197. Springer, Cham (2020). https://doi.org/10.1007/978-3-030-58112-1_13

20. Ulmer, H., Streichert, F., Zell, A.: Evolution strategies with controlled model assistance. In: Proceedings of the 2004 Congress on Evolutionary Computation, vol. 2, pp. 1569–1576 (2004). https://doi.org/10.1109/CEC.2004.1331083
21. Yang, J., Arnold, D.V.: A surrogate model assisted (1+1)-ES with increased exploitation of the model. In: Proceedings of the Genetic and Evolutionary Computation Conference (GECCO), pp. 727–735. ACM Press (2019). https://doi.org/10.1145/3321707.3321728

Author Index

J. Correia et al. (Eds.): EvoApplications 2023, LNCS 13989, pp. 815–817, 2023.
https://doi.org/10.1007/978-3-031-30229-9

Printed in the United States
by Baker & Taylor Publisher Services